Bob Weir

ENVIRONMENTAL MODELING WITH GIS

Michael F. Goodchild
Bradley O. Parks
Louis T. Steyaert

New York Oxford
OXFORD UNIVERSITY PRESS
1993

Oxford University Press

Oxford New York Toronto
Delhi Bombay Calcutta Madras Karachi
Kuala Lumpur Singapore Hong Kong Tokyo
Nairobi Dar es Salaam Cape Town
Melbourne Auckland Madrid

and associated companies in
Berlin Ibadan

Published by Oxford University Press, Inc.,
200 Madison Avenue, New York, New York 10016

Library of Congress Cataloging-in-Publication Data
Environmental modeling with GIS / [edited by] Michael F. Goodchild,
Bradley O. Parks, Louis T. Steyaert.
p. cm. Includes bibliographical references.
ISBN 0-19-508007-6
1. Environmental engineering–Data processing.
2. Environmental protection–Data processing.
3. Geographic information systems.
I. Goodchild, Michael F. II. Parks, Bradley O.
III. Steyaert, L. T. (Louis T.)
TD153.E58 1993 628'.0285–dc20 92-27454

9 8 7 6 5 4 3 2
Printed in the United States of America
on acid-free paper

PREFACE

This book brings together a collection of invited interdisciplinary perspectives on the topic of environmental modeling with geographic information systems (GIS). GIS by definition is a technology designed to capture, store, manipulate, analyze, and visualize the diverse sets of georeferenced data that are required to support accurate modeling of the Earth's environmental processes. Chapters by leading authorities introduce GIS technology and a broad range of environmental simulation models, while additional chapters illustrate current levels of integration and suggest opportunities for future research. The focus is on contemporary modeling in natural science as related to global change research, land and water resource management, and environmental risk assessment. In general, the environmental simulation modeling and GIS concepts in this book are relevant to each of these three types of uses, as well as for applications that can include local, regional, and global scales.

Our goal in this book is to facilitate the integration of GIS and environmental simulation models in the areas of scientific research, quantitative resource assessment, and risk analysis. Our specific purposes are: (1) to improve the interdisciplinary understanding of GIS technology and selected types of simulation models in natural science; (2) to enhance cross-disciplinary communication; (3) to identify requirements and opportunities for integration; and (4) to generate the enthusiasm to meet these challenges.

We believe that there are substantial opportunities in the integration of GIS and advanced environmental simulation models. For example, detailed consideration of landscape properties and spatially distributed processes at the land surface are fundamental to global climate and mesoscale models, watershed and water resource assessment models, and various types of ecological models involving landscape, population, and community development processes. The issues of multiple space and time scales are fundamental to "integrated systems" modeling, a highly cross-disciplinary modeling approach exemplified by the suite of models for land–atmosphere interactions research. In addition to the issue of spatial processes operating at multiple time and space scales, GIS and environmental simulation models share converging interests in terms of spatial data. The availability of spatial data from many sources including land cover characteristics based on multitemporal satellite remote sensing is growing rapidly. In fact, the advanced simulation models require a rich variety of multidisciplinary land surface characteristics data of many types. Such data are needed to investigate environmental processes that are functions of a complex terrain and heterogeneous landscape. Finally, both GIS and quantitative models are increasingly recognized as vital to natural resource management, environmental risk assessment, and major research programs such as global climate change issues at local and regional scales.

Thus, this overall modeling environment seems ideally suited for GIS as a tool to support integrative modeling, to conduct interactive spatial analysis across multiple scales for understanding processes, and to capture complex land surface properties to drive the models based on innovative thematic mapping of multitiered land database products. Despite their many functional strengths, however, GIS are often limited by static representations of dynamic space/time processes, use of simple logical operations to explore complex relationships, nonstochastic treatment of uncertain events, quasi-three-dimensional (surface or perspective) treatments of truly three-dimensional properties, and insufficient emphasis on continuous, differentiable surfaces of ratio scale data. The entire area of spatial statistical analysis is largely undeveloped. Thus, these "tool level" limitations add to the challenge of using GIS for conceptualizing and modeling complex, dynamic environmental processes. To meet modeling requirements, the repertoire of analytical tools and methods available in GIS must be expanded. The challenge is to clarify the role of GIS in scientific research and environmental modeling, and to influence the evolution or integration of modeling functionality within the GIS domain.

The book is subdivided into six parts: Introduction, GIS, Environmental Simulation Models, Risk Assessment, Spatial Data, and Spatial Statistics. The Introduction focuses on the scientific demands of environmental issues ranging from local to global scales, the status of environmental simulation models and GIS technology, the need for integration, and the role of the GIS software vendor. Part II summarizes GIS in terms of its scope, current analytical functions, spatial data models, error analysis issues, and contemporary trends.

In Part III, individual sections feature state of the art environmental simulation models that are representative of atmospheric, hydrological, biological/ecological, land/subsurface processes, and cross-disciplinary integrated systems modeling. Within each modeling section, an overview paper and a series of case studies illustrate contemporary applications including the use of GIS. The chapters in this section begin with examples of state of the art R&D models and gradually shift in emphasis to examples of the more qualitative issues concerning risk assessment modeling for policy input. Because these R&D models will eventually become the "next generation" of resource assessment and risk analysis models, the issue of GIS integration with these models has added relevance.

Part IV moves from modeling to policy by addressing the assessment of risk, and includes a commentary on the use of models in policy formulation. In Part V, the focus is on the spatial data that underlie much environmental modeling, and the chapters in this section cover data sources and some of the issues that arise in constructing large, multithematic databases. Finally Part VI covers the important area of spatial statistics and includes reviews of spatial interpolation and spatial sampling, both significant problems for environmental modeling with GIS.

The book was planned as a coherent whole, and the authors were asked to fit their pieces into that overall design. Almost inevitably the results vary widely in length, but the editors felt that it was appropriate to allow chapter length to reflect the progress that has been made in each field. There is also some degree of overlap in content; for example, the issues of GIS integration in hydrologic modeling clearly share some content with those in land surface/subsurface modeling. However, as editors we have not been overly concerned with removing such overlap, believing that it is often useful to have more than one perspective on an issue.

Not all environmental processes are covered in the book—we have not, for example, included reviews of physical oceanography or the modeling of processes in the upper atmosphere. The map metaphor has played a key role in the development of GIS to date, and the current technology has a strong emphasis on static, two-dimensional views of the world that favors processes occurring at or near the land surface. We hope that in the future GIS will be more successful at handling fully three-dimensional, time-dependent perspectives, but the emphasis in the book is on areas where GIS currently has the most to offer.

The chapters on spatial statistics in Part VI are mathematically and conceptually difficult. Unfortunately the extension of statistical methods to the spatial case—including explicit treatment of spatial dependence—has proved far from straightforward. However, we have included this section not only because we believe that models must be verified and analyzed using statistical methods that reflect the realities of the spatial context, but also because GIS seems to offer a way around the conceptual difficulties of spatial statistics through its emphasis on visualization.

GIS is certainly not a panacea, and its proponents have often been accused of overselling the capabilities of an immature and limited technology. The reader will quickly realize that the chapters of this book are often concerned as much with the limitations of current GIS as with its capabilities. However, we are convinced that some degree of systematic concern for the problems of handling digital spatial data is essential if environmental models are ever to have a significant influence on the development of environmental policy. Few policy-makers will ever be competent modelers or mathematicians, and most will have to rely on computational tools that can present the results of modeling in effective, honest ways. We use the term GIS to refer to such tools, but our purpose in doing so is not to make claims about the capabilities of current technology, but to identify the improvements that need to be made.

Finally, the book covers a lot of ground. The Earth's environment is incredibly complex, and no one can claim to understand the current state of modeling of all of its processes. Yet that is precisely what is needed in the development of effective environmental policy. We see only one possible solution to this fundamental dilemma—the development of computer systems that include well-formulated models based on high-quality data, and allow the user to work with them through understandable user interfaces. Modern computing technology, of which GIS is a part, has made great progress in the past decade at programming systems that allow users to analyze and model complex processes effectively, yet without demanding a complete comprehension of all of the system's components.

The foundation for this book was established at the First International Conference/Workshop on Integrating GIS and Environmental Modeling, which was held in Boulder, Colorado, during the period September 15–18, 1991. The Conference/Workshop was organized by the National Center for Geographic Information and Analysis, US Environmental Protection Agency, US Geological Survey, Department of Energy, and National Oceanic and Atmospheric Administration. Researchers from the spatial data and geoprocessing communities were introduced to advanced environmental simulation models by internationally recognized scientists representing the atmospheric, hydrologic, biologic, ecologic, and related Earth sciences.

The response to the conference was very positive, with more than 600 people in attendance. Participants included representatives from government, university, and the private sector communities. There was an effective mixture of persons from a broad range of environmental modeling specialties, as well as persons involved with researching, developing, or using geoprocessing technologies. The conference presentations were made by the authors in this book. They were selected because of their status as recognized leaders in the integration of geoprocessing and modeling tools, as well as their expertise in the use of scientific models for environmental problem solving.

The generous contributions of the agencies and organizations that provided conference support is appreciated. The conference sponsors included:

U.S. Environmental Protection Agency
U.S. Department of the Interior/Geological Survey
U.S. Department of Energy/Oak Ridge National Laboratory
U.S. Army Corps of Engineers
U.S. Department of the Interior/Fish and Wildlife Service
U.S. Department of Agriculture/Forest Service
U.S. Department of Commerce/NOAA
U.S. Department of the Interior/Minerals Management Service
National Aeronautics and Space Administration
National Institute for Global Environment Change
IBM Corporation
National Center for Geographic Information and Analysis
National Science Foundation

The success of the conference and the realization of this book were made possible only through the very dedicated efforts of the conference Steering Committee. This unique group of individuals worked closely together as a team to design, plan, coordinate, and conduct the conference. The Committee was also instrumental in providing valuable assistance to the editing of these papers. The second conference in the series will take place in Breckenridge, Colorado, September 26–29, 1993. Its program will review progress in integrating GIS and environmental modeling in each of the major areas of concern identified in this book. Practical demonstration of integration efforts will be a major feature of the conference.

Sincere appreciation is extended to the staff of NCGIA—Sandi Glendinning, Judith Parker, and Cassie Taylor—for dedicated support of the conference and this book. The contribution of Sandi, who provided the essential coordination of both projects with unfailing energy and good humor, is particularly appreciated. Special thanks go to Franz Schubert for his timely and efficient preparation of the camera-ready copy for this manuscript. The editors would also like to thank the authors, and particularly the authors of the overview chapters, for their efforts in providing a series of extensive reviews of the state of the field.

The National Center for Geographic Information and Analysis (NCGIA) is supported by the National Science Foundation, through cooperative agreement SES 88-10917 with the University of California, Santa Barbara. Reference to hardware and software products in this volume does not imply endorsement by the authors, editors, or sponsoring organizations. The names of many of these products are registered trade marks.

STEERING COMMITTEE
Mike Crane—USGS, Denver, Colorado
Michael Goodchild—NCGIA, Santa Barbara, California
Dave Hastings—NOAA, Boulder, Colorado
Mason Hewitt—EPA, Las Vegas, Nevada
Carolyn Hunsaker—ORNL, Oak Ridge, Tennessee
Brad Parks—AScI Corp, EPA-LLRS, Grosse Ile, Michigan
Dave Rejeski—EPA, Washington, D.C.
Lou Steyaert—USGS, Reston, Virginia
Denis White—METI, EPA, Corvallis, Oregon

Santa Barbara, Calif. M.F.G.
Grosse Ile, Mich. B.O.P.
Reston, Va. L.T.S.
September, 1992

CONTENTS

THE EDITORS

Michael F. Goodchild
Professor of Geography at the University of California, Santa Barbara, and Director, National Center for Geographic Information and Analysis. He received his BA degree from Cambridge University in Physics in 1965 and his PhD in Geography from McMaster University in 1969. After 19 years at the University of Western Ontario, including three years as Chair, he moved to Santa Barbara in 1988. His research interests center on geographic data models, and the accuracy of geographic databases.

Department of Geography
University of California
Santa Barbara, CA 93106-4060
USA

Louis T. Steyaert
Research physical scientist within the National Mapping Division of the U.S. Geological Survey. The focus of his USGS research is on land surface characterization requirements for land-atmosphere interactions models as part of the USGS global change research program. His previous USGS projects have included research on water and energy exchange modeling and the integration of GIS, remote sensing and environmental modeling technologies. Dr. Steyaert received his PhD in atmospheric science from the University of Missouri-Columbia in 1977. He is a former US Naval Line Officer (1965-1971) and received his AB in mathematics from the University of Missouri in 1965.

US Geological Survey
521 National Center
Reston, VA 22092
USA

Bradley O. Parks
GIS Coordinator for Great Lakes Research at the U.S. Environmental Protection Agency's Large Lakes Research Station, Grosse Ile, Michigan, where he is an environmental scientist with AScI Corporation, a scientific contractor. His research interests span the breadth of GIS applications to natural resources management and environmental problem solving. He has a particular interest in the development of water management tools which integrate GIS, modeling and spatial decision support techniques. Dr. Parks has also worked extensively with spatial analytic applications to problems in forest pest management, particularly Gypsy Moth risk assessment and vegetation resources management in urban and developing environments. He received his PhD and MS in natural resources management from Michigan State University in 1986 and 1983, respectively.

AScI Corporation
US EPA
9311 Groh Road
Grosse Ile, MI 48138
USA

THE AUTHORS

Luc Anselin
Professor of Geography and Economics at the University of California, Santa Barbara, California, and Associate Director of the National Center for Geographic Information and Analysis (NCGIA). His research deals with spatial data analysis and its integration into GIS, applied to regional modeling, on which he has published widely.

Department of Geography
University of California
Santa Barbara, CA 93106
USA

Roni Avissar
Professor of Atmospheric Science at Rutgers University, New Brunswick, New Jersey. His major research interest is in land-atmosphere interactions at various scales, and the parameterization of subgrid-scale processes in atmospheric numerical models. He has published over fifty papers dealing with these issues.

Department of Meteorology and Physical Oceanography
Rutgers University
New Brunswick, NJ 08903
USA

Barry B. Baker
Systems Ecologist for Great Plains Systems Research Unit, USDA-ARS, Fort Collins, Colorado. Current research direction is in the construction and use of simulation models combined with GIS to examine the potential effects of climate change on rangelands. Specific research interests are in plant-herbivore dynamics.

U.S. Department of Agriculture
Agricultural Research Service
Great Plains Systems Research
Fort Collins, CO 80522
USA

William L. Baker
Assistant Professor in the Department of Geography and Recreation, University of Wyoming, Laramie, Wyoming. Research interests in the application of GIS to biogeography, landscape ecology, and spatial modeling.

Department of Geography
University of Wyoming
Laramie, WY 82071
USA

Lawrence E. Band
Associate Professor in the Department of Geography, University of Toronto, Ontario, Canada. Has been active in the area of distributed hydrological and ecological modeling of watershed processes, including the use of digital terrain data and remote sensing to characterize watershed structure and land surface conditions.

Department of Geography
University of Toronto
Toronto, Ont M5S 1A1
Canada

William A. Battaglin
Hydrologist for the U.S. Geological Survey, Water Resources Division. Currently working on studies that use geographic information systems to investigate the fate and transport of agricultural chemicals in the midwestern United States, and the effects of climate change on water resources in the Gunnison River Basin, Colorado.

U.S. Geological Survey
Water Resources Division
Denver Federal Center
Lakewood, CO 80225
USA

Joseph K. Berry
Principal in Berry and Associates and the Director of Training for GIS World, Inc. He has authored over one hundred papers on the analytic capabilities of GIS technology, and presented workshops on GIS to over five thousand professionals. Special Faculty Member of Colorado State University, author of the Academic Map Analysis Package used by universities worldwide, and co-developer of the Professional Map Analyais Package software by Spatial Information Systems Corporation.

Berry & Associates
19 Old Town Square
Fort Collins, CO 80524
USA

Daniel B. Botkin
Professor of Biological Sciences and Environmental Studies at the University of California at Santa Barbara. Research interests related to a GIS environment include ecological modeling, ecosystem dynamics, simulation modeling, vegetation mapping, and biomass estimation. His research is interdisciplinary, involving biological sciences, geography (remote sensing), and statistics.

Department of Biological Sciences
University of California
Santa Barbara, CA 93106
USA

Joan A. Browder
Systems Ecologist at the Southeast Fisheries Science Center, National Marine Fisheries Service, NOAA, Miami, Florida. Her principal research activity is the determination of quantitative relationships between fish and their habitat. In this pursuit, she integrates catch and effort data from fisheries with information obtained from satellite imagery. She presently is assigned to the Oceanic Pelagics Division of the Miami Laboratory, where her work supports the longterm monitoring of such species as bluefin tuna, blue and white marlin, and swordfish.

National Marine Fisheries Service
National Oceanic and Atmospheric Administration
Miami, FL 33149
USA

Ingrid C. Burke
Ecosystem Ecologist with the Department of Forest Sciences and the Natural Resource Ecology Laboratory at Colorado State University. Her research has focused on spatial patterns in soil organic matter dynamics. She is currently directing a program in regional assessment of

ecosystem processes in U.S. grasslands by linking simulation models and GIS.

Department of Forest Science
Colorado State University
Fort Collins, CO 80522
USA

Michael P. Crane
Director of the Central Region Geographic Information Systems Laboratory of the U.S. Geological Survey in Denver, Colorado. This interdivisional facility makes available to Survey scientists a variety of spatial/analytic tools to help accomplish the Earth science mission of the agency.

Central Region GIS Lab
U.S. Geological Survey Denver Federal Center
Denver, CO 80225
USA

Noel A. Cressie
Professor of Statistics at Iowa State University, Ames, Iowa. His research interests are in statistics for spatial data and its application to the environmental sciences, in empirical Bayes methods, and in goodness of fit.

Department of Statistics
Iowa State University
Ames, IA 50011
USA

Jack Dangermond
President of Environmental Systems Research Institute, Redlands, California. He has over twenty years experience in developing geographic information systems for multi-agency data management. Recent projects include two studies for the U.S. Air Force: the design, development, and application of geographic information systems to support geotechnical/siting evaluation; and the environmental assessment of the Small ICBM Program.

Environmental Systems Research Institute
580 New York St
Redlands, CA 92373
USA

Robin L. Dennis
Physicist by training, but has spent most of his research career in air quality modeling. A key interest is environmental problem solving while making use of the best information available. At the U.S. Environmental Protection Agency he is involved in regional modeling and is in charge of the evaluation of the Regional Acid Deposition Model, RADM, and its application to assessments and environmental problem solving. He is a co-leader of EPA's HPCC Program.

Atmospheric Sciences Modeling Division
Air Resources Laboratory
National Oceanic and Atmospheric Administration
Research Triangle Park, NC 27711
USA

John A. Eddy
Vice-President of Research and Chief Scientist for the Consortium for International Earth Science Information Network (CIESIN). His research interests include Earth systems science, solar physics, and the history of science.

CIESIN
2250 Pierce Road
University Center, MI 48710
USA

Bernard A. Engel
Associate Professor in the Purdue University Agricultural Engineering Department. His research and teaching interests include systems engineering, decision support systems, expert systems, soil and water conservation engineering, watershed modeling, GIS applications for environmental and natural resources protection, integration of GIS and environmental modeling capabilities, and multimedia.

Department of Agricultural Engineering
Purdue University
West Lafayette, IN 47907-1146
USA

Evan J. Englund
Working on geostatistical methods for site characterization and mediation, with emphasis on contaminated soils. He is currently doing research and development on geostatistical simulation as a tool for integrated optimization of remediation plans, that is, finding the most cost effective combination of sampling network, sampling and analytical methods, QA/QC, interpolation method, and decision criteria for a specified remediation objective. Responsible for developing the Geo-EAS software package.

U.S. Environmental Protection Agency
Environmental Monitoring Systems Laboratory
Las Vegas, NV 89193-3478
USA

Kurt Fedra
An expert in environmental computing applications, and as project leader of the Advanced Computer Applications Project team at the International Institute for Applied Systems Analysis (IIASA), responsible for the design and development of information and decision support systems in the area of environmental management, development planning and risk analysis for international governmental and industrial clients. Has authored more than 80 articles and reports, and has contributed to several books on environmental systems analysis and related computer applications.

Advanced Computer Applications
International Institute for Applied Systems Analysis
A-2361 Laxenburg
Austria

Thomas R. Fisher
Senior Staff Scientist in the Austin, Texas offices of Radian Corporation, a major environmental engineering and technical services firm. He is responsible for Geographic Information Systems and Services and a specialist in 3D modeling and visualization for environmental applications.

Radian Corporation
Geographic Information Systems and Services
Computer Applications Department
Austin, TX 78720-1088
USA

Xiaogang Gao
A PhD student in the Department of Hydrology and Water Resources at the University of Arizona. He is now working on the large scale landsurface hydrology parameterization in a General Circulation Model (GCM), especially using GIS techniques to improve the spatial variability representation in the GCM subgrid scale.

Department of Hydrology and Water Resources
University of Arizona
Tucson, AZ 85621
USA

David C. Goodrich
Research Hydraulic Engineer with the USDA-Agricultural Research Service, Southwest Weatershed Research Center in Tucson, Arizona, and Adjunct Assistant Professor in the Department of Hydrology and Water Resources at the University of Arizona. Current research interests include GIS and hydrologic model integration, hydrologic response as a function of scale as well as model representation of hydrologic parameter variability.

Agricultural Research Service
U.S. Department of Agriculture
Tucson, AZ 85719
USA

Sumant Gupta
Over 35 years of surface and subsurface hydrology research and technology development experience which includes planning, regulatory issue resolution and strategy development, project manager, and three-dimensional finite-element code development. Since 1987, his work has focused on the characterization, and remedial investigations/feasibility study of several large, highly contaminated Superfund sites.

CH2M Hill
Santa Ana, CA 92705
USA

Jon D. Hanson
An Ecologist with USDS-Agricultural Research Service in Ft. Collins, Colorado. His work involves the development and testing of theories relating to the effect of environment and management on grassland ecosystems. Interest in spatial processes has led to Hanson's interest in geographic information systems.

Agricultural Research Service
Great Plains Systems Research
U.S. Department of Agriculture
Fort Collins, CO 80522
USA

Jonathan Harris
Currently manages CH2M Hill's Groundwater Resources Group in Southern California. His experience includes application of a variety of numerical methods to hydrogeologic and geologic problems that typically involve large spatially-dependent databases. Has been Site Manager of the San Gabriel Superfund site since 1989.

CH2M Hill
Santa Ana, CA 92705
USA

Lauren E. Hay
A Hydrologist for the U.S. Geological Survey, Water Resources Division. She is presently working on studies that examine the effects of climate change on water resources in mountainous regions.

U.S. Geological Survey
Water Resources Division
Denver Federal Center
Lakewood, CO 80225
USA

Ann Henderson-Sellers
Director of the Climatic Impacts Centre and Professor of Physical Geography in the School of Earth Sciences at Macquarie University. Trained as a mathematician with a PhD in meteorology. In 1988 the Royal Geographical Society presented her with the Gill Award for contributions to climatology and she is the Australian Meteorological and Oceanographic Society's 1990 R.H. Clarke Lecturer.

Climatic Impacts Centre
School of Earth Sciences
Macquarie University North Ryde, NSW 2109
Australia

Mason J. Hewitt III
Manager and Chief Scientist for the U.S. Environmental Projection Agency GIS Research and Development Program. He leads an interdisciplinary team in developing GIS applications tailored to the EPA's environmental protection mission.

U.S. Environmental Protection Agency
Las Vegas, NV 89193-3478
USA

Carolyn T. Hunsaker
Staff Ecologist in the Environmental Sciences Division of Oak Ridge National Laboratory in Oak Ridge, Tennessee. Her research interests include landscape ecology and regional ecology risk assessment, especially the analysis of water quality and aquatic ecology data for large spatial and temporal scales.

Environmental Sciences Division
Oak Ridge National Laboratory
Oak Ridge, TN 37831-6038
USA

Michael Hutchinson
Currently a Senior Fellow at the Centre for Resource and Enviornmental Studies of the Austrlaian National University. His main research interests are in the development and application of techniques for spatial and temporal analysis and modeling of climate and topography, particularly at continental scale.

Centre for Resource and Environmental Studies
Australian National University
Canberra, ACT 2601
Australia

Henriette I. Jager
Research Associate in the Environmental Sciences Division of Oak Ridge National Laboratory. She and Dr. Scott Overton have been interested in the role of spatial processes in environmental survey design and analysis since cokriging stream ANC with elevation for the USEPA National Stream Survey in 1988.

Oak Ridge National Laboratory
Oak Ridge, TN 37831-6036
USA

Susan K. Jenson
Physical scientist with the National Mapping Division of the U.S. Geological Survey at the EROS Data Center in Sioux Falls, South Dakota where her primary areas of research have been image classification, surface generation, and applications of digital elevation models.

U.S. Geological Survey
EROS Data Center
Sioux Falls, SD 57198
USA

Carol A. Johnson
Directs the Natural Resources Geographic Information System Laboratory at the University of Minnesota, Duluth, where she uses GIS in her research on the influence of beavers and wetlands on landscape ecology. She serves on the Minnesota Governor's Council on Geographic Information.

Natural Resources Research Institute
University of Minnesota
Duluth, MN 55811
USA

Norman L. Jones
Associate Professor of Civil Engineering at Brigham Young University in Provo, Utah. His research interests include the application of computational geometry and scientific visualization techniques to terrain and subsurface modeling.

Department of Civil Engineering
Brigham Young University

Provo, UT 84602
USA

Karen K. Kemp
Previously Coordinator of Education Programs for the NCGIA, and now a member of the academic staff of the Technical University of Vienna. Her recently accepted PhD dissertation was entitled "Environmental Modeling and GIS: Strategies for Dealing with Spatial Continuity."

Abteilung Landesvermessung
Technische Universitaet Wien
Gusshausstrasse 27-29/127/1
A-1040 Wien
Austria

John J. Kineman
Principal Investigator for the Global Ecosystems Database Project, a five-year inter-agency project sponsored by the U.S. EPA and NOAA, and was formerly the lead scientist for the IGBP-DIS Global Change Diskette Project for Africa. He is currently employed as a Physical Scientist and Data Manager at the NOAA National Geophysical Data Center and World Data Center-A in Boulder, Colorado. He has published on the design and application of GIS and spatial databases for studies of global change.

John L. King
Professor of Information and Computer Science, University of California, Irvine, and widely known for his work on the impact of electronic data processing on organizations. His publications include *The Dynamics of Computing*, with Kenneth L. Kraemer, published by Columbia University Press in 1985.

Department of Information and Computer Science
Graduate School of Management
University of California
Irvine, CA 92717
USA

Tim G.F. Kittel
Assistant Project Scientist for the Climate System Modeling Program, University Corporation for Atmospheric Research, and Research Associate for the Natural Resource Ecology Laboratory, Colorato State University at Fort Collins, Colorado. His research interests include global change (biosphere-climate interactions), climatology, and vegetation geography.

Climate System Modeling Program
University Corporation for Atmospheric Research
Boulder, CO 80307-3000
USA

Kenneth L. Kraemer
Director of the Center for Research on Information Technology and Organizations, and professor in the Graduate School of Management and the Department of Information and Computer Science at the University of California, Irvine. His research interests include: the management of information systems in organizations; the organizational, social, and public policy implications of computing; and the strategic and political uses of computing.

Department of Information and Computer Science
Graduate School of Management
University of California
Irvine, CA 92717
USA

David C.L. Lam
Chief, Environmental Synthesis and Prediction, at the National Water Research Institute, Environment Canada, Burlington, Ontario. He is also a professor (part-time) at the Civil Engineering Department, McMaster University, Hamilton, Canada. His research interests are in expert systems, GIS, and modeling.

Canada Center for Inland Waters
National Water Research Institute
Burlington, Ont L7R 4A6
Canada

George H. Leavesley
Employed since 1973 by the Water Resources Division of the U.S. Geological Survey. He is presently Project Chief of the Precipitation-Runoff Moedling Project developing hydrologic models to simulate the effects of climate and land-use change.

U.S. Geological Survey
Water Resources Division
Denver Federal Center
Lakewood, CO 80225
USA

Tsengdar J. Lee
Current a Research Associate at Colorado State University. He has been working on atmospheric circulations induced by different landscape and landuse surface characteristics. A soil-vegetation-atmosphere transport

model has been developed for his research work. The impact of man's agricultural practice on local weather and climate is one of his major research interests.

Department of Atmospheric Science
Colorado State University
Fort Collins, CO 80523
USA

Thomas R. Loveland
Manager of Land Sciences at the U.S. Geological Survey, National Mapping Division's EROS Data Center in Sioux Falls, South Dakota. He manages EDC's land science research program, and is responsible for research in the use of remote sensing and GIS technology for land cover characterization. Mr. Loveland has been engaged in applications research programs at EDC for over 12 years. In addition, he has held GIS and remote sensing applications and management positions with state governments in South Dakota and Arizona.

U.S. Geological Survey
EROS Data Center
Sioux Falls, SD 57198
USA

Dennis J. Lytle
National Leader for Soil Geography and Information Systems at the National Soil Survey Center. he has US federal leadership responsibilities for soils digital spatial data. He has held soil scientist positions in Maine, Wyoming, and California during his 17 year career with the Soil Conservation Service.

US Department of Agriculture
National Soil Survey Center
Lincoln, NE 68508
USA

David R. Maidment
Professor of Civil Engineering at the University of Texas at Austin, and editor of Journal of Hydrology, and Editor in Chief of the Handbook of Hydrology.

Department of Civil Engineering
University of Texas
Austin, TX 78712
USA

Ian Moore
Adjunct Professor of Geography in the Earth Sciences Department of Montana State University, Missoula, and holder of the Jack Beale Chair of Water Resources at the Centre for Resource and Environmental Studies, the Australian National University in Canberra.

Center for Resource and Environmental Studies
Australian National University
GPO Box 4, Canberra
ACT 2601, Australia

Donald E. Myers
Professor of Mathematics at the University of Arizona. His research specialty is multivariate spatial statistics and applications in hydrology, mining engineering, soil sciences, and environmental monitoring and assessment.

Department of Mathematics
University of Arizona
Tucson, AZ 85721
USA

Ramakrishna Nemani
Research Assistant Professor at the University of Montana. His research interests include ecosystem modeling, climatology, and remote sensing. His current research activity is in the integration of ecosystem models with remote sensing and geographic information systems for regional and global ecosystem analysis.

School of Forestry
University of Montana
Missoula, MT 59801
USA

Robert A. Nisbet
Acting Professor in the Department of Biological Sciences at University of California, Santa Barbara. Currently he is working on the linkage of a geographic information system to a forest growth model to analyze effects of global warming on regional forest growth.

Department of Biological Sciences
University of California
Santa Barbara, CA 93106
USA

Joan H Novak
Chief of the Modeling Systems Analysis Branch of the Environmental Protection Agency's Atmospheric Research and Exposure Assessment Laboratory in Research Triangle Park, North Carolina. She also serves as the EPA's representative on the High Performance Computing and Communications Information Technology (HPCCIT) subcommittee of the Federal Coordinating Council for

Science, Engineering, and Technology.

Atmospheric Sciences Modeling Division
Air Resources Laboratory
National Oceanic and Atmospheric Administration
Research Triangle Park, NC 27711
USA

Timothy L. Nyerges
Associate Professor of Geography at the University of Washington in Seattle. His research interests include spatial problem solving with GIS and spatial decision support systems, particularly concerning transportation planning and environmental management applications. Special focus is on linking transportation and environmental models to GIS architectures.

Associate Professor of Geography
Department of Geography, DP-10
University of Washington
Seattle, WA 98195
USA

Donald O. Ohlen
Senior Scientist with Hughes STX Corporation at the EROS Data Center in Sioux Falls, South Dakota. His research interests include image analysis and database development for spatial environmental analysis of global land processes.

Hughes STX Corporation
EROS Data Center
Sioux Falls, SD 57198
USA

W. Scott Overton
Professor of Statistics at Oregon State University in Corvallis, Oregon. His areas of specialty are sampling theory, statistic ecology, and systems ecology. He is currently investigating several aspects of sampling design and parameter estimation in spatial environmental surveys.

Department of Statistics
Oregon State University
Corvallis, OR 97331
USA

Randolph S. Parker
Project Chief for the US Geological Survey's Water Resources Division, on their precipitation-runoff modeling project developing hydrologic models to simulate the effects of climate and land-use change.

US Geological Survey
Water Resources Division
MS 412, Box 25046
Denver Federal Center
Lakewood, CO 80225
USA

David A. Parrish
Geographic Information Systems Coordinator with the US Environmental Protection Agency, working with water quality monitoring, emergency response, and information resources management.

US Environmental Protection Agency
Region 6, Environmental Services Division
1445 Ross, Suite 1200
Dallas, TX 75202-2733
USA

Emma J. Pearson
Researcher with Perot, Inc. in the UK. Has extensive experience in applying GIS to environmental science problems.

NERC Unit for Thematic Information Systems
Department of Geography
University of Reading Whiteknights, PO Box 227
Reading RG6 2AS
UK

Roger A. Pielke
Professor at the Department of Atmospheric Sciences, Colorado State University, Fort Collins, Colorado. Formerly employed with NOAAs Experimental Meteorology Lab, and the University of Virginia.

Department of Atmospheric Sciences
Colorado State University
Fort Collins, CO 80523
USA

A. J. Pitman
Lecturer in Atmospheric Science at the School of Earth Sciences, Macquarie University. He has his doctorate from Liverpool University's Department of Geography.

Climatic Impacts Centre
School of Earth Sciences
Macquarie University
North Ryde, NSW 2109
Australia

David Rejeski
Program Systems Division, PM-218B
US Environmental Protection Agency
401 "M" St., S.W.
Washington, DC 20460
USA

Chris Rewerts
Doctoral candidate in the Agricultural Engineering Department at Purdue University. His research focuses on integration of watershed simulation and geographic information systems for nonpoint source pollution management and assessment.

Department of Agricultural Engineering
Purdue University
West Lafayette IN 47907-1146
USA

Hannah R. Rhodes
Doctorate in Applied Mathematics at the University of Arizona, with her dissertation on multivariate and geostatistical analysis of Phase I Eastern Lake Survey Data.

Cray Research Inc.
655-F Lone Oak Dr.
Egon, MN 55121
USA

David R. Richards
Chief of the Estuarine Simulation Branch of the Hydraulics Laboratory at the US Army Corps Engineer Waterways Experiment Station in Vicksburg, Mississippi. He has been active in the hydraulic modeling field for more than ten years, and is responsible for the maintenance and development of the TABS numerical modeling system at WES.

Hydraulics Laboratory WESHE-S
Army Corps of Engineers
Waterways Experiment Station
Vicksburg, MS 39180
USA

Steven Running
Professor in Forest Ecology at the University of Montana. His research interests include simulation of carbon, water and nitrogen budgets of coniferous forests, and mountain microclimatology. He is also currently working as a member of the NASA EOS/MODIS team on the integration of simulation models with satellite data for estimating carbon and water fluxes at global scales.

School of Forestry
University of Montana
Missoula, MT 59812
USA

David S. Schimel
Project Scientist for the Climate Systems Modeling at the University Corporation for Atmospheric Research (UCAR) at Boulder, Colorado.

Climate System Modeling Program
University Corporation for Atmospheric Research
Box 3000
Boulder CO 80307-3000
USA

Piers J. Sellers
Goddard Space Flight Center
National Aeronautical and Space Administration
Greenbelt, MD 20771

W. Chris Skelly
Doctoral candidate in the School of Earth Sciences, Macquarie University. His research focuses on spatial evaluation of general circulation model results.

Climatic Impacts Centre
School of Earth Sciences
Macquairie University
North Ryde, NSW 2109
Australia

Soroosh Sorooshian
Professor and Head of the Department of Hydrology and Water Resources at the University of Arizona, Tucson. He is also Chief Editor of the Water Resources Research Journal, published by the American Geophysical Union. His research interests are in surface hydrology, and applications for remote sensing in hydrology and climate studies.

Department of Hydrology and Water Resources
University of Arizona
Tucson, AZ 85621
USA

Raghavan Srinivasan
Agricultural Engineer and Associate Research Scientist with Texas Agricultural Experiment Station, Temple, TX. He is currently working in integrating multimedia tech-

niques with geographic information systems and integrating GIS with basic scale environmental simulation models (SWRRB/ROTO). He has worked to develop techniques for the assessment of environmental impact using nonpoint source pollution models such as integration of the distributed parameter models (Agricultural NonPoint Source) with GIS.

Department of Agricultural Engineering
Purdue University
West Lafayette, IN 47907-1146
USA

Monica G. Turner
Research scientist in the Environmental Sciences Division of Oak Ridge National Laboratory in Tennessee. Her research interest include the effects of spatial heterogeneity on ecological processes at broad scales and integrates theory and empirical studies. Currently, her landscape studies include the spatial spread of disturbance, plant re- establishment following large-scale fire, the movement and foraging dynamics of large ungulates, and the interactions between socioeconomic and ecological factors in creating landscape mosaics.

Environmental Sciences Division
Oak Ridge National Laboratory
PO Box 2008
Oak Ridge, TN 37831-6038
USA

A. Keith Turner
Long-standing interest in computer applications to geological and environmental studies. His research in Europe led to the 1989 Directorship for a NATO Advanced Research Workshop on 3-dimensional modeling using geoscientific information systems. He is currently applying these methods to Southern Nevada studies for the US Geological Survey.

Department of Geology and Geological Engineering
Colorado School of Mines
Golden, CO 80401-1887
USA

Jay M. Ver Hoef
Biometrician for the Alaska Department of Fish and Game. He is also Affiliate Assistant Professor at the University of Alaska, Fairbanks. His research interests include spatial statistics and ecology.

Wildlife Conservation Division
Alaska Department of Fish and Game
1300 College Rd.
Fairbanks, AK 99701
USA

Margrit von Braun
Assistant Professor in the Chemical Engineering Department at the University of Idaho, and Principal and Vice-President of Terra Graphics Environmental Engineering, Inc. She teaches courses in hazardous waste management and risk assessment. Projects have included the reconstruction of ground water contaminant exposures to a population, development of environmental remedial strategies at an abandoned lead smelter and numerous hazardous waste site characterizations.

Department of Chemical Engineering
Buchanan Engineering Lab 308
University of Idaho
Moscow, ID 83843
USA

G. Wadge
Geologist with the University of Reading Department of Geography. Research experience in volcanology and the geology of the Caribbean who now applied computer-based techniques to geological problems. Current research includes Expert-GIS, natural hazard simulation and geological map creation from imaging spectrometry and radar data.

NERC Unit for Thematic Information Systems
Department of Geography
University of Reading
Whiteknights, PO Box 227
Reading RG6 2AS
UK

John F. Weaver
Research meteorologist with the National Environmental Satellite Data and Information Service (NOAA). His research interests include remotely-sensed climatological studies and forecast techniques for tornadoes and severe hailstorms.

Cooperative Institute for Research in the Atmosphere
Colorado State University
Fort Collins, CO 80523
USA

Denis White
Research geographer with Mantech Environmental Technology, Inc. at the US. EPA Environmental Research

Laboratory, Corvallis, Oregon. His research interests are spatial analysis, cartography and regionalization applied to environmental science. Formerly he was at the Laboratory fore Computer Graphics and Spatial Analysis at the Harvard University Graduate School of Design.

Environmental Protection Agency
Environmental Resources Laboratory
1600 S.W. Western Blvd.
Corvallis, OR 97337
USA

John P. Wilson
Associate Professor of Geography and Director of the Geographic Information and Analysis Center of Montana State University. He teaches classes in physical geography, quantitative methods, GIS, and environmental modeling. His recent research emphasizes the measurement and modeling of surface hydrology, soil erosions, and waterquality.

Department of Earth Sciences
Montana State University
Bozeman, MT 58717-0348
USA

Anton P. Wislocki
Mining engineer and mine manager with an interest in hazard mapping and risk assessment. He is currently developing risk maps from engineering geology databases for use in development planning within an urban environment.

NERC Unit for Thematic Information Systems
Department of Geography
University of Reading
Whiteknights, PO Box 227
Reading RG6 2AB
USA

Greg Woodside
Groundwater hydrologist for CH2M HILL. His interests include ground water modeling, and data management and evaluation using geographic information systems. He has worked on numerous large-scale modeling investigations of ground water flow and contaminant transport using GIS.

CH2M Hill
2510 Redhill Ave.
Santa Ana, CA 92705
USA

Yongkang Xue
Research scientist for the Department of Meteorology at the University of Maryland. His research interests focus on land surface modeling, land surface-atmosphere interaction, and remote sensing.

Center for Ocean-Land-Atmosphere Interactions
2213 Computer and Space Science Building
Department of Meteorology
University of Maryland
College Park, MD 20742-2425
USA

I

Six Introductory Perspectives

1

Environmental Research: What We Must Do

JOHN A. EDDY

Some years ago I met with a number of astronauts who were preparing to man the SpaceLab module on one of the U.S. space shuttle flights. In the group was a young German astronaut, named Ulf Merbold. Later, after their successful mission, he had this to say, of what he saw, when he first looked down on the Earth from the vantage point of space:

> For the first time in my life I saw the horizon as a curved line. It was accentuated by a thin seam of dark blue light—our atmosphere. Obviously, this was not the ocean of air I had been told it was so many times. I was terrified by its fragile appearance.

Science in recent years has focused more and more on the Earth as a planet. And one that for all we know is unique: where a thin blanket of air, a thinner film of water, and the thinnest veneer of soil combine to support a web of life of wondrous diversity and continuous change.

What is more, living and nonliving parts of the system are by nature entwined and interconnected: "When one pulls up any part of it," wrote the naturalist John Muir, "he finds its roots entangled with all the rest."

The mystic poet Francis Thompson may have said it better in 1893:

> All things by immortal power near or far hiddenly to each other linked are. That thou canst not stir a flower without troubling of a star.

No one doubts that this is the way the environment works, least of all any of the readers of this book.

This book tries to join Geographic Information Systems and Environmental Modeling—a connection that offers the hope of predicting environmental changes at local, regional, and global scales. These predictions are needed to allow sensible responses to impending changes in climate and other aspects of the environment. They include greenhouse warming, land use, and a number of other concerns that today come under the general heading of "global change."

How do the changes of today compare with others of the past? Need we be concerned?

GLOBAL CHANGES

What are the most drastic environmental changes that the Earth has known in its long history? For the long run, my own list would include:

- The acquisition of an atmosphere and oceans of water.
- The birth and spread of life.
- The repeated annihilations of much of it, through repeated extinctions of species, as at the Cretaceous-Tertiary boundary 65 million years ago.
- The drifting and gnashing of the continents that give us our present geography and all the action in geology today.

For the last one or two million years, there would be only two on my list:

- The recurrent ice ages that signalled a drastic change in tempo of the Pleistocene epoch.
- The rise of man, whose whole history has been played out in the shadow of the ice ages.

From all of these life on Earth has recovered, apparently through interactive processes that seem to provide resilience to drastic change.

What about the past 100 years? What is the most drastic global change of this period? Although the Earth has warmed in this time by about 0.6C, that global warming cannot qualify as a drastic change. Nor can wars. It is surely instead the complete ascendancy of man on the planet, now numbering about 5 billion souls, or about 34 people per square kilometer of the land surface of the Earth.

We get an erroneous impression of the twentieth century crowding in this country, where in a very large country the population density—about 25 people per square kilometer—is less than the world average.

- In China, where one quarter of the people live, it is now 108.
- In the U.K. and India, where one seventh of the Earth's population live, there are over 200 people per square kilometer.
- In Japan and the Netherlands, over 300.
- In Taiwan, 513, and in Hong Kong, 5000.

The marks of this many people are now indelibly written on the face of the planet: marks that were first subtle, and for millennia insignificant in any global view. That is no longer the case, and never will be again.

We see them most easily at night, as in the montage of modern dark-Earth views from space assembled several years ago by Woody Sullivan, already out of date. The curved arc of the aurora borealis seen there, and caught by chance above Norway, is natural, a reminder of how the world was but a century ago. The world at night is now so bright that cities on the Northeast coast of the U.S. are indistinguishable in the growing blur of megapolitan BosWash. Puerto Rico is an island of light, as is the whole south of India and Southeast Asia, where so many now and will reside.

- About 2% of the Earth is now urban.
- More than 10% of the land surface is now under cultivation.
- More than 30% is today under active management for the purposes of mankind.

More pervasive, and more troubling in 1991, are the changes we have wrought in but the last seconds of geologic time in the essential chemistry of the planet. These changes are best seen in records, now about 30 years long, of the global abundance of atmospheric carbon dioxide and methane, and other important trace gases, and the extension of these into the past as recovered from polar ice cores.

As every reader will know, these gases are all rising at an accelerating rate of roughly 1% per year, which is extremely fast. Carbon dioxide, one of the slower ones, has increased more than 10%, globally, in the time since any of you was born! Methane is rising today at almost 2% per year. You also recognize the seasonal, high-frequency modulation in the ascending CO_2 record that reflects the seasonal cycles of growth and decay of plants—largely trees—in the northern hemisphere. Trees are the lungs of the planet, and we see them breathing out and breathing in, reducing the atmosphere's load of carbon dioxide in the summer of their growth, and giving it back in the autumn and winter as products of the decay of leaves and vegetable matter.

Both carbon dioxide and methane are long-lived constituents, and because of that they are well mixed over the globe. The amount of carbon dioxide in the air in Boulder is the same that one would measure in the air of Nairobi or Nagasaki or Buenos Aires, or in Hawaii where these measurements were made. The graph tells a global story, and a permanent one. The carbon dioxide that I put into the atmosphere today will remain there until the end of the next century, long after any of us is here. Half of all the carbon dioxide released through burning of fossil fuels since the start of the Industrial Revolution—before Charles Dickens was born—is still there today.

Inadvertently, we have initiated a global experiment whose ultimate outcome is at present unknown. And in doing so we may have driven Ulf Merbold's fragile seam of dark blue light—in which is all of life—beyond the range of natural repair, at a time when we must prepare the planet for a doubling of the world's population to about 10 billion people, in but a few decades. Almost all of the growth will take place in the less developed world, where the local and regional environment is already the most stressed.

It is this, coupled with concerns of loss of species diversity, degradation of soils, the loss of tropical forests, and the burgeoning burden of modern waste, that prompted *Time* magazine a few years ago to declare the endangered Earth the planet of the year. The Earth is changing, at a rate faster than it has ever known: in terms of climate, in terms of chemistry, in terms of the number of species of animals and of plants, and in terms of land use, which is probably the most dramatic change; and at a time when a change in any part of the environment affects all the rest.

There is something else that is new, and possibly troubling, about the current predictions of future climate change, now defined by the broadest possible international consensus of scientists through the Intergovernmental Panel on Climate Change: It is the first time in human history that we as individuals or a society have been allowed to look clearly at the face of the future. Never before have we known, tens of years in advance, what awaits us. There are no precedents as to what we should do, or how, as peoples of the world, we should do it.

SIX IMPERATIVES

I think it clear that the Earth faces an environmental crisis. And as in the case of a city under siege, there is a need to organize and mobilize, if not to put out fires, at least to understand the nature and extent of environmental threat. What is it now that we must do? I have a personal list of six actions that seem to me so essential as to constitute a set of imperatives for science:

1. *We need to put the Earth in intensive care* to monitor the vital signs of the planet.

We must put in place systems to detect global changes in what are, to our best knowledge, the most vital signs of the planet. We must continue to track them, indefinitely, into the future. This we owe to future generations, and it is a task neither started nor yet fully defined. What we now know best of long-term global changes concerns trace gas abundances and, to a lower level of certainty, surface temperatures. The scattered readings of other parameters now in hand are not enough to diagnose a system as complex and interconnected as the Earth.

The analogy with medicine is a good one: We are physicians in a hospital emergency room. A patient, our Earth, comes in with signs of illness. It may be serious; it may be hypochondria. What is it we should monitor, besides temperature? We cannot measure everything all of the time. What would you measure, if you could only choose five, or ten? In defining which Earth parameters are to be monitored, we should use Francis Bretherton's 20-year test: What is it that in 20 years time we shall most want to know of what has changed?

Artificial satellites, which will carry most of the load of this endeavor, have been around for 34 years: How sad it is that this task of monitoring the planet was not started earlier. But we can start it now, and some of it will be with the Earth Observing System that is now in the President's budget and in NASA's long-term strategy. That is why that element of the U.S. Global Change Research budget is so dominated by this necessary space investment.

2. *We need to begin a crash effort to recover the past history* of climate and other significant environmental changes, including sea-level and vegetation changes.

There is no other way to put present change in perspective, and we have at present only the sketchiest histories with which to compare. What is the history of drastic land-use change through the last 2000 years, and what were the impacts? How irreparable are the damages we are doing today? Was there, or was there not, a Little Ice Age, and what was it like on the regional level here, or in China, or Brazil? What happened the last time that surface temperatures reached the levels anticipated for the next 50 years? How did the natural Earth respond to CO_2 or CH_4 levels as high, or almost as high, as levels of today? Which came first at the time of the last glacial-to-interglacial transitions, changes in atmospheric chemistry, or changes in surface temperature? On the answer to that question ride policy decisions involving trillions of dollars.

In recovering the past, we must learn to read Nature's secret diaries, in trees and sediments and rocks, and we need to be prepared for surprises, for human experience is a very poor indicator of climate change, or any other environmental shift. A single life is too short to detect very much at all, least of all slow and surreptitious changes, however drastic, and by benevolent design we forgive and forget and adjust to change.

The greatest change that the Earth has known since the time of man has been the coming and going of the major ice ages, yet these were not known or suspected until scarcely more than 100 years ago when the Swiss geologist, Louis Agassiz, found evidence of them in the placement of rocks at the base of the Aare glacier in Switzerland. Cro Magnon man was there when the ice came down, and when it retreated, but they remained a secret for more than 10,000 years after they were gone. The record of past environmental changes give us our only clues to how the Earth works as a system, and the only fodder for validating climate models or other environmental models.

3. *We need to develop an Earth system science,* to extend what we now know in classical disciplines, of elements of the whole.

For 100 years we have been specializing in science, from what was in the nineteenth century "natural philosophy" to today's list of specialties within specialities. This process of divide and conquer has greatly expanded what we know of the Earth, but it has come at the expense of knowing enough about the system itself.

Because of these prejudices, we have been surprised when we have learned, for example, that the recovered curves of surface temperature and of the abundance of atmospheric greenhouse gases, drawn from Arctic and Antarctic ice, have followed the same ups and downs through at least the past 160,000 years of Earth history; that there is a "spring bloom" each year of surface organisms in the North Atlantic, like the burst of spring flowers in the Rockies, that is somehow tied to ocean circulation and the flow of nutrients; or that carbon and phosphorus and nitrogen recycle themselves through various parts of the system as though to ignore the academic divisions of water or soil or biota or air, with the same atoms abiding awhile in living forms, then in nonliving forms, to return again in a cyclic flow that allows and even regulates life on the planet. None of these phenomena belongs to atmospheric science, or biology, or hydrology, or geology, but to all of them.

This need for general practitioners must not take from the need for specialists, or for continued strong disciplinary studies, but must instead supplement them. The need has been particularly driven in the past few years by the development of coupled Earth system models, or coupled environmental models that operate at the local and regional scales. A broader understanding of how the Earth works has come from repeated failures in matching, for example, climate *data* with climate *models* that failed to include the effects of leaves or ice or topography. It has been intensified by the demands put on science by policy-makers who ask very practical questions and who are not satisfied by our customary, qualified answers, or our requests that they wait a while longer for answers.

These very practical and appropriate questions of what is going to happen, regionally or globally, and what can be done to mitigate or adapt to it have put science in an extremely awkward position. It has been described by Wally Broecker as how you or I would feel were we the new manager of, say, a large and complex oil refinery where things are going wrong, or thought to be awry. Into our large office with its large mahogany desk comes a stream of complaints and worries, one after the other: We are told that what used to come out clear and cold in vat #4 is now brackish and hot; fires keep breaking out in Building 9; the lights keep going out in the storage area; and so on. "Calm down," we would say; "just bring me the *plans* for the refinery and we shall trace the flows and deduce the problem and find out what to do." Those around us pause and shake their heads: "But... there *are* no plans. We *don't know how it works.* We have electricians, and pipefitters, and carpenters, who know about parts of it, but how it fits together no one knows." We need to reconstruct, from observations, process studies, and modeling efforts, that missing set of plans for how the Earth works.

4. *We need to begin now to develop Earth system models* built upon models of smaller scale that incorporate significant elements of the system in realistic ways.

There are today elaborate atmospheric global circulation models and ocean models and forest succession models and economic models and a host of others. But the aim here is to couple them to describe environmental processes on local, regional, or global scales. What might be called "toy models" of this sort have already been devised, and are being improved upon. The goal should be operable models capable of test within 10 years and of models capable of reliable prediction within 20 years, by 2011. If this seems like a long time away, think what a short a time ago was 1981 or 1971. To me they seem like yesterday.

Analytical models can guide thinking and provide ways to test hypotheses, and we can hope, if the system is indeed predictable, they can be used for pragmatic forecasts of how the system will respond to anticipated changes in forcing functions, and even predictions of what these forcing functions will be. A particular need is for models that take anticipated global conditions to describe regional impacts.

The fact that so wide a variety of models are discussed in a single conference reveals one of the roles that modeling plays in the science of today. Modeling has become the common language of environmental science, one that is spoken in the same words by hydrologists or biologists and by natural scientists and social scientists or scientists and engineers. Computers and modeling are sledgehammers and headache balls that have broken down most of the walls that in the past separated the disciplines and, I think, inhibited their progress. And as policy-makers become in the future more computer and model oriented, I suspect it will make it easier for science, as it must, to communicate better with them.

5. *We must put in place a global data and information system* that makes environmental data, past and current, available to all who need it, in a form that they can use.

I do not think we can even imagine what it would do for environmental studies—locally, regionally, or globally—were scientific information on everything available easily and instantly to all who need it. The data system that is needed must make biological data available to meteorologists, hydrologists, sociologists, or whomever, and in any country. And it needs to be in a form and format to be standardized and user friendly.

We can hope that the political developments that have shaken the world in the past 3 years will in time wash away some of the barriers that have inhibited the exchange of many types of scientific data in the past. It was a Russian writer who many years ago said:

> There is no such thing as national science. Just as there is no national multiplication table.

Given the modern need to address environmental problems that are global in extent, maybe the day will come when this is true, as it is true, for example, today in medicine. The dream of a breakthrough environmental data system seems possible because of advances in computers and even more propitiously, the advent of geographic information systems. Both have reached the stage where we can rightly wonder how we ever got along without them.

In providing a standardization, geographic information systems provide another big benefit to data systems: Data of different kinds can exist in a wide network of storage nodes. It is a frightening thought to think of a single or only a few data storage depots—like the Egyptian library in Alexandria of long ago. Computer access and standardization allow the user to know the contents and limitations of data. The greatest legacy of the national and international environmental research programs that now are underway, including the IGBP and WCRP and HDGEC, will surely be the revolutionary changes they will make and leave behind in the access and use and dissemination of data.

6. *Finally, we must enlist and train a new army* of scientists who are needed to create the set of plans that God never gave us.

We have made promises in science in order to win support for practical programs of environmental research like the U.S. Global Change Research Program. Like Babe Ruth at the plate, we have pointed to where in deep center field a hit will now be placed, inviting a world of watchers to see that it is done. Promised are not singles or doubles, but home runs, pragmatic assessments and predictions at the global and regional level, sufficiently specific to be of use to policy-makers. To do this will require very challenging research across a very broad front, the creation of new methods of synthesizing and modeling and of handling and disseminating vast stores of data, and the equally challenging task of holding on to the hot wire of the policy process. Most of these steps are as yet untried.

It will also require that we inspire young scientists to work in these fields. We must build the cadre of scientists and technicians needed to keep the promises we have made, through efforts in education in both the developed and the less developed world. The challenge is to recruit and educate a new and larger generation of scientists to work in the environmental and related sciences, and in the process to strengthen disciplinary as well as interdisciplinary capability, for one will be as needed as the other. We must also involve, much more than we have, the talents of scientists from the undeveloped world, for two reasons: They are needed, and, if policy answers are our goal, we can never expect policy interest from nations that are not involved from the start in the process.

In particularly short supply at the moment are hydrologists who work at scales larger than watershed, and regional and global ecologists. The situation in these fields is not unlike that in the early days of radio astronomy, just after the Second World War, when it became apparent that one could explore the sky with radio waves. In those days there were very few who worked in that new field. At that time the saying was that whenever two radio astronomers got together the first thing they would do is try to hire each other! We also need hermaphroditic scientists who can work with one foot in the natural sciences and the other in the social sciences, to address what is probably the biggest unknown in the whole system: Which is the human element of environmental change, both in terms of causative factors and in terms of the responses that can be expected, if either of these is predictable.

THE PROMISE

The good news is that many of these imperatives are now underway, or at least started, and the better news is that two of these are being addressed in even more powerful form in this volume, as a combination of two of them that could increase the power of both. Moreover there is support, here and abroad, for the kind of science that is addressed here.

Not everyone in science likes organized research programs. In today's climate, in my view, they offer the best hope for addressing regional or global problems. They also offer funds to do it. The U.S. Congress has authorized, in an era of severe budgetary constraints, almost a billion dollars in the current fiscal year for "focused" research in this country's Global Change Research Program. In the view of many of us who have worked for many years towards that end, this year's $954 million is not yet enough for what must be done, but it is still a lot, when measured in terms with which we are more familiar.

Toward the close of a thought-filled book, *The Farther Shore*, Don Gifford puts such incomprehensible figures into a more personal perspective: If you or I were given $954 million in the year that Christ was born, invested none of it, and spent it at the profligate rate of $1000 each day, through all the weeks and months and years of history that have since passed by, we would have enough left today to last another 621 years. A more appropriate comparison may be found in the fractional increases granted to the program over the last three years, confirming that Congress is willing to give more than lip service to a solid program aimed at an eminently pragmatic end. The global change budget has risen in the past 3 years very steeply, from about $100 million in 1989 to more than $600 million in 1990, to $954 million in 1991.

It is my hope that what is addressed in this volume will somehow be heard and seen around the world, as a fusing of two front-line scientific technologies at a time when so much depends upon each of them.

2

The State of GIS for Environmental Problem-Solving

MICHAEL F. GOODCHILD

WHAT IS GIS?

The use of the term *Geographic* (or Geographical) *Information System* dates back to the mid-1960s, where it seems to have originated in two quite different contexts. In Canada, it was devised to refer to the use of a mainframe computer and associated peripherals (notably a scanner) to manage the mapped information being collected for the Canada Land Inventory, and to process it to compute estimates of the area of land available for certain types of uses. A rigorous analysis was used to show that a computer was the only cost-effective means of producing the vast numbers of measurements of area required by the project, even with the primitive and expensive nature of digital technology at the time, because manual measurement of area remains an inaccurate and labor-intensive task. Much of the proposed analysis was concerned with measuring areas simultaneously on two maps, to answer questions like "How much area is class 1 agricultural land and not currently used for agriculture?" The ability to overlay two or more maps for analysis (in this case a map of soil capability for agriculture with a map of land use) has always been a strong argument for GIS, because it is so cumbersome by hand.

Almost at the same time, researchers in the U.S. were struggling with the problems of accessing the many different types of data required by the large-scale transportation models then in vogue, and conceived of a GIS as a system capable of extracting appropriate data from large stores, making them available for analysis, and presenting the results in map form (Coppock and Rhind, 1991). Such models combined information on population distributions with other spatially distributed information on places of employment and transportation routes, and required access to data in a variety of formats.

Almost 30 years later, these same arguments are still among the most frequently heard justifications for the use of GIS, particularly in environmental modeling and policy development. GIS is seen as a general-purpose technology for handling geographic data in digital form, and satisfying the following specific needs, among others:

- The ability to preprocess data from large stores into a form suitable for analysis, including such operations as reformatting, change of projection, resampling, and generalization.
- Direct support for analysis and modeling, such that forms of analysis, calibration of models, forecasting, and prediction are all handled through instructions to the GIS.
- Postprocessing of results, including such operations as reformatting, tabulation, report generation, and mapping.

In all of these operations, the typical GIS user now expects to be able to define requirements and interact with the system through a "user-friendly," intuitive interface that makes use of such contemporary concepts as graphic icons and desktop metaphors (Mark and Gould, 1991).

GIS has evolved dramatically since the early days of mainframe computing, particularly in the past 12 to 15 years. Its first commercial successes came in the early 1980s, primarily in resource management, but more recently large markets for GIS software have developed in local government, utility companies, and a host of activities that use geographic data or manage geographically distributed facilities. It is estimated that the global GIS industry grosses over $1 billion annually (for general, popular reviews of the GIS industry see Bylinski, 1989; The Economist, 1992; for a comprehensive industry overview see GIS World, Inc., 1991; for a review of GIS as a whole see Maguire, Goodchild, and Rhind, 1991).

GIS applications now span a wide range, from sophisticated analysis and modeling of spatial data to simple inventory and management. Since the latter account for the lion's share of the commercial market for software, they also dictate the development directions of much of the industry. However, several vendors have chosen to concentrate on the niche market of environmental applications, and to emphasize support for environmental modeling, and GRASS is significant as a public-domain GIS developed by a branch of the military (the U.S. Army Corps of Engineers' Construction Engineering Research Laboratory) and having substantial capabilities for environmental modeling.

Growth has brought confusion, notably to the meaning of the term GIS itself. The 371 software products listed in the GIS World, Inc. (1991) survey include a vast range of capabilities, and run on platforms from the Macintosh to mainframes. Some products are offered on platforms as diverse as the PC and the IBM 3090, while others focus on Unix workstations. All handle geographic data in some form and provide capabilities for input and output. But there is currently no consensus on the minimum set of functional capabilities required to qualify as a GIS. This diversity is illustrated when one compares the relatively focused approach taken by Berry in Chapter 7 with the much broader perspective of Nyerges in Chapter 8.

Besides its collection of tools, GIS now also has a broadly based community of interest, drawn together by a common concern for the computerized handling of geographic data. It includes established disciplines like surveying, remote sensing, geodesy, and cartography, which see GIS as another valuable digital technology with capabilities that augment those of GPS (global positioning systems), image processing, digital cartography etc. In some senses GIS is the common ground between all of these, the broad technology that attempts to integrate data from a number of acquisition systems, and provide it to the user with appropriate analytic tools.

The GIS community also includes specialists in various application fields: local government officials, urban and regional planners, land records administrators, the oil and gas industry, and many others. It includes geographers, planners, resource managers, environmental modelers, geologists, epidemiologists, soil scientists, and representatives of the disciplines that work with geographical data.

The technology that supports this community is complex, but at the same time primitive in the eyes of most of its users. Geographical reality is enormously complex, and it can be represented in digital form in a rich variety of ways. Moreover, the set of GIS functions is long and growing, as uses are found for a greater and greater variety of forms of spatial analysis. Yet ideally, all of this complexity should be presented to the user in a friendly, intuitive manner. The human eye and mind are incredibly powerful processors of two-dimensional data, and compared to them even supercomputers sometimes appear impossibly clumsy. At the same time, the computer is much more efficient at primitive operations like the measurement of length and area and the combination of data from different sources.

This chapter presents an introduction to some of the principles of GIS and the issues surrounding its use. The later chapters on GIS in this book, Chapters 7–9, provide greater depth. Some of the argument in this chapter was published in a previous article (Goodchild, 1991).

PRINCIPLES OF GIS DATA MODELING

Standard models

Many geographical distributions, such as those of soil variables, are inherently complex, revealing more information at higher spatial resolution apparently without limit (Mandelbrot, 1982). Because a computer database is a finite, discrete store, it is necessary to sample, abstract, generalize or otherwise compress information. "Geographical data modeling" is the process of discretization that converts complex geographical reality into a finite number of database records or "objects." Objects have geographical expression as points, lines, or areas, and also possess descriptive attributes. For example, the process of sampling weather-related geographic variables such as atmospheric pressure at weather stations creates point objects and associated measured attributes.

GIS technology recognizes two distinct modeling problems, depending on the nature of the distributions being captured. When the distributions in reality are spatially continuous functions or "fields," such as atmospheric pressure or soil class, the database objects are creations of the data modeling process. The set of objects representing the variation of a single variable is termed a "layer", and the associated models are "layer models" or "field models." However, there are numerous instances where the database objects are defined a priori, rather than as part of the modeling process. The object "Lake Ontario" is meaningful in itself, and has an identity that is independent of any discretization of a binary water/land variable over North America. We refer to these as "object models." In a field model every location by definition has a single value of the relevant variable, whereas in the object model there would be no particular problem in allowing a location to be simultaneously occupied by more than one object. For example the "Bay of Quinte" is also in "Lake Ontario." The term "planar enforcement" is often used to reflect the fact that objects in a field model may not overlap; planar enforcement clearly is not relevant to the object models.

A major difficulty arises in the case of the object models when a well-defined object has no equally well-defined location. For example, the spatial extent of Lake Ontario would most likely be defined by some notion of average elevation, but this is not helpful in deciding when Lake Ontario ends and the St Lawrence River begins. Many geographical objects have inherently fuzzy spatial extents. One common solution to this problem is to allow objects to have "multiple representations"—spatial extents that vary with scale. A river, for example, might be a single line at scales smaller than 1:50,000, but a double line at larger scales. Both geometric and topological expressions vary in this case as the object changes from line to area.

The field models

The purpose of field models is to represent the spatial variation of a single variable using a collection of discrete objects. A spatial database may contain many such fields or layers, each able in principle to return the value of one variable at any location (x, y) in response to a query, and fields may be associated with variables measured on either continuous or discrete scales. Because information is lost in modeling, the value returned may not agree with observation or with the result of a ground check; so accuracy is an important criterion in choosing between alternative data models. We define the accuracy of a field measured on a continuous scale as $E(z - z')^2$ where z is the true value of the variable, as determined by ground check, and z' is its estimated value returned from the database; for a field measured on a discrete scale, accuracy will likely be defined by the probability that the class recorded at a randomly chosen point is indeed the class at that point on the ground (further discussion of the measurement of accuracy can be found in Chapter 9). Note that the true value may be inherently uncertain because of definition or repeated measurement problems.

Six field models are in common use in GIS:

1. Irregular point sampling: the database contains a set of tuples $<x, y, z>$ representing sampled values of the variable at a finite set of irregularly spaced locations (e.g., weather station data).
2. Regular point sampling: as (1) but with points regularly arrayed, normally on a square or rectangular grid (e.g., a Digital Elevation Model).
3. Contours: the database contains a set of lines, each consisting of an ordered set of $<x, y>$ pairs, each line having an associated z value; the points in each set are assumed connected by straight lines (e.g., digitized contour data).
4. Polygons: the area is partitioned into a set of polygons, such that every location falls into exactly one polygon; each polygon has a value that is assumed to be that of the variable for all locations within the polygon; boundaries of polygons are described as ordered sets of x,y pairs (e.g., the soil map).
5. Cell grid: the area is partitioned into regular grid cells; the value attached to every cell is assumed to be the value of the variable for all locations within the cell (e.g., remotely sensed imagery).
6. Triangular net: the area is partitioned into irregular triangles; the value of the variable is specified at each triangle vertex, and assumed to vary linearly over the triangle (e.g., the Triangulated Irregular Network, or TIN, model of elevation) (Weibel and Heller, 1991).

Other possibilities, such as the triangular net (6) with nonlinear variation within triangles (Akima, 1978), have not received much attention in GIS to date.

Each of the six models can be visualized as generating a set of points, lines, or areas in the database. Models (2) and (5) are commonly called "raster" models, and (1), (3), (4), and (6) are "vector" models (Peuquet, 1984); storage structures for vector models must include coordinates, but in raster models locations can be implied by the sequence of objects. Models (3) and (6) are valid only for variables measured on continuous scales.

Models (4), (5), and (6) explicitly define the value of the variable at any location within the area covered. However, this is not true of models (1), (2), and (3), which must be supplemented by some method of spatial interpolation before they can be used to respond to a general query about the value of z at some arbitrary location. For example, this is commonly done in the case of continuous-scaled variables in model (2) by fitting a plane to a small 2x2 or 3x3 neighborhood. However, this need for a spatial interpolation procedure tends to confound attempts to generalize about the value of models (1), (2), and (3).

In practice, model (6) is reserved for elevation data, where its linear facets and breaks of slope along triangle edges fit well with many naturally occurring topographies (Mark, 1979). It would make little sense as a means of representing other variables, such as atmospheric pressure, since curvature is either zero (within triangles) or undefined (on triangle edges) in the model. Models (2) and (4) are frequently confused in practice, since the distinction between point samples and area averages is often unimportant. Models (1) and (3) are commonly encountered because of the use of point sampling in data collection and the abundance of maps showing contours, respectively, but are most often converted to models (2), (4), (5), or (6) for storage and analysis. The ability to convert between data models, using various algorithms, is a key requirement of GIS functionality.

The object models

Objects are modeled as points, lines, or areas, and many implementations make no distinction in the database between object and field models. Thus a set of lines may represent contours (field model) or roads (object model), both consisting of ordered sets of x,y pairs and associated attributes, although the implications of intersection, for example, are very different in the two cases.

Object models are commonly used to represent man-made facilities. An underground pipe, for example, is more naturally represented as a linear object than as a value in a layer. Pipes can cross each other in object models, whereas this would cause problems in a field model. Most manmade facilities are well defined, so the problems of fuzziness noted earlier are likely not important. Another common use of object models is in capturing features from maps.

Object models are also commonly used to capture aspects of human experience. The concept "downtown" may be very important in building a database for vehicle routing or navigation, forcing the database designer to confront the issue of its representation as a geographical object. In an environmental context, McGranaghan (1989a,b) has shown the importance of this for handling the geographical referrents used in herbarium records.

Finally, object models can be conceptualized as the outcome of simple scientific categorization. The piecewise approximation inherent in field model (4) assigns locations to a set of discrete regions, in the geographical equivalent of the process of classification. In geomorphology, the first step in building an understanding of the processes that formed a given landscape is often the identification of "landforms" or "features," such as "cirque" or "drumlin." Band (1986), among others, has devised algorithms for detecting such objects from other data. Frank and Mark (1991) have discussed the importance of categories in the GIS context, and there is growing interest in understanding the process of object definition and its effects. For most purposes, environmental data modeling is dominated by the field view, and its concept of spatially continuous variables. But the object view is clearly important, particularly in interpreting and reasoning about geographical distributions.

Network models

Both field and object models have been presented here as models of two-dimensional variation. An important class of geographic information describes continuous variation over the one-dimensional space of a network embedded in two-dimensional space. For example, elevation, flow, width, and other parameters vary continuously over a river network, and are not well represented as homogeneous attributes of reaches. Models (1), (2), (4), (5), and (6) can all be implemented in one-dimensional versions, but none is supported in this form in any current GIS.

Choosing data models

In principle, the choice of data model should be driven by an understanding of the phenomenon itself. For example, a TIN model will be an appropriate choice for representing topography if the Earth's surface is accurately modeled by planar facets. Unfortunately other priorities also affect the choice of model. The process of data collection often imposes a discretization, the photographic image being a notable exception. The limitations of the database technology may impose a data model, as, for example, when a "raster" GIS is used and the choices are reduced to field models (2) and (4), or when a "vector" GIS is used and a cell grid must be represented as polygons. Finally convention can also be important, particularly in the use of certain data models to show geographic variation on maps. For example, digitized contours are used in spatial databases not because of any particular efficiency—in fact accuracy in a field sense is particularly poor—but because of convention in topographic map-making.

Relationships

A digital store populated by spatial objects—points, lines, and areas—would allow the user to display, edit, or move objects, much as a computer-aided design (CAD) system. However, spatial analysis relies heavily on interactions between objects, of three main forms:

- Relationships between simple objects, used to define more complex objects (e.g., the relationships between the points forming a line);
- Relationships between objects defined by their geometry (e.g., containment, adjacency, connectedness, proximity); and
- Other relationships used in modeling and analysis.

Examples of the third category of relationships not determined by geometry alone include "is upstream of" (connectedness would not be sufficient to establish direction of flow, and a sink and a spring may not be connected by any database object). In general, a variety of forms of interaction may exist between the objects in the database. In order to model these, it is important that the database implement the concept of an "object pair," a virtual object that may have no geographical expression but may nevertheless have attributes such as distance or volume of flow.

Recent trends in data modeling

Recently there has been much discussion in the GIS community over the value of "object orientation," a generic term for a set of concepts that have emerged from theoretical computer science (see, for example, Egenhofer and Frank, 1988a,b). Unfortunately the debate has been confused by the established usages of "object" in GIS, both in the sense of "spatial object" as a point, line, or area entity in a database, and also "object model" as defined here.

Three concepts seem particularly relevant. "Identity" refers to the notion that an object can possess identity that is largely independent of its instantaneous expression, with obvious relevance to the independence of object identity and geographic expression in GIS. "Encapsulation" refers to the notion that the operations that are possible on an object should be packaged with the object itself in the database, rather than stored or implemented independently. Finally, "inheritance" refers to the notion that an object can inherit properties of its parents, or perhaps its component parts. As a geographical example, the object "airport" should have access to its component objects—runway, hangar, terminal—each of which is a spatial object in its own right.

Of the three concepts, inheritance seems the most clearly relevant, particularly in the context of complex objects, and in tracking the lineage of empirical data. It seems increasingly important in the litigious environment that surrounds many GIS applications to track the origins and quality of every data item.

Encapsulation seems to present the greatest problems for modeling using GIS. In a modeling context, the operations that are permissible on an object are defined by the model, and are therefore not necessarily treatable as independent attributes of the object. This issue seems particularly important in the context of the discussion of object orientation in location/allocation modeling by Armstrong, Densham, and Bennett (1989). For example, one can rewrite the shortest path problem by treating each node in the network as a local processor, making it possible to encapsulate the operations of a node with the object itself. It is possible that this process of rewriting may lead to useful insights in other models as well.

A related debate is that over procedural and declarative languages: A user should be able to declare "what" is required (declarative), and not have to specify "how" it should be done (procedural). But are these largely distinguishable in a modeling context, and do they imply that the modeler should somehow surrender control of the modeling process to the programmer?

The role of data models in environmental analysis and modeling is clearly complex. Models written in continuous space, using differential equations, are independent of discretization. But for all practical purposes modeling requires the use of one or more of the data models described here. Perhaps the greatest advantage of GIS is its ability to handle multiple models, and to convert data between them.

GIS AND ENVIRONMENTAL PROBLEM-SOLVING

One of the strongest and most successful application areas for GIS has been in addressing problems of the environment. GIS were acquired by many forest resource companies and regulatory agencies in the early 1980s, and subsequently by other environmental agencies such as EPA (U.S. Environmental Protection Agency), the National Parks Service, and the Bureau of Land Management. Virtually all North American resource management agencies now have some form of GIS program and associated policies. However, this pattern gives no impression of the diversity of GIS applications within agencies. The following subsections provide a categorization of different types of applications within the broad area of environmental problem-solving.

Mapping

In principle, it is possible to make a clear distinction between GIS and digital cartography. The latter deals with map features and with associated attributes of color, symbology, name or annotation; provides capabilities for input, editing, and output; deals with such cartographic features as legends, neatlines, and north arrows; and includes algorithms for projection and scale change. GIS, on the other hand, frequently allows geographic entities to have multiple attributes, frequently includes capabilities for storing and handling relationships between entities, and includes the capabilities of digital cartography in its input and output subsystems.

Of particular importance to this discussion is the manner in which digital cartography and GIS handle notions of continuous variation. Consider the elevation of the Earth's surface as an example of a variable conceived as varying continuously across geographic space and having a unique value everywhere in that space. Cartographically, such continuous variation must be represented in the form of map features, typically contour lines and irregularly spaced spot heights, and digital cartography provides capabilities for processing these as line and point objects. From a GIS perspective, however, numerous other methods exist for creating a digital representation of continuous variation. The objects used in that representation should be viewed as artifacts, and hidden from the user so as not to obscure the latter's concept of a continuum.

In practice, however, many applications of GIS technology turn out to be little more than digital cartography. The map is a very persuasive form of data display, and a computer-drawn map carries the authority of a powerful technology. Many GIS contain user-friendly map editing and formatting subsystems, and allow the user with little cartographic training to make a convincing product quickly and efficiently. But cartographers have accumulated centuries of knowledge and experience about the effective visual communication of geographic information, and GIS may be doing its users a disservice if it encourages them to ignore this (Buttenfield and Mackaness, 1991).

Data preprocessing

Because there are so many ways of representing geographic variation, and such diversity in data structures between GIS products and databases, most available GIS include substantial facilities for input and output of data in different formats. Some of these are sanctioned by official standards: the major U.S. federal effort at standardization has produced a Spatial Data Transfer Standard (SDTS) that has been accepted (Federal Information Processing Standard (FIPS) 173 (DCDSTF, 1988)).

Other useful preprocessing functions include the ability to extract information in a user-defined window; scale and projection change, including knowledge of such coordinate systems as UTM (Universal Transverse Mercator) and SPC (State Plane Coordinates); and capabilities for resampling. All of these help to give GIS an important role as a manager of data, particularly if data must come from diverse sources in mutually inconsistent formats.

The chapters of this book provide strong support for this argument for GIS as a data integrator. Models for specific environmental processes, such as non-point-source pollution, have often been developed without reference to GIS, and without the ability to interface with GIS software. But models that integrate more than one process, or attempt to provide links to policy development, such as those described in Chapters 26–34 of this book, are much more likely to rely on GIS for preprocessing of data. Moreover, it is clear that in areas like global climate modeling, the need to integrate data on a variety of themes, such as soil or land cover, is leading to increasing concern with GIS (see Chapters 10–13).

GIS in modeling

Many applications of GIS in environmental modeling have followed the scheme outlined in the previous sections, where a GIS is used to preprocess data, or to make maps of input data or model results. We refer to this mode as "loose coupling," implying that the GIS and modeling software are coupled sufficiently to allow the transfer of data, and perhaps also of results in the reverse direction. In many instance loose coupling requires the development of a linking module of specialized code, if the GIS is not capable of providing data in the format(s) required by the modeling software.

Closer forms of coupling are possible if the GIS and modeling module share the same data structures, obviating the need for a linking module and allowing both systems to interact with the same database. Nyerges discusses issues of coupling in more detail in Chapter 8. At the highest level of sophistication, models are calibrated and run directly in the GIS, using the GIS command language. Berry describes the concept of cartographic modeling in Chapter 7, and illustrates the use of a GIS to perform complex analysis of layered data. However, the current generation of GIS command languages falls well short of satisfying the requirements of this style of environmental modeling. To support modeling directly in the GIS's command language, it would be necessary for the GIS's data model to match that of the environmental model. For example, if an atmospheric model requires that space be represented as a set of square finite elements, it would be necessary for the GIS database also to represent spatial variation in terms of these same square finite elements. In other words, the GIS data model must match the needs of environmental modeling.

In practice, GIS data models tend to have more in common with maps than with the finite elements of environmental models. GIS databases have often been constructed from mapped information, and maps are a common form of output. Thus elevation is often represented by digitized contours, since this is the preferred method of map representation, as noted earlier. Soil or land cover maps are often represented as collections of digitized polygons. The space of environmental models is often conceived as continuous, and discretized as finite elements only for the purpose of numerical analysis. In GIS, however, the finite elements tend to be the basis of conceptualization, and frequently it is the cartographic model that dominates the choice of finite elements. Based on these arguments, the following are suggested as the minimum requirements for an environmental modeling language interface to GIS:

- In the first instance, the user should be able to work with symbolic representations of continuous geographic variation (e.g., T and h as representations of the continuous variation of temperature and elevation, respectively).
- The language should allow such continuous variations (fields) to be combined symbolically (e.g., $T = 20 - h/100$ to compute ground surface temperature by approximating the adiabatic lapse rate).
- As far as possible, the language should hide issues of discrete digital representation from the user. This is straightforward when T and h are both evaluated for the same set of sample points, as in the examples and language presented by Berry in Chapter 7. Where operations change the discrete representation, the user should provide the necessary information. For example, $T(100)=T$ might resample T to a 100 m sampling interval, using a default method of resampling, and assuming that the sampling interval of T was much smaller than 100 m.
- The language should include all of the common primitives of environmental modeling. For example, although both discrete and continuous scalar fields are handled, in current GIS products there is no support for the concept of a vector field. Related operators, such as grad or the dot product, are not supported, and neither are simple methods of display.

GIS in policy

Some of the greatest interest in the use of GIS for environmental problem-solving has come from those who would apply the technology to translate the results of environmental modeling into policy. Postprocessing is essential if the results of a spatially distributed model are to be used for policy development. Results must often be aggregated by administrative unit, or brought into consistency with social and economic data for comparison and correlation. Displays must be developed to present the results of modeling in convincing form. Finally, increasing use is being made of the paradigm of spatial decision support, in which the technology is made available directly to decision-makers for scenario development, rather than being confined to use by analysts.

As the chapters of this book progress from models to policy, the need for a technology that can cross the gap between rigorous science and responsible policy formulation will become increasingly clear. We spend vast sums on collecting raw geographic information with technologies such as remote sensing, and on modeling environmental processes, and yet it often seems that the biggest problem of all is the translation of this knowledge into useful and effective policy.

FUNCTIONALITY FOR ENVIRONMENTAL ANALYSIS

The statistical packages such as S, SAS, or SPSS are integrated software systems for performing a wide variety of forms of analysis on data. By analogy, we might expect GIS to integrate all reasonable forms of spatial analysis. However, this has not yet happened, for several reasons. First, while the analogy between the two systems may be valid, there are important differences. The statistical packages support only one basic data model—the table—with one class of records, whereas GIS must support a variety of models with many classes of objects and relationships between them. Much of the functionality of GIS must therefore be devoted to supporting basic housekeeping and transformation functions that would be trivial in the statistical packages.

Second, spatial databases tend to be large, and difficult and expensive to create. This is particularly true of imagery, where a single analysis might require gigabytes (10^9 bytes) or even terabytes (10^{12} bytes) of data, and of representations of geographical variation in three dimensions. While many users of statistical packages input data directly from the keyboard, it is virtually impossible to do anything useful with a GIS without devoting major effort to database construction. Recently there have been significant improvements in this situation, with the development of improved scanner and editing technology.

Third, while there is a strong consensus on the basic elements of statistical analysis, the same is not true of spatial analysis. The literature contains an enormous range of techniques (for examples, see Serra, 1982; Unwin, 1981; Upton and Fingleton, 1985), few of which could be regarded as standard.

Because of these issues and the diversity of data models, GIS has developed as a loose consortium, with little standardization. While ESRI's ARC/INFO and TYDAC's SPANS are among the most developed of the analytically oriented packages, they represent very different approaches and architectures. Among the most essential features to support environmental modeling are:

- Support for efficient methods of data input, including importing from other digital systems;
- Support for alternative data models, particularly models of continuous spatial variation, and conversions between them using effective methods of spatial interpolation;
- Ability to compute relationships between objects based on geometry (e.g., intersection, inclusion, adjacency), and to handle attributes of pairs of objects;
- Ability to carry out a range of standard geometric operations (e.g., calculate area, perimeter length);
- Ability to generate new objects on request, including objects created by simple geometric rules from existing objects (e.g., Voronoi polygons from points, buffer zones from lines);
- Ability to assign new attributes to objects based on existing attributes and complex arithmetic and logical rules;
- Support for transfer of data to and from analytic and modeling packages (e.g., statistical packages, simulation packages).

Because of the enormous range of possible forms of spatial analysis, it is clearly absurd to conceive of a GIS as a system to integrate all techniques, in contrast to the statistical packages. The last requirement listed proposes that GIS should handle only the basic data input, transformation, management, and manipulation functions, leaving more specific and complex modeling to loosely coupled packages. Whereas the statistical packages are viable because they present all statistical techniques in one consistent, readily accessible format, GIS is viable for environmental modeling because it provides the underlying support for handling geographical data, and the "hooks" needed to move data to and from modeling packages, at least until command languages can be developed that are more suitable for environmental modeling.

CONCLUSIONS

GIS is a rapidly developing technology for handling, analyzing, and modeling geographic information. To those sciences that deal with geographic information it offers an integrated approach to data handling problems, which are often severe. As with all GIS applications, the needs of environmental modeling are best handled not by integrating all forms of geographic analysis into one GIS package, but by providing appropriate linkages and hooks to allow software components to act in a federation.

Models that lack a spatial component clearly have no use for GIS, but increasing concern for geographically distributed models of the environment, particularly as inputs to environmental policy, has led to increasing interest in GIS in the past decade in the environmental modeling community, and this book describes much of that interest. The ability of GIS to integrate spatial data from different sources, with different formats, structures, projections, or levels of resolution is a powerful aid to spatially distributed models, particularly those models that integrate more than one process. Finally, the need to integrate environmental information with administrative, political, social, and economic data in developing environmental policy and regulating the use of the environment provides another powerful incentive to remove some of the impediments to greater integration of GIS and environmental modeling.

This chapter has provided a brief overview of GIS and some of the issues involved in its use in environmental problem-solving. Many of its themes are explored in greater detail later in this book in Chapters 7–9.

REFERENCES

Armstrong, M.P., Densham, P.J., and Bennett, D.A. (1989) Object oriented locational analysis. *Proceedings, GIS/LIS '89*, Bethesda, MD: ASPRS/ACSM, pp. 717–726.

Band, L.E. (1986) Topographic partition of watersheds with digital elevation data. *Water Resources Research* 22(1): 15–24.

Buttenfield, B.P., and Mackaness, W.A. (1991) Visualization. In Maguire, D.J., Goodchild, M.F., and Rhind, D.W. (eds.) *Geographical Information Systems: Principles and Applications* (2 volumes), London: Longman, and New York: Wiley, pp. 427–443.

Bylinski, G. (1989) Managing with electronic maps. *Fortune* (April 24): 237.

Coppock, J.T., and Rhind, D.W. (1991) The history of GIS. In Maguire, D.J., Goodchild, M.F., and Rhind, D.W. (eds.) *Geographical Information Systems: Principles and Applications* (2 volumes), London: Longman, and New York: Wiley, pp. 21–43.

DCDSTF (Digital Cartographic Data Standards Task Force) (1988) The proposed standard for digital cartographic data. *The American Cartographer* 15(1): 9–140.

Economist, The (1992) The delight of digital maps. *The Economist* (March 21): 69–70.

Egenhofer, M.J., and Frank, A.U. (1988a) Object-oriented databases: database requirements for GIS. *Proceedings, IGIS: The Research Agenda*, Washington DC: NASA, Vol. 2, pp. 189–211.

Egenhofer, M.J., and Frank, A.U. (1988b) Designing object-oriented query languages for GIS: human interface aspects. *Proceedings, Third International Symposium on Spatial Data Handling*, Columbus, OH: International Geographical Union, 79–96.

Frank, A.U., and Mark, D.M. (1991) Language issues for GIS. In Maguire, D.J., Goodchild, M.F., and Rhind, D.W. (eds.) *Geographical Information Systems: Principles and Applications* (2 volumes), London: Longman, and New York: Wiley, pp. 147–163.

GIS World, Inc. (1991) *1991–92 International GIS Sourcebook*, Fort Collins, CO: GIS World, Inc.

Goodchild, M.F. (1991) Integrating GIS and environmental modeling at global scales. *Proceedings, GIS/LIS '91*, Bethesda, MD: ASPRS/ACSM, Vol. 1, pp. 117–127.

Maguire, D.J., Goodchild, M.F., and Rhind, D.W. (eds.) (1991) *Geographical Information Systems: Principles and Applications* (2 volumes), London: Longman, and New York: Wiley.

Mandelbrot, B.B. (1982) *The Fractal Geometry of Nature*, San Francisco: Freeman.

Mark, D.M. (1979) Phenomenon-based data structuring and digital terrain modeling. *GeoProcessing* 1: 27–36.

Mark, D.M., and Gould, M.D. (1991) Interacting with geographic information: a commentary. *Photogrammetric Engineering and Remote Sensing* 57(11): 1427–1430.

McGranaghan, M. (1989a) Incorporating bio-localities in a GIS. *Proceedings, GIS/LIS '89*, Bethesda, MD: ASPRS/ACSM, pp. 814–823.

McGranaghan, M. (1989b) Context-free recursive-descent parsing of location-descriptive text. *Proceedings, AutoCarto 9*, Bethesda, MD: ASPRS/ACSM, pp. 580–587.

Peuquet, D.J. (1984) A conceptual framework and comparison of spatial data models. *Cartographica* 21(4): 66–113.

Serra, J.P. (1982) *Image Analysis and Mathematical Morphology*, New York: Academic Press.

Unwin, D.J. (1981) *Introductory Spatial Analysis*, New York: Methuen.

Upton, G.J.G., and Fingleton, B. (1985) *Spatial Data Analysis by Example*, New York: Wiley.

Weibel, R., and Heller, M. (1991) Digital terrain modelling. In Maguire, D.J., Goodchild, M.F., and Rhind, D.W. (eds.) *Geographical Information Systems: Principles and Applications* (2 volumes), London: Longman, and New York: Wiley, pp. 269–297.

3

A Perspective on the State of Environmental Simulation Modeling

LOUIS T. STEYAERT

Computer-based, mathematical models that realistically simulate spatially distributed, time-dependent environmental processes in nature are increasingly recognized as fundamental requirements for the reliable, quantitative assessment of complex environmental issues of local, regional, and global concern. These environmental simulation models provide diagnostic and predictive outputs that can be combined with socioeconomic data for assessing local and regional environmental risk or natural resource management issues. Such assessments may involve air and water quality; the impact of man's activities on natural ecosystems; or conversely, the effects of climate variability on water supplies, agriculture, ecosystems, or other natural resources.

More recently, the importance of scientific models for the assessment of potential global environmental problems, including regional response to global change, has been illustrated by the National Research Council (NRC) (1986, 1990), Earth System Sciences Committee (ESSC) (1986, 1988), International Council of Scientific Unions (ICSU) (1986), International Geosphere–Biosphere Programme (IGBP) (1990), and Committee on Earth Sciences (CES) (1989, 1990). Eddy (Chapter 1) has discussed several environmental issues and suggested various courses of action including the need for environmental models. Mathematical models are needed to help understand the current behavior and to project the future state of the complex Earth system processes. These modeling tools are necessary to help differentiate between environmental changes that are due to natural variability in the environmental system versus possible changes due to human interactions. Although much progress has been made, research is still needed to understand and model environmental processes in natural science.

This chapter highlights some of the general research themes that tend to characterize the state of environmental simulation modeling within, and particularly across, natural science disciplines. The goal of these remarks is to help facilitate the understanding of the models subsequently discussed within this volume, as well as to establish further potential links between environmental modeling and geographic information systems (GIS). These remarks are not intended to address detailed scientific questions, such as the pros and cons of various approaches for modeling or parameterizing some process in an atmospheric or ecosystem dynamics model. This is left to the respective modeling communities.

The focus is on state-of-the-art dynamical simulation models that are evolving through global climate change research programs, specifically as noted by the Committee on Earth and Environmental Sciences (CEES) (1991), NRC (1990), and IGBP (1990). The emphasis is on the terrestrial component of biosphere–atmosphere interactions modeling at the local and regional scales. These types of advanced models are also applicable to issues within the areas of land and water resource management, environmental risk assessment, and other applications. This chapter addresses the status of research on the development and testing of models, not the use of models in an applications or an assessment mode.

There are four major sections in this chapter. First, the concept of environmental simulation modeling is introduced to help understand modeling research themes. Second, the role of global change research programs as major contributors to progress in the overall advancement of the environmental simulation modeling field is discussed. Although there is increasing awareness, global change research also means modeling at local and regional scales of interest in order to evaluate the cumulative effect of these smaller-scale features on the global environment properly. Third, several research themes that help to characterize the status of environmental simulation modeling are examined. This section also introduces the overview and case study papers within this volume. Specifically, an attempt is made to integrate these contributions, especially overview chapters by Lee et al. (Chapter 10), Maidment (Chapter 14), Hunsaker et al. (Chapter 22), and Schimel and Burke (Chapter 26). Fourth, potential links between environmental simulation models and GIS are examined.

ENVIRONMENTAL SIMULATION MODELS

General concepts

The environmental processes in the real world are typically three dimensional, time dependent, and complex. Such complexity can include nonlinear behavior, stochastic components, and feedback loops over multiple time and space scales. There may be significant qualitative understanding of a particular process, but the quantitative understanding may be limited. The ability to express the physical process as a set of detailed mathematical equations may not exist, or the equations may be too complicated to solve without simplifications.

Furthermore, computer limitations or the manner in which mathematical equations are converted for numerical processing on a grid (discretization) lead to the parameterization of subgrid small-scale complex processes that cannot be explicitly represented in the model. In some cases, these sets of equations can be viewed as a collection of hypotheses, concerning physical processes, whose inputs and outputs are linked. This set of parameterized equations represents the modeler's best approach to account for these processes, given these collective constraints. (This concept is illustrated in the subsection on Simulation Models.)

Therefore, it is important to recognize that environmental models are usually, at best, just a simplification of real world processes (IGBP, 1990). Reality is only approximated by the model. In spite of all these qualifications, models in the hands of a skilled user do, in fact, provide useful information of scientific and applied interest.

The concept of an *environmental simulation model* is used in this chapter as a general term to characterize the types of models under discussion in this volume. There is no one definition or all-encompassing term that will adequately describe this type of modeling. In fact, most of these models are more commonly referred to as an Earth science model, atmospheric or hydrological model, ecosystem dynamics model, or some other type of model. The term by which a model is named is frequently associated with the particular scientific discipline, spatial and temporal scale, application, or type of bureaucratic program. In fact, such terms as "land–atmosphere interactions" or "biosphere–atmosphere interactions" modeling stem from global change programs.

However, the phrase "environmental simulation model" can be dissected and examined word by word to help understand these types of models. In fact, other key terms such as "parameterization" or "discretization" permit even more understanding of the modeling process, as well as the current state of the models.

Three references provide the framework for the subsequent discussion. First, Law and Kelton (1982) introduce the overall topic of simulation modeling. Frenkiel and Goodall (1978) provide an excellent summary of early use of simulation models for environmental problem solving. The topic of modeling is addressed by several global change documents (ESSC, 1988; IGBP, 1990; NRC, 1990).

Types of models

Models can be classified into three major categories: scale, conceptual, and mathematical. An example of a scale model would be a scaled-down replica of a mountain range or an airplane wing for use in wind tunnel experiments. There are also analog scale models such as a topographic map. (Note: GIS deal with analog models, hence integration implies links between scale models and mathematical models.) Conceptual models are frequently used in the modeling process in block diagrams that show major systems, processes, and qualitative interrelationships between subsystems.

Mathematical models can be further classified as either deterministic or statistical (i.e., nondeterministic). Statistical or probability models contain at least one stochastic process represented by one or more random variables. A deterministic model does not have any random variables. Thus, a stochastic model has output data that are also random variables, and a deterministic model has a unique set of output data for a given input set (Law and Kelton, 1982). (In this categorization, deterministic models are associated with environmental processes, and statistical models are based on empirical analysis of observations.)

Both deterministic and statistical models can be further subdivided into either "steady-state" or "dynamic" models, where dynamic models contain at least one term that is a function of time. In the case of deterministic relationships, steady-state models can be represented by algebraic expressions for diagnostic study. Similarly, the deterministic-dynamic model is represented by differential equations that include at least one time derivative or by algebraic relationships that include a time term. Both total and partial differential equations are used.

Diagnostic models may represent the interrelationships within a system that is in a static or steady-state condition (that is no temporal component) or for some fixed point in time given a quasi-steady-state assumption. Prognostic models are used for prediction and depict a dynamic system that is a function of time (IGBP, 1990).

Simulation models

An environmental simulation model may be defined as a computer-based technique to imitate, or *simulate*, the operations of various kinds of real-world processes (Law and Kelton, 1982). Examples of real-world environmental processes are hydrodynamic fluid flow, radiative and heat transfer, biological growth mechanisms, and ecological development. Physically based laws describing these processes may not be known.

To study these types of processes, either individually or as part of a system, physically based laws (e.g., Newton's laws of motion) or other types of *assumptions* are usually made on how the processes actually work. These laws or assumptions can be expressed as mathematical or logical relationships; collectively they represent a *model.*

There are no fixed rules, but typically the environmental simulation model will include time-dependent partial differential equations and algebraic equations. The model is a dynamic model because of the time dependency. In some cases, these may be based on assumption or may be derived empirically through statistical analysis of observed data. For complex processes, there may be an extensive number of equations. Frequently, the goal is to make prognostic projections based on the simultaneous solution of a set of equations (with the same number of unknowns), as discussed by Lee et al. (Chapter 10). Frequently, assumptions must be made to simplify the set of equations.

For a very simple deterministic model, it may be possible to calculate an exact solution for idealized conditions (e.g., couette flow). However, real-world processes are typically so complex and nonlinear that an analytic solution is not possible. In such cases, the model is converted so that a numerical solution can be calculated on a computer.

In the case of numerical solutions, the system of mathematical equations that make up a model is usually converted to run on a two- or three-dimensional grid. The methodology for restructuring a system of equations to run on a grid is termed *discretization.* There may be small-scale processes, termed *subgrid processes*, that must be accounted for at the grid level, usually by a method that is termed *parameterization.*

Parameterization may be viewed as a method for scaling subgrid processes up to the grid level. However, the concept of parameterization is also used to link models across space and time scales. For example, detailed data and models near the process level at small scales are used to parameterize relationships at the next higher scales. Such an approach is of interest to scaling instantaneous biophysical data at the plant leaf level (evapotranspiration and photosynthesis) to annual regional estimates of net primary production or evapotranspiration (Running, 1991) through multiple parameterizations.

These parameterizations can be quite elaborate, for example, the land surface parameterization for a global climate model. (To complicate matters, the terms "model" and "parameterization" are sometimes used interchangeably.)

Additional information is provided by Lee et al. (Chapter 10) for the case of meteorological numerical models. Because the equations are discretized, only a portion of the equations are based on fundamental physical quantities (for example, the pressure gradient and advective terms in the equation of motion). The remainder of the terms are parameterized and generally are not based on fundamental concepts. Lee et al. (Chapter 10) provide examples of such parameterizations for meteorological models: cloud physics, long- and short-wave radiative flux divergence, subgrid turbulent fluxes, and soil and vegetation effects.

There are several reasons for uncertainty in model results: (1) Only a limited number of processes can be treated; (2) processes may not be well understood or, for some other reason, may be treated inadequately; and (3) the spatial and temporal resolution is inadequate (IGBP, 1990). Also, the solutions may be very sensitive to initial conditions if the interactions in the models are sufficiently nonlinear (e.g., Lorenz, 1963).

Finally, the term "modeling" may be referred to as the research process that leads to the development of a "model." The modeling process typically involves the development and testing of complex, interrelated hypotheses as part of the scientific method. This process may include the collection and analysis of observations that may be used to formulate a model or to test the hypothesis. The modeling process can include steps to develop, test and evaluate, validate, and apply the model.

GLOBAL CHANGE RESEARCH PROGRAMS

The challenge of Earth system modeling

Global change research programs have grown out of the need for scientifically valid assessments of critical environmental issues to support the policy decision process. Such requirements stem, in part, from widespread concerns about the potential for global environmental change due to such issues as depletion of stratospheric ozone, rising concentrations of CO_2 and other greenhouse gases in the atmosphere, tropical deforestation, and other activities leading to potential anthropogenic-induced changes to the Earth system.

As a consequence of these concerns, policy-level questions have been raised about the scope of the threat or the degree of risk for such global change. A fundamental issue involves the distinction between anthropogenic-induced variability versus change due to natural variability. Some of the questions concerning these issues include "Will there be a change? If so, when and how much? At what rates will change occur?" and "What can be done to mitigate or adapt to the potential change if it should occur?" Regional response to global forcing and regional

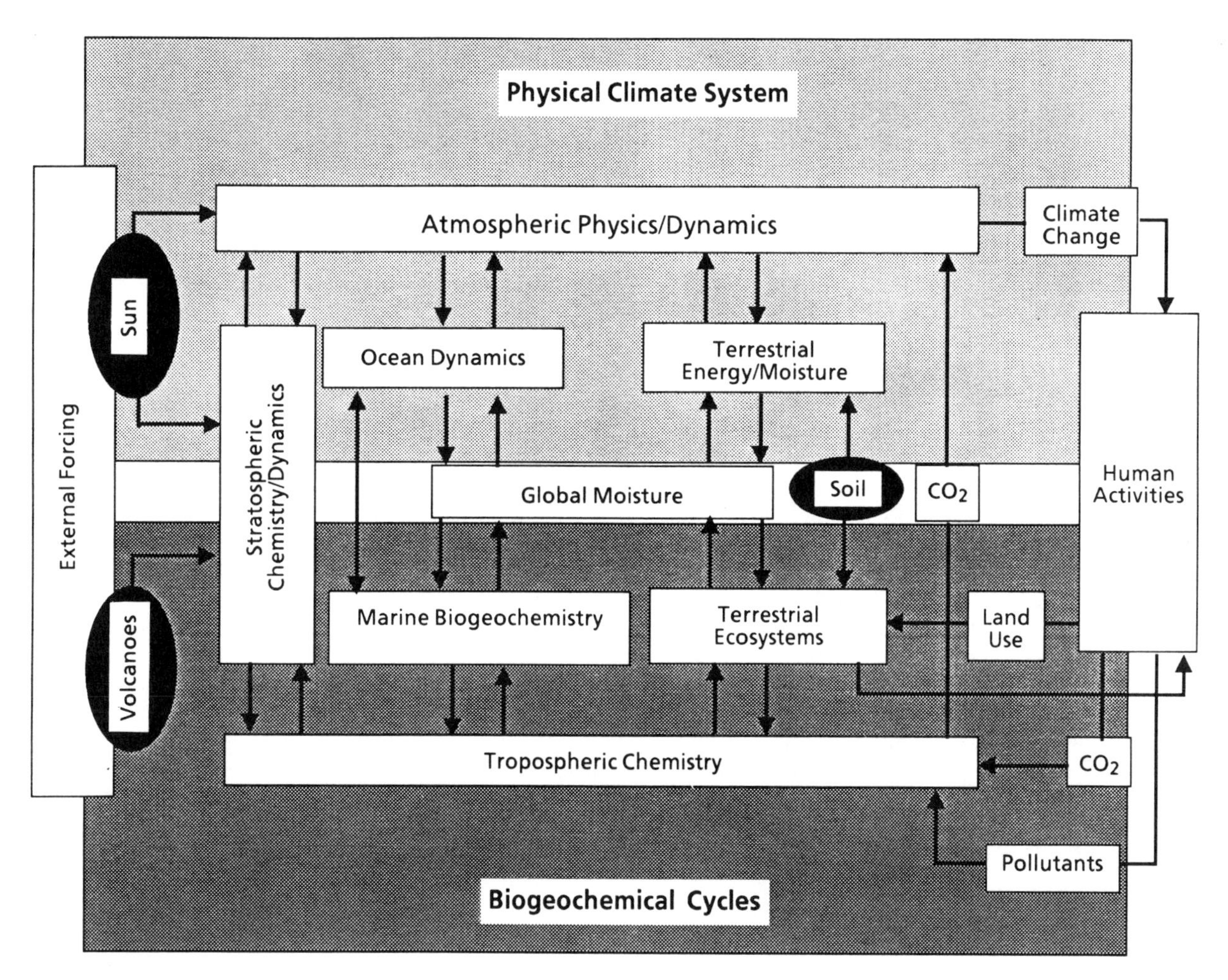

Figure 3-1: Conceptual model of the Earth system process operating on time scales of decades to centuries (Source: ESSC, 1986).

feedbacks to the climate system are central issues in this research.

By the mid-1980s, research strategies were developed to provide the scientific foundation to help answer these challenging questions. As outlined in the NRC (1986) and ICSU (1986) reports, the overall goal of proposed global change research was to describe and understand the interactive physical, chemical, and biological processes that regulate the Earth's unique environment for life, the changes that are occurring in this system, and the manner in which they are influenced by human actions.

In concert with this goal and as part of the Earth System Science concept, the development of an integrated model of the entire Earth system was proposed by ESSC (1986, 1988). Such a model is a necessary requirement to understand the Earth system and make reliable, long-term projections of future climate conditions. Under this proposed global change research plan, a network of Earth satellites, the proposed Earth Observation System, would provide the essential data in combination with in situ observations to understand the environmental processes to be modeled.

Figure 3-1 shows the conceptual model of the Earth system as originally discussed by the ESSC (1986, 1988). Such a modeling endeavor requires understanding of the physical climate system and biogeochemical cycles (the upper and lower sections of Figure 3-1, respectively). Processes within the atmospheric, oceanographic, cryospheric, and terrestrial components of the total Earth system must be modeled, as well as the exchanges across the interfaces, for example, land–atmosphere interactions.

The major components of the *physical climate system* include the atmospheric physics/dynamics, ocean dynamics, terrestrial energy/moisture/trace gas (CO_2 and methane) exchange at the land–atmosphere interface, and global moisture. The physical climate system is driven by variations of solar heating with latitude and the modes of energy partitioning at the surface, which leads to differential patterns of circulation, precipitation, evaporation, surface conditions, vegetation, and climate (ESSC, 1988). Differential heating between the equator and the poles leads to atmospheric and oceanic circulations that help to redistribute heat, momentum, and moisture.

Biogeochemical cycles are essential to the maintenance of life on Earth (ESSC, 1988). Key chemical constituents, such as carbon, nitrogen, sulfur, and oxygen, cycle through the atmosphere, lithosphere, and hydrosphere on varying time scales. This cycling has led to the evolution of the Earth's atmospheric constituents through the release of trace gases, such as methane, ammonia, nitrous oxide, and carbon dioxide, historically from natural processes, but more recently because of human activities. The understanding of the fluxes and reservoirs of these biogeochemical products, including the role of biological systems across several time scales, is essential.

This block diagram (Figure 3-1) of the global Earth system processes has become one of the most widely recognized symbols of global change research, as well as a challenge to the scientific community. In fact, the ESSC (1988) and NRC (1990) suggest that the development of a fully coupled, dynamical model of the Earth system by the year 2000, as conceptually described by Figure 3-1, is a realistic goal. Since these pioneering studies, global change research programs have evolved into focused research efforts at both international and national levels. The international component includes the IGBP (1990) under the auspices of the International Council of Scientific Unions (ICSU). The IGBP complements and is closely coordinated with the World Climate Research Program (WCRP) (1990), which is conducted jointly by the World Meteorological Organization and the ICSU. These two programs are complementary in that the IGBP focus is on the biological and biogeochemical components of the Earth system at local and regional scales, and the focus of the WCRP is more on the physical aspects of the climate system, including climate predictability and the influence of man on the climate system.

The U.S. Global Change Research Program (USGCRP), a national program, is most recently documented by the CEES (1991). The central goal of the USGCRP is to establish the scientific basis to support national and international policy-making relating to natural and human-induced changes in the global Earth system by:

- Establishing an integrated, comprehensive, long-term program of documenting the Earth system on a global scale;
- Conducting a program of focused studies to improve our understanding of the physical, geological, chemical, biological, and social processes that influence Earth system processes and trends on global and regional scales; and
- Developing integrated conceptual and predictive Earth system models.

All of these programs provide the conceptual framework for an advanced systems approach to research based on data collection, focused studies to understand the underlying processes, and the development of quantitative Earth system models for diagnostic and prognostic analyses. Concepts such as Earth system science (ESSC, 1986), global geosphere–biosphere modeling (IGBP, 1990), and proposed integrated modeling and "coupled systems" modeling at multiple scales (NRC, 1990) have emerged.

The terrestrial components

Modeling of terrestrial processes at local and regional scales is a large component of global change research. The term "atmosphere–terrestrial" subsystem is used by the NRC (1990) to describe the environmental processes that occur continuously at the land surface and that interact with the lower atmosphere. This interface zone near the land surface and lower atmosphere is also termed the terrestrial biosphere because of the predominance of life and biological processes.

The atmosphere–terrestrial subsystem is represented by the terrestrial ecosystems box on the biogeochemical cycles side and by the terrestrial energy/moisture box on the physical climate system side of Figure 3-1, respectively. Terrestrial ecosystems are linked to biogeochemical cycles through tropospheric chemistry, and terrestrial energy/moisture is linked with atmospheric physics/dynamics. There are fundamental links among terrestrial ecosystems, soils, and terrestrial energy/moisture components, also termed the "soil–plant–atmosphere continuum." Thus, there is a biosphere–atmosphere coupling in terms of tropospheric chemistry and physics/dynamics.

One of the scientific challenges lies in determining the quantitative relationships between the Earth's biosphere and the physical and chemical components of the global climate system (Sellers and McCarthy, 1990). Some of the features and components for this biosphere–atmosphere coupling are illustrated in Figure 3-2.

The major components illustrated by Figure 3-2 include the atmosphere, soils, surface physiology and hydrology (analogous to terrestrial energy/moisture in Figure 3-1), community composition structure (analogous to terrestrial ecosystems in Figure 3-1), biogeochemical and hydrological cycles, and anthropogenic activities. (Figure 3-2 applies to terrestrial biosphere interactions and does not consider effects or feedbacks due to the oceanic or ice components of the Earth system or due to external forcing from solar influences or volcanic activity such as illustrated in Figure 3-1.)

Figure 3-2 also shows how the coupling between the atmosphere and biosphere is characterized by multiple time scales, ranging from seconds to centuries. For example, there are two-way interactions involving atmospheric forcing on the biosphere (precipitation, wind, solar radi-

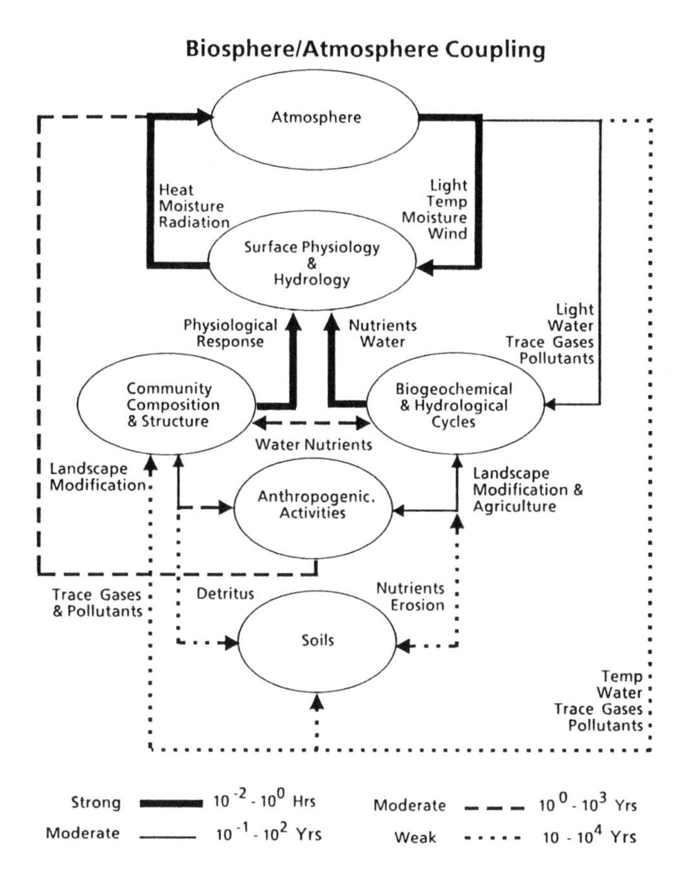

Figure 3-2: Major elements in the coupling between the atmosphere and terrestrial ecosystems. This diagram emphasizes the time scales involved in the various couplings. The lower atmosphere and surface vegetation are coupled with a fast-response loop through partitioning of incident solar radiation at the land surface and the subsequent circulation of moisture and heat in the lower atmosphere, which affects the physiology of the surface vegetation. The atmosphere is also coupled through weaker responses at longer time scales by climate modifications to biogeochemical and hydrological cycles, soils and community, and composition and structure (Source: Hall et al., 1988).

ation, temperature, and humidity) and the biospheric response at multiple time scales. These processes and feedback loops are integral to the discussion in the next section on modeling themes.

MODELING RESEARCH THEMES

Research on the terrestrial–atmosphere subsystem of global change programs is contributing to the development of environmental simulation models. This modeling is usually focused on biosphere–atmosphere coupling or land–atmosphere interactions, which includes the regional response to large-scale atmospheric forcing.

As a guide to understanding some of the approaches for this state-of-the-art modeling, the research activities are categorized according to six major research themes. These research themes are:

1. Cross-disciplinary models.
2. Models at multiple time scales.
3. Models at multiple space scales.
4. Physically based models of spatially distributed processes across complex terrain and heterogeneous landscape environments.
5. Integrated or coupled systems models.
6. Field experiments to develop and test models with ground-based, aircraft, and various remote sensing systems.

These themes are not necessarily distinct. Perhaps the dominant theme with the highest research priority is the development of integrated systems models, also referred to as coupled systems models. This research and modeling encompasses most of the other themes, especially those involving multiple time and space scales. Of major importance is the need for multiscale field experiments based on a wide variety of in situ, aircraft, and remote sensing technologies.

Many of the chapters in this volume include discussion of one or more of these research and modeling themes.

Cross-disciplinary models

Environmental processes themselves and associated feedback loops cross several disciplines, including atmospheric, hydrologic, soils, geomorphologic, biologic, and ecologic sciences (see Figures 3-1 and 3-2). To understand and model these processes, a cross-disciplinary research focus is needed.

This focus is illustrated by the modeling components of the hydrological cycle at the land–atmosphere interface. Specifically, the modeling of water and energy exchange mechanisms proceeds within the classic concept of the soil–plant–atmosphere continuum, which crosses several disciplines. However, even this concept must be viewed in a more cross-disciplinary mode for, as Lee et al. (Chapter 10) point out, there must be parameterizations to account for transfers of "water and energy exchange" between the land surface and the "free" atmosphere. Another example of this cross-disciplinary focus is biogeochemical cycling within plants, soil, water, and atmosphere.

This cross-disciplinary research focus is explicitly built into the global change research activities of the USGCRP (CEES, 1991), IGBP (1990), and WCRP (1990). For example, the science elements of the USGCRP include climate and hydrological systems, biogeochemical dynamics, ecological systems and dynamics, Earth system history, human interactions, solid earth processes, and solar influences.

The terrestrial components of the IGBP core projects include terrestrial biosphere–atmosphere chemistry interactions, biospheric aspects of the hydrological cycle, and effects of climatic change on terrestrial ecosystems. These are complemented by WCRP research at larger scales, such as the global energy water experiment (GEWEX) and climate modeling. Because processes including feedbacks and cross-disciplinary links are not well understood, the research approach includes strong links among data observation, environmental monitoring, process field studies conducted at multiple space scales, and model development activities. For example, in the recommended strategy for the USGCRP, the NRC (1990) emphasized focused studies on water–energy–vegetation interactions, as well as terrestrial trace gas and nutrient fluxes. These studies are to be linked to integrated or coupled modeling of the atmosphere-terrestrial subsystem. These cross-disciplinary links are basic to major field experiments conducted by the International Satellite Land Surface Climatology Project (ISLSCP) and the GEWEX Continental-Scale International Project (GCIP).

In response to the IPCC (1990), the USGCRP will emphasize four integrating themes: climatic modeling and prediction, global water and energy cycles, global carbon cycle, and ecological systems and population development. Furthermore, links with the IGBP and WCRP studies are planned.

Models at multiple time scales

As indicated in Figure 3-2, environmental processes for biosphere-atmosphere interactions occur across a wide spectrum of time scales. These include biophysical exchanges (fast response times); biological and biogeochemical processes (intermediate to long time scales); and changes in community, composition, and structure of the ecosystem (longer time scales).

In another approach to the time scale issue, the NRC (1990) organized current models of biosphere–atmosphere interactions according to three time step intervals based on the model structure. These three characteristic time step intervals are: (1) seconds to days, (2) days to weeks, and (3) annual increments. In addition to recommending focused research to improve models at each of these three time steps, the NRC (1990) also proposed a conceptual framework to link these three types of models. The coupling of these models across time is one of the central challenges in global change research. The NRC (1990) classification of models is used as a general guide to discuss the papers in this volume.

Seconds to days

At the shortest time scale, biophysical interactions between the biosphere at the land surface and the free atmosphere are dominant. Basic processes occur at time step intervals of seconds as part of the diurnal cycle. These processes include exchanges of solar and infrared radiation, momentum (i.e., wind), and heat fluxes, as well as associated plant biological processes such as CO_2 exchange and plant transpiration through stomatal cells during photosynthesis. The results of these exchanges are reflected in the land surface energy balance, the soil moisture budget, soil temperatures, and plant behavior. The very fast loop in Figure 3-2 shows the atmospheric forcing (light, temperature, and wind) and water and energy exchange feedbacks from the land surface as determined by the physiological, nutrient, and hydrologic responses from the underlying ecosystem and soil.

Figure 3-3 illustrates some of the major land surface processes that are characterized by time scales on the order of seconds to days. The concept of the soil–plant–atmosphere continuum is integral to the modeling of water and energy exchange processes, as well as associated biological processes such as photosynthesis. The energy exchanges in these biophysical processes are a function of land surface characteristics, such as albedo, vegetation type, and surface roughness. Modern atmospheric models, such as atmospheric general circulation models (GCMs) and regional mesoscale meteorological models, include various types of land surface parameterizations to account for exchanges between the atmosphere and these biophysical processes and land surface characteristics. Henderson-Sellers and McGuffie (1987) introduce basic concepts for GCMs, while Pielke (1984) provides an in-depth discussion on mesoscale meteorological models.

The overview chapter on atmospheric science by Lee et al. (Chapter 10) introduces atmospheric modeling concepts and provides a detailed comparison of three state-of-the-art land surface parameterizations for use in these types of atmospheric models. They discuss the Land Ecosystem–Atmosphere-Feedback Model (LEAF) (Lee, 1992), the Biosphere–Atmosphere Transfer Scheme (BATS) (Dickinson et al., 1986; Dickinson, 1983; 1984), and the Simple Biosphere Model (SiB) (Sellers et al., 1986; Xue et al., 1991; Xue and Sellers, Chapter 27).

The purpose of these models is to determine realistically the exchange of radiative, momentum, and sensible and latent heat fluxes between the land surface and the lower atmosphere. The models take into account atmospheric forcing (solar radiation, temperature, humidity, wind, and precipitation) and, in particular, the role of vegetation and other land surface properties in determining these exchanges. Figure 3-3 was used by Dickinson et

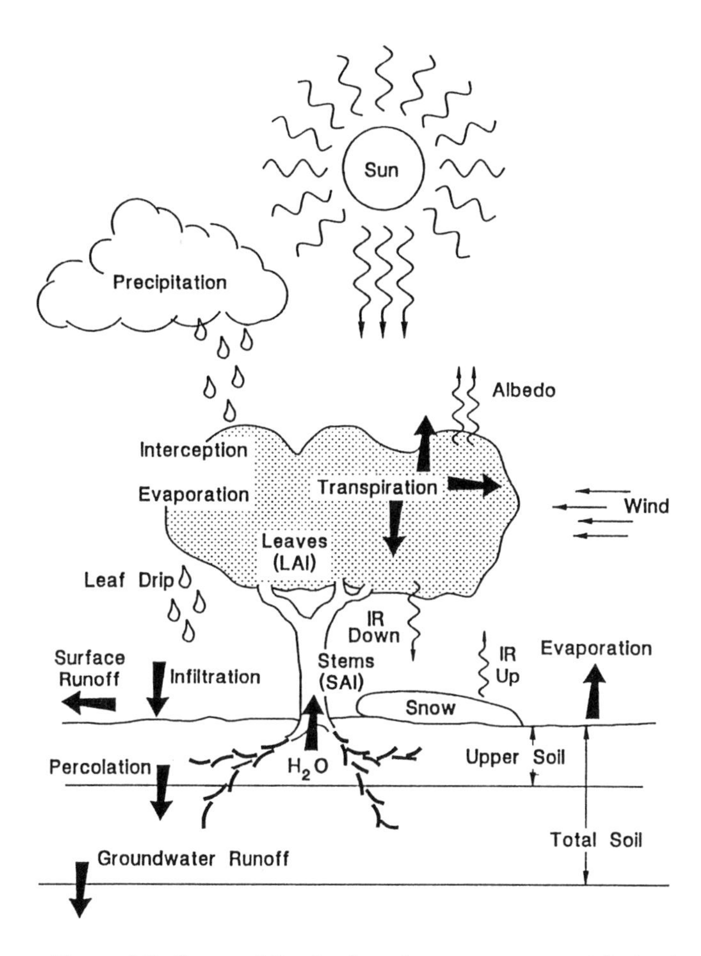

Figure 3-3: Some of the land surface processes used in land surface parameterizations (Source: Dickinson et al., (1986).

al. (1986) to illustrate land surface processes parameterized in the BATS.

Lee et al. (Chapter 10) illustrate how the LEAF Model is used with the Regional Atmospheric Modeling System (RAMS), a mesoscale meteorological model (see Pielke, 1984), to investigate how landscape properties and structural patterns influence mesoscale weather circulations.

Two other chapters (Avissar, Chapter 11; Skelly et al., Chapter 12) outline the use of similar land surface parameterization models to investigate appropriate methods to aggregate land surface properties within grid cells for atmospheric models. The issue of spatial aggregation of data for input to models is a scaling problem that remains unresolved.

Xue and Sellers (Chapter 27) illustrate the use of the SiB Model to investigate desertification in Africa. Models such as the SiB and BATS have been used to investigate how tropical deforestation potentially affects regional and global climate.

Days to weeks

The second class of models operates at intermediate time steps ranging from days to weeks. However, it must be noted that the models may be iterated by these intermediate time steps to get annual accumulations and multiyear simulations out to decades and centuries.

These types of models account for biological and plant physiological processes (such as nutrient cycles and growth processes), as well as some soil biogeochemical processes including decomposition, release of trace gases, and soil organic formulation at longer time scales. Thus, the energy exchanges at the short time scales drive the biology and biogeochemistry, which dominate this intermediate time scale.

The NRC (1990) characterizes models in this class as capturing weekly to seasonal dynamics of plant phenology, carbon accumulation, and nutrient uptake and allocation. Furthermore, the primary outputs are carbon and nutrient fluxes, biomass, leaf area index, and canopy height and roughness.

In general, ecosystem dynamic simulation models based on plant processes and biogeochemical cycling models operate in the range of this time scale. The diversity of modeling approaches is illustrated by three key models that are featured in this volume: Nemani et al. (Chapter 28), Hanson and Baker (Chapter 29), and Schimel and Burke (Chapter 26).

Nemani et al. discuss the Forest-Biogeochemical Cycle (BGC) as part of the Regional Hydro-Ecological Simulation System (RHESSys). The Forest-BGC is an ecosystem process model that calculates the carbon, water, and nitrogen cycles in a forest ecosystem (Running and Coughlan, 1988). The Forest-BGC model has been used to simulate the annual hydrologic balance and net primary production in a variety of studies involving climatic change scenarios (Running and Nemani, 1991; Nemani and Running, 1989; Running and Coughlan, 1988).

The model has daily and annual components. The daily time step accounts for the hydrologic balance, canopy gas exchanges, and photosynthetic partitioning to respiration and primary production. The model is simplified for regional scale analysis by the use of the leaf area index (LAI) to quantify forest structure important for water and energy exchange processes. The model is driven by daily meteorological data: air temperature, radiation, precipitation, and humidity. As described by Running et al. (1987), daily meteorological data are extrapolated from weather stations into microclimate estimates in mountainous terrain by an algorithm called MT-CLIM.

Annual summations for photosynthesis, evapotranspiration, and respiration are passed from the daily component to the yearly submodel. The yearly processes are carbon partitioning, growth respiration, litterfall, and decomposition (Running and Coughlan, 1988). The yearly submodel partitions available carbon passed from the daily model as an annual summation into leaves, stems, and roots.

As another example of a daily model for seasonal analysis, Hanson and Baker (Chapter 29) describe the Simulation and Utilization of Rangelands (SPUR) model. The model is a general grasslands ecosystems simulation model (Hanson et al., 1985) composed of seven basic components: soils, hydrology, plants, insects, management, animals, and economics. The purpose of the model is to simulate and investigate how these factors interrelate to determine grasslands behavior. There are regional watershed and field level versions of the SPUR model.

The plant component of the SPUR model is a multispecies model for rangeland plant communities described by Hanson et al. (1988). The model is driven by daily inputs of meteorological data, which are fed into the soils/hydrology component. In the plant component, carbon and nitrogen cycling account for standing green, standing dead, live roots, dead roots, seeds, litter, and soil organic matter (Hanson and Baker, Chapter 29). The model can be used to study how daily processes over the growing season are affected by other factors, such as management practices.

Schimel and Burke (Chapter 26) discuss the Century model, the final example in this section. This model simulates the temporal dynamics of soil organic matter (SOM) formulation and plant production (Parton et al., 1987) for the study of biogeochemical cycling in grasslands ecosystems. As discussed by Schimel et al. (1991), the model is used for climate change studies involving interactions of terrestrial biogeochemical cycles with atmospheric and hydrological processes.

The Century model is driven by monthly climatic data (precipitation and temperature), surface-soil physical properties, plant nitrogen and lignin contents, and land use to calculate plant production and the cycling of carbon, nitrogen, sulfur, and phosphorus through the soil (Burke et al., 1991). The model simulates decomposition and soil organic matter formulation for active SOM (1–2 years), intermediate SOM (20–40 years), and passive SOM (1000 years) (Schimel et al., 1991). Thus, the model is intended for multiyear simulations.

Annual time step

The annual time step is used for various types of ecological models. Hunsaker et al. (Chapter 22) review historical modeling approaches for terrestrial, freshwater, and marine ecosystems. Johnston (Chapter 25) discusses quantitative methods and modeling in community, population, and landscape ecology. A more comprehensive treatment of quantitative methods in landscape ecology can be found in Turner and Gardner (1991).

One major component of this research is on the biomass component of terrestrial ecosystems, specifically in terms of the development of community, composition, and structure. The specific focus is on the modeling of forest growth and the succession over periods of several decades to centuries due to natural causes, stochastic processes, or potential human-induced global change. In this context, the NRC (1990) characterizes this class of ecological models by annual changes in biomass and soil carbon (i.e., net ecosystem productivity and carbon storage). These types of ecological models address the community, composition, and structure of terrestrial ecosystems. As indicated by Figure 3-2, there is a very slow response by terrestrial ecosystems to changes in climate forcing, but the physiological characteristics of existing community, composition, and structure directly determine the fast response of biosphere/atmosphere coupling for heat, moisture, and radiation. Thus, the models such as LEAF or SiB for fast response processes are dependent on vegetation characteristics including regional landscape composition, structure, and function.

Many types of ecological models are characterized by the annual time step. Most notable are the forest dynamics models that simulate growth and forest succession. Two well-known examples are the JABOWA forest growth model (Botkin et al., 1972) and the class of FORET forest succession models (Shugart, 1984), descendants of the JABOWA modeling approach. These are termed Gap Models because they follow the birth, growth, and death of individual trees according to prescribed rules.

Nisbet and Botkin (Chapter 23) discuss the latest version of the JABOWA model, in particular the extension of the modeling to include complex spatial interactions. Forest growth is modeled in terms of annual changes in tree diameter as a function of available light, temperature, soil moisture, and available nitrogen in the soil.

Shugart (1984) discusses FORET for simulation of forest succession dynamics of complex mixed species and mixed age forests. Competition of individual trees on small plots in terms of annual growth is the basis for the simulation. Forest succession dynamics are estimated by the average behavior of 50 to 100 of these small plots.

As an example of an empirical approach, Hall et al. (1991) investigated large-scale patterns of forest succession based on remote sensing data. Landsat images for 1973 and 1983 were used to infer the spatial pattern of, and transition rates between, forest ecological states.

Across time scales

Environmental simulation models are categorized by three characteristic time steps in this section: seconds to days, days to weeks, and annual time steps. In general, these time steps correspond to the processes illustrated by Figure 3-2. The focus of much ongoing research is on the understanding of processes at each of these time slices. However, there is also effort to "couple" models across time, for example, as in the proposed strategy for integrated modeling within the NRC (1990). A subsequent section focuses on coupled systems models.

Models at multiple space scales

Several chapters in this volume illustrate some of the spatial scaling issues of importance to environmental simulation modeling. Spatial scaling issues are closely connected with other research themes, such as temporal scaling or coupled systems modeling themes. Basic concepts and examples are illustrated in this section.

Fundamentally, environmental processes operate at multiple space and time scales. Multiple spatial scales are evident in the fields of atmospheric science (hemispheric long-wave patterns, synoptic waves, mesoscale circulations, surface layer fluxes, and eventually the dissipation of microscale turbulence fluctuations near the land surface), hydrology (networks of small watersheds scaling up to large river basins), or ecology (patches, landscapes, and biomes).

One of the basic goals of land–atmosphere interactions modeling research is to be able to move up and down spatial scales, within disciplines and across disciplines. The use of remote sensing and GIS technologies to extend and extrapolate local results to the regional level is one approach. Another is the use of nested modeling concepts, sometimes in combination with remote sensing and GIS.

In fact, these represent the types of scientific issues recently addressed at two global change workshops. Running (1991) outlines the proposed use of a nested modeling strategy to extrapolate detailed ground-based results to large areas as part of the Boreal-Ecosystem Atmosphere Study (BOREAS) project. This proposed use of nested models to design, monitor, and assess a wide variety of field experiment observations is part of the strategy. Pielke et al. (1992) outline research strategies including nested modeling techniques for: (1) scaling global forcing down to the regional levels to determine regional response, (2) scaling up regional processes including land surface fluxes to determine their influence on global climate, and (3) determining the data requirements of various types of environmental simulation models to be linked in both space and time. The overlapping concepts of extrapolation and nested modeling are integral strategies to spatial scaling issues.

Extrapolation

One of the basic modeling challenges is to generate meaningful estimates at the regional level based on the appropriate extrapolation of detailed results derived at the local level. This process typically involves the use of a model developed at the plot or field site level to extrapolate across an entire region. Schimel and Burke (Chapter 26) discuss extrapolation techniques for the Century model, and Schimel et al. (1991) describe techniques involving the use of GIS and remote sensing technologies. The field of landscape ecology is fundamentally about the study of spatial patterns and ecological processes in terms of scaling from small areas to regional landscapes (Chapter 25).

To illustrate, Burke et al. (1991) describe their approach for regional analysis of potential climate effects on net primary production and the carbon balance of grasslands ecosystems in the central Great Plains. The three-step process for extrapolating site level information involved: (1) conducting process-level research at field sites to identify the major driving variables for ecosystem structure and function; (2) developing and testing simulation models, such as the Century model, that represent these variables; and (3) generating regional patterns and dynamics based on input of regional data into the simulation model.

As another example, Nemani et al. (Chapter 28) discuss their approach for extrapolation of daily weather station data throughout a complex terrain mountain environment. The MT-CLIM algorithm is used to obtain terrain-adjusted inputs of daily weather data for the Forest-BGC model.

In fact, extrapolation of results across broad spatial scales is one of the primary objectives of the ISLSCP field experiments, for example, the First ISLSCP Field Experiment (FIFE) (Sellers et al., 1988) and the planned BOREAS project. Parameterizations and other methods are developed on the basis of results from detailed process-level studies, and then regional extrapolations are made on the basis of the combined use of these methods with satellite and other remote sensing data.

Nested models

The nested modeling concept is receiving widespread attention as a means to address both complex spatial and temporal scaling problems. Some spatial aspects are illustrated here.

In one approach, the output of a large-scale model such as a GCM is used as the input for a smaller-scale, higher-resolution model. For example, Dickinson et al. (1989) used a nested modeling approach to simulate more realistic rainfall distributions from winter storm systems over the Western United States. Hemispheric wave and synoptic features for major weather systems, as simulated by the coarse-resolution GCM, were used as input to a higher-resolution mesoscale meteorological model. The mesoscale model, in combination with more detailed terrain and other land surface properties, was able to simulate more realistic precipitation distribution patterns. Such capabilities are important to water resource assessment and regional response of terrestrial ecosystems. Hay et al. (Chapter 16) describe a similar approach for nesting a mesoscale meteorological model in a GCM for hydrologic basin assessment. They also use an alternative approach in which GCM output of atmospheric winds and humidity are used as input to a detailed orographic precipitation model. The terrain features, including the orientation with respect to the direction of moist, low-level atmospheric flow, are a major factor in determining precipitation distributions through the orographic uplift process. Hay et al. (Chapter 16) and Leavesley et al. (in preparation) also address spatial scaling issues in the development of hydrologic basin characteristics.

Running (1991), reporting on the results of a Pre-BOREAS Ecological Modeling Workshop, described a proposed nested modeling approach for extrapolation to larger space scales. The intent is to use nested modeling to integrate land surface processes made at different spatial and temporal scales (ground, towers, aircraft, and remotely sensed observations). The plan calls for using data and models at one scale to develop parameters and validation criteria for the next higher scale.

As outlined by Running (1991), hierarchical or nested modeling is needed "to connect fine spatial and temporal scales, where many biophysical processes are well measured and understood, to coarser scales amenable to aircraft flux and remote sensing measurements." In this proposed approach, multitiered, complex canopy models of instantaneous radiation and photosynthesis are calibrated against daily observations at the tree level. These observations are used to define and calibrate simple canopy models of daily photosynthesis, such as Forest-BGC, which can be run over the entire growing season for multihectare stands. The goal is then to define and calibrate very simple annual models relating absorbed photosynthetically active radiation (APAR) to net primary production (NPP). These annual models can be run for very large regions based on satellite data input.

These same concepts are also applicable to the scaling of evapotranspiration estimates from the microscale to large regions. Jarvis and McNaughton (1986) provide an excellent overview on uncertainties concerning basic processes and other issues relating to the scaling of plant transpiration from the stomatal cell level to the leaf, plant, canopy, and region. The rationale for this nested modeling approach is outlined by Running (1991). Nemani et al. (Chapter 28) discuss the use of RHESSys model to investigate multiscale evapotranspiration processes within various sizes of watersheds within a complex terrain environment.

Spatial aggregations

Many aspects of the parameterization of subgrid processes or the aggregation of landscape properties up to a grid level estimate have been previously addressed. Important issues on this topic are discussed by Lee et al. (Chapter 10), Avissar (Chapter 11), and Skelly et al. (Chapter 12). These issues are critical to the parameterization of land surface processes for input to meteorological models.

Physically based models of spatially distributed processes

For many reasons, the issue of using environmental simulation models in a complex terrain and heterogeneous landscape environment has become one of the top research priorities. Elements of this issue are reflected in the scaling themes. Almost all chapters in this volume address some aspect of this topic, which really is about land surface characterization and its role in environmental simulation modeling.

In a sense, two diverse groups are the main drivers. These are the atmospheric modeling community with their large GCM grid cells at one extreme, and the ecologists and other researchers with most of their investment on small-scale plot or field studies. The atmospheric modeling community wants more detailed information on land surface processes, including higher resolutions for enhanced modeling results. The ecological community wants to extrapolate their results across the landscape to regional estimates. Previous discussion on spatial scaling issues is relevant to these two groups.

The importance of the landscape environment and land surface properties to atmospheric modeling has been demonstrated by the GCM modeling community (Chapter 27; Pielke et al., 1991). As discussed by Lee et al. (Chapter 10), several land-atmosphere transfer schemes such as LEAF, BATS, and SiB have been developed. These models require detailed land surface data. Spatial scaling issues have not been resolved (Chapter 11; Chapter 12). However, the importance of landscape to biosphere–atmosphere modeling has become more evident based on recent research results at the mesoscale level (Pielke and Avissar, 1990; Pielke et al., 1992; Avissar and Pielke, 1989). Novak and Dennis (Chapter 13) discuss the importance of land surface characteristics for regional oxidant modeling and biogenic emissions modeling.

There is also the trend away from using "lump sum" or "black box" types of models in favor of using more physically based, distributed parameter models within a complex terrain and heterogeneous landscape environment (Chapter 28). This type of approach is at the forefront of research on hydrological watershed models. The potential benefits are obvious in modeling the land–atmosphere interactions, as well as the runoff, infiltration, and groundwater flow processes in terms of realistic, detailed hydrologic basin and landscape properties (Hay et al., Chapter 16; Gao et al., Chapter 17). This distributed, physically based approach contrasts with more limited models such as "black box" methods applied in a "lump sum" approach to estimate the water budget within the watershed or the classic bucket model for soil moisture budgeting.

Research on the nature of heterogeneous landscape environments is fundamental to ecological modeling, including landscape ecology (Hunsaker et al., Chapter 22; Johnston, Chapter 25). Determining the factors that lead to forest succession in terms of changes in community, composition, and structure is basic to global change modeling. This research includes the use of remote sensing to investigate spatial patterns of change and disturbance (Hall et al., 1991).

Integrated or coupled systems models

The issues raised within the preceding discussion on each of these modeling themes help to illustrate the challenges in the development of an integrated coupled systems model of the Earth system as shown in Figure 3-1. Such environmental simulation modeling involves a mixture of all these themes. The importance of cross-disciplinary modeling across multiple time and space scales is evident.

Figure 3-2 provides insight for the portion of the Earth system model that involves biosphere–atmosphere interactions. The preceding discussion has examined the types of cross-disciplinary processes, the multiple scale relationships, and some of the research efforts to improve models at each scale. The NRC (1990) suggests a conceptual strategy to link the models across time scales, where processes range from seconds to centuries. One of the challenges is to model the feedbacks. For example, if climate forces a change in vegetation (community, composition, and structure), then the surface characteristics that control fast response processes must also change. The NRC (1990) suggests the need for translator and integrator modules between the GCM and the land surface parameterizations, such as BATS and SiB, as well as between the GCM and the ecosystem dynamics models. Nested modeling approaches are integral to this system integration.

Several chapters in this volume address coupled systems integration issues. For example, the land surface parameterizations (LEAF, BATS, and SiB) are designed to make a first-order coupling between the land surface and the atmosphere. Hay et al. (Chapter 16) illustrate nested modeling approaches that couple a GCM with a mesoscale meteorological model, as well as an orographic rainfall model for hydrological research. Nemani et al. (Chapter 28) describe how the RHESSys modeling approach is designed to provide nested ecosystems modeling within a complex terrain and landscape environment.

Several papers in this volume illustrate the initial steps for integrated modeling across multiple time scales. These include: (1) fast response time models such as LEAF (Lee et al., Chapter 10) and SiB (Xue and Sellers, Chapter 27), (2) intermediate time step modeling approaches such as RHESSys (Nemani et al., Chapter 28), SPUR (Hanson and Baker, Chapter 29) and Century (Schimel and Burke, Chapter 26), and (3) longer-term time scales involving forest successional issues (Nisbet and Botkin, Chapter 23).

Field experiments to develop and test models

The development and validation of integrated coupled systems models, even at regional levels, require a major effort to understand complex environmental processes. To help meet this need, the overall research approach of global change programs incorporates strong links between data observation, environmental monitoring, process-oriented field studies conducted at multiple space scales, and model development activities. Remote sensing technology in combination with in situ observations is essential to understanding processes, to building the models, and to developing the data sets for input to the models.

The ISLSCP projects such as FIFE and the planned BOREAS project illustrate the types of process-oriented field experiments that are needed. These projects involve a wide variety of ground-based, aircraft, and remote sensing systems to investigate cross-disciplinary environmental processes such as previously outlined and shown in Figure 3-2. The preceding discussion on the proposed use of a nested modeling approach illustrates the importance of combining field-level data, model results, and remote sensing technology for maximum benefit.

CONCLUDING REMARKS ON LINKS TO GIS

The integrated systems approach for developing and testing environmental simulation models suggests potential links to GIS technology. In conceptual terms, GIS seem well suited to address data and modeling issues that are associated with a modeling environment that includes multiscale processes, all within a complex terrain and heterogeneous landscape domain. GIS can help address data integration questions associated with multiscale data from ground-based and remote sensing sources. GIS could potentially support exploratory analysis of complex spatial patterns and environmental processes. Finally, these advanced environmental simulation models require detailed spatial data, which provides an opportunity for innovative thematic mapping and error analyses with a GIS. However, there is much to be done to meet the needs of the modeling community.

GIS can contribute to spatial data issues. To illustrate, land surface characteristics data are needed by the scientific research community. These data include land cover, land use, ecoregions, topography, soils, and other properties of the land surface to help understand environmental processes and to develop environmental simulation models (Loveland et al., 1991). These advanced land surface process models also require data on many other types of land surface properties, for example, albedo, slope, aspect, LAI, potential solar insolation, canopy resistance, surface roughness, soil information on rooting depth and water holding capacity, and the morphological and physiological characteristics of vegetation.

These land surface properties are complex. For example, land surface properties such as albedo, canopy conductance, LAI, and surface roughness have distinct temporal components that are critical to realistic models. Frequently, there is uncertainty in how to define requirements for these types of land surface properties for input to advanced land surface process modeling. Such uncertainty often stems from incomplete knowledge about fundamental processes (e.g., the plant transpiration and photosynthetic process at the stomatal cell level), scaling from small- to large-area estimates, methods for integrating and aggregating data in space and time, and the interrelationship of data sets in space and time.

GIS can potentially help meet these requirements and provide the flexibility for the development, validation, testing, and evaluation of such innovative data sets that have distinct temporal components. Capabilities are needed to convert existing data sets into derivative data sets with provisions for flexible scaling, multiple parameterizations and classifications, grid cell resolutions, or spatial aggregations and integrations. In summary, the integration of GIS and environmental systems process models can proceed along at least two major themes: database issues and analysis and modeling issues.

ACKNOWLEDGMENTS

Permission to reproduce Figures 3-1, 3-2, and 3-3 from Dr. Francis Bretherton of the University of Wisconsin and formerly Chairman of the NASA Earth System Science Committee, Dr. Robert Dickinson of the University of Arizona and formerly of the National Center for Atmospheric Research, and Dr. Forrest Hall of NASA, respectively, is acknowledged and sincerely appreciated. Permission from the editor of *Landscape Ecology* to reproduce Figure 3-3 is appreciated. The many helpful comments and suggestions provided by Dr. Roger A. Pielke, Department of Atmospheric Science, Colorado State University, and Dr. Forrest G. Hall, Biospheric Sciences Branch, NASA/Goddard Space Flight Center, are sincerely appreciated. This research was conducted in conjunction with U.S. Geological Survey approved research projects GIS/Modeling (RGEG 88-4) and Land Surface Energy Balance (RGEG 81GC 1791).

REFERENCES

Avissar, R., and Pielke, R.A. (1989) A parameterization of heterogeneous land surfaces for atmospheric numerical models and its impact on regional meteorology. *Monthly Weather Review*, 117(10): 2113–2136.

Botkin, D.B., Janak, J.F., and Wallis, J.R. (1972) Rationale, limitations, and assumptions of a Northeastern forest growth simulator. *IBM Journal of Research and Development* 16: 101–116.

Burke, I.C., Kittel, T.G.F., Lauenroth, W.K., Snook, P., Yonker, C.M., and Parton, W.J. (1991) Regional analysis of the Central Great Plains. *Bioscience* 41(10): 685–692.

Committee on Earth and Environmental Sciences (1991) *Our Changing Planet: The FY 1992 U.S. Global Change Research Program*, Washington, DC: Federal Coordinating Council for Science, Engineering, and Technology, Office of Science and Technology Policy.

Committee on Earth Sciences (1989) *Our Changing Planet: A U.S. Strategy for Global Change Research.* Washington, DC: Federal Coordinating Council for Science, Engineering, and Technology, Office of Science and Technology Policy.

Committee on Earth Sciences (1990) *Our Changing Planet: The FY 1990 Research Plan*, Washington, DC: Federal Coordinating Council for Science, Engineering, and Technology, Office of Science and Technology Policy.

Dickinson, R.E. (1983) Land surface processes and climate-surface albedo energy balance. *Advances in Geophysics* 25: 305–353.

Dickinson, R.E. (1984) Modeling evapotranspiration for three-dimensional climate models. In Hansen, J.E., and Takahashi, T. (eds.) *Climate Processes and Climate Sensitivities*, American Geophysical Union, Vol. 29, pp. 58–72.

Dickinson, R.E., Errico, R.M., Giorgi, F., and Bates, G.T. (1989) A regional climate model for the Western United States. *Climatic Change* 15: 383–422.

Dickinson, R.E., Henderson-Sellers, A., Kennedy, P.J., and Wilson, M.F. (1986) Biosphere-Atmosphere Transfer Scheme (BATS) for the NCAR Community Climate Model. *NCAR Technical Note NCAR/TN-275+STR*, 69 pp.

Earth System Sciences Committee (1986) *Earth System Science Overview: A Program for Global Change*, Washington, DC: National Aeronautics and Space Administration.

Earth System Sciences Committee (1988) *Earth System Science: A Closer View*, Washington, DC: National Aeronautics and Space Administration.

Frenkiel, F.N., and Goodall, D.W. (eds.) (1978) *Simulation Modeling of Environmental Problems*, New York: John Wiley and Sons.

Hall, F.G., Botkin, D.B., Strebel, D.E., Woods, K.D., and

Goeta, S.T. (1991) Large-scale patterns of forest succession as determined by remote sensing. *Ecology* 72(2): 628–640.
Hall, F.G., Strebel, D.E., and Sellers, P.J. (1988) Linking knowledge among spatial and temporal scales: vegetation, atmosphere, climate and remote sensing. *Landscape Ecology* 2(1): 3–22.
Hanson, J.D., Parton, W.J., and Innis, G.S. (1985) Plant growth and production of grasslands ecosystems: a comparison of modeling approaches. *Ecological Modelling* 29: 131–144.
Hanson, J.D., Skiles, J.W., and Parton, W.J. (1988) A multi-species model for rangeland plant communities. *Ecological Modelling* 32: 89–123.
Henderson-Sellers, A., and McGuffie, K. (1987) *A Climate Modeling Primer*, New York: John Wiley and Sons.
Intergovernmental Panel on Climatic Change (1990) *Climate Change: The IPCC Scientific Assessment*, Cambridge University Press.
International Council of Scientific Unions (1986) The International Geosphere–Biosphere Program: a study of global change. *Report No. 1, Final Report of the Ad Hoc Planning Group, ICSU Twenty-first General Assembly, September 14–19, 1986, Bern, Switzerland.*
International Geosphere–Biosphere Programme (1990) The International Geosphere–Biosphere Programme: a study of global change. The initial core projects. *Report No. 12*, Stockholm: IGBP Secretariat.
Jarvis, P.G., and McNaughton, K.G. (1986) Stomatal control of transpiration: scaling up from leaf to region. *Advances in Ecological Research* 15: 1–49.
Law, A.M., and Kelton, W.D. (1982) *Simulation Modeling and Analysis*, New York: McGraw-Hill Book Company.
Leavesley, G.H., Branson, M.D., and Hay, L.E. (in preparation) Coupled orographic-precipitation and distributed-parameter hydrologic model for investigating hydrology in mountainous regions.
Lee, T.J. (1992) The impact of vegetation on the atmospheric boundary layer and convective storms. *PhD Dissertation*, Fort Collins, CO: Department of Atmospheric Science, Colorado State University (in preparation).
Lorenz, E.N. (1963) Deterministic nonperiodic flow. *Journal of Atmospheric Science* 20: 131–141.
Loveland, T.R., Merchant, J.W., Ohlen, D., and Brown, J.F. (1991) Development of a land-cover characteristics database for the conterminous U.S.. *Photogrammetric Engineering and Remote Sensing* 57(11): 1453–1463.
National Research Council (1986) *Global Change in the Geosphere–Biosphere, Initial Priorities for an IGBP*, Washington, DC: U.S. Committee for an International Geosphere-Biosphere Program, National Academy Press.
National Research Council (1990) *Research Strategies for the U.S. Global Change Research Program*, Washington, DC: Committee on Global Change (U.S. National Committee for the IGBP), National Academy Press.
Nemani, R.R., and Running, S.W. (1989) Testing a theoretical climate–soil–leaf area hydrologic equilibrium of forests using satellite data and ecosystem simulation. *Agricultural and Forest Meteorology* 44: 245–260.
Parton, W.J., Schimel, D.S., Cole, C.V., and Ojima, D.S. (1987) Analysis of factors controlling soil organic levels in Great Plains grasslands. *Journal, Soil Science Society of America* 51: 1173–1179.
Pielke, R.A. (1984) *Mesoscale Meteorological Modeling*, Orlando, FL: Academic Press.
Pielke, R.A., and Avissar, R. (1990) Influence of landscape structure on local and regional climate. *Landscape Ecology* 4: 133–155.
Pielke, R.A, Bretherton, F., Schimel, D., and Kittel, T.G.F. (1992) Global Forcing and Regional Interactions Workshop report of the Coupled Systems Modeling Program (CSMP) (submitted for publication).
Pielke, R.A., Dalu, G.A., Snook, J.S., Lee, T.J., and Kittel, T.G.F. (1991) Nonlinear influence of mesoscale land use on weather and climate. *Journal of Climate* 4: 1053–1069.
Running, S.W. (ed.) (1991) *Proceedings, Pre-BOREAS Ecological Modeling Workshop, Flathead Lake, MT, August 18–24, 1991.*
Running, S.W., and Coughlan, J.C. (1988) A general model of forest ecosystem processes for regional applications. I: hydrologic balance, canopy gas exchange and primary production processes. *Ecological Modelling* 42: 125–154.
Running, S.W., and Nemani, R.R. (1991) Regional hydrologic and carbon balance responses of forests resulting from potential climate change. *Climatic Change* 19: 349–368.
Running, S.W., Nemani, R.R., and Hungerford, R.D. (1987) Extrapolation of synoptic meteorological data in mountainous terrain and its use for simulating forest evapotranspiration and photosynthesis. *Canadian Journal of Forest Research* 17: 472–483.
Schimel, D.S., Kittel, T.G.F., and Parton, W.J. (1991) Terrestrial biogeochemical cycles: global interactions with the atmosphere and hydrology. *Tellus* 43AB: 188–203.
Sellers, P.J., Hall, F.G., Asrar, G., Strebel, D.E., and Murphy, R.E. (1988) The First ISLSCP Field Experiment (FIFE). *Bulletin, American Meteorological Society* 69: 22–27.
Sellers, P.J., and McCarthy, J.J. (1990) Planet Earth: part III, biosphere interactions. *EOS* 71(52): 1883–1884.
Sellers, P.J., Mintz, Y., Sud, Y.C., and Dalcher, A. (1986) A Simple Biosphere Model (SiB) for use within general circulation models. *Journal of Atmospheric Science* 43(6): 505–531.

Shugart, H.H. (1984) *A Theory of Forest Dynamics: The Ecological Implications of Forest Succession Models*, New York: Springer-Verlag, 278 pp.
Turner, M.G., and Gardner, R.H. (eds.) (1991) *Quantitative Methods in Landscape Ecology: The Analysis and Interpretation of Landscape Heterogeneity*, New York: Springer-Verlag, 536 pp.
World Climate Research Program (1990) *Global Climate Change: A Scientific Review*, Geneva: World Meteorological Organization and International Council of Scientific Unions.
Xue, Y., Sellers, P.J., Kinter, J.L., and Shukla, J. (1991) A simplified biosphere model for global climate studies. *Journal of Climate* 4: 345–364.

4

The Need for Integration

BRADLEY O. PARKS

It is tempting to presume that better integration of GIS and environmental modeling is demanded by the opportunity of combining ever-increasing computational power, more plentiful digital data, and more advanced models. However, while important scientific and technological opportunities at the frontiers of research do exist, and while timely supporting applications for these tools can and will be found, such opportunities do not present the most important reasons for more tightly coupling these tools. Opportunities alone will not enable practitioners and technocrats with limited means and closely defined responsibilities to incorporate GIS/modeling tools in their work nor will they necessarily encourage the best implementation of new and better "hybrid" tools, as is advocated here. And in any case, such integration is certainly not foreordained or self-propelled, as was evidenced by the large and enthusiastic support for the innovative conference from which this volume springs.

There are more compelling, and perhaps more commonplace, reasons to recognize and obtain the benefits of joining GIS and environmental modeling. These have to do with the current limitations of the tools themselves and with the constraints imposed on their potential users, many of whom need to respond to problems in ways not typical of research endeavors. Such problems are often expressed through multiple social or institutional filters and frequently demand brief evaluation of fragmentary, poorly understood, strongly contested, and sometimes rapidly changing data with inadequate resources committed to resolving the problem for a waiting public. This is probably indicative of the manner in which research-level environmental issues will increasingly evolve into difficult daily management problems requiring simple but much more imaginative analytical and interpretive tools.

Actually, one could make an argument against the trouble and expense of coupling these tools if it were to mean that each would remain largely unaffected by the other, simply becoming joined "tail to nose," or that their combinations would prove inaccessible, inflexible, or unfriendly to nonresearch users. Strong consideration is needed in the beginning for the ways in which these tools might cross-fertilize and mutually reinforce each other and for the new ways in which they could be designed to serve additional users—including those who must inform or participate in common decision-making processes.

What is described in this volume, and was explored in the conference that preceded it, is not a new idea as such, but rather the further co-adaptation of existing tools, and methods. It is worth remembering that models have long been employed with GIS and they are generally useful whether they are simply schematic descriptions for the procedures used in combining spatial datasets, or whether they are more rigorous quantitative exercises deriving new data sets based upon empirical or theoretical relationships, knowledge of trends, and information about human needs and preferences.

Likewise, numerically oriented models have long made use of spatial and graphical representations of data, physical properties, and interesting phenomena, even if these representations have sometimes seemed crude by current standards. Moreover, modeling has formed the core of a great deal of research focusing on inherently geographical aspects of our environment, and has led to understanding of distributions and spatial relationships in everything from astronomy to microbiology and chemistry. In fact, many scientific endeavors have made and continue to make sophisticated use of tools that have strong functional relationships to GIS but that are not celebrated as equivalent or distinct technologies and that may not have the same ancestors as GIS. As the chapters in this volume attest, many persons, especially those active in research fields, already make effective use of GIS and modeling techniques on a parallel or serial basis wherein GIS becomes a pre/postprocessor for model input and output data, an ancillary data exploration tool, or a means for parameterizing, calibrating, or otherwise adjusting models. These models, in turn, serve as "engines" for anticipating alternate environmental outcomes that are valuable because they help to define our available choices. Given all this, one might ask, what is unique, urgent, or particular in the call to intensify efforts to couple GIS and modeling techniques?

DEFINING NEEDS FOR INTEGRATION

The need for integration is driven by our need to make environmental choices. Research to increase knowledge can inform the act of making choices, with hopes of producing good decisions. But the co-adaptation of GIS and environmental modeling to support practical decision processes can simultaneously, without self-contradiction, and with little extra effort, be made to support research needs as well. The reverse would not necessarily be true.

GIS and environmental modeling originated in and still represent substantially different domains of expertise (see the preceding perspectives in Chapters 2 and 3, by Goodchild and Steyaert, respectively), yet their complementarity will benefit environmental problem-solving most if the tools themselves, and the methods for their use, can be made to interpenetrate. Success with that task will then have the greatest benefit if results can be made accessible and productive for large numbers of users not specifically trained in both domains, or users whose primary purpose lies outside of either domain. Other fields and methods such as artificial intelligence and expert systems may need to be tapped to develop the means to guide nonexpert users in the appropriate handling of these hybrid tools.

It is unlikely, given fundamental differences, that environmental modeling would converge naturally with GIS in a meaningfully short period of time and without significant intervention by proponents of the idea. Despite historical affinity seen between GIS and other now-allied technologies and the progress made by users who have already been forced to harness GIS and modeling in tandem, unaided gravitation of these tools toward a common software environment would likely be a protracted process producing little of enduring value. The alternative is to formalize first the need and then the method for achieving a desirable blend and then to stimulate discussion and development of the required tools. Three primary reasons for integration suitable to begin such discussion are given here.

1. Spatial representation is critical to environmental problem solving, but GIS currently lack the predictive and related analytic capabilities necessary to examine complex problems.

Environmental problem-solving, ostensibly the goal of environmental modeling and environmental applications of GIS, requires that spatial aspects of problems be examined explicitly and that solutions incorporate such knowledge. GIS currently excels in such tasks so long as it relies on its ability to map what is manifest within a largely two-dimensional universe where all things seem certain, or equally uncertain, and where change with time is typically not a limiting consideration. Current GIS are typically limited to analytic compromises that include: static representations of dynamic space/time processes; use of simple logical operations to explore complex relationships; nonstochastic treatments of uncertain events; and quasi-two-dimensional or perspective treatments of inherently three-dimensional properties. Unfortunately GIS can become the implement of very simplistic thinking and the victim of perpetual data inadequacies insofar as they do not include or cannot easily facilitate the use of numerical models, highly time-variant relationships, true three-dimensional objects, and robust spatial statistical functions.

Models particularly could allow GIS to function outside the limits of a static, planar domain where complex, often nonlinear relationships can be usefully expressed and where change and uncertainty can be addressed in a direct way. Whether models are used to diagnose or predict, it is their ability to help us understand or anticipate change in complex or dynamic systems that makes them so useful. GIS are currently very limited in their ability to examine any dynamical processes unless they are "wired up" in advance by analysts having singular objectives and very good understanding of both the technical aspects of GIS and the operation of a given model. This probably presumes more knowledge than many users possess. Similarly, the toolbox approach to GIS design whereby functionality can be extended by adapting a kit of primitive operators to perform new tasks also demands too much specialized knowledge on the part of users and presumes too much of the vendor's motivation.

The capabilities to represent true three-dimensional data and to make statistical inferences about data, models, and results are supporting functions that are perhaps not as central as the issue of model compatibility, but they are no less important. Most environmental features or phenomena have inherent three-dimensional properties, the complexity of which may need to be preserved and represented as volumes of space with interiors. This additional complexity must of course be made tractable by appropriate analytical tools as it varies in both time and space, and any uncertainty or error introduced by data, tools, or processes needs to be examined and accounted for. All of this requires additional intelligent tools or suites of such tools that interact more or less seamlessly.

We need to encourage the development of GIS beyond that of a map-based data viewer capable of limited logical analytic operations. Although these functions are clearly useful, they are inadequate to support fully the evaluation and decision-making processes required to manage complex environmental problems.

2. Modeling tools typically lack sufficiently flexible GIS-like spatial analytic components and are often inaccessible to potential users less expert than their makers.

Implementations of physically based process models have often relied on specialized graphical user interfaces and limited spatial operators developed specifically to structure the model's use or to help reveal spatial relationships discovered in data or model output. GIS have not generally been amenable to this task without considerable effort to customize them. In recent years, simpler models have been linked closely to GIS and some more complex ones have begun to make extensive, although more remote, use of GIS.

What GIS could offer to modeling, in part, is a flexible environment with a standardized array of spatial operators based on mathematical principles that describe the motion, dispersion, transformation, or other meaningful properties of spatially distributed entities. Sufficient intelligence could be built into such a tool to prevent some forms of misuse by the uninformed. Such an approach offers obvious economies in addition to the benefits of a potentially common analytic medium in which more comparability would be possible and through which it might be possible to improve communications among modelers working in different disciplines. But for this to happen will require that modeling linkages be designed in. Whether such functions are to be supported entirely within or outside GIS software is less important than the engineering of workable linkages. Many examples of such linkages are included in this volume. A related problem is that models whose value is not strictly limited to the research sphere or that are already widely accepted are often not easily implemented by the variety of users who might benefit from them. At the operational level it is often true that persons without training and experience identical to that of the maker of a model will not necessarily know how to make proper use of it. Those without any such specialized training are more constrained in their ability to make intelligent use of models that could help solve or inform debate about environmental problems. On a more abstract level, nontechnical users will typically not be conversant with the mathematical methods or statistical reasoning upon which a given model may rely. And most fundamentally, such nonexpert users will likely not adequately understand the fundamental physical processes that govern all other aspects of a model's use.

It should be possible, through a properly organized effort, to begin to adapt at least well-tested models for a broader group of users. This process could begin with the identification of any common intellectual basis for the mathematical, physical, statistical, or other processes in use within or across disciplines. From this beginning design criteria could be developed to exploit the functionality thought feasible within GIS maximally. Those functions best supported externally could then be made the focus of efforts to build effective links and user guidance systems.

3. Modeling and GIS technology can both be made more robust by their linkage and coevolution.

Regardless of where new GIS/modeling co-functionality comes to reside, in or out of a GIS or a modeling software environment, the effort to combine the strengths of these tools will be mutually beneficial. Naturally it is impossible to foresee all such benefits, but one can speculate with confidence about several that already have a basis in experience.

GIS (and their 3D and spatial statistical counterparts) could grow from what many believe to be comparative immaturity by being examined relative to the increased functionality required to enable many more users to execute process modeling tasks in support of environmental decision-making. Modeling, on the other hand, would benefit by the better engagement of the visual senses in evaluating the assumptions, operations, and results of models. Doubtless many readers could recount experiences wherein the better mapping/visualization of spatial properties brought new and sometimes startling understanding to those previously confident that they fully understood a target system and methods for its analysis. Similar revelations could await new users of models that become integrated with GIS.

Moving beyond the first stages of integration to achieve a co-evolution of GIS and environmental modeling will benefit both, as fundamental assumptions are reevaluated and as new knowledge and techniques are incorporated, eventually producing replacement tools that will transcend current issues of integration.

CONCLUSION

It is simple enough to imagine synergy by combining the spatial representations of GIS with the predictive capabilities of environmental models and adding more adequate three-dimensional and temporal representation and the abilities to characterize uncertainty and error. It is more difficult and more challenging to ask how these tools should be made to become more interdependent and interactive. It is clear that to solve pressing environmental problems we will need different tools that work together in an intelligent way and are easy to use but that can be employed in a flexible manner to handle complicated problems. These seemingly conflicting objectives will need to be harmonized in order to make rapid progress.

Without strong intervention, the allure of constantly improving technologies will continue to draw both GIS and environmental modeling along behind it. Without formalization of an effort to achieve integration, only the very able and the very fortunate will be able to incorporate the benefits of modeling and GIS in their work because only they will have sufficient understanding or resources to overcome the difficulty of coupling tools that, despite certain affinities, remain quite dissimilar.

5

GIS and Environmental Modeling

KURT FEDRA

Many human activities, such as large-scale industrial, energy, construction, water resources, or agricultural projects driven by increasing resource consumption with increasing affluence and population numbers, considerably affect the natural environment. Growing concern about these impacts and their immediate, as well as long-term, consequences, including risk involved with technological systems and the inherent uncertainty of any forecast, makes the prediction and analysis of environmental impacts and risks the basis for a rational management of our environment, a task of increasing global importance.

Environmental modeling, as one of the scientific tools for this prediction and assessment, is a well-established field of environmental research. International conferences, monographs, and dedicated journals illustrate a mature field.

Most environmental problems do have an obvious spatial dimension. Within the domain of environmental modeling this is addressed by spatially distributed models that describe environmental phenomena in one (for example, in river models), two (land, atmospheric, and water-quality models, models of population dynamics), or three dimensions (again air and water models). The increasing development and use of spatially distributed models replacing simple spatially aggregated or lumped parameter models is, at least in part, driven by the availability of more and more powerful and affordable computers (Loucks and Fedra, 1987; Fedra and Loucks, 1985).

On the other hand, geographical information systems are tools to capture, manipulate, process, and display spatial or geo-referenced data. They contain both geometry data (coordinates and topological information) and attribute data, that is, information describing the properties of geometrical spatial objects such as points, lines, and areas.

In GIS, the basic concept is one of location, of spatial distribution and relationship, and basic elements are spatial objects. In environmental modeling, by contrast, the basic concept is one of state, expressed in terms of numbers, mass, or energy, of interaction and dynamics; the basic elements are "species," which may be biological, chemical, and environmental media such as air, water, or sediment.

The overlap and relationship is apparent, and thus the integration of these two fields of research, technologies, or sets of methods, that is, their paradigms, is an obvious and promising idea.

This chapter will first present an argument for the integration of the two fields as a paradigm change or rather extension. It will then try to summarize the state of the respective arts and current trends in these two fields, drawing on the available overview papers of this volume. This will be followed by a more detailed analysis of the ways and means of integration from a technical, a modeler's perspective. Finally, the idea of integration will be expanded to cover other areas such as expert systems, scientific visualization, or multimedia systems, and to discuss integration from a perspective of users and uses, that is, institutional aspects of integrated environmental information systems.

INTEGRATING FIELDS OF ENQUIRY: MERGING OF PARADIGMS

Merging of fields of research, or adding a new technology to an established and mature field, usually leads to new and exciting developments. Adding the telescope to astronomy (or astrology, for that matter), the portable clock and the sextant to cartography, the microscope to classical biology (i.e., anatomy and morphology), space probes to astrophysics, x rays or the laser to medicine, are just a few arbitrary examples. But they all have had profound effects on the respective fields of enquiry.

Adding the computer to environmental sciences is yet another one of these possibly fruitful mergers. It is not only the technology that allows us to do things better and faster, it is new concepts and ideas, or a new paradigm, that leads us to do different things.

Normal science à la Kuhn (Kuhn, 1962) tends to organize itself into research programs (Lakatos, 1968). They are supposed to anticipate novel facts and auxiliary theory, and as opposed to pedestrian trial and error (the hallmark of immature science according to Lakatos), they have heuristic power. In other words, they tell us what to do, how to do it, and what to look for. Journals, monographs, textbooks, and peer groups see to that. However, in his Consolations for the Specialist, Feyerabend (1970) states that "Everybody may follow his inclinations, and science, conceived as a critical enterprise, will profit from such an activity".

Central to normal science is the paradigm: It has a sociological notion related to the researchers rather than the research; that is, it can function even if an all-encompassing theory or any amount of theoretical underpinning, for that matter, is not there; it is a puzzle-solving device rather than a metaphysical world view; it has to be a concrete picture used analogically, a way of seeing. As something concrete or even crude, a paradigm may literally be a model or a picture, or an analogy-drawing story, or some combination (Masterman, 1970). Thus, a paradigm is something that works, and makes people work, in practical science.

Clearly, this could describe environmental modeling and probably GIS as a field of research, often enough a puzzle-solving activity short on theory with its own way of seeing (and displaying) things. It is exactly this way of seeing things that gets changed, or enlarged, when paradigms are merged and thus at least shifted if not revolutionized. Language, concepts, and tools of different fields can certainly enrich each other.

In GIS, the basic concept is maybe one of location, of spatial distribution, and relationship. Spatial objects such as areas, lines, or points and their usually static properties are the basic units. Interaction is more or less limited to being at the same location, or maybe in close proximity to each other.

By contrast, in environmental modeling the basic concept is one of system state, of mass and energy conservation, of transformation and translocation, of species and individuals' interaction and dynamics. Populations and species, environmental media such as air, water, and soil, and environmental chemicals are the basic units. Since all the basic units, or better, actors, in environmental modeling do have a spatial distribution, and this distribution does affect the processes and dynamics of their interactions considerably, GIS has a lot to offer to environmental modeling. At the same time, an enriched repertoire of object interactions and more explicit dynamics can make GIS a more attractive tool as well.

GIS AND ENVIRONMENTAL MODELING: STATE OF THE ART

Both environmental modeling and GIS are well-established methods and fields of research. Their integration, however, seems at best an emerging field. A simple analysis of a DIALOG computer search in a number of relevant databases seems to indicate just that. Using Geographical Information System or GIS as a key, individual files like Water Resources Abstracts or Enviroline would yield maybe 100 to 200 entries. Using Environment, several thousand entries would typically be found. Combining the GIS and Environment keys, just a few publications could be identified.

Datafile	GIS	ENV	GIS + ENV
SocSciSearch	181	9,898	6
SciSearch	143	6,118	7
Enviroline	121	25,310	34
Water Resources Abstracts	165	44,696	56
Computer DB	501	21,933	81
INSPEC	1,711	105,781	266

Another piece of circumstantial evidence might be the following: In a hefty volume on Computerized Decision Support Systems for Water Managers (Labadie et al. 1989) a conference proceedings of close to 1000 pages, GIS is not mentioned once (at least according to the subject index). In contrast, and three years later, at a session of the 1991 General Assembly of the European Geophysical Society, dedicated to Decision Support Systems in Hydrology and Water Resources Management, more than half the papers discuss GIS as a component of the research method (EGS, 1991).

And here is another nugget of corroboration: in an article in the Government Computer News, Weil (1990) quotes an assistant administrator of EPA as singling out "GIS and modeling as cornerstones for achieving EPA's goals for the 1990s."

While this literature search was neither systematic nor exhaustive, I believe it is certainly indicative: The integration of environmental modeling and GIS is a new and emerging field, and thus, full of opportunities.

GIS

GIS are computer-based tools to capture, manipulate, process, and display spatial or geo-referenced data. They contain both geometry data (coordinates and topological information) and attribute data, that is, information describing the properties of geometrical objects such as points, lines, and areas. Berry (Chapter 7), Nyerges (Chapter 8), and Goodchild (Chapter 9) summarize everything you ever wanted to know about GIS.

GIS are quite common and generally accepted in surveying and mapping, cartography, geography, and urban and regional planning (Smyrnew, 1990; Scholten and Stillwell, 1990) and land resources assessment, with conferences, journals, and monographs documenting an established field (e.g., Burrough, 1986).

For environmental applications in a rather loose sense, including land management, there is considerable tradition in the field, for example, in Canada under the header of land modeling in the Lands Directorate of Environment Canada (e.g., Gélinas, Bond, and Smit, 1988). There are also major initiatives to build up or integrate geographical and environmental databases in many countries worldwide (for example, Kessell, 1990), and in most European countries (e.g., van Est and de Vroege, 1985; Jackson, James, and Stevens, 1988; Sucksdorff, Lemmelä, and Keisteri, 1989) at the European Community level (Wiggins et al., 1986) or at the global level within the UN framework with systems such as GRID (Witt, 1989) or GEMS (Gwynne, 1988).

While many of these systems have explicit environmental components and functions, they are not usually integrated with any modeling capabilities in the sense of simulation models, that is, transport or process and fate models, models of population development, etc. The idea of this integration, however, is obvious and discussed frequently (Granger, 1989; Tikunov, 1989; Fedra, 1990b; Lam and Swayne, 1991).

Recent overview papers on environmental GIS, and more generally, information technology for environmental applications include Jackson, James, and Stevens (1988), or Woodcock, Sham, and Shaw (1990); Moffat, (1990); Bishop, Hull, and Bruce (1991). There are also critical appraisals of the field (e.g., Arend, 1990), but often enough critical question like: GIS, useful tool or expensive toy? are rhetorical in nature.

Current trends in GIS include a better integration between raster- and vector-based systems; in the GIS World survey of 1988 (Parker, 1988), 17 vector-based, 7 raster-based, and 12 supporting both raster and vector formats were listed, with the ArcInfo/ERDAS combination as one of the more widely used ones (Tilley and Sperry, 1988). A recent discussion of hybrid systems is given by Fedra and Kubat (forthcoming). There also is increasing emphasis on remote sensing data as a valuable source of environmental data (Woodcock and Strahler, 1983; Welch, Madden Remillard, and Slack, 1988).

Another interesting line of development is the integration of GIS and expert systems (Maidment and Djokic, 1990; Lam and Swayne, 1991; Davis and Nanninga, 1985; Davis et al. 1991). This aims at a more flexible and complex analysis of maps and map overlays, based on rules and logical inference. Alternatively, GIS can provide spatial data to rule-based assessment systems (Fedra and Winkelbauer, 1991).

Related is the explicit use of GIS as decision support systems (Fedra and Reitsma, 1990; Parent and Church, 1989).

Adding dynamics or temporality in spatial databases (Armstrong, 1988; Langran, 1988) gets them closer to spatially distributed dynamic models. The integration with video technology for fast animation adds yet another feature for better visualization and presentation (Gimblett and Itami, 1988).

One of the more frequently discussed issues is the extension of traditional 2D GIS into full 3D systems (Turner, 1991).

Technically, moving from proprietary systems to open systems, embracing window environments (Gardels, 1988), and experiments with distributed systems (Seaborn, 1988; Ferreira and Menendez, 1988) are notable trends. And Dangermond (Chapter 6) in his report on the GIS developer or vendor's perspective on GIS and model integration provides an overview of what is in store in the commercial sector.

Environmental modeling

Environmental modeling has a considerable history and development. A number of analytical approaches applied to biological and ecological problems date back to Lotka (1924), and fields like hydrology can look to more than a hundred years of modeling history (Maidment, Chapter 14).

With the advent of digital computers, numerical simulation models became feasible. Early linear models (Patten, 1971), applications of system dynamics to ecological problems (Forrester, 1982), and ever more complex multi-compartment models like CLEANER and MS. CLEANER (Park et al., 1974, 1979) were developed. An overview of environmental systems process modeling is given by Steyaert (Chapter 13).

None of these approaches had explicit spatial dimensions yet. In limnology, oceanography, and plant sociology concepts such as patchiness were discussed, however, not in relation to a fixed coordinate system as used in GIS.

More powerful computers allow more complex models to be developed and run. Spatial distribution and increased dimensionality and resolution is one straightforward way of "improving" models. And spatially distributed models can interact with GIS.

The following sections, based mainly on the overview papers of this volume and the literature survey, summarize current developments and trends in environmental modeling, and emphasize examples of GIS integration.

Atmospheric systems

In modeling the atmospheric environment, the relationship to geographical data should be self-evident. For more complex models that go beyond the classical Gaussian plume models, topographic relief, surface roughness, and surface temperatures are important input parameters. Sources of pollution are spatially distributed, and may be point sources such as large industrial stacks or power plants, line sources such as highways, and area sources, such as urban areas.

And for the impact and exposure calculations, land use and population distribution are required, again spatial data that a GIS could well handle. And what is true for local and regional air and air quality models is certainly the case for global models in climate change research.

A recent "prototypical" application of atmospheric modeling with GIS integration is Zack and Minnich (1991), who applied a diagnostic wind field model for forest fire management. A GIS was used both for input data preparation (DEM and meteorological stations) and the display and analysis of model results. And case studies at IIASA of air pollution management for the City of Vienna and for Northern Bohemia also apply a tight coupling between air quality simulation models, optimization models for the design of pollution control strategies, and GIS functions (see the following).

Larger-scale models, and in particular, general circulation models (GCM) at a global scale, and their GIS connections are discussed by Lee et al. (Chapter 10). The chapter describes different surface modeling schemes to represent the interface to the GCMs. Coupling with GIS here is mainly seen as a way for proper input characterization, that is, landscape and land-use data preparation.

The coupling between terrestrial and atmospheric systems, and in particular, the role of vegetation in shaping weather and climate, influencing the hydrological cycle, and as sources and sinks of greenhouse gases, is discussed in Schimel's overview paper (Chapter 26).

Hydrological systems

Maidment (Chapter 14) summarizes the state of the art of hydrological modeling. Hydrological modeling deals with two major topics, namely the quality and quantity of water. Quality concerns have undergone a change in emphasis from oxygen to eutrophication to toxics, following both improvements in treatment technology and analytical chemistry.

Spatial elements are important in marine systems, lake models, and groundwater problems, which have obvious 2 or 3D structure. Finite element and finite difference models provide a well-established discretization of space for these models. River models, in contrast, usually operate in a one-dimensional representation of a sequence of reaches or cells, and networks such as canals or pipes can be represented by a graph with nodes defined in 2D and arcs with the necessary connectivity information.

Maidment sees a major role of GIS in hydrological modeling in its capability to assist explicit treatment of spatial variability. It is important to note, however, that a GIS is not a source of information, but only a way to manipulate information. Unless appropriate spatially distributed input data exist, even a state of the art GIS coupled with a 3D model will not guarantee reasonable results.

GIS coupling and linkages are described for hydrologic assessment, that is, mapping of hydrologic factors using qualitative or semi-quantitative index-based assessment, for example, of groundwater contamination potential.

The estimation of hydrologic parameters is another area of GIS application: Watershed parameters such as slope, soils, land cover, and channel characteristics can be used for terrain models and simple flow descriptions. These parameters are, of course, of central importance for the land surface and sub-surface process models (Moore et al., Chapter 19).

Recent applications include work on runoff and erosion models (De Roo, Hazelhoff, and Burrough, 1989; Oslin and Morgan, 1988), river basin management (Goulter and Forrest, 1987; Hughes, 1991), surface water modeling (Arnold, Datta, and Haenscheid, 1989; Andreu and Capilla, 1991; White, 1991; Wilde and Drayton, 1991), or groundwater modeling (Steppacher, 1988; Fedra and Diersch, 1989; Fedra, Diersch and Härig, forthcoming; Hedges, 1991; Nachtnebel et al., 1991).

Land surface and subsurface processes

Watershed models, erosion and non-point modeling, and groundwater modeling are areas of environmental modeling that have an obvious and explicit spatial dimension. Distributed models, and the use of finite difference and finite element schemes, provide a natural opportunity for GIS coupling for both input data preparation as well as for the display and further analysis of model results.

Moore et al. (Chapter 19) discuss GIS and land surface-subsurface process modeling. An important concern in their analysis is one of scale. There are several scale-related issues and problems identified for spatially distributed models and GIS applications: the grid or polygon size, the method used to derive attribute values such as slope, aspect, soil type, for these elements, merging data of different resolution, and the scale differences between model representation and the observational methods used to derive a priori parameter values. A related issue is the concern that by moving from a lumped parameter model to a spatially distributed one, the interpretation of parameters may have to differ.

Biological and ecological models

Spatial patterns have a considerable history in plant sociology, forestry, and plankton studies. Until recently, any consideration of patterns or spatial distribution, however, was statistical in nature rather than explicit, that is, connected to absolute location with an *X,Y* coordinate system. An earlier spatially distributed ecological model is the famous spruce budworm exercise (Holling, 1978). And in the 1982 state of the art conference in ecological modeling (to pick one more or less at random), only very few examples of spatially distributed models, mainly in the river and lake modeling areas can be found (Lauenroth, Skogerboe, and Flug, 1983).

Hunsaker et al. (Chapter 22) also review a rich literature with numerous examples of studies that include GIS methods in the various applications fields. Again, the review demonstrates a movement from point or spatially lumped models towards distributed models, a development that in part seems to be made possible if not motivated by increasingly affordable computer resources. The trend is clearly towards more spatial resolution, and linkage with GIS as a ready-made technology to handle spatial information.

Application examples include Johnston (1989); Johnston and Naiman (1990); Johnston and Bonde (1989); Johnston et al. (1988); Lindenmayer et al. (1991); and Johnson (1990).

Problems identified by Hunsaker et al. (Chapter 22) include the software engineering problems of tight linkage and data requirements; related to the need for large volumes of data are problems of their effective storage, although this is a largely technical constraint that is changing rapidly as computer technology advances. Problems of scale and uncertainty come up again (see Moore et al., Chapter 19)

Approaches to integration are found to range from GIS as pre- and post-processors to complex "intelligent" GIS with built-in modeling capabilities or expert systems integration (Lam and Swayne, 1991), high-level application, and modeling languages.

Risk and hazards

The mapping of risk, as a rather abstract concept, makes it much easier to communicate. And elements of environmental and technological risk, from its sources to the recipient, are spatially distributed. Exposure analysis as an overlay of sources and receptors is an almost classical GIS application.

Rejeski (Chapter 30), in his analysis of GIS and risk, emphasizes the cultural dimensions and problems of plural rationalities. Believability, honesty, decision utility, and clarity are major issues he addresses. GIS have the ability to integrate spatial variables into risk assessment models, and maps are powerful visual tools to communicate risk information. A major concern, since risk analysis is a risky business, is again uncertainty.

Recent applications include Best et al. (1990) or the XENVIS system developed at IIASA (see the following).

Modeling in policy-making

Models are built for a purpose, and scientific research as an end in itself and better understanding is, although noble, increasingly insufficient to get research funded. Direct responsiveness to society's actual and perceived needs is important as well.

Modeling for decision support or model-based decision support systems for environmental problems have been discussed and advocated for a considerable time (Holcomb Research Institute, 1976; Bachmat et al., 1980; Andelin and Niblock, 1982; Loucks, Kindler, and Fedra, 1985; de Wispelaere, Schiermeier, and Gillani, 1986; Labadie et al., 1989; Fedra and Reitsma, 1990; Fedra, 1991). Success stories of actual use in the public debate and policy-making process are somewhat more rare.

In his overview chapter, King (Chapter 34) presents a view of models in what he calls the datawars of public policy-making. Implementation, that is, putting a model into an institution for a political purpose, is the key concept, and a consequently partisan rather than "value free science" approach is advocated.

The specific role of environmental models integrated with GIS would largely be in their ability to communicate effectively, using maps as a well-understood and accepted form of information display, generating a widely accepted and familiar format for a shared information basis.

Summary

Every field of environmental modeling is increasingly using spatially distributed approaches, and the use of GIS methods can be found everywhere. With ever more powerful and affordable computer technology, spatial distribution and increasing resolution for dynamic environmental models become feasible.

A repeated concern of modelers, however, is in the area of uncertainty, scale, and data availability. Powerful tools can be tempting, and distributed models without good distributed data are at best expensive interpolation tools, and at worst subject to the GIGO (garbage in–garbage out) syndrome. Linkage with GIS is frequently found, but in the majority of cases, GIS and environmental models are not really integrated, they are just used together. GIS are frequently used as pre-processors to prepare spatially distributed input data, and as post-processors to display and possibly analyze model results further. Alternatively, modeling approaches directly built into GIS appear rather simple and restrictive.

LEVELS OF INTEGRATION

Integration, trans and multidisciplinary, hybrid, embedded, etc., are recurring keywords in today's modeling lit-

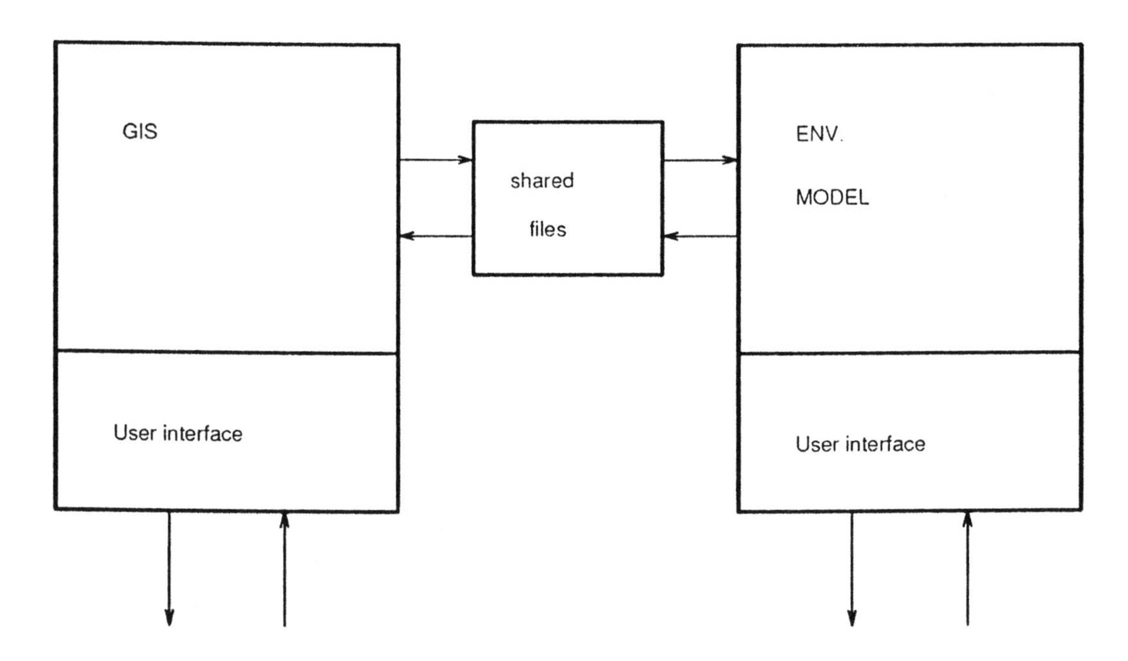

Figure 5-1: Linkage of separate programs through common files.

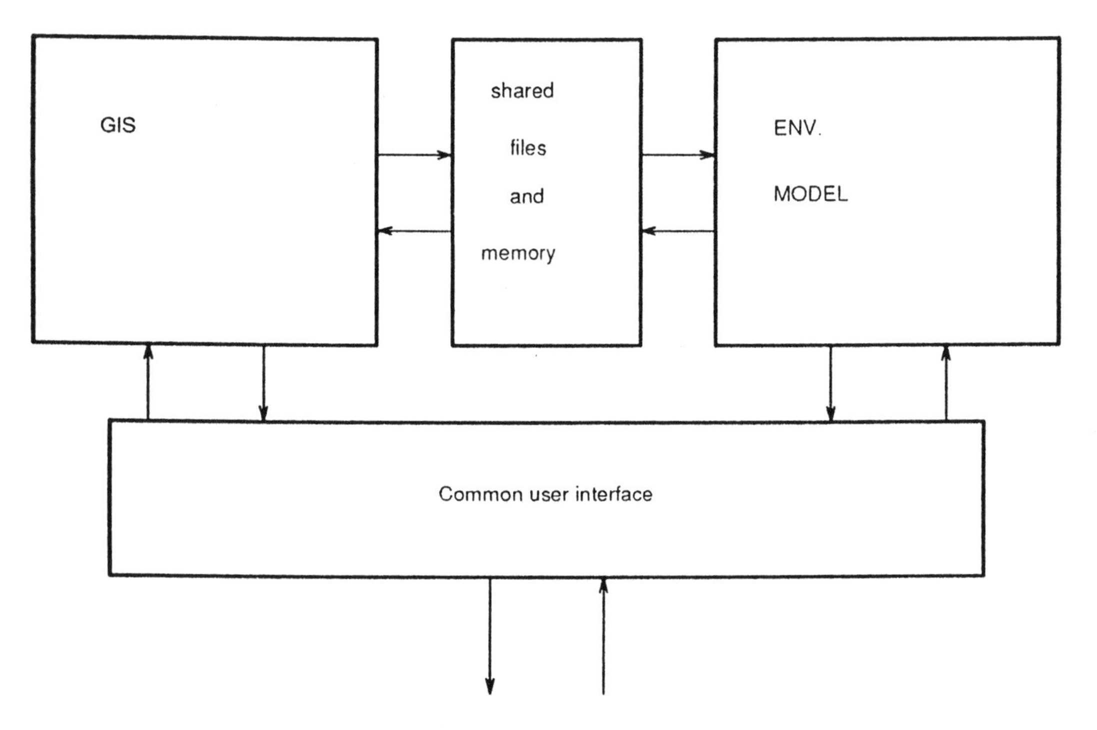

Figure 5-2: Integration within one program with a common interface.

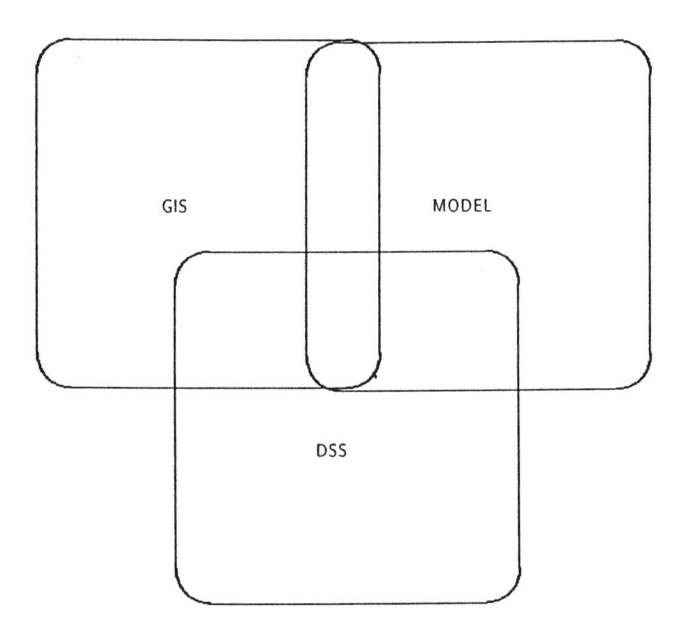

Figure 5-3: Partial functions overlap in a dedicated system.

erature. The integration of GIS and environmental models can come in many forms. In the simplest case, two separate systems, the GIS and the model, just exchange files: The model may read some of its input data from GIS files, and produce some of its output in a format that allows processing and display with the GIS (Figure 5-1).

This seems to be a rather common approach, since it requires little if any software modifications. Only the file formats and the corresponding input and output routines, usually of the model, have to be adapted.

Depending on the implementation, however, a solution based on files shared between two separate applications is cumbersome and error prone. Deeper integration provides a common interface and transparent file or information sharing and transfer between the respective components (Figure 5-2).

One possible way is the use of a higher-level application language or application generators built into the GIS. An alternative is the use of tool kits that provide both GIS functionality as well as interface components for simulation models, and, in the worst case, there is always assembler programming.

A recent example of integration that draws together GIS, models, spreadsheet, and expert systems in a programmable system is RAISON (Lam and Swayne, 1991). Application generators and modeling capabilities with commercial GIS also offer the possibility of tight integration within the limits of the respective package's options.

Any integration at this level, however, requires a sufficiently open GIS architecture that provides the interface and linkages necessary for tight coupling.

For a problem-specific information and decision support system rather than a generic tool, only a subset of the functions a GIS supports may be required for a given application. Functions such as data capture and preprocessing and final analysis can conveniently be separated: They support different users with different time frames. This subset of functionality concept (Figure 5-3) applies equally to models: For the analysis stage, for example, we would assume that the model has already been successfully calibrated. Calibration is an important and often time-consuming and difficult task, but it can be separated from the interactive decision support use of a model.

Parallel to this technical level of coupling, there are different conceptual levels of integration. In the simplest case, the GIS is used to store and manipulate and maybe also analyze distributed model input data; alternatively, the GIS is used to present and maybe further analyze modeling results. A majority of applications found in the literature represent this approach.

A deeper level of integration would merge the two approaches, such that the model becomes one of the analytical functions of a GIS, or the GIS becomes yet another option to generate additional state and output variables in the model, and to provide additional display options.

This requires, however, tools that are sufficiently modular, so that the coupling of software components within one single application with shared memory rather than files and a common interface becomes possible (Figure 5-4). Obviously, this most elegant form of integration is also the most costly one in terms of development effort.

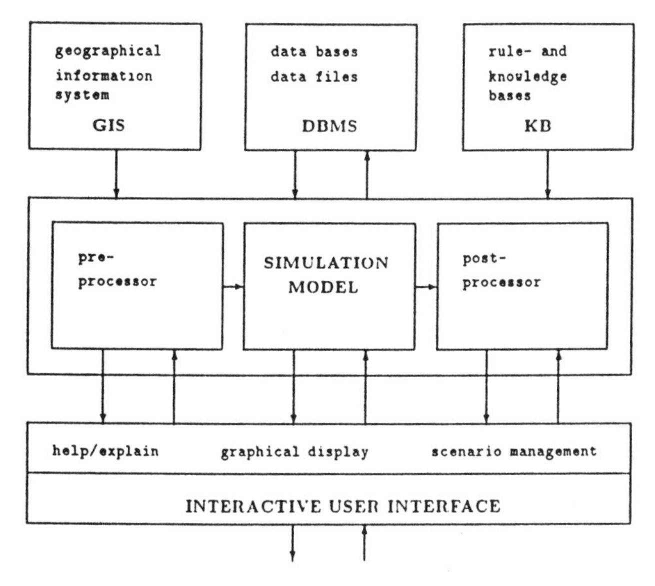

Figure 5-4: Interactive modeling in an integrated framework: a model oriented perspective.

However, if the ultimate goal is not only to develop a better research tool, in the form of more powerful models and analysis software, but also to aid the environmental planning and policy-making process, more than the integration of environmental models and GIS technology will be required to integrate these methods successfully into the policy- and decision-making process.

TOWARDS BETTER INTEGRATED ENVIRONMENTAL INFORMATION SYSTEMS

Given this overall objective of institutional integration for practical application, the task then is to construct and apply better tools for better results. This includes not only collecting more and better data of ever increasing resolution and precision and developing better models and tools for analysis, but also providing more effective interfaces in a technical as well as procedural, organizational, and institutional sense of our efforts with the policy- and decision-making process.

Integrated environmental information and decision support systems, built around one or more coupled models, numerical simulation models or rule-driven inference models, and integrated with GIS, feature:

- An interactive, menu-driven user interface that guides the user with prompt and explain messages through the application. No command language or special format of interaction is necessary. The computer assists the user in its proper use; help and explain functions can be based on hypertext and possibly include multi-media methods to add video and audio technology to provide tutorial and background information.
- Dynamic color graphics for the model output and a symbolic representation of major problem components that allow easy and immediate understanding of basic patterns and relationships. Rather than emphasizing the numerical results, symbolic representations and the visualization of complex patterns support an intuitive understanding of complex systems behavior; the goal is to translate a model's state variables and outputs into the information requirements of the decision-making process.
- The coupling to one or several databases, including GIS, and distributed or remote sources of information in local or wide area networks that provide necessary input information to the models and the user. The user's choice or definition of a specific scenario can be expressed in an aggregated and symbolic, problem-oriented manner without concern for the technical details of the computer implementation.
- Embedded AI components such as specific knowledge bases allow user specifications in allowable ranges to be checked and constrained, and ensure the consistency of an interactively defined scenario.
- They are, wherever feasible, built in direct collaboration with the users, who are, after all, experts in the problem areas these systems address.

In summary, integrated information systems are designed for easy and efficient use, even in data-poor situations, and cater to the user's degree of computer expertise. The "intelligent" interface and its transparent pre- and post-processing functions free the user from the time-consuming and error-prone tasks of data file preparation, the mechanics of model runs, and finally the interpretation and translation of numerical results into meaningful terms that are adequate to the problem. This not only allows the user to employ the models more freely in a more experimental and interesting way, it also allows the analyst to concentrate on the more important tasks he can do best, that is, the recognition of emerging patterns, the comparative evaluation of complex alternatives, and the entire institutional aspects of any environmental impact assessment rather than its technicalities.

The models, and their interfaces, are representations of the problems they address as much as of the planning and decision-making processes they are designed to support. In the latter field, if not also in the former, their users are the real experts. Thus, their expertise and experience needs to be included in the systems. As a consequence, the user must be involved in the design and development, so that he can accept responsibility and ownership for the software system.

Institutional integration also must look at aspects such as user training, data entry, maintenance issues of keeping systems current and operational, providing adaptations and updates, etc. Any complex information system has more than one user at more than one level of technical competence and with different roles within an institution. Different users have different requirements that need to be supported: Flexibility and adaptability are therefore important features. Systems must be able to grow with their users. Therefore, the institutional commitment and technical infrastructure to keep a system alive and evolving are as important as the scientific and technical quality of the original software system.

It is, however, important to recognize that there is a price to be paid for the ease of use and all the features of these systems: Not only are they more expensive to build—after all, all the information that makes them smart has to be compiled and included at some stage—they are also much less flexible than their more conventional, general-purpose siblings. Only by restricting the range of applications are we able to build more application-specific knowledge into the systems and thus make them appear smart. There is no such thing as the general-purpose problem solver, or a generic model or decision support system that is easy to use. Mastery of a problem area comes at the price of increasingly narrow specialization.

Certainly the expert systems approach, or any computerized decision support system for that matter, is not a replacement for the human expert in such a complex problem domain; it still requires a knowledgeable and responsible person to use it, to interpret and apply the results. However, the system will take care of the more mundane tasks of data handling, freeing the analyst to concentrate on the real problems that require human creativity.

Integrated software systems, whether they are expert systems or based on simulation or optimization models, organize the planning or decision-making process; they provide structure, ensure completeness, and may even ascertain plausibility. It is the easy to use "smart" interface, the fast and efficient operation, and the apparent intelligence of the programs that makes them attractive. Based not only on the organized collection of experience from numerous experts, and international literature, but also on various guidelines, regulations, and environmental law, a system's knowledge base, with or without one or more numerical models in its core, may indeed provide intelligent advice to any individual user.

For the specific model and GIS coupling, this means that their respective functions are fully and transparently integrated. Imagine a system that is structured not in terms of state variables and parameters, or spatial objects and attributes, but in terms of problems and problem owners, intentions and objectives, constraints and regulations, options and decision alternatives, facts and assumptions, preferences and perceptions.

A problem is represented by a set of descriptors, some of which may be spatially distributed or derived from spatially distributed descriptors, within a context of facts and assumptions. The user can now manipulate and analyze his problem situation in terms of decisions or assumptions, and explore the behavior of his system in response to any of his specifications. This may involve queries to databases, browsing through a hypertext system, or running simulation or optimization models, using expert systems, and of course GIS functions for both mapping as well as spatial analysis. In most practical situations, it will involve all of the above and more. For the user it is immaterial which method is used to generate the answer to his questions, to provide insights or arguments, help structure his thinking and communicate information within a group. In fact, it will usually be the combination of several "methods" or tools that are required.

If the problem is spatial in nature, and most if not all practical environmental problems are, the distinction between GIS and spatial model disappears. The system provides a coordinated set of functions or tools that cooperate in a common environment, within a single integrated system.

Application examples

To illustrate the concept of integrated environmental information systems better, I would like to present a few examples, first prototypes that illustrate, or rather experiment with, some of these concepts. I apologize for drawing them all from our own work at IIASA's Advanced Computer Applications Project, but, for obvious reasons, these are the ones I have most information on. A typical example system is an environmental information system for the city of Hanover. It combines a GIS component with simulation models for specific problems, such as groundwater or air pollution.

The GIS forms a central framework and integrating component, providing a variety of map types for use in the system. Maps or overlays include simple line features, such as the city boundaries or complex topical maps as background for the spatially distributed models, including model input data sets. Examples would be a landuse map, the geological map, or a biotope map, stored in vector format, or groundwater head and groundwater recharge, stored as grid cell files. Similar to the model input files, model results, (i.e., computed groundwater heads or concentration fields of pollutants from air, ground- or surface water models) can also be stored as grid cell files.

Another raster format integrated in the GIS is a SPOT satellite image of the city area. The satellite image is stored and treated as a "true" raster, that is, only color numbers, the attribute data, are stored rather than the original multi-spectral data.

While most of the maps fully cover the entire area, an interactive map editor allows one to select individual features from a given map for an overlay. For example, from the full area coverage of the landuse map, only the road and rail network, or the area above a certain threshold value of pollution as computed by a model, can be extracted as an overlay for the biotope map, for example, for transportation corridor analysis. A color editor offers the possibility to adjust the display color and style of a given feature so that any arbitrary combination of features and overlays will result in a well-designed display, highlighting the important features.

From a user point of view, all these different maps, including model input data and model results, are equivalent; the user is not necessarily aware of their structural differences. Composite maps can be generated by overlaying the various maps or subsets of features of the maps (Plate 5-1), and the GIS offers the possibility of zooming into any subarea down to the limits of the resolution of the database. Here, the differences between vector and raster formats become obvious. One of the models in the system is a finite-difference-based groundwater flow and transport model, using a particle tracking scheme (Fedra, Diersch, and Härig, forthcoming). The model uses spatially distributed input data, such as initial head, porosity, or groundwater recharge, that are also available as overlay planes in the GIS. The GIS functions can be called directly from the model so that the various data sets can be viewed and analyzed. At the same time, a problem relevant background map such as the geological map or the biotope map, or combined selected features can be prepared interactively.

Model output (groundwater head, flows, and pollutant concentration) is displayed dynamically over the background map. The user can modify the display at any point, stop or rerun the model with alternative scenario assumptions, etc. (Plate 5-2). Model output can also be stored as a map overlay, and thus passed to the GIS for further analysis in conjunction with other overlay planes such as landuse, etc. However, since the GIS functions are directly accessible from the model interface, there are no differences for the user between model or GIS functions. They both serve to analyze his problems and help him to design a meaningful representation and display.

To this end, other features include a built-in expert system for parameter estimation and input feasibility checking. This will, for example, advise the user on reasonable pumping rates for wells introduced in the simulation (Plate 5-3) or check the proper design and location for remediation strategies such as hydraulic barriers or interception pumping. Another feature is a hypertext-based help and explain system, which can provide background information on models, data, assumptions, and explain parameters as well as the results from the expert system's use.

A similar approach is used in a series of air quality models, both on a local and a regional scale (Plate 5-4). Again the models are operated in a GIS context, with a map background and data such as the emission sources, topography, surface roughness, and temperature managed by the GIS.

In addition to the dynamic simulation model, optimization tools allow one to design cost-effective pollution abatement strategies. Here objective functions including human health criteria, such as exposure, are derived from the spatially distributed model output (on a regular grid), superimposed on administrative units (polygons) with associated population and age distribution data (tabular). Land use, analyzed for an environmental impact criterion, is derived from satellite imagery.

It is interesting to note that, from the GIS perspective, many of the "maps" or "overlays" are not stored as data, but are dynamically generated and regenerated by a model, or a set of rules from an expert system. There is, however, a trade-off between computation times and storage space. In other examples, dynamic data may be stored rather than computed on request, simply for reasons of efficiency. An example is CLIMEX, an expert system for climate impact assessment with a global GIS. Monthly climate data and GCM results and population data are stored in the respective formats, and can be seen in an animated display. These data are then used for regional rule-based impact assessment.

Another reason for storing selected model results is the ability to compare different scenarios and generate, for example, the delta of two pollution concentration fields resulting from alternative abatement policies. But even in this case, parallel or distributed processing may provide the computer power to run several scenarios simultaneously.

Other applications of integrated environmental information systems with various technologies combined include MEXSES, an expert system for environmental impact assessment that includes both GIS and dynamic simulation models (Fedra and Winkelbauer, 1991). The inference engine in processing the rules to assess environmental impacts can use more rules to infer facts. It can, if appropriate, get data from the GIS (examples would be soils and slopes, vegetation, land use, etc. for a given project location), derive them from a simulation model, or ask the user. Where the information required comes from is more or less transparent for the user. And the strategy (i.e., which source of information to try first) is controlled by the knowledge base of the system and can be modified dynamically, based on context and state of the system.

REPLACE is a spatial expert system for site suitability analysis (Reitsma, 1990). Implemented in PROLOG and with a graphical user interface, it matches the spatial requirements of "activities" such as industrial plants or hospitals with spatial properties such as physiography, infrastructure, or environmental constraints (Plate 5-5). REPLACE, rather than being a spatial model, models space. By sharing data with a number of related simulation and optimization models as well as statistical and geographical databases, it is an integrated component of a modular regional information and decision support system (Fedra et al., 1987).

XENVIS is a national level environmental information system implemented for the Netherlands, that incorporates a GIS, a water quality model for simulation of spills of toxics into the Rhine, a transportation risk analysis model, and an interface to a fault-tree-based risk assessment system for process industries. Model output, including risk contour plots based on plant safety characteristics, weather data, and population distribution, is displayed over an interactively constructed map (Plate 5-6). Designed for risk analysis and risk communication, the system also includes a noise analysis module for railways and a number of interrelated databases, implemented in a hypertext structure, covering topics such as hazardous installations or chemicals.

These and similar applications are described by Fedra (1991), Fedra (1990a,b), and Fedra and Reitsma (1990).

Uses and users

Advocating integrated environmental information systems as a central theme for environmental research, research in GIS, and in the coupling of GIS and environmental models is based on a few personal political science premises.

First, that what scientists do, or should do, is to ultimately assist societal decision making-processes; that research priorities are set, or should be set, in response or better anticipation of societal needs and problems; and third, that a sustainable development of life on this planet and the generation or maintenance of an enjoyable environment for future generations is, or should be, one of the basic goals of our societies.

A very similar credo, by the way, was formulated by E.W. Manning from the Lands Directorate in Canada, in the context of land modeling (Manning, 1988). Like any other tools, environmental models with integrated GIS, or the other way around, are built for a purpose, for users.

Like many computer-based models and methods, integrated environmental information systems and their components, such as simulation models and GIS, are potentially useful. A large amount of formal, mathematical, and computational methods have been developed in the area of environmental planning and management, and the field has a considerable history in the use of computers. However, to turn a potentially useful method into one actually used requires a number of special features as well as an approach that takes psychological and institutional aspects as well as scientific and technical ones into account.

Tools that are easy to use, equipped with a friendly user interface, use problem-adequate representation formats and a high degree of visualization, are customized for an institution and its specific view of problems and are developed in close collaboration with the end user, stand a better chance of being used than tools that are based on "only" good science.

Good science is a necessary, but certainly not sufficient, condition for a useful and usable information and decision support system; there are definite advantages to increased user participation, with consideration of questions of maintenance and the update of information requirements from the very beginning, but also questions of control and ownership, responsibility, and credibility.

All science is propaganda, to paraphrase Paul Feyerabend again, and a strong argument along this line is provided by King (Chapter 34) on modeling in the policy process. I must hasten to add, however, that while I find his arguments most convincing, I cannot follow all his conclusions. Decades of neopositivist brainwashing in academe (and in Vienna circles) have led me to believe that indeed, and if only in the long run, truth wins (sometimes). You can probably cheat your way out of a hearing with the fancier model, but you cannot cheat thermodynamics (not with cold fusion) and evolution in the long run. Having real practical use is important for a model. Being close to reality (which of course includes the policy-making process as part of the overall environmental system) is at least equally important.

Having said that, I can agree, however, that in most decision-making situations the model is not so much a model of reality, and thus subject to all our scientific aspirations. It is rather a tool to help organize a learning or bargaining exercise, where it is more important that it provides a framework, a mirror for our thinking, stimulation or excuses, or justification for compromise. However, models and information systems are used at various levels in the policy making process, and the research level is certainly one of them. Uses and user requirements differ considerably at these levels, and the challenge is to provide a smooth and credible connection between these different levels of abstraction, detail, and interpretation.

Advanced information technology provides the tools to design and implement smart software, where in a broad sense, the emphasis is on the man-machine interface. Integration, interaction, intelligence, visualization, and customization are key concepts that are briefly discussed below.

Integration implies that in any given software system for real-world applications, more than one problem representation form or model, several sources of information or databases, and finally a multi-faceted and problem-oriented user interface ought to be combined in a common framework to provide a useful and realistic information base. The integration of environmental modeling and GIS is one step in this direction.

Interaction is a central feature of any effective man-machine system: a real-time dialogue allows the user to define and explore a problem incrementally in response to immediate answers from the system; fast and powerful systems with modern processor technology can offer the possibility to simulate dynamic processes with animated output, and they can provide a high degree of responsiveness that is essential to maintain a successful dialogue and direct control over the software.

Intelligence requires software to be "knowledgeable," not only about its own possibilities and constraints, but also about the application domain and about the user, that is, the context of its use. Defaults and predefined options in a menu system, sensitivity to context and history of use, learning, or alternative ways of problem specification, can all be achieved by the integration of expert systems technology in the user interface and in the system itself.

Visualization provides the bandwidth necessary to understand large amounts of highly structured information, and permits the development of an intuitive understanding of processes and interdependencies, of spatial and temporal patterns, and complex systems in general. Many of the problem components in a real-world planning or management situation are rather abstract: Representing them in a symbolic, graphical format that allows visual inspection of systems behavior and symbolic interaction with the machine and its software is an important element in friendly and easy to use computer-based systems.

Customization is based on the direct involvement of the end user in systems design and development. It is his view of the problem and his experience in many aspects of the management and decision-making process that the system is designed to support. This then must be central to a system's implementation to provide the basis for user acceptance.

Software and computer-based tools are designed to make things easier for the human user, and they improve the efficiency and quality of information processing tasks. In practice, only very few programs do that. They make things possible that would not be possible without the computer, but they rarely make it easy on the user.

As with the better mousetrap, one would expect to see demand for such techniques. However, simply doing things faster—once all the input has been painstakingly collected and entered, or solving a more complex version of the problem—and then leaving it to the user to extract the meaning from a flood of output and translate into his problem description language may not rate as a better mousetrap in the eye of the practitioner. All tools, and models in particular, have to become integrated parts in a much more complex information processing and decision-making procedure, which involves not only running the model, but certainly preparing its inputs over and over again, interpreting and communicating its results, and making them fit the usually rather formalized framework of the existing procedures.

There are several important aspects that need to be addressed. Computer-based tools, information and decision support systems as a rule imply a change in personal work habits, institutional procedures, and thus, institutional culture. While they may or may not change what is done, they most certainly change the way things are done—if they are used.

There is a tradeoff between the efficiency and ease of use and the flexibility of a system. The more options are predetermined and available from a menu of choices, the more defaults are provided, the easier it becomes to use a system for an increasingly smaller set of tasks.

There also is a tradeoff between the ease of understanding and the (at least numerical) precision of results. Providing a visual graphical or symbolic representation changes the quality of the information provided from a quantitative and thus at least apparently precise format, to a qualitative format. The latter, however, certainly is more appropriate to display patterns and complex interdependencies.

Finally, the easier a system is to use for some, the harder it is to make, and possibly also to maintain. Predefined options need to be defined at some point, and a knowledge base must be well developed and tested to work reliably. Automatic downloading of data and defaults requires that these data and defaults have been compiled and prepared in the first place. Thus, use has to be understood in a much wider sense, including problems of data collection and preparation, keeping data current, communicating and using the output within the institutional framework and communication channels, adapting the system to changing requirements, training new users, etc.

Ceterum censeo

The integration of environmental models and GIS is an obvious, challenging, and promising development in environmental research.

The need for better tools to handle ever more critical environmental problems is obvious, and the rapidly developing field of information technology provides the necessary machinery. To exploit the full potential of this integration, however, I believe it is important to try to really merge and combine modeling and GIS, rather than just using them together. The challenge is in merging the respective paradigms to create a new field of integrated environmental information systems that goes beyond models and GIS. Problem-oriented but scientifically based, with the computer as one of the most versatile tools and technologies as its basis, integrated environmental information systems should find their place both in academic research as well as in public policy-making.

REFERENCES

Andelin, J., and Niblock, R.W. (1982) *Use of Models for Water Resources Management, Planning, and Policy. Office of Technology Assessment Report*, Washington, DC: Congress of the United States.

Andreu, J., and Capilla, J. (1991) A computer aided support system for complex water resources schemes management. *Annales Geophysicae.* Supplement to Volume 9. XVI General Assembly Wiesbaden. 22–26 April, 1991, European Geophysical Society, p. C476.

Anthony, S.J., and Corr, D.G. (1987) Data structures in an integrated geographical information system. *ESA Journal*, 11/12: 69–72.

Arend, R.B. (1990) GIS: useful tool or expensive toy? *Journal of Surveying Engineering* 116(2): 131–138.

Armstrong, M.P. (1988) Temporality in spatial databases. *GIS/LIS '88, Accessing the World. Proceedings, Third annual international conference, San Antonio*, Vol 2, pp. 880–889.

Armstrong, M.P., and Densham P.J. (1990) Database organization strategies for spatial decision support systems. *Int. J. Geographical Information Systems* 4(1): 3–20.

Arnold, U., Datta, B., and Haenscheid, P. (1989) Intelligent geographic information systems (IGIS) and surface water modeling. *New Directions for Surface Water Modeling (Proceedings of the Baltimore Symposium)*, pp. 407–416.

Bachmat, Y., Bredehoeft, J., Andrews, B., Holtz, D., and Sebastian, S. (1980) Groundwater management: the use of numerical models. *Water Resources Monograph 5*, Washington, D.C.: American Geophysical Union.

Best, R.G., Guber, A.L., Kliman, D.H., Kowalkowski, S.M., Rohde, S.L., and Tinney, L.R. (1990) Emergency response and environmental applications of geographic information systems. *Transactions of the American Nuclear Society* 62: 42–43.

Bishop, I.D., Hull I.V., and Bruce, R. (1991) Integrating technologies for visual resource management. *Journal of Environmental Management* 32: 295–312.

Burrough, P.A. (1986) *Principles of Geographical Information Systems for Land Resources Assessment.* Monographs on Soil and Resources Survey, Oxford: Clarendon Press, 193pp.

Davis, J.R., Cuddy, S.M., Laut, P., Goodspeed, M.J., and Whigham, P.A. (1991) Canberra: Division of Water Resources, CSIRO.

Davis, J.R., and Nanninga, P.M. (1985) GEOMYCIN: towards a geographic expert system for resource management. *Journal of Environmental Management* 21: 377–390.

De Roo, A.P.J., Hazelhoff, L., and Burrough P.A. (1989) Soil erosion modelling using 'Answers' and geographical information systems. *Earth Surface Processes and Landforms* 14: 517–532.

De Wispelaere, D., Schiermeier, F.A., and Gillani, N.V. (eds.) (1986) *Air Pollution Modeling and Its Application V. NATO Challenges of Modern Society*, Plenum Press, 773pp.

EGS (1991) *XVI General Assembly, Wiesbaden, 22–26 April 1991. Annales Geophysicae*, Supplement to Volume 9.

Fedra, K. (1990a) A computer-based approach to environmental impact assessment. *RR-91-13*, Laxenburg, Austria: International Institute for Applied Systems Analysis.

Fedra, K. (1990b) Interactive environmental software: integration, simulation and visualization. *RR-90-10*, Laxenburg, Austria: International Institute for Applied Systems Analysis.

Fedra, K. (1991) Smart software for water resources planning and management. *Decision Support Systems. NATO ASI Series, Vol. G26. Proceedings of NATO Advanced Research Workshop on Computer Support Systems for Water Resources Planning and Management. September 24–28, 1990.* Eiriceiria, Portugal, Springer, pp. 145–172.

Fedra, K., and Diersch, H.J. (1989) Interactive groundwater modeling: color graphics, ICAD and AI. *Proceedings of International Symposium on Groundwater Management: Quantity and Quality. Benidorm, Spain October 2–5, 1989. IAHS Publication No 188*, pp. 305–320.

Fedra, K., Diersch, H.J., and Härig, F. (forthcoming) *Interactive Modeling of Groundwater Contamination: Visualization and Intelligent User Interfaces. Advances in Environmental Science: Groundwater Contamination Series*, New York: Springer Verlag.

Fedra, K., and Kubat M. (forthcoming) Hybrid geographical information systems. *Paper presented at the EARSeL Workshop on Relationship of Remote Sensing and Geographic Information Systems. Hanover, Germany. September 1991.*

Fedra, K., Li, Z., Wang, Z., and Zhao, C. (1987) Expert systems for integrated development: a case study of Shanxi Province, The People's Republic of China. *SR-87-001*, Laxenburg, Austria: International Institute for Applied Systems Analysis.

Fedra, K., and Loucks, D.P. (1985) Interactive computer technology for planning and policy modeling. *Water Resources Research* 21/2: 114–122.

Fedra, K., and Reitsma, R.F. (1990) Decision support and geographical information systems. *RR-90-9*, Scholten, H.J., and Stillwell, J.C. H. (eds.) *Geographical Information Systems for Urban and Regional Planning*, Dordrecht: Kluwer Academic Publishers, pp. 177–186.

Fedra, K., and Winkelbauer, L. (1991) MEXSES: An expert system for environmental screening. *Proceedings Seventh IEEE Conference on Artificial Intelligence Applications. February 24–28, 1991, Miami Beach, Florida*, Los Alamitos, California: IEEE Computer Society Press. pp. 294–298.

Ferreira, J., Jr. and Menendez, A. (1988) Distributing spatial analysis tools among networked workstations. *URISA '88—Mapping the Future*, Washington, DC: Urban and Regional Information Systems Association, pp. 200–215.

Feyerabend, P. (1970) Consolations for the specialist. In: Lakatos, I., and Musgrave, A. (eds.) *Criticism and the Growth of Knowledge*, Cambridge: Cambridge University Press, pp. 197–230.

Fischer, D. (1988) CAD thematic mapping as part of an environmental information system. In Kovacs, P., and Straub, E. (eds.) *Proceedings of the IFIP TCB Conference on Governmental and Municipal Information Systems*, The Netherlands: North-Holland, Amsterdam, pp. 23–29.

Forrester, J.W. (1982) *Principles of Systems. Second Preliminary Edition*, Cambridge, MA: MIT Press.

Gardels, K. (1988) GRASS in the X-Windows environment. Distributing GIS data and technology. *GIS/LIS '88, Accessing the World. Proceedings, Third annual international conference, San Antonio*, Vol 2, pp. 751–758.

Gélinas, R., Bond, D., and Smit B. (eds.) (1988) *Perspectives on Land Modelling. Workshop Proceedings, To-*

ronto, Ontario. November 17–20, Montreal: Polyscience Publications Inc., 230pp.

Gimblett, R.H., and Itami, R.M. (1988) Linking GIS with video technology to simulate environmental change. *GIS/LIS '88, Accessing the World. Proceedings, Third annual international conference, San Antonio*. Vol 1, pp. 208–219.

Goulter, I.C., and Forrest, D. (1987) Use of geographic information systems (GIS) in river basin management. *Water Science and Technology* 19(9): 81–86.

Granger, K.J. (1989) Process modelling and geographic information systems: breathing life into spatial analysis. *Proceedings of Eighth Biennial Conference and Bushfire Dynamics Workshop*, Canberra, ACT, Australia: Australian National University. pp. 37–341.

Gwynne, M.D. (1988) The Global Environment Monitoring System (GEMS): Some recent developments. *Environmental Monitoring and Assessment* 11: 219–223.

Hedges, P.D. (1991) Evaluation of the effects of groundwater drawdown on vegetation using GIS. *Annales Geophysicae. Supplement to Volume 9. XVI General Assembly Wiesbaden. 22–26 April, 1991*, European Geophysical Society, p. C481.

Holcomb Research Institute (1976) Environmental modeling and decision making. The United States experience. *A report by the Holcomb Research Institute Butler University for the Scientific Committee on Problems of the Environment*, Praeger Publishers. 152pp.

Holling, C.S. (ed.) (1978) *Adaptive Environmental Assessment and Management*, John Wiley and Sons, 377pp.

Hughes, D.A. (1991) The development of a flexible, PC based hydrological modelling system. *Annales Geophysicae. Supplement to Volume 9. XVI General Assembly Wiesbaden. 22–26 April, 1991*, European Geophysical Society, p. C481.

Jackson, M.J., James, W.J., and A. Stevens (1988) The design of environmental geographic information systems. *Phil. Trans. R. Soc. Lond.* A 324: 373–380

Johnson, L.B. (1990) Analyzing spatial and temporal phenomena using geographical information systems: a review of ecological applications. *Landscape Ecology* 4(1): 31–43.

Johnston, C.A. (1989) Ecological research applications of geographic information systems. *GIS/LIS'89 Proceedings, Orlando, Florida. November 26–30*, Vol. 2, pp. 569–577.

Johnston, C.A., and Bonde, J. (1989) Quantitative analysis of ecotones using a geographic information system. *Photogrammetric Engineering and Remote Sensing* 55(11): 1643–1647.

Johnston, C.A., Detenbeck, N.E., Bonde, J.P., and Niemi, G.J. (1988) Geographic information systems for cumulative impact assessment. *Photogrammetric Engineering and Remote Sensing* 54(11): 1609–1615.

Johnston, C.A, and Naiman, R.J. (1990) The use of a geographic information system to analyze long-term landscape alteration by beaver. *Landscape Ecology*, 4(1): 5–19.

Kessell, S.R. (1990) An Australian geographical information and modelling system for natural area management. *International Journal of Geographical Information Systems* 4(3): 333–362.

Khan, M. Akram, and Liang, T. (1989) Mapping pesticide contamination potential. *Environmental Management* 13(2): 233–242.

Kuhn, T.S. (1970 and 1962) *The Structure of Scientific Revolutions, 2nd Edition*, Chicago University Press.

Labadie, J.W., Brazil, L.E., Corbu, I., and Johnson, L.E. (1989) Computerized decision support systems for water managers. *Proceedings of the 3rd Water Resources Operations Workshop, Colorado State University, Fort Collins, CO, June 27–30, 1988*, New York: ASCE, 978pp.

Lakatos, I. (1968) Criticism and the methodology of scientific research programmes. *Proceedings of the Aristotelian Society* 69: 149–186.

Lam, D.C.L., and Swayne, D.A. (1991) Integrating database, spreadsheet, graphics, GIS, statistics, simulation models and expert Systems: experiences with the Raison system on microcomputers. *NATO ASI Series, Vol. G26*, Heidelberg: Springer, pp. 429–459.

Langran, G. (1988) Temporal GIS design tradeoffs. *GIS/LIS '88, Accessing the World. Proceedings, Third annual international conference, San Antonio*, Vol 2., pp. 890–899.

Lauenroth, W.K., Skogerboe, G.V., and Flug, M. (eds.) (1983) Analysis of Ecological Systems: State-of-the-Art in Ecological Modelling. *Proceedings of a Symposium 24–28 May 1982 at Colorado State University, Fort Collins, Colorado,* Elsevier Scientific Publishing Company, 992pp.

Lindenmayer, D.B., Cunningham, R.B., Nix, H.A., Tanton, M.T., and Smith, A.P. (1991) Predicting the abundance of hollow-bearing trees in montane forests of Southeastern Australia. *Australian Journal of Ecology* 16: 91–98.

Lotka, A.J. (1924) *Elements of Physical Biology*, Baltimore: Williams and Wilkins Co. Inc., 460pp.

Loucks, D.P., and Fedra, K. (1987) Impact of changing computer technology on hydrologic and water resource modeling. *Review of Geophysics*, 25(2).

Loucks, D.P., Kindler, J., and Fedra, K. (1985) Interactive water resources modeling and model use: an overview. *Water Resources Research* 21/2: 95–102.

Maidment, D.R., and Djokic, D. (1990) Creating an expert geographic information system: the Arc-Nexpert interface. *Internal paper*, Austin, TX: Dept. of Civil Engineering, University of Texas.

Manning, E.W. (1988) Models and the decision maker. In: Gélinas, R., Bond, D., and Smit, B. (eds.) *Perspectives on Land Modelling. Workshop Proceedings, Toronto, Ontario, November 17–20, 1986*, Montreal: Polyscience

Publications, pp. 3–7.
Masterman, M. (1970) The nature of a paradigm. In: Lakatos, I., and Musgrave, A. (eds.) *Criticism and the Growth of Knowledge*, Cambridge: Cambridge University Press, pp. 59–89.
Moffatt, I. (1990) The potentialities and problems associated with applying information technology to environmental management. *Journal of Environmental Management* 30: 209–220.
Nachtnebel, H.P., Fürst, J., Girstmair, G., and Holzmann, H. (1991) Integration of a GIS with a groundwater model to assist in regional water management. *Annales Geophysicae. Supplement to Volume 9. XVI General Assembly Wiesbaden. 22–26 April, 1991*, European Geophysical Society, p. C484.
Oslin, A.J., and Morgan, D.S. (1988) Streams: A basin and soil erosion model using CADD remote sensing and GIS to facilitate ratershed management. *Proceedings of NCGA's Mapping and GIS System '88*, Fairfax, VA: National Computer Graphics Association.
Parent, P., and Church, R. (1989) Evolution of geographic information systems as decision making tools. In Ripple, W.J. (ed.) *Fundamentals of Geographic Information Systems: A Compendium*, Bethesda, MD: American Society for Photogrammetry and Remote Sensing, pp. 9–18.
Park, R.A., Collins, C.D., Leung, D.K., Boylen, C.W., Albanese, J.R., de Caprariis, P., and Forstner H. (1979) The aquatic ecosystem model MS.CLEANER. In Jorgensen, S.E. (ed.) *State-of-the-Art in Ecological Modelling*, Copenhagen: International Society of Ecological Modelling.
Park, R.A., O'Neill, R.V., Bloomfield, J.A., Shugart, Jr., H.H., Booth, R.S., Koonce, J.F., Adams, M.S., Clesceri, L.S., Colon, E.M., Dettman, E.H., Goldstein, R.A., Hoopes, J.A., Huff, D.D., Katz, S., Kitchell, J.F., Kohberger, R.C., LaRow, E.J., McNaught, D.C., Peterson, J.L., Scavia, D., Titus, J.E., Weiler, P.R., Wilkinson, J.W., and Zahorcak, C.S. (1974) A generalized model for simulating lake ecosystems. *Simulation* 23: 33–50.
Parker, H.D. (1988) A comparison of current GIS software: the results of a survey. *GIS/LIS '88, Accessing the World. Third annual international conference, San Antonio*, Vol. 2, pp. 911–914.
Patten, B.C. (ed.) (1971) *Systems Analysis and Simulation in Ecology. Volume I*, Academic Press.
Phillips, J.D. (1990) A saturation-based model of relative wetness for wetland identification. *Water Resources Bulletin* 26(2): 333–342.
Piwowar, J.M., LeDrew, E.F., and Dudycha, D.J. (1990) Integration of spatial data in vector and raster formats in a geographic information system environment. *Int. J. Geographical Information Systems* 4(4): 429–444.
Reitsma, R.F. (1990) Functional classification of space. Aspects of site suitability assessment in a decision support environment. *RR-90-2*, A-2361 Laxenburg, Austria: International Institute for Applied Systems Analysis, 301pp.
Reitsma, R.F., and Behrens, J.S. (1991) Integrated river basin management (IRBM): a decision support approach. *Proceedings of Second International Conference on Computers in Urban Planning and Urban Management*, Akron, OH: Institute for Computer Aided Planning, University of Akron.
Scholten, H.J., and Stillwell, J.C.H. (eds.) (1990) *Geographical Information Systems for Urban and Regional Planning*, Dordrecht: Kluwer, 261pp.
Seaborn, D.W. (1988) Distributed processing and distributed databases in GIS—separating hype from reality. *GIS/LIS '88, Accessing the World. Proceedings, Third annual international conference, San Antonio*, Vol 1., pp. 141–144.
Smyrnew, J.M. (1990) Trends in geographic information system technology. *Journal of Surveying Engineering*, 116(2) 105–111.
Steppacher, L. (1988) Ground water risk management on Cape Cod: the applicability of geographic information system technology. *Proceedings of FOCUS Conference on Eastern Regional Ground Water Issues. September 27–29, Stamford, Connecticut*, Dublin, OH: National Water Well Association.
Sucksdorff, Y., Lemmelä, R., and Keisteri T. (1989) The environmental geographic information system in Finland. *New Directions for Surface Water Modeling (Proceedings of the Baltimore Symposium)*. May, pp. 427–434.
Tikunov, V.S. (1989) Modern research tools on the 'Society-Natural Environment' System in Soviet geography. *Journal of Environmental Systems* 19(1): 59–69.
Tilley, S., and Sperry, S.L. (1988) Raster and vector integration. *Computer Graphics World* (August 1988): 73–76.
Turner, A.K. (ed.) (1991) *Three-dimensional Modeling with Geoscientific Information Systems*, Dordrecht: Kluwer Academic Publishers, 443pp.
van Est, J.V., and de Vroege, F. (1985) Spatially oriented information systems for planning and decision making in the Netherlands. *Environment and Planning B: Planning and Design* 12: 251–267.
Weil, U. (1990) Info technology helps with environmental cleanup. *Government Computer News* (November 26): 50.
Welch, R., Madden Remillard, M., and Slack, R.B. (1988) Remote sensing and geographic information system techniques for aquatic resource evaluation. *Photogrammetric Engineering and Remote Sensing* 54(2): 177–185.
White, S.M. (1991) The use of a geographic information system to automate data input to the Systĕme Hydrologique Europĕen (SHE-UK). *Annales Geophysicae. Supplement to Volume 9. XVI General*

Assembly Wiesbaden. 22–26 April, 1991, European Geophysical Society, p. C488.

Wiggins, J.C., Hartley, R.P., Higgins, M.J., and Whittaker, R.J. (1986) Computing aspects of a large geographic information system for the European Community. *Conference Proceedings of Auto Carto London*, Vol. 2, pp. 28–43.

Wilde, B.M., and Drayton, R.S. (1991) A geo-information system approach to distributed modelling. *Annales Geophysicae. Supplement to Volume 9. XVI General Assembly Wiesbaden. 22–26 April, 1991*, European Geophysical Society, p. C488.

Witt, R.G. (1989) An overview of GRID: a working global database. In Zirm, K.L., and Mayer, J. (eds.) *Envirotech Vienna 1989—Computer-Based Information Systems in Environmental Protection and Public Administration (Workshop). February 20–23, 1989*, Vienna/Hofburg, Austria: International Society for Environmental Protection.

Woodcock, C.E., Sham, C.H., and Shaw, B. (1990) Comments on selecting a geographic information system for environmental management. *Environmental Management* 14(3): 307–315.

Woodcock, C.E., and Strahler, A.H. (1983) PROFILE: remote sensing for land management and planning. *Environmental Management* 7(3): 223–238.

Zack, J.A., and Minnich, R.A. (1991) Integration of geographic information systems with a diagnostic wind field model for fire management. *Forest Science* 37(2): 560–573.

6

The Role of Software Vendors in Integrating GIS and Environmental Modeling

JACK DANGERMOND

This chapter presents the perspective of a major developer of GIS software, a firm that has been in the GIS field for more than 20 years. Nevertheless, it does not represent the view of other firms in the field, certainly not of all such firms. The scope of this chapter is quite narrow, at the request of the editors of this book. For broader perspectives other chapters will need to be consulted. It is particularly important to understand that this chapter presents a point of view rather than an exhaustive survey or a position paper on the topic. This approach is probably valid because the response of the GIS software vendor community to requests for the development of interfaces between GIS and environmental modeling is as likely to be based on the industry's perceptions as it is on more tangible evidence.

THE GIS MARKET

The chief driving force that determines what kinds of GIS software products will be developed by software vendors is the market as it is perceived by the vendor organizations. GIS vendors generally develop new products because they believe that the products may be successful and return a profit. For many of these vendors the market must be, at least potentially, rather large; for example, for a mature GIS software product selling for under $10,000 the market would often need to be one in which hundreds or thousands of copies of a particular software product would be sold. As the price of some GIS software products falls to under $500, a truly mass market will be required in which thousands and perhaps tens of thousands of copies will need to be sold in order to support software development and marketing costs of the vendors. The implication is clear: GIS products for which only a small market exists are much less likely to be developed by GIS software vendors.

Products for which only a small market exists are, nevertheless, still created by GIS software vendors, but they are likely to be developed on a time and materials basis as a discrete contracted project for a fixed profit. This is the method by which a good deal of customized software is developed for particular clients or client organizations, and it is the means by which a good deal of software for use by government agencies was originally developed. Of course this approach requires an agency that can provide the funds for the development contract. Equally important, for the product to be useful, the development work must be performed by competent programmers. Unlike products that are tested and proven by the market, where the buyer can select among competing products and obtain the one most suited to her or his needs, and where products are usually in a constant state of continuing development, products of custom programming are only as good as the programming that goes into them, and are only improved to the extent that the contract provides. Vendors may also attempt to create a market for a new product for which no known market exists, but in this case the vendors must first be convinced that either the potential market exists or that the appropriate product can create such a market.

Vendors also are likely to believe that, over some period of time, they know what the market is since their representatives are in constant contact with potential buyers. For a large vendor organization, with more than a hundred sales representatives, this means thousands of contacts with potential customers every week. Vendors are also in the business of distinguishing casual interest from real possibilities for sales. Given these kinds of considerations, vendors have some reason for believing that they know what the real market for GIS software is better than anyone else.

All these factors inevitably tend to reduce the likelihood of vendors developing some kinds of GIS software products, chiefly those that are perceived to have only a very limited market appeal, and those that would be very expensive to develop, especially in the absence of a large market. Products that can be developed inexpensively, or that might be the product of a single entrepreneurial programmer or of a very small company, might still be developed by those hoping to found a company on the basis of a unique, new product. In the present context, the question is where products integrating GIS software and environmental modeling fit: What kind of market is there for them, or what kind of market might there be if such products were created?

CHARACTERISTICS OF FUTURE GIS SOFTWARE PRODUCTS

If it is assumed that ideal products for integrating GIS and environmental modeling are not now available, then the question of their development may also be approached by a consideration of the characteristics of those GIS software products that are most likely to be developed for the market in the future. It might be, for example, that the future development of such products is inevitable because of the direction in which the GIS software market is going. Of course, different GIS software vendors see their markets differently, or attempt to serve different segments of the overall market for GIS software. Inevitably, these remarks represent part of the views of one vendor organization.

It is our view that the future market for GIS software will increasingly be dominated by software that runs on open hardware systems, that is, systems in which the hardware of different vendors is interconnected. Increasingly, also, some of these systems will be networks, either local area (small) networks or much larger, even global, networks. GIS is now a sufficiently important software product that it will be developed for the latest hardware products of all kinds, both central processing units and peripherals. Many kinds of GIS products will be developed for personal computers, especially as these become more powerful. GIS software will be interfaced to a growing number of database management systems and will be increasingly capable of dealing with very large databases; it is already the case for some GIS software that the only limitations on database size have to do with hardware capacities. GIS software will increasingly be able to interface with a widening range of data types: CAD data, images, remote sensing, text, elevational data, and so on. An inevitable result of this is that the types of disciplines that make use of GIS will continue to increase.

Role of GIS users in vendor development of future GIS products

It is important to recognize the important role that GIS users play in the development of new GIS software, particularly in the development of what might loosely be called "applications software." In earlier years, when there were many fewer users of GIS technology, the development of new GIS software capabilities was chiefly in the hands of the software vendors and their programmers, partly because they knew more about GIS software than anyone else. However, as the number of GIS users and GIS-using organizations has rapidly increased in recent years, the users have become more knowledgeable and their buying power has increasingly influenced the development plans of GIS software vendors. This is especially the case for those users who have software maintenance contracts with the vendors and who are organized into users groups.

At the present time GIS users continually bombard the Environmental Systems Research Institute (ESRI) and other GIS vendors with requests for new or expanded GIS capabilities and applications. These requests come in many forms, some of them very precise, others rather vague. It is physically impossible for GIS vendor programmers to satisfy all these requests. Inevitably those requests that promise the most return on the programming investment, or those that come closest to meshing with our plans or those of other vendors for software development, are likely to be satisfied first.

In the case of applications, it is now fairly clear that expansion in the range of applications is chiefly in the hands of GIS users rather than GIS vendors. This is so because of the large and growing number of GIS users, the diversity of the users, and the relative constancy in the number of vendor organizations. The GIS field continues to be dominated by a handful of GIS vendors while the number of GIS users is rapidly increasing into the tens of thousands. One consequence is that vendor organizations are directly involved in the development of less and less of the total range of GIS applications.

Other vendor activities that promote GIS software development

Quite apart from the formal business of software development itself, GIS software vendors also conduct other activities, or support the conduct of activities, that may foster the development of GIS software that integrates GIS and environmental modeling. Some vendor organizations, for various reasons, support university research in various ways. They may supply funding for research, supply software or subsidize its purchase, or encourage hardware suppliers to adopt similar policies. Vendors may conduct their own long-range research (as distinct from product development efforts) on GIS technology. Vendors also are a significant source of employment for skilled programmers and other experts on GIS software; inevitably these persons find the means of carrying on research efforts of their own and in the process may expand the capabilities of GIS software. Such efforts are sometimes encouraged by vendor company policies of various kinds. GIS software vendors, as do others in the computer field, also donate software and other GIS-related technology to nonprofit environmental and conservation organizations, some of which then use these gifts to develop additional GIS capabilities. Vendors also contribute importantly to the scientific literature of the field and to the education of people about GIS technology. These efforts may also foster the integration of GIS technology with environment modeling.

CURRENT ENVIRONMENTAL MODELING USING GIS TECHNOLOGY

While the other chapters in this volume discuss this topic in considerably greater detail, it is appropriate to include here a few words, from the point of view of a GIS software vendor, about what the results of GIS vendor efforts have been as reflected in the current use of GIS technology for environmental modeling. Many of these are efforts that software vendors have supported in various ways, including application development.

At present, GIS technology is used extensively in the environmental field. GIS technology is also used extensively for modeling. However, as Kurt Fedra also points out in Chapter 5, GIS technology is not used for modeling in the environmental field either as extensively as it might be used or as extensively as it ought to be used. From ESRI's vantage point we have seen GIS technology used extensively as a means to take inventory of information about the environment and planning and managing actions that have environmental impact. For example, perhaps the majority of U.S. federal agencies having to do with the environment now make use of GIS technology in their efforts. GIS technology is also extensively used throughout the world by governments at all levels, as well as by private sector organizations and educational institutions, in environmentally related efforts of many kinds.

GIS technology is also used extensively for modeling of various kinds. This modeling ranges from simple use of GIS as conceptual models of spatial reality, through modeling efforts involving GIS manipulative capabilities (like overlaying, buffering, network analysis, and the like), to such uses as gravity modeling, traffic modeling, emergency evacuation modeling, and the like.

GIS technology is used extensively in such environmentally related efforts as forest management, management of toxic sites and toxic waste disposal, environmental monitoring and management, monitoring of rainforests and old growth forests, planning and management of interconnected water resources, siting of waste disposal sites including nuclear wastes, and the like. Modeling in various forms has been a component of most of these efforts.

GIS technology has also been used in some quite comprehensive models of environments and environmental processes. Among the efforts that ESRI has supported has been the construction of hierarchical environmental models of environments in various parts of Germany using a combination of GIS technology and various continuous system modeling software. J. Schaller, M. Sittard, and others from the ESRI Gesellschaft für Systemforschung und Umweltplanung, and W. Haber, W.-D. Grossman, and others from the Lehrstuhl für Landschaftsökologie der Technischen Universität München, are among those who have worked in this area over the years. Individual models of various types of environments (meadows, lakes, cultivated land, differing types of woodlands, etc.) have been created and coupled with one another. Basic data about the existing real situation have been accessed in a GIS and supplied as initial data for these models. The models are then stepped through time with their output updating the GIS. Model compartments have been coupled with one another, and coupled atmospheric/terrestrial modeling has been done. Economic and social data have been incorporated into the modeling process. Moreover, the results of these modeling efforts have been used to provide policy and planning guidelines for real application. Unfortunately, this sort of modeling seems to be the exception rather than the rule; the few examples are striking because they are unusual. An evaluation of the field as a whole suggests that the use of GIS technology for environmental modeling has just begun.

SOME PROBLEMS IN INTEGRATING EXISTING GIS TECHNOLOGY WITH ENVIRONMENTAL MODELING

While not all the technical problems inhibiting the integration of GIS technology with environmental modeling can be discussed here, a few will be mentioned from the perspective of software vendors interested in developing modeling applications for users.

Static data

For many purposes, for many processes, for many time periods, and for many kinds of spatial data, such as land use, surficial geology, vegetation, etc., GIS can safely represent data as invariant with respect to time. Those using the data make judgments as to how frequently and thoroughly their data require updating, often deciding that only yearly updates or even less frequent updates are sufficient for their purposes; city planners, tax assessors, and the like fit this category. Most present GIS users probably employ and are satisfied with this kind of relatively invariant data. The technical capacity to store even very large amounts of time series data in GISs now exists, and this capacity will continue to increase rapidly due to coming developments in computer hardware and in algorithms for data compression. While storage capacity may have been a primary limiting technical factor in the past, that is changing very quickly. Static GIS data are probably used by many other GIS users only because they cannot afford to update their data or they lack available technical means to do so. Data gathering continues to be a very costly exercise, and no immediate solution to that problem is at hand. Indeed, availability of the needed data, time sequenced or not, may be the chief limiting factor in integrating GIS with environmental modeling, as it probably is for both GIS and environmental modeling as separate activities. Where repetitive remote sensing can provide the required data, it may partially resolve the data problem, but automated image interpretation is not yet effective for many kinds and scales of data, and manual image interpretation requires time and its costs have not been reduced. For many kinds of data that modelers want, appropriate sensors and data capture systems do not exist, are extremely expensive, or are available only in such low numbers that the data coverage is very sparse. All these limitations suggest that modelers who want to use GIS will continue to be well advised to work with geographical areas where rich, frequently updated data already exist.

Because of various technical developments (e.g., much faster, much less expensive central processing units; and much larger, much less expensive memories with faster search times), GIS users are beginning to give more serious thought to including time series data. But to do so, technical problems beyond storage capacity and processing speed must be overcome. Many changes of interest occurring in the real world are likely to occur at differing rates and require differing scales and means of data representation. Capturing, storing, and representing widely differing rates of change in a GIS presents technical problems with which GIS technology has little experience. How, for example, should geological change, the movement of automobile traffic, succession in a forest stand, and the movement of a toxic plume through an aquifer be represented for the same geographical region? GIS have usually displayed geographic data as single maps or a short series of maps for comparison purposes. Where time series data are to be displayed, some form of full animation will likely be required, and automatic means of generating interprocess comparison data will probably be wanted. Modelers and others, for example, are sometimes likely to want to see what the relationship is between two processes that occur simultaneously but that may occur at widely differing rates, over overlapping regions, and that may not (seem to) be closely coupled. How should data about these interactions be provided? Probably good solutions to such problems will only evolve over time, though vendors are now beginning to look into them seriously.

Data sharing

Even if it is assumed that suitable data exist (e.g., time series data including appropriate variables for a region of interest), the data still may not be available to those who want to use them. This might be so for many reasons: secrecy, unavailability in digital form, too high a cost for acquisition, and so on. A good deal of attention is being given by at least some vendors to the means by which existing data can be more widely shared, including questions relating to how the public may access government data. This problem involves a range of issues: data collection, data control and security, charges, cost recovery, public/private relationships, GIS data publication, technical means of delivering stored data, technical means of transmitting large quantities of data over networks, and the like. There are some successful models for cost-effective data gathering and subsequent data sharing, but they have not been widely copied. This is an area in which a good deal of interinstitutional organization work will have to be done, and more organizations must be found that are willing to take all the risks of being pioneers in data sharing efforts.

GIS standards

A large GIS operating in the U.S. at the present time might be appropriately subject to as many as a thousand different hardware, software, communications, and other kinds of standards. One implication of this is that even were mechanisms in place to facilitate data sharing, or networking of modelers (in something like collaboratories), additional technical hurdles, in the form of standards, might still have to be cleared before users, a GIS, and a model could be brought together. This is an area in which GIS are especially driven by GIS users. When enough users conform to a standard and require that a standard be adopted by GIS vendors, the adoption usually takes place.

Representation of functional relationships

At present, in the experience of many software vendors, most users of GIS represent functional relationships between variables in quite simple ways, ways they feel fit their requirements and are also appropriate to their understanding of the relationships. Thus in considering environmental influences on the choice of building sites, slope, underlying geology, and soil type are very likely to be considered, but many economic and most sociological, cultural, ecological, visual, and other relationships are more likely to be ignored, or taken into account only in the planner's or designer's thought processes. Moreover, the relationship between slope and suitability in a construction suitability model is likely to be represented in the form of straightforward tables relating acceptable slope classes to construction suitability. Many GIS users (engineers, planners, foresters, and so on) are simply not prepared to make use of more speculative environmental models in making decisions. This inevitably inhibits the use of environmental modeling in cases in which the processes seem to be extremely complicated, interrelated, and not well understood. It may be that only practical successes in model use, or widely recognized improvements in environmental models, will overcome this problem.

Deterministic representation of stochastic variables

Since so much of the use of GIS technology is based on either static or quasi-steady-state conditions, issues related to the causation of data variability arise rather infrequently. When they do arise, as when models are stepped through time, it is probably the case that most of the variation is now treated in a deterministic way. Facilities for allowing stochastic modeling using GIS do exist; however, the stochastic variation is generally injected into the model from the simulation languages used with the GIS rather than from the GIS software system itself; there are particular exceptions to this general rule (for example, traffic modeling using network GIS capabilities).

CONCLUSION: THE FUTURE OF GIS AND ENVIRONMENTAL MODELING FROM A VENDOR'S PERSPECTIVE

Environmental modeling and simulation are among the most important applications of GIS technology. They are, in one vendor's view, chiefly hindered at present by a lack of suitable modeling tools, a lack of sound theoretical understanding of how environmental systems work, and a lack of data about such systems. GIS vendors are perhaps most responsible for providing the modeling tools that will work with a GIS.

GIS and environmental modeling

In the future, ESRI and other GIS software vendors are likely to supply GIS technology that is considerably more user friendly, requires less GIS-specific technical knowledge, is more tailored to each user's needs (e.g., specialized GIS workstations for planners, hydrologists, foresters, etc.), is less expensive to acquire, more graphical, easier to use, and more powerful. Artificial intelligence, initially used for routine functions (map design, data selection, etc.), will unobtrusively support users. Useful forms of data will be easier to import into and export from the GIS of the future. As computer processors increase in power and fall in cost, and as computer memory becomes orders of magnitude less expensive, the multidimensional (time, space, other parameters) nature of environmental systems will be easier to represent in GIS. In dynamic modeling, maps and other graphics will display, in continuous animation, the changes occurring in environmental systems as events proceed. Data collection devices, including a more fully instrumented universe, will provide modelers and others with a richer and more continuous flow of information. One likely means for providing this instrumented universe is the use of nanotechnology. (Nanotechnology includes the device fabrication methods now used in computer chip manufacture to produce complete ultraminiature machines, such as environmental sensing and measuring devices. Strain gauges and velocity meters are already produced by such means; other environmental sensors are likely to follow in the years just ahead. Produced as cheaply as computer chips and in very large numbers, such sensors could be very widely distributed through the environment, providing greatly enriched data resources for modelers.)

There is also a clear trend in ESRI's own software towards closer integration within a single software system of what were once separate and disparate software capabilities. Thus CAD capabilities and GIS capabilities have been integrated into a single system, image processing and GIS capabilities have been integrated into a single system, raster and grid capabilities have been integrated into a single system, and so on. It is likely that this trend will continue. Thus, as environmental modeling becomes more popular, modelers can probably look forward to the integration of at least some modeling capabilities into GIS software systems (or vice versa).

Communication between GIS vendors and environmental modelers

There is obviously some community of interests between those working in the GIS field, including software vendors, and those doing environmental modeling. If GIS and environmental modeling are to be more closely and effectively coupled, communication between the two groups would be useful; moreover, if modelers are to be heard by GIS software vendors, they may need to join together and agree as much as possible on the messages they want to send. An effective way to do this might be patterned on existing professional associations, special interest groups (SIGs), or organizations of computer software users.

SUMMARY

The function of science is to understand, predict, and control; as GIS-based environmental models become accepted, they will become part of the decision support systems for environmental managers and decision-makers, and environmental models will be used to predict, and perhaps ultimately control, both the consequences for the environment of human actions and the impact on humans of environmental processes. The chapters in this volume testify to the progress being made in this direction. This is also a direction Environmental Systems Research Institute has supported from its founding, and one I believe other software vendors will also increasingly support in the years ahead.

II

GIS for Modelers

MICHAEL F. GOODCHILD

The three chapters of this section provide three distinct perspectives on GIS, and expand on the introduction of Chapter 2. Like word processing, GIS is now a widely installed class of software, but unlike word processing, its functionality and objectives remain complex and broad-based. Is GIS a set of spatial data handling capabilities that are broadly useful in preprocessing data for environmental modeling, or a shortcut to high-quality cartographic display, or is it closer to a language for spatially distributed models? Elements of all three can be found in the chapters of Part III of this book.

Berry's Chapter 7 comes closest to presenting GIS as a language, in his discussion of cartographic modeling. This tradition, which has its roots in the work of McHarg (1969) and has recently been described at length by Tomlin (1990), views a spatial database as a collection of layers, each representing the distribution of one spatial variate as a coregistered raster. The language of cartographic modeling includes commands to transform and combine layers, convolve them with filters, measure them in various ways, and perform assorted simple analyses. As a language, it bears some similarity to image algebras (e.g., Serra, 1982). The cartographic modeling approach can be found imbedded in several widely available GIS, such as the MAP package described by Berry.

In Chapter 8, Nyerges expands the discussion substantially with a broad perspective on the current state of GIS as a whole, with its plethora of approaches to the digital representation of variation in space. Two examples—the contamination of an urban estuary and the modeling of a river network—are used to illustrate differences in data models and associated GIS functionality. Nyerges argues that three different perspectives are important in defining and understanding the nature of GIS: functional, procedural, and structural. GIS can be defined as a set of manipulations on spatial data and their associated objectives; as a set of procedures for solving specific problems; or as a software architecture assembled from assorted pieces. In the final sections of the chapter, Nyerges looks at the problems of coupling current GIS with software for environmental modeling, and at recent trends in the GIS industry.

The quality of any computer-based analysis is ultimately constrained by the quality of its input, and unfortunately spatial data are particularly prone to uncertainty and inaccuracy. In the final chapter of this section, Goodchild (Chapter 9) examines the question of data quality in GIS, and the impact of inaccuracy and error on GIS output. Like all measurements, maps are approximations to the truth, and often no more than subjective interpretations of real spatial variation. Yet the ease of access and manipulation provided by GIS, and its inherent precision, encourage the user to see spatial data as accurate, and to lose touch with the uncertainties of the measurement and capture processes. Thus a major research priority is the incorporation within GIS of a suite of techniques for describing error, and propagating it through the modeling process into confidence limits on results.

The three chapters provide an overview of GIS and draw attention to some of its basic issues. They cannot, however, provide a complete and comprehensive introduction to GIS. At the end of Chapter 7, Berry gives a short bibliography of basic texts and other sources. Other references to the current state of GIS are scattered throughout the book.

REFERENCES

McHarg, I.L. (1969) *Design with Nature*, New York: Doubleday.

Serra, J.P. (1982) *Image Analysis and Mathematical Morphology*, New York: Academic Press.

Tomlin, C.D. (1990) *Geographic Information Systems and Cartographic Modeling*, Englewood Cliffs, NJ: Prentice-Hall.

7

Cartographic Modeling: The Analytical Capabilities of GIS

JOSEPH K. BERRY

Geographic Information Systems (GIS) technology is more different from than it is the same as traditional mapping and map analysis. Mathematics provides a useful starting point in discussing the differences. In fact, GIS is based on a mathematical framework of primitive map analysis operations analogous to those of traditional statistics and algebra. From this perspective, GIS forms a "toolbox" for processing maps that embodies fundamental concepts and stimulates creative applications. This toolbox is as flexible as conventional mathematics in expressing relationships among variables—but with a GIS, the variables are entire maps.

The map analysis toolbox helps resource managers define and evaluate spatial considerations in land management. For example, a map of unusually high concentrations of lead in the soil can be generated from a set of soil sample data. These hazardous areas can then be combined with population data to determine high-risk areas. Or, the map analysis tools can be used to identify the downwind neighbors of either a point or a nonpoint source of air pollution. In a similar fashion, surface and groundwater flows can be modeled using effective distance measures. Such applications of GIS technology are still in the early development phases and are the focus of this volume.

BASIC FUNCTIONS OF A GIS

The main purpose of a geographic information system is to process spatial information. The data structure can be conceptualized as a set of "floating maps" with common registration, allowing the user to "look" down and across the stack of maps. The spatial relationships of the data can be summarized (database inquiries) or manipulated (analytic processing). Such systems can be formally characterized as "internally referenced, automated, spatial information systems, designed for data mapping, management, and analysis."

All GIS packages contain hardware and software for data input, storage, processing, and display of computerized maps. The processing functions of these systems can be broadly grouped into three functional areas: computer mapping, spatial database management and cartographic modeling.

Computer mapping, also termed automated cartography, involves the preparation of map products. The focus of these operations is on the input and display of computerized maps. *Spatial database management* procedures focus on the storage component of GIS, efficiently organizing and searching large sets of data for frequency statistics and coincidence among variables. The database allows rapid updating and examining of mapped information. For example, a spatial database can be searched for areas of silty-loam soil, moderate slope, and Ponderosa Pine forest cover. A summary table or map of the results can then be produced.

These mapping and database capabilities have proved to be the backbone of current GIS applications. Aside from the significant advantages of processing speed and ability to handle tremendous volumes of data, such uses are similar to those of manual techniques. Here is where parallels to mathematics and traditional statistics may be drawn. From this perspective *map analysis* is the natural extension of scalar statistics and algebra into a spatial format.

SPATIAL STATISTICS

In light of these parallels, a "map-ematics"-based GIS structure provides a framework for discussing the various data types and storage procedures involved in computer mapping and data management. It also provides a foundation for advanced analytic operations involving spatial statistics and map algebra.

The dominant feature of GIS technology is that spatial information is represented numerically, rather than in

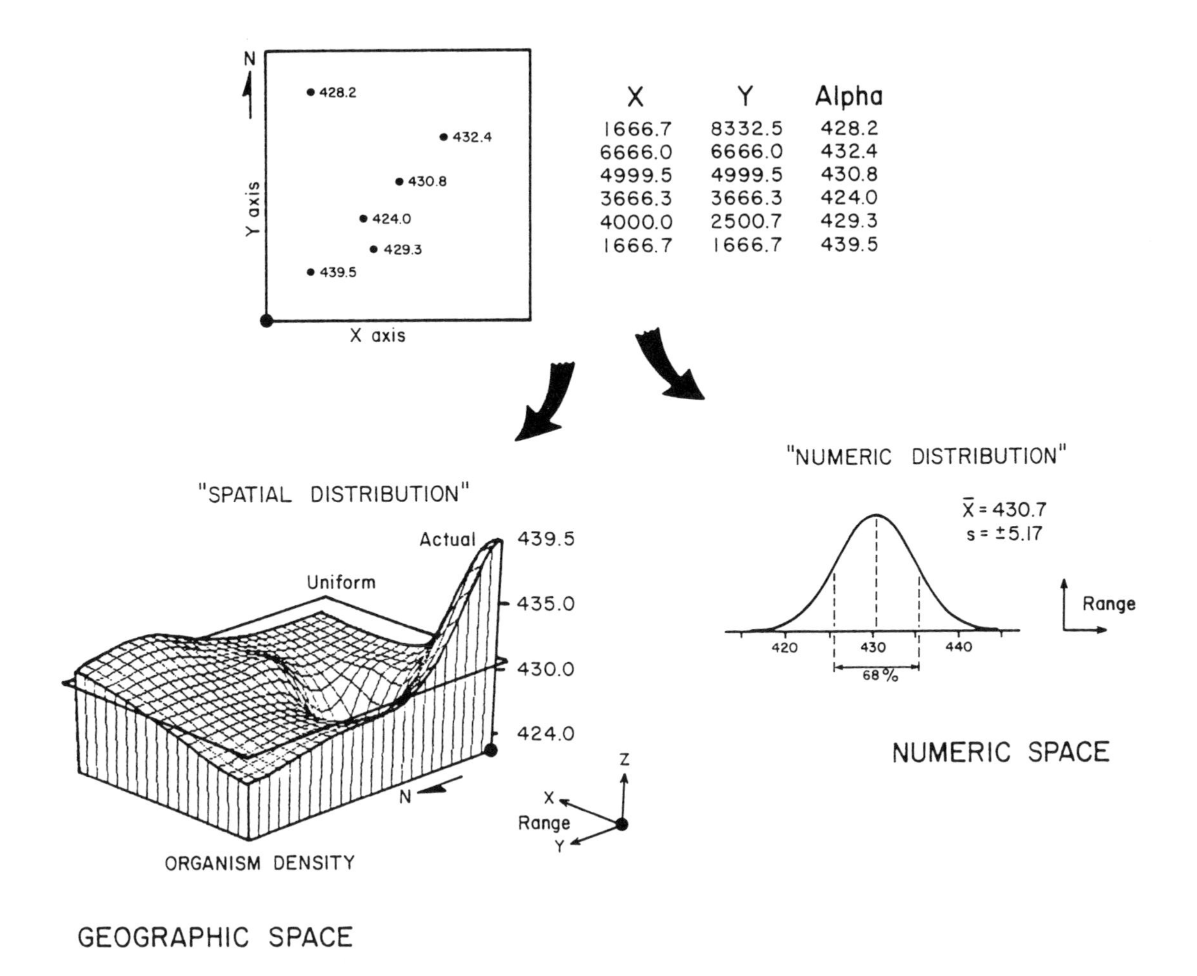

Figure 7-1: Spatially characterizing data variation. Traditional statistics identifies the typical response and assumes this estimate to be distributed uniformly in geographic space. Spatial statistics seeks to characterize the geographic distribution, or pattern, of mapped data.

analog fashion as in the inked lines of a map. Because of the analog nature of map sheets, manual analytic techniques are principally limited to nonquantitative processing. Digital representation, on the other hand, has the potential for quantitative as well as qualitative processing.

The growing field of GIS has stimulated the development of spatial statistics, a discipline that seeks to characterize the geographic distribution or pattern of mapped data. Spatial statistics differs from traditional statistics by describing the more refined spatial variation in the data, rather than producing typical responses assumed to be uniformly distributed in space.

An example of spatial statistics is shown in Figures 7-1 and 7-2. Figure 7-1 depicts density mapping of a microorganism for a portion of a lake as determined from laboratory analysis of surface water samples. A plot and a listing of the data are shown at the top of the figure. The *x,y* values identify the geographic coordinates (termed locational attribute) of the sampled densities (termed thematic attribute) of the microorganism during the "alpha" time period. When analyzed in numeric space, the locational information is disregarded and an average concentration of 430.7 (±5.17) is assumed to be true everywhere in the lake during that period.

Spatial statistics, on the other hand, uses both the positioning and the concentration information to interpolate a map of the microorganism densities spatially. The lower-left three-dimensional plot in Figure 7-1 shows the density surface. Note that both the highest and lowest densities occur in the southwest portion of the lake. The uniform plane identifies the average density.

Figure 7-2 depicts the natural extension of multivariate statistics for the spatial coincidence between the two time periods. Traditional statistics describes the coincidence between the two periods by their joint mean and covariance matrix. Again, this typical coincidence is assumed to be uniformly distributed in geographic space. The coincidence plot on the right shows the difference between the

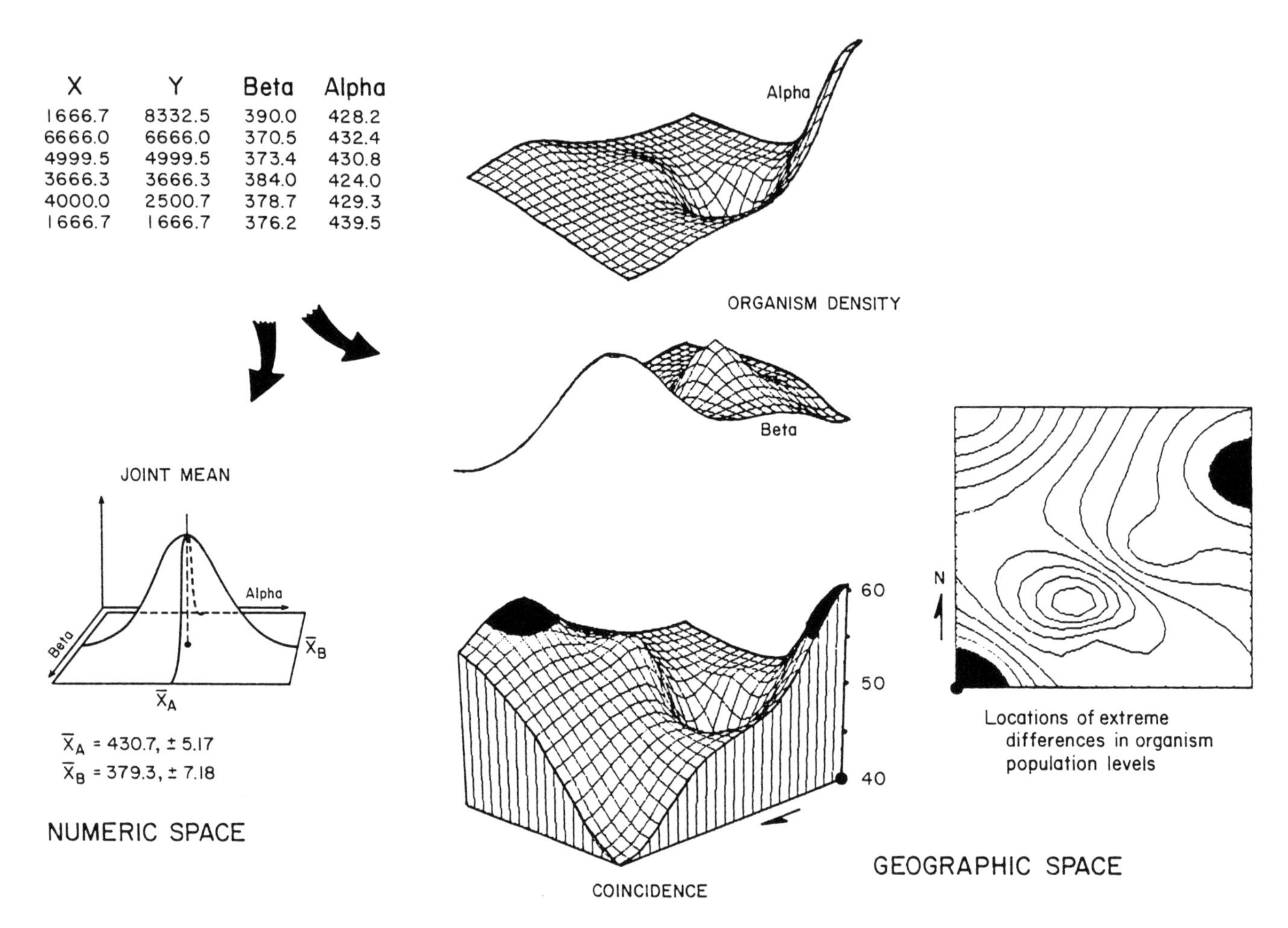

Figure 7-2: Assessing coincidence among mapped data. Maps characterizing spatial variation among two or more variables can be compared and locations of unusual coincidence identified.

density surfaces for the alpha and beta time periods. Note the darkened areas in the southwest and northeast depicting locations of extreme differences in organism population levels. Traditional and spatial statistics complement each other for environmental modeling. In both statistical approaches, the nature of the data is crucial. For analyses similar to those in Figures 7-1 and 7-2, thematic values ("what" information) must identify variables that form gradients in both numeric and geographic space.

Many forms of mapped data exist, including digital maps that, coupled with traditional mapping considerations—scale, projection, registration, resolution—can help environmental modelers understand the potentials and pitfalls of GIS applications. The quantitative nature of digital maps provides the foundation for a mathematics of maps.

MAP ALGEBRA

Just as a spatial statistics has been developed by extending concepts of conventional statistics, a spatial mathematics has evolved. This map algebra uses sequential processing of mathematical primitives to perform complex map analyses. It is similar to traditional algebra, in which primitive operations (add, subtract, exponentiate) are logically sequenced on variables to form equations, but in map algebra, entire maps represent the variables.

Most of the traditional mathematical capabilities plus an extensive set of advanced map-processing primitives emerge. Transpose, inverse, and diagonalize are examples of new primitives based on the nature of matrix algebra. Within map analysis, the spatial coincidence and juxtapositioning of values among and within maps create new operators such as masking, proximity, and optimal paths. This set of map analysis operators can be flexibly combined through a processing structure similar to conventional mathematics. The logical sequence involves retrieval of one or more maps from the database, processing those data as specified by the user, creation of a new map containing the processing results, and the storage of the new map for subsequent processing.

The cyclical processing is similar to "evaluating nested parentheticals" in traditional algebra. Values for the "known" variables are first defined, then they are manipulated by performing the primitive operations on those numbers in the order prescribed by the equation.

For example, in the equation $A = (B+C)/D$ the variables B and C are first defined and then added, with the sum stored as an intermediate solution. This intermediate value, in turn, is retrieved and divided by the variable D to derive the value of the unknown maps. The numbers contained in a solution map (in effect, solving for A) are a function of the input maps and the primitive operations performed.

CARTOGRAPHIC MODELING

This mathematical structure forms a conceptual framework easily adapted to a variety of applications in a familiar and intuitive manner. For example, % change = [(new value – old value)/old value] × 100 is a general equation for calculating the percent change in any variable. If applied to the microorganism data in Figures 7-1 and 7-2, the equation calculates the percent change in organism density between the alpha and beta time periods. Using the joint means, % change = [(379.3 – 430.7)/430.7] × 100 = –11.9 percent. The decrease is assumed to occur everywhere in the lake. In a similar manner a map of percent change can be expressed in such commands as:

COMPUTE ALPHA.MAP MINUS BETA.MAP FOR DIFFERENCE.MAP
COMPUTE DIFFERENCE.MAP DIVIDEDBY ALPHA.MAP TIMES 100 FOR %CHANGE.MAP

or simply,

COMPUTE ((ALPHA.MAP – BETA.MAP) / BETA.MAP) * 100 FOR %CHANGE.MAP

[The pMAP (Professional Map Analysis Package) software used in the derivation and display of all of the examples in this paper is a PC-based system developed by Spatial Information Systems, Inc., 6907 Sprouse Court, Springfield, Virginia, 22153, phone (703) 866–9458.]

This process first calculates the "difference" map shown in Figure 7-2, then normalizes the difference to the alpha density for each map location. The result is a map depicting the geographic distribution of the percent change in organism density.

Within this simple model, field data are collected and interpolated into computerized maps. These data are evaluated as shown to form a solution map of percent change. The simple model might be extended to provide coincidence statistics, such as:

CROSSTAB DEPTH.MAP WITH %CHANGE.MAP

for a table summarizing the spatial relationship between the depth map and the change in organism density. Such a table would show which depth zones experienced the greatest change in density.

The basic model might also be extended to include such geographic searches as

RENUMBER %CHANGE.MAP FOR BIG-CHANGE.MAP ASSIGNING 1 TO –200 THRU –20 ASSIGNING 0 TO –19 THRU 200

for a map isolating all locations that experienced a 20% or more decrease in organism density.

RASTER AND VECTOR DATA STRUCTURES

The main purpose of a geographic information system is to process spatial information. In doing so, it must be capable of four things:

- Creating digital abstractions of the landscape (ENCODE);
- Efficiently handling these data (STORE);
- Developing new insights into the relationships of spatial variables (ANALYZE); and
- Creating "human-compatible" summaries of these relationships (DISPLAY).

The data structure used for storage has far reaching implications in how we encode, analyze, and display digital maps. It has also fueled a heated debate as to the "universal truth" in data structure since the inception of GIS. In truth, there are more similarities than differences in the various approaches. All GIS packages are "internally referenced," which means they have an automated linkage between the data (or thematic attribute) and the whereabouts (or locational attribute) of that data.

There are two basic approaches used in linkage. One approach (vector) uses a collection of line segments to identify the boundaries of point, linear, and areal features. The alternative approach (raster) establishes an imaginary grid pattern over a study area, then stores values identifying the map characteristic occurring within each grid space.

Although there are significant practical differences in these data structures, the primary theoretical difference is that the *raster structure stores information on the interior of areal features*, and implies boundaries, whereas, the *vector structure stores information about boundaries*, and implies interiors. This fundamental difference determines, for the most part, the types of applications that may be addressed by a particular GIS.

It is important to note that both systems are actually gridbased; it is just in practice that line-oriented systems use a very fine grid of "digitizer" coordinates. Point features, such as springs or wells on a water map, are stored the same for both systems—a single digitizer *x,y* coordinate pair or a single "column,row" cell identifier. Similarly, line features, such as streams on a water map, are stored the same—a series of *x,y* or "column,row" identifiers. If the same gridding resolution is used, there is no theoretical difference between the two referencing schemes, and, considering modern storage devices, only practical differences in storage requirements. Yet, it was storage considerations that fueled most of the early debate about the relative merits of the two data structures. Demands of a few, or even one, megabyte of storage were considered a lot in the early 1970s. To reduce storage, very coarse grids were used in early grid systems. Under this practice, streams were no longer the familiar thin lines assumed a few feet in width, but represented as a string of cells of several acres each. This, coupled with the heavy reliance on pen-plotter output, resulted in ugly, "saw-toothed" map products when using grid systems. Recognition of any redeeming qualities of the raster data form were lost to the unfamiliar character of the map product.

DATA STRUCTURE IMPLICATIONS

Consideration of areal features presents significant theoretical differences between the two data structures. A lake on a map of surface water can be described by its border (defined as a series of connecting line segments which identify the shoreline), or its interior (defined by a set of cells identifying open water). This difference has important implications in the assumptions about mapped data.

In a line-based system, the lines are assumed to be "real" divisions of geographic space into homogeneous units. This assumption is reasonable for most lakes if one accepts the premise that the shoreline remains constant. However, if the body of water is a flood-control reservoir the shoreline could shift several hundred meters during a single year—a fat, fuzzy line would be more realistic.

A better example of an ideal line feature is a property boundary. Although these divisions are not physical, they are real and represent indisputable boundaries that, if you step one foot over the line, often jeopardize friendships and international treaties alike.

On the other hand, consider the familiar contour map of elevation. The successive contour lines form a series of long skinny polygons. Within each of these polygons the elevation is assumed to be constant, forming a layercake of flat terraces in three-dimensional data space. For a few places in the world, such as rice paddies in mountainous portions of Korea or the mesas of New Mexico, this may be an accurate portrayal. However, the aggregation of a continuous elevation gradient discards much of the information used in its derivation.

An even less clear example of a traditional line-based image is the familiar soil map. The careful use of a fine-tipped pen in characterizing the distribution of soils imparts artificial accuracy at best. At worst, it severely limits the potential uses of soil information in a geographic information system. As with most resource and environmental data, a soil map is not "certain" (as contrasted with the surveyed and legally filed property map). Rather the distribution of soils is probabilistic—the lines form artificial boundaries presumed to be the abrupt transition from one soil to another. Throughout each of the soil polygons, the occurrence of the designated soil type is treated as equally likely. Most soil map users reluctantly accept the "inviolately accurate" assumption of this data structure, as the alternative is to dig soil pits everywhere within a study area.

A more useful data structure for representing soils is gridded, with each grid location identified by its most probable soil, a statistic indicating how probable, the next most probable soil, its likelihood, and so on. Within this context, soils are characterized as a continuous statistical gradient—detailed data, rather than an aggregated, human-compatible image.

Such treatment of map information is a radical departure from the traditional cartographic image. It highlights the revolution in spatial information handling brought about by the digital map. From this new perspective, maps move from images describing the location of features to mapped information quantifying a physical or abstract system in prescriptive terms—from inked lines and colorful patterns to mapped variables affording numerical analysis of complex spatial interrelationships.

THE NATURE OF SPATIAL DATA

In Figure 7-3, a dichotomy is drawn based on the nature of the mapped data and the intended application. If you are a fire chief, your data consist of surveyed roads, installed fire hydrants and tagged street addresses (lines real and data certain). Whenever a fire is reported, you want to query the database to determine the best route to the fire and the closest fire hydrants. You might also check the hazardous materials database if toxic chemicals are stored nearby. Such applications are *descriptive*, involving computer mapping and spatial database management, and are identified by a field of GIS termed "AM/FM" for Automated Mapping and Facilities Management.

Contrast this use with that of a resource manager working with maps of soils, vegetation, and wildlife habitat (lines are artificial and data probabilistic) to identify the

GEOGRAPHIC INFORMATION SYSTEMS

AM-FM (Inventory)	DSS (Analysis)
• Lines Real	• Lines Artificial
• Data Certain	• Data Probabilistic
• Descriptive Processing	• Prescriptive Processing
• Mapping	• Statistics
• Database Management	• Modeling

Figure 7-3: Two groups of GIS users can be identified by differences in the nature of the mapped data and the intended application.

optimal timber harvesting schedule. Such applications are *prescriptive*, involving spatial statistics and modeling, and are identified by a field of GIS termed DSS for Decision Support Systems.

Most often, the fire chief's application is best solved with a "vector"-based data structure considering precise locations with minimal analysis. Think of the consequences of a "raster" data structure—"a 0.85 probability that a fire hydrant is located in this hectare grid cell." The resource manager, on the other hand, has an entirely different circumstance. Most often, her application is best solved with a "raster" structure with complex analyses and minimal spatial precision. Think of the "vector" consequence—"a 1.00 probability that a quarter-acre parcel is Ponderosa Pine on Cohassett soil." Both results are equally inappropriate.

So, what can be concluded as the implication of data structure? The only firm conclusion is that the "best" data structure depends on the nature of the data used and the analytic tools required by the application. The user is the individual closest to the data and the application, and therefore, holds "the answer." This puts GIS principles squarely in the lap of the user. It requires the user to be well versed in the nature of spatial data, the implications of data structure, and map analysis capabilities.

MAP ANALYSIS CAPABILITIES

The development of a generalized analytic structure for map processing is similar to those of many other nonspatial systems. For example, the popular dBASE III package contains fewer than twenty analytic operations, yet may be used to create models for such diverse applications as address lists, payroll, inventory control, or commitment accounting. Once developed, these logical dBASE sentences are fixed into menus for easy end-user applications.

A flexible analytic structure provides for dynamic simulation as well as database management. For example, the Lotus spreadsheet package allows users to define interrelationships among variables. A financial model of a company's production process may be established by specifying a logical sequence of primitive operations and map variables. By changing specific values of the model, the impact of several fiscal scenarios can be simulated. The advent of database management and spreadsheet packages has revolutionized the handling of nonspatial data. The potential of computer-assisted map analysis promises a similar revolution for spatial data processing.

From this perspective, four classes of primitive operations can be identified, viz., those that

- Reclassify map categories,
- Overlay two or more maps,
- Measure distance and connectivity, and
- Characterize cartographic neighborhoods.

Reclassification operations merely repackage existing information on a single map. Overlay operations, on the other hand, involve two or more maps and result in delineations of new boundaries. Distance and connectivity operations generate entirely new information by characterizing the juxtaposing of features. Neighborhood operations summarize the conditions occurring in the general vicinity of a location. The reclassifying and overlaying operations based on point processing are the backbone of current GIS applications, allowing rapid updating and examination of mapped data. However, other than the significant advantage of speed and ability to handle tremendous volumes of data, these capabilities are similar to those of manual processing. The category-wide overlays, distance, and neighborhood operations identify more advanced analytic capabilities.

RECLASSIFYING MAPS

The first, and in many ways the most fundamental, class of analytical operations involves the reclassification of map categories. Each operation involves the creation of a new map by assigning thematic values to the categories of an existing overlay. These values may be assigned as a function of the initial value, the position, the contiguity, the size, or the shape of the spatial configuration of the individual categories. Each of the reclassification operations involves the simple repackaging of information on a single overlay and results in no new boundary delineations. Such operations can be thought of as the purposeful "recoloring" of maps.

Figure 7-4 shows the result of simply reclassifying a map as a function of its initial thematic values. For display, a unique symbol is associated with each value. In the figure, the COVERTYPE map has categories of Open Water, Meadow, and Forest. These features are stored as thematic values 1, 2, and 3, respectively, and displayed as separate graphic patterns. Color codes could be used in place of the black and white patterns for more elaborate

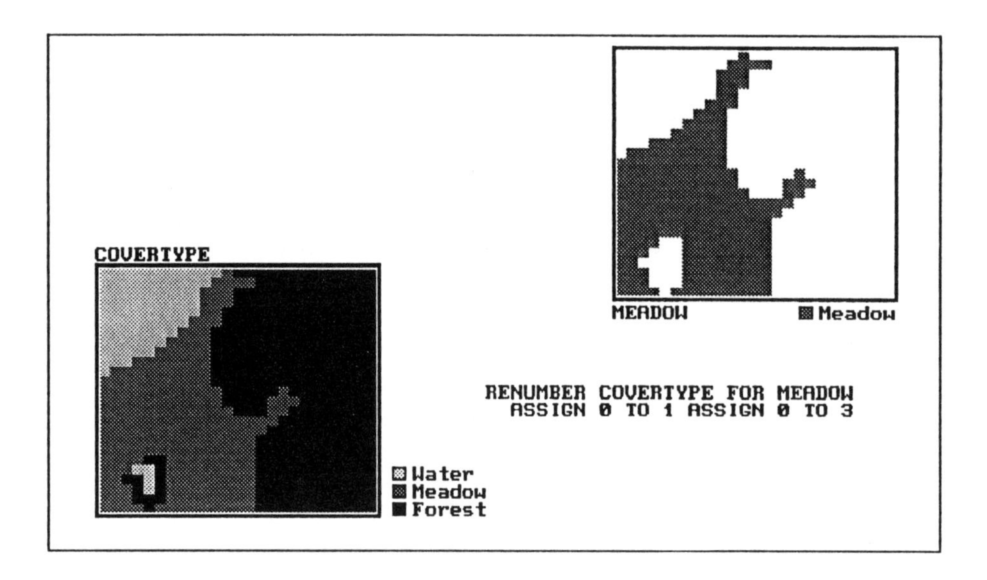

Figure 7-4: Reclassification of map categories can be based on initial thematic value, as in this example. The COVERTYPES of open water and forest are renumbered to the value zero, which is displayed as a blank. The resulting map isolates the MEADOW region.

graphic presentation. A binary map that isolates the meadow can be created by simply assigning 0 to the open and forested areas, and displaying a blank symbol wherever this value occurs.

A similar reclassification operation might involve the ranking or weighing of qualitative map categories to generate a new map with quantitative values. A map of soil types, for example, might be assigned values that indicate the relative suitability of each soil type for residential development. Quantitative values may also be reclassified to yield new quantitative values. This might simply involve a specified reordering of map categories (e.g., given a map of soil moisture content, generate a map of suitability levels for plant growth). Or, it could involve the application of a generalized reclassifying function, such as "level slicing," which splits a continuous range of map category values into discrete intervals (e.g., derivation of a contour map from a map of topographic elevation values).

Other quantitative reclassification functions include a variety of arithmetic operations involving map category values and a specified or computed constant. Among these operations are addition, subtraction, multiplication, division, exponentiation, maximization, minimization, normalization, and other scalar mathematical and statistical operators. For example, a map of topographic elevation expressed in feet may be converted to meters by multiplying each map value by the appropriate conversion factor of 3.28083 feet per meter.

Reclassification operations can also relate to locational, as well as purely thematic attributes associated with a map. One such characteristic is position. An overlay category represented by a single "point" location, for example, might be reclassified according to its latitude and longitude. Similarly, a line segment or areal feature could be reassigned values indicating its center or general orientation.

A related operation, termed "parceling," characterizes category contiguity. This procedure identifies individual "clumps" of one or more points having the same numerical value and being spatially contiguous (e.g., generation of a map identifying each lake as a unique value from a generalized water map representing all lakes as a single category).

Another locational characteristic is size. In the case of map categories associated with linear features or point locations, overall length or number of points might be used as the basis for reclassifying those categories. Similarly, an overlay category associated with a planar area might be reclassified according to its total acreage or the length of its perimeter. For example, an overlay of surface water might be reassigned values to indicate the area of individual lakes or the length of stream channels. The same sort of technique might also be used to deal with volume. Given a map of depth to bottom for a group of lakes, for example, each lake might be assigned a value indicating total water volume based on the areal extent of each depth category.

In addition to the value, position, contiguity, and size of features, shape characteristics may also be used as the basis for reclassifying map categories. Categories represented by point locations have measurable "shapes" insofar as the set of points imply linear or areal forms (i.e., just as stars imply constellations). Shape characteristics associated with linear forms identify the patterns formed by

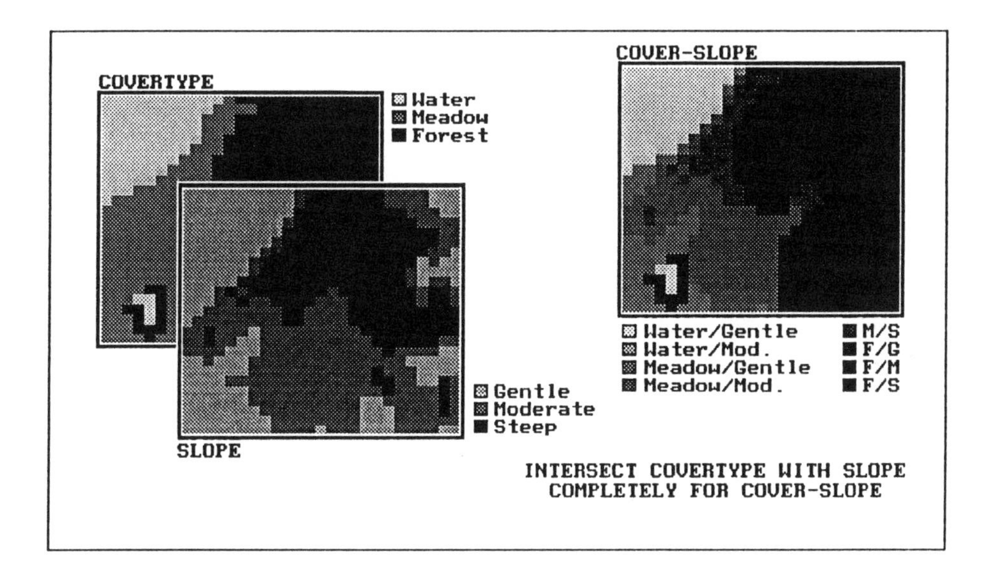

Figure 7-5: Point-by-point overlaying operations summarize the location specific coincidence of two or more maps. In this example, each map location is assigned a unique value identifying the COVERTYPE and SLOPE conditions at that location.

multiple line segments (e.g., dendritic stream pattern). The primary shape characteristics associated with areal forms include feature integrity, boundary convexity, and nature of edge.

Feature integrity relates to "intactness" of an area. A category that is broken into numerous fragments and/or contains several interior holes is said to have less spatial integrity than ones without such violations. Feature integrity can be summarized as the Euler number, which is computed as the number of holes within a feature less one short of the number of fragments that make up the entire feature. An Euler number of zero indicates features that are spatially balanced, whereas larger negative or positive numbers indicate less spatial integrity.

The other shape characteristics, convexity and edge, relate to the configuration of boundaries of areal features. Convexity is the measure of the extent to which an area is enclosed by its background, relative to the extent to which the area encloses this background. The convexity index for a feature is computed by the ratio of its perimeter to its area. The most regular configuration is that of a circle that is totally convex and, therefore, not enclosed by the background at any point along its boundary. Comparison of a feature's computed convexity with that of a circle of the same area results in a standard measure of boundary regularity. The nature of the boundary at each point can be used for a detailed description of boundary configuration. At some locations the boundary might be an entirely concave intrusion, whereas others might be at entirely convex protrusions. Depending on the "degree of edginess," each point can be assigned a value indicating the actual boundary convexity at that location. This explicit use of cartographic shape as an analytic parameter is unfamiliar to most GIS users. However, a nonquantitative consideration of shape is implicit in any visual assessment of mapped data. Particularly promising is the potential for applying quantitative shape analysis techniques in the areas of digital image classification and wildlife habitat modeling. A map of forest stands, for example, might be reclassified such that each stand is characterized according to the relative amount of forest edge with respect to total acreage and the frequency of interior forest canopy gaps. Those stands with a large proportion of edge and a high frequency of gaps will generally indicate better wildlife habitat for many species.

OVERLAYING OPERATIONS

Operations for overlaying maps begin to relate to spatial coincidence, as well as to the thematic nature of cartographic information. The general class of overlay operations can be characterized as "light-table gymnastics." These involve the creation of a new map on which the value assigned to every point, or set of points, is a function of the independent values associated with that location on two or more existing overlays. In location-specific overlaying, the value assigned is a function of the point-by-point coincidence of the existing overlays. In category-wide compositing, values are assigned to entire thematic regions as a function of the values on other overlays that are associated with the categories. Whereas the first overlay approach conceptually involves the vertical spearing of a set of overlays, the latter approach uses

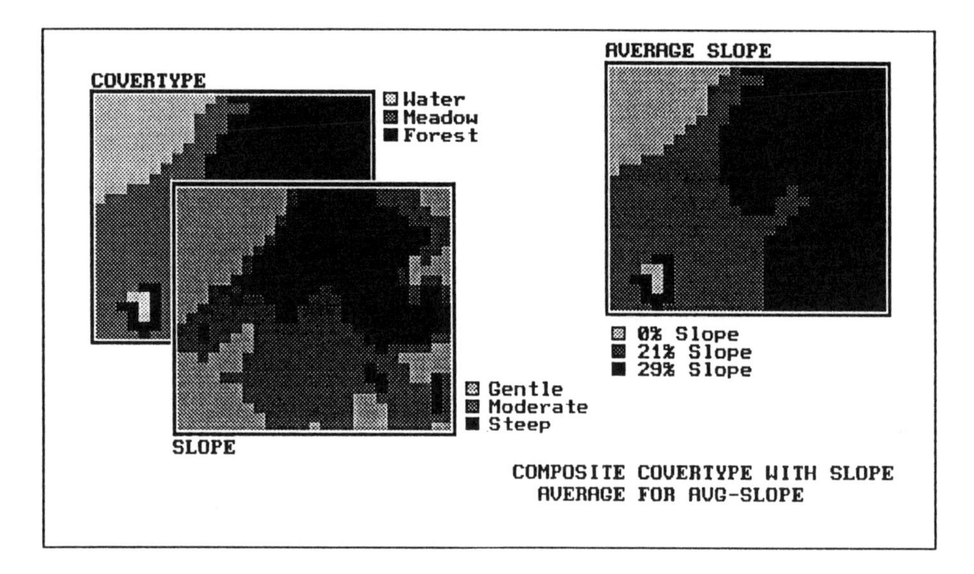

Figure 7-6: Category-wide overlay operations summarize the spatial coincidence of map categories. In this example, each of the three COVERTYPES (open water, meadow, and forest) is assigned a value equal to the average of the SLOPE values occurring within its boundaries.

one overlay to identify boundaries by which information is extracted from other overlays.

Figure 7-5 shows an example of location-specific overlaying. Maps of COVERTYPE and topographic SLOPE are combined to create a new map identifying the particular cover/slope combination at each location. A specific function used to compute new map category values from those of existing maps being overlaid may vary according to the nature of the data being processed and the specific use of that data within a modeling context. Environmental analyses typically involve the manipulation of quantitative values to generate new values that are likewise quantitative in nature. Among these are the basic arithmetic operations such as addition, subtraction, multiplication, division, roots, and exponentiation. Functions that relate to simple statistical parameters such as maximum, minimum, median, mode, majority, standard deviation, or weighted average may also be applied in this manner. The type of data being manipulated dictates the appropriateness of the mathematical or statistical procedure used. For example, the addition of qualitative maps such as soils and land use would result in mathematically meaningless sums, since their thematic values have no numerical relationship. Other map overlay techniques include several that may be used to process either quantitative or qualitative data and generate values that can likewise take either form. Among these are masking, comparison, calculation of diversity, and permutations of map categories (as depicted in Figure 7-5). More complex statistical techniques may also be applied in this manner, assuming that the inherent interdependence among spatial observations can be taken into account. This approach treats each map as a variable, each point as a case, and each value as an observation. A predictive statistical model can then be evaluated for each location, resulting in a spatially continuous surface of predicted values. The mapped predictions contain additional information over traditional nonspatial procedures, such as direct consideration of coincidence among regression variables and the ability to spatially locate areas of a given level of prediction.

An entirely different approach to overlaying maps involves category-wide summarization of values. Rather than combining information on a point-by-point basis, this group of operations summarizes the spatial coincidence of entire categories shown on one map with the values contained on another map(s). Figure 7-6 contains an example of a category-wide overlay operation using the same input maps as those in Figure 7-5. In this example, the categories of the COVER type map are used to define an area over which the coincidental values of the SLOPE map are averaged. The computed values of average slope are then assigned to each of the cover type categories.

Summary statistics that can be used in this way include the total, average, maximum, minimum, median, mode, or minority value; the standard deviation, variance, or diversity of values; and the correlation, deviation, or uniqueness of particular value combinations. For example, a map indicating the proportion of undeveloped land within each of several counties could be generated by superimposing a map of county boundaries on a map of land use and computing the ratio of undeveloped land to the total land area for each county. Or a map of ZIP code bound-

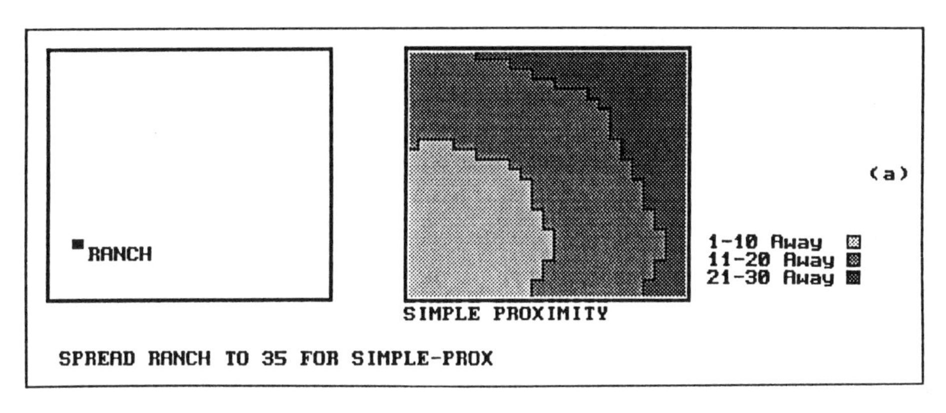

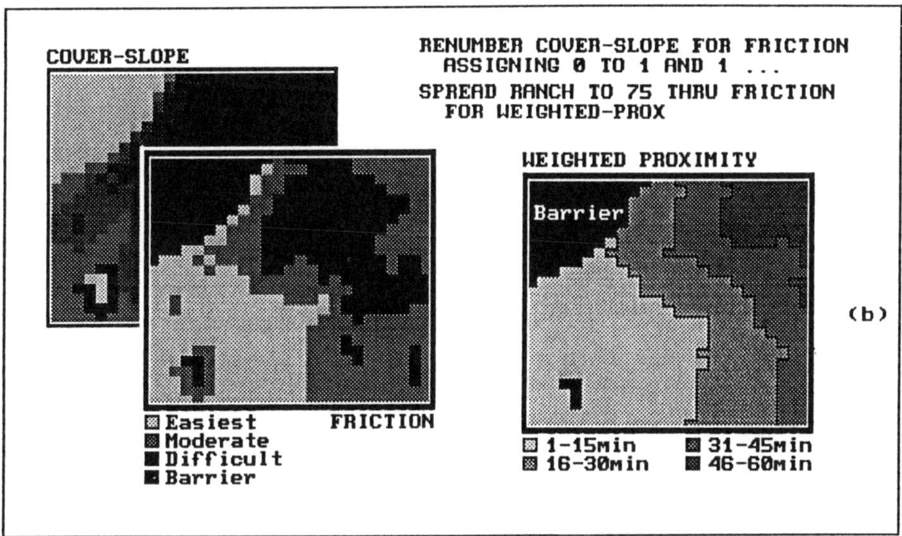

Figure 7-7: Distance between locations can be determined as simple distance or a function of the effect of absolute and relative barriers. In this example, inset (a) identifies equidistance zones around the RANCH. Inset (b) is a map of hiking traveltime from the RANCH. It was generated by considering the relative ease of travel through various COVER and SLOPE conditions (see Figure 7-5) where flat meadows are the fastest to traverse, steep forested are intermediate, and flat water is an absolute barrier to hiking.

aries could be superimposed over maps of demographic data to determine the average income, average age, and dominant ethnic group within each ZIP code.

As with location-specific overlay techniques, data types must be consistent with the summary procedure used. Also of concern is the order of data processing. Operations such as addition and multiplication are independent of the order of processing. Other operations, such as subtraction and division, however, yield different results depending on the order in which a group of numbers is processed. This latter type of operations, termed noncommutative, cannot be used for category-wide summaries.

MEASURING DISTANCE AND CONNECTIVITY

Most geographic information systems contain analytic capabilities for reclassifying and overlaying maps. These operations address the majority of applications that parallel manual map analysis techniques. However, to integrate spatial considerations into decision-making more fully, new techniques are emerging. The concept of distance has been historically associated with the "shortest straight-line distance between two points." While this measure is both easily conceptualized and implemented with a ruler, it is frequently insufficient in an environmental modeling context. A straight-line route may indicate the distance "as the crow flies," but offers little information for a walking or hitchhiking crow, or other flightless creature. Equally important to most travelers is to have the measurement of distance expressed in more relevant terms, such as time or cost. The group of operations concerned with measuring this effective distance is best characterized as "rubber rulers."

Any system for the measurement of distance requires two components—a standard measurement unit and a measurement procedure. The measurement unit used in most computer-oriented systems is the "grid space" implied by the superimposing of an imaginary grid over a geographic area. A ruler with its uniform markings implies such a grid each time it is laid on a map. The measurement procedure for determining actual distance from any location to another involves counting the number of intervening grid spaces and multiplying by the map scale. If the grid pattern is fixed, the length of the hypotenuse of a right triangle formed by the grid is computed.

This concept of point-to-point distance may be expanded to one of proximity. Rather than sequentially computing the distance between pairs of locations, concentric equidistant zones are established around a location, or set of locations. This is analogous to the wave pattern generated when a rock is thrown into a still pond.

Insert (a) in Figure 7-7 is an example of a simple proximity map indicating the shortest, straight-line distance from the ranch to all other locations. A more complex proximity map would be generated if, for example, all housing locations were considered target locations—in effect, throwing a handful of rocks. The result would be a map of proximity indicating the shortest straight-line distance to the nearest target area (i.e., house) for each nontarget location.

Within many application contexts, the shortest route between two locations may not always be a straight line. And even if it is straight, the Euclidean length of that line may not always reflect a meaningful measure of distance. Rather, distance in these applications is best defined in terms of movement expressed as travel time, cost, or energy that may be consumed at rates that vary over time and space. Distance-modifying effects may be expressed cartographically as barriers located within the space in which the distance is being measured. Note that this implies that distance is the result of some sort of movement over that space and through those barriers.

Two major types of barriers can be identified as to how they affect the implied movement. Absolute barriers are those which completely restrict movement, and, therefore, imply an infinite distance between the points they separate, unless a path around the barrier is available. A river might be regarded as an absolute barrier to a nonswimmer. To a swimmer or a boater, however, the same river might be regarded as a relative rather than an absolute barrier. Relative barriers are ones that are passable, but only at a cost which may be equated with an increase in physical distance.

Insert (b) of Figure 7-7 shows a map of hiking time around the target location identified by the ranch. The map was generated by reclassifying the various cover/slope categories (see Figure 7-5) in terms of their relative ease of foot travel. In the example, two types of barriers are used. The lake is treated as an absolute barrier, completely restricting hiking. The land areas, on the other hand, represent relative barriers, which indicate varied hiking impedance of each point as a function of the cover/slope conditions occurring at that location.

In a similar manner, movement by automobile may be effectively constrained to a network of roads (absolute barriers) of varying speed limits (relative barriers) to generate a riding travel-time map. Or, from an even less conventional perspective, weighted distance can be expressed in such terms as accumulated cost of powerline construction from an existing trunkline to all other locations in a study area. The cost surface that is developed can be a function of a variety of social and engineering factors, such as visual exposure and adverse topography, expressed as absolute and/or relative barriers.

The ability to move, whether physically or abstractly, may vary as a function of the implied movement as well as the static conditions at a location. One aspect of movement that may affect the ability of a barrier to restrict that movement is direction. A topographic incline, for example, will generally impede hikers differently depending on whether their movement is uphill, downhill, or across slope. Another possible modifying factor is accumulation. After hiking a certain distance, molehills tend to become disheartening mountains, and movement is more restricted.

A third attribute of movement that might dynamically alter the effect of a barrier is momentum, or speed. If an old car is stopped on a steep hill, it may not be able to resume movement, whereas if it were allowed to maintain its momentum (i.e., green traffic light), it could easily reach the top. Similarly, a highway impairment that effectively reduces traffic speeds from 55 to 40 miles per hour, for example, would have little or no effect during rush hour, when traffic is already moving at a much slower speed.

Another distance-related class of operations is concerned with the nature of connectivity among locations on an overlay. Fundamental to understanding these procedures is the concept of an accumulation surface. If the thematic value of a simple proximity map from a point is used to indicate the third dimension of a surface, a uniform bowl would be formed. The surface configuration for a weighted proximity map would have a similar appearance; however, the bowl would be warped with numerous ridges and pinnacles. Also, the nature of the surface is such that it cannot contain saddle points (i.e., false bottoms). This bowl-like topology is characteristic of all accumulation surfaces and can be conceptualized as a football stadium with each successive ring of seats identifying concentric, equidistant halos.

The bowl need not be symmetrical, however, and may form a warped surface responding to varying rates of accumulation. The three-dimensional insert in Figure 7-8 shows the surface configuration of the hiking travel time map from the previous Figure 7-7. The accumulated distance surface is shown as a perspective plot in which the ranch is the lowest location and all other locations are

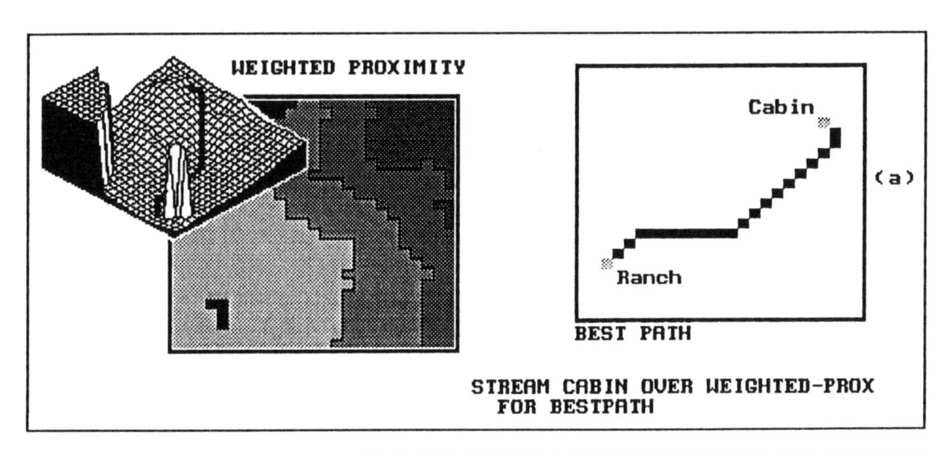

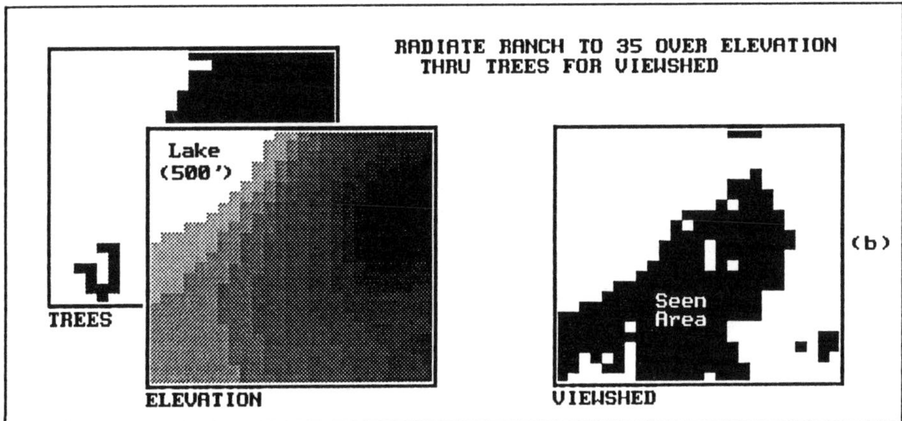

Figure 7-8: Connectivity operations characterize the nature of spatial linkages among locations. Insert (a) delineates the shortest (i.e., the least time) hiking route between the CABIN and the RANCH. The route traces the steepest downhill path along the traveltime surface derived in Figure 7-7 (also shown as a three-dimensional plot). Inset (b) identifies the viewshed of the RANCH. The ELEVATION surface and trees act as absolute barriers when establishing visual connectivity.

assigned progressively larger values of the shortest distance, but not necessarily straight, to the ranch. When viewed in perspective, this surface resembles a topographic surface with familiar valleys and hills. However, in this case the highlands indicate areas that are effectively farther away from the ranch.

In the case of simple distance, the delineation of paths, or connectivity, locates the shortest straight-line path between two points considering only two dimensions. Another technique traces the steepest downhill path from a location over a complex three-dimensional surface. The steepest downhill path along a topographic surface will indicate the route of surface runoff. For a surface represented by a travel-time map, this technique traces the optimal (e.g., the shortest or quickest) route. Insert (a) of Figure 7-8 indicates the optimal hiking path from a nearby cabin to the ranch, as the steepest downhill path over the accumulated hiking-time surface shown in Figure 7-7.

If an accumulation cost surface is considered, such as the cost surface for powerline construction described, the minimum cost route will be located. If powerline construction to a set of dispersed locations were simultaneously considered, an optimal path density map could be generated that identifies the number of individual optimal paths passing through each location from the dispersed termini to the trunkline. Such a map would be valuable in locating major feederlines (i.e., areas of high optimal path density) radiating from the central trunkline.

Another connectivity operation determines the narrowness of features. The narrowness at each point within a map feature is defined as the length of the shortest line segment (i.e., chord) that can be constructed through that point to diametrically opposing edges of the feature. The result of this processing is a continuous map of features with lower values indicating relatively narrow locations. For a narrowness map of forest stands, low thematic values indicate interior locations with easy access to edges.

Viewshed characterization involves establishing intervisibility among locations. Locations forming the

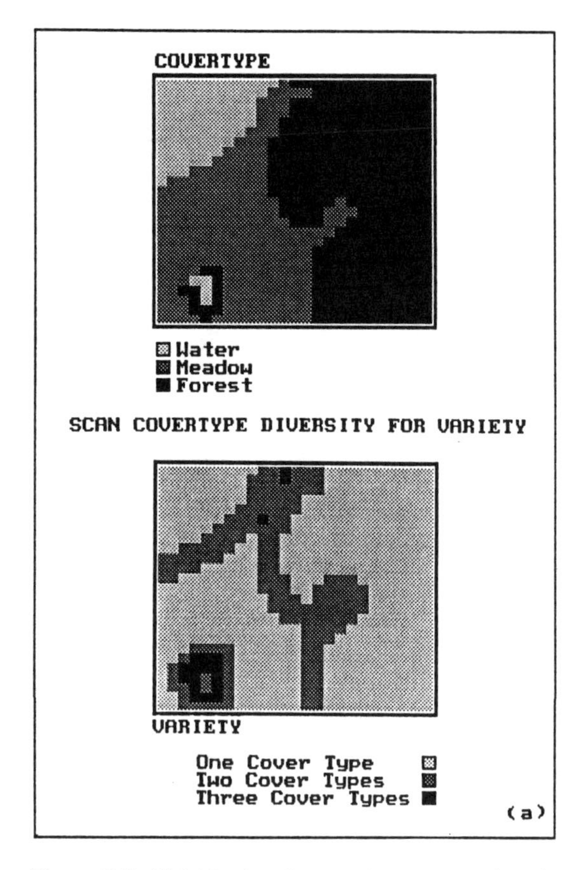

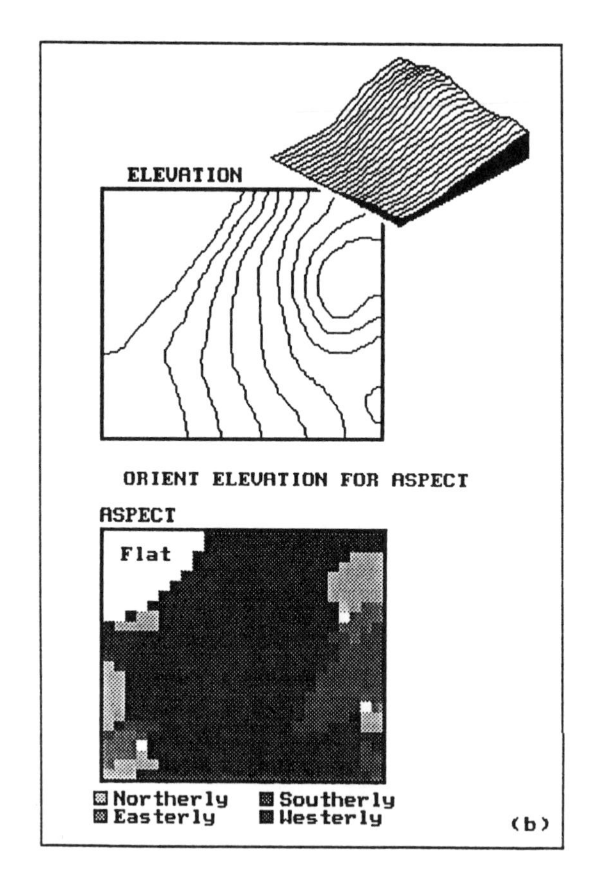

Figure 7-9: Neighborhood operations summarize the attributes occurring in the vicinity of each location. Inset (a) is a map of COVERTYPE diversity generated by counting the number of different COVERTYPES in the immediate vicinity of each map location. Inset (b) is a map of topographic aspect generated by successively fitting a plane to neighborhoods of adjoining ELEVATION values.

viewshed of an area are connected by straight rays in three-dimensional space to the viewer location, or a set of viewers. Topographic relief and surface objects form absolute barriers that preclude connectivity. If multiple viewers are designated, locations within the viewshed may be assigned a value indicating the number or density of visual connections. Insert (b) of Figure 7-8 shows a map of the viewshed of the ranch, considering the terrain and forest canopy height as visual barriers.

CHARACTERIZING NEIGHBORHOODS

The fourth and final group of operations includes procedures that create a new map where the value assigned to a location is computed as a function of independent values surrounding that location (i.e., its cartographic neighborhood). This general class of operations can be conceptualized as roving windows moving throughout the mapped area. The summary of information within these windows can be based on the configuration of the surface (e.g., slope and aspect) or the mathematical summary of thematic values.

The initial step in neighborhood characterization is the establishment of neighborhood membership. The members are uniquely defined for each target location as the set of points that lie within a specified distance and direction around that location. In most applications, the window has a uniform geometric shape and orientation (e.g., a circle or square). However, as noted in the previous section, the distance may not necessarily be Euclidean nor symmetrical, such as a neighborhood of downwind locations within a quarter mile of a smelting plant. Similarly, a neighborhood of the 10-minute drive along a road network could be defined.

The summary of information within a neighborhood may be based on the relative spatial configuration of values that occur within the window. This is true of the operations that measure topographic characteristics, such as slope, aspect, or profile from elevation values. One such technique involves the least-squares fit of a plane to adjacent elevation values. This process is similar to fitting a linear regression line to a series of points expressed in two-dimensional space. The inclination of the plane denotes terrain slope, and its orientation characterizes the aspect. The window is successively shifted over the entire elevation map to produce a continuous slope or aspect map. Insert (a) of Figure 7-9 shows the derived map of aspect for the area.

Note that a slope map of any surface represents the first derivative of that surface. For an elevation surface, slope depicts the rate of change in elevation. For an accumulation cost surface, its slope map represents the rate of change in cost (i.e., a map of marginal cost). For a travel-time overlay, its slope map indicates relative change in speed and its aspect map identifies direction of travel at each location. Also, the slope map of an existing topographic slope map (i.e., second derivative) will characterize surface roughness.

The creation of a profile map uses a window defined as the three adjoining points along a straight line oriented in a particular direction. Each set of three values can be regarded as defining a cross-sectional profile of a small portion of a surface. Each line of data is successively evaluated for the set of windows along that line. This procedure may be conceptualized as slicing a loaf of bread, then removing each slice and characterizing its profile (as viewed from the side) in small segments along its upper edge. The center point of each three-member neighborhood is assigned a value indicating the profile form at that location. The value assigned can identify a fundamental profile class (e.g., inverted "V" shape indicates a ridge or peak), or it can identify the magnitude, in degrees, of the "skyward angle" formed by the intersection of the two line segments of the profile. The result of this operation is a continuous map of a surface's profile as viewed from a specified direction. Depending on the resolution of an elevation map, its profile map could be used to identify gullies or valleys running east/west (i.e., "V" shape as viewed from the east or west profile) or depressions (i.e., "V" shape as viewed from both the east and west).

The other group of neighborhood operations are those that summarize thematic values. Among the simplest of these is the calculation of summary statistics associated with the map categories occurring within each neighborhood. These statistics might include, for example, the maximum income level, the minimum land value, or the diversity of vegetation types within a 1/8-mile radius (or perhaps, a 5-minute walking radius) of each target point. Insert (b) of Figure 7-9 shows the cover type diversity occurring within the immediate vicinity of each map location. Other thematic summaries might include the total, the average, or the median value occurring within each neighborhood; the standard deviation or variance of those values; or the difference between the value occurring at a target point itself and the average of those surrounding it.

Note that none of the neighborhood characteristics described so far relates to the amount of area occupied by the map categories within each neighborhood. Similar techniques might be applied, however, to characterize neighborhood values that are weighted according to areal extent. One might compute, for example, total land value within 3 miles of each target point on a per-acre basis. This consideration of the size of the neighborhood components also gives rise to several additional neighborhood statistics including mode (i.e., the value associated with the greatest proportion of neighborhood areas), minority value (i.e., the value associated with the smallest proportion of a neighborhood area), and uniqueness (i.e., the proportion of the neighborhood area associated with the value occurring at the target point itself).

Another locational attribute that might be used to modify thematic summaries is the cartographic distance from the target point. While distance has already been described as the basis for defining a neighborhood's absolute limits, it might also be used to define the relative weights of values within the neighborhood. Noise level, for example, might be measured according to the inverse square of the distance from surrounding sources.

The azimuthal relationship between neighborhood location and the target point may also be used to weight the value associated with that location. In conjunction with distance weighting, this gives rise to a variety of spatial sampling and interpolation techniques. For example, weighted nearest-neighbor interpolation of lake bottom temperature data assigns a value to an unsampled location as the distance-weighted average temperature of a set of sampled points within its vicinity.

GENERALIZED CARTOGRAPHIC MODELING APPROACH

As an example of some of the ways in which fundamental map processing operations might be combined to perform more complex analyses, consider the cartographic model outlined in Figure 7-10. Note the format uses boxes to represent encoded and derived maps and lines to represent primitive map processing operations. The flowchart structure indicates the logical sequencing of operations on the mapped data that progresses to the desired final map.

The simplified cartographic model shown depicts the siting of the optimal corridor for a highway considering only two criteria: an engineering concern to avoid steep slopes, and a social concern to avoid visual exposure. Implementation of the model using the pMAP system requires fewers than 25 lines of code. The execution of the entire model using the database in the previous figures requires less than 3 minutes in a 386SX notebook personal computer that fits in a briefcase. Such processing power was the domain of a mainframe computer just a decade ago, which effectively made GIS technology out of reach for most users.

Given a map of topographic elevation values and a map of landuse, the model allocates a minimum-cost highway

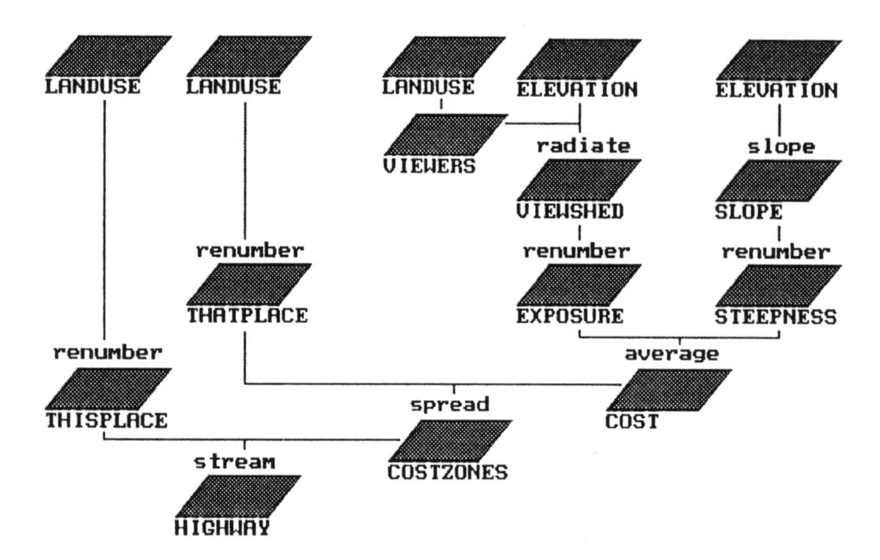

Figure 7-10: This simplified cartographic model depicts the siting of an optimal corridor for a highway with reference to only two criteria: an engineering concern to avoid steep slopes and a social concern to avoid visual exposure. In a manner similar to conventional algebra, this process uses a series of map operations (indicated by lines) to derive intermediate maps (indicated by boxes), leading to a final map of the optimal corridor.

alignment between two predetermined termini. Cost is not measured in dollars, but in terms of locational criteria. The right portion of the flowchart develops a discrete cost surface in which each location is assigned a relative cost based on the particular steepness/exposure combination occurring at that location. For example, those areas that are flat and not visible from houses would be assigned low values (good places for a road), whereas areas on steep slopes and visually exposed would be assigned high values (bad places for a road).

Similar to the hiking example described in Figures 7-7 and 7-8, the discrete cost surface is used as a map of relative barriers for establishing an accumulated cost surface from one of the termini to all other locations within the mapped area. The final step locates the other terminus on the accumulated cost surface and identifies the minimum cost route as the steepest downhill path along the surface from that point to the bottom (i.e., the other endpoint).

In addition to the benefits of efficient data management and of automating cartographic procedures, the modeling structure of computer-assisted map analysis has several other advantages. Foremost among these is the capability of dynamic simulation (i.e., spatial "what if" analysis). For example, the highway siting model could be executed for several different relative weightings of the engineering and social criteria. What if the terrain steepness is more important? Or what if the visual exposure is twice as important? Where does the minimum cost route change, or just as important, where does it not change? From this perspective, the model replies to user inquiries, rather than answering them, providing information, rather than tacit decisions.

Another advantage to cartographic modeling is its flexibility. New considerations may be easily added and existing ones refined. For example, the nonavoidance of open water bodies in the highway model is a major engineering oversight. In its current form, the model favors construction on lakes, as they are flat and frequently not visually exposed. This new requirement can be readily incorporated by identifying open water bodies as absolute barriers (i.e., infinite cost) when constructing the accumulation cost surface. The result will be routing of the minimal cost path around these areas of extreme cost.

Finally, cartographic modeling provides an effective structure for communicating both specific application considerations and fundamental processing procedures. The flowchart structure provides a succinct format for communicating the logic, assumptions, and relationships embodied in an analysis. Such a presentation encourages decision-makers' active involvement in the analytic process.

CARTOGRAPHIC VERSUS SPATIAL MODELING

Environmental modelers are pushing at the frontier of GIS technology. To the environmental modeler, a model is an abstraction of reality used to gain conceptual clarity and improve understanding of a system. GIS by definition is an abstraction of landscape complexity. The merging of these two fields is inevitable, but the form of the merger is not yet clear.

The *cartographic model* of the land planner described in the previous section is different from many environmental modelers' concept of a model. It is basically an implementation of procedures that would be used in a manual map analysis. In this sense, the model serves as a recipe for siting alternative highway corridors. It is the subjective expression of map variables responding to various weighting factors. It is a nonprocess model with minimal mathematical rigor, in which the GIS acts as a conceptual blackboard for decision-making.

Contrast this type of model with one determining the surface water flow over the landscape. In this application, extensive mathematical equations have been developed to describe the physical process and closely track the cause and effect of a system. It is a process model whose empirical relationships are rigorously expressed in mathematical terms.

For the process-oriented modeler, the mathematical relationships are usually parameterized with spatially aggregated values, such as the average slope and dominant soil of an entire watershed. A modeler might say "If it is steep over here, but flat over there, I guess it is moderately sloped everywhere," and use average slope in the execution of a hydrologic model. From this perspective, maps are used to generalize equation parameters using traditional drafting aids, such as dot grids, planimeters, and light-table overlays. Based on such experiences with maps, the modeler initially views GIS as an automated means of deriving model parameters that are passed as input to existing models.

An alternative perspective is the *spatial model*, and its implied "map-ematics." In this context, the GIS provides not only the definition of model parameters, but the engine of the model itself. The result is a full integration of GIS and a mathematical model. A major advantage of this approach is that input variables are defined as continuous surfaces, and areas different from the average can be recognized in the model. Another advantage is that spatially dependent operators, such as effective distance, can be incorporated. A third advantage is the ability to deal with error propagation in models.

The 1960s and 1970s saw the development of computer mapping and spatial database management. The extension of these fields to a robust set of analytical capabilities and cartographic modeling applications became the focus of the 1980s. Spatial modeling is a direct extension of cartographic modeling and will likely dominate research attention in the 1990s. The two modeling approaches can be seen as the extremes of a continuum in the mathematical rigor used in modeling applications.

To move towards spatial modeling and the full implementation of "map-ematics," several conditions must be met. On one front, the GIS community must become familiar with the process modeler's requirements and incorporate the more mathematical functionality in their GIS products. On the other front, the modeling community must see GIS as more than an electronic planimeter and become familiar with the conditions, considerations and capabilities of the technology. This dialogue between modelers and GIS specialists has begun in earnest and will be a focal point for GIS research in the 1990s.

SUMMARY AND CONCLUSION

The preceding discussion developed a conceptual framework for computer-assisted map analysis, described a set of independent analytical operations common to a broad range of applications and suggested GIS's role in environmental modeling. By systematically organizing these operations, the basis for a modeling approach is identified. This approach accommodates a variety of analytic procedures in a common, flexible, and intuitive manner, analogous to the mathematical structure of conventional algebra.

Environmental modeling has always required spatial information as its cornerstone. However, procedures for fully integrating this information into the modeling process have been limited. Traditional statistical approaches seek to constrain the spatial variability within the data. Sampling designs involving geographic stratification attempt to reduce the complexity of spatial data. However, it is the spatially induced variation of mapped data and their interactions that most often concern environmental modelers. This approach retains the quantitative aspects of the data necessary for most models, but lacks spatial continuity. On the other hand, the drafting approach is spatially precise, but limited by both its nonquantitative nature and its laborious procedures. In most instances, a final drafted map represents an implicit decision considering only a few possibilities, rather than a comprehensive presentation of information.

Computer-assisted map analysis, on the other hand, involves quantitative expression of spatially consistent data. In one sense, this technology is similar to conventional map processing, involving traditional maps and drafting aids, such as pens, rub-on shading, rulers, planimeters, dot grids, and acetate sheets for light-table overlays. In another sense, these systems provide advanced analytic capabilities, enabling modelers to address complex systems in entirely new ways. Within this analytic context, mapped data truly become spatial information for inclusion in environmental models. The GIS itself becomes an integral part of the model by providing both the mapped variables and the processing environment.

REFERENCES

Berry, J.K. (1986) Learning computer-assisted map analysis. *Journal of Forestry* 39–43.

Berry, J.K. (1987a) A mathematical structure for analyzing maps. *Journal of Environmental Management* 11(3): 317–325.

Berry, J.K. (1987b) Fundamental operations in computer-assisted map analysis. *International Journal of Geographical Information Systems* 1(2): 119–136.

GENERAL REFERENCES

Textbooks

Aronoff, S. (1989) *Geographic Information Systems: A Management Perspective*, Ottawa: WDL Publications.

Burrough, P.A. (1986) *Principles of Geographical Information Systems for Land Resources Assessment*, Oxford: Clarendon.

Huxhold, W.E. (1991) *An Introduction to Urban Geographic Information Systems*, New York: Oxford University Press.

Maguire, D.J., Goodchild, M.F., and Rhind, D.W. (eds.) (1991) *Geographical Information Systems: Principles and Applications* (2 volumes), London: Longman, and New York: Wiley.

Star, J.L., and Estes, J.E. (1990) *Geographic Information Systems: An Introduction*, Englewood Cliffs, NJ: Prentice-Hall.

Tomlin, C.D. (1990) *Geographical Information Systems and Cartographic Modeling*, Englewood Cliffs, NJ: Prentice-Hall Publishers.

Tutorial materials

Berry, J.K. (1991) *Map Analysis Tutor (maTUTOR)*, Springfield, VA: Spatial Information Systems, Inc.

Periodicals

GIS World, Fort Collins, CO: GIS World Inc.

GeoInfo Systems, Eugene, OR: Aster Publishing.

International Journal of Geographical Information Systems, London: Taylor and Francis.

Other sources

American Society for Photogrammetry and Remote Sensing (ASPRS) and the American Congress on Surveying and Mapping (ACSM), 5410 Grosvenor Lane, Bethesda, MD 20814; phone (301) 493–0290.

Earth Sciences Information Center (ESIC), U.S. Geological Survey, 507 National Center, Reston, VA 22092.

Urban and Regional Information Systems Association (URISA), 900 Second Street, N.E., Washington, DC 20002; phone (202) 289–1685.

AM/FM (Automated Mapping and Facilities Management) International, 8775 East Orchard Road, Englewood, CO 80111.

8

Understanding the Scope of GIS: Its Relationship to Environmental Modeling

TIMOTHY L. NYERGES

THREE PERSPECTIVES FOR UNDERSTANDING GIS

This chapter serves to introduce the nature of a geographic information system (GIS), particularly in regard to environmental issues, to those not intimate with the GIS literature. You have probably heard it many times—just what is a geographic information system? In answering that question, not all introductory comments are presented in this paper. This chapter focuses on how people come to understand the nature of GIS from three basic perspectives: functional, procedural, and structural. Each of these perspectives is treated in turn. Berry has already discussed GIS from the perspective of cartographic modeling in Chapter 7, and Goodchild earlier introduced some general GIS issues in Chapter 2.

The term GIS means many things to many people, and as such one can be confident in saying that GIS has a considerable potential for influencing environmental modeling. Although such a statement does not suffice for an introduction to the topic, it does imply that the flexible nature of a GIS has made it a useful information processing environment for many applications, including those in subjects such as resource management, highway transportation, and electrical utilities. In describing GIS it is important to remember that it will continue to evolve to suit the needs of those who are interested in applying it to fit their field of expertise, whether the problem calls for small-scale or large-scale investigation, and/or static or dynamic data analysis. It will become what we make of it.

Any description of something, whether it be an elephant or a GIS, stems from some perspective. Cowen (1988) identified several ways to define a GIS found in the literature. All seemed lacking, and he proposed to describe a GIS as a "decision support system"—although this phrase did not receive much elaboration. Definitions often arise due to perspective on a topic. Fundamentally, three perspectives are important for GIS:

1. A functional perspective concerning what applications a GIS is used for—the nature of GIS use;
2. A procedural perspective concerning how a GIS works with regard to the various steps in the process to perform this work—the nature of GIS work flow; and
3. A structural perspective concerning how a GIS is put together with regard to various components—the nature of GIS architecture.

All three perspectives add a different insight into the nature of a GIS, and any comprehensive definition should incorporate all three. Although no single definition says it all, a definition that combines a procedural and structural perspective and has had some discussion and agreement is: "A system of hardware, software, data, people, organizations, and institutional arrangements for collecting, storing, analyzing, and disseminating information about areas of the Earth," particularly in this case for understanding environmental processes (Dueker and Kjerne, 1989, p. 7).

Such a definition is in keeping with the most popular rendition as given by those with considerable experience in the field (cf. Marble, 1990), but is slightly broader in scope because it admits of the institutional considerations that give flavor to a GIS. A single definition of GIS is too limited to offer a full, functional description. Consequently, each of the three perspectives will be treated in turn in the following overview to provide a comprehensive treatment.

Looking to the future, there are several reasons why we should strive for integrating GIS and spatial process models. There comes a stage in the development of science when issues mature and one needs to be more realistic with the data sets, to try out ideas, to explore from a new perspective. The principal benefit to modeling is to be able to deal with large volumes of spatially oriented data that geographically anchor processes occurring across space. The principal benefit to GIS is the in-depth injection involving temporal and attribute issues to make GIS models more realistic. Both will undoubtedly gain from the other.

Definitions evolve like most other things that are exposed to dynamic environments. In keeping with the topic at hand that deals with integrating GIS and modeling, it is easy to envision a much broader sense of GIS due to the influences of model-based approaches (Goodchild, 1987; Burrough et al., 1988; Birkin et al., 1990; Nyerges, 1991a). Whether the term eventually is model-based GIS, spatial decision support system, geographic information modeling system, geographic modeling system, or something else, perhaps does not matter in the short term. What does matter is that GIS will evolve to suit the needs of those who take an active interest in its evolution.

GIS IN BRIEF: PAST, PRESENT, AND FUTURE

GIS has a longer history than most realize. Depending of what lineage one traces, one can find hints of what was to come in GIS perhaps some 25 or more years ago, through computer-assisted cartography (Tobler, 1959), in civil engineering (Horwood et al., 1962), and solid vestiges in geography (Dacey and Marble, 1965, cited by Wellar and Graf, 1972). There was clearly some sense of what a GIS was in 1972 as indicated by the publication of the Proceedings from the Second Symposium on Geographical Data Handling (Tomlinson, 1972). Tomlinson and Petchenik (1988) edited a series of articles that treat computer-assisted cartography as one of the core lineages, particularly analytical cartography (Tobler, 1977) with its focus on the transformation of spatial data geometries.

The Canada Geographic Information System developed between 1960 and 1969 has often been called the first production GIS (Tomlinson, 1988). Industrial-strength commercial GIS first took root in the early 1980s when spatial data managers were teamed with relational data managers to provide for spatial and attribute data management. The design for spatial data manipulation in the Odyssey system developed in the mid-1970s at the Harvard Laboratory for Computer Graphics and Spatial Analysis is the predecessor design for many of these (Morehouse, 1978).

Remote sensing has had a significant impact on GIS, but in remote sensing the major focus has been on the development of image processing systems. The use of remote sensing data in geographic information systems was discussed as early as 1972 by Tomlinson (1972). The link between image processing systems and geographic information systems has also been maturing for quite some time (Marble and Peuquet, 1983).

GIS as applied to environmental topics has been a long-standing concern. As in any field, topics mature with continued effort and focus. The theme of the 1979 AutoCarto IV Conference held in Crystal City, Virginia was on environmental issues. Many remote sensing conferences over the years such as the PECORA VII Conference in 1981 held in Sioux Falls, South Dakota have highlighted the linkage between GIS and remote sensing environmental issues.

Recent developments in GIS are too voluminous to review in this limited space. However, an indication of recent growth and development is given by the number of software packages in the commercial marketplace, and the number of GIS textbooks and reference volumes of collected papers published recently. No fewer than 55 commercial products are available worldwide according to the GIS Sourcebook (GIS World, 1990). Overall, the data, software, and hardware market is estimated to be in the multi-billions of dollars.

The state of accomplishments and knowledge in a field of study is often documented by texts in that field. GIS has now become a field so large that no single, basic reference source has it all. Thus, the state of knowledge must be taken as an aggregate of references. Several books have appeared that are useful for environmental problems, each with a particular perspective on GIS. Burrough's (1986) text focuses on GIS for land resource assessment, with a large portion dedicated to database issues. Starr and Estes' (1990) text deals with GIS from a remote sensing perspective, but emphasizing the importance of satellite imagery as well as point, line, and polygon vector-structured data. Tomlin's (1990) text treats GIS from an analytical, map algebra perspective, dealing principally with data analysis rather than data entry or map product displays. Clarke's (1990) text, although focusing on analytical cartography, provides a very valuable perspective on the algorithmic transformations of space.

Basic readers as collections of articles have also appeared. Peuquet and Marble (1990) provide an introductory collection of papers, and Ripple (1987) provides a compendium of natural resource management topics. Such a compendium is currently warranted for environmental issues. As of early 1992, the most ambitious reference on the general principles and applications of GIS is by Maguire, Goodchild, and Rhind (1991). The two-volume reference is indicative of both the wealth of knowledge and the maturation of certain GIS topics, but certain topics like human factors issues and network GIS remain largely untreated in texts.

Although it is true that there has been a tremendous growth in knowledge related to GIS, there is still considerable room for improvement. Improvements will occur on several fronts, including issues related to institutional arrangements, data, software and hardware (Frank, Egenhofer, and Kuhn, 1991). A better understanding of how GIS facilitates (or hinders) information flow within organizations is a principal concern facing institutions implementing GIS. Characterizing the quality of data, and using various qualities appropriate for certain decisions, is an important consideration for data improvements. Being able to represent multiple scales of data in a database is an important consideration for software improvements. Central processing unit size and cost of a workstation has always been a standard concern for hardware improvement. Many other improvements make GIS even more efficient and effective in the future.

Another major part of those improvements, dealing with the study of the environment, will involve the linking of GIS and environmental process models. Although analytical models have been a part of GIS for some time (Wheeler, 1988) and have been linked to GIS for such topics as resource assessment (Burrough et al., 1988), the cross-fertilization between environmental process models and GIS is only getting started. As such, part of the future of GIS is represented in a significant way by the theme of the conference that motivated this book—integrating environmental models and GIS. Much of the current experience in environmental modeling comes from hydrological modeling, particularly as performed by the Water Resources Division of USGS. Environmental modeling as a broad topic can place some of the heaviest demands on GIS because of its need to handle many kinds of data in both space and time in a more dynamic way than has been demonstrated to date. These requirements will help GIS evolve, perhaps in ways not currently envisioned. The current state of GIS should not be viewed so much as a hindrance, but as an opportunity that needs both direction and steps for progress. Realizing this opportunity involves putting GIS to use in creative ways.

FUNCTIONAL PERSPECTIVE—THE NATURE OF GIS USE

There are many dimensions that can be considered in the use of GIS. There have been recent attempts through the National Center for Geographic Information and Analysis (NCGIA) at developing a taxonomy of use (Obermeyer, 1989). The framework for such a taxonomy can include many dimensions, such as (Onsrud, 1989):

- Type of task: resource inventory, assessment, management and development, etc.;
- Application area: environmental, socioeconomic, etc.;
- Level of decision: policy, management, operations;
- Spatial extent of problem: small, medium, or large study area size; and
- Type of organization: public, private, or not-for-profit.

All five dimensions are pertinent for any particular use of a GIS. A generic treatment of use can only hope to focus on a couple of dimensions. Of particular interest here are the dimensions that concern tasks and environmental applications.

Environmental tasks include those for inventorying, assessing, managing, and predicting the fate of environmental resources. An environmental inventory is an accounting of the state of a resource environment, what exists and what does not in particular areas. This can be accomplished with a GIS at different levels of spatial, temporal, and thematic resolution as deemed necessary by the group interested in the inventory. Such an accounting is a very traditional, descriptive approach to "doing geography." An assessment can be used to determine what has been lost due to environmental influences. The relationship between what exists and what no longer exists is a difference in two maps over time. The difference is computed through use of a change detection technique. Environmental management is a task requiring certain policy controls to determine what resources should receive protected use. Environmental prediction is a task requiring a thorough understanding of the causal mechanisms of change, and is the most difficult of all tasks as it requires assumptions about many unknowns.

Another of the dimensions regarding use concerns applications. A GIS can provide support for several different environmental modeling applications. These include: atmospheric modeling (Chapter 10), hydrological modeling (Chapter 14), land surface–subsurface modeling (Chapter 19), ecological systems modeling (Kessell, 1990; Chapter 22), plus integrated environmental models (Chapter 26), as well as policy considerations for risk/hazard assessment (Hunsaker et al., 1990; Chapter 30) involving these models.

Another of the dimensions of use is the decision-making level. Different levels of decision making are supported by different GIS processing environments. Until recently, a single GIS could not provide support at all levels since human–computer interactions were rather different. Now, tools for customizing interfaces are beginning to appear that allow a single GIS to take on many faces, depending on the requirements of the decision-maker. However, since the tasks differ among policy, management, and operations decision-makers, it is likely that the use of the GIS also will differ.

GIS use in terms of tasks, applications, and levels of decision-making takes on particular meanings when we understand who the users are. Several types of users can be identified: scientists, managers, technical specialists/analysts, clerks, and the general public. At the current time the largest group of users are mainly specialists with a background in both GIS jargon and their own disciplinary jargon. The question has been asked many times—Why don't others use it?—meaning more scientists, managers, clerks, and the general public. One of the responses often given is: It is a matter of language. That is, the use of GIS requires an understanding of principles underlying GIS, and unfortunately this means an understanding of GIS jargon that communicates those principles. Current research is investigating ways to make GIS more useful (Mark and Gould, 1991). At some time in the future, an even bigger user group is expected to be either clerks or the general public. However, for the time being the next most obvious user group could be scientists, as long as GIS addresses needs appropriately, particularly in regard to mode of use.

Three primary modes of GIS use can be identified: map, query, and model (Nyerges, 1991a). The map mode provides referential and browse information, when a user wishes to see an overview of a spatial realm, and needs to get a sense of what is there, sometimes in order to clarify issues at hand. The map mode is usually satisfied by having hardcopy visuals on hand, but pan (as a scan across a display screen) and zoom (as in a photographic blowup) have been useful operations for browsing a computer graphics display. With these operations one has access to much more geographic area on demand than with hardcopy maps.

A query mode is used to address specific requests for information posed in two ways. One is that the user could specify a location and request information on phenomena surrounding that location or nearby. A second is that a user could specify a kind of phenomenon (or phenomena) and request to see all locations where the phenomenon occurs. However, there are several other renditions of these basic questions (cf. Nyerges, 1991a). Questions dealing with "when" and "how much" can be added to the "what" and "where" in queries about the geographic phenomenon under study.

Model invocation is the third mode of use. After having prepared the nature of the inputs to be retrieved for a model, the model is run and an answer is computed. More realistic data with a locational character do have an impact on model results. In addition, geographical displays interactively depicting the nature of the sensitivity of certain parameters can be very useful in support of model parameterization. Examining scale effects can be accomplished by changing interactively the nature of the data aggregation. The model brings together the locational, temporal, and thematic aspects of phenomena in a geographic process characterization.

Oftentimes a combination of the three modes of use can be very beneficial in a problem-solving situation. How to incorporate these three modes of use is a topic of current research dealing with human-interaction considerations (Mark and Gould, 1991; Nyerges, 1991d). The goal here is interactive modeling for problem solving. This goal in many respects requires that we consider the operating mode of a GIS before expectations rise too high.

GIS operate in all conceivable modes for information processing. There are systems that are standalone, and there are systems that are wholly integrated. In the standalone mode the GIS is the entire workhorse for problem solving, whereas with integrated systems GIS is only one part of a comprehensive solution (Nyerges, 1991c). To consider such issues more thoroughly, we can view GIS from a procedural perspective.

PROCEDURAL PERSPECTIVE—THE NATURE OF GIS WORK FLOW

A GIS workflow process consists primarily of four steps: (1) problem definition (and system setup if needed); (2) data input/capture (with subsequent data storage/management); (3) data manipulation/analysis; and (4) data output/display (see Figure 8-1). Data storage/management functions support all the others.

As in any problem-solving environment, a problem definition follows from some goal, no matter how well or ill defined the goal may be. Each of the steps is taken in turn. Some overlap might exist in terms of what is done when, and some amount of iteration among steps occurs depending on the results of each step.

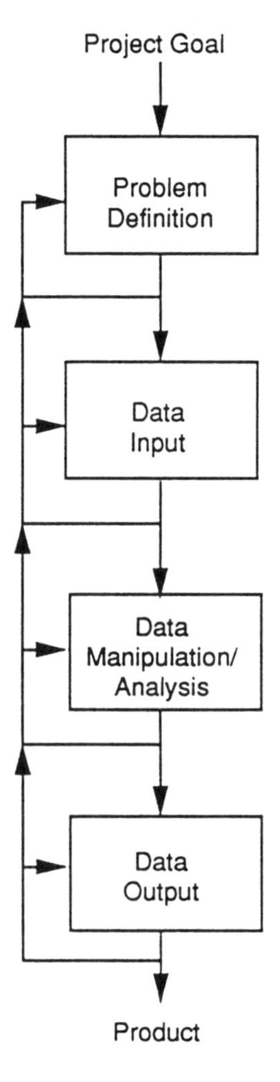

Figure 8-1: GIS as a work-flow process from a procedural perspective.

In the problem definition step in GIS, just as in any scientific study, one examines what must be done with regard to an understanding of the problem and the needs for information processing. We describe the nature of the problem, doing the best that can be done with the information at hand. Obviously, after we know more we can describe the problem better. The information that is needed to solve the problem at hand is the basis for the data required for processing.

Tasks and applications that solve problems require functions (sometimes called operations) to make use of a GIS. The major functions that support use of a GIS include data entry/capture, data storage/management, data manipulation/analysis, and data display/output.

Data entry/capture functions support all other processing steps. There is no question that data entry/capture functions can support environmental modeling. In fact, this is the most obvious of the support functions, and unfortunately sometimes the only function that is recognized. Examples of data entry functions useful for river network modeling and urban estuary contaminant modeling are provided in Tables 8-1a and 8-1b, respectively. Although substantial similarity exists in the listing of those functions, there are differences due to the nature of the spatial, thematic, and temporal character of the phenomena, hence data representations, that are maintained in the data records.

In the data entry step, data are either converted to digital form from hardcopy sources or they are acquired from digital sources and reformatted as appropriate for use. The digitizing process involves the use of an instrument to capture spatial coordinates in computer-compatible form for a sample or phenomenon of interest on the map. Both vector digitizing and raster scanning are used for data capture. Maps may or may not be of suitable form and resolution to be digitized. That decision is among the first in a series of informed decisions based upon the needs of the project (Marble, Lauzon, and McGranaghan, 1990). Alternatively, acquiring digital data is usually a much less expensive undertaking than digitizing data. Software to support importation of several different file formats is needed at the current time, until the national spatial data transfer standard (U.S. Geological Survey, 1991) comes into wider use. Making sure a digital data set is exactly what is needed is probably the most important task as part of this process (Marble, Lauzon, and McGranaghan, 1990). It requires data about the characteristics of the data set (i.e., metadata) (Nyerges, 1987). Using the metadata to make an informed decision is an undertaking well worth the effort. Many GIS installations store metadata as part of the data being managed.

Data management functions directly support all tasks. Example lists of GIS data management functions to support inventories of phenomena with regard to river network modeling and urban estuary contaminant modeling appear in Tables 8-2a and 8-2b, respectively. The functions focus on characterizing the state of the data environment for each context. Although substantial similarity exists in the listing of those functions, there are differences due to the nature of the spatial, thematic, and temporal character of the phenomena, hence data representations, that are maintained in the data records, the principal difference being that river networks are horizontally segmented in a linear manner, whereas estuaries are often horizontally segmented in an areal manner.

Complete data records for geographic phenomena would include an observation of thematic attribute character, an observation of location, and an observation of time. The tendency has been to characterize the present, forgetting about the past. In many instances new data replace old data, with the record of old data being eliminated (except for database backup and archive). As a result of this view of the value of old data, current systems can only handle the character of phenomena as time slices for past, present, and future observations. Current inventories may or may not constrain analyses, depending on the functions available for transforming data into a suitable form for analysis. Langran and Chrisman (1988) describe the value of retaining old data, and some of the conceptual issues underpinning design requirements for on-line spatio-temporal inventories.

Functions to support GIS data manipulation focus on preparing data for the analysis phase of processing. Lists of data manipulation functions to support river network modeling and urban estuary contaminant modeling are provided in Tables 8-3a and 8-3b, respectively. Manipulation functions prepare data for further processing. Conversion from one structure to another is often necessary to support spatial data analysis.

Functions to support GIS data analysis focus on developing and synthesizing spatial relationships in geographic data to provide answers. They range from simple models developed wholly within a GIS context to elaborate coupled models linked into a GIS environment. Examples of such functions for river network modeling and urban estuary contaminant modeling appear in Tables 8-4a and 8-4b, respectively. These functions produce answers that take the form of single numbers or words, perhaps in response to a query, or they take the form of several thousand numbers or words as the basis of a map display. Many of the current concerns with analytical functions are with the design tradeoffs for integrating and/or linking spatial process models into a GIS context, that is, the model structuring function in Tables 8-4a and 8-4b.

The display function in a GIS has a close lineage with efforts in cartography. Examples of display/output functions appear in Tables 8-5a and 8-5b, respectively. Output can be generated in either softcopy or hardcopy form. Softcopy output to the computer monitor is useful for interactive problem solving. Hardcopy output is useful for presentation to a large group of people over an extended period of time. Using a GIS to present the outputs from

Table 8-1a: GIS data entry/capture functions to support river network modeling

Spatial Data Entry

Manual/automatic digitize: digitize station point, river line and basin polygon data with or without topology, using table, state plane, and latitude/longitude coordinates.
Set default: set default values for river description.
Duplicate entries: duplicate entries for sample stations if desired.
Topological build: creating topology, both manual and automatic approaches for chain to area; area to chain; chain to node; node to chain; node to area; area to node; generate polygon from chains.
Assign feature labels/codes: either while digitizing or as separate process assign labels/codes to river reaches and basins.
Object move: interactive insertion, deletion, changing, moving singly or in a group parts of network.
Snap: snap lines to nodes; snap lines to lines for river network.
Close polygon: automatic polygon closure for basin and subbasin development.
Centroid: manual and automatic polygon centroid creation for subbasins.
Data checking: automatic topological error checking for river connectivity, graphic display of errors, and facility for correction; checking overshoots and undershoots of line intersections, with interactive and batch mode correction.
Valid codes: check for valid names/codes of river reaches while digitizing or batch mode.

Attribute Data Entry

Enter attributes: interactive or batch entry of attributes describing river water quality.
Linkage: develop linkage by feature name; location identifier; manual and/or automatic for attaching spatial to attribute data records.
Insert: manual or automatic insertion of area, perimeter, length, width, depth, statistics as attribute values.
Interactive change: insertion, deletion, change, and move of names/codes.
Valid values: check for valid attribute values or combination of values; check for missing names/codes; format checking, range checking.
Query update: query select function for updating groups of attribute values.

Spatial and Attribute Data Entry

Import data: import spatial and attribute data in various formats, for example, Arc, DXF, DLG-3, user-defined ASCII.
Export data: export spatial and attribute data in various formats, e.g., Arc, DXF, DLG-3, user defined ASCII.

Table 8-1b: GIS data entry/capture functions to support urban estuary contaminant modeling

Spatial Data Entry

Manual/automatic digitize: digitize point samples, contour lines and sediment and parcel polygons etc. with or without topology, using table, state plane, and latitude/longitude coordinates.
Set default: set default values for sediment, water column, bioassay description.
Duplicate entries: duplicate entries for points, lines, areas if desired.
Datum choice: chose from multiple vertical datums, for example, terrain data shoreline and bathymetric data shoreline.
Topological build: creating topology, both manual and automatic approaches for chain to area; area to chain; chain to node; node to chain; node to area; area to node; generate polygon from chain.
Complex polygon: generate complex polygons for sediment volume characterization.
Snap: snap lines to nodes; snap lines to lines.
Close polygon: automatic polygon closure.
Centroid: manual and automatic polygon centroid creation (both visual and geometrically weighted).
Assign feature labels/codes: either while digitizing or as separate process.
Data checking: automatic topological error checking, graphic display of errors and facility for correction for drainage network and abutting land parcel character; checking overshoots and undershoots of line intersections, with interactive and batch mode correction.
Object change: interactive insertion, deletion, changing, moving singly or in a group.
Valid codes: check for valid names/codes of bathymetric data, drainage network, etc. while digitizing or batch mode.

Attribute Data Entry

Enter attributes: interactive or batch entry of attributes for parcel land use, drainage flow, sediment contaminant concentrations, water column concentrations, etc.
Linkage: develop data linkage for spatial and attribute data by feature name; location identifier using manual and/or automatic approach.
Insert: manual or automatic insertion of area, perimeter, length, width, statistics as attribute values.
Interactive change: insertion, deletion, change, and move of names/codes for estuary features.
Valid values: check for valid attribute values or combination of values for concentrations; check for missing names/codes for estuary features; format checking, range checking of attribute data.
Query update: query select function for updating groups of attribute values to change from one value to another.

Spatial and Attribute Data Entry

Import data: import spatial and attribute data in various formats, for example, Arc, DXF, DLG-3, user-defined ASCII.
Export data: export spatial and attribute data in various formats, for example, Arc, DXF, DLG-3, user-defined ASCII.

Table 8-2a: GIS data management functions to support river network modeling

Spatial Data Management

Spatial data description: construction of point features for stations, link/node topology with shape records, chain encoding of polygon boundaries. Surface coverages may be useful. Complex features may be useful.

Network data model: Node-based network or link-based network created as an option.

Global topology: global network topology for any geographic domain and any set of data categories to be defined by system administrator at the request of users.

Locational reference: use of absolute referencing such as latitude/longitude, state plane or UTM coordinate reference system or linear referencing such as river station point, reach number, or other specialized system depending on problem orientation.

Locational cross-reference: cross-reference of locational reference systems such as coordinate system with linear locational reference.

Spatial sorting: sort (reorder) the data on x or y or a combination of both.

Spatial indexing: after data are sorted then build an index.

Attribute Data Management

Attribute data description: construction of attribute fields to qualitatively and quantitatively describe data categories.

Attribute sort: sort data for any attribute.

Attribute indexing: create an index for quick retrieval by attribute value.

Both Spatial and Attribute Data Management

Map area storage/retrieval: continuous geographic domain for any area or group of areas can be stored and retrieved as a single database.

Object store/retrieval: storage/retrieval of objects described as points, lines, or areas and the attributes that describe them.

Browse facility: retrieval of any and all data categories.

Access and security: multiuser or single-user access with read/write protection.

Roll-back facility: supports restoration of database state in the event of system failure. Minimal data redundancy.

Subschema capability: select parts of a corporate-wide database for special management.

Database size: No limitation on the number of points, lines, or areas per map, maps per database, or coordinates per line or area should exist for logical storage of elements within the capacity of physical storage.

Data definition: software to manage descriptions and definitions of the data categories and the spatial and attribute data descriptors of these categories.

Catalog: create a catalog of all data by spatial or attribute data.

Table 8-2b: GIS data management functions to support urban estuary contaminant modeling

Spatial Data Management

Point-based data model: both vector and grid data models for samples of sediment chemistry.

Network-based data model: for drainage network.

Area-based data model: for drainage basin, concentration polygons.

Volume-based data model: for volumes of sediments, bioassay, water column.

Spatial sort: locational coordinate values, for example, x and/or y.

Spatial index: build quadtree or other for spatial data files after sort or for temporary reasons.

Attribute Data Management

Attribute index: sequential, random, and keyed index to attribute data files.

Calculate values: new attribute values using arithmetic expressions or table lookups in related tables.

Attribute sort: sort on attribute data values.

Spatial and Attribute Data Management

Data representation: ability to handle geodata with a 0D, 1D, 2D, or 3D spatial character, and with varied thematic attribute and temporal character.

Data dictionary: for defining database contents.

Metadata functions: for entering data quality information, such as lineage, positional accuracy, attribute accuracy, logical consistency, completeness.

Cataloging: automatically cataloging or indexing all data in the database, including data quality, contents, location, and date last maintained, and provide status report to this effect.

Journaling: for tracking transactions.

Record access: capability to set read/write/access authority for spatial and attribute data; multiple access for read only; single access for write.

File access: full add, delete, modify of user-created work files, but only authorized user ability to modify database.

Database size: capability to add points, lines, polygons, grid points constrained only by disk size.

Data organization: organize spatial files by location, project, theme, geographic area.

Data selection: selection of a specific data category; of spatial data by any shape polygon window; of attribute data with conditions by any shape polygonal window; of spatial and/or attribute data by data category name or groups of names; of geographic objects by Boolean condition on attributes.

Locate: geographic object by pointing, by specifying coordinate.

Table 8-3a: GIS data manipulation functions to support river network modeling

Spatial Data Manipulation

Structure conversion: conversion of vector to raster, raster to vector, quadtrees to vector, network to network in coordination with locational referencing.
Object conversion: point, line, area, cell, or attribute conversion to point, line, area, cell, or attribute.
Coordinate conversion: map registration, "rubber sheet" transformations, translation, rotation, scaling, map projection change, or image warping.
Spatial selective retrieval: retrieval of data based on spatial criteria such as rectangular, circular, or polygonal window, river station reference, point proximity, or feature name.
Locational classification: grouping of data values to summarize the location of an object such as calculations of river basin area centroids.
Locational simplification: coordinate weeding (line thinning) of lines.
Locational aggregation: grouping of spatial objects into a superordinate object such as river reaches into branches.
Locational disaggregation: subdivide a branch into reaches based on assumptions of continuous character of river reach.
Spatial clustering: interactive creation of areas that impact river reaches, based on clustering of areas with homogeneous characteristics. Clustering algorithm interfaces.
Micro-macro network hierarchy: different levels of the network should be supported for different aspects of a model(s) simultaneously allowing input from one level to another (as smaller branches into larger branches).

Attribute Data Manipulation

Attribute selective retrieval: retrieval of data based on thematic criteria such as attribute of river, and Boolean combination of attributes.
Attribute classification: grouping of attribute data values into classes.
Class generalization: grouping data categories into the same class based on characteristics of those categories.
Attribute aggregation: creation of a superordinate object based on attribute characteristics of two or more other objects such as reaches of similar character.
Arithmetic calculation: calculate an arithmetic value based on any other values (add, subtract, multiply, divide).

Spatial and Attribute Data Manipulation

Node/link attribution: distances and flow drag (friction) functions associated with node/reaches used to compute reach attribute values dynamically from attribute and/or spatial data; hence loading values onto confluence/reach in real time.
Variable length segmentation: segmentation of a river in a linear manner based on the homogeneity of attribute values along feature. Could be done dynamically for best results to support display and network overlay for different water quality characteristics.

Table 8-3b: GIS data manipulation functions to support urban estuary contaminant modeling

Spatial Data Manipulation

Data conversion: raster to vector and vector to raster, with user selectable priority for point, line, or area object; convert spatial object types from surface to polygon, polygon to line, point to line, and reverse.
Merging: interactive or automatic merging of geometrically adjacent spatial objects to resolve gaps or overlaps with user-specified tolerance.
Compress: or decompress raster formatted data such as run length or quadtree.
Resample: modify grid cell size through resampling.
Weed: weed coordinate detail, retain topology, sinuosity.
Smoothing: smooth line data to extract general sinuosity.
Coordinate geometry: protraction of parallel lines, generate boundaries from survey data; intersect lines, bisect lines.
Adjustments: mathematical adjustment of coordinate data using rotation, translation scaling in 4 parameters for X and Y, or 6 parameters for X or Y, least-squares adjustment.
Coordinate recovery: recover geographic ground coordinates from photographic data using single photo resection/intersection together with digital elevation data.
Location transformation: transform of ground survey bearing and distance data to geographic coordinates using least-squares adjustment of traverse to ground control; for example, from lat/long to Washington state plane and the reverse.
Spatial clustering: supervised or unsupervised clustering.

Attribute Data Manipulation

Attribute selective retrieval: retrieval of data based on thematic criteria such as attribute of sediment, and Boolean combination of attributes.
Attribute classification: grouping of attribute data values into classes.
Class generalization: grouping data categories into the same class based on characteristics of those categories.
Attribute aggregation: creation of a superordinate object based on attribute characteristics of two or more other objects such as concentration of contaminant in sediment volume.
Arithmetic calculation: calculate an arithmetic value based on any other values (add, subtract, multiply, divide).

Spatial and Attribute Manipulation

Contours: generate contours from irregularly or regularly spaced data; constrain contour generation by break lines, for example, shoreline, faults, ridges.
Grid: generate grid data from TIN; generate a grid from contours.
Rescale: rescaling of raster data values, for example, contrast stretching.
Variable area segmentation: segmentation of an area based on the homogeneity of attribute values among subareas of the study area. Could be done dynamically for best results to support display and overlay.

Table 8-4a: GIS data analysis functions to support river network modeling

Spatial Data Analysis

Spatial intraobject measurement: individual object calculations for line length, polygon area, and surface volume, polygon perimeter, percent of total area.
Spatial interobject measurement: interobject calculations for distance and direction point to point, point to line, polygon perimeter, percent of total area, percentiles, range, midrange.
Descriptive spatial statistics: centroid (weighted, geometric).
Inferential spatial statistics: trend analysis.

Attribute Data Analysis

Descriptive nonspatial statistics: frequency analysis, measures of dispersion (variance, standard deviation, confidence intervals, Wilcoxon intervals), measures of central tendency (mean, median, mode), factor analysis, contingency tables.
Inferential nonspatial statistics: correlation, regression, analysis of variance, discriminant analysis.

Spatial and Attribute Data Analysis

Overlay operators: point, line, area object on/in point, line, area object.
Network indices: compute network indices for connectivity, diameter, and tree branching structures.
Significance tests: t-test, chi-square, Mann-Whitney, runs.
Simulation: test the interaction of flows on the network over time. Test modal choice over time.
Model structuring: a model structuring environment that provides linkages between parts of models and the GIS environment through a special language interface.

Table 8-4b: GIS data analysis functions to support urban estuary contaminant modeling

Spatial Data Analysis

Spatial intraobject measurement: individual object calculations for line length, polygon area, and surface volume, polygon perimeter, percent of total area.
Spatial interobject measurement: interobject calculations for distance and direction point to point, point to line, polygon perimeter, percent of total area, percentiles, range, midrange.
Descriptive spatial statistics: centroid (weighted, geometric).
Inferential spatial statistics: trend analysis, autocorrelation.
Buffers: compute distances from point, line, polygon.
Slope/aspect: compute slope, aspect, and sun intensity.
Azimuth: compute azimuth, bearings, and geographic point locations.
Traverses: define open and closed traverses.
Nearest neighbor: compute closest geographic phenomenon to each phenomenon of a particular kind.

Attribute Data Analysis

Descriptive nonspatial statistics: frequency analysis, measures of dispersion (variance, standard deviation, confidence intervals, Wilcoxon intervals), measures of central tendency (mean, median, mode), factor analysis, contingency tables, percentile, range, midrange.
Inferential nonspatial statistics: correlation, regression, analysis of variance, discriminant analysis.
Significance tests: t-test, chi-square, Mann-Whitney, runs.
Inference intervals: confidence intervals, Wilcoxon intervals.

Spatial and Attribute Data Analysis

Variogram: variogram and cross-variogram calculations.
Covariance: calculation of generalized covariances in nonstationary cases.
Kriging samples: kriging in 1D, 2D, and 3D for scattered points, grids of points or blocks of points, irregular grids, and irregular blocks.
Kriging computation: ordinary, cokriging, universal, and disjunctive kriging for 1D, 2D, and 3D; stationary and nonstationary kriging.
Simulation: conditional simulation in 2D and 3D.
Raster and vector overlay: data operators for Boolean AND, OR, XOR, NOT; for point in polygon, point on line, line in polygon, and polygon in polygon.
Weighting: weight features by category in the overlay process.
Superimpose: superimpose one feature on another with replacement.
Merging: merge attribute information automatically or manually as a result of composite process; two attributes associated with same area.
Assignment: assign binary (1/0) and discrete (0-32767) or real continuous data values to cells in raster data set.
Group mathematical operations: on two or more data categories perform add, subtract, multiply, divide, minimum, maximum on the grid cell.
Single mathematical operations: on single data category perform exponential, logarithm, natural logarithm, absolute value, sine, cosine, tangent, arcsine, arccosine, arctangent on the grid cell.
Replacement: replace cell values with new value reflecting mathematical combination of neighborhood cell values: average, maximum, minimum, total, most frequent, least frequent, mean deviation, standard deviation.
TIN production: TIN modeling with fault specification.
Network tracing: shortest path source to multiple destinations in the drainage network; multiple sources to a destination.
Feature identification: automatically identify drainage networks, watersheds, viewshed in sub-basins.
Cut/fill: compute cut/fill volumes for sediments.
Profile: generate profiles for concentrations.
Model structuring: a model structuring environment that provides linkages between parts of models and the GIS environment through a special language interface.

Table 8-5a: GIS data display/output functions to support river network modeling

Spatial and Attribute Data Display/Output

Zoom: enlarge the map scale of basin display area, perhaps show more detail of the river network.
Pan: move across the display area to show a different part of the river basin.
Screen dump: make a hardcopy of what is on screen.
Mosaic output: hardcopy display can be produced in strips from plotter/printer.
Flat map: create birdseye view map of any part of the river basin area (3D orthographic display).
Contours: 2D contouring display with attribute highlighting.
Design composites: user-defined map design displays, and default output.
Composition variables: compose a map using location, size, scale, orientation, multiple views on a single display.
Graduated point-symbol map: user-defined graduated symbols for different magnitudes of point source pollution in basin.
Line map: 2D line maps of the river network.
Scatter diagrams: relationship between two variables.
Line graphs: two variable graphs.
Histogram: any number of data categories for nominal, ordinal or interval/ratio data.
Report generator: create reports using tabular formatting capabilities such as: lines breaks, page breaks, calculation of subtotals, totals, pagination, page and column headings, multiple line displays from single records.
Reference format: neat line, graticule, tick marks in lat/long and/or state plane.
Create symbolization: default and user-defined libraries of symbolization for point symbols, line styles, polygon area fill, and volume fill for the entire river or any portion thereof.
Assign symbolization: Assign symbolization by specifying feature name, group of names, feature display color, attribute or group of attributes.
Fonts: large selection of fonts and user-defined ability for fonts with size, slant, width, orientation, curved aspect.
Design object position: create, name, store, retrieve and interactively position the map title, legend, bar scale, representative fraction, north arrow, text blocks.
Text strings: automatic or manual positioning of text strings.

Table 8-5b: GIS data display/output functions to support urban estuary contaminant modeling

Spatial and Attribute Data Display/Output

Zoom: enlarge the map scale of display area, perhaps show more detail of parts of the estuary and surrounding area.
Pan: move across the display area to show a different area.
Screen dump: make a hardcopy of what is on screen.
Mosaic output: hardcopy display can be produced in strips from plotter/printer.
Perspective display: create 3D solids in 3D 1-point or 2-point perspective display with color highlighting of contaminant concentrations in the estuary bottom and the drainage into the estuary.
Flat map: create birdseye view map (3D orthographic display) of the entire area or parts thereof.
Contours: 2D contouring display with attribute highlighting for certain ranges of values, for example, contaminant concentration thresholds.
Design composites: user-defined map design displays, and default output for any part of the study area.
Composition variables: compose a map using location, size, scale, orientation, multiple views on a single display as in desktop publishing applications.
Dot maps: create dot maps for dot density and dot percentage using various symbols for population density or other ratio level data.
Graduated point-symbol map: user-defined graduated symbols for different magnitudes for showing possible sources of contaminants, for example, outfalls.
Choropleth map: user-defined classification; fixed interval defined; shading defined by user for showing land use differences.
Line map: 2D line maps for showing the drainage network.
Scatter diagrams: relationship between two variables
Line graphs: for showing relationships between any two variables.
Histogram: depicting any number of data categories for nominal, ordinal, or interval/ratio data.
Report generator: create reports using tabular formatting capabilities such as: lines breaks, page breaks, calculation of subtotals, totals, pagination, page and column headings, multiple line displays from single records.
Reference format: neat line, graticule, tick marks in lat/long and/or state plane.
Create symbolization: default and user-defined libraries of symbolization for point symbols, line styles, polygon area fill, and volume fill.
Assign symbolization: assign symbolization by specifying feature name, group of names, feature display color, attribute, or group of attributes.
Symbol fills: area and volume fill defined by color, line style, rotation angle, spacing of any kind of primitive symbolization for showing magnitudes of data.
Fonts: large selection of fonts and user-defined ability for fonts with size, slant, width, orientation, curved aspect.
Design object position: create, name, store, retrieve, and interactively position the map title, legend, bar scale, representative fraction, north arrow, text blocks.
Text strings: automatic or manual positioning of text strings.

a model is one of the most obvious functions. In this sense GIS has contributed to visual analysis. However, using GIS to support exploratory analysis with analytical techniques is still on the horizon, and should prove as useful, if not more useful, in the future (MacEachren and Ganter, 1990).

The advantage of a GIS is that it provides interactive data processing support. Consequently, the four steps described can occur in rather rapid succession. However, for some projects, days rather than minutes are required to proceed through all of the steps. The ease with which these necessary steps are carried out is influenced by GIS architecture.

STRUCTURAL PERSPECTIVE— THE NATURE OF GIS ARCHITECTURE

Another way to describe the general nature of a GIS is to examine the nature of the software architecture. As mentioned in the definition, a GIS is composed of data, software, and hardware used by personnel to develop and disseminate geographic information. In a sense, each of the components of data, software, and hardware has an architecture, a framework for how it is put together. For this discussion, and from a systems design point of view, perhaps the most fundamental of these three is the software architecture. Of course, a GIS would not be a GIS without data and hardware, and in most instances not people, but the software, provides for effectiveness and efficiency in many solutions to problems.

GIS architecture can be described in terms of generic subsystems (See Figure 8-2). Many authors (Calkins and Tomlinson, 1977; Clarke, 1986; Guptill, 1988) list the following subsystems that are of significance: input/capture, management, manipulation/analysis, output/display. Not surprisingly, these subsystems are a reflection of the use and processing activities in a GIS.

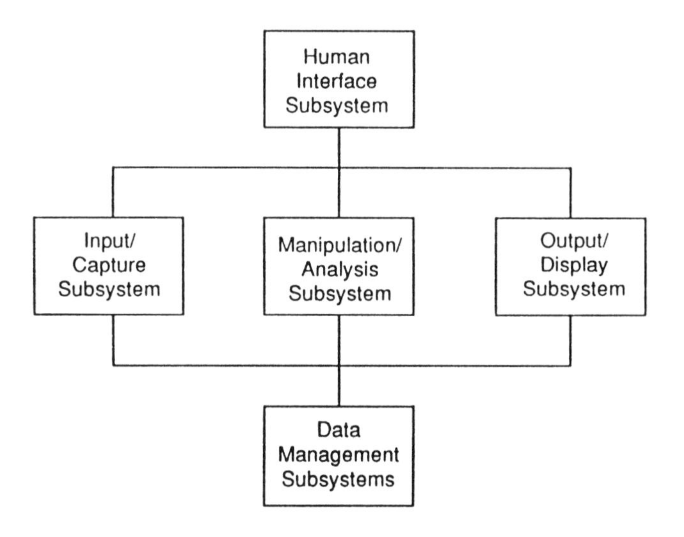

Figure 8-2: GIS as an architecture of subsystems from a structural perspective.

Each of the subsystems has been described in terms of functions that the respective subsystem performs. The data capture subsystem provides operational functions for acquiring data. The data management subsystem stores and retrieves the data elements. The manipulation and analysis subsystem handles the transformation of data from one form to another and the derivation of information from data. The data output and display subsystem provides a way for the user to see the data (information) in the form of diagrams, maps, and/or tables, etc.

The architecture of the data management subsystem determines the design of the descriptive constructs used for data storage. As such it is the fundamental mechanism for determining the nature of data representation as presented to applications, whether these be integrated functions in the GIS or models linked to the GIS. The architecture of the subsystem is based on the types of data models used. A data model determines the constructs for storage, the operations for manipulation, and the integrity constraints for determining the validity of data to be stored (Nyerges, 1987). The data model concept is often misunderstood, many authors including only the data construct aspect of this concept. Goodchild (1992) reviews the different data structuring approaches used in GIS data models. Together with the numerous functions described in the previous section and the integrity constraints for keeping data valid, these aspects of a GIS data model determine the architecture of a system.

Among the more common GIS data models are the layer (field), object, and network data models (Goodchild, 1992), together with a relational data model (see Figure 8-3). Layer or field models consist of spatial data constructs (location samples) with attached attribute data constructs. Object models consist of culturally defined attribute descriptions (attribute data constructs) with attached spatial data constructs. The principal difference between layer models and object models is that layer models do not bundle spatial and attribute data for data management, where object models do bundle spatial and attribute data for management. Network models can be developed like layer-based or object models, but with the additional stipulation that the linear geometry along the phenomenon must be part of the spatial description. The network model in Figure 8-3 indicates this through the use of the linear address construct, which can be thought of as a sequence of river sampling stations. Relational data models carry the attribute information for layer-based, object, and network models. Consequently, the difference in the data modeling approach is whether the spatial, attribute, and temporal characteristics are bundled as described phenomena or not. Bundling characteristics

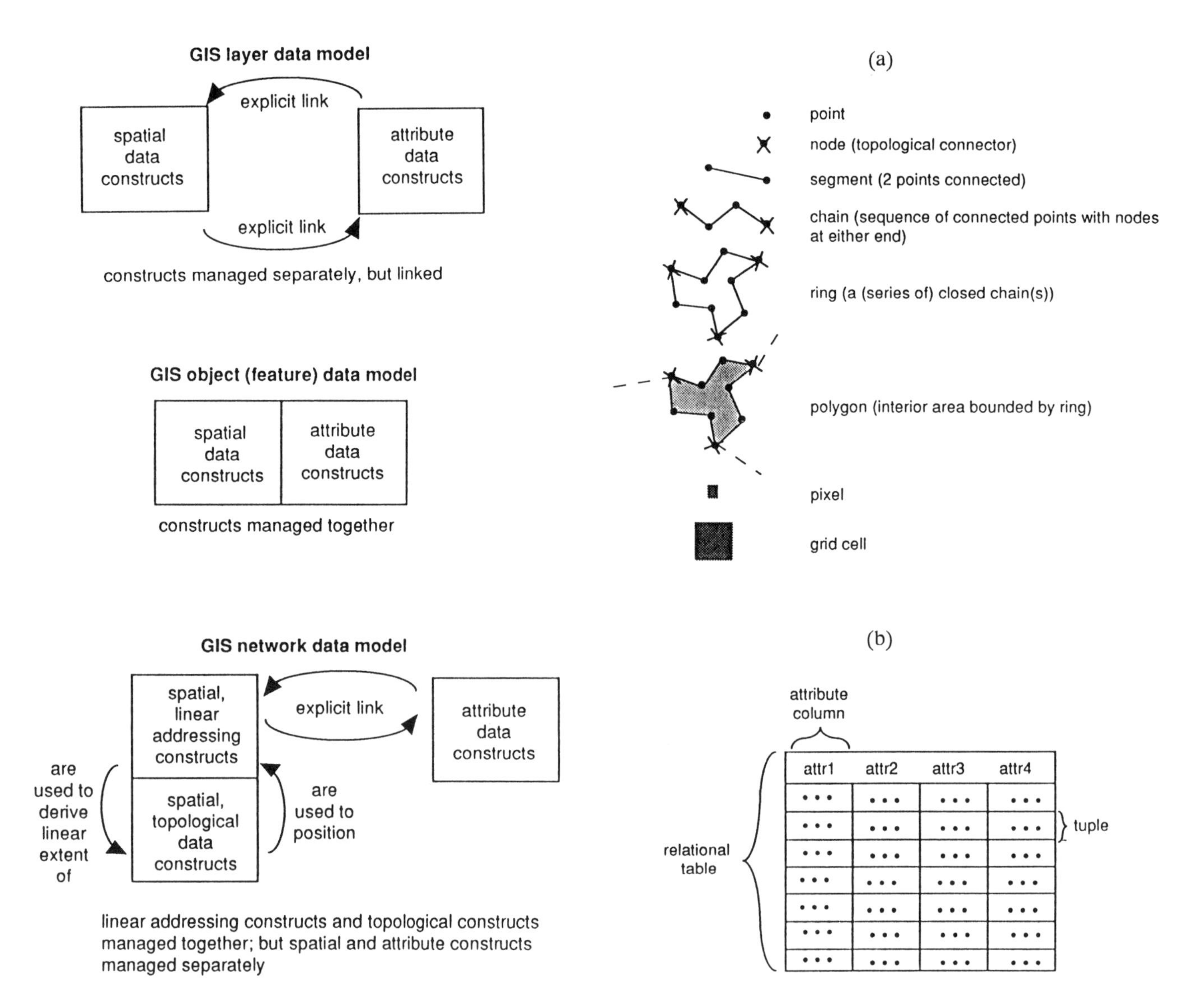

Figure 8-3: Common GIS models.

Figure 8-4: (a) Graphic depiction of spatial data constructs (after USGS 1991). (b) Graphic depiction of attribute data constructs.

allows for easier data manipulation. Not bundling them allows for easier data management.

The primitive, spatial data constructs in layer, object, and network models include: point, node, segment, chain, ring, polygon, cell, and table (see Figure 8-4a). The primitive data constructs in the relational data model include: table (relation), tuples, and columns (see Figure 8-4b). These basic constructs are combined in various ways to form the layer, object, and relational models.

Six different types of layer models include: irregular or regular sampled points, contours, polygons, grid cells, and triangular nets. The irregular or regular sampled points are composed usually of points in X, Y, (Z) space with an attached attribute. Contours are (X,Y) points with implicit segments, the aggregate sequence of points having an attached attribute, usually for terrain elevation or water depth. The polygon model is composed of topological chains (also called arcs or edges in some systems) that form a ring (closed boundary) around individual areas, with all areas together exhaustively covering a spatial extent, having no overlaps or gaps between polygons. The grid cells are areas (usually of the same rectangular size) exhaustively covering a surface. Triangular nets are (usually) irregularly spaced points that form vertices of triangles, whose sizes are optimized to cover a surface (with no gaps or overlaps) for the most effective representation. The layer models are discussed in more detail by Goodchild (1992).

Table 8-6: Issues and strategies for coupling GIS and models

	Coupling Strategy			
	Data transfer	Reference/ cross index	Federated	Integrate/ embedded
Issue				
data object types	• unresolved object differences	• resolved object differences	• resolved object differences	• resolved object differences
	• stored off-line	• stored off-line	• stored off-line	• stored off-line
	• manual resolution	• manual resolution	• computer-assisted resolution	• automatic resolution
data handling (software functions)	• different	• different	• different	• as one
human interface (hardware functions)	• different	• same	• different	• same
organi-zational environment (personnel)	• different	• same	• different	• same

The object model includes categorization by feature name and objects of well-defined cultural significance. Network models include: node (node star), link, node–link, and node–link–chain with named features. A recent advance in spatial data management software is the explicit linking of vector and raster representations so that data can be provided for analysis in a direct manner, without having to spend lots of time converting them from one form to the other.

The data management subsystem is connected directly to the data/capture, manipulation/analysis, and output/display subsystems (Figure 8-2). It provides the clearinghouse function to pass data between the various subsystems. One of the major reasons why GIS has been so useful is because it provides a mechanism to manage large volumes of spatially related data in a systematic fashion. Originally the graphics were acceptable, but computer-assisted cartography was better. The analytical capability was acceptable, but spatial models (in many different disciplines) provided for better analysis. For this reason a GIS often had been referred to as a database engine. With the advances in graphics and analysis incorporated into many GIS in recent years, this view is changing. Some organizations look at a GIS as a graphics engine, while others look at it as an analytical engine. Nonetheless, a GIS cannot do everything; it would be too expensive if it did.

One of the functions that it still does not do well (if it ever will) is to perform process modeling. For that reason, linking models into a GIS is seen as an area of interest to many researchers. This problem is part of the larger problem that deals with systems integration and linking applications to each other.

CONSIDERATIONS FOR EFFECTIVE COUPLING OF GIS AND MODELS

The architecture of a system, that is, the GIS data model, determines how easy or difficult it is to couple GIS and environmental process models. Any one of the subsystems of the architecture can be used as a focal point for coupling, but the most natural is through the data management subsystem. With this in mind, Smyth (Nyerges and Smyth, 1983) developed a data management system as the basis of a geographic modeling system architecture. Many of the same requirements were described a few years later by Dangermond (1987) and Goodchild (1987) at an International Conference on GIS research. Dangermond (1987) called the GIS that could easily integrate models a geographic information modeling system. Since then there have been various attempts at defining the nature of GIS and model coupling.

Burrough et al. (1988) offered a list of important concerns for effective linking of GIS and land resource assessment models. From the modeling perspective they see the following as important:

- What are the basic assumptions and methods?
- At what scale or organizational level is the model designed to work?
- What kinds of data are needed for control parameters?
- What kinds of data are needed to feed the model?
- Under which conditions are certain control parameters or input data more important than others?
- How are errors propagated through the model?

From a GIS perspective, Burrough et al. (1988) see the following as important concerns:

- Are the right input data available at an appropriate spatial resolution?
- Do sufficient data exist for creating an appropriate substrate?
- Are data available for calibrating and validating the model?
- If data are not available, can surrogates be used, and how should they be transformed?
- How should a user be made aware of the intrinsic quality of the results of modeling?
- Is information available on data quality and errors?
- If the results are not good enough, should the GIS suggest alternative data or alternative models to the user?

Coupling a model with a GIS is an information integration problem, somewhat like coupling one GIS to another for data transfer purposes (Nyerges, 1989). This is more than just a software systems integration problem. In keeping with the definition of GIS in this paper, it is a challenge that involves data, software, hardware, and personnel issues (Table 8-6). When coupling addresses all of these issues, major institutional arrangements (another aspect of the definition) come into play in considering such integration. Access to appropriate information in the long run requires that technical and managerial members of an organization work together to produce effective institutional arrangements.

Focusing on software architecture, Nyerges (1990, 1991c) and Lewis (1990) have discussed the linkage of GIS and network-based transportation models that may have some relevance to environmental modeling based on networks, for example, hydrologic modeling in drainage basins or toxic waste outfalls into water bodies. They describe different coupling environments based on the nature of the models and the GIS involved. Coupling environments can range from loose to tight coupling depending on the compatibility of the data constructs and the software operations used to process them (see Table 8-6). A loose coupling involves a data transfer from one system to another. A tight coupling is one with integrated data management services. The tightest of couplings is an embedded or integrated system, where the GIS and models rely on a single data manager. Embedded systems have been shown to be either too superficial for solving problems or too complex in their development. Since embedded systems require a substantial amount of effort, and are developed for selected user groups, they tend to be rather expensive, and constraining when changes are desired.

In light of the coupling issues mentioned by Burrough et al. (1988), Nyerges (1990), and Lewis (1990), identifying the compatible data constructs between the GIS data model and the modeling system data model is the fundamental issue; a second consideration only after the data constructs is with processing support. These considerations are important regardless of a loose or tight coupling. Some environmental models such as those used in atmospheric (Chapter 10) or ecological systems modeling (Kessell, 1990) use a grid-based spatial construct to describe space. This has a direct parallel in GIS data models. Sometimes environmental models such as those in hydrological modeling (Chapter 14) use a network-based description of space. These, as mentioned, have a parallel in GIS data models. The 3D environmental models that use time do not currently have a direct parallel in GIS data models. Such models can be loaded with data from a GIS, but of course, this would require more information processing effort—remembering that 2D GIS data models with an integral temporal component are still an active research area (Langran and Chrisman, 1988).

Technical solutions to problems are commonly guided by the information processing needs and institutional constraints in an organization. Therefore, no single solution is likely to suit all organizations, and every approach to a solution should build some amount of flexibility into the implementation. Consequently, this brings to light the continued effort that is needed to characterize the nature of the GIS and model coupling problem, and approaches that might provide solutions.

FRONTIERS FOR GIS AND MODELING

Nyerges (1991b) reviews several trends in data, software, and hardware that collectively represent frontiers of GIS development. Whether or not these developments are fostered by frontiers of GIS use remains to be seen.

One significant development involves raster and vector processing. There was once a debate among researchers as to which of the two data structures, raster or vector, would predominate in the architecture of systems (Chrisman, 1978). After some time, a calm came over the debate when both groups recognized that the other would find a niche for processing. Now we understand that both are necessary because each has advantages and disadvantages (cf. Burrough, 1986). In following up, hybrids of raster and vector structures have been discussed for some time (Peuquet, 1984), but commercial implementations of anything like a true hybrid are few and far between (GIS World, 1990). Side by side raster and vector structures are useful, but become more useful when they reference each other. Such systems are now beginning to appear. This has probably been the single most important development in GIS in the past several years. Goodchild (1987) and Nyerges (1987, 1991a) suggest that advancements in GIS data models are required to support better integration with spatial models. Although data representation is not the entire answer to an environmental problem expressed in a data model, advances in data representation are a significant start. To be able to go either way with raster or vector and then change as data and/or processing requirements change is certainly a new frontier to support GIS problem solving with models. It is hoped that more flexibility will result in more creativity.

Another significant development in GIS in the recent past has been the development of data structuring for complex features. Some GIS are no longer constrained to simple point, line, and area representations. Complex features allow a database designer to build in more phenomenological complexity than has been possible before. GIS has resisted this for some time because simple representations provide for flexible database development and use. However, when processes are reasonably well known, such that the details of phenomena are better understood, then this understanding can be designed into the database.

With GIS spatial data models becoming more flexible, another significant task for researchers to tackle will be the temporal aspects, both for data representation and processing. Recent efforts (Armstrong, 1988; Langran and Chrisman, 1988; Langran, 1990) have produced valuable contributions, but concepts that form the basis of design suggestions for software architecture are still needed. When these developments are incorporated into commercial GIS then we will see substantial progress in pushing back the GIS frontier as related to environmental modeling. Such progress could in fact be almost as significant as the maturing of space-only based GIS. Coupling process models to a GIS is required at the current time because no GIS currently has the data representation flexibility for space and time, together with the algorithmic flexibility to build process models internally. Coupling models to GIS depends on the compatibility of the software architectures. One-off coupling attempts have been the order of business, because few organizations have had expertise in both sophisticated models and GIS, and because the immediate solutions are more important than generic attempts. A good example of an integrated environment is discussed in Chapter 13. However, a more systematic and thus easier approach to such coupling is needed for the future in order to reduce the overall effort for linking models and GIS.

Smyth (1991) has been working to formalize a description of integration frameworks, and toward this end produced a glossary of terms that describes the various concepts involved. He calls these "GIS application integration frameworks." Some of the work has been borrowed from recent successful attempts to standardize issues of the same nature in the CAD/CAM/CIM area. Object-oriented database (OODB) management potentially plays a major role in developing the frameworks. Results for GIS will only be forthcoming with a more widespread, concerted effort to formalize these issues in terms of a reference model for GIS application integration. A reference model identifies issues that are mature for standardization, while at the same time identifies cogent issues in need of further research.

Sorting out the nature of coupling will certainly take more research. Some productive avenues on the horizon involve a mix of both technical and human issues. One productive direction appears to be a focus on model management (in addition to data management) with so-called model management systems (Geoffrion, 1987). Jankowski (1989) has applied the idea of model management to water quality modeling in a stream using ideas from system entity structuring (Ziegler, 1987). Once these ideas become tightly coupled into GIS then perhaps as Birkin et al. (1990) suggest these systems will become known as model-based GIS rather than data-based GIS.

Another productive direction involves studies of how we use information. Through such studies we can gain insight into the actual benefits of enhanced problem solving and decision-making. Benefits in coupling GIS and environmental models are based on the assumption that indeed the two will provide a better information processing environment than either could separately. These benefits need to be verified by undertaking systematic studies of GIS and model use. From the GIS side, systematic investigation of the decision-making process appears to be a rather fruitful undertaking to determine the value of geographic information in this process (Dickinson, 1990). Carried one step further, studies in cognitive work analysis (Rasmussen et al., 1990) will help us design systems to better suit the needs of geographic problem-solving and decision-making environments, as such studies attempt to uncover cognitive goals, needs, strategies, and constraints in particular problem-solving contexts. Together, these systematic methodologies should be able to help with studying the combinations of computer-based models and GIS.

RECOMMENDATIONS

The way in which a model is coupled with a GIS depends on the compatibility of the architectures. A description of the data transformations required between the data representation constructs is the first step. Specifying software to export and import between the constructs, without going through an intermediate structure, is the next step. Determining whether that software can run without human intervention follows. Being able to set up the transfer as bidirectional is the next step. An interprocess utility that shuttles data between the modeling software and the GIS data management subsystem is a common kind of linkage. Figure 8-5 depicts an interprocess utility solution common for a loose coupling of software architectures. This solution can be implemented using an off-line data transfer approach or on-line operating system approach. The former are better when programming staff is not available. The on-line operating systems approach is difficult to develop and costly to change, but provides the best performance. Notice the human interface subsystems are separate, as are the analytical and management subsystems. If these are integrated to provide a tight coupling, then the solution becomes a custom implementation, hence rather costly. Consequently, a coupling solution depends on the particular needs of an organization in regards to timeliness of the solution, constrained by economic considerations.

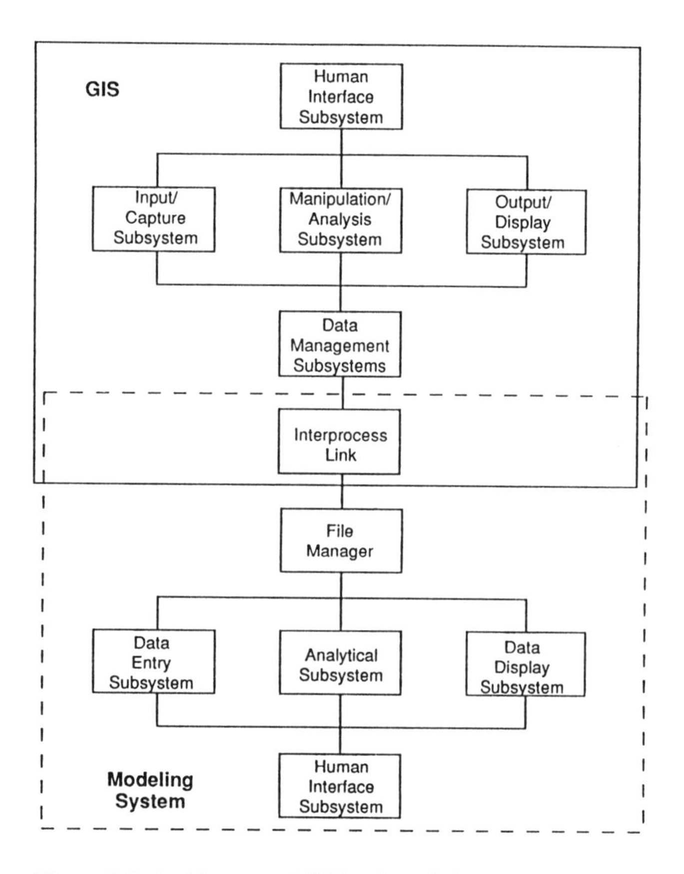

Figure 8-5: Architecture of GIS and modeling system coupling.

From a slightly broader perspective, access to information is related to policy that in turn is related to how societal goals are addressed (Chrisman, 1987; Clapp et al., 1989). The same concerns about whether and how computer-based models are used in policy formation (Chapter 34) occur with GIS. Consequently, the same concerns arise when we consider a linkage of computer-based models and GIS. Information generated from the coupled solution will not get used unless it addresses, and is intimately associated with, the policy issues at hand. Policy sets the direction for technical development whether this be at local, regional, or national levels.

One-off technical solutions are fine for a single organization. Perhaps they meet the policy objectives of the organizations sponsoring the effort. Such solutions tend to be the least expensive option in the short run, assuming alternatives are unknown, but such solutions are often expensive in the long run because knowledge of "making the coupling" is not reusable. In the long run, a systematic approach to coupling that relies on a generic framework is probably going to be less expensive. Unfortunately, the overhead for development of such a framework is prohibitive for any single organization.

To make systematic headway with the coupling problem, what is required is a multiple path effort. The conference that motivated this book was only a beginning of an understanding the problem. A larger concerted effort is required. Convening a group sponsored by multiple organizations, having interest in the integration of models and GIS from a general policy perspective, which can work together to develop a reference model for the integration problem, should be an important consideration. Members of such a group would include those with managerial as well as technical concerns. The National Institute of Standards and Technology (NIST) can provide insight into how this might come about. The proposed National Spatial Data Transfer Standard distributed by USGS (U.S. Geological Survey, 1991) and sponsored by NIST is a start in regards to data representation. The glossary by Smyth (1991) is a first published attempt at addressing the several technical aspects of the integration problem.

CONCLUSIONS

Models are not new to GIS. For some time various organizations have been trying to link operational and engineering-oriented spatial models to GIS environments. When this has been accomplished as a commercial endeavor, it has been called an application module. These simple models have sometimes satisfied operational and management uses of GIS in many local, state, and federal agencies. At the same time that GIS has matured to where it is useful in many applications areas with simple models, there has been a recognition that it might not be what it could be in other applications areas, for example, scientific uses of GIS, whether these be environmental science, natural science, or social science endeavors. Many scientific undertakings by their very nature have been rather narrow outlooks on the world. This is a result of an intense attempt to cope with the model complexity of a portion of the world.

Developing models in a GIS context is something that occurs daily across the world, but the kinds of models developed are not in the category we call process models. The mathematics of process models, and the representation requirements, are more complex than any vendor might wish to deal with because of a lack of marketplace potential. Developing effective process models within a GIS context depends on the nature of the architecture that fosters such development. Developing process models within a GIS can be accomplished if the tools are there. However, the tools get developed only if the marketplace or a benefactor act to spur on that development. Goodchild (1987) sees a dilemma with such development for advanced social science oriented systems, as such system use suffers from a lack of marketplace and benefactor—the social scientific community is not large enough and not focused enough. The same is probably true of the environmental science community, meaning the scientific marketplace is not now large enough, but it is growing. The alternative is to take advantage of what models exist and couple them to a GIS.

Making a concerted effort to specify a framework for coupling models and GIS can produce concrete results. The opportunity exists at the current time because both GIS and environmental models are matured sufficiently that proponents of each have an idea what they want from the proponents of the other. Now is the time to make significant headway in integrating GIS and environmental models.

REFERENCES

Armstrong, M. (1988) Temporality in spatial databases. *Proceedings, GIS/LIS '88*, Falls Church, VA: ASPRS/ACSM, Vol. 2, pp. 880–889.

Birkin, M., Clarke, G., Clarke, M., and Wilson, A.G. (1990) Elements of a model-based GIS for the evaluation of urban policy. In Worrall, L. (ed.) *Geographic Information Systems: Developments and Applications*, London: Belhaven Press, pp. 133–162.

Burrough, P.A. (1986) *Principles of Geographical Information Systems for Land Resource Assessment*, Oxford: Oxford University Press.

Burrough, P.A., van Deursen, W., and Heuvelink, G. (1988) Linking spatial process models and GIS: a marriage of convenience of a blossoming partnership? *Proceedings, GIS/LIS '88*, Falls Church, VA: ASPRS/ACSM, Vol. 2, pp. 598–607.

Calkins, H.W., and Tomlinson, R.F. (1977) *Geographic Informations Systems: Methods and Equipment for Land Use Planning*, Ottawa: IGU Commission on Geographical Data Sensing and Processing.

Chrisman, N.R. (1978) Concepts of space as a guide to cartographic data structures. In Dutton, G. (ed.) *Harvard Papers on Geographic Information Systems*, Cambridge, MA: Harvard University Laboratory for Computer Graphics and Spatial Analysis, Vol. 7.

Chrisman, N.R. (1987) Design of geographic information systems based on social and cultural goals. *Photogrammetric Engineering and Remote Sensing* 53(10): 1367–1370.

Clapp, J.L., McLaughlin, J.D., Sullivan, J.G., and Vonderohe, A.P. (1989) Toward a method for the evaluation of multipurpose land information systems. *Journal of the Urban and Regional Information Systems Association* 1(1): 39–45.

Clarke, K.C. (1986) Advances in geographic information systems. *Computers, Environment and Urban Systems* 10(3/4): 175–184.

Clarke, K.C. (1990) *Analytical and Computer Cartography*, Englewood Cliffs, NJ: Prentice Hall.

Cowen, D. (1988) GIS versus CAD versus DBMS: what are the differences? *Photogrammetric Engineering and Remote Sensing* 54(11): 1551–1555.

Dacey, M., and Marble, D.F. (1965) Some comments on certain aspects of geographic information systems. *Technical Report No. 2*, Evanston, IL: Department of Geography, Northwestern University.

Dangermond, J. (1987) The maturing of GIS and a new age for geographic information modeling (GIMS). *Proceedings, International Geographic Information Systems Symposium: The Research Agenda*, Washington, DC: NASA, Vol. II, pp. 55–66.

Dickinson, H.J. (1990) Deriving a method for evaluating the use of geographic information in decision making. *Technical Paper 90-3*, Santa Barbara, CA: National Center for Geographic Information and Analysis.

Dueker, K., and Kjerne, D. (1989) *Multipurpose Cadastre: Terms and Definitions*, Bethesda, MD: American Congress on Surveying and Mapping.

Frank, A.U., Egenhofer, M.J., and Kuhn, W. (1991) A

perspective on GIS technology in the nineties. *Photogrammetric Engineering and Remote Sensing* 57(11): 1431–1436.

Geoffrion, A.M. (1987) An introduction to structured modeling. *Management Science* 33(5): 547–588.

GIS World (1990) *GIS Sourcebook*, Boulder, CO: GIS World, Inc.

Goodchild, M.F. (1987) A spatial analytical perspective on geographical information systems. *International Journal of Geographical Information Systems* 1(4): 327–334.

Goodchild, M.F. (1992) Geographical data modeling. *Computers and Geosciences* 18(4): 401–408.

Guptill, S. (ed.) (1988) A process for evaluating geographic information systems. *Open File Report 88-105*, Reston, VA: U.S. Geological Survey.

Horwood, E.M., Rogers, C.D., Rom, A.R.M., Olsonski, N.A., Clark, W.L., and Weitz, S. (1962) Using computer graphics in community renewal. *Community Renewal Program Guide No. 1*, Washington, DC: Urban Renewal Administration, Housing and Home Finance Agency.

Hunsaker, C.T., Graham, R.L., Barnthouse, L.W., Gardner, R.H., O'Neill, R.V., and Suter II, G.W. (1990) Assessing ecological risk on a regional scale. *Environmental Management* 14: 325–332.

Jankowski, P. (1989) Knowledge-based structured modeling: an application to stream water quality management. *PhD dissertation*, Seattle, WA: Department of Geography, University of Washington.

Kessell, S.R. (1990) An Australian geographical information and modeling system for natural area management. *International Journal of Geographical Information Systems* 4(3): 333–362.

Langran, G. (1990) Temporal GIS design trade-offs. *Journal of the Urban and Regional Information Systems Association* 2(2): 16–25.

Langran, G., and Chrisman, N.R. (1988) A framework for temporal geographic information. *Cartographica* 25(3): 1–14.

Lewis, S. (1990) Use of geographical information systems in transportation modeling. *ITE Journal* (March): 34-38.

MacEachren, A., and Ganter, J.H. (1990) A pattern identification approach to cartographic visualization. *Cartographica* 27(2): 64–81.

Maguire, D.J., Goodchild, M.F., and Rhind, D.W. (1991) *Geographical Information Systems: Principles and Applications*, London: Longman.

Marble, D.F. (1990) Geographic information systems: an overview. In Peuquet, D.J., and Marble, D.F. (eds.) *Introductory Readings in Geographic Information Systems*, London: Taylor and Francis, pp. 8–17.

Marble, D.F., Lauzon, J.P., and McGranaghan, M. (1990) Development of a conceptual model of the manual digitizing process. In Peuquet, D.J., and Marble, D.F. (eds.) *Introductory Readings in Geographic Information Systems*, London: Taylor and Francis, pp. 341–352.

Marble, D.F., and Peuquet, D.J. (eds.) (1983) Geographic information systems and remote sensing. In Colwell, R.N. (ed.) *Manual of Remote Sensing*, Bethesda, MD: ASPRS.

Mark, D.M., and Gould, M.D. (1991) Interacting with geographic information: a commentary. *Photogrammetric Engineering and Remote Sensing* 57(11): 1427–1430.

Morehouse, S. (1978) The Odyssey file system. In Dutton, G. (ed.) *Harvard Papers on Geographic Information Systems*, Cambridge, MA: Harvard University Laboratory for Computer Graphics and Spatial Analysis, Vol. 7.

Nyerges, T.L. (1987) GIS research issues identified during a cartographic standards process. *Proceedings, International Symposium on Geographic Information Systems*, Washington, DC: NASA.

Nyerges, T.L. (1989) Information integration for multipurpose land information systems. *Journal of the Urban and Regional Information Systems Association* 1(1): 28–39.

Nyerges, T.L. (1990) Development of a state-route digital highway network for Washington State. *Final Report on Task 3, Contract JSA 88-28*, Seattle, WA: University of Washington.

Nyerges, T.L. (1991a) Analytical map use. *Cartography and Geographic Information Systems* 18(1): 11–22.

Nyerges, T.L. (1991b) Directions and issues in the future of GIS. Paper presented at the Third North Carolina GIS Conference, Raleigh, NC, January 11, 1991.

Nyerges, T.L. (1991c) GIS data handling support for spatial transportation modeling. Paper presented at AAG Annual Convention, Miami, April 16, 1991.

Nyerges, T.L. (1991d) Thinking with geographic information: user interface requirements for GIS. Position paper presented to Initiative 13, National Center for Geographic Information and Analysis, Buffalo, NY, June, 1991.

Nyerges, T.L., and Smyth, C.S. (1983) IGDMS—an integrated geographic data management system. Paper presented at the International Symposium for Automated Cartography, AUTO-CARTO VI, Ottawa, Canada, October, 1983.

Obermeyer, N. (1989) A systematic approach to the taxonomy of geographic information use. *Proceedings, GIS/LIS '89*, Falls Church, VA: ASPRS/ACSM, Vol. 2, pp. 421–429.

Onsrud, H. (1989) Understanding the uses and assessing the value of geographic information. *Proceedings, GIS/LIS '89*, Falls Church, VA: ASPRS/ACSM, Vol. 2, pp. 404–411.

Peuquet, D.J. (1984) A conceptual framework and comparison of spatial data models. *Cartographica* 21: 66–113.

Peuquet, D.J., and Marble, D.F. (eds.) (1990) *Introductory Readings in Geographic Information Systems*, London: Taylor and Francis.

Rasmussen, J.A., Pejtersen, A.M., and Schmidt, K. (1990) Taxonomy for cognitive work analysis. *Technical Report Riso-M-2871*, Denmark: Riso National Laboratory.

Ripple, W.J. (ed.) (1987) *GIS for Resource Management: A Compendium*, Bethesda, MD: American Society for Photogrammetry and Remote Sensing.

Smyth, C.S. (1991) *GIS Application Integration Frameworks*, Bellevue, WA: Spatial Data Research, Inc.

Starr, J.L., and Estes, J.E. (1990) *An Introduction to Geographic Information Systems*, Englewood Cliffs, NJ: Prentice-Hall.

Tobler, W.R. (1959) Automation and cartography. *Geographical Review* 49: 526–534.

Tobler, W.R. (1977) Analytical cartography. *The American Cartographer* 3(1): 21–31.

Tomlin, C.D. (1990) *Geographic Information Systems and Cartographic Modeling*, Englewood Cliffs, NJ: Prentice-Hall.

Tomlinson, R.F. (1972) *Geographical Data Handling*, Ottawa: International Geographical Union Commission on Geographical Data Sensing and Processing.

Tomlinson, R.F. (1988) The impact of the transition from analogue to digital cartographic representation. *The American Cartographer* 15(3): 249–261.

Tomlinson, R.F., and Petchenik, B.B. (eds.) (1988) Reflections on the revolution: the transition from analogue to digital representations of space, 1958–1988. *The American Cartographer* 15(3).

U.S. Geological Survey (1991) *Spatial Data Transfer Standard*, Reston, VA: U.S. Geological Survey, National Mapping Division.

Wellar, B., and Graf, T. (1972) Bibliography on urban and regional information systems: focus on geographic perspectives. *Exchange Bibliography 316, 317*, Montecello, IL: Council of Planning Libraries.

Wheeler, D.J. (1988) A look at model building with geographic information systems. *Proceedings, GIS/LIS '88*, Falls Church, VA: ASPRS/ACSM, pp. 580–589.

Ziegler, B.P. (1987) Hierarchical, modular discrete-event modeling in an object-oriented environment. *Simulation* 49(5): 219–230.

9

Data Models and Data Quality: Problems and Prospects

MICHAEL F. GOODCHILD

GIS can bring enormous benefits because of its ability to handle spatial data in ways that are precise, rapid, and sophisticated. The previous two chapters have provided perspectives on the current state of GIS functionality and on the range of data types currently recognized by GIS. Many of these have direct relevance to the needs of environmental modeling and policy formulation, as later chapters in this book demonstrate. However, much of the data that have found their way into spatial databases over the past two decades have originated from map documents, and the uses to which the data are put in a GIS may differ substantially from the traditional uses of maps. Indeed, GIS may be leading to uses for which the data were never intended when they were collected and mapped.

Like all measurements, spatial data have levels of accuracy that are limited by the measuring devices used. For example, the spatial resolution of an imaging sensor is limited by its pixel size. Similarly, the spatial resolution of mapped data is determined by the inherent limitations of mapmaking technology. It is hard to draw a line of less than 0.2 mm width, and to prevent paper documents from stretching and shrinking with changes in humidity. However, such limitations have a way of disappearing in the digital domain. It is easy to print single-precision real numbers with seven or eight significant digits, even though these may be entirely unjustified by the quality of the input data. Spatial databases have no explicit scale, and rarely inform the user of their accuracies in any useful manner.

So while digital processing of geographic data brings immense benefits in the form of rapid, precise and sophisticated analysis, at the same time it reveals in data weaknesses that may not otherwise be apparent. Computers are very precise machines, and errors and uncertainties in data can lead to serious problems, not only in the form of inaccurate results but also in the consequences of decisions made on the basis of poor data, and increasingly in legal actions brought by affected parties.

In this chapter we first review the dimensions of the accuracy problem, and discuss the premises and assumptions on which the remainder of the chapter is based. The second section reviews the models that are available for analyzing and understanding errors in spatial data. The chapter concludes with a summary of the current state of the art and an agenda for future research and development, and draws heavily on arguments presented by Goodchild (1991). An annotated bibliography and taxonomy of the literature in spatial data error and uncertainty have been provided by Veregin (1989a,b).

The terms accuracy, precision, resolution, and scale are used almost interchangeably in reference to spatial data, but we will need to establish more exact definitions. *Accuracy* refers to the relationship between a measurement and the reality it purports to represent. *Precision* refers to the degree of detail in the reporting of a measurement or in the manipulation of a measurement in arithmetic calculations. Finally, the *resolution* of a data set defines the smallest object or feature that is included or discernable in the data. Scale and resolution are intimately related because there is a lower limit to the size of an object that can be usefully shown on a paper map. This limit is often assumed to be 0.5 mm as a rule of thumb; so the effective resolution of a 1:1000 map is about 1000 x 0.5 mm, or 50 cm, although the standards of most mapping agencies are substantially better. The effective resolutions of some common map scales, assuming 0.5 mm resolution, are shown in Table 9-1.

As noted earlier, because digital spatial data sets are not maps, and have no similar physical existence, they do not have an obvious or explicit scale, and it is common to think of such data as essentially scale free. But this is to oversimplify. If the data were obtained by digitizing, their resolution is that of the digitized map. If they were obtained from imagery, their resolution is that of the pixel size of the imagery. In spite of appearances to the contrary, then, no spatial data are ever resolution free. The term *scale* is often used loosely as a surrogate for database resolution.

Table 9-1: Scale and resolution for some common map scales

Scale	Effective resolution	Minimum resolvable area
1:1,250	62.5 cm	
1:10,000	5 m	
1:24,000	12 m	
1:50,000	25 m	0.0625 ha
1:100,000	50 m	0.25 ha
1:250,000	125 m	1.56 ha
1:500,000	250 m	6.25 ha
1:1,000,000	500 m	25 ha
1:10,000,000	5 km	2500 ha

DIMENSIONS OF THE PROBLEM

Precision

One of the widely accepted design principles of spatial data handling systems is that they should be able to process data without significant distortion. It is important when calculations are carried out on coordinates, for example, that the results should not be affected by lack of precision in the machine. We have seen that a figure of 0.5 mm can be assumed to be typical for the resolution of an input map document. 0.5 mm is also roughly the accuracy with which an average digitizer operator can position the crosshairs of a cursor over a point or line feature. Most digitizing tables have a precision of 0.25 mm or better, thus ensuring that the table itself does not introduce additional distortion as points are captured.

If we take the size of the source map sheet on the digitizing table to be 100 cm by 100 cm, then a precision of 0.5 mm represents an uncertainty of 0.05% relative to the size of the sheet, or an uncertainty in the fourth significant digit. So in order not to introduce distortion, internal storage and processing of coordinates must be correct to at least four digits. However, seven or eight decimal digits are commonplace in arithmetic calculations in spatial data handling systems, and many GIS operate at double precision, in effect allowing GIS users to ignore the possibility of arithmetic distortion and to assume that internal processing is carried out with infinitely high precision. Double precision of 15 decimal digits in latitude and longitude is sufficient to resolve location to the size of a single atom on the surface of the Earth; single precision of 7 decimal digits resolves to about a meter.

Despite this, arithmetic precision, or lack of it, is sometimes a problem in spatial data processing. One common situation occurs when point locations are specified in a global coordinate system. For example, the UTM system assigns coordinates with respect to an origin on the equator, so locations in the mid-Northern latitudes will have one of their coordinates in the millions of meters. Processing of data taken from large-scale maps, perhaps 1:1000, will therefore require precision in the sixth or seventh significant digit. This can be particularly problematic in performing operations such as calculation of the intersection point of two lines, or the centroid of a parcel of land, where differences in the products of six or seven digit numbers become important.

Another artifact of high precision is well known in connection with polygon overlay operations. It is common to find that two coverages of different themes for the same area share certain lines. For example the shoreline of a lake will appear on both a map of soils and a map of vegetation cover. When soils and vegetation cover are overlaid, the two versions of the shoreline should match perfectly. In practice this is never the case. Even if the two lines matched perfectly on the input documents, errors in registration and differences in the ways in which the digitizer operators captured the lines will ensure different digital representations in the database. Moreover, since the overlay operation is carried out to high precision, the differences are treated as real and become small sliver polygons. Table 9-2, from Goodchild (1979), shows the results of overlaying five layers from the Canada Geographic Information System, and the enormous numbers of very small polygons produced.

Several approaches have been taken to deal with the spurious polygon problem in polygon overlay, which has the annoying property that greater accuracy in digitizing merely leads to greater numbers of slivers. Some systems attempt to distinguish between spurious and real polygons after overlay, using simple rules, and to delete those which are determined to be spurious. A suitable set of rules to define spurious polygons might be:

Table 9-2: Polygons by area for five CGIS layers and overlays

Acres	1	2	3	4	5	1+2+5	1+2+3+5	1+2+3+4+5
0–1	0	0	0	1	2	2640	27566	77346
1–5	0	165	182	131	31	2195	7521	7330
5–10	5	498	515	408	10	1421	2108	2201
10–25	1	784	775	688	38	1590	2106	2129
25–50	4	353	373	382	61	801	853	827
50–100	9	238	249	232	64	462	462	413
100–200	12	155	152	158	72	248	208	197
200–500	21	71	83	89	92	133	105	99
500–1000	9	32	31	33	56	39	34	34
1000–5000	19	25	27	21	50	27	24	22
>5000	8	6	7	6	11	2	1	1
Totals	88	2327	2394	2149	487	9558	39188	90599

- Small in area;
- Long and thin in shape (high ratio of perimeter to square root of area);
- Composed of two arcs only (an arc is the common boundary between two polygons, or a topological edge of the boundary network);
- Both nodes have exactly four incident arcs.

In addition, the attributes of spurious polygons are characteristic. Suppose our soil map line marks the boundary between soils A and B, and our vegetation map shows the boundary between classes 1 and 2. Ideally, when overlaid the result should be a line with soil A and vegetation 1 on one side, and soil B and vegetation 2 on the other. Sliver polygons will be either soil B and vegetation 1, or soil A and vegetation 2. Moreover, the attributes will alternate from one sliver to the next, as the soil and vegetation boundaries cross and recross. So the detection of one spurious polygon provides additional evidence that nearby polygons are also spurious.

Another approach to dealing with the sliver problem is in effect to reduce the precision of the overlay operation, by requiring the user to establish a tolerance distance. Any lines lying within the tolerance distance of each other are assumed to be the same. A similar approach is often taken in digitizing, where two lines are assumed to join or "snap" if the distance between them is less than some user-established tolerance distance.

Because a tolerance distance is in effect a reduced level of spatial resolution, all objects or features smaller than the tolerance become ambiguous or indistinguishable. In a raster representation, the pixel size has the same effect of establishing an upper limit to the spatial resolution of GIS processing operations; thus there is no equivalent of the spurious polygon problem in raster-based systems. But in vector-based systems, the ambiguity that arises when the processor has a level of spatial resolution that is inconsistent with the positional accuracy of the data leads to problems where the known topological properties of the data conflict with geometric properties.

The potential conflict between topology and geometry is a consistent theme in spatial data handling. Franklin (1984; Franklin and Wu, 1987) has written about specific instances of conflict, and about ways of resolving them. For example, a point may be given an attribute indicating that it lies inside a given polygon, which might be a county, but geometrically, because of digitizing error or other problems, the point may lie outside the polygon (Blakemore, 1984). The conflict may be resolved by moving the point inside the polygon, or by moving the polygon boundary to include the point, or by changing the point's attributes. The first two are examples of allowing topology to correct geometry, and the third of allowing geometry to correct topology. In all cases, and particularly in the second, the implications of a change on other objects and relationships must be considered; if the polygon's boundary is moved, do any other points now change properties? Unfortunately the propagation of changes of this type can have serious effects on the overall integrity of the database.

In summary, the average user will approach a GIS or spatial data handling system with the assumption that all internal operations are carried out with infinite precision. High precision leads to unwanted artifacts in the form of spurious polygons. Moreover in reducing effective precision in such operations as digitizing and polygon closure, it is common for unwanted effects to arise in the form of conflicts between the topological and geometrical properties of spatial data. The ways in which system designers choose to resolve these conflicts are important to the effective and data-sensitive application of GIS technology.

Accuracy

Thus far we have used examples of the distortions and errors introduced into spatial data by digitizing and processing. However, if we are to take a comprehensive view of spatial data accuracy, it is important to remember that accuracy is defined by the relationship between the measurement and the reality it purports to represent. In most cases reality is not the source document but the ground truth that the source document models. The map or image is itself a distorted and abstracted view of the real world, interposed as a source document between the real world and the digital database.

We will use two terms to distinguish between errors in the source document and in the digitizing and processing steps. Source errors are those that exist in the source document, and define its accuracy with respect to ground truth. Processing errors are those introduced between the source document and the GIS product, by digitizing and processing. We will find that in general processing errors are relatively easier to measure, and smaller than source errors.

All spatial data without exception are of limited accuracy, and yet it is uncommon to find statements of data quality attached to such data, whether in the form of source maps, images, or digital databases. The accuracy of much locational or attribute data is limited by problems of measurement, as for example in the determination of latitude and longitude, or elevation above mean sea level. In other cases accuracy is limited by problems of definition, as when a soil type is defined using terms such as "generally," "mostly," "typically," and thus lacks precise, objective criteria. Locations defined by positions on map projections are limited by the accuracy of the datum, and may change if the datum itself is changed.

More subtle inaccuracies result from the way in which a source document models or abstracts reality. For example, an extended object such as a city may be shown at some scales as a point, or a census reporting zone may be located by its centroid or some other representative point (Bracken and Martin, 1989). The transition between two soil types, which occurs in reality over some extended zone, may be represented on the source document and in the database by a simple line or sharp discontinuity. The attributes assigned to a polygon may not in fact apply homogeneously to all parts of the polygon, but may be more valid in the middle and less valid toward the edges (Mark and Csillag, 1989), and the reverse is also sometimes observed to be true. All of these errors result from representing spatial variation using objects (Chapter 2); the objects are either of the wrong type for accurate representation, in the case of the city represented by a point, or used to model what is in fact continuous variation. Models such as these allow complex spatial variation to be expressed in the form of a comparatively small number of simple objects, but at a corresponding cost in loss of accuracy. Unfortunately the process of modeling has usually occurred in the definition of the source document, long before digitizing or processing in a GIS, and rarely is any useful information on levels of accuracy available.

Separability

Most digital spatial data models distinguish clearly between locational and attribute information. From an accuracy perspective, errors in location and attributes are often determined by quite different processes, and in these cases we term the two types of error separable. For example, a reporting zone such as a census tract may be formed from segments that follow street center lines, rivers, railroads, and political boundaries. The processes leading to error in the digitizing of a river-based boundary are quite different from those typical of street center lines. Errors in imagery are similarly separable, attribute errors being determined by problems of spectral response and classification, while locational errors are determined by problems of instrument registration. The same is true to a lesser extent for topographic data, since errors in determining elevation at a point are probably only weakly dependent on errors in locating the point horizontally.

However, in maps of vegetation cover, soils, or geology, where polygon objects are commonly used to model the continuous variation of fields, the two types of error are not separable, as object boundaries are derived from the same information as the attributes they separate (Goodchild, 1989). The transition zone between types A and B may be much less well defined than the transition between types A and C. The process by which such maps are created gives useful clues to the likely forms of error they contain. A forest cover map, for example, will be derived from a combination of aerial imagery and ground truth. Polygons are first outlined on the imagery at obvious discontinuities. Ground surveys are then conducted to determine the attributes of each zone or polygon. The process may be iterative, in the sense that the ground survey may lead to relocation of the interpreted boundaries, or the merger or splitting of polygons.

The forest example illustrates the relationship between different error processes particularly well. In forest management the homogeneous zones identified in the mapping process are termed *stands*, and become the basic unit of management. If a stand is cut and reseeded or allowed to regenerate, the processes determining its attributes may no longer be related to the processes determining its boundaries, and thus attribute and locational errors will become separable. Similarly, even in the initial map obtained from imagery and ground survey there will be boundary segments that follow roads, rivers or shorelines and again are subject to error processes that are independent of those affecting attributes.

The ideal

The arguments in the preceding sections lead to a clear idea of how an ideal GIS might be structured. First, each object in the database would carry information describing its accuracy. Depending on the type of data and source of error, accuracy information might attach to each primitive object, or to entire classes of objects, or to entire datasets. Accuracy might be coded explicitly in the form of additional attributes, or implicitly through the precision of numerical information.

Every operation or process within the GIS would track error, ascribing measures of accuracy to every new attribute or object created by the operation. Uncertainty in the position of a point, for example, would be used to determine the corresponding level of uncertainty in the distance calculated between two points, or in the boundary of a circle of specified radius drawn around the point.

Finally, accuracy would be a feature of every product generated by the GIS. Again it might be expressed either explicitly or implicitly in connection with every numerical or tabular result, and means would be devised to display uncertainty in the locations and attributes of objects shown in map or image form.

Of course the current state of GIS technology falls short of this ideal in all three areas. We currently lack comprehensive methods of describing error, modeling its effects as it propagates through GIS operations, and reporting it in connection with the results of GIS analysis. Moreover there seems little doubt that if such a system were available, it would reveal frightening levels of uncertainty in many GIS analyses and much GIS-based modeling. In Chapter 43, Englund shows the consequences of simulating actual levels of uncertainty in analyses based on kriging, using conditional simulation techniques. Very few techniques exist for displaying known levels of uncertainty on maps (Beard et al., 1991); so such information is commonly hidden from the user. Finally, it seems clear that the nonlinear characteristics of many of the models described in this book, and the ability of many of them to propagate between scales, must lead to high and perhaps disastrous sensitivity to uncertainty in spatial data inputs.

The remaining sections of this chapter describe the current state of knowledge and such techniques as currently exist for achieving a partial resolution of the error problem in GIS.

MODELS OF ERROR

Background

With the basic introduction to accuracy and error in the previous section, we can now consider the specific issue of quality in spatial data. This is not a simple and straightforward extension; as we will see, spatial data require a somewhat different and more elaborate approach.

The objective is to find an appropriate model of the errors or uncertainties that occur in spatial data. A model is taken here to mean a statistical process whose outcome emulates the pattern of errors observed in real digital spatial data. The first subsection considers models of error in the positioning of simple points; the second section looks at the ways this can be extended to more complex line or area features. For simple scalar measurements, conventional error theory is based on the Gaussian distribution. In effect, what we seek is the equivalent of the Gaussian distribution for the ensembles of measurements referred to here as spatial databases.

Points

The closest analogy between the classical theory of measurement and the problem of error in spatial data concerns the location of a single point. Suppose, for example, that we wished accurately to determine the location, in coordinates, of a single street intersection. It is possible to regard the coordinates as two separate problems in measurement, subject to errors from multiple sources. If the coordinates were in fact determined by the use of a transparent roamer on a topographic sheet, then this is a fairly accurate model of reality. Our conclusion might be that both coordinates had been determined to an accuracy of 100 m, or 2 mm on a 1:50,000 sheet.

If the errors in both coordinates are represented by Gaussian or normal distributions, then we can represent accuracy in the form of a set of ellipses centered on the point. If the accuracies are the same in each coordinate and if the errors are independent, then the ellipses become circles, again centered on the point. This gives us the circular normal model of positional error, as the two-dimensional extension of the classic error model. The error in position of a point can be visualized as a probability density forming a bell-shaped surface centered over the point. Using the model we can compute the probability that the true location lies within any given distance of the measured location, and express the average distortion in the form of a standard deviation. A commonly used summary measure is the Circular Map Accuracy Standard, defined as the distance within which the true location of the point lies 90% of the time, and equal to 2.146 times the standard deviation of the circular normal distribution.

These indices have been incorporated into many of the widely followed standards for map accuracy. For example, the NATO standards rate maps of scales from 1:25,000 to 1:5,000,000 as A, B, C, or D as follows:

A: CMAS = 0.5 mm (e.g., 12.5 m at 1:25,000)
B: CMAS = 1.0 mm (e.g., 25 m at 1:25,000)
C: CMAS determined but greater than 1.0 mm
D: CMAS not determined.

Many other mapping programs use CMAS as the basis for standards of horizontal or planimetric accuracy, and use similar values. Note, however, that CMAS is meaningful only for well-defined point features. Fisher (1991a) reviews many map accuracy standards, and Chrisman (1991) describes the background to the proposed U.S. Federal Spatial Data Transfer Standard (DCDSTF, 1988).

Lines and areas: the Perkal band

In vector databases lines are represented as sequences of digitized points connected by straight segments. One solution to the problem of line error would therefore be to model each point's accuracy, and to assume that the errors in the line derived entirely from errors in the points. Unfortunately this would be inadequate for several reasons. First, in digitizing a line a digitizer operator tends to choose points to be captured fairly carefully, selecting those that capture the form of the line with the greatest economy. It would therefore be incorrect to regard the points as randomly sampled from the line, or to regard the errors present in each point's location as somehow typical of the errors that exist between the true line and the digitized representation of it.

Second, the errors between the true and digitized line are not independent, but instead tend to be highly correlated (Keefer, Smith, and Gregoire, 1988). If the true line is to the east of the digitized line at some location along the line, then it is highly likely that its deviation immediately on either side of this location is also to the east by similar amounts. Much of the error in digitized lines results from misregistration, which creates a uniform shift in the location of every point on the map. The relationship between true and digitized lines cannot therefore be modeled as a series of independent errors in point positions.

One commonly discussed method of dealing with this problem is through the concept of an error band, often known as the Perkal epsilon band (Perkal, 1956, 1966; Blakemore, 1984; Chrisman, 1982). The model has been used in both deterministic and probabilistic forms. In the deterministic form, it is proposed that the true line lies within the band with probability 1.0, and thus never deviates outside it. In the probabilistic form, on the other hand, the band is compared to a standard deviation, or some average deviation from the true line. One might assume that a randomly chosen point on the observed line had a probability of 68% of lying within the band, by analogy to the percentage of the normal curve found within one standard deviation of the mean.

Some clarification is necessary in dealing with line errors. In principle, we are concerned with the differences between some observed line, represented as a sequence of points with intervening straight-line segments, and a true line. The gross misfit between the two versions can be measured readily from the area contained between them, in other words the sum of the areas of the spurious or sliver polygons. To determine the mismatch for a single point is not as simple, however, since there is no obvious basis for selecting a point on the true line as representing the distorted version of some specific point on the observed line. Most researchers in this field have made a suitable but essentially arbitrary decision, for example, that the corresponding point on the true line can be found by drawing a line from the observed point that is perpendicular to the observed line (Keefer, Smith, and Gregoire, 1988). Using this rule, we can measure the linear displacement error of any selected point, or compute the average displacement along the line.

Although the Perkal band is a useful concept in describing errors in the representation of complex objects and in adapting GIS processes to uncertain data, it falls short as a stochastic process model of error. An acceptable model would allow the simulation of distorted versions of a line or polygon, which the Perkal band does not, and would provide the basis for a comprehensive analysis of error.

Consider, for example, the problem of placing confidence limits on the area of a polygon, and assume that uncertainty in the position of the boundary is described by an epsilon band. It is easy to compute an estimate of the polygon's area, and to place upper and lower limits on area by taking the area inside the inner and outer edges of a band of width epsilon on either side of the boundary. However, these limits will greatly overestimate the true uncertainty, because in reality the observed position of the boundary will sometimes be inside the true position, and sometimes outside; the two types of distortion tend to cancel in a calculation of area. In order to estimate confidence limits correctly, we would need information on the pattern of distortions, such as a measure of their spatial dependence.

Alternatives to the Perkal band

The standard method of measuring accuracy in a classified image is to compare the classes assigned to a sample of pixels to the true classes on the ground ("ground truth") and express the result in the form of a table, the rows representing the assigned class and the columns the ground truth. Pixels that fall on the diagonal of the table are correctly classified; pixels that fall off the diagonal in a column are termed "errors of omission," since the true value is omitted from the assigned class; and pixels that fall off the diagonal in a row are termed "errors of commission," since they appear as false occurrences of a given class in the data. The contents of the table can be summarized in statistics such as the percentage of cells correctly classified, or using Cohen's kappa statistic, which allows for correct classification by chance (see Congalton, Oderwald, and Mead, 1983; van Genderen and Lock, 1977; van Genderen, Lock, and Vass, 1978; Greenland, Socher, and Thompson, 1985; Mead and Szajgin, 1982; Rosenfield, 1986; Rosenfield and Fitzpatrick-Lins, 1986).

Goodchild, Sun, and Yang (1992) describe a stochastic process that might be used to model error in soil, vegetation, or forest cover maps, and to propagate its effects. The model is based on a raster representation, in which every cell is assumed to have a vector of probabilities of membership in each of n classes. Such information might come from the use of a fuzzy classifier on a remotely sensed image, or from the misclassification matrix described above, or from the legend of a soil map from statements such as "class A is a mixture of 90% sand and 10% clay."

Fisher (1991b) has simulated the uncertainty implied by this model. If we interpret the probabilities in each pixel as indicative of the proportion of times a given pixel would be assigned a given class by the mapping process, then different realizations correspond to the range of uncertainty inherent in the process. However, it is unrealistic to assume that outcomes are independent in each pixel, since pixels are an artifact of the mapping process. Goodchild, Sun, and Yang (1992) introduce a level of spatial dependence between neighboring outcomes, and by varying the level are able to simulate a range of realistic scenarios regarding the spatial distribution of uncertainty. They use the model to propagate error from the database into confidence limits on the measurement of area, and the results of overlay.

Although suitable estimates of the probability vectors required by the model are readily available from fuzzy classifiers and many types of land cover maps, the estimation of the spatial dependence parameter presents more of a problem. While a single value of the parameter may be sufficient in some circumstances, in others it is likely that the parameter varies by class, and even regionally. Moreover, there is no simple intuitive interpretation of the parameter, and thus it seems unreasonable to expect the average GIS user to participate in its estimation. However, the parameter has a simple visual expression in the texture of each realization. Suppose, for example, that a large patch is predominantly class A, but is known to have inclusions of class B. At low values of spatial dependence, class B will appear in the realizations of this process as small, pixel-sized patches; as spatial dependence increases, the inclusions of B become larger and larger; and in the limit the entire patch is uniformly A in most realizations, but sometimes B.

This visual connection to texture provides a simple way of obtaining crude estimates of the value of the spatial dependence parameter(s). It also illustrates a more general point, that the visual nature of GIS may provide an intuitive access to some of the more sophisticated ideas of spatial statistics discussed in Part VI of this book.

Models for dot and pixel counting

One of the traditional methods of measuring area from a map involves placing an array of grid cells or dots over the area to be measured, and counting. In principle this process is similar to that of obtaining the area of a patch from an image by counting the pixels that have been classified as belonging to or forming the patch. The accuracy of the area estimate clearly depends directly on the density of dots or the pixel size, but so does the cost of the operation; so it would be useful to know the precise relationship in order to make an informed judgment about the optimum density or size.

The literature on this topic has been reviewed by Goodchild (1980). In essence two extreme cases have been analyzed, although intermediates clearly exist. In the first the area consists of a small number of bounded patches, often a single patch; in the second, it is highly fragmented so that the average number of pixels or dots per fragment of area is on the order of one or two. We refer to these as Cases A and B, respectively. Case A was first analyzed by Frolov and Maling (1969), and later by Goodchild (1980). The limits within which the true area is expected to lie 95% of the time, expressed as a percentage of the area being estimated, are given by:

$$1.03(kn)^{1/2}(SD)^{-3/4} \qquad (9\text{-}1)$$

where n is the number of patches forming the area; k is a constant measuring the contortedness of the patch boundaries, as the ratio of the perimeter to 3.54 times the square root of area ($k \sim 1$ for a circle); S is the estimated area; and D is the density of pixels per unit area.

The results for Case B follow directly from the standard deviation of the binomial distribution, since this case allows us to assume that each pixel is independently assigned. The limits within which the true area is expected to lie 95% of the time, again as a percentage of the area being estimated, are given by:

$$1.96(1 - S/S_0)^{1/2}(S/D)^{1/2} \qquad (9\text{-}2)$$

where S_0 is the total area containing the scattered patches.

Error rises with the 3/4 power of cell size in Case A, but with the 1/2 power in Case B. Thus a halving of cell size will produce a more rapid improvement in error in Case

A, other things being equal, because only cells that intersect the boundary of the patch generate error in Case A, whereas all cells are potentially sources of error in Case B, irrespective of cell size.

PROSPECTS

As we have seen, techniques for dealing with uncertainty in spatial data are much better developed in some areas than others. Uncertainties in point locations are comparatively easy to deal with, and a significant proportion of geodetic science is devoted to dealing with the issue of accuracy in control networks. Substantial progress has been made in modeling the error introduced by digitizing, and in predicting its effects on the measures of perimeter and area of objects (Griffith, 1989; Chrisman and Yandell, 1988). But the more general problem of modeling the accuracy of spatial data with respect to ground truth remains. In part this is because of the complexity of the processes involved, and in part because many ground truth definitions are themselves uncertain. But in addition, the manner in which spatial variation is represented in GIS makes a significant difference. Descriptors of uncertainty such as the Perkal band are convenient and intuitive, but are unable to satisfy the requirements of error models, and thus do not lead to methods for assessing uncertainty in analyses or models based on GIS. For example, an epsilon band description cannot be used to estimate uncertainty in the area of a polygon, because it contains no information on the spatial dependence of errors around the polygon's boundary.

In general, certain GIS data models are much more suitable than others as bases for the modeling of uncertainty. Digitized contours are highly unsuitable, because it is virtually impossible to connect the information on the accuracy of contour positioning provided by standard descriptions of topographic map quality with the parameter(s) of a suitable error model. Similarly, we have seen above how the Perkal epsilon band description of positional uncertainty in the polygon model fails to provide the basis of an adequate error model. In general, it appears that the raster data models are much better bases for error modeling in GIS. Representation of spatial variation through the use of irregular objects acts in many cases to make error modeling more difficult, simply because of the nature of these data models. But unfortunately choices among data models for spatial data representation are bound in strong traditions, implying that progress on the accuracy issue will necessarily be slow.

Because of interest in error models for spatial data, there has been substantial progress in recent years in understanding the propagation of uncertainty that occurs in spatial data handling. Newcomer and Szagin (1984) and Veregin (1989c) have discussed error propagation during simple Boolean overlay; Lodwick (1989; Lodwick and Monson, 1988) and Heuvelink et al. (1989) have analyzed the sensitivity of weighted overlay results to uncertainty in the input layers and weights; and Arbia and Haining (1990) have analyzed a general model of uncertainty in raster data. However, overlay is perhaps the most simple of the primitive spatial data handling operations. Much more research is needed on the effects of buffering (dilating) uncertain objects, and of changing from one data model to another (e.g., raster/vector conversion).

If traditional map data models tend to give a misleading impression of the reliability of data, it is not surprising that many cartographic products fail to make the user aware of uncertainties. The move to standards of data quality (see Chapter 35) is a significant step toward greater awareness of the reliability of spatial data, and includes concepts of lineage, consistency, and completeness as well as the more statistical issues discussed in this chapter. It is particularly important as GIS users continue to apply spatial data to purposes for which they may not have been designed.

This chapter has covered a range of material broadly connected with the issue of accuracy in spatial databases. As we noted at the outset, there are good reasons for believing that GIS technology will enhance our sensitivity to the nature of spatial data, and will increase the need to develop comprehensive and precise models of its errors and distortions. This enhanced understanding is likely to come not from study of the abstract objects that populate our maps and databases, but from the processes that created spatial variation in the first place and by which that variation was interpreted and modeled. In this sense GIS technology will help to fill many of the gaps that currently exist in our understanding of spatial differentiation and the processes that produce it.

It seems unlikely that the ideal accuracy-sensitive GIS that was outlined earlier in this chapter will ever become a reality. Its analog in simple, aspatial measurement, the statistical analysis system that tracks uncertainty as an attribute of each of its data elements and through each of its procedures, is also readily justified but equally far from implementation at this time. On the other hand, standard errors of estimate and confidence intervals are a common product of statistical systems, and it is not difficult to imagine similar products from GIS procedures.

One of the results of the information age is an erosion of the role of the central data collection agencies, and society has not yet begun to come to grips with a growing need to reorganize the entire system of public data provision. It is now possible to determine position at low cost and at any location using satellite receivers, and to do so with an accuracy that exceeds that of the USGS 1:24,000 map series or other readily available and authoritative sources. Population counts from local agencies are increasingly being used as alternatives to those of the Bureau of the Census. Accuracy is now increasingly seen as a variable attribute of spatial information, and as a component of its value to decision-makers. In some areas its associated costs can be dramatic, when disputes based on spatial information involve litigation. The complex relationships between accuracy and cost in the GIS area will be a challenging area for research for many years to come.

The measurement of data quality in spatial data products is dealt with in numerous publications and standards. Perhaps the most comprehensive view of quality standards is that developed in the U.S. by the Digital Cartographic Data Standards Task Force (DCDSTF, 1988), which provides a fivefold model for describing data quality, although it sets no precise numerical thresholds in any of the five categories and defines no standard measurement procedures.

This chapter has taken the view that measurement of data quality for the digital databases now being developed to support GIS is substantially different from measurement of the accuracy of cartographic features. In fact the two questions "Is this feature correct?" and "What is at this place?" require entirely different strategies. Some of these already exist, particularly in the feature-oriented domain, because they have been developed in support of traditional map-making. Others, particularly in the GIS-oriented domain, are still very limited and inadequate. Despite recent advances, and with certain very limited exceptions, current GIS still regard the contents of the database as perfectly accurate, and make no attempt to estimate the uncertainties that are inevitably present in their products.

To summarize the state of the art, we have good techniques for describing and measuring:

- The accuracy of location of a single point;
- The accuracy of a single measured attribute;
- The probability that a point at a randomly chosen location anywhere on a map has been misclassified;
- The effects of digitizing error on measures of length and area;
- The propagation of errors in raster-based area class maps through GIS operations such as overlay; and
- The uncertainty in measures of area derived from dot counting.

However, many of these are still in the domain of research literature, and we are in a very early stage of implementing them in standard practice.

From a practical point of view, the coming decade will see a greater awareness of the problems caused by uncertainty in the manipulation and analysis of geographical data, driven largely by the increasing availability of GIS. There will be increasing awareness also of the distinction between accuracy of cartographic features, and accuracy of GIS databases and products. We will see the emergence of systems equipped with tools for tracking uncertainties and the lineage of data through complex series of operations, allowing the user to visualize and interact directly with such information through graphic interfaces. And finally, we will see increasing pressure on the providers of data to supply information on data quality.

REFERENCES

Arbia, G., and Haining, R.P. (1990) Error propagation through map operations. *Unpublished paper*, Sheffield: University of Sheffield.

Beard, M.K., Buttenfield, B.P., and Clapham, S. (1991) Initiative 7 specialist meeting: visualization of spatial data quality. *Technical Paper 91-26*, Santa Barbara, CA: National Center for Geographic Information and Analysis.

Blakemore, M. (1984) Generalization and error in spatial databases. *Cartographica* 21: 131–139.

Bracken, I., and Martin, D. (1989) The generation of spatial population distributions from census centroid data. *Environment and Planning A* 21(8): 537–544.

Chrisman, N.R. (1982) Methods of spatial analysis based on errors on categorical maps. *Unpublished PhD Dissertation*, Bristol: Department of Geography, University of Bristol.

Chrisman, N.R. (1991) The error component in spatial data. In Maguire, D.J., Goodchild, M.F., and Rhind, D.W. (eds.) *Geographical Information Systems: Principles and Applications*, London: Longman, pp. 165–174.

Chrisman, N.R., and Yandell, B. (1988) A model for the variance in area. *Surveying and Mapping* 48: 241–246.

Congalton, R.C., Oderwald, R.G., and Mead, R.A. (1983) Assessing Landsat classification accuracy using discrete multivariate analysis statistical techniques. *Photogrammetric Engineering and Remote Sensing* 49(12): 1671–1678.

DCDSTF (Digital Cartographic Data Standards Task Force) (1988) The proposed standard for digital cartographic data. *The American Cartographer* 15(1): 9–140.

Fisher, P.F. (1991a) Spatial data sources and data problems. In Maguire, D.J., Goodchild, M.F., and Rhind, D.W. (eds.) *Geographical Information Systems: Principles and Applications*, London: Longman, pp. 175–189.

Fisher, P.F. (1991b) Modeling soil map-unit inclusions by Monte Carlo simulation. *International Journal of Geographical Information Systems* 5(2): 193–208.

Franklin, W.R. (1984) Cartographic errors symptomatic of underlying algebra problems. *Proceedings, International Symposium on Spatial Data Handling, Zurich*, University of Zurich, pp. 190–208.

Franklin, W.R., and Wu, P.Y.F. (1987) A polygon overlay system in PROLOG. *Proceedings, AutoCarto 8*, Falls Church, VA: ASPRS/ACSM, pp. 97–106.

Frolov, Y.S., and Maling, D.H. (1969) The accuracy of

area measurements by point counting techniques. *Cartographic Journal* 6: 21–35.

van Genderen, J.L., and Lock, B.F. (1977) Testing land-use map accuracy. *Photogrammetric Engineering and Remote Sensing* 43(9): 1135–1137.

van Genderen, J.L., Lock, B.F., and Vass, P.A. (1978) Remote sensing: statistical testing of thematic map accuracy. *Remote Sensing of Environment* 7(1): 3–14.

Goodchild, M.F. (1979) Effects of generalization in geographical data encoding. In Freeman, H., and Pieroni, G.G. (eds.) *Map Data Processing*, New York: Academic Press, pp. 191–206.

Goodchild, M.F. (1980) Fractals and the accuracy of geographical measures. *Mathematical Geology* 12: 85–98.

Goodchild, M.F. (1989) Modeling error in objects and fields. In Goodchild, M.F., and Gopal, S. (eds.) *Accuracy of Spatial Databases*, New York: Taylor and Francis, pp. 107–114.

Goodchild, M.F. (1991) Issues of quality and uncertainty. In Muller, J.-C. (ed.) *Advances in Cartography*, New York: Elsevier, pp. 113–139.

Goodchild, M.F., Sun, G., and Yang, S. (1992) Development and test of an error model for categorical data. *International Journal of Geographical Information Systems* 6(2): 87–104.

Greenland, A., Socher, R.M., and Thompson, M.R. (1985) Statistical evaluation of accuracy for digital cartographic data bases. *Proceedings, AutoCarto 7*, Falls Church, VA: ASPRS/ACSM, pp. 212–221.

Griffith, D.A. (1989) Distance calculations and errors in geographic databases. In Goodchild, M.F., and Gopal, S. (eds.) *Accuracy of Spatial Databases*, New York: Taylor and Francis, pp. 81–90.

Heuvelink, G.B.M., Burrough, P.A., and Stein, A. (1989) Propagation of errors in spatial modelling with GIS. *International Journal of Geographical Information Systems* 3(4): 303–322.

Keefer, B.J., Smith, J.L., and Gregoire, T.G. (1988) Simulating manual digitizing error with statistical models. *Proceedings, GIS/LIS '88*, Falls Church, VA: ASPRS/ACSM, pp. 475–483.

Lodwick, W.A., and Monson, W. (1988) Sensitivity analysis in geographic information systems. *Paper RP886*, Denver, CO: Department of Mathematics, University of Colorado at Denver.

Lodwick, W.A. (1989) Developing confidence limits on errors of suitability analyses in geographical information systems. In Goodchild, M.F., and Gopal, S. (eds.) *Accuracy of Spatial Databases*, New York: Taylor and Francis, pp. 69–78.

Mark, D.M., and Csillag, F. (1989) The nature of boundaries on area-class maps. *Cartographica* 21: 65–78.

Mead, R.A., and Szajgin, J. (1982) Landsat classification accuracy assessment procedures. *Photogrammetric Engineering and Remote Sensing* 48(1): 139–141.

Newcomer, J.A., and Szajgin, J. (1984) Accumulation of thematic map errors in digital overlay analysis. *The American Cartographer* 11(1): 58–62.

Perkal, J. (1956) On epsilon length. *Bulletin de l'Academie Polonaise des Sciences* 4: 399–403.

Perkal, J. (1966) On the length of empirical curves. *Discussion Paper No. 10*, Ann Arbor, MI: Michigan Inter-University Community of Mathematical Geographers.

Rosenfield, G.H. (1986) Analysis of thematic map classification error matrices. *Photogrammetric Engineering and Remote Sensing* 52(5): 681–686.

Rosenfield, G.H., and Fitzpatrick-Lins, K. (1986) A coefficient of agreement as a measure of thematic map accuracy. *Photogrammetric Engineering and Remote Sensing* 52(2): 223–227.

Veregin, H. (1989a) Accuracy of spatial databases: annotated bibliography. *Technical Paper 89-9*, Santa Barbara, CA: National Center for Geographic Information and Analysis.

Veregin, H. (1989b) A taxonomy of error in spatial databases. *Technical Paper 89-12*, Santa Barbara, CA: National Center for Geographic Information and Analysis.

Veregin, H. (1989c) Error modeling for the map overlay operation. In Goodchild, M.F., and Gopal, S. (eds.) *Accuracy of Spatial Databases*, New York: Taylor and Francis, pp. 3–18.

III

ENVIRONMENTAL SIMULATION MODELING

Louis T. Steyaert

The chapters within this section on environmental modeling fall into two broad categories. First, selected examples of environmental simulation models that illustrate state of the art modeling approaches within atmospheric, hydrologic, and ecologic sciences are presented. These within-discipline models are supplemented by four chapters that explicitly focus on the cross-disciplinary theme of "integrated/coupled systems" modeling. A good example of this integrated approach is the land–atmosphere interactions modeling at local and regional scales within global change research programs.

Most chapters illustrate the theme of spatially distributed environmental simulation models used within the context of a complex terrain and heterogeneous landscape environment. This theme represents a key link between environmental models and GIS technology. The models are just beginning to deal with spatial complexity. In contrast, the second set of chapters is focused on the subset of "land surface/subsurface" process modeling that is traditionally conducted strictly within the GIS as terrain analysis or in special 3D subsurface analysis environments. The issue is how to bring the two approaches together.

Many characteristics of these environmental simulation models are described in Chapter 3 (Steyaert) of the Introductory Section to this book. In general, the focus is on the mathematical modeling of dynamic, spatially distributed terrestrial processes. The conversion of these models for computer-based, numerical simulations by discretization and parameterization creates special scaling and subgrid process considerations. The models discussed in this section have potential for broad application to problem solving in land and water resource management, environmental risk, or special research programs involving global environmental change.

For the most part, this book emphasizes potential 2D connections between these environmental simulation models and GIS. These models were chosen because of their overall importance, yet they represent just a sampling of the ongoing types of modeling within the natural science disciplines. Furthermore, many important modeling topics were not addressed in order to limit the scope of this book. Models of natural ecosystems are given greater emphasis than more specialized models that incorporate human activities such as agriculture. In addition, the potential links between GIS and other important environmental modeling topics associated with upper atmospheric, oceanographic, coastal, or polar region processes were excluded, again partly to limit the scope of the book, and also because these processes may be less readily coupled to GIS.

The atmospheric models (Chapters 10–13) are by design focussed on interactions between the land surface and the free atmosphere, involving processes with time scales (or rates of change) on the order of seconds to hours. For example, three of the chapters focus on advanced models, termed land surface parameterizations, which account for quantitative water and energy exchanges between the land surface and the atmosphere in mesoscale meteorological models and atmospheric general circulation models. These processes include atmospheric forcing at the land surface—for example, solar radiation, precipitation, wind, temperature, and humidity, as well as the exchange of heat, momentum, and radiation fluxes.

The overview chapter (Lee et al., Chapter 10) provides a comparison of three advanced land surface parameterizations. The importance of land surface characteristics (for example, albedo and surface roughness) as major determinants of water and energy exchange processes are highlighted. Critical spatial scaling issues of relevance to GIS are also illustrated in the overview chapter, as well as in two case study examples (Avissar, Chapter 11; Skelly et al., Chapter 12). The final case study (Novak and Dennis, Chapter 13) describes the U.S. EPA regional air quality and acid deposition models in terms of spatial analysis requirements, results of a pilot GIS project, and the functional requirements for GIS and visualization within this type of modeling.

A broad range of hydrological modeling approaches is presented in Chapters 14–18, to illustrate the significant progress that has been made in linking GIS and hydrological models. Maidment (Chapter 14) provides a very detailed and in-depth review of the state of hydrological modeling, current modeling issues, current hydrological modeling with GIS, and opportunities for further integration. The integrated use of GIS and a three-dimensional finite element model for groundwater flow analysis is discussed by Harris et al. in Chapter 15. Richards et al. (Chapter 18) describe the development and selected applications of a graphical user interface that is used as an alternative to current GIS capabilities to support two-dimensional, unsteady, surface water flow analysis by the U.S. Army Corps of Engineers.

Two chapters in this section address the topic of water resource and watershed modeling. Both feature extensive use of GIS. Hay et al. (Chapter 16) describe the development of a prototype coupled systems modeling approach for the assessment of potential climatic change effects on water resources in the Gunnison River Basin of Colorado. This advanced modeling approach features the coupled use of a distributed parameter watershed model to account for water and energy exchange processes, the estimation of enhanced regional precipitation based on an orographic precipitation model, and the development of climate scenarios based on nesting these models within various atmospheric models. The use of GIS for data management, display, hydrologic response unit delineation, interactive water balance analysis, model output verification, investigation of scale effects, and project coordination is discussed. Gao et al. (Chapter 17) discuss the role of GIS as a tool to support a physically based distributed parameter model for smaller-spatial-scale watershed processes (precipitation, evapotranspiration, runoff production, and subsurface flow as a function of land surface properties). GIS is used to support parameterization and calibration of the models, integrate diverse spatial datasets, and visualize model output processes.

The chapters on atmospheric and hydrological modeling illustrate the contemporary theme of physically based, distributed parameter modeling in a complex terrain and heterogeneous landscape environment. GIS is not the driving force for these types of models. However, Chapters 19–21 (Moore et al., Engel et al., and Fisher, respectively) address this same topic, but from the perspective of land surface/subsurface modeling generally based on GIS and 3D analysis tools. Specifically, the overview chapter (Moore et al., Chapter 19) provides a comprehensive review of the existing interfaces between GIS and various hydrological models. The emphasis is on the role of terrain in modeling and analysis with GIS. The links between GIS and such models are examined in terms of model and database structures, the effects of scale and accuracy on GIS databases, scaling issues with models, and the requirements for adequate soil information. The role of GIS for database development, model parameterization, spatial analysis, and display as part of the AGricultural Non-Point Source Pollution (AGNPS) prediction modeling system for agricultural watersheds is described by Engel et al. (Chapter 20). GIS is effectively used to analyze model output results as part of a decision support system designed to assist with the management of runoff, erosion, and nutrient movement from agricultural watersheds. Several recommendations to expand links between GIS and the AGNPS model are suggested. The final chapter by Fisher addresses important 3D GIS related issues for subsurface analysis of hazardous waste sites. The current approaches for 3D GIS are reviewed, several applications illustrating the need for 3D are presented, and the requirements for time-dependent subsurface analysis tools are outlined.

Selected topics in ecological modeling are discussed in Chapters 22–25 by Hunsaker et al., Nisbet and Botkin, Lam, and Johnston. Chapter 22 (Hunsaker et al.) is designed to provide a very broad overview of ecological modeling and to lay the foundation for the integration of spatial ecological models with GIS. The chapter only addresses ecological modeling issues that involve some type of spatial dependency, in terms of three broad themes: terrestrial, fresh water aquatic, and marine ecosystems models. Each of these themes is examined in terms of historical modeling approaches and the current state of spatial modeling in ecology. Recommendations for enhanced integration of GIS and ecological models center on the need for extended GIS capabilities for data management including remote sensing data inputs, interactive landscape level modeling, sensitivity and error analysis, and environmental risk assessment. Johnston (Chapter 25) complements the overview chapter by presenting an excellent summary discussion on quantitative methods and modeling in community, population, and landscape ecology including current usage of GIS. Two case study applications are presented. Nisbet and Botkin (Chapter 23) describe the historical roots, analytical approach, and recent advances in the JABOWA forest growth model. A conceptual approach to incorporate GIS into dynamic spatial interactions modeling with JABOWA is proposed to test hypotheses involving windshed effects, pollution plume effects, and horizontal transport of materials by water, wind, or biological vectors. In Chapter 24, Lam presents a case study that illustrates the combined use of ecological modeling, GIS, and expert systems to study the effects of acid deposition on lake alkalinity, pH, and fish species richness.

The final set of chapters in this section is explicitly designed to highlight elements of integrated/coupled systems modeling, in particular advanced land–atmosphere interactions modeling at multiple time and space scales. Chapter 27 by Xue and Sellers illustrates the short time scales of seconds to hours as related to the Simple Biosphere Model (SiB). The use of this land surface model to study desertification and reforestation in the Sahel, Amazon deforestation experiments, and the U.S. 1988 drought is highlighted. In general, these fast response water and energy exchange processes help to drive biological and some soil biogeochemical processes characterized by intermediate time steps (or rates of change) of days to seasons. These intermediate time scale processes are illustrated by three different simulation models. First, the Forest–BioGeochemical Cycles model, a forest ecosystem dynamics model, is discussed by Nemani et al. (Chapter 28) in terms of the Regional Hydrological Ecosystems Simulation System (RHESSys). The Simulation and Utilization of Rangelands (SPUR) model, a general grasslands simulation model, is described in Chapter 29 by Hanson and Baker. Chapter 26, by Schimel and Burke, describes the Century model, which simulates the temporal dynamics of soil organic matter formulation and plant production for the study of biogeochemical cycling in grasslands ecosystems. These models are complemented by the JABOWA forest growth simulation model (Nisbet and Botkin, Chapter 23), which illustrates longer-term time scales associated with forest succession and the coupled systems modeling illustrated for watershed assessment in Chapter 16 by Hay et al.

These overview and case study papers provide insight into the types of environmental simulation models, their data and analysis requirements, their interrelationships to each other, the current usage of GIS with these models, and some of the opportunities for further integration. GIS has an important role to play in terms of the development of tailored data sets, including model calibration and parameterization. There are many important spatial analysis and scaling issues including opportunities for selective interactive analysis. The use of distributed parameter, multiscale models within a complex terrain and heterogeneous landscape environment seems very promising for GIS. As a final comment, GIS is not suggested as a panacea for this type of modeling, but GIS and allied technologies such as remote sensing do have an important role.

10

Atmospheric Modeling and Its Spatial Representation of Land Surface Characteristics

T. J. LEE,
R. A. PIELKE,
T. G. F. KITTEL, AND
J. F. WEAVER

Atmospheric modeling is a technique used to study atmospheric phenomena in a mathematical and physical framework. The behavior of the atmosphere can be described using a set of differential equations that describe external forcing to the system and the response of the atmosphere to that forcing. In order to solve the equation set, initial and boundary conditions must be provided. However, there is currently no comprehensive geographical data set of land surface characteristics (e.g., canopy structural and radiometric properties, soil hydrological properties) that can provide surface boundary conditions for these models. GIS have become a natural choice for the reconciliation and storage of such data available from different sources with different projections and spatial resolutions.

In the following sections, we briefly describe atmospheric models and provide an overview of three current land surface parameterization schemes and their data requirements. Finally, using observations and atmospheric model simulations, we demonstrate the need for accurate characterization of the land surface.

ATMOSPHERIC MODELS

Atmospheric models range in spatial scales from the entire globe to one kilometer or less. The time period of model simulations range from on the order of an hour for the smallest domain size simulations to centuries for global simulations. Each of these models, regardless of domain size, is based on conservation laws of physics. As summarized in Pielke (1984), these relations are expressed as:

- Conservation equation for velocity;
- Conservation equation for heat;
- Conservation equation for air;
- Conservation equations of water substance in its three phases;
- Conservation equations of other chemical constituents.

The conservation relationship for velocity is obtained from Newton's second law of motion, while the conservation equation for heat is from the first law of thermodynamics. The conservation equations permit an accounting of the changes in each of the variables in time, and are expressed as a simultaneous set of partial differential equations.

Mathematically these equations can be expressed as

$$\frac{d\varphi}{dt} = \frac{\partial \varphi}{\partial t} + \vec{V} \cdot \vec{\nabla}\varphi = S_{\varphi} \qquad (10\text{-}1)$$

where φ can represent dependent variables such as velocity, potential temperature, water substance, etc. In this equation, $\vec{V}$ is the three-dimensional velocity vector. The source–sink term S_{φ} represents the nonconservative physical mechanisms by which these dependent variables can be changed. For velocity, for example, the pressure gradient force and gravitational force represent terms in S_{φ}. For potential temperature, radiative flux divergence, heat changes due to phase changes of water substance, etc. represent S_{φ}.

In the creation of a numerical model, this simultaneous set of partial differential equations is integrated over grid volumes corresponding to the spatial separation within the model. The result is a set of partial *difference* equations that have correlation terms to represent subgrid-scale processes. Mathematically, this integration can be represented as

$$\frac{\partial \overline{\varphi}}{\partial t} = -\vec{\nabla} \cdot \vec{\nabla}\overline{\varphi} - \left(\overline{\overline{\vec{V}} \cdot \vec{\nabla}\varphi''} + \overline{\vec{V}'' \cdot \vec{\nabla}\overline{\varphi}} + \overline{\vec{V}'' \cdot \vec{\nabla}\varphi''}\right) \tag{10-2}$$

where the overbar represents the grid volume average and the terms in parentheses represent the subgrid-scale fluxes. The dependent variable with the double primes represents the subgrid-scale values of the variable.

It is important to recognize that in meteorological numerical models, because the equations are discretized, only a portion of these models are explicitly based on fundamental physical quantities. These are:

- The pressure gradient force;
- Advection;
- Gravitational force.

The Coriolis force is also a fundamental concept, but it only appears because the meteorological equations are defined with respect to the rotating earth.

The quantities that are *parameterized* (and, in general, are not necessarily based on fundamental concepts) are:

- Cumuliform clouds;
- Stratiform clouds;
- Long- and short-wave radiative flux divergence;
- Subgrid-scale turbulent and coherent circulation fluxes;
- Soil and vegetation effects.

There are specific named subsets of atmospheric numerical models. These include, from smaller to larger domain size:

- Large eddy simulation (LES) models;
- Boundary layer (BL) models;
- Cumulus cloud models;
- Cumulus field models;
- Mesoscale models;
- Regional models;
- Hemispheric models;
- Global models.

The smallest-scale models (LES, BL, and cumulus cloud and cumulus field models) usually use the more general version of the conservation equations. Mesoscale and large-scale models often utilize the hydrostatic equation to substitute for the vertical velocity conservation equation. Regional, hemispheric, and global models are often used routinely for numerical weather prediction (NWP). For instance, the U.S. National Weather Service currently creates prognosis meteorological fields twice daily using a regional model referred to as the Nested Grid Model (NGM). The European Centre for Medium Range Weather Forecasting (ECMWF) performs model forecasts out to 10 days or so using the ECMWF global forecast model. General circulation models (GCMs) are a form of global models that up to the present have been integrated using coarser model resolution than the NWP global models. A recent summary of GCM modeling techniques is presented in Randall (1992).

The term *resolution* has been used in a wide variety of peer-reviewed publications to refer to the grid increment used in a model. For example, general circulation models (GCMs) are said to have a resolution of about 400 km by 400 km when that scale more appropriately refers to the horizontal grid mesh.

From sampling theory it is well known, however, that at least two grid increments are required to represent data. Real information at scales smaller than two grid increments is erroneously aliased to larger scales. An illustration of this is presented in Pielke (1984, Figure 10.7). Models such as GCMs, however, require additional grid resolution to simulate meteorological processes adequately as a result of serious computational inaccuracies at scales less than four grid increments (e.g., see Tables 10.1, 10.2, and 10.3 in Pielke, 1984). Some investigators suggest even more grid increments are needed for adequate simulations.

Using these clarifications, resolution within a numerical model should refer to at least four times the grid interval. For instance, a GCM with 400 km by 400 km horizontal grid increments would have a resolution of no less than 1600 km by 1600 km. Diagnostic data (e.g., terrain) with sampling at a 400 km interval would have a resolution of no better than 800 km. Information that has spatial resolution smaller than the resolvable scales, however, is not completely lost. Using statistical techniques, some of this information can still be captured. This is commonly referred to as "subgrid-scale parameterization," as briefly discussed earlier in this section. Subgrid-scale land surface heterogeneities, for instance, should be parameterized so that the integral effect on the atmosphere of these subgrid-scale land surface features can be accounted for.

Atmospheric motion is fueled by energy received from the sun, and a majority of this energy is first absorbed by the surface and then transferred to the atmosphere through surface turbulent exchange processes. Consequently, atmospheric processes are sensitive to surface characteristics. For example, a simple sensitivity analysis (Pielke et al., 1993) showed that a small change of surface albedo can result in a change of equilibrium atmospheric temperature as large as the proposed greenhouse warming effect. Given human ability to alter surface characteristics, we expect that landscape changes will have a large impact on global climate and weather. In the next section we overview three current land surface parameterizations that are used in atmospheric models to describe surface exchange processes. In the long term, we expect to upgrade these surface schemes to simulate the response of land cover (i.e., vegetation) to changes in atmospheric conditions.

OVERVIEW OF LAND PARAMETERIZATION SCHEMES USED IN NUMERICAL MODELS

Since the Earth's surface is the only natural boundary of the atmosphere, to include a surface representation scheme in numerical weather/climate prediction models is not a new idea. In his famous work on numerical weather prediction, Richardson (1922) noted:

> The atmosphere and the upper layers of the soil or sea form together a united system. This is evident since the first meters of ground have a thermal capacity comparable with 1/10 that of the entire atmospheric column standing upon it, and since buried thermometers show that its changes of temperature are considerable. Similar considerations apply to the sea, and to the capacity of the soil for water.

Richardson then went on to discuss the possible treatments of three principal surface covers, namely the sea, bare soil surface, and vegetation covered surface. For land surface, Richardson considered the motion of water in soil, the transfer of heat in soil, and evapotranspiration. Analogous to the electric conductance, the rate of transpiration is proportional to the stomatal conductance and the vapor pressure difference between the intercellular space and the canopy air. He described the physics in the soil–vegetation–atmosphere continuum as:

> Leaves, when present, exert a paramount influence on the interchanges of moisture and heat. They absorb the sunshine and screen the soil beneath. Being very freely exposed to the air they very rapidly communicate the absorbed energy to the air, either by raising its temperature or by evaporating water into it. A portion of rain, and the greater part of dew, is caught on foliage and evaporated there without ever reaching the soil. Leaves and stems exert a retarding friction on the air.

It has been 70 years since Richardson published his book on numerical weather prediction, and the idea of treating the exchange processes in the soil–vegetation–atmosphere continuum is still the same. A similar modeling concept is still widely used except that many of the detailed physical and biophysical processes have become understood over the years. New model parameters have been introduced. For example, the relation between the leaf area index (LAI) and the stomatal conductance (Jarvis and McNaughton, 1986) and between the root zone water stress and the evapotranspiration (ET) (Kramer, 1949) are parameterized in recent land surface models. It should be noted that even the ideas of introducing the LAI and the root zone water stress were briefly mentioned in Richardson's book.

Three land surface parameterization schemes, which are currently used in numerical weather/climate prediction models, are overviewed in this paper. The simple biosphere model (SiB; Sellers et al., 1986) and the biosphere–atmosphere transfer scheme (BATS; Dickinson et al., 1986) are used in general circulation models (GCMs). The land ecosystem–atmosphere feedback model (LEAF; Lee, 1992) is used in a smaller-scale model (the regional atmospheric modeling system; RAMS).

Model structure

As mentioned in the previous section, each of the land surface parameterization schemes currently used in atmospheric models have virtually the same model structure. Following Richardson's "vegetation-film" and Deardorff's (1978) "big-leaf" concept, the three models (BATS, LEAF, and SiB) each introduce a layer of vegetation that interacts with the atmosphere. Notice that although SiB has a separate layer of ground cover vegetation, it is treated as part of the ground (no prognostic equations are introduced to describe the thermal and hydrological properties of this layer). Besides, the recent simplified simple biosphere model (SSiB; Xue et al., 1991) has eliminated the second layer of vegetation and still obtains nearly the same results as SiB.

For this layer of vegetation, averaged quantities, such as wind speed, thermal capacity, exchange coefficients, and radiative extinction coefficient, are utilized so that the detailed flow structure and the interception of radiation in the canopy are not resolved. However, prescribed wind profile and radiation distributions are used to calculate these averaged quantities. This type of model has been classified as a "greenhouse canopy" model by Goudriaan (1989) so that, in addition to the "big leaf," the impact of the canopy air is also parameterized. Figure 10-1 shows schematically the structure of the three models. In this figure and the following sections, the notation used in SiB (Sellers et al., 1986) is also used for BATS and LEAF for ease of discussion. The variables H and λE are sensible and latent heat fluxes, T is temperature, e is water vapor pressure, Ψ is water potential, and r indicates a resistance function. Subscripts r, a, g, c, and s denote variables at different locations, namely the atmospheric reference level, the canopy air at the zero plane displacement height, the ground-level vegetation coverage, the canopy, and the soil surface, respectively. Three aerodynamic resistance functions are used: $\bar{r}_b$ is a bulk boundary layer resistance,

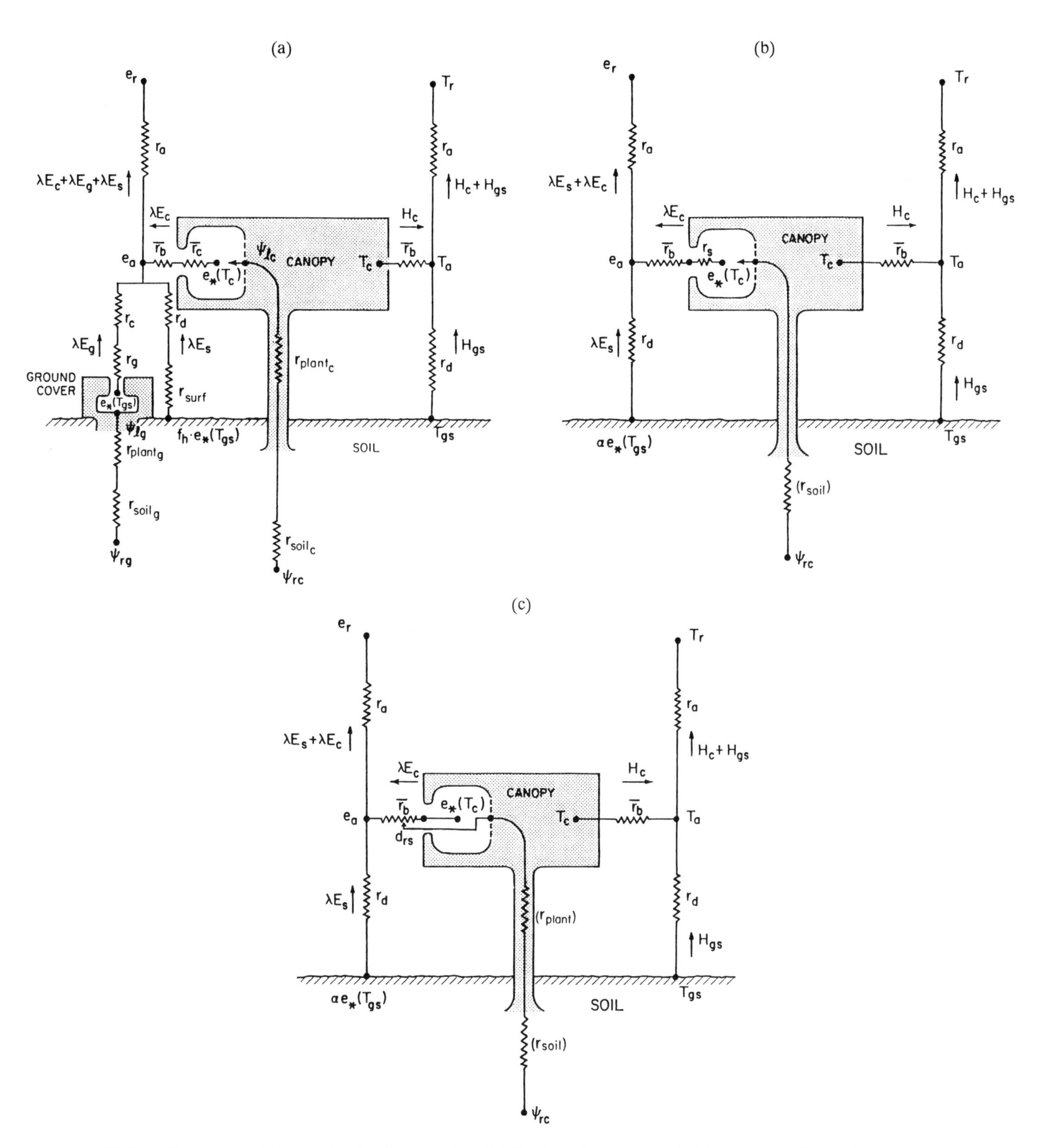

Figure 10-1:(a) Framework of the simple biosphere(SiB) model. The transfer pathways for latent and sensible heat flux are shown on the left- and right-hand sides of the diagram, respectively (from Sellers et al., 1986). Symbols are described in the text. (b) same as (a) but for the biosphere–atmosphere transfer scheme (BATS). (c) same as (a) but for the land ecosystem atmosphere feedback model (LEAF).

which is a resistance function between leaves and the canopy air, r_a is a resistance function between canopy air and the atmospheric reference level, and r_d is a resistance between the ground and the canopy air. In Figures 10-1b and c, the resistance functions in parentheses indicate the functions are implicitly embedded in other functions.

For example, in Figure 10-1c the resistance functions for water transport between the roots and the canopy, that is, r_{plant} and r_{soil}, implicitly exist in the formulation of relative stomatal conductance d_{rs}. The bulk stomatal resistance is noted as $\bar{r}_c$ in these figures.

It is not too difficult to see that these models have a similar structure, as shown in Figure 10-1. The canopy and the ground first exchange sensible and latent heat with the surrounding air and then the canopy air exchanges heat with the boundary layer atmosphere. We view this two-step exchange process as a major improvement to the "big-leaf" model in that the surface heat and moisture fluxes from the vegetation to the atmosphere are regulated by the heat and moisture capacity of the canopy air. Also, the water flow from the root zone to the surface of the leaves is regulated by the soil resistance, the plant resistance, and the bulk stomatal resistance. Although the three models show a similar model structure, it is the parameterization of these resistance functions that are different. A detailed comparison between the models is presented in the following sections.

Aerodynamic resistance

The primary goal of the existing soil–vegetation models is to provide a realistic boundary forcing to the atmosphere. This is accomplished through parameterizing the momentum, sensible heat, and latent heat fluxes. A common practice in parameterizing these fluxes is to utilize the resistance formulation:

$$\text{flux} = \frac{\text{potential difference}}{\text{resistance}} \tag{10-3}$$

Employing the Monin–Obukhov surface-layer similarity theory, the momentum flux (or the shear stress) in all three models is in the form of:

$$\tau = \rho C_D u^2 = \rho u_*^2 = \rho \left[\frac{ku}{\ln\left(\frac{z-d}{z_0}\right) + \Psi_M} \right]^2 \tag{10-4}$$

where τ is the shear stress, ρ is the density of air, u is the wind speed at a reference height z within the surface layer, C_D is the drag coefficient, u_* is the friction velocity, k is von Karman's constant, d is the zero plane displacement height, z_0 is the surface aerodynamic roughness, and Ψ_M is a stability adjustment function for momentum transport. In this formulation, we have assumed that the average flow speed at the displacement height is zero. The displacement height, d, and the surface roughness length, z_0, are prescribed in LEAF and BATS and are calculated in SiB by using "K theory." An additional assumption must be made, in order to calculate the displacement height, about the variation of the momentum transfer coefficient in the canopy. Sellers et al.(1986) noted that using "K theory" in the canopy may not be realistic and it also introduces new parameters. However, additional flexibility is obtained by using the SiB formulation so that the displacement height and surface roughness are actually varying with the density of the canopy (i.e., the LAI). This is important especially in climate simulations since the LAI varies with season. It may not be physically consistent to hold d and z_0 fixed while varying the LAI, such as applied, for example, in BATS. For short-range weather forecasts, where LEAF has been used, this may not be a problem, and we can supply these variables as model parameters and hold them fixed during the period of the model simulation (usually no more than several days).

For the sensible and latent heat fluxes, the boundary layer aerodynamic resistance function, r_a, is used in LEAF and SiB, so that:

$$H = \rho c_p \frac{T_r - T_a}{r_a} \tag{10-5}$$

$$\lambda E = \rho \frac{c_p}{\gamma} \frac{e_r - e_a}{r_a} \tag{10-6}$$

$$\frac{1}{r_a} = \frac{k^2 u}{\left[\ln\left(\frac{z-d}{z_0}\right) + \Psi_M\right]\left[\ln\left(\frac{z-d}{z_0}\right) + \Psi_H\right]}. \tag{10-7}$$

In these formulas, the variable Ψ_H is a stability adjustment function for heat and moisture transport, λ is the latent heat of vaporization, and γ is the psychrometric constant. The stability adjustment functions can be found, for example, in Paulson (1970), Businger et al., (1971), and Louis (1979). Paulson's scheme is used in SiB, while Louis' scheme is used in LEAF and SSiB due to the fact that the latter scheme is noniterative. A slightly different form for the boundary layer resistance function is used in BATS where

$$\frac{1}{r_a} = u \left[\frac{k}{\ln\left(\frac{z-d}{z_0}\right)} \right]^2 \Phi, \tag{10-8}$$

in which Φ is a stability correction function suggested by Deardorff (in Dickinson et al., 1986).

The bulk boundary layer resistance, $\bar{r}_b$, should depend upon the morphology of the vegetation and take into account the bluff-body effect of air flow around the leaves. Following the work by Goudriaan (1977), LEAF and SiB use the form:

$$\frac{1}{\bar{r}_b} = \int_0^{\mathrm{LAI}} \frac{\sqrt{u_f}}{C_s P_s} d\,(\mathrm{LAI}) \qquad (10\text{-}9)$$

where u_f is the wind speed in the canopy, C_s is a transfer coefficient that depends on the shape of the leaves, and P_s is a shelter coefficient. As noted by Sellers et al.(1986), the major difficulty in using this formula is the determination of the shelter coefficient, P_s. This coefficient is highly dependent on the morphology of the vegetation and can only be determined empirically (e.g., Thom, 1972). A much simpler form is used in BATS, where

$$\frac{1}{\bar{r}_b} = \mathrm{LAI}\; C_f \left(\frac{u_f}{D_f}\right)^{\frac{1}{2}}, \qquad (10\text{-}10)$$

in which C_f is an exchange coefficient and D_f is the dimension of the leaves in the wind flow. Although this formula is quite similar to the one used by LEAF and SiB, the effect of P_s on the bulk aerodynamic coefficient is not taken into account. This might be a drawback of this formula; however, on the other hand, the shelter coefficients, P_s, is one of the most ill-defined coefficient in LEAF and SiB due to the lack of measurements.

Knowing the bulk boundary layer resistance, the resultant sensible heat flux from the leaves to the canopy air can be written as:

$$H_c = \rho c_p \frac{T_a - T_c}{\bar{r}_b}, \qquad (10\text{-}11)$$

where T_a and T_c are the temperature of the canopy air and the leaves, respectively. The resultant latent heat flux will be discussed later with the stomatal resistance function.

The surface aerodynamic resistance, r_d, should also depend on the morphology of the vegetation. For example, a constant stress profile may be able to be used in a hardwood forest, where the wind profile may be logarithmic near the surface and below the elevated canopy. However, there may be no turbulent exchange of heat and moisture between the soil and the canopy air beneath a dense grass canopy since the wind reduces to zero in the canopy. For sparse canopy, it is even more difficult to parameterize this effect. The relative contribution from the soil should be very important when the density of the canopy is small.

Assuming the logarithmic wind profile is valid beneath an elevated canopy layer, Sellers et al., (1986) used the following form in SiB:

$$r_d = \frac{C}{u_f \varphi_H} \qquad (10\text{-}12)$$

where C is a surface-dependent constant and u_f is an average wind speed between the soil surface and the displacement height, and φ_H is a stability correction factor. The variable C can be calculated from a prescribed wind profile law. As mentioned, this formulation is designed for hardwood forests, but would be expected to fail when applied to other vegetation cover where the canopy is not elevated.

Due to the fact that this soil-to-canopy exchange process is not yet clearly understood, BATS chooses to use a simple formula to represent this effect without a fundamental physical base. Following Deardorff's (1978) work, a linear interpolation of resistance values between bare and vegetated surfaces is used, so that:

$$\frac{1}{r_d} = \frac{1}{r_a}\left[\,(1 - \sigma_f) + \frac{u}{u_f}\sigma_f\right] \qquad (10\text{-}13)$$

where σ_f is the fractional coverage of vegetation. Recently, Dickinson (1991, personal communication) indicated that a change has been adopted to make the transition between vegetation and bare soil smoother. It is observed, in this formula, that r_d approaches r_a when vegetation coverage is small and $r_a u_f / u$ when the coverage is large. Following Shuttleworth and Wallace (1985), a similar form is used in LEAF, with some variation:

$$r_d = r_{\mathrm{bare}} \max\left(\left(1 - \frac{\mathrm{LAI}}{3}\right), 0\right) + r_{\mathrm{close}} \min\left(\frac{\mathrm{LAI}}{3}, 1\right) \qquad (10\text{-}14)$$

where r_{bare} is the resistance function when the surface is bare and r_{close} is the resistance when the surface is covered by closed canopy. Instead of vegetation cover, LEAF uses LAI as an indicator of the coverage. It is assumed that there is no soil contribution when LAI is larger than 3. The advantage of using this formula is that it uses a realistic wind profile law to calculate the resistance functions, r_{bare} and r_{close}. The resultant turbulent heat fluxes from the soil are:

$$H_{gs} = \rho c_p \frac{T_a - T_{gs}}{r_d}, \qquad (10\text{-}15)$$

$$\lambda E_{gs} = \rho \frac{c_p}{\gamma} \frac{e_a - \alpha e_*(T_{gs})}{r_d}, \qquad (10\text{-}16)$$

where $e_*(T_{gs})$ is the saturation water vapor pressure immediately above the surface with a surface temperature T_{gs}. The variable α is an adjustment factor (or surface resistance) in determining the soil surface water vapor pressure.

Stomatal resistance

The major difference between vegetated and bare soil surface is the access to water in the soil. Over a bare soil surface, water is available for evaporation only from the top soil layers. In the presence of vegetation, water is also available from deep soil layers where roots are present. Although the transfer of water in the plant is mostly passive (meaning that water is not directly used by the photosynthesis process), it is responsible for the transport of nutrition from the root zone to the leaves where the photosynthesis process is taking place. The result of this transport process is that water is lost to the atmosphere through the opening (stomata) on leaves. It is known that water vapor pressure is at its saturation value in the intercellular space (Rutter, 1975). However, the vapor pressure at the surface of the leaves is regulated by the size of opening of the stomata, which is, in turn, a function of the environmental variables (e.g., photosynthetically active radiation, water stress, temperature, and CO_2 concentration).

Assuming this stomatal opening can be parameterized by a single resistance function, r_s, the water vapor flow from the intercellular space to the canopy is a two-step process. First, water vapor is transferred to the leaf surface:

$$\lambda E_1 = \rho \frac{c_p}{\gamma} \frac{[e_{\text{sfc}} - e_*(T_c)]}{\bar{r}_c} \tag{10-17}$$

where $\bar{r}_c$ is a canopy resistance function, or bulk stomatal resistance function, e_{sfc} is the water vapor pressure at the surface of the leaves, and $e_*(T_c)$ is the saturation water vapor pressure at the intercellular space with the temperature of the leaves, T_c. Following Jarvis and McNaughton (1986), $\bar{r}_c$ is defined as:

$$\frac{1}{\bar{r}_c} = \int_0^{\text{LAI}} \frac{1}{r_s} d(\text{LAI}) \tag{10-18}$$

Second, water vapor is transferred to the canopy air:

$$\lambda E_2 = \rho \frac{c_p}{\gamma} \frac{(e_a - e_{\text{sfc}})}{\bar{r}_b}. \tag{10-19}$$

If we further assume there is no accumulation of water vapor at the surface of the leaves (i.e., $E_1 = E_2$), we obtain:

$$\lambda E_c = \rho \frac{c_p}{\gamma} \frac{(e_a - e_*)}{\bar{r}_b + \bar{r}_c}. \tag{10-20}$$

In this equation E_c is the evaporation rate from the intercellular space to the canopy air. This equation is used in both BATS and SiB and is an absolute approach, where the magnitude of the stomatal resistance function is parameterized.

LEAF, on the other hand, adopted a relative approach, where the "potential evaporation" (maximum evaporation rate when there is no stomatal resistance) is evaluated first and then adjusted by a "relative stomatal conductance." This approach has been referred to as the "threshold concept," so that:

$$\lambda E_c = \rho \frac{c_p}{\gamma} \frac{d_{rs}}{\bar{r}_b} (e_a - e_*), \tag{10-21}$$

where d_{rs} is the relative stomatal conductance. This can be conceptually seen in Figure 10-1c, where the actual boundary aerodynamic resistance $\bar{r}_b$ is adjusted by a dial (i.e., d_{rs}).

Obviously, the stomatal resistance/conductance is still to be determined. Since transpiration is controlled by the stomata, a realistic parameterization of the stomatal opening is necessary in order to estimate the amount of latent heat flux correctly. Past studies showed that the stomata opening is affected by environmental variables (Allaway and Milthorpe, 1976; Jarvis, 1976; Avissar et al., 1985), and is parameterized in these models with the following forms:

BATS:
$$r_s = r_{s\,\text{min}} f_R f_S f_M f_V, \tag{10-22}$$

LEAF:
$$d_{rs} = \frac{d_{\text{min}} + (d_{\text{max}} - d_{\text{min}}) f_R f_T f_V f_\Psi}{d_{\text{max}}}, \tag{10-23}$$

SiB:
$$r_s = \left(\frac{a}{b + f_{\text{PAR}}} + c \right) \frac{1}{f_L f_T f_V}, \tag{10-24}$$

where a, b, and c are plant-related constants, f_{PAR} is a environmental adjustment factor for photosynthetically active radiation (PAR), f_T is an adjustment factor for leaf temperature, f_V is an adjustment factor for vapor pressure deficit, f_R is an adjustment factor for total solar radiation, f_Ψ is an adjustment factor for soil water potential, f_S is an adjustment factor for seasonal temperature, and f_M is an adjustment factor for soil water availability. The subscripts min and max indicate the minimum and maximum values.

It is evident that both LEAF and SiB use functions of leaf temperature, soil water potential, solar radiation, and vapor pressure deficit. BATS considered radiation, water availability, seasonal temperature, and vapor pressure deficit. Note that BATS uses only one plant-dependent parameter, $r_{s\,\text{min}}$, while LEAF and SiB use several. Although LEAF and SiB are more realistic in describing the stomatal response to environmental variables, a major difficulty in applying these models is to define these plant-related functions. Recently, Sellers et al. (1992) have modified the formulation of the stomatal conductance function to account for the rate of photosynthesis explicitly.

Model equations

If we can relate the resistance function to some environmental variables, as shown in the previous two sections, and assume the atmospheric condition is predicted, then the only unknown becomes the corresponding surface value. For example, the wind speed, air temperature, and water vapor pressure at the displacement height are required in order to estimate the amount of momentum, sensible heat, and latent heat fluxes. Since the surface roughness length and the displacement height are either calculated (SiB) or prescribed (LEAF and BATS), the momentum flux is immediately obtained by employing surface layer similarity theory. Sensible and latent heat fluxes, on the other hand, are still to be resolved. As mentioned before, the three models prescribe wind profiles in the canopy, which can either be a constant or varying with height. This will leave two variables to be determined, namely the water vapor pressure and the temperature of the canopy air (i.e., e_a and T_a in Figure 10-1). Assuming the canopy air has minimum heat or moisture storage, the turbulent heat fluxes gained from the soil and vegetation must be balanced by the loss to the atmosphere, so that:

$$H = H_c + H_{gs}, \text{ and} \tag{10-25}$$

$$\lambda E = \lambda E_c + \lambda E_{gs}. \tag{10-26}$$

Substituting Equations (10-11), (10-15), (10-16) and (10-20) or (10-21) into Equations (10-25) and (10-26), we can solve for the temperature (T_a) and the water vapor pressure (e_a) of the canopy air. However, we still need to determine the surface temperatures T_c and T_{gs}. This is done by solving the surface energy budget equation. Assuming a very small heat capacity in the canopy layer and the top soil layer, LEAF and SiB use a prognostic equation for the energy balance:

$$C_c \frac{\partial T_c}{\partial t} = Rn_c + H_c + \lambda E_c, \tag{10-27}$$

$$C_{gs} \frac{\partial T_{gs}}{\partial t} = Rn_{gs} + H_{gs} + \lambda E_{gs} + G, \tag{10-28}$$

where C_c and C_{gs} are heat capacity, in J m^{-2} K^{-1}, of the canopy and the top soil layer respectively. The variables Rn_c and Rn_{gs} are net radiation absorbed in the canopy and by the soil, and G is the ground heat flux to the deep soil layers. BATS, on the other hand, uses the balance equation at the vegetation surface:

$$0 = Rn_c + H_c + \lambda E_c, \tag{10-29}$$

and solves for the surface temperature, iteratively. For the soil surface, a "force–restore" method is used so that:

$$\frac{\partial T_{gs}}{\partial t} = \frac{c_1 G}{C_{gs}} - \frac{c_2 (T_{gs} - T_{gz})}{\tau_1} \tag{10-30}$$

where c_1 and c_2 are constants, τ_1 is a time scale for soil heat transfer, and T_{gz} is a deep soil temperature. Solving the prognostic equations, as in LEAF and SiB, has the advantage of saving computation time but is less accurate and also introduces new parameters, namely the heat capacities C_c and C_{gs}. Fortunately these heat capacities are usually small so that the prognostic equations can still simulate the fast response of the surface temperature to the radiative forcing, especially if a very small time step (< 10 sec) is used as an example in LEAF. Using the iterative scheme not only increases the computational time requirement, but the model can fail to converge especially because the coupling between the surface and the atmosphere is a highly nonlinear process. It is also very well documented that soil heat and moisture transfers respond to temperature and soil moisture content in a very nonlinear manner (Clapp and Hornberger, 1978). The use of an iterative scheme with a soil model increases the chance of model failure.

Radiation fluxes

From Equations (10-27) and (10-28) or (10-29) and (10-30), it is obvious that the surface energy budget is mainly forced by radiation. A correct representation of the radiative flux in the canopy is necessary. The optical properties of the canopy are summarized by Sellers et al. (1986) so that three radiation bands are considered in SiB:

1. Visible or PAR (0.4 – 0.72 μm): most of the energy in this region is absorbed by the leaves.
2. Near infrared (0.72 – 4 μm): radiation is moderately reflective in this region.
3. Thermal infrared (> 4 μm): leaves behave like a black body in these wavelengths.

For the visible and near infrared regions, SiB treats direct and diffuse radiation separately. This is because the radiative transfer in the canopy for these short waves is highly dependent on the angle of the incident flux. For this reason, SiB also considers the change of surface albedo with solar angle, and the values are higher in the morning and evening when the solar angle is low. However, it might not be necessary to consider the variation of surface albedo since the optical depth also increases when albedo increases and the error in the net radiation should not be too large.

The surface albedo, in BATS and LEAF, is not varying with time, and there is no distinction between direct and diffuse radiation.

Soil representation

The major difference between the "greenhouse canopy" and the "big-leaf" model is that vegetation is treated separately from the ground surface in the first approach. In order to close the surface energy budget equations in the previous section [e.g., Equations (10-27), (10-28))], a soil model must be used to obtain the soil surface temperature T_{gs} and also the soil heat flux G. BATS and SiB use

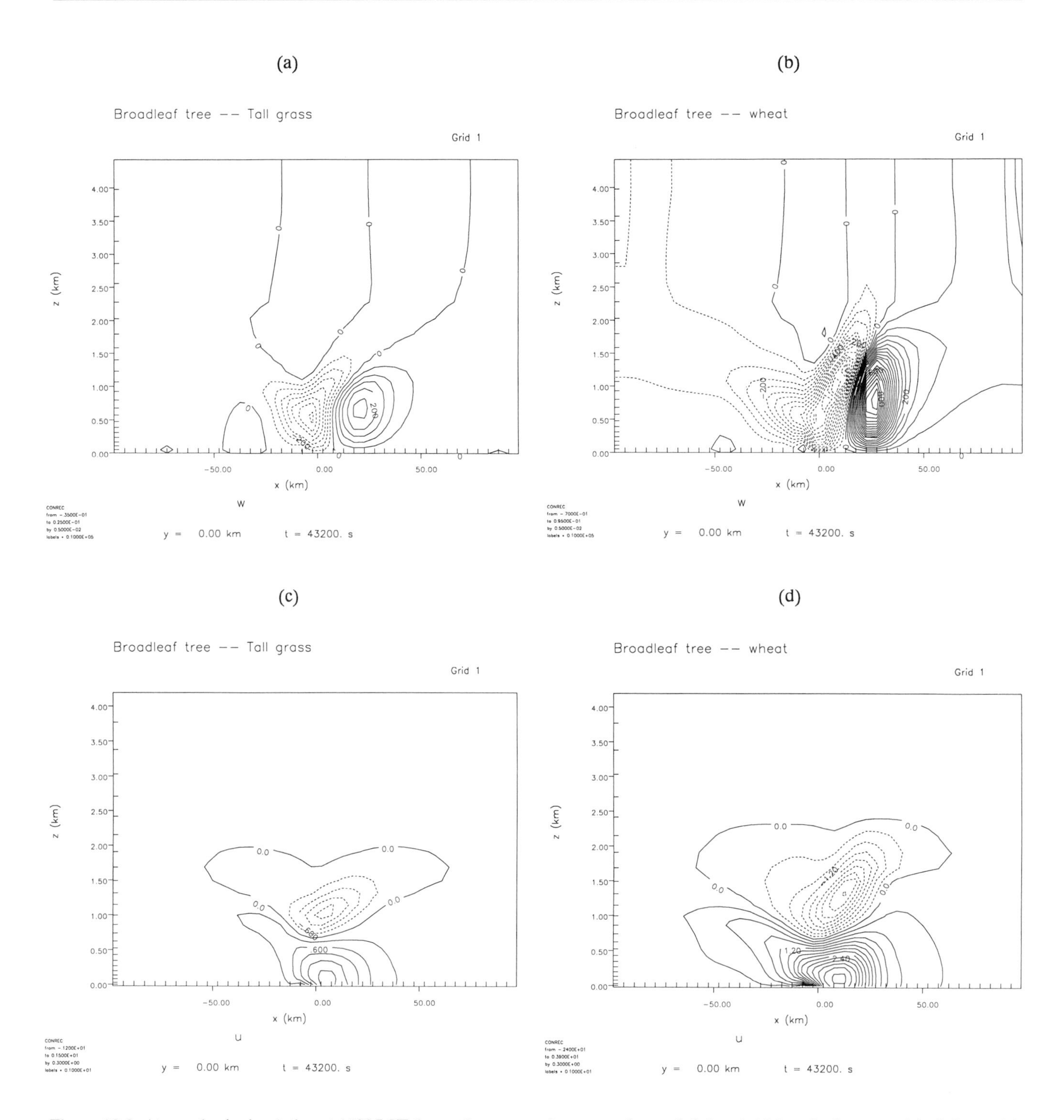

Figure 10-2: Atmospheric circulation at 1400 LST due to the contrast between a forest (left-hand side) and tallgrass prairie (left panels) and a forest and wheat farming (right panels). Contour intervals are (a) 0.5 cm s^{-1} for W; (b) 0.3 cm s^{-1} for U; (c)1 K for θ; and (d)1 g kg^{-1} for mixing ratio.

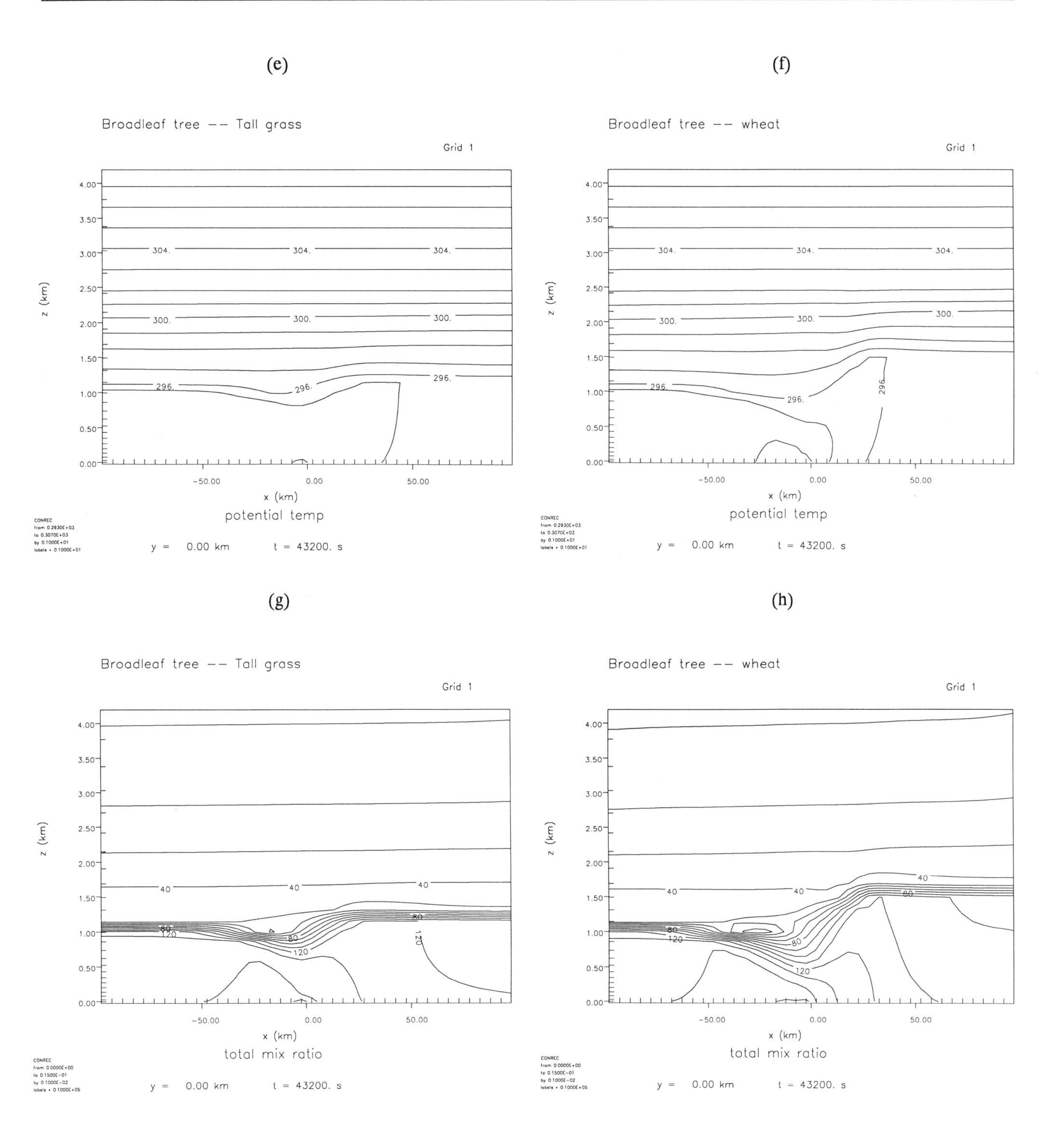
(e)
Broadleaf tree -- Tall grass
Grid 1
z (km)
x (km)
potential temp
y = 0.00 km
t = 43200. s
(f)
Broadleaf tree -- wheat
Grid 1
z (km)
x (km)
potential temp
y = 0.00 km
t = 43200. s
(g)
Broadleaf tree -- Tall grass
Grid 1
z (km)
x (km)
total mix ratio
y = 0.00 km
t = 43200. s
(h)
Broadleaf tree -- wheat
Grid 1
z (km)
x (km)
total mix ratio
y = 0.00 km
t = 43200. s

a force–restore method (Deardorff, 1978) that has three layers; LEAF uses a multilayer soil model (McCumber and Pielke, 1981) and has a detailed description for the transfer of moisture and temperature (Tremback and Kessler, 1985). The force–restore method is especially appropriate for climate models since it requires very little computer time. However, it fails to resolve the strong gradient of temperature and soil moisture potential close to the surface. Originally there were two soil layers used in the force–restore method such that the roots would be in the top soil layer with an averaged soil water content provided. However, the soil water potential can vary over several orders of magnitude from wet to dry so that it is very difficult to describe the appropriate average thermal and hydrological behaviors of the soil. It is also difficult to represent the soil water stress on vegetation correctly using this method. Due to the fact that this averaging process may lose important information on the water stress on vegetation, LEAF has chosen the use of a multilayer soil model. However, it is extremely difficult to initialize the multilayered soil model because of the lack of observations. Special measurements must be performed to obtain the needed information on soil moisture and temperature profiles.

Data requirement

In order to integrate an atmospheric model [e.g., Equation (10-1)] forward in time, initial and boundary conditions must be provided. For the initial conditions, an atmospheric model usually requires the state of atmosphere, which includes winds, temperature, pressure, humidity, and the state of soil, which includes soil moisture and temperature. For the surface boundary conditions, surface roughness and thermal and hydrological properties are essential. Soil texture, vegetation leaf area index, land-use, and land-cover data are necessary in order to estimate the surface thermal and hydrological properties. The percentages of surface coverage by different vegetation and landscapes are useful when a subgrid-scale land surface parameterization is used. DEM (digital elevation model) data can be used to define terrain characteristics. Clearly, it is the lower boundary condition information that GIS can contribute the most. In the future, it is expected that GIS can be linked to atmospheric modeling and analysis packages.

In this section, the major building blocks of three land surface parameterizations have been discussed. Many other model details are not covered in this chapter, for example, the treatment of dew, soil water flow, interception of precipitation by leaves, and evaporation from wet surfaces. Due to the fact that BATS and SiB are designed for use in GCMs, a complete hydrological cycle is available in these models. Both BATS and SiB can handle snow, ice, and dripping of water from leaves to the ground, while LEAF cannot. Generally SiB is more sophisticated and more realistic in the treatment of vegetation than BATS and LEAF. However, given the inhomogeneities in plant distribution and biophysical states within a model grid box (usually 400 × 400 km in GCMs and 20 × 20 km in RAMS), there is perhaps no need to use such a sophisticated model. Recently, Avissar (1992) proposed a statistical-dynamical approach in which the resistance functions are described by a distribution function, as an alternative to be used in numerical models.

THE IMPACT OF LANDSCAPE AND LAND USE ON ATMOSPHERIC CIRCULATIONS

Using these surface modeling schemes, the role of landscape type on planetary boundary layer structure and associated atmospheric circulations can be studied. Of specific interest is the difference in the partitioning of sensible and latent turbulent heat flux as a function of land use and landscape. Over irrigated areas and other areas of unstressed vegetation, boundary layer structure in the lower troposphere can be enhanced sufficiently to result in more vigorous cumulonimbus convection. Even slight differences in vegetation type, due to their different stomatal conductance and albedo characteristics, can cause substantial changes in the atmospheric response.

It has been shown in modeling (e.g., Ookouchi et al., 1984; Segal et al., 1988; Avissar and Pielke, 1989; Pielke and Avissar, 1990; Pielke and Segal, 1986) and observational studies (e.g., Segal et al., 1989; Pielke and Zeng, 1989; Pielke et al., 1990a) that the partitioning of sensible and latent heat fluxes into different Bowen ratios as a result of spatially varying landscape can significantly influence lower boundary layer structure and result in mesoscale circulations as strong as a sea breeze. Over and adjacent to irrigated land in the semiarid west, for example, enhanced cumulonimbus convection can result, as reported by Pielke and Zeng (1989). Schwartz and Karl (1990) document how the appearance of transpiring leaves on vegetation in the spring has the effect of substantially cooling (and thus moistening) the lower atmosphere. In their observational study, Rabin et al., (1990) demonstrate the effect of landscape variability on the formation of convective clouds. Dalu et al., (1991) evaluated, using a linear model, how large these heat patches must be before they generate these mesoscale circulations, while Pielke et al. (1991) present a procedure to represent this spatial landscape variability as a subgrid-scale parameterization in general circulation models.

These landscape variations result from a variety of reasons including:

1. Man-caused variations

   - Agricultural practice (e.g., crop type, land left fal-

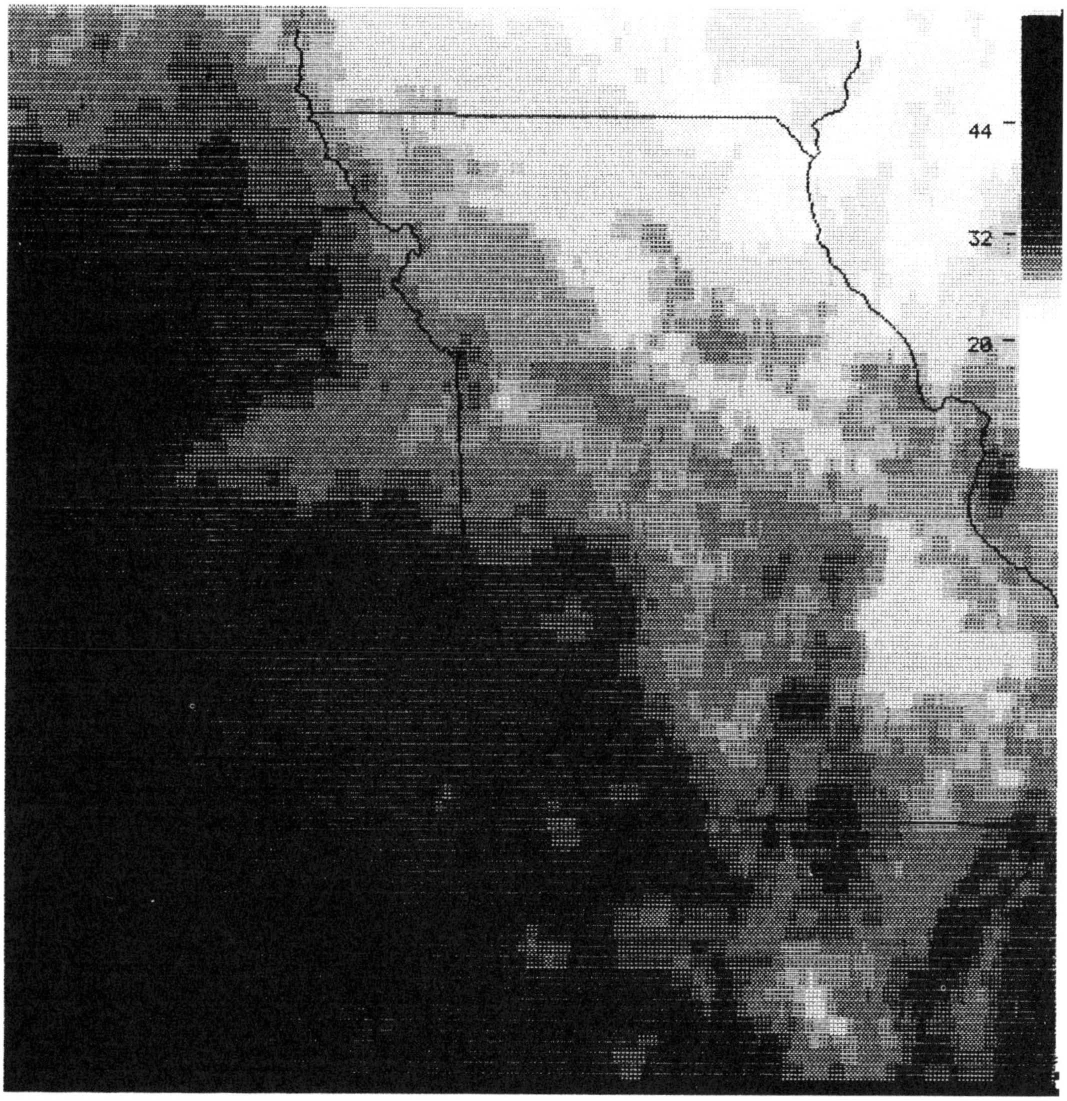

Figure 10-3: Plot of the highest measured surface skin temperature irradiance in °C as measured by the GOES geostationary satellite from July 24–August 31, 1986 for a region centered on eastern Kansas and western Missouri.

low, deforestation);

- Political practices of land subdivision (e.g., housing developments)
- Forest management (e.g., clearcutting);
- Irrigation.

2. Natural variations

- Fire damage to prairies and forests;
- Insect infestation and resultant damage to vegetation;
- Drought.

Plates 10-1 and 10-2 illustrate observed variations in photosynthetically active vegetation, as measured by NDVI (normalized difference vegetation index) satellite imagery over the Great Plains of the U.S. Presumably, active vegetation is transpiring efficiently during the daytime, while the other areas, with very low vegetation cover or vegetation under water stress, have most of their turbulent heat flux in sensible heat transfer. These two plates illustrate the large spatial and temporal variability of photosynthetically active vegetation and therefore suggest large corresponding variability in sensible heat flux (Pielke et al., 1991). GIS stored data must be of a sufficient spatial and temporal resolution to monitor these variations.

To illustrate the influence of landscape variations on weather, we have performed modeling and observational studies that demonstrate the major importance of vegetation and its spatial gradients on planetary boundary layer structure and mesoscale atmospheric circulations. Figure 10-2, for example, shows the results of a numerical model simulation with zero synoptic flow for the early afternoon in the summer for (1) a region in which a tallgrass prairie is adjacent to a forest region; and (2) the same as (1) except the tallgrass prairie is replaced by wheat.

For both simulations, the vertical velocity, east–west velocity, potential temperature, and mixing ratio fields are shown. Among the important results is the generation of a wind circulation as a result of the juxtaposition of the two vegetation types, and the change in the intensity of this circulation when the prairie is replaced by wheat. Higher transpiration over the forest, in conjunction with the thermally forced circulation, which can advect the elevated low-level moisture into the resultant low-level wind circulation, can be expected to result in enhanced convective rainfall when the synoptic environmental conditions are favorable. Changes in convective rainfall resulting from the conversion of the natural prairie to wheat also seem possible.

Satellite observations support the existence of large gradients in atmospheric conditions across a forest–grassland boundary in the United States, as illustrated in Figure 10-3 where the highest satellite-measured surface skin temperature irradiances are presented for a 5 week period in 1986 (July 24–August 31) as measured by the GOES geostationary satellite. Temperatures are over 10°C cooler over the forest as contrasted with prairie regions even short distances away (~30 km), as suggested to be the case by the modeling study.

CONCLUSIONS

Land-use and landscape patterns strongly influence atmospheric boundary layer structures and mesoscale circulations. For example, vegetation types as similar as tallgrass prairie and wheat cropland result in different atmospheric responses. This suggests that human modification of the land surface has had a major role in local climate, and, since such modifications have occurred worldwide, a global response to land-use changes should be expected. Through atmospheric modeling techniques, impacts of land-use change and its spatial heterogeneity on weather and climate systems can be studied and monitored. Consequently, there is a need for accurate characterization of the land surface for use as boundary conditions in atmospheric models. Important biophysical data for various vegetation species and hydrological data for soil states are necessary for correctly initializing these models. Clearly, given natural and human-induced spatial and temporal variability of land surface properties, model resolution needs to be greatly increased in order to include these features. GIS represents an important advancement in the refinement of global and regional land surface data sets that are necessary model inputs. Through improvements in model techniques and through the development of geo-referenced data sets, our future ability to model the global climate response to human-induced environmental change will be greatly enhanced.

ACKNOWLEDGMENTS

This work was sponsored under NSF Grant ATM-8915265 and NOAA/ERL Grant NA85RAH-05045. The authors are very grateful to Drs. R. E. Dickinson and P. J. Sellers for their helpful comments on our original draft. The authors would also like to thank Jeff Eidenshink from the EROS Data Center in South Dakota for his assistance. T. G. F. Kittel was supported by grants from NASA/FIFE, NSF Long-Term Ecological Research Program, and DOE Theoretical Ecology Program. The editorial preparation was ably handled by Dallas McDonald. The last section of this paper appeared in Pielke et al., (1990b).

REFERENCES

Allaway, W.G., and Milthorpe, F.L. (1976) Structure and functioning of stomata. In Kozlowski, T.T. (ed.) *Water Deficits and Plant Growth*, New York: Academic Press, pp. 57–102.

Avissar, R. (1992) Conceptual aspects of a statistical-dynamical approach to represent landscape subgrid-scale heterogeneities in atmospheric models. *Journal of Geophysical Research* 97: 2729–2742.

Avissar, R., Avissar, P., Mahrer, Y., and Bravdo, B.A. (1985) A model to simulate response of plant stomata to environmental conditions. *Agricultural and Forest Meteorology* 34: 21–29.

Avissar, R., and Pielke, R.A. (1989) A parameterization of heterogeneous land surfaces for atmospheric numerical models and its impact on regional meteorology. *Monthly Weather Review* 117: 2113–2136.

Businger, J.A., Wangaard, J.C., Izumi, Y., and Bradley, E.F. (1971) Flux profile relationships in the atmospheric surface layer. *Journal of the Atmospheric Sciences* 28: 181–189.

Clapp, R.B., and Hornberger, G.M. (1978) Empirical equations for some soil hydraulic properties. *Water Resources Research* 14: 601–604.

Dalu, G.A., Pielke, R.A., Avissar, R., Kallos, G., Baldi, M. and Guerrini, A. (1991) Linear impact of subgrid-scale thermal inhomogeneities on mesoscale atmospheric flow with zero synoptic wind. *Annales Geophysicae* 9: 641–647.

Deardorff, J.W. (1978) Efficient prediction of ground surface temperature and moisture, with inclusion of a layer of vegetation. *Journal of Geophysical Research* 83(C4): 1889–1903.

Dickinson, R.E., Henderson-Sellers, A., Kennedy, P.J., and Wilson, M.F. (1986) Biosphere–atmosphere transfer scheme for the NCAR community climate model. *NCAR Technical Note, NCAR/TN-275+STR, Boulder, CO*, 69 pp.

Goudriaan, J. (1977) *Crop Micrometeorology: A Simulation Study.* Wageningen, The Netherlands: Pudoc, 249 pp.

Goudriaan, J. (1989) Simulation of micrometeorology of crops, some methods and their problems, and a few results. *Agricultural and Forest Meteorology* 47: 239–258.

Jarvis, P.G. (1976) The control of transpiration and photosynthesis by stomatal conductance found in canopies in the field. *Philosophical Transactions of the Royal Society of London, Series B* 273: 593–610.

Jarvis, P.G., and McNaughton, K.G. (1986) Stomatal control of transpiration: scaling up from leaf to region. *Advances in Ecological Research* 15: 1–49.

Kramer, P.J. (1949) *Plant and Soil Water Relationships.* New York: McGraw-Hill, 349 pp.

Lee, T.J. (1992) The impact of vegetation on the atmospheric boundary layer and convective storms. *Ph.D. Dissertation*, Fort Collins, CO: Department of Atmospheric Science, Colorado State University, 137 pp.

Louis, J.-F. (1979) A parametric model of vertical eddy fluxes in the atmosphere. *Boundary-Layer Meteorology* 17: 187–202.

McCumber, M.C., and Pielke, R.A. (1981) Simulation of the effects of surface fluxes of heat and moisture in a mesoscale numerical model. Part I: Soil layer. *Journal of Geophysical Research* 86: 9929–9938.

Ookouchi, Y., Segal, M., Kessler, R.C., and Pielke, R.A. (1984) Evaluation of soil moisture effects on the generation and modification of mesoscale circulations. *Monthly Weather Review* 112: 2281–2292.

Paulson, C.A. (1970) Mathematical representation of wind speed and temperature profiles in the unstable atmospheric surface layer. *Journal of Applied Meteorology* 9: 857–861.

Pielke, R.A. (1984) *Mesoscale Meteorological Modeling*, New York: Academic Press, 612 pp.

Pielke, R.A., and Avissar R. (1990) Influence of landscape structure on local and regional climate. *Landscape Ecology* 4: 133–155.

Pielke, R.A., Dalu, G., Snook, J.S., Lee, T.J., and Kittel, T.G.F. (1991) Nonlinear influence of mesoscale landuse on weather and climate. *Journal of Climate* 4: 1053–1069.

Pielke, R.A., Lee, T.J., Weaver, J., and Kittel, T.G.F. (1990b) Influence of vegetation on the water and heat distribution over mesoscale sized areas. *Preprints, 8th Conference on Hydrometeorology, Kananaskis Provincial Park, Alberta, Canada, October 22–26, 1990*, pp. 46–49.

Pielke, R.A., Schimel, D.S., Lee, T.J., Kittel, T.G.F., and Zeng, Z. (1993) Atmosphere–terrestrial ecosystem interactions: implications for coupled modeling. *Ecological Modelling* (in press).

Pielke, R.A., and Segal, M. (1986) Mesoscale circulations forced by differential terrain heating. In Ray, P. (ed.) *Mesoscale Meteorology and Forecasting*, AMS, Chapter 22, 516–548.

Pielke, R.A., Weaver, J., Kittel, T., and Lee, J. (1990a) Use of NDVI for mesoscale modeling. *Proceedings Workshop on the "Use of Satellite-Derived Vegetation Indices in Weather and Climate Prediction Models"*, Camp Springs, Maryland. February 26–27, 1990, pp. 83–85.

Pielke, R.A., and Zeng, X. (1989) Influence on severe storm development of irrigated land. *National Weather Digest* 14: 16–17.

Rabin, R.M., Stadler, S., Wetzel, P.J., Stensrud, D.J., and Gregory, M. (1990) Observed effects of landscape variability on convective clouds. *Bulletin of the American Meteorological Society* 71: 272–280.

Randall, D.A. (1992) Global climate models: what and how. In Levi, B. G., Mafemeister, D., and Scribner, R. (eds.) *Global Warming: Physics and Facts,* American Physical Society, pp. 24–35.

Richardson, L.F. (1922) *Weather Prediction by Numerical Process.* London: Cambridge University Press, 236 pp.

Rutter, A.J. (1975) The hydrological cycle in vegetation. In Monteith, J.L. (ed.) *Vegetation and the Atmosphere. V. I: Principle*, New York: Academic Press, pp. 111–154.

Schwartz, M.D., and Karl, T.R. (1990) Spring phenology: nature's experiment to detect the effect of "green-up" on surface maximum temperatures. *Monthly Weather Review* 118: 883–890.

Segal, M., Avissar, R., McCumber, M.C., and Pielke, R.A. (1988) Evaluation of vegetation effects on the generation and modification of mesoscale circulations. *Journal of the Atmospheric Sciences* 45: 2268–2292.

Segal, M., Schreiber, M., Kallos, G., Pielke, R.A., Garratt, J.R., Weaver, J., Rodi, A., and Wilson, J. (1989) The impact of crop areas in northeast Colorado on midsummer mesoscale thermal circulations. *Monthly Weather Review* 117: 809–825.

Sellers, P.J., Berry, J.A., Collatz, G.J., Field, C.B., and Hall, F.G. (1991) Canopy reflectance, photosynthesis and transpiration, III: a reanalysis using enzyme kinetic-electron transport models of leaf physiology. *Remote Sensing of Environment* 42: 187–217.

Sellers, P.J., Mintz, Y., Sud, Y.C., and Dalcher, A. (1986) A simple biosphere model (SiB) for use within general circulation models. *Journal of the Atmospheric Sciences* 43: 505–531.

Shuttleworth, W.J., and Wallace, J.S. (1985) Evaporation from sparse crops—an energy combination theory. *Quarterly Journal of the Royal Meteorological Society* 111: 839–855.

Thom, A.S. (1972) Momentum, mass and heat exchange of vegetation. *Quarterly Journal of the Royal Meteorological Society* 98: 124–134.

Tremback, C.J., and Kessler, R. (1985) A surface temperature and moisture parameterization for use in mesoscale models. *Proceedings Seventh Conference on Numerical Weather Prediction*, Boston, MA: AMS, pp. 355–358.

Xue, Y., Sellers, P.J., Kinter, J.L., and Shukla, J. (1991) A simplified biosphere model for global climate studies. *Journal of Climate* 4: 345–364.

11

An Approach to Bridge the Gap Between Microscale Land-Surface Processes and Synoptic-Scale Meteorological Conditions Using Atmospheric Models and GIS: Potential for Applications in Agriculture

RONI AVISSAR

The Earth's surface absorbs over 70% of the energy absorbed into the climate system, and many physical, chemical, and biological processes take place there. Of particular importance are the exchanges of mass (notably water), momentum, and energy between the surface and the atmosphere. Also, the bulk of the biosphere lives there and plays a major role in many biogeochemical cycles, some of which may be affecting the chemical composition of the atmosphere and thereby the climate. The production and transport of atmospheric pollutants from various human activities is but one aspect of this question. Other natural processes also contribute to the modification of the radiative properties of the atmosphere: For example, most aerosols originate at the surface (Avissar and Verstraete, 1990).

The role of Earth's surface processes on climate and weather has been demonstrated with both general circulation models (GCMs) (e.g., Shukla and Mintz, 1982; Dickinson and Henderson-Sellers, 1988; Henderson-Sellers et al., 1988; Shukla et al., 1990) and mesoscale atmospheric models (e.g., Mahfouf et al., 1987; Segal et al., 1988; Avissar and Pielke, 1989; Pielke and Avissar, 1990; Avissar, 1991, 1992a). While GCMs simulate the entire Earth's atmosphere with a horizontal resolution (i.e., the horizontal domain represented by one element in the numerical grid) of 100,000 to 250,000 km^2, mesoscale models are used to simulate regional and local atmospheric phenomena with typical horizontal resolutions of 25 to 2,500 km^2. For instance, Shukla et al. (1990) showed with a GCM that turning the Amazonian tropical forests into agriculture could reduce precipitation and lengthen the dry season in this region. As a result, serious ecological implications would be expected. In this chapter, it will be demonstrated with a mesoscale model that the development of an agricultural area in an arid region or in a forested area (e.g., the Amazonian tropical forest) can produce atmospheric circulations stronger than sea breezes, and eventually convective clouds and precipitation. These apparently contradictory results produced by the two types of model emphasize the importance of scaling in land-atmosphere interaction studies.

These global and regional processes are due mainly to the fact that the input of solar and atmospheric radiation at the Earth's surface is redistributed very differently on dry and wet land. Indeed, on bare dry land, the absorption of this energy results in a relatively strong heating of the surface, which usually generates high sensible and soil heat fluxes. Typically, the soil heat flux represents 10 to 20% of the sensible heat flux. In that case, there is no evaporation (i.e., no latent heat flux). On wet land, however, the incoming radiation is used mainly for evaporation, and the heat flux transferred into the atmosphere and conducted into the soil is usually much smaller than the latent heat flux. When the ground is covered by a dense vegetation, water is extracted mostly from the plant root zone by transpiration. In that case, latent heat flux is dominant even if the soil surface is dry, but as long as there is enough water available in the plant root zone and plants are not under stress conditions. As a result, the characteristics of the atmosphere above dry land and wet (vegetated) land are significantly different. The differential heating of the atmosphere produced by the distinct surfaces creates a pressure gradient at some distance above the ground, which generates a circulation (e.g., Avissar, 1991, 1992a). Pielke and Avissar (1990) summarized a few observational evidences of such mesoscale circulations.

Hence, it is widely agreed that the parameterization of the Earth's surface is one of the more important aspects of climate modeling, and that it should be treated with care to ensure successful simulations with these models.

PARAMETERIZATION OF LAND SURFACE IN ATMOSPHERIC MODELS

Several parameterizations of the land surface have been suggested for application in atmospheric models. As recently reviewed and discussed by Avissar and Verstraete (1990), these parameterizations have improved from the prescription of surface potential temperature as a periodic heating function (e.g., Neumann and Mahrer, 1971; Pielke, 1974; Mahrer and Pielke, 1976) to more realistic formulations based on the solution of energy budget equations applied to the soil surface (e.g., Pielke and Mahrer, 1975) and, when present, vegetation layers (e.g., McCumber, 1980; Yamada, 1982; Sellers et al., 1986; Dickinson et al., 1986; Avissar and Mahrer, 1988a).

For instance, McCumber (1980) introduced a refined parameterization of the soil layer (based on the solution of governing equations for soil water and heat diffusion) and a bulk layer of vegetation following the approach suggested by Deardorff (1978). Yamada (1982) suggested a multilayer vegetation canopy representation, and Sellers et al. (1986) proposed a two-layer vegetation module to account for two different types of vegetation (e.g., trees and shrubs) that may be found in a single grid element of their GCM. These parameterizations all are based on the concept of "big leaf–big stoma," which implies that land is homogeneously covered by one (or more) big leaf within a grid element of the numerical atmospheric model. This big leaf usually has a single stoma that is sensitive, in the most sophisticated parameterizations, to the environmental conditions known to have an effect on the mechanism of stomata (i.e., solar radiation, temperature, humidity, carbon dioxide, and soil water potential in the root zone). This stoma controls the plant transpiration and, as a result, the development of the atmospheric planetary boundary layer (Avissar, 1992a). Thus, this stoma affects the climate system strongly.

At the scale of resolution of GCMs and mesoscale models, however, continental surfaces are very heterogeneous. This can be readily appreciated, for instance, by looking at maps of soil, vegetation, topography, or land use patterns (which are already generalizations of the truth!). The combined effect of this type of heterogeneity with the heterogeneity of precipitation results in large variations of water availability for evapotranspiration (ET) and, consequently, of surface energy fluxes. For instance, Wetzel and Chang (1988) estimated that for numerical weather prediction models with a grid spacing of 100 km or greater, the expected subgrid-scale variability of soil moisture may be as large as the total amount of potentially available water in the soil. Mahrer and Avissar (1985) demonstrated with a two-dimensional numerical soil model and with field observations that temperature differences as large as 10°C occur between the north-facing and south-facing slopes of small furrows produced by agricultural tillage. In the same study, they also showed that while the south-facing slope of the furrows dried up quickly, the north-facing slope remained wet. Even larger gradients can be expected between slopes of larger topographical features with the same inclination because no significant diffusion is possible between these slopes, in contrast to small furrows. Avissar and Pielke (1989) showed that this type of heterogeneity can generate circulations as strong as sea breezes under appropriate synoptic-scale atmospheric conditions.

Furthermore, the microenvironment of each leaf of most plant canopies is different from that of the other leaves in the same canopy and it varies constantly. This is due to the different position (i.e., inclination, orientation, level within the canopy, and shading) of the different leaves, which, as a result, absorb different amounts of solar radiation and, consequently, have different temperature, vapor pressure, and heat exchange (sensible and latent) with the canopy ambient air. In addition, the leaf orientation and inclination is constantly modified by the wind (which varies through the canopy) and by physiological stimuli, such as the tracking of the sun (i.e., heliotropism). Obviously, because of a rapid reaction of the stomata to the micro-environment (e.g., Squire and Black, 1981), a large variability of stomatal conductance and microenvironmental conditions is expected, even in otherwise homogeneous canopies, generating gradients of temperature and humidity within the canopy.

This expectation was confirmed by Avissar (1992b), who measured about 1,150 stomatal conductances in a homogeneous (i.e., spatially distributed in a similar configuration) potato field. Measurements were collected within a small area (about 30 m^2) in the middle of the field and under a range of atmospheric conditions that seemed to affect stomatal aperture in a similar way. He characterized the stomatal conductance variability measured in this field by a distribution with an average of 0.340 cm s^{-1}, a standard deviation of 0.242 cm s^{-1}, a positive skewness of 1.381, and a kurtosis of 4.752. The frequency distribution resulting from these observations was best represented by a log-normal distribution. Naot et al. (1991) reproduced the field experiment of Avissar (1992b) on a turf grass, and obtained a frequency distribution somewhat similar to that obtained in the potato field, which was also best represented by a log-normal distribution.

In addition to the problem of heterogeneity, state-of-the-art "big leaf–big stoma" parameterizations contain a large number of settable constants. For instance, McNaughton (1987) criticized the simple biosphere (SiB) scheme developed by Sellers et al. (1986) based on the fact that this scheme requires 49 settable constants, plus a leaf angle distribution function, to characterize the land surface. Obviously, even an extremely sophisticated micrometeorological field experiment allowing complete control of the plant environment could not provide exact values for these constants. Also, because these constants might be different for different growing periods and for the various plants found in the domain represented by a single grid element of the models, it is not clear how these constants could be combined to provide grid-scale "representative" values that could account for the nonlinearity of the involved processes. Obviously, this could represent an impossible task for the geographical information system scientific community.

Thus, "big leaf–big stoma" parameterizations cannot represent the subgrid-scale spatial variability that is found in the real world (e.g., McNaughton, 1987; Avissar and Pielke, 1989). To describe exactly the multiple processes occurring at the land surface, one would ideally build a model able to simulate each type of soil, each topographical feature, each leaf of each plant, etc., within the domain of interest. Unfortunately, we are still far removed from mastering the appropriate scientific knowledge and from having the computing facilities and appropriate data sets necessary for such a model to be used at the regional or global scale. Therefore, an alternative, more realistic method should be adopted.

In developing a parameterization of land-surface processes for atmospheric models, one wants to account for the contribution of the landscape heterogeneity, but in a global form that can be used at the grid scale of the model. Therefore, a parameterization based on a statistical-dynamical approach seems conceptually appropriate (Avissar and Pielke, 1989; Entekhabi and Eagleson, 1989; Avissar, 1991, 1992a; Famiglietti and Wood, 1991). With such an approach, probability density functions (pdfs) are used to represent the spatial variability of the various parameters of the soil–plant–atmosphere system that affect the input and redistribution of energy and water at the land surface. For instance, water availability for ET, surface roughness, topography, and leaf area index are probably the most important parameters to be represented by pdfs. They must be implemented in the model equations to produce, at the grid scale, the energy fluxes that force the atmosphere at its lower boundary. But since these pdfs are likely to be correlated, their covariance must be taken into consideration. Clearly, this presents a major complication, with computing time implications, that will need to be addressed before such a parameterization could be adopted in an operational model.

Conceptually, the statistical-dynamical approach presents several possible advantages (and a few shortcomings) for atmospheric numerical models (Avissar, 1992a):

1. In theory, each heterogeneity contributes to the entire pdf and, therefore, is represented at the grid scale. It affects the surface energy fluxes at the grid scale according to its probability of occurrence and in a nonlinear way. In practice, however, one is limited by the possibility to identify these heterogeneities.
2. The pdf of most land-surface characteristics is expected to be relatively stable. Indeed, it is reasonable to assume that the pdfs of topography and soil texture are constant. The vegetation leaf area index usually varies on a seasonal basis and, of course, so does its pdf. The surface aerodynamic roughness, which is affected by the vegetation cover, can be somehow associated with the leaf area index and, therefore, its pdf should vary on a seasonal basis as well.

   However, the pdf of surface wetness, which expresses the water availability for ET and, as emphasized, is one of the most important variables of the system, is unlikely to be stable at the local scale. Yet it is likely that this pdf could be characterized reasonably well as a function of topography, vegetation leaf area index, and grid-scale precipitation. For instance, Famiglietti and Wood (1991) found a good relation between topography and soil moisture, based on an analysis of observations collected as part of FIFE.
3. There is probably no need to represent in great detail the interacting mechanism of the soil–plant–atmosphere system at the microscale. This is because a fluctuation of the surface fluxes at one particular location within the grid might be counterbalanced by a fluctuation of these fluxes, in an opposite direction, somewhere else in the same grid.

   To illustrate this process, it is convenient to consider two identical patches of bare land located within the same grid. If these patches are distant enough from each other, at a particular time, one of them might receive some precipitation while the other might remain dry, generating completely different surface heat fluxes. At a later time, the patch that was dry might now get precipitation while the patch that was wet might have dried up. In order to represent the precipitation and the soil drying mechanism at the patch scale, one would need to use a sophisticated model that includes appropriate land–atmosphere interacting mechanisms. From a (large) grid-scale statistical perspective, however, we just need to know that during the time period considered, one land patch will produce a high latent heat flux, and another one will produce a high sensible heat flux. Where and how exactly the process occurs in the grid is probably not important.

   Obviously, this is a problem of grid scale. The larger the domain represented by a single grid, the more likely this process can be expected. But, on the other hand, the larger the grid spacing, the more complicated the parameterization of important atmospheric dynamical processes. The optimal scale has yet to be found.

   This aggregation procedure can save a tremendous amount of computing resources and avoid the parameterization of complex processes. For instance, Avissar (1992b) discussed the effect of integrating the plant stomatal conductance from the leaf to the canopy level. While it is well admitted that in order to describe the mechanism of plant stomata at the leaf level it is necessary to evaluate the impact of at least five environmental conditions (i.e., solar radiation, leaf temperature, vapor pressure difference between the leaf and the air, carbon dioxide concentration, and soil water potential in the root zone), he found that the daytime distribution of stomatal conductance within a potato canopy remained approximately constant on a monthly basis. Thus, even though the conductance of each stomata within the canopy fluctuates many times during such a time period, a complex model that simulates in detail these fluctuations is probably not necessary to describe the global canopy behavior and, a fortiori, to describe the vegetation behavior at the grid scale of atmospheric models.

It is important to emphasize that a stable pdf of land characteristics does not mean that the fluxes are constant. Clearly, the input of energy at the land surface that affects the intensity of the surface heat fluxes varies a lot, even on a diurnal basis.

4. Most pdfs are characterized by two to four parameters (e.g., mean, standard deviation, skewness, and kurtosis) and, therefore, could be identified relatively easily, as compared to the large number of settable parameters required in a big leaf scheme.
5. The shape of some of the pdfs might be used to provide valuable information for estimating subgrid-scale convective activity. Indeed, extended contrast of sensible heat flux might generate, under appropriate synoptic-scale conditions, local circulations (e.g., breezes) and convective clouds. For instance, if the pdf of surface wetness has a "U" shape, this indicates that a large part of the grid is dry and another large part of the grid is wet. As a result, thermally induced circulations are expected to develop and increase the convective activity within the grid, depending on the synoptic-scale wind speed forcing, the specific humidity, etc. If, on the other hand, the surface wetness distribution is characterized by a peak value, which indicates homogeneity of surface wetness within the grid, then not much subgrid-scale convective activity will be expected. There is obviously a problem of spatial scale of dry and wet land associated to this point. A systematic analysis of the impact of horizontal contrast of dry and wet land could provide an important contribution to the parameterization of this process.
6. A land-surface parameterization based on this approach produces the entire distribution of surface energy fluxes, temperatures, and humidities within the grid. This information might be of great interest for ecologists, agronomists, and others who use climate models for predictions of biospheric changes. Often, in such studies, knowledge about extreme environmental conditions is more valuable than the average conditions. This is because the biosphere tends to be more affected by extreme conditions.
7. The simulation of energy fluxes with a statistical-dynamical scheme requires the numerical integration over the various pdfs of land-surface characteristics (including their covariances). This integration is likely to cause an additional burden on the computing system, even though some of the processes simulated with big leaf schemes might not need to be represented any more.

Avissar (1992a) demonstrated the potential of this approach by comparing the heat fluxes produced by a primary version of a statistical-dynamical parameterization, to the corresponding fluxes produced by a "big leaf" parameterization. Large absolute and relative differences were obtained between the two schemes for many combinations of stomatal conductance pdfs and environmental conditions. Based on 1280 such comparisons (including five different pdfs of stomatal conductance), he found differences in sensible heat flux and latent heat flux as high as 167 W m^{-2} and 186 W m^{-2}, respectively, corresponding to relative differences of 2290% and 23%!

APPLICATIONS IN AGRICULTURE

Since the soil–plant–water system interacts strongly at all scales with the atmosphere, it is particularly interesting to evaluate the impact of extended agricultural activity on atmospheric processes, as well as the impact of extreme weather conditions on agriculture. It is important to emphasize that the case studies to be presented here are just a few selected examples from many potential applications of atmospheric models in agriculture.

Impact of agriculture on the atmosphere

The simulations presented in this section were produced with the mesoscale hydrostatic, incompressible, three-dimensional numerical model originally developed by Pielke (1974) and subsequently improved by Mahrer and Pielke (1978), McNider and Pielke (1981), Mahrer (1984), Avissar and Pielke (1989), and Avissar (1992a). This model is hereafter referred to as "the mesoscale model."

The mesoscale model was used to simulate the effects of the development of an agricultural area in (1) an arid region that consists essentially of bare, dry lands; and (2) a densely forested tropical area that has been cleared. The former case has been observed at various locations (e.g., Northern Colorado, the Sacramento Valley, and the area adjacent to the Salton Sea in California; along the Nile River in Egypt and the Sudan; Lake Chad in Africa; the northern Negev in Israel), and the latter illustrates the process of deforestation as it can be observed in the Amazon Basin in Brazil and is also frequent in midlatitudes. For instance, Plate 11-1 shows a satellite-derived image of such agricultural developments in the Amazon Basin.

The model was initialized with the parameters summarized in Tables 11-1 to 11-4. For both the arid and the tropical cases, the agricultural area was assumed to consist of a large number of randomly distributed patches of agricultural fields, built-up areas, bodies of water, and wastelands that cover 55%, 20%, 10%, and 15% of the total area, respectively. The simulated region was 240 km wide and was represented in the model by 40 grid elements with a grid resolution of 6 km. The agricultural area was 60 km wide and was located in the middle of the domain. It was flanked at its eastern and western boundary by a 90 km wide region of either arid land or forests, depending on the simulated case. The dynamics of the atmosphere was simulated up to a height of 10 km (i.e., roughly the tropopause).

The numerical integration for the simulations was started at 0600 a.m. (local standard time), which corresponds to the time when the sensible heat flux becomes effective in the development of the convective planetary boundary layer on sunny summer days. The spatial distribution of surface heat fluxes and surface temperature obtained after 8 hours of simulation (i.e., at 0200 p.m.) are presented in Figure 11-1a for the arid case and in Figure 11-1b for the tropical case. Figure 11-2 presents the meteorological fields (u, the west-east horizontal component of the wind parallel to the domain; w, the vertical component of the wind, θ, the potential temperature, and q, the specific humidity) obtained at the same time for both the arid (Figure 11-2a) and the tropical (Figure 11-2b) cases. The agricultural area is indicated in these figures by a

Table 11-1: Soil properties and initial conditions of the sandy loam soil type that was adopted for the simulations, as provided by the United States Department of Agriculture textural classes. The hydraulic properties are soil water content at saturation (η_s), soil water potential at which water content (η) departs from saturation (ψ_{cr}), hydraulic conductivity at saturation (K_{hs}), and b is an exponent in the function that relates soil water potential and water content.

Soil Property	Value
Density	1250 kg m^{-3}
Roughness	0.01 m
η_s	0.435 $m^3 m^{-3}$
ψ_{cr}	–0.218 m
K_{hs}	2.95 m day^{-1}
b	4.90
Albedo	0.20
Emissivity	0.95

Initial Profiles

	Arid Area		Tropical Area	
Soil Depth (m)	Temperature (K)	Wetness ($m^3 m^{-3}$)	Temperature (K)	Wetness ($m^3 m^{-3}$)
0.00	303	0.04	298	0.04
0.01	303	0.04	298	0.04
0.02	303	0.04	298	0.07
0.03	303	0.04	298	0.10
0.04	303	0.04	298	0.15
0.05	303	0.04	298	0.25
0.075	303	0.04	298	0.25
0.125	303	0.04	298	0.25
0.20	303	0.04	298	0.25
0.30	303	0.04	298	0.25
0.50	303	0.04	298	0.25
0.75	303	0.04	298	0.25
1.00	303	0.04	298	0.25

Table 11-2: Vegetation input parameters

Property	Forest	Agricultural Fields
Roughness	1.0 m	0.1 m
Leaf Area Index	5	3
Height	10.0 m	1.0 m
Albedo	0.15	0.20
Transmissivity	0.031	0.125
Emissivity	0.98	0.98
Initial Temperature	298 K	298 K

Root Distribution

Soil Depth (m)	Root Fraction
0.00	0.00
0.01	0.00
0.02	0.00
0.03	0.00
0.04	0.00
0.05	0.09
0.075	0.12
0.125	0.14
0.20	0.18
0.30	0.16
0.50	0.16
0.75	0.15
1.00	0.00

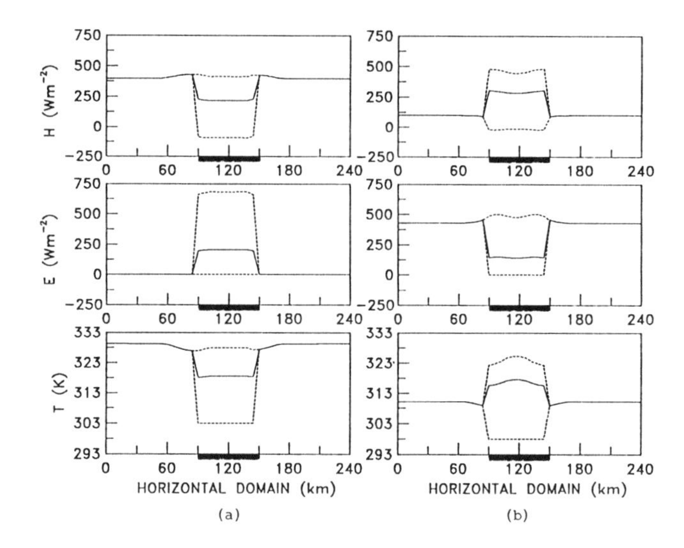

Figure 11-1: Horizontal variation of (i) sensible heat flux (H); (ii) latent heat flux (E); and (iii) temperature (T) at the surface of the simulated domain at 1400LST. The 60 km wide agricultural area (indicated by the dark underbar) consists of agricultural field (55%), built-up areas (20%), bodies of water (10%), and wastelands (15%). The surrounding areas are (a) bare and dry; and (b) covered by a dense unstressed vegetation. Dashed lines in the rural area delimit the range of values obtained from the different patches (from Avissar, 1991).

black underbar, and only the lower 3 km of the atmosphere (i.e., roughly the planetary boundary layer) are displayed in Figure 11-2.

The regional circulations depicted by the horizontal and vertical components of the wind in these simulations result from the differential heating produced by the different partitioning of the net radiative energy absorbed in the agricultural area and in the surrounding areas. This is illustrated by the relatively high sensible heat flux obtained in the arid regions (Figure 11-1a) as well as in the built-up areas and wastelands of the agricultural areas (Figures 11-1a and 11-1b), and by the low sensible heat flux and high evapotranspiration obtained in the vegetated part of the domains.

It is interesting to note that even though similar agricultural areas were considered for both simulations, the sensible heat flux distribution in the deforestation simulation exhibits higher values than those obtained in the simulation of the arid region (dashed lines in Figure 11-1 indicate maximum and minimum values obtained from the different land patches). This is due mainly to a stronger wind intensity and a lower atmospheric temperature (generating a higher temperature gradient between the land surface and the atmosphere) in the deforestation case (Figure 11-2). Reciprocally, the dry and warm atmosphere in the arid region (Figure 11-2a) generates a strong gradient of specific humidity between the land surface and the atmosphere, resulting in a very strong latent heat flux from the agricultural fields in the rural area (Figure 11-1a). For comparison, the sensible heat flux from the built-up patches (maximum values in Figure 11-1) in the tropical region exceeds by about 100 W m^{-2} that from the same built-up patches in the arid region. The latent heat flux from the agricultural fields (maximum values in Figure 11-1) in the arid region exceeds by about 200 W m^{-2} that from the same fields in the tropical region. These results emphasize the importance of the interactions between the land surface and the atmosphere as well as the benefits of the statistical-dynamical land-surface parameterization.

The faster heating rate of the dry land surface generates a vigorous turbulent mixing and an unstable, stratified planetary boundary layer, as can be seen from the field of potential temperature in Figure 11-2a. On the contrary, the slower heating rate of the transpiring forest in the deforestation simulation limits the development of the planetary boundary layer, which remains shallow above the forest (Figure 11-2b). The different patches in the agricultural area produce an intermediate heating. This creates a pressure gradient with the surrounding land-

Table 11-3: Atmospheric input parameters and initial conditions

Input Parameter	Value
Number of vertical grid points	19
Number of horizontal grid points	40
Horizontal grid resolution	6 km
Synoptic-scale background flow	0.5 m s^{-1}
Initial potential temperature lapse	3.5 K/1000 m

Specific Humidity (g kg–1)		
Height (m)	Arid Area	Tropical Area
5	10.0	25.0
15	10.0	25.0
30	10.0	25.0
60	10.0	25.0
100	10.0	25.0
200	10.0	25.0
400	10.0	25.0
700	10.0	25.0
1000	8.0	22.0
1500	6.0	17.0
2000	2.5	12.0
2500	1.5	7.0
3000	1.0	4.0
3500	0.5	3.0
4000	0.5	2.0
5000	0.5	1.0
6000	0.5	0.5
8000	0.5	0.5
10,000	0.5	0.5

Table 11-4: General model input parameters

Input Parameter	Arid Area	Tropical Area
Latitude	32°N	0°
Day of the year	August 5	September 21
Initialization time	0600 LST	
Integration time step	20 s	

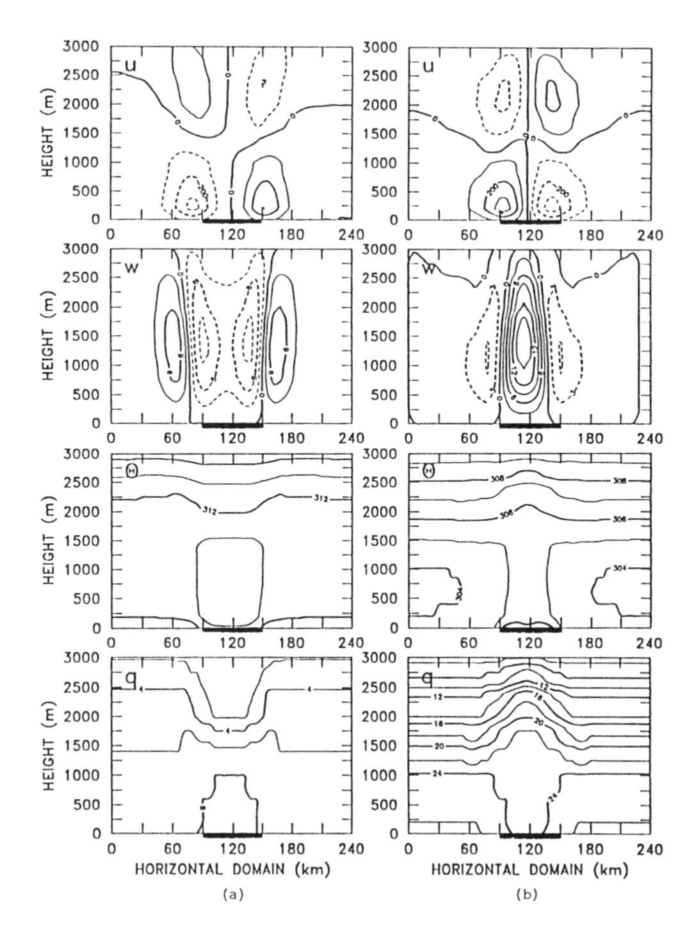

Figure 11-2: Vertical cross-section of the simulated domain at 1400 LST for (i) the horizontal wind component parallel to the domain (u) in cm s^{-1}, positive from left to right; (ii) the vertical wind component (w) in cm s^{-1}, positive upward; (iii) the potential temperature (θ) in K; and (iv) the specific humidity (q) in g kg^{-1}, resulting from the implementation of a 60 km wide agricultural area (indicated by the dark underbar) which consists of agricultural field (55%), built-up areas (20%), bodies of water (10%), and wastelands (15%), in (a) an arid area; and (b) a tropical area. Solid contours indicate positive values; dashed contours indicate negative values (from Avissar, 1991).

scape at some distance above the ground that generates the circulation from the relatively cold to the relatively warm area.

In the deforestation simulation, the transpiration from the forest provides a supply of moisture that significantly increases the amount of water in the shallow planetary boundary layer. This moisture is advected by the thermally induced flow, which strongly converges toward the agricultural area, where it is well mixed within the relatively deep convective boundary layer. This process may eventually generate convective clouds and precipitation under appropriate synoptic-scale atmospheric conditions.

Impact of atmosphere on agriculture

Radiative frost is one of the most severe weather conditions that affects agriculture in many parts of the world. Since various protective methods to reduce frost impact are available, refinements of frost forecasting methodologies should provide economical benefits (Avissar and Mahrer, 1988a,b). In the past, topoclimatologic surveys of air minimum temperature at 0.5 m height above the soil surface have been used to map relatively cold and relatively warm areas in an agricultural region of interest. These surveys were based on a statistical analysis of min-

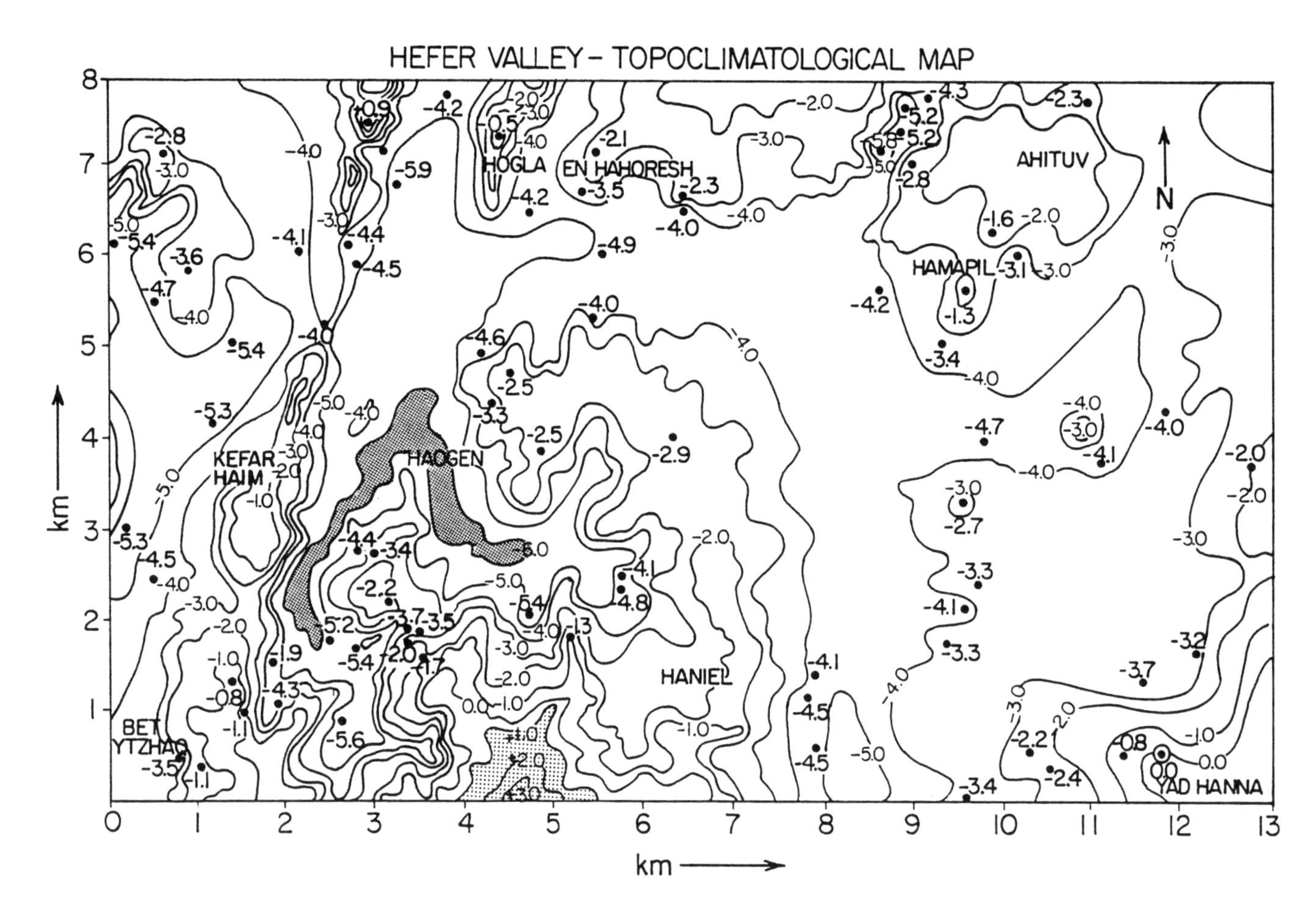

Figure 11-3: Topoclimatologic map of Hefer Valley (Israel) derived from that produced by the Meteorological Service of Israel, based on observations from 1965 to 1969 (from Avissar and Mahrer, 1988b).

imum temperatures measured at several locations within the studied area during various frost events. Therefore, these surveys required an extensive campaign of observations over a period of several years and, consequently, were tedious and expensive (e.g., Lomas and Gat, 1971).

Figure 11-3 illustrates the results of a topoclimatologic survey, which was carried out in the Hefer Valley (Israel) by the Meteorological Service of Israel. This is a map of the difference of the minimum temperature from the main station (Yad Hanna, at the lower right corner of the map) expected during radiative frost events. Dots indicate the location of the measurements, and the numbers next to them indicate the mean minimum temperature difference with the main station measured at these locations. Isotherms (i.e., the contours of temperature between the points of measurement) are drawn according to the variation of topography. A detailed description of the methodology and technique used for this survey is given in Lomas and Gat (1971) and summarized in Avissar and Mahrer (1988b).

Avissar and Mahrer (1988b) pointed out several problems associated with this technique. For instance, they demonstrated that topography was not clearly correlated with minimum temperature (R^2=0.13), thus questioning the reliability of the map contouring procedure. They pointed out that other land-surface characteristics, such as vegetation coverage, soil type, soil wetness, and wind speed, should have been considered for the extrapolation between points of measurement. Also, they emphasized that large temperature differences could be found between the air temperature at 0.5 m above the ground and leaf temperature, in particular under clear, calm, and cold nights, as is the case during radiative frosts. Of course, from an agricultural practical perspective, leaf temperature is more relevant than air temperature.

Avissar and Mahrer (1988a) suggested a numerical modeling approach as an alternative method to topoclimatologic surveys of air minimum temperature at 0.5 m above the soil surface. The mesoscale numerical model that they developed for this purpose is physically

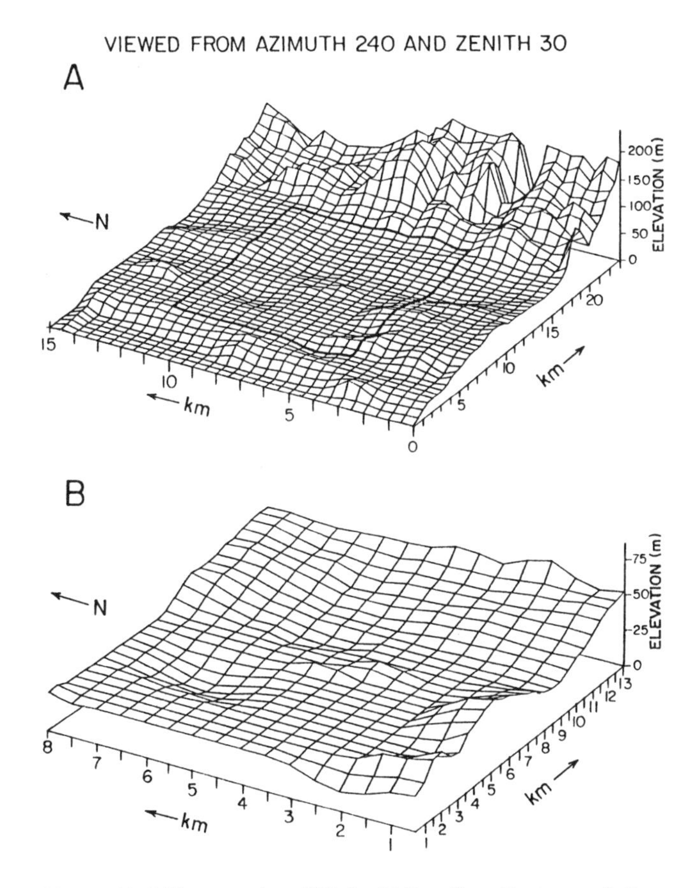

Figure 11-4: Topography of Hefer Valley (Israel) at a resolution of 500 m, (a) general view of the region, and (b) domain of the survey (from Avissar and Mahrer, 1988b).

and conceptually similar to the mesoscale model used in the previous section. It should be mentioned, however, that the two models use different numerical schemes and engineering concepts. In order to emphasize the potential application of this approach, Avissar and Mahrer (1988b) produced with their model a numerical topoclimatologic survey of Hefer Valley, which they compared to the survey produced by the Meteorological Service of Israel (Figure 11-4).

The model was initialized with the topography, soil types, and soil covers of Hefer Valley, which are presented in Figures 11-4, 11-5, and 11-6, respectively. The resolution of these maps corresponds to the resolution of the model (i.e., 500 by 500 m).

Figure 11-7 shows predicted versus observed differences of minimum air temperatures (at a height of 0.5 m above the soil surface) from the main station. A relatively low correlation was found between them ($R^2=0.42$). The discrepancies obtained between predicted and observed values were explained mainly by the fact that observations were always taken in open areas, even when most of the area around the station was covered with dense vegetation. Since during radiative frost events the atmosphere is very stable and advective effects are generally negligible due to weak winds, the temperature measured at the station is only representative of the immediate vicinity of the thermometer and, unlike the numerical model, does not give a mean value of the considered grid. Thus, the divergence between predicted and observed minimum temperature increases with the density of the vegetation.

To check the validity of this assumption, Avissar and Mahrer (1988b) produced another numerical

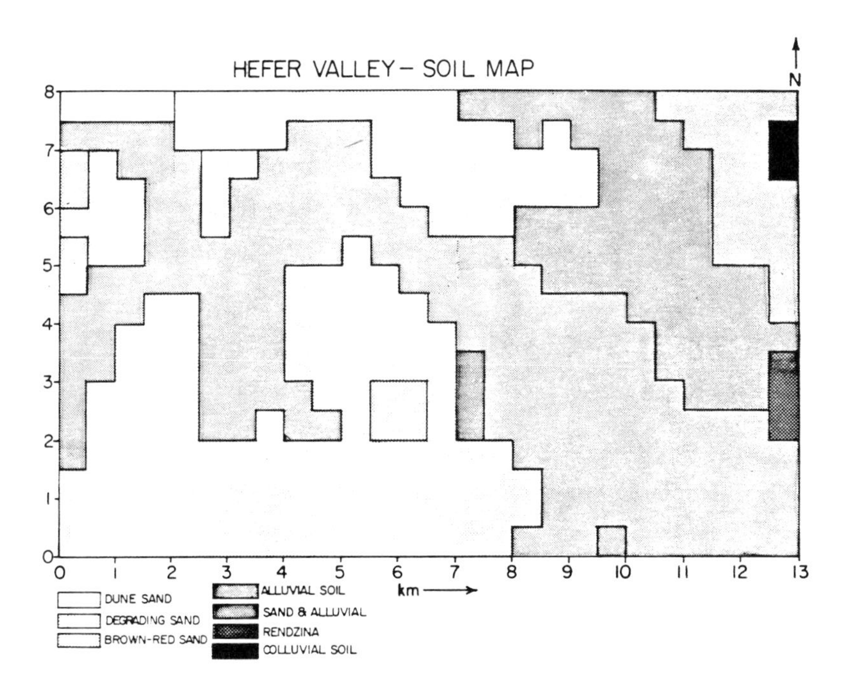

Figure 11-5: Soil map of Hefer Valley (Israel) at a resolution of 500 m (from Avissar and Mahrer, 1988b).

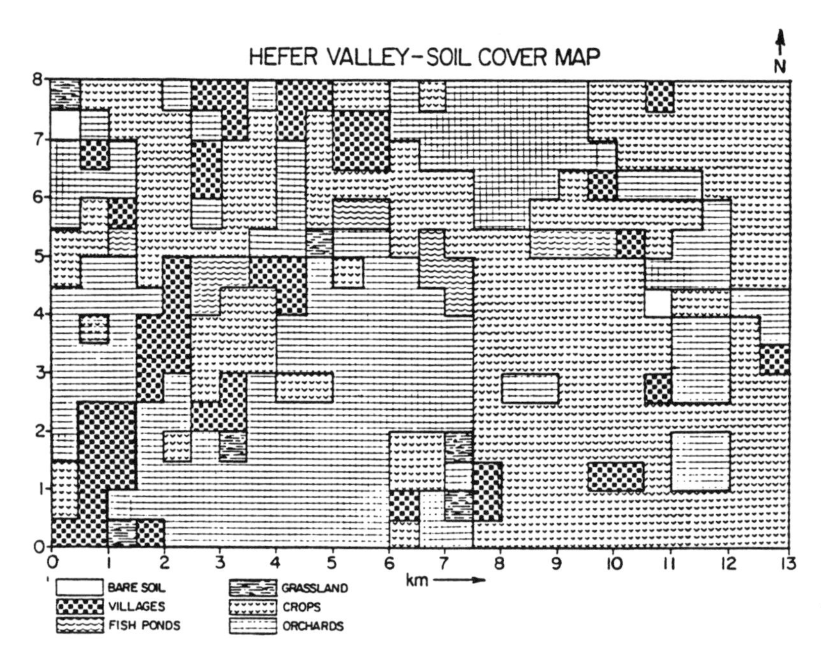

Figure 11-6: Soil cover map of Hefer Valley (Israel) at a resolution of 500 m (from Avissar and Mahrer, 1988b).

topoclimatologic survey of Hefer Valley, but assuming that the entire valley was bare (i.e., without considering the effects of vegetation, fish ponds, and villages). As illustrated in Figure 11-8, a better correlation (R^2=0.58) was obtained between predictions and observations.

The two numerical topoclimatologic surveys of Hefer Valley (i.e., when considering and ignoring the soil covers) are presented in Figures 11-9 and 11-10, respectively. The comparison between these two figures emphasizes the significant influence of dense vegetation and water ponds on the minimum temperature near the ground surface. The largest discrepancies between the two surveys are

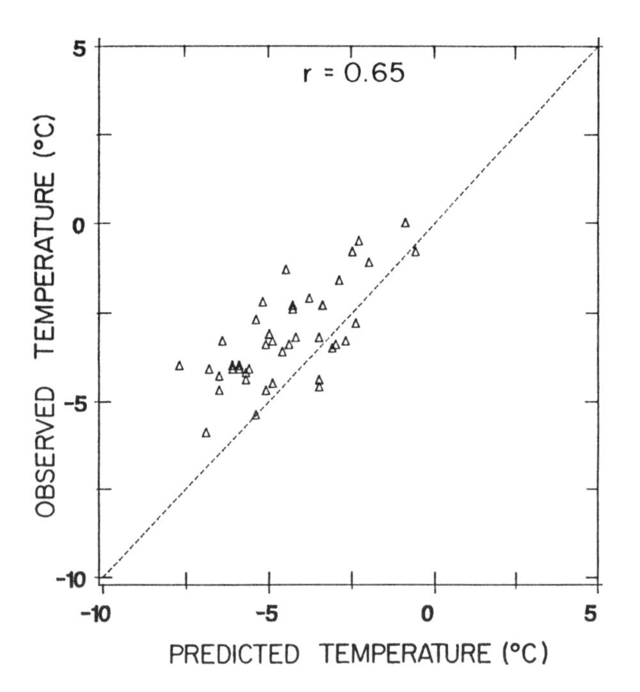

Figure 11-7: Predicted versus observed differences of minimum temperatures in Hefer Valley (Israel) during radiative frosts (from Avissar and Mahrer, 1988b).

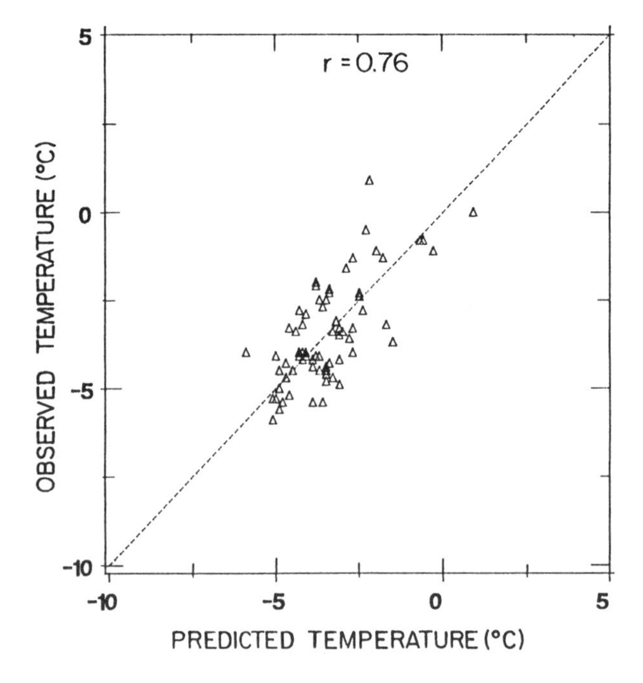

Figure 11-8: Same as Figure 11-7, but assuming Hefer Valley (Israel) is bare (from Avissar and Mahrer, 1988b).

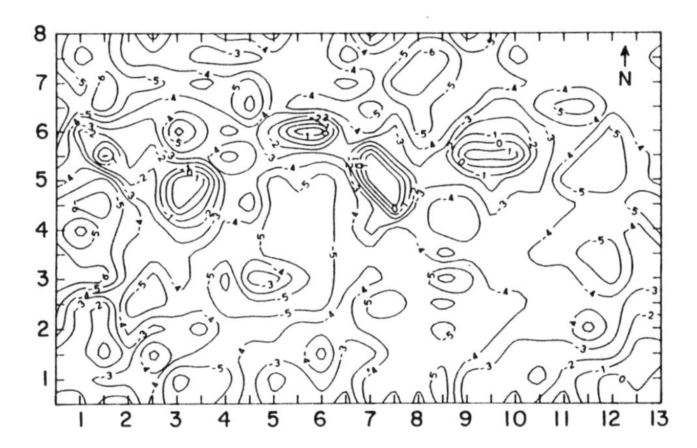

Figure 11-9: Numerical topoclimatologic survey of minimum temperature in Hefer Valley (Israel) (from Avissar and Mahrer, 1988b).

obtained at the water ponds, where the temperature during the night is much higher than near the bare soil surface or within the plant canopy.

IMPLICATIONS FOR GIS

The various simulations presented in this chapter emphasize the importance of land-surface characteristics on land-atmosphere interactions and, thereby, climate and weather predictions. Unfortunately, so far only limited studies have been carried out to understand which of these characteristics is important for atmospheric modeling purposes, and at which scale they should be provided to the model. These two questions challenge the atmospheric modeling community. When addressed, they should provide an essential information for the GIS scientific community.

ACKNOWLEDGMENTS

This research was supported by the National Science Foundation under grants ATM-9016562 and EAR-9105059, by the National Aeronautics and Space Administration under grant NAGW-2658, and by the New Jersey Agricultural Experiment Station under grant NJ13105. Computations were partly performed using the National Center for Atmospheric Research CRAY supercomputer (NCAR is supported by NSF). The author wishes to thank C.A. Nobre (Center for Weather Prediction and Climate Studies, Brazil) for providing the satellite image shown in Plate 11-1.

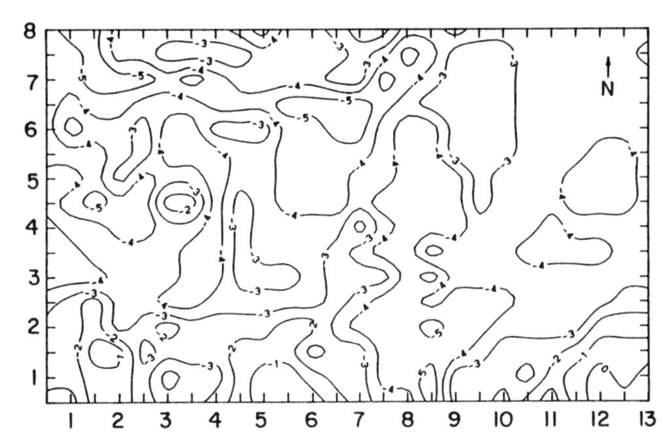

Figure 11-10: Same as Figure 11-9. but assuming Hefer Valley (Israel) is bare (from Avissar and Mahrer, 1988b).

REFERENCES

Avissar, R. (1991) A statistical-dynamical approach to parameterize subgrid-scale land-surface heterogeneity in climate models. In Wood, E. F. (ed.) *Land Surface–Atmosphere Interactions for Climate Models: Observations, Models, and Analyses*, Kluwer, pp. 155–178.

Avissar, R. (1992a) Conceptual aspects of a statistical-dynamical approach to represent landscape subgrid-scale heterogeneities in atmospheric models. *Journal of Geophysical Research* 97: 2729–2742.

Avissar, R. (1992b) Observations of leaf stomatal conductance at the canopy scale: an atmospheric modeling perspective. *Boundary-layer Meteorology* (in press).

Avissar, R., and Mahrer, Y. (1988a) Mapping frost-sensitive areas with a three-dimensional local scale model. Part I: Physical and numerical aspects. *Journal of Applied Meteorology* 27: 400–413.

Avissar, R., and Mahrer, Y. (1988b) Mapping frost-sensitive areas with a three-dimensional local scale numerical model. Part II: Comparison with observations. *Journal of Applied Meteorology* 27: 414–426.

Avissar, R., and Pielke, R.A. (1989) A parameterization of heterogeneous land surface for atmospheric numerical models and its impact on regional meteorology. *Monthly Weather Review* 117: 2113–2136.

Avissar, R., and Verstraete, M.M. (1990) The representation of continental surface processes in atmospheric models. *Reviews of Geophysics* 28: 35–52.

Deardorff, J.W. (1978) Efficient prediction of ground surface temperature and moisture, with inclusion of layer of vegetation. *Journal of Geophysical Research* 83: 1889–1903.

Dickinson, R.E., and Henderson-Sellers A. (1988) Modelling tropical deforestation, a study of GCM land-surface parameterizations. *Quarterly Journal of the Royal Meteorological Society* 114: 439–462.

Dickinson, R.E., Henderson-Sellers, A., Kennedy, P.J., and Wilson, M.F. (1986) Biosphere-atmosphere transfer scheme (BATS) for the NCAR community climate model. *NCAR Technical Note: NCAR/TN-275+STR*, 69 pp.

Entekhabi, D., and Eagleson, P.S. (1989) Land-surface hydrology parameterization for atmospheric general circulation models including subgrid-scale spatial variability. *Journal of Climate* 2: 816–831.

Famiglietti, J.S., and Wood, E.F. (1991) Evapotranspiration and runoff from large land areas: land surface hydrology for atmospheric general circulation models. In Wood, E.F. (ed.) *Land Surface–Atmosphere Interactions for Climate Models: Observations, Models, and Analyses*, Kluwer, pp. 179–197.

Henderson-Sellers, A., Dickinson, R.E., and Wilson, M.F. (1988) Tropical deforestation, important processes for climate models. *Climatic Change* 13: 43–67.

Lomas, J., and Gat, Z. (1971) Methods in agrotopoclimatologic surveys—low temperatures. *Agron. Rep. No. 1*, Bet Dagan, Israel: Israel Meteorological Service.

Mahrer, Y. (1984) An improved numerical approximation of the horizontal gradients in a terrain following coordinate system. *Monthly Weather Review* 112: 918–922.

Mahrer, Y., and Avissar, R. (1985) A numerical study of the effects of soil surface shape upon the temperature and moisture profiles. *Soil Science* 139: 483–490.

Mahrer, Y., and Pielke, R.A. (1976) Numerical simulation of the air flow over Barbados. *Monthly Weather Review* 104: 1392–1402.

Mahrer, Y., and Pielke, R.A. (1978) A test of an upstream spline interpolation technique for the advective terms in a numerical mesoscale model. *Monthly Weather Review* 106: 818–830.

McCumber, M.C. (1980) A numerical simulation of the influence of heat and moisture fluxes upon mesoscale circulations. *Ph.D. Dissertation*, Charlottesville: University of Virgina,

McNaughton, K.G. (1987) Comments on “Modeling effects of vegetation on climate”. In Dickinson, R.E. (ed.) *The Geophysiology of Amazonia: Vegetation and Climate Interactions*, Wiley & Sons, 339–342.

McNider, R.T., and Pielke, R.A. (1981) Diurnal boundary-layer development over slopping terrain. *Journal of the Atmospheric Sciences* 38: 2198–2212.

Mahfouf, J.F., Richard, E., and Mascart, P. (1987) The influence of soil and vegetation on the development of mesoscale circulation. *Journal of Climate and Applied Meteorology* 26: 1483–1495.

Neumann, J., and Mahrer, Y. (1971) A theoretical study of the land and sea breeze circulations. *Journal of the Atmospheric Sciences* 28: 532–542.

Naot, O., Otte, M.J., Collins, D.C., and Avissar, R. (1991) A statistical-dynamical approach to evaluate evapotranspiration from agricultural fields. *Proc. 20th Conference on Agricultural and Forest Meteorology, Salt Lake City, Utah, September 9–13, 1991*, pp. 54–57.

Pielke, R.A. (1974) A three-dimensional numerical model of the sea breezes over south Florida. *Monthly Weather Review* 102: 115–139.

Pielke, R.A., and Avissar, R. (1990) Influence of landscape structure on local and regional climate. *Landscape Ecology* 4: 133–155.

Pielke, R.A., and Mahrer, Y. (1975) Technique to represent the heated-planetary boundary layer in mesoscale models with coarse vertical resolution. *Journal of the Atmospheric Sciences* 32: 2288–2308.

Segal, M., Avissar, R., McCumber, M., and Pielke, R.A. (1988) Evaluation of vegetation effects on the generation and modification of mesoscale circulations. *Journal of the Atmospheric Sciences* 45: 2268–2292.

Sellers, P.J., Mintz, Y., Sud, Y.C., and Dalcher, A. (1986) A simple biosphere (SiB) for use within general circulation models. *Journal of the Atmospheric Sciences* 43: 505–531.

Shukla, J., and Mintz, Y. (1982) Influence of land-surface evapotranspiration on the Earth's climate. *Science* 215: 1498–1501.

Shukla, J., Nobre, C., and Sellers, P. (1990) Amazon deforestation and climate change. *Science* 247: 1322–1325.

Squire, G.R., and Black, C.R. (1981) Stomatal behaviour in the field. In: Jarvis, P.G., and Mansfield, T.A. (eds.) *Stomatal Physiology*, Cambridge University Press, pp. 248–279.

Wetzel, P.J., and Chang, J.T. (1988) Evapotranspiration from nonuniform surfaces: A first approach for short-term numerical weather prediction. *Monthly Weather Review* 116: 600–621.

Yamada, T. (1982) A numerical model simulation of turbulent airflow in and above canopy. *Journal of the Meteorological Society of Japan* 60: 439–454.

12

Land Surface Data: Global Climate Modeling Requirements

W. C. SKELLY,
A. HENDERSON-SELLERS, AND
A. J. PITMAN

Global atmospheric modeling, which developed in the 1950s, gained momentum in the 1960s and 1970s, spawning a rapid development of 3D numerical weather prediction and climate models (Washington and Parkinson, 1986). The increasing public awareness of the greenhouse effect in the 1980s propelled climate modeling into the fore of climate change research (Bolin et al., 1986). Climate models are currently our most valuable means of understanding the climate system, its potential for change, and the role of radiatively active gases within the atmosphere (Mitchell, 1989).

Atmospheric general circulation models (AGCMs) are valuable tools for investigating the effects of large-scale perturbations (e.g., desertification, Charney et al., 1977; increasing greenhouse gases, Mitchell et al., 1987; deforestation, Dickinson and Henderson-Sellers, 1988; Lean and Warrilow, 1989; and volcanic dust in the atmosphere, Hansen and Lacis, 1990) on the Earth's climate. Although AGCMs were specifically designed for global-scale climate simulations, they are now being used to evaluate climatological and hydrological quantities at or near the land surface and at subcontinental scales (Rind, 1982; Shukla and Mintz, 1982; Sud and Smith, 1985; Wilson et al., 1987; Sato et al., 1989).

As the climate modeling community takes its first steps towards assessing the potential impacts from climatic change, the incorporation of increasingly realistic continental surfaces is becoming more important. While climate modeling has provided insight into the functioning of the atmosphere and the role of radiatively active gases, this has been accomplished with little consideration of the Earth's land surface. Even though the ocean, by virtue of its greater role in latent heat transfer, is undeniably more important to the general circulation than is the land surface, a series of experiments over the past decade have shown that the atmosphere is sensitive to the parameterization of the land surface (Mintz, 1982). For instance, changes in the state of the land surface (e.g., soil moisture, Charney, 1975; albedo, Chervin, 1979; roughness length, Sud and Fennessy, 1984; soil texture, Wilson et al., 1987) have all been shown to affect the simulation of the Earth's climate by AGCMs.

The improvement of land-surface–atmospheric interaction schemes, increasing spatial resolution of global and mesoscale models and the desire to use AGCMs to study the Earth's land surface have created a need for a better spatial and temporal representation of the land surface. Verstraete and Dickinson (1986) indicate that there are four basic reasons why the land surface, and in particular vegetation, is important and needs to be incorporated into AGCMs. These are: radiative interactions (canopies have low albedos in general); water balance (canopies intercept a large but rather variable fraction of incoming precipitation while transpiration moves large quantities of water from within the soil to the atmosphere); energy balance (the leaves comprise a large area in contact with the atmosphere enhancing energy exchange); and roughness (plants are aerodynamically rough and are responsible for much of the variability in the surface roughness).

The realization that the atmosphere and land surface are closely linked led to a number of models being developed for AGCMs. The most important models are the biosphere atmosphere transfer scheme (BATS; Dickinson et al., 1986) and the simple biosphere model (SiB; Sellers et al., 1986, Xue et al., 1991). More recently, Pitman et al. (1991) developed the bare essentials of surface transfer (BEST) model, which attempts to simplify the parameterization of the surface–atmosphere interaction as far as possible, while still retaining an adequate description of land surface processes (Cogley et al., 1990). These "sub" models, which are nested within AGCMs, are currently driving the development of global land surface data for climate studies. How climatic change may impact the land surface, as well as how changes in the land surface may affect the climate, are now active areas of research. We therefore believe that there is an immediate need to re-evaluate the use and manipulation of global land surface data. It has been repeated many times, but nowhere is there a better example of the need for multidisciplinary cooperation than at the current juncture of climate modeling, climate change, and impact assessment research.

Integrating GIS technology with land-surface–atmosphere interaction schemes is going to be handled in two stages: (1) utilizing the GIS as a data management, analysis, and display tool upon which the land-surface–atmosphere model operates; and (2) fully integrating the land-surface–atmosphere software into the GIS. Our research is currently in the first stage. The second stage comprises what we would call a "self-parameterizing" database in which a physical representation of the Earth in terms of vegetation, soil, topography, etc. is used to update the land surface parameterizations needed for AGCMs automatically.

The specific objective of this chapter, however, is to expose one particularly urgent and problematic spatial data requirement—the aggregation of land surface data to different climate model resolutions. At present, even the same location on the Earth's surface is likely to be parameterized by different AGCMs in different ways (Grotch and MacCracken, 1991). This presents problems in comparing predicted regional impacts from different AGCMs. Common land surface data sets and standard aggregation techniques will reduce the uncertainty that is associated with modeling groups employing very different representations of the Earth's surface.

LAND SURFACE DATA

Climate models need estimates of energy, mass, and momentum fluxes in both space and time. In AGCMs the Earth's surface exists primarily as a flux interface. Thus it would be incorrect to suggest that climate models need to know the actual type of vegetation and soil at the surface. Energy, mass, and momentum fluxes are estimated indirectly by specifying observable parameters known to play a part in these land-surface–atmosphere fluxes. The number of parameters is dependent upon the complexity of the land-surface–atmospheric interaction scheme, but some of the basic parameters include albedo (α), roughness length (Z_0), and leaf area index (LAI). Providing these exchange flux parameterizations presents two immediate problems: (1) Special data sets must be prepared for each of these quantities; and (2) multiple data sets would need to be created to simulate this seasonal variability (Matthews, 1985; Henderson-Sellers et al., 1986).

In addition to the lack of observational data from which these required data sets would need to be drawn, it is not well understood how many of these parameters vary in space and time. What are known, classified, and more readily available are vegetation and soils distributions around the globe. While some of these data are of suspect accuracy, vegetation and soils maps provide the starting point from which a variety of parameters can be derived from the known physical relationships between the needed parameters and the vegetation–soil associations.

Global vegetation and soils data are available from several sources (e.g., Olson et al., 1983; Matthews, 1984; Wilson and Henderson-Sellers, 1985; Zobler, 1986) at 0.5 degrees (Olson et al., 1983) or 1 degree (all others) latitude–longitude resolution. Climate model resolutions are invariably larger than that of the original data (AGCM resolutions are about 3 degrees latitude × 4 degrees lon-

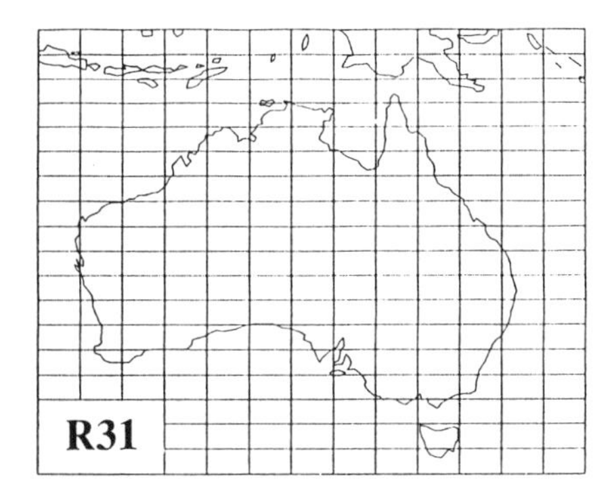

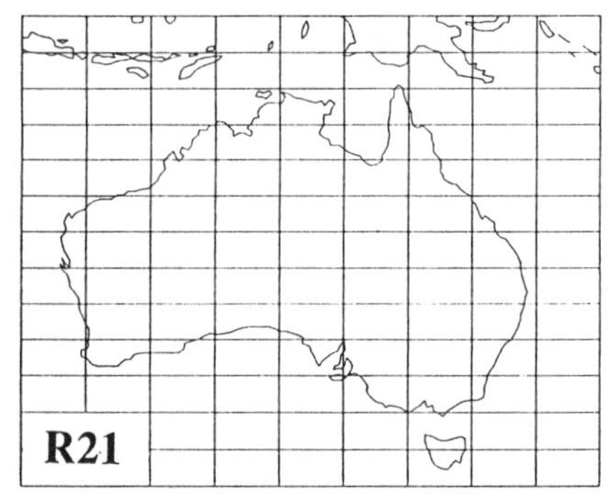

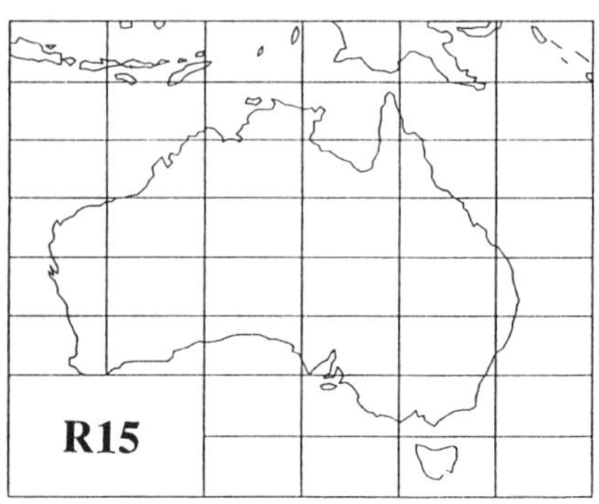

Figure 12-1: Three commonly used spectral atmospheric general circulation model resolutions, R31, R21, and R15.

gitude or larger), necessitating aggregation. Future regional studies using nested mesoscale models will be performed at much less than one degree resolution, which will necessitate the application of appropriate areal interpolation techniques.

Having come full circle from what is really needed (land-surface–atmosphere exchange fluxes) to what it is possible to obtain (vegetation and soil cover), the now classified land surface is used to derive the parameters needed to estimate the fluxes. The fluxes ultimately need to be provided at AGCM resolutions. To accomplish this, one can aggregate the classified land surface (vegetation and soils), the parameters (α, Z_0, LAI, etc.), or the fluxes themselves.

Aggregating the "classified land surface" has been the most common of the three methods used, and usually the most common vegetation or soil classification within the required AGCM resolution is used. The second method, which is used in this chapter, is to aggregate the parameters derived from the best available global vegetation and soils data sets using various numerical aggregation procedures. In this chapter we use a simple arithmetic averaging procedure. A third method, involving far more computation, is to aggregate the exchange fluxes themselves, which are calculated at the finest land surface data resolution available (Giorgi, 1991). This method, which is not straightforward, is made even more difficult by the absence of suitable areal flux observations (Mahrt, 1987).

SURFACE ROUGHNESS PARAMETERIZATION

Surface roughness, for purposes of climate modeling, is determined by vegetation distribution, and, particularly, height. It is parameterized as Z_0, the height above the ground surface where the average wind speed is zero, for different vegetation types or associations. Generally, Z_0 ranges from 0.001 to about 2 m (Wieringa, 1991a,b). Except for urban areas, which are not resolved at scales used by AGCMs, only forested areas have a Z_0 of 1 m or greater. Further, most vegetation types and/or associations that can be resolved at AGCM resolutions will probably have a Z_0 of 0.1 m or less because of the heterogeneity of such large areas (Wieringa, 1991b).

Three resolutions of particular interest to our group are known as the R31, R21, R15 resolutions (Figure 12-1). These resolutions are so termed because they specify the type ("R"=rhomboidal) and magnitude (e.g., "31"=31st wave) of a wave truncation in a spectral AGCM (Henderson-Sellers and McGuffie, 1987). The rhomboidal truncation after the thirty-first wave "R31" has a resolution of approximately 3.7 degrees long. × 2.2 degrees lat. R21 and R15 have resolutions of 5.6 degrees long. × 3.3 degrees lat. and 7.5 degrees long. × 4.5 degrees lat. respectively. R15, historically the "workhorse" for most climate simulations, is often being replaced with the R21 resolution in climate studies, although for long simulations R15 is still used (Houghton et al., 1990). The R31 resolution is not often used. The computational overhead at this resolution is still too high for most climate modeling purposes.

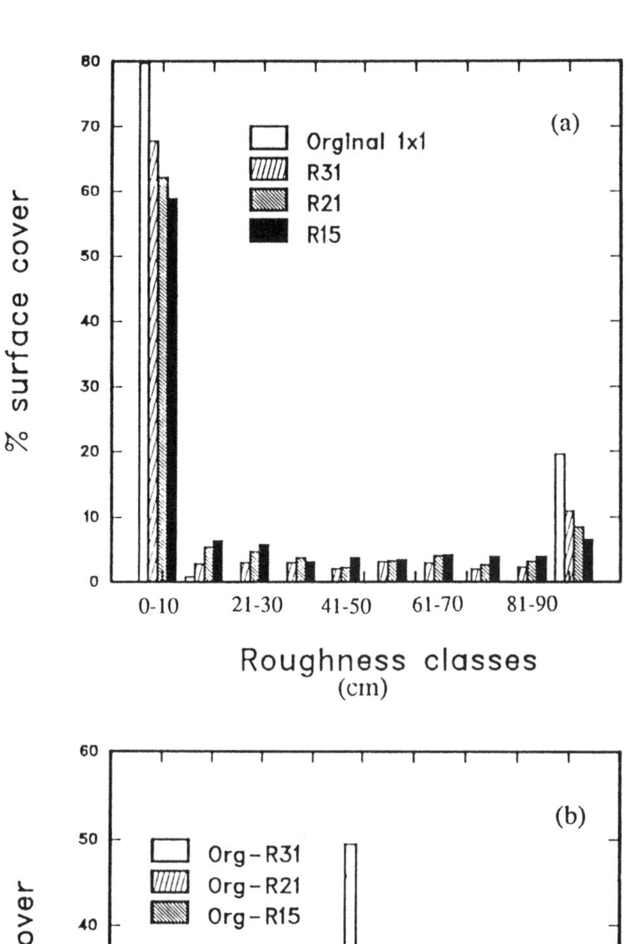

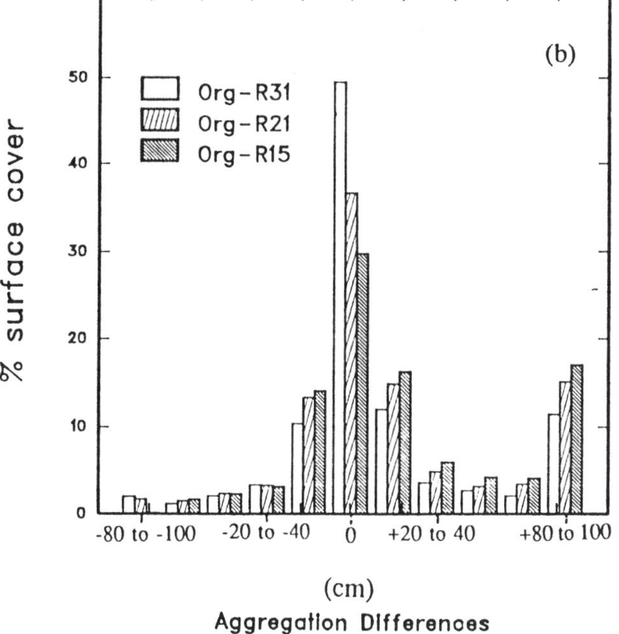

Figure 12-2: The distribution of the Global Roughness parameterization (a) and the aggregation differences (b), in 10 and 20 cm classes, respectively.

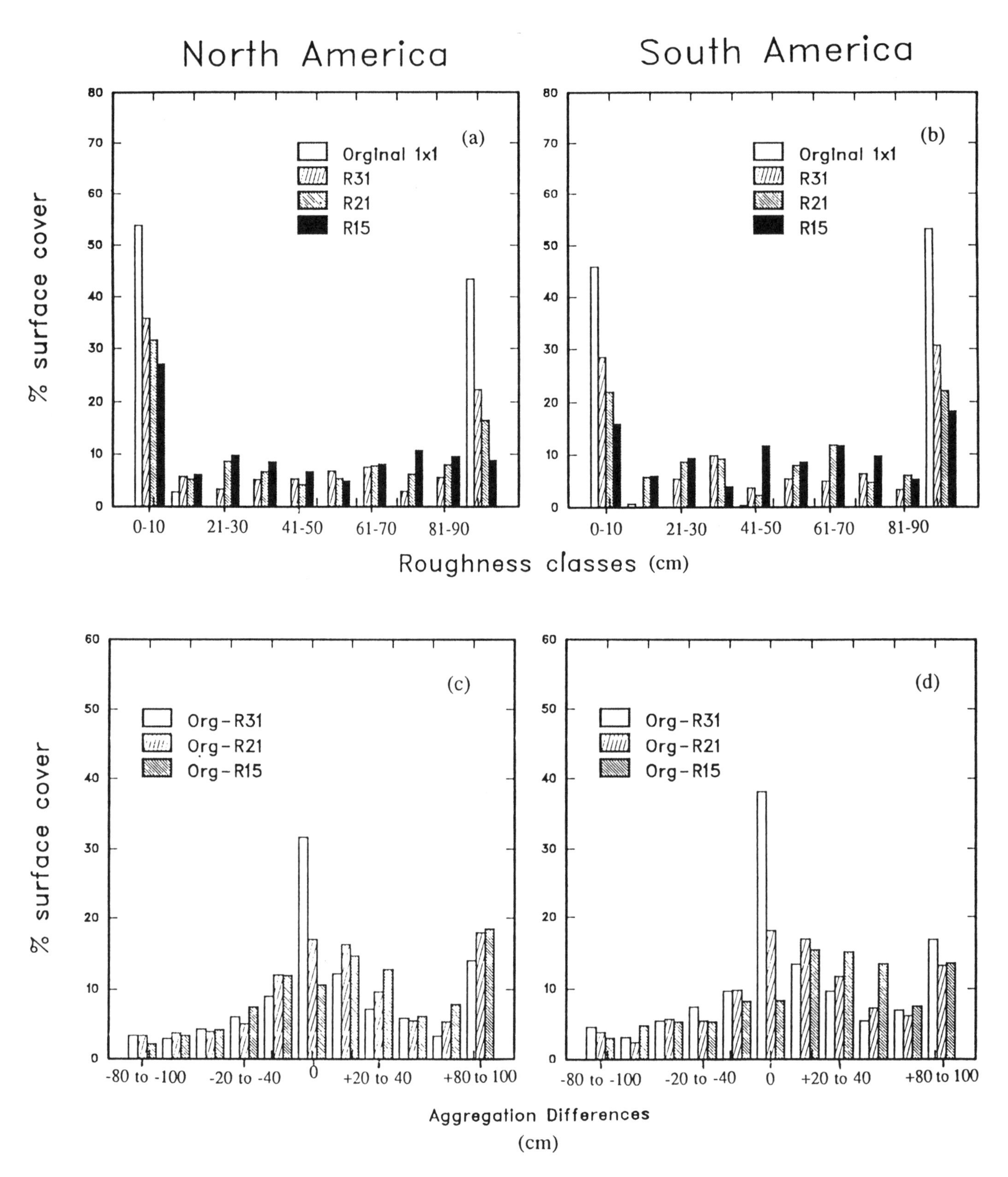

Figure 12-3: The distribution of North (a, c) and South (b, d) American roughness parameterization and aggregation differences, in 10 and 20 cm classes, respectively.

Global roughness distribution

The Wilson and Henderson-Sellers (1985) primary vegetation database is used to provide the global distribution of the Z_0 parameter. At a global scale Z_0 is bimodally distributed reflecting the forest/nonforest gap in surface roughness parameterizations for resolutions coarser than 1 degree × 1 degree (Figure 12-2a). Aggregating the 1 degree × 1 degree parameterizations to AGCM resolutions creates "new" Z_0 classes, and effectively makes forested areas smoother and some areas with initially small Z_0 rougher. This is shown by looking at the aggregation differences in which the original 1 degree × 1 degree data are coarsened to the R31, R21, and R15 AGCM resolutions (Figure 12-2b). The aggregation differences range from the most negative (most smoothed) to the most positive (most roughened). It is readily evident that the coarser the resolution used, the less of the Earth's surface will remain unchanged. In the case of the most coarse resolution used, R15, less than 30% of the continental land surface retains the same roughness parameterization. Interestingly, except for large aggregation differences (Figure 12-2b), the changes in roughness length due to aggregation are normally distributed, implying that the net effect of aggregation makes the Earth as a whole neither rougher nor smoother. The aggregation differences in the +80 to +100 cm range are an anomaly arising because the coarser resolution identifies grid cells as land that was formerly water at the original resolution. This flaw in the aggregation process was not corrected here because AGCMs can cope with grid cells identified as land that should be water by means of a simple respecification, although the reverse confusion causes more difficulty because there are no parameters available to characterize the soil and vegetation.

South and North American roughness distributions

The distribution of surface roughness on the South and North American continents is found to be bimodal, as was the global distribution (Figure 12-3a,c). As well, the creation of "new" roughness classes is also evident. However, the parameterization of surface roughness on both continents is divided more evenly between "forest" roughness and nonforest roughness than was the global distribution. Additionally, a greater relative decrease in the extent of "forestlike" surface roughness on these two continents is seen as compared to the global-scale coverage. The aggregation differences further illustrate the discrepancies between global- and continental-scale roughness distributions (Figure 12-3b,d). It appears that at resolutions coarser than R31, both South and North America are preferentially smoothed; that is, the net effect is to smooth both continents.

LAND-SURFACE–ATMOSPHERE MODELING

The effects of aggregation on sensible and latent heat fluxes are shown using the BEST model (Figure 12-4a,b). We simulate an AGCM grid cell as a combination of nine finer-resolution cells consisting of conifer forest, C, and tundra, T. This is accomplished by arithmetically averaging the parameters (i.e., Z_0, α, LAI, etc.). It is evident that there is a nonlinear relationship between the assigned parameters and the simulated fluxes. That is, as the number of tundra cells is increased (i.e., Z_0 decreased, albedo

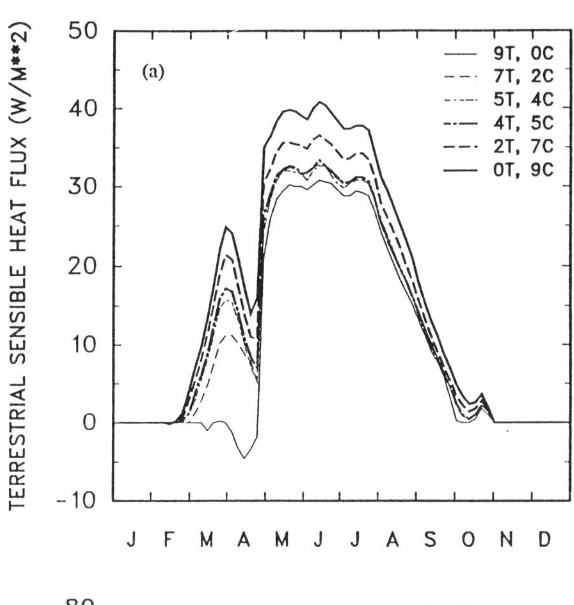

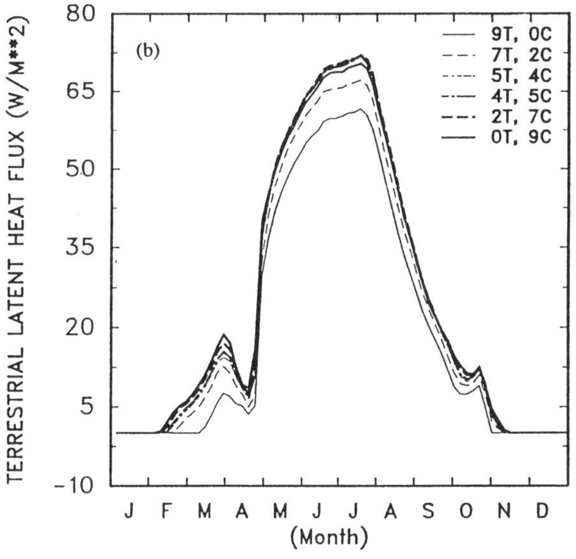

Figure 12-4: The sensible (a) and latent (b) heat fluxes as simulated using the BEST model for varying combinations of conifer forest C and tundra T.

increased, LAI decreased, etc.) and the number of conifer forest cells is consequently decreased, there is not a corresponding linear change in the turbulent energy fluxes (Figure 12-4). The roughness of a grid cell obtained by averaging 5 tundra cells, $5T$, and 4 conifer forest cells, $4C$, is not much different from the roughness obtained by averaging 4T and 5C together—and neither is either of the energy fluxes (Figures 12-4a,b). However, had it been decided to aggregate by taking the most "common" vegetation type, where:

$5T > 4C = T$ ($9T$ & $0C$, Figure 12-3a,b)
$5C > 4T = C$ ($0T$ & $9C$, Figure 12-3a,b)

then the resulting energy fluxes of the coarsened cell would differ greatly at times during the simulated year.

DISCUSSION

These preliminary results indicate that while the global distribution of surface roughness does not appear to be greatly affected by aggregation typically required for current GCMs, the continental distributions do appear to be affected. Indeed, aggregation-induced changes in the parameterization of the Earth's surface may be as great as those used in climate modeling studies that aim to investigate the effect of a changed land surface on the climate. The investigation of this very specific concern, that is, the Z_0 parameterization, has hinted at the wider implications of the management of spatial land surface data for global climate modeling. It suggests the value of integrating spatial data management techniques (i.e., GIS) within the overall development of land-surface–atmosphere interaction schemes.

The first stage of GIS and land-surface–atmosphere model integration will prove valuable in the intercomparison of the land surface climatologies from different AGCMs (e.g., Grotch and MacCracken, 1991). We already have some indication of the necessity for standardizing the representation of the land surface, as it is already evident that some AGCMs have grassland where others have forest. Until AGCMs agree on what exists where (and it seems probable that agreement is more important, in the first instance, than actually obtaining the "correct" distribution), intercomparisons of the climates predicted for the land surface cannot be expected to provide similar results.

As mentioned previously, we believe that the integration of GIS with land-surface–atmosphere interaction models needs to be taken further than the current stage of using GIS to simply "manage" a land-surface–atmosphere model's spatial data. In the very near future we will have integrated a GIS and a land-surface–atmosphere model at the software level. This will provide the basis for the development of a "self-parameterizing" database, in which we will use the land-surface–atmosphere model to derive flux parameters based on the best available information regarding the physical relationships between vegetation, soils, and the needed flux parameters.

ACKNOWLEDGMENTS

WCS holds both an ELCOM Scholarship and a Macquarie Overseas Postgraduate Research Award.

REFERENCES

Andre, J.C. and Blondin, C. (1986) On the effective roughness length for use in numerical three-dimensional models. *Boundary Layer Meteorology* 35: 231–245.
Bolin, B., Doos, B., Jager, J., and Warrick, R.A. (1986) *SCOPE 29, The Greenhouse Effect, Climatic Change, and Ecosystems*, Chichester: John Wiley and Sons, 541 pp.
Charney, J. (1975) Dynamics of deserts and drought in the Sahel. *Quarterly Journal, Royal Meteorological Society* 101: 193–202.
Charney, J., Quirk, W.J., Chow, S-H., and Kornfield, J. (1977) A comparative study of the effects of albedo change on drought in semi arid regions. *Journal of Atmospheric Science* 34: 1366–1385.
Chervin, R.M. (1979) Responses of the NCAR General Circulation Model to a changed land surface albedo. In *Report to the JOC Study Conference on Climate Models: Performance, Intercomparison and Sensitivity Studies*, Washington, DC: GARP Publications, Ser. 22, Vol. 1, pp. 563–581.
Cogley, J.G., Pitman, A.J., and Henderson-Sellers, A. (1990) A land surface for large scale climate models. *Trent University Technical Note 90-1*, Peterborough, Ontario: Trent University.
Dickinson, R.E., and Henderson-Sellers, A. (1988) Modeling tropical deforestation: a study of GCM land-surface parameterizations. *Quarterly Journal, Royal Meteorological Society* 114(b): 439–462.
Dickinson, R.E., Henderson-Sellers, A., Kennedy, P.J., and Wilson, M.F. (1986) Biosphere atmosphere transfer scheme (BATS) for the NCAR community climate model. *NCAR Technical Note TN275+STR*, Boulder, CO: National Center for Atmospheric Research, 69 pp.
Giorgi, F. (1991) Personal communication.
Grotch, S.L., and MacCracken, M.C. (1991) The use of general circulation models to predict regional climatic change. *Journal of Climate* 4: 286–303.
Hansen, J.E., and Lacis, A.A. (1990) Sun and dust versus greenhouse gases: an assessment of their relative roles in global climate change. *Nature* 346: 713–719.
Henderson-Sellers, A., and McGuffie, K. (1987) *A Cli-*

mate Modeling Primer, Chichester: John Wiley and Sons, 217 pp.

Henderson-Sellers, A., Wilson, M.F., Thomas, G., and Dickinson, R.E. (1986) Current global land surface data sets for use in climate related studies. *NCAR Technical Note TN-272+STR*, Boulder, CO: National Center for Atmospheric Research, 110 pp.

Houghton, J.T., Jenkins, G.J., and Ephraums, J.J. (eds.) (1990) *Climate Change: The IPCC Scientific Assessment*, Cambridge University Press, 364 pp.

Lean, J., and Warrilow, D.A. (1989) Simulation of the regional climatic impact of Amazon deforestation. *Nature* 342: 411–413.

Mahrt, L. (1987) Grid-averaged surface fluxes. *Monthly Weather Review* 115: 1550–1560.

Mason, P.J. (1988) The formation of areally-averaged roughness lengths. *Quarterly Journal, Royal Meteorological Society* 114: 399–420.

Matthews, E. (1984) Vegetation, land-use and seasonal albedo data sets: documentation of archived data tape. *NASA Technical Memorandum 86107*, Washington, DC: National Aeronautics and Space Administration, 12 pp.

Matthews, E. (1985) Atlas of archived vegetation, land-use and seasonal albedo data sets. *NASA Technical Memorandum 86199*, Washington, DC: National Aeronautics and Space Administration.

Mintz, Y. (1982) The sensitivity of numerically simulated climates to land surface conditions. In Eagleson, P.S. (ed.) *Land Surface Processes in Atmospheric General Circulation Models*, Cambridge: Cambridge University Press, pp. 109–111.

Mitchell, J.F.B. (1989) The "greenhouse" effect and climate change. *Reviews of Geophysics* 27: 115–139.

Mitchell, J.F.B., Wilson, C.A., and Cunnington, W.M. (1987) On CO_2 sensitivity and model dependence of results. *Quarterly Journal, Royal Meteorological Society* 113: 293–322.

Olson, J.S., Watts, J.A., and Allison, L.J. (1983) Carbon in live vegetation of major world ecosystems. *DOC/NBB Report No. TR004*, Oak Ridge, TN: Oak Ridge National Laboratory, 152 pp.

Pitman, A.J., Yang, Z-L., Cogley, J.G., and Henderson-Sellers, A. (1991) Description of bare essentials of surface transfer for the Bureau of Meteorology Research Centre AGCM. *Bureau of Meteorology Research Report*, Melbourne: BMRC (in press).

Rind, D. (1982) The influence of ground moisture conditions in North America on summer climate as modelled in the GISS AGCM. *Monthly Weather Review* 110: 1487–1494.

Sato, N., Sellers, P.J., Randall, D.A., Schneider, E.K., Shukla, J., Kinter III, J.L., Hou, Y.-T., and Albertazzi, E. (1989) Effects of implementing the simple biosphere model in a general circulation model. *Journal of Atmospheric Science* 46: 2757–2782.

Sellers, P.J., Mintz, Y., Sud, Y.C., and Dalcher, A. (1986) A simple biosphere model (SiB) for use within general circulation models. *Journal of Atmospheric Science* 43: 505–531.

Shukla, J., and Mintz, Y. (1982) Influence of land surface evapotranspiration on Earth's climate. *Science* 215: 1498–1501.

Sud, Y.C., and Fennessy, M.J. (1984) Influence of evaporation in semi-arid regions on the July circulation: a numerical study. *Journal of Climatology* 4: 383–398.

Sud, Y.C., and Smith, W.E. (1985) Influence of local land-surface processes on the Indian Monsoon: a numerical study. *Journal of Climatology and Applied Meteorology* 24: 1015–1036.

Taylor, P.A. (1987) Comments and further analysis on effective roughness lengths for use in numerical three-dimensional models. *Boundary Layer Meteorology* 39: 403–418.

Verstraete, M.M., and Dickinson, R.E. (1986) Modeling surface processes in atmospheric general circulation models. *Annales Geophysicae* 4: 357–364.

Washington, W.M., and Parkinson, C.L. (1986) *An Introduction to Three-Dimensional Climate Modeling*, University Science Books, 422 pp.

Wieringa, J. (1991a) Representative roughness parameters for homogeneous terrain. Submitted to *Boundary-Layer Meteorology*, 54 pp.

Wieringa, J. (1991b) Updating the Davenport roughness classification. *8th International Conference on Wind Engineering, London, Canada, July 1991*.

Wilson, M.F., and Henderson-Sellers, A. (1985) A global archive of land cover and soil data sets for use in general circulation climate models. *Journal of Climatology* 5: 119–143.

Wilson, M.F., Henderson-Sellers, A., Dickinson, R.E., and Kennedy, P.J. (1987) Sensitivity of the biosphere–atmosphere transfer scheme (BATS) to the inclusion of variable soil characteristics. *Journal of Climatology and Applied Meteorology* 26: 341–362.

Xue, Y., Sellers, P.J., Kinter, J.L., and Shukla, J. (1991) A simplified biosphere model for climate studies. *Journal of Climate* 4: 345–364.

Zobler, L. (1986) A world soil file for global climate modeling. *NASA Technical Memorandum 87802*, Washington, DC: National Aeronautics and Space Administration, 32 pp.

13

Regional Air Quality and Acid Deposition Modeling and the Role for Visualization

JOAN H. NOVAK AND
ROBIN L. DENNIS

Airborne pollutants are responsible for a variety of environmental problems in the U.S. from violations of the health-based National Ambient Air Quality Standards (NAAQS) for ozone to acidification of lake and streams. EPA uses mathematical models to simulate atmospheric and chemical processes leading to air pollution and acidic deposition. Because atmospheric processes are often nonlinear, that is, a 50% reduction in source emissions most likely will result in somewhat less than a 50% reduction in related ambient air concentrations, detailed and complicated models are required. These models require input data related to the location and strength of pollutant sources, and meteorological parameters such as precipitation, and wind speed and direction, and are primarily used to simulate the quality of the air and the amount of acidic deposition expected to occur in future years (i.e., 2010, 2030) if different levels of source emissions controls are required by legislation. Verification and interpretation of model predictions require a significant amount of graphical and statistical analysis. Thus, software tools with spatial analysis functionality, such as GIS, are very useful to both the researchers and policy analysts.

This chapter begins with a brief summary of EPA's regional air quality and deposition models with some thoughts on the researchers' need for hands-on data analysis capabilities. Next a pilot project using a GIS for spatial analysis related to regional air quality model interpretation is discussed along with lessons learned and remaining obstacles to overcome. Finally a plan to develop an integrated modeling and analysis framework is presented, followed by a summary of recommendations to ensure that GIS functionality is directly accessible by the researchers.

REGIONAL AIR POLLUTION MODELS

Regional acid deposition model

The regional acid deposition model (RADM) (Chang et al., 1987) is a complex mathematical model that simulates the dominant physical and chemical processes in the troposphere related to acidic deposition. The RADM is used both to assess the relative effectiveness of various options for controlling pollutant source emissions and to understand the interrelationships among transport, chemical transformations, and deposition of acidic pollutants (Dennis et al., 1990). The processes characterized in RADM include 3D advection and vertical diffusion, gas and aqueous phase chemistry, dry deposition, convective and stratiform cloud effects, and both manmade and natural source emissions. These processes are mathematically represented by a general system of conservation equations solved using finite difference methods. Thus the model domain, which typically extends over an area of 3000 × 3000 km, is divided into small three-dimensional cells of 80 km on a side with varying vertical resolution. The RADM extends up to 16 km with the vertical cell dimension increasing logarithmically with height to form six to fifteen vertical layers, as needed, to represent cloud processes and vertical transport adequately. The system of differential equations is solved in each cell with time steps varying from seconds to 1 hour, for periods on the order of 5 days. Approximately 30 of these 5 day episodes are aggregated with a statistically derived weighting scheme based on the frequency of occurrence of stratified groupings of meteorological conditions to produce annual sulfur and nitrogen deposition estimates across the current domain, which covers the eastern half of the U.S. and extends northward into southern portions of Canada. RADM is also being adapted to include aerosol chemistry and dynamics to provide additional functionality to address regional particulate and visibility issues critical to western portions of the U.S.

Regional oxidant model

The regional oxidant model (ROM) (Lamb, 1983) is also a complex mathematical model designed to simulate the photochemical production and transport of ozone and associated precursors. ROM has been used extensively (Lamb, 1986, 1988; Possiel et al., 1991) to evaluate the effectiveness of regional control strategies for ozone in the Northeastern U.S. The process and associated mathematical representations in ROM are similar to those described for RADM: horizontal and vertical transport and diffusion, gas-phase chemistry (Gery et al., 1988), dry deposition, terrain and convective cloud effects, and manmade and natural emissions. However, ROM is limited to fair weather processes conducive to ozone formation, whereas RADM must deal with both wet and dry processes contributing to acidic deposition. A typical study domain extends 1000×1000 km with a horizontal cell resolution of 18.5 km. This finer resolution is necessary to capture the gradients in emissions and time constraints of chemical conversions that lead to steep gradients typically found in regional ozone patterns. ROM has three predictive vertical layers designed to change both in space and time to adjust to locally varying physical parameters, and a diagnostic surface layer incorporating effects of sub-grid-scale source emission distributions. ROM produces hourly estimates of ozone and ozone precursors for periods up to 15 days. And as in RADM, internal time steps vary considerably in areas with steep emissions gradients. ROM is currently being applied in the northeastern, southeastern, and midwestern regions of the U.S.

Other air pollution models

Several other air quality models are being used and/or developed to handle related air pollution problems. The urban airshed model (UAM) (Morris and Myers, 1990), a fine-resolution grid-based photochemical oxidant model, is used by the states and local air pollution control groups to perform ozone attainment demonstrations for urban areas as mandated by the Clean Air Act. Since UAM is very sensitive to boundary conditions, the ROM is used to provide those conditions for a variety of regional control strategies. This is a prime example of where the integration of regional and urban scale approaches is necessary to model the overall environmental assessment properly.

The regional Lagrangian model of air pollution (RELMAP) (Eder et al., 1986) is designed to estimate monthly-to-annual ambient concentrations of wet and dry deposition of SO_2, SO_4, and particulates over the Eastern U.S. This relatively simple linear chemistry model is used primarily as a screening tool prior to application of the more complex, nonlinear RADM.

The user's dilemma

These models have been developed independently to address issues related to different pollutants, and different temporal and spatial scales. Even the linkage of the UAM and ROM came after the individual models were fully developed. Only the regional particulate model development is directly building on the existing RADM model. Thus each model has its own collection of input/output data processors and analysis software that requires a team of specialized computer programmers or scientists for effective utilization. Unfortunately the researchers and policy analysts who are the best qualified to analyze the data have limited access to either the data or the models. They routinely receive a standard set of output graphics for review, but have little to no capability for hands-on exploratory data analysis.

GIS PROTOTYPE DEVELOPMENT EXPERIENCE

Pilot GIS project

Recognizing the need for better spatial analysis capabilities for the users, several groups within the EPA Research Triangle Park location performed a cooperative pilot project (Birth et al., 1990) to evaluate the usefulness of a GIS for air pollution modeling research and assessment activities. The groups included the Atmospheric Research and Exposure Assessment Laboratory (AREAL), the Office of Air Quality Planning and Standards (OAQPS), and the Health Effects Research Laboratory (HERL). Three main goals for the project were: (1) to provide the end user (researcher, policy analyst) with easy, hands-on access to spatial display and analysis capabilities available in ARC/INFO; (2) to provide the end-user transparent interface between ARC/INFO and several agency databases typically not available outside a small user community; and (3) to provide a systematic approach to database management for numerous input/output and map files.

The study area for the pilot project was the ROM domain covering the northeastern region of the United States. Several types of data were prepared for use with ARC/INFO: ROM predicted pollutant concentrations, manmade point and area source emissions, natural hydrocarbon emissions, air quality measurements, land use data, and census data. A pilot Interactive Display for Environmental Analysis System (IDEAS) was developed using ARC/INFO to enable the user to study relationships between (1) measured and predicted air pollution concentrations, (2) point and area source emissions, (3) land use, and (4) health effects. A series of pull-down menus was implemented to allow point and click interface to ARC/INFO features and user databases.

Positive lessons

At the completion of the pilot study, we assessed the capabilities of ARC/INFO for our specific environmental data analysis needs. The users realized that hands-on access to data display and analysis capabilities is a necessity for a productive analysis environment. The user interface to the analysis capabilities must be easy to use and provide flexibility to the users to pursue their own direction for analysis. ARC/INFO provides several advantages for spatial display and overlay of multiple data types. It allows for selective display of attribute information and supports spatial allocation/gridding. Since much of the work of preparing emissions data for input to models involves spatial distribution of emissions sources, human activities, active emissions control legislation, and land use and land cover, a GIS provides an excellent means to design emissions control strategies and estimate the resultant emissions inputs to air quality models. We are currently designing an Emissions Preprocessor and Analysis Module (EPAM) in an ARC/INFO environment for performing quality control, speciation, temporal and spatial allocation, and emissions projection and control of man-made and natural emissions. We are also preparing preliminary estimates of toxic emissions for the Great Lakes area using a combination of SAS and ARC/INFO.

The user's dilemma

Viewing spatial data over time is critical in understanding the evolution of pollutant formation and transport within a region. However, ARC/INFO and other GIS approaches lack the ability to deal effectively with the element of time in spatial data analysis. Animation or rapid display of time sequences of spatial data is difficult within ARC/INFO and other geographical information systems because (1) features are not provided within the systems to manipulate or manage a large number of files or images easily; and (2) the ARC/INFO software executes very slowly in the shared VAX cluster computing environment currently supported by EPA. A dedicated workstation environment provides the needed computing power but introduces data access and transfer problems that increase demands for network bandwidth and transparent data access methods such as the Network File Server (NFS). In addition, the cost of GIS software can severely limit the number of users who can have direct access to these GIS applications.

GIS applications often involve passing data to and from the GIS portion of the application. For example, sorts of large toxic emissions data sets are so slow, even on individual workstations, that we pass large data sets to SAS sort routines on an IBM mainframe and then return them to ARC/INFO for completion of the spatial analysis portion of the task. ARC/INFO does not provide a simple feature to assist in data conversion and transfer activities; specific code must be written. Also air quality and deposition models perform numerous computations on a grid cell (spatial) basis. Theoretically a GIS could be used for many of these computations if more advanced computational capabilities were supported, such as integration or finite difference operations on nonuniform rectangular grid cells. Obviously computational speeds would be critical with these types of operations.

Air quality models predict three-dimensional fields of ambient air concentrations. Current GIS software has difficulty in adequately representing and exploring those three-dimensional fields. Thus to perform a routine analysis of a typical air quality model simulation, the "user" must access a variety of high- and low-level language software tools such as ARC/INFO, Graphic Kernel System (GKS), SAS, the National Center for Atmospheric Research (NCAR) graphics routines, Numerical Designs Limited (NDL) volume rendering and animation routines, Dynamics Graphics volume analysis tools, etc. Since most of these tools require extensive experience and training for effective use, the researcher or policy analyst must rely on skilled programmers to schedule and carry out selected portions of the analysis activity. The "real user" loses direct access to the analysis tools.

THIRD GENERATION MODELING AND ANALYSIS FRAMEWORK

Presented with a similar dilemma for access to both the models and the analysis tools, and in response to the Federal High Performance Computing and Communications (HPCC) program, we have initiated a project to design a third-generation modeling and analysis framework (Models-3) (Dennis and Novak, 1991) specifically to address many of the problems discussed previously. These include: (1) limited and difficult user access, (2) numerous modeling and analysis tools with overlapping functionality, (3) difficult transition to emerging high-performance computer hardware, software, and network systems that can provide relief from existing performance limitations. The main goal of the Models-3 project is to develop a new generation of air quality models that are more functional and directly accessible to the users for solving critical environmental issues of the decade.

The Models-3 framework is currently envisioned to have five major components: user interface and system manager, core models—modules and processors, database management subsystem, input/output subsystem, and analysis subsystem. This framework will provide a common architecture for many air quality models to allow interchangeability of process modules such as transport, chemistry, clouds, and aerosols. The design includes a flexible grid system capable of variable spatial resolution to enable integrated simulations from urban to regional scales. This can be achieved using multiple levels of nesting and ultimately adaptive gridding techniques. The Models-3 framework will incorporate algorithms developed under EPA's HPCC program to take advantage of massively parallel computing hardware both for scientific module execution and for visualization/analysis activities including remote distributed visualization.

A proof of concept prototype was recently demonstrated for the user interface component of the system. The prototype was developed to show the feasibility of using a graphical user interface for (1) collection of inputs and execution of an air quality model, and (2) extraction and visualization of model predictions. The prototype was built using a commercially available software package, Application Visualization System (AVS). Through point and click mouse operations the user combines a variety of standard AVS software modules to form an executable network flow diagram, in essence "the program." When the flow network is activated, each module in turn performs its designated operations on input data and produces desired outputs which flow to the next module in the process. Modules can execute in parallel, on different hardware, as designated by the flow network. In the prototype, data extractions were performed on a supercomputer, while rendering and display were performed on a graphics workstation. Data flow between hardware was handled transparently through standard AVS modules. Significant gains in productivity were achieved through this prototype automation process. A picture of the AVS customized user interface is presented in Figure 13-1. Our experience with this proof of concept prototype leads us to believe we are on the right track to resolving our dilemmas with this approach.

A few implementations of visual programming environments (Upson, 1990) have been available on a wide variety of hardware platforms for several years, and some new approaches were announced at SIGGRAPH '91. These approaches provide spatial analysis capabilities such as animation over time sequences and 3D visualization that are weak within current GIS implementations. Recent advances in database systems capable of storing and accessing features extracted from maps and imagery offer functionality for typical GIS applications within a visual programming environment. The ultimate goal is to provide each "user" with an affordable desktop workstation network connected to a variety of specialized and conventional high-performance computing resources that act together, transparent to the user, to satisfy a wide variety of computational and visualization needs.

SUMMARY AND RECOMMENDATIONS

The atmospheric modeling community has a definite need for spatial analysis capabilities and must currently use a wide variety of software tools to achieve their data analysis objectives. We have identified several areas where enhancements in GIS functionality would be extremely useful: (1) animation of spatial fields over time, (2) three-dimensional volume images for exploratory analysis over time, (3) more sophisticated computational functionality within a GIS, (4) simplified, flexible "user" interface, and (5) transparent data flow between GIS and other functional statistical and graphical modules.

At least two conceptual approaches exist to attempt to obtain the full range of functional capabilities required to support environmental assessment activities. First, the users can communicate specific needs to GIS software vendors with the hope that enough users have similar needs to influence priority on the development of new functionality to be offered with new releases of vendor products. Or, second, the introduction of application building packages such as AVS, aPE, Khoros, Explorer, etc. provides the GIS vendors an opportunity to supply basic modular GIS functionality within a generalized framework that attempts to standardize user interface and data flow to support a variety of functional capabilities beyond what is currently found in geographical information systems.

EPA's atmospheric modeling community has chosen to explore the use of AVS as a user interface not just for data analysis, but as a tool to provide the "real users" hands-on access to model applications, large quantities of data, and a variety of graphical and statistical analysis packages. AVS holds the promise of providing the flexibility to extend its capabilities to emerging, distributed high-performance computing environments with minimum impact to the user's application.

DISCLAIMER

This paper has been reviewed in accordance with the U.S. Environmental Protection Agency's peer and administrative review policies and approved for presentation and publication. Mention of trade names or commercial products does not constitute endorsement or recommendation for use.

REFERENCES

Birth, T., Dessent, T., Milich, L.B., Beatty, J., Scheitlin, T.E., Coventry, D.H., and Novak, J.H. (1990) *A Report on the Implementation of a Pilot Geographic Information System (GIS) at EPA, Research Triangle Park, NC.*

Research Triangle Park, NC: U.S. Environmental Protection Agency.

Chang, J.S., Brost, R.A., Isaksen, I.S.A., Madronich, S., Middleton, P., Stockwell, W.R., and Walcek, C.J. (1987) A three-dimensional Eulerian acid deposition model: physical concepts and formulation. *Journal of Geophysical Research* 92(D12): 14,681–14,700.

Dennis, R.L., Binkowski, F.S., Clark, T.L., McHenry, J.N., Reynolds, S.J., and Seilkop, S.K. (eds.) (1990) Selected applications of the Regional Acid Deposition Model and Engineering Model. Appendix 5F(Part 2) of NAPAP SOS/T Report 5. In *National Acid Precipitation Assessment Program: State of Science and Technology*, Washington, DC: National Acid Precipitation Assessment Program, Vol. 1.

Dennis, R.L., and Novak, J.H. (1991) EPA's Third Generation Modeling System (Models-3): an overview. *Proceedings of the Air & Waste Management Association Specialty Conference on Tropospheric Ozone and the Environment II: Effects, Modeling and Control and The Response of Southern Commercial Forests to Air Pollution, Atlanta, Georgia, November 4–7, 1991*, Atlanta, GA: Air & Waste Management Association.

Eder, B.K., Coventry, D.H., Clark, T.L., and Bollinger, C.E. (1986) RELMAP: a Regional Lagrangian Model of Air Pollution user's guide. *EPA/600/8-86/013*, Research Triangle Park, NC: U.S. Environmental Protection Agency.

Gery, M.W., Whitten, G.Z., and Killus, J.P. (1988) Development and testing of the CBM-IV for urban and regional modeling. *EPA-600/3-88-012*, Research Triangle Park, NC: U.S. Environmental Protection Agency.

Lamb, R.G. (1983) A regional scale (1000 km) model of photochemical air pollution. Part 1. Theoretical formulation. *EPA-600/3-83-035*, Research Triangle Park, NC: U.S. Environmental Protection Agency.

Lamb, R.G. (1986) Numerical simulations of photochemical air pollution in the Northeastern United States. ROM1 applications. *EPA/600/3-86/038*, Research Triangle Park, NC: U.S. Environmental Protection Agency.

Lamb, R.G. (1988) Simulated effects of hydrocarbon emissions controls on seasonal ozone levels in the Northeastern United States: a preliminary study. *EPA/600/3-88/017*, Research Triangle Park, NC: U.S. Environmental Protection Agency.

Morris, R.E., and Myers, T.C. (1990) User's guide for the Urban Airshed Model, Vol. I: user's manual for UAM (CB-IV). *EPA-450/4-90-007A*, Research Triangle Park, NC: U.S. Environmental Protection Agency.

Possiel, N.C., Milich, L.B., and Goodrich, B.R. (eds.) (1991) Regional Ozone Modeling for Northeast Transport (ROMNET). *EPA-450/4-91-002a*, Research Triangle Park, NC: U.S. Environmental Protection Agency.

Upson, C. (1990) Tools for creating visions. *UNIX Review* 8(8): 39–47.

14

GIS AND HYDROLOGIC MODELING

David R. Maidment

GIS provides representations of the spatial features of the Earth, while hydrologic modeling is concerned with the flow of water and its constituents over the land surface and in the subsurface environment. There is obviously a close connection between the two subjects. Hydrologic modeling has been successful in dealing with time variation, and models with hundreds or even thousands of time steps are common, but spatial disaggregation of the study area has been relatively simple. In many cases, hydrologic models assume uniform spatial properties or allow for small numbers of spatial subunits within which properties are uniform. GIS offers the potential to increase the degree of definition of spatial subunits, in number and in descriptive detail, and GIS–hydrologic model linkage also offers the potential to address regional or continental-scale processes whose hydrology has not been modeled previously to any significant extent.

The goal of this chapter is to outline an intellectual basis for the linkage between GIS and hydrologic modeling. Its specific objectives are to present a taxonomy of hydrologic modeling; to understand the kinds of models that are used and what they are used for; to indicate which kinds of models could be incorporated within GIS and which are best left as independent analytical tools linked to GIS for data input and display of results; to examine the object-oriented data model as an intermediate link between the spatial relational model inherent in GIS and the data models used in hydrology; and to look at some future directions of hydrologic models that have not been possible before but that might now be feasible with the advent of GIS. The scope is limited to a fairly abstract discussion looking over the field as a whole rather than to one or other types of models within the field.

BRIEF HISTORICAL REVIEW

A hydrologic model can be defined as a mathematical representation of the flow of water and its constituents on some part of the land surface or subsurface environment. In this sense, hydrologic modeling has been going on for at least 150 years. Darcy's Law (the fundamental equation governing groundwater flow) was discovered in 1856, the St. Venant equations describing unsteady open channel flow were developed in 1871, and a steady stream of analytical advances in description of the flow of water has occurred in the succeeding decades. Transport of constituents in natural waters was sparsely treated before about 1950; after that time, first for transport in pipes, and then later in rivers, estuaries, and groundwater systems, transport issues gradually assumed a greater prominence and are now a major factor in hydrologic modeling. Computer models began to appear in the middle 1960s, first for surface water flow and sediment transport, then in the 1970s for surface water quality and groundwater flow, then in the 1980s for groundwater transport. There are literally hundreds of public domain computer programs for hydrologic modeling; however, the most frequently used models in the United States are produced or endorsed by the Federal government, and these are much fewer in number (not more than a few dozen in total). Table 14-1 summarizes some of the models commonly used in hydrology.

Hydrologic modeling depends on a representation of the land surface and subsurface because this is the environment through which water flows. There are also interactions with biological and ecological modeling because the transport of constituents in natural waters is influenced by biological activity that may increase or decrease the amount of constituents in water, and because the amount and condition of water flow can affect habitats for fish, plants and animals. The degree of saturation of the soil is a time-varying hydrologic parameter that impacts biological and ecological processes. Hydrology is closely tied to weather and climate, so, in principle, modeling of atmospheric processes should be linked to hydrologic modeling, but in practice a close linkage between these two types of models is difficult to achieve because the large, grid-square spatial scale of atmospheric modeling, especially global climate modeling, is so much larger than the watershed or aquifer scale normally used for hydrologic models.

Table 14-1: A summary of some commonly used computer codes for hydrologic modeling, and the sources from which they can be obtained.

Model	Source
SURFACE WATER HYDROLOGY MODELS	
(1)Single-event rainfall-runoff models	
HEC-1	U.S. Army Corps of Engineers, Davis, California
TR-20	Soil Conservation Service, U.S. Dept. of Agriculture, Washington DC
Illudas	Illinois State Water Survey, Champaign, Illinois
DR3M	U.S. Geological Survey, Reston, Virginia
(2) Continuous streamflow simulation	
SWRRB	Agricultural Research Service, U.S. Dept. of Agriculture, Temple, Texas
PRMS	U.S. Geological Survey, Reston, Virginia
SHE	Institute of Hydrology, Wallingford, England
(3) Flood hydraulics	
Steady flow:	
HEC-2	U.S. Army Corps of Engineers, Davis, California
WSPRO	U.S. Dept. of Transportation, Washington, DC
Unsteady flow:	
DMBRK	U.S. National Weather Service, Silver Spring, Maryland
DWOPER	U.S. National Weather Service, Silver Spring, Maryland
(4) Water quality	
SWMM	University of Florida Water Resources Center, Gainesville, Florida
HSPF	USEPA Environmental Research Laboratory, Athens, Georgia
QUAL2	USEPA Environmental Research Laboratory, Athens, Georgia
WASP	USEPA Environmental Research Laboratory, Athens, Georgia
SUBSURFACE WATER HYDROLOGY MODELS	
(1) Groundwater flow	
PLASM	International Groundwater Modeling Center (IGWMC), Colorado School of Mines, Golden, Colorado
MODFLOW	U.S. Geological Survey, Reston, Virginia
AQUIFEM-1	Geocomp Corporation
(2) Groundwater contaminant transport	
AT123D	IGWMC, Golden, Colorado
BIO1D	Geotrans, Inc., Herndon, Virginia
RNDWALK	IGWMC, Golden, Colorado
USGS MOC	U.S. Geological Survey, Reston, Virginia
MT3D	S.S. Papadopulis and Associates, Inc; National Water Well Association
MODPATH	U.S. Geological Survey, Reston, Virginia
(3) Variably saturated flow and transport	
VS2D	U.S. Geological Survey, Reston, Virginia
SUTRA	U.S. Geological Survey, Reston, Virginia; National Water Well Association

CURRENT AND FUTURE ISSUES IN HYDROLOGY

There are three basic issues: *pollution control* and mitigation for both groundwater and surface water; *water utilization* for water supply for municipalities, agriculture, and industry and the competing demands for instream water use and wildlife habitat; and *flood control* and mitigation. Most hydrologic modeling is directed at solving problems in one of these three areas.

A pressing current issue is the cleanup of contaminated soil and groundwater. Thousands of contaminated sites exist and billions of dollars are being poured into cleanup efforts in the U.S. The chief point sources of contamination are dump sites of old industrial chemicals and wastes, leakage of petroleum products from storage tanks and washoff of working areas, and leakage of sewage from municipal septic tanks and sewer systems. The chief non-point source of groundwater contamination is drainage of pesticides and fertilizers used in agriculture.

STATE OF HYDROLOGIC MODELING INDEPENDENT OF GIS

All of the waters of the Earth can be classified into three types: atmospheric water, surface water, and subsurface water. Atmospheric water includes water vapor in the atmosphere and also water and ice droplets being carried by clouds or falling as precipitation. Surface water is water flowing on the land surface or stored in pools, lakes, or reservoirs. Subsurface water is water contained within the soil and rock matrix beneath the land surface. Hydrologic modeling is largely concerned with surface and subsurface water. When modeling of atmospheric water processes is considered, such as precipitation or evaporation, it is generally to supply input information needed for some aspect of the surface or subsurface water balance. The dynamics of atmospheric processes are so complex that models of precipitation and evaporation are more analogies to the real processes rather than precise physical descriptions.

In modeling the flow of water, the main issue is to determine the disposition of rainfall: how much of it

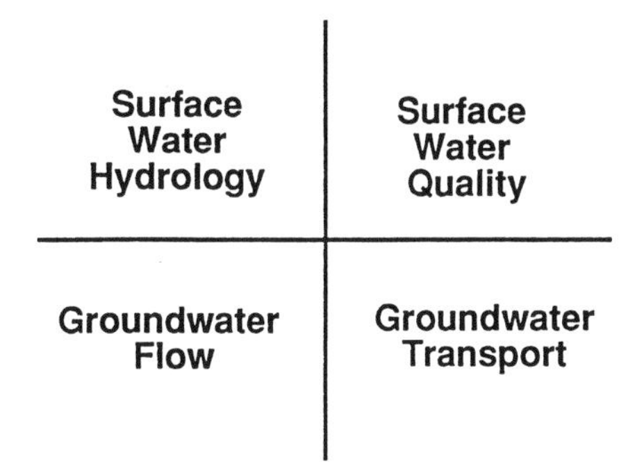

Figure 14-1: A classification of hydrologic models.

becomes runoff, infiltration, groundwater recharge, evaporation, and water storage? Once the discharge of water is determined at particular point, the hydraulics of flow are sometimes also considered, such as the flow velocity and water surface elevation in a channel, or the Darcy flux and piezometric head field in an aquifer. Transport issues are also important. These include transport of material floating or suspended in water such as immiscible oils or sediment; constituents dissolved in water, such as toxic chemicals or pesticides; and biological constituents in water, whose effect is measured by their consumption of dissolved oxygen. The objective in modeling pollutant flows is to be able to predict how far and how fast a pollutant will travel in a water body, what will happen to it as it travels, and what will be its ultimate fate. If remedial activities are being undertaken to clean up pollution, modeling of the extraction of polluted waters is also sometimes undertaken, such as in the design of pumping schemes to extract contaminated groundwater. Obviously, physical processes such as adsorption of the pollutant onto the soil or rock matrix, or chemical processes, such as the oxidation or reduction of chemical species, are important in assessing the pollutant fate and transport.

In a general way, hydrologic modeling can be divided into four components, as shown in Figure 14-1: Vertically, there is a distinction between modeling surface or subsurface waters; sequentially, there is a distinction between modeling flow and transport, or equivalently between modeling water quantity and water quality. Thus, one can speak of a surface water flow model or a surface water quality model, and likewise a subsurface flow or transport model. It is necessary to model or map the flow field before modeling the transport of constituents, since their motion is driven by the motion of the flow field.

One can also distinguish three major variables for which hydrologic models are constructed: for flow or discharge, Q (units of volume per unit time), for water surface elevation or piezometric head, h (units of length), and for constituent concentrations, C (units of mass per unit volume of water). Once the time and space distribution of these variables has been determined, a hydrologic modeling exercise is usually complete. It may be noted that while h and C are scalar variables, their gradients are vectors. The discharge Q is also a vector and is oriented in the direction of declining head gradient. The product of concentration and discharge produces the constituent or contaminant loading (in units of mass per unit time) that is being carried by the flow, such as the sediment load of a river.

Spatial components in hydrologic modeling

There are four basic spatial components used in hydrologic models: surface watersheds, pipes or stream channels, subsurface aquifers, and lakes or estuaries. Each of these components is a three-dimensional object, but they can be approximated satisfactorily for many purposes with a model of lesser dimensions.

Watersheds

It is most common to treat watersheds as *lumped* systems; that is, their properties are spatially averaged, and no attempt is made to describe the topology of the watershed and its stream network. A lumped model is like a zero-dimensional representation of spatial features. When more detail is desired, a *linked–lumped system* is developed in which the watershed is divided into subwatersheds that are each represented as lumped systems connected by stream links, rather in the sense of a schematic diagram, as shown in Figure 14-2, which illustrates how a watershed is treated in the HEC-1 rainfall–runoff model. The linked–lumped system used in this schematic diagram is widely employed in surface water hydrology. In GIS terms, its topology has connectivity but not area definition or contiguity because no explicit attempt is made to represent the outline of the watershed or the stream system beyond a schematic representation.

Some attempts have also been made to treat watersheds as *distributed* systems; that is, their surface terrain is explicitly described using sets of (x,y,z) points, and some progress has been made on constructing hydrologic analysis making use of digital terrain models of the land surface. But this is still a nascent field in which GIS has a significant potential for making a contribution.

Pipes and stream channels

Pipes and stream channels are linear elements that can be represented by arcs in a point–arc–polygon model of topology. In GIS databases, such as the USGS Digital Line

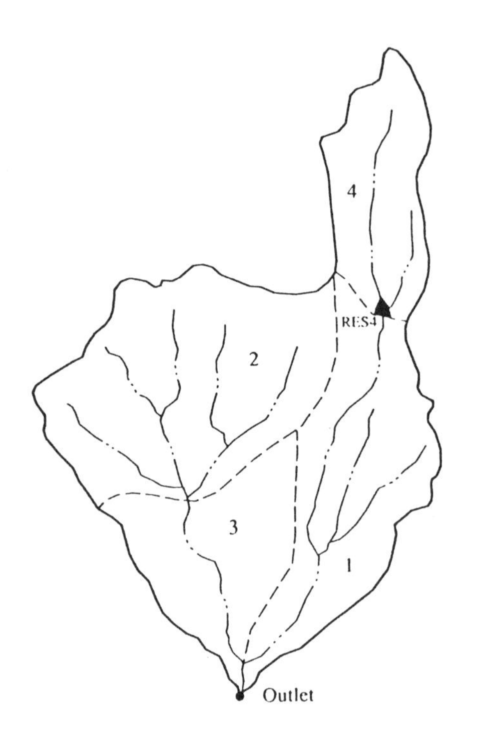

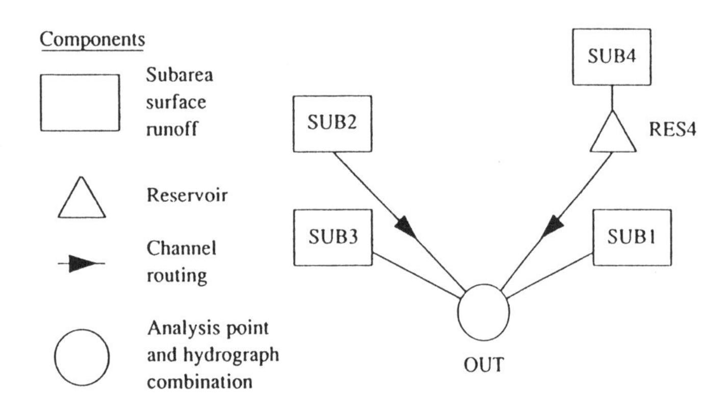

Figure 14-2: A lumped representation of a watershed system as used in the HEC-1 model of surface runoff (Source: Chow, Maidment, and Mays, 1988).

Graph files, streams are shown as lines located on a plan view of the area. The length of the line is known, and sometimes the order of the stream is determined using the Horton stream ordering system in which the smallest recognizable stream is assigned order 1; when two streams of order 1 join, an order 2 stream is created, and so on.

If a particular point on a stream is selected and the properties of the stream as a flow environment are required, four additional items are needed: bed slope, channel roughness (generally indicated by the Manning's n coefficient), cross-sectional area, and the hydraulic radius, which is the cross-sectional area divided by the wetted perimeter of the cross section. The latter two variables, cross-sectional area and hydraulic radius, are depth-dependent quantities; so to describe them one needs cross-sectional profiles of the channel at that point. Thus, a complete description of a stream as a flow environment consists of a plan view showing the location of the center line of the stream (the x dimension), a series of (y,z) cross-section profiles whose location is indicated on the plan view, and a longitudinal profile of the stream bed generally constructed by connecting the lowest points on the cross-sectional profiles. Sometimes, these profiles include areas of the flood plain as well as the stream channel to allow for flow over the stream banks.

For water quality modeling, it is common to use a segment or tank model in which the stream is considered to be a sequence of segments through which water passes at a velocity determined by the flow properties of the stream. Within a segment, the constituents in the water undergo physical, chemical, and biological transformations, and mass balances are performed for each constituent to determine the output concentration of this constituent in the flow going to the next segment (Thomann and Mueller, 1987).

For pipe systems, the cross-sectional area is constant and known from the pipe size; so a plan view of the pipe locations and a vertical profile of the pipe slopes is sufficient. Thus, for a pipe network, the database representation requires explicit location of the end points of the pipes in (x,y,z) coordinates, and connectivity information so that the linkage of pipes from one to the next is known.

Aquifers

Describing aquifers requires maps of the geologic strata, both in the horizontal plane and also vertical cross sections in strategic directions, such as parallel and perpendicular to the direction of groundwater flow. In most cases, however, the spatial extent of the aquifer in the horizontal plane is much greater than in the vertical plane. In Figure 14-3 are shown the horizontal and vertical cross sections of a plume of contaminated water flowing in an aquifer in Massachusetts, the contamination being derived from leakage of sewage effluent from wastewater treatment lagoons. The plume is approximately 2 miles long, half a mile wide and 100 ft thick, so the width-to-depth ratio is approximately 20:1 and the length-to-depth ratio is approximately 120:1. It is thus reasonable as a first approximation to represent the plume as a 2D planar body having depth attributes, rather than as a true 3D body. Hence, aquifers and contaminated plumes of water within them can reasonably be represented within 2D planar GIS systems. In computing groundwater flow, it is common to assume that the flow is horizontal even though it is actually going downhill at a very slight slope of about 1:1000.

Groundwater hydrologists make considerable use of vertical cross sections of an aquifer. The magnitude of the Darcy flux, or flow per unit area of an aquifer system, is computed as the product of the hydraulic conductivity and the slope of the water table or piezometric head surface. The specific discharge, or discharge per unit width of the aquifer system, is found as the product of the Darcy flux and the aquifer thickness at the point in question. The properties of aquifer thickness and slope of the piezometric surface are easier to see on a vertical cross section view of the aquifer than they are on a planar view. The planar view is very helpful in planning pumping schemes because the effect on the piezometric head surface of different spatial patterns of pumping from a network of wells can be readily ascertained.

Lakes and estuaries

When considering what spatial representation is needed for lakes and estuaries, the depth of the water is a critical item. If the water body is shallow, say less than 10–20 feet deep, water currents and wind action on the surface keep the water well mixed between the surface and the bed so that flow properties can be depth averaged in a reasonable fashion. One thus needs to compute the velocity of water and transport rates of contaminants only in the horizontal plane; so the GIS spatial planar representation is quite adequate as a basis for modeling. If the water body is deeper, however, a pronounced temperature profile develops, with the surface in contact with the air fluctuating in temperature and the deep waters having fairly uniform, generally cooler temperatures. This thermal stratification of the water body has profound effects on the lake limnology that need to be considered. One can do this with finite difference or finite element modeling in 3D or with resort to the segment type of models discussed earlier in application to water quality in rivers.

Some lakes, particularly reservoirs created by damming rivers, have a shape determined by the river valley in which they reside; so their flow behaves somewhat like a slow river flow with water perhaps 100 feet deep. Flow and transport in such lakes can be computed by viewing the channel as a linear flow element possessing 1D flow, with adjacent bays and tributaries contributing or receiving water mass from the main channel but not affected by the momentum of the flow in that channel.

Flow and transport processes in hydrology

In each of the four flow systems previously described, the same physical principles govern water flow and transport and similar equations are used to describe these phenomena. The most fundamental principle is the conservation of mass—water or constituent mass cannot be created or destroyed. This principle is expressed as the continuity equation, which states that the rate of change of storage in a system is equal to the inflow minus the outflow. To complete the description of water motion, an additional principle is usually employed—either the momentum principle (for fairly rapidly varying flow with time steps of the order of hours or days) or the energy principle (for long-term studies with time steps of the order of months or years). The momentum principle is contained in Newton's second law of motion, which states that if a body is acted upon by an unbalanced external force, its motion will change in proportion to the magnitude of the force and in the direction of the external force. The energy principle is contained in the first law of thermodynamics, which states that the change of energy of a body is equal to the amount of heat input it receives minus the amount of work it does, work being measured by the product of force applied by the body and the distance through which it moves.

Transport processes are characterized by the rate of advection, dispersion, and transformation of the transported constituent. Advection refers to the passive motion of a transported constituent at the same velocity as the flow. This is the simplest motion that one can conceive, but it is a reasonable approximation, particularly in groundwater flow, where this approximation is used to determine how long it will take leakage from a contaminant source to flow through the groundwater system to a strategic point, such as a drinking water well, a lake, or a river.

Dispersion refers to the spreading of the constituent because of mixing in the water as it flows along. In surface water flow, dispersion occurs because of eddy motion mixing masses of water containing different concentrations of the constituent; in groundwater flow, dispersion occurs because of molecular diffusion of constituents in the water (which occurs even when the water is not moving) and because of differential advection of water moving at different velocities through variously sized flow paths in the aquifer. Dispersion occurs in all three directions: longitudinally (in the direction of flow) and laterally, spreading the constituent both horizontally and vertically around the main flow direction. In the groundwater plume shown in Figure 14-3, lateral dispersion is obviously much greater in the horizontal plane than in the vertical plane. Transport in both surface and groundwater is commonly described by the advection–dispersion equation, which combines mass balance of the constituent with its advection and dispersion in one or more directions (Freeze and Cherry, 1979).

The transformation of constituents by degradation, adsorption, or chemical reaction can change the concentration from that predicted by the advection–dispersion equation, usually resulting in decreased concentrations as the constituent leaves the solution.

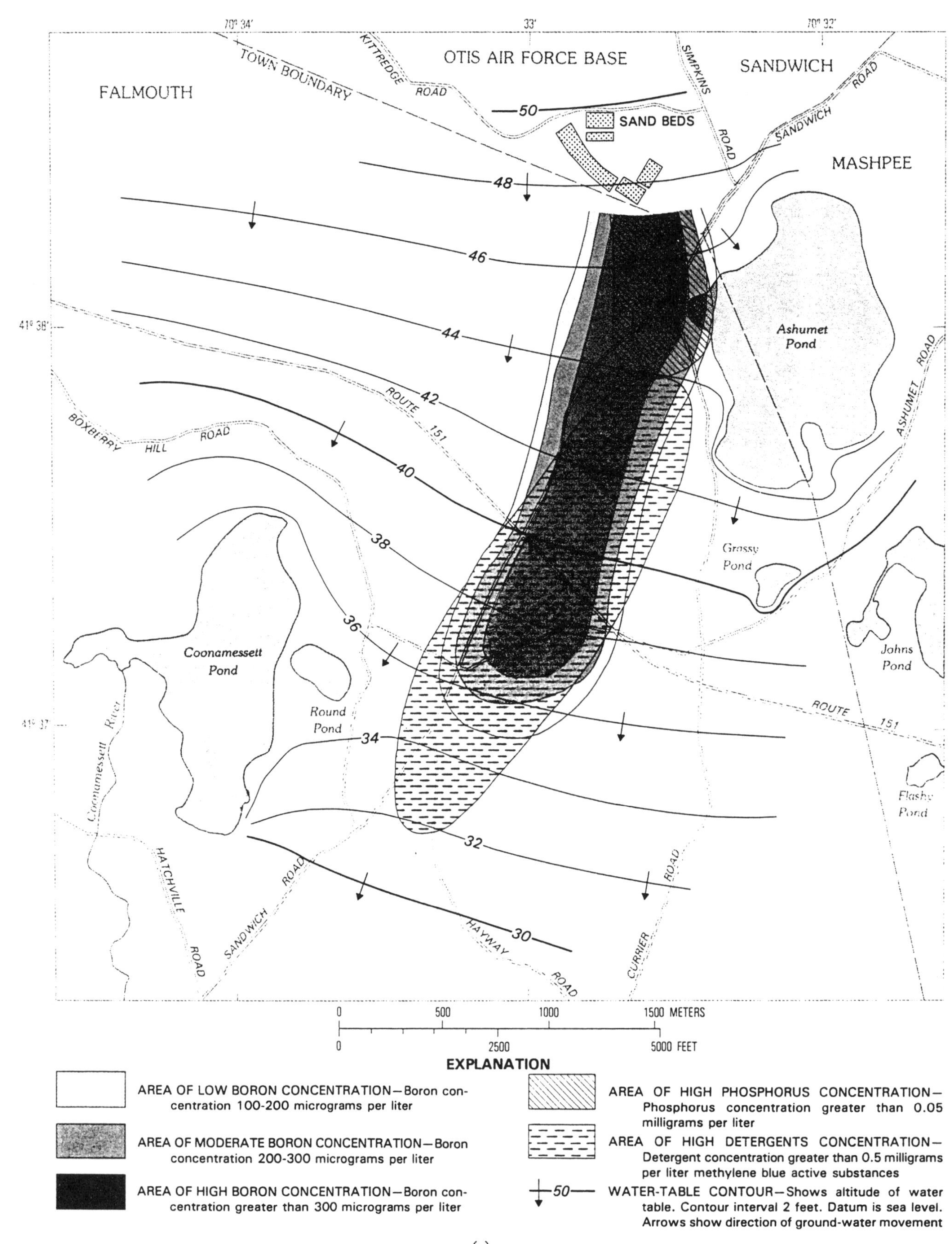
70° 34′
33′
70° 32′
OTIS AIR FORCE BASE
SANDWICH
FALMOUTH
MASHPEE
TOWN BOUNDARY
KITTREDGE ROAD
SIMPKINS ROAD
SANDWICH ROAD
SAND BEDS
50
48
46
44
42
40
38
36
34
32
30
Ashumet Pond
ASHUMET ROAD
ROUTE 151
BOXBERRY HILL ROAD
Grassy Pond
Johns Pond
Coonamessett Pond
Round Pond
Coonamessett River
Flashy Pond
41° 38′
41° 37′
HATCHVILLE ROAD
SANDWICH ROAD
HAYWAY ROAD
CURRIER ROAD
0
500
1000
1500 METERS
0
2500
5000 FEET
EXPLANATION
AREA OF LOW BORON CONCENTRATION—Boron concentration 100-200 micrograms per liter
AREA OF MODERATE BORON CONCENTRATION—Boron concentration 200-300 micrograms per liter
AREA OF HIGH BORON CONCENTRATION—Boron concentration greater than 300 micrograms per liter
AREA OF HIGH PHOSPHORUS CONCENTRATION—Phosphorus concentration greater than 0.05 milligrams per liter
AREA OF HIGH DETERGENTS CONCENTRATION—Detergent concentration greater than 0.5 milligrams per liter methylene blue active substances
50
WATER-TABLE CONTOUR—Shows altitude of water table. Contour interval 2 feet. Datum is sea level. Arrows show direction of ground-water movement

(a)

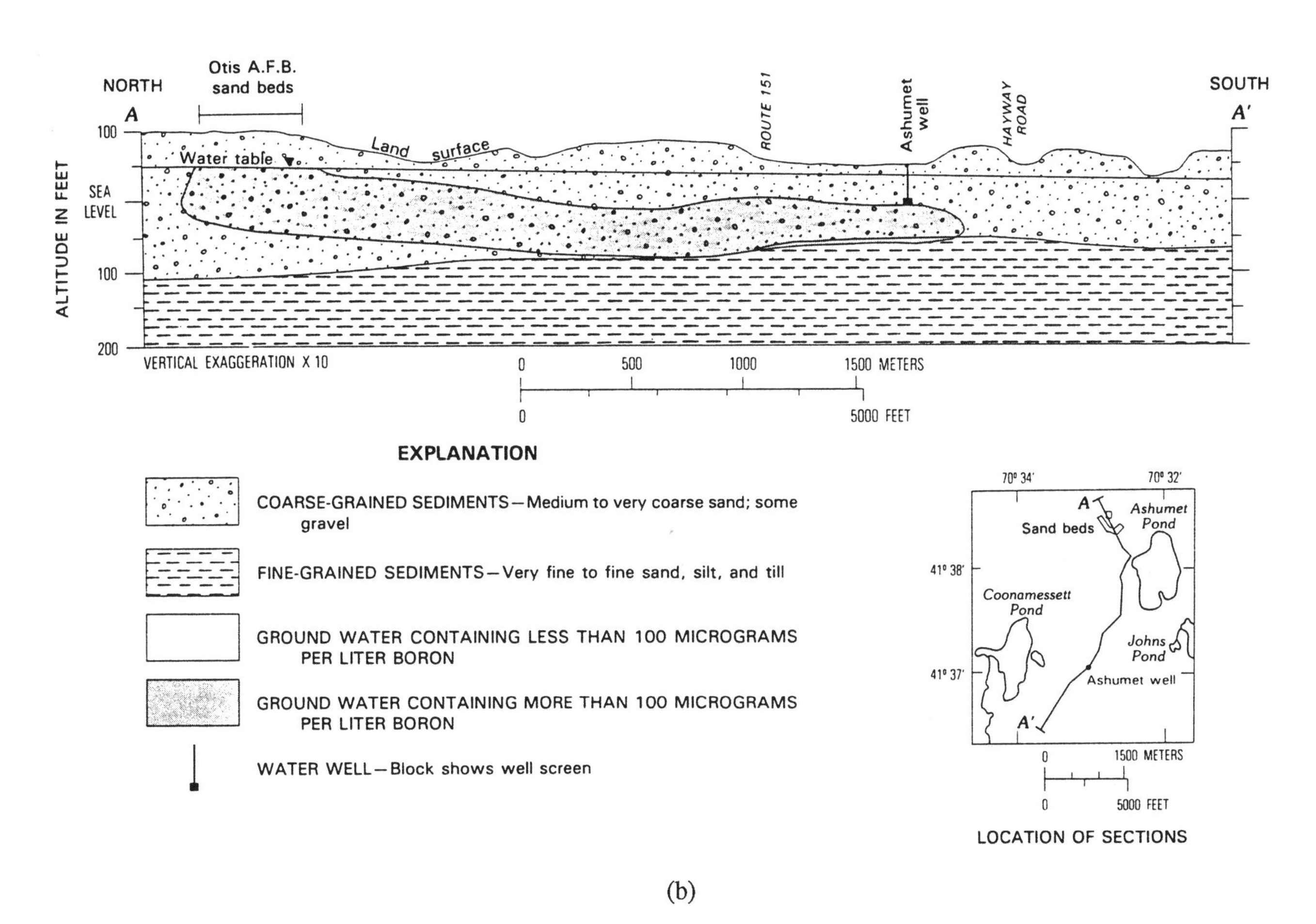

(b)

Figure 14-3: (a) A plan view of a plume of contaminated groundwater produced by leakage from sewage lagoons. The length-to-width ratio of the plume is about 6:1 (Source: LeBlanc, 1984). (b) A vertical cross-section of the same contaminant plume as in (a) with the vertical scale exaggerated by a factor of 10. The actual length-to-depth ratio of the plume is about 120:1 (Source: LeBlanc, (1984).

Taxonomy of hydrologic models

From the previous discussion, it is apparent that similar physical principles and various spatial representations can be used for hydrologic modeling. It is thus possible to propose a taxonomy of hydrologic modeling that classifies available models. There are a number of ways such taxonomies can be built up, but the one that follows serves as a basic framework within which more detailed classifications can be made.

Hydrologic phenomena vary in all three space dimensions, in time, and are random or uncertain because they are driven by rainfall and because many of the properties of the flow domain are unknown, especially for subsurface flow. There are thus five sources of variation that one can consider in a hydrologic model: time, three space dimensions, and randomness. All hydrologic models can be classified according to the assumptions made about these three factors, shown in Figure 14-4. The simplest case is a deterministic (no randomness), lumped (processes are assumed spatially uniform), steady flow model (no variation in time). This is typical of steady, uniform flow in a pipe, channel, or aquifer system, for which the mechanics are well understood and modeled.

Allowing any one of the factors (time, one space dimension, or randomness) to be accounted for explicitly, is also commonly handled, and good models with computer programs for this purpose have existed for about 20 years (e.g., steady, nonuniform flow in rivers, radial flow towards wells in groundwater, regression modeling, and statistical analysis of extreme floods); modeling variation with respect to any two of these factors taken as independent variables has become possible with standardized computer programs within the last 5–10 years (e.g., unsteady, nonuniform flow in rivers, two-dimensional steady flow and transport in groundwater systems; time series and geostatistical analysis of hydrologic data). Allowing any three of these factors to be explicitly varying is really still largely in the research realm (e.g., two-dimensional, unsteady flow in rivers and estuaries; three-dimensional steady groundwater flow models; geostatistical studies with randomness explicitly characterized in two dimensions on a spatial plane). Models with four or five independent variables are even more in the research realm.

The state of flow modeling is more advanced than that of transport modeling, both for surface and subsurface waters. Often the flows are computed first, and then transport is modeled using the precomputed flow field. In subsurface waters, flow in saturated groundwater below the water table is better understood than is unsaturated flow above the water table, especially near the soil surface.

GIS can make a contribution to hydrologic modeling by solidifying the treatment of spatial variation. In so doing, it is likely that models with three independent variables will become capable of general application during the next decade, and those with one or two independent variables will become more accurate and less costly to implement than they are now.

Eulerian and Lagrangian views of motion

In modeling motion of any kind, there are two views that can be taken, the Eulerian view or the Lagrangian view.

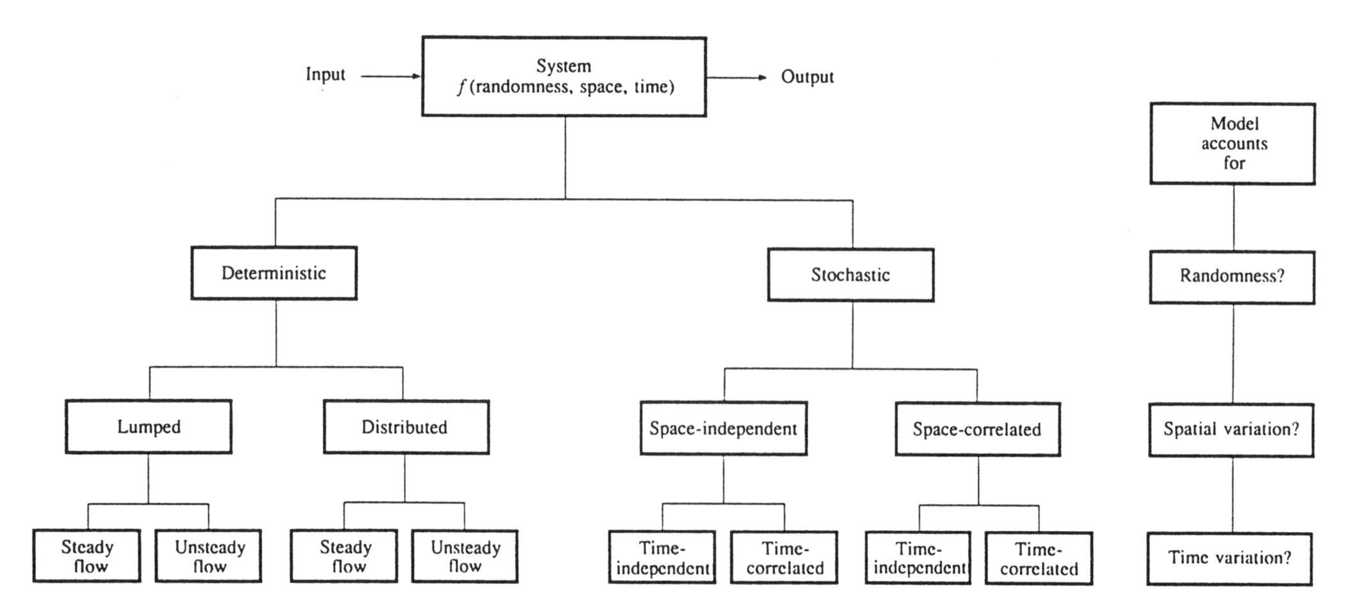

Figure 14-4: A taxonomy of hydrologic models based on the way they represent the space, time, and uncertainty of hydrologic systems (Source: Chow, Maidment, and Mays, 1988).

In the Lagrangian view of motion, the focus is on a moving object, such as a rocket blasting off into the sky—our eyes follow the rocket, and we see the sky pass by as a background. In the Eulerian view of motion, the focus is on a fixed frame in space through which the motion occurs, such as when watching a movie. In modeling traffic flow, for example, a Lagrangian model assumes the general traffic flow has been established and one wishes to follow the motion of a single car, while an Eulerian model would focus on one section of the highway and study the motion of all the cars passing through that section.

This distinction is important in hydrologic modeling because the modeling of fluid flow is nearly always done using the Eulerian view of motion, that is, focusing on a fixed frame in space through which the fluid passes. The mass, momentum, and energy principles governing the rate of flow can all be written in a general form called the Reynolds Transport Theorem, which applies to a control volume. A control volume is a volume in space surrounded by a boundary, somewhat like a polygon surrounded by lines or a line bounded by nodes. One can readily envisage fluid flow modeling being done in GIS using the Eulerian view of motion because GIS presents fixed frames in space through which the fluid passes, as required by the Eulerian view.

The only analysis technique in current use in GIS is network flow modeling, which is normally based on Dykstra's algorithm for traffic flow routing. This algorithm assumes the general flow environment has been established and operates by determining the optimal or least impedance path for a single vehicle through the network. Thus, network flow modeling, as currently practiced in GIS, uses the Lagrangian view of motion, and it is unlikely that fluid motion as it occurs in hydrology will naturally be captured by this type of spatial analysis method. Different network algorithms based on the Eulerian view of motion are needed.

The Lagrangian view of motion is appropriate, however, for describing environmental dispersion, which can be conceived as a process of tracing the motion of the centroid of a dispersing cloud of contaminant particles.

STATE OF MODELING COUPLED WITH GIS

Several levels of hydrologic modeling in association with GIS can be distinguished: hydrologic assessment, hydrologic parameter determination, hydrologic modeling inside GIS, and linking GIS and hydrologic models.

Hydrologic assessment

By hydrologic assessment is meant the mapping in GIS of hydrologic factors that pertain to some situation, usually as a means of risk assessment. The best known example of this is DRASTIC mapping of groundwater contamination potential, which is quite popular for measuring the relative likelihood that groundwater will become contaminated if a waste source is placed on the surface at one point in one region as compared to another. This type of model does not have any explicit physical laws imbedded in its relationships but is a weighted and summed index of the influence of various factors.

Hydrologic parameter determination

Hydrologic parameter determination is currently the most active area in GIS related to hydrology. The goal is to determine the parameters that go into hydrologic models by analysis of terrain and land cover features. Thus land surface slope, channel lengths, land use, and soil characteristics of a watershed are starting to be extracted from both raster and vector GIS systems, with most work up to this time being in raster systems. Three forms of terrain representation are used: rectangular grids, triangulated irregular networks (TIN), and topographic contours. Contour lines are the most common terrain representation in GIS coverages, but they have to be converted to grids or TINs to have automated analysis. Alternatively, grids or TINs can be built up by direct data input, such as by using the USGS Digital Elevation Models for grids or land survey data for TINs.

Once a grid or TIN is available, automated routines are available for watershed and stream network delineation and for some simple flow descriptions assuming flow proceeds in the line of steepest descent. Using the watersheds and stream networks so determined, characteristics can be extracted for hydrologic modeling in external codes. The grid, contour, and TIN models are also applicable for subsurface modeling in much the same way as for surface modeling, and they have been extensively developed in the petroleum industry for that purpose. These terrain models are suitable for describing the motion of an inanimate object rolling on the land surface, but they do not really capture the full complexities of water flow, which considers four forces: gravity, friction, pressure, and inertia. The GIS terrain models of this type consider only gravity and in a few cases gravity and friction together. These models cannot account for the dynamics of turning a stream of water by a wall or gutter or other type of flow obstruction.

A simple way for a GIS to supply hydrologic parameters is through linkage to a library of georeferenced parameter values. The SWRRB model (Arnold et al., 1990) for simulation of the water resources of rural basins has a library of weather parameters defined for about 100 weather stations in the United States, so that in application of the model without local weather data, the analyst chooses parameters from the weather station closest to the study area. Likewise, for soils information, SWRRB has detailed information on soil properties for hundreds of soils classified on county soil maps. If these maps were likewise stored in GIS, then the linkage of these parameters as attributes of the soil polygons would be a straightforward matter. In Canada a similar georeferenced parameter library is used for modeling pesticide movement in soils. A linkage between GIS and the EarthInfo CD-ROM data on the hydrology and climatology of the United States would also be useful in this regard.

Hydrologic modeling within GIS

It is possible to do some hydrologic modeling directly within GIS systems, so long as time variability is not needed. This is the case when considering annual averages of variables, such as annual average flow or pollutant loadings from a watershed. One can then implement spreadsheet-type models in which flows or loadings are computed as flow or load per unit area times the area; one can also capture some more complex equations, such as those for pollutant loadings derived from the regression, where the independent variables in the regression equations are mapped in coverages and then the loadings are worked out based on a mathematical combination of coverage data.

Another way of eliminating time as a variable is to take a snapshot at the peak flow condition and model that by assuming the discharge is at its peak value throughout the system. It is thus possible to route water through GIS networks using analogies to traffic flow routing in which each arc is assigned an impedance measured by flow time or distance and flow is accumulated going downstream through the network. A limitation of this type of modeling is that for water flow the impedance to flow in the arc is related to the amount of flow; that is, there is a chicken and egg situation where one cannot specify the impedance without first knowing the flow and cannot calculate the flow without first knowing the impedance. But this problem has been known in hydrology for many years, and the well-used rational formula for storm sewer design gets around this by working successively downstream, so that the amount of flow is being accumulated as the computations proceed and thus the impedance to be assigned to the next arc can be determined.

It seems theoretically possible that one-dimensional and two-dimensional steady flow computations could be done explicitly based on a GIS database of river channels (for 1D flow) or shallow lakes and estuaries or aquifer systems (for 2D flow). That is not done in commonly used GIS systems at this time, however. Incorporating these flow descriptions for groundwater models within GIS would best be done using the analytical solutions which exist for many different types of groundwater configurations.

Linking GIS and hydrologic models

This is an active area of research, especially in connection with groundwater flow. Two-dimensional finite difference and finite element codes have been linked to GIS for spatial data input to the flow computation and for display of results, such as piezometric head surfaces or contaminant plumes. These models are spatially distributed and rely fairly heavily on the topological representation of the flow domain in GIS as compared to a lighter emphasis on the descriptive attributes attached to the spatial features. Finite difference and finite element codes are extremely complex entities on their own, and solving them is not a straightforward matter. Even if a steady-state solution is ultimately sought, it is generally approached by solving a series of time-varying conditions until an equilibrium is reached. Finite difference and finite element methods are more difficult to solve in surface water flow than in slowly flowing groundwater, because the velocities of flow are much greater and the numerical instabilities generated as part of the solution procedure are more difficult to remove. There does not seem to have been as much work with surface water flow codes of this type linked to GIS as there has been with groundwater flow codes. Good potential for real-time tracking of oil spills and other contaminant spills in surface water appears possible if remotely sensed data on the initial extent of the spill could be quickly funnelled through GIS to the hydrologic model.

Object-oriented linkage

Figure 14-5 shows a concept of GIS representation of the real world, proposed by Scott Morehouse of ESRI. He suggests that beginning with image processing, building through raster GIS, then through vector GIS, one obtains raster and tabular representations of reality. He indicates that the next step beyond tabular modeling is semantic modeling, in which one attempts to capture the functional interrelationship of the things located on the land surface. Semantic modeling is a very familiar concept for hydrologists, as it is the prime way that the data structures for lumped hydrologic models are organized, as shown in Figure 14-2. In such models, geography serves a starting point for building up the semantic representation of the system, but once it is constructed, the semantic representation becomes the primary mental model that the hydrologist uses when thinking about the system.

Building semantic models in hydrology requires an abstraction of spatial features into spatial objects to form a schematic representation of the flow system. An object-oriented model, such as that used in some expert system programs, appears to be a good linkage between GIS and hydrologic models for these types of models. Each line in the input code of the hydrologic model can be conceived of as an object whose properties are the various fields in

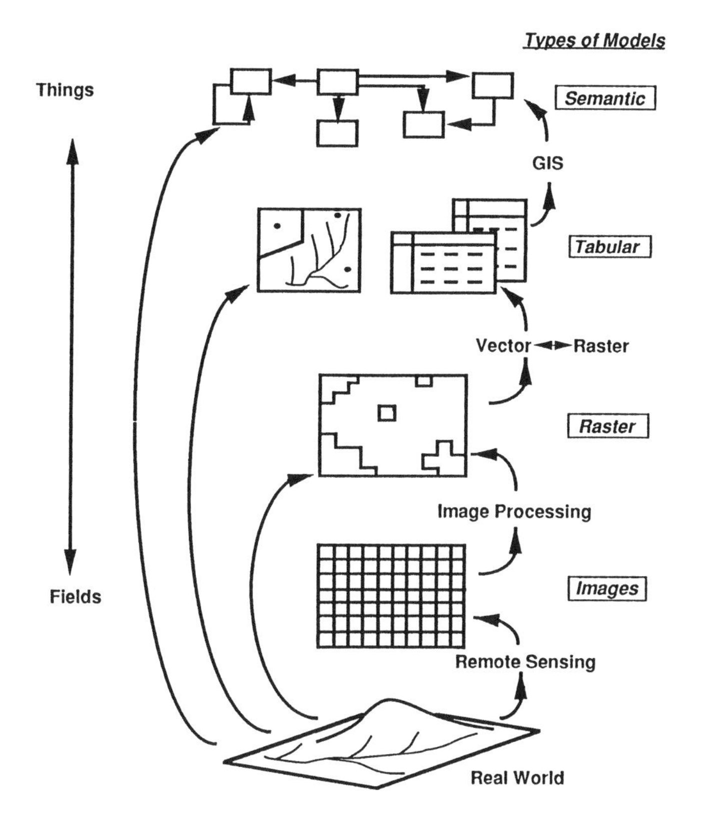

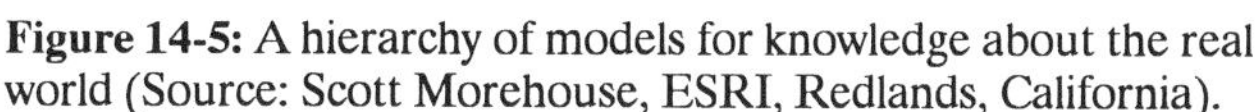
Figure 14-5: A hierarchy of models for knowledge about the real world (Source: Scott Morehouse, ESRI, Redlands, California).

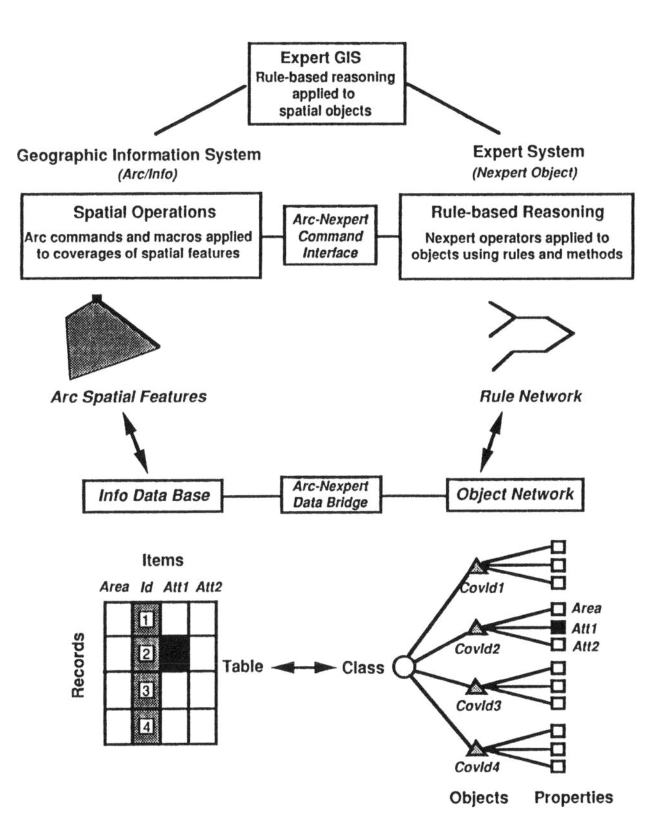

Figure 14-6: An object-oriented representation of GIS data formed by linking the ARC/INFO GIS with the Nexpert Object expert system.

that line. By querying GIS, the values of these properties can be filled in with a direct expert GIS linkage and with linkage to other databases for nonlocationally dependent data, such as physical/chemical reaction rates. Then the input file for the hydrologic model can be written from the expert system or, if it is not very complex, directly from the GIS model itself. Using a back transformation, the output from the hydrologic model could be sent back to the GIS for display and representation. We are currently conducting some research along these lines at the University of Texas at Austin using a linkage depicted in Figure 14-6 between the ARC/INFO GIS and the expert system Nexpert Object (Djokic and Maidment, 1991; Djokic, 1991).

Six common data structures

If two computer programs are going to communicate with one another, it is essential that there be a one-to-one mapping between the portions of their data structures that share common data. In linking GIS and hydrologic modeling, these common data structures must involve the representation of the land surface and subsurface. It is suggested that GIS possesses six fundamental data structures, including three basic structures (point, line, and polygon), and three derived structures [grid, triangulated irregular network (TIN), and network]. Each of these data structures is well understood in hydrologic modeling, where they have been used for many years as a basis for flow modeling.

Indeed, one might interpret coverages of many land cover features in GIS as "flow systems" because their hydrologic properties are usually well understood. For example, a coverage of storm drainage inlets is a flow system of point sinks from the land surface where the discharge into the inlet is proportional to the depth of water over it; a streets coverage has well-defined flow properties because of the regular geometry of streets (though flow at an intersection is a more complex matter); a buildings coverage has simple flow properties—flow does not pass through a building!

Let us examine each of the six data structures to see what it represents in GIS and in hydrology.

Points

A point in GIS is the simplest kind of geographic feature—it has no dimensions. Its sole spatial property is location. In hydrology, points represent water wells, stormwater inlets, or point sources of contamination such as an industrial outfall into a river. But points have a more general utility as well—in GIS an annotation point is used to locate the labeling of a polygon, and the annotation point can be anywhere inside the polygon. In hydrology, a lumped system model reduces geographic features to a point or node (without spatial dimensions) to which are attached properties representative of the corresponding land feature, such as a watershed or stream reach. But the connectivity of these nodes is important because that determines the order in which hydrologic operations will be performed on water flowing through the system.

Lines

In GIS, a line or arc is a set of connected line segments beginning and ending at a node. In hydrology, this concept represents pipes or channels that can be subdivided into segments in much the same way. Water flow and transport along a line is a distributed process; that is, it can be defined as a function of distance along the line, and differential equations written to describe the motion of water and its constituents as they move along the line. In open channel flow, standardized computer programs exist to define the steady-flow water surface profile in an open channel (e.g., HEC-2 from the U.S. Army Corps of Engineers), and these programs are widely used for flood plain delineation.

For unsteady flow in a river, the discharge and water surface elevation are mapped out on a mesh of points called the "space–time domain," as shown in Figure 14-7. The horizontal axis represents distance along the channel, and the vertical axis represents time. Discharge and elevation are computed on "time lines," which are the horizontal lines representing all the distance points at a particular point in time—the equations of motion are written in finite difference form so that values of discharge and water surface elevation on the current time line are computed from the corresponding values of those variables on the previous time line. If the computation proceeds one node at a time, it is called an explicit finite difference solution, while if all points on a given time line are solved simultaneously, this is an implicit solution. Implicit solutions are numerically more stable and are most often used. The U.S. National Weather Service has a standard unsteady, one-dimensional finite difference code called DMBRK, which performs these types of computations on flow in channels, including the effect of structures such as dams, reservoirs, and bridges.

The thinking behind these programs can be illustrated by the following simplified representation. The continuity equation, expressing the conservation of mass, can be written for a cross section in an open channel as:

$$\partial Q/\partial x + \partial A/\partial t = q \tag{14-1}$$

where Q is the discharge at time t at a point located at distance x along the channel where the cross-sectional area of the water flow is A, which is receiving a lateral discharge q of water from land bordering the channel (measured in units of flow per unit length of channel). What Equation (14-1) states is that the change in discharge as a function of distance, $\partial Q/\partial x$, plus the change of the water cross-sectional area with respect to time, $\partial A/\partial t$, is equal to the lateral inflow to the channel, q. The partial differential symbol ∂ is used in this equation to indicate that the changes with respect to x and t are both being considered in the equation but that they are computed separately, rather than jointly, that is, $\partial Q/\partial x$ refers to the change in Q with respect to x at a particular instant of time. In this equation, there are two dependent variables, Q and A, and two independent variables, x and t, and the solution for Q and A is computed at a mesh of points in the space–time domain for particular values of x and t.

It is not possible to solve a single equation for two unknown values; so a second equation is required, in this case the momentum equation, which can be written in the form:

$$A = aQ^b \tag{14-2}$$

where a and b are coefficients determined by the shape, slope, and roughness of the channel from a standard formula called Manning's equation. By differentiating Equation (14-2) with respect to time, and substituting the result into Equation (14-1), the following result is derived:

$$\partial Q/\partial x + abQ^{b-1}\partial Q/\partial t = q \tag{14-3}$$

which now has only one unknown variable Q to be computed as a function of x and t. Equation (14-3) is the kinematic wave equation applicable for determining the motion of a flood wave in a river. It is solved for $Q(x, t)$ by approximating the partial derivatives $\partial Q/\partial x$ and $\partial Q/\partial t$ by the finite differences $\Delta Q/\Delta x$ and $\Delta Q/\Delta t$ between the values of Q at successive mesh points in distance, and in time, respectively, as illustrated in Figure 14-7. In this way, a grid of values of discharge Q as a function of distance and time is computed. By substituting these values into Equation (14-2), the corresponding values of the water cross-sectional area A are found at each point, and from them, water surface elevation or water depth can also be determined.

The finite difference method just described for solving the kinematic wave equations in an open channel is applied in different forms for solving many other problems in surface and groundwater. A partial differential equa-

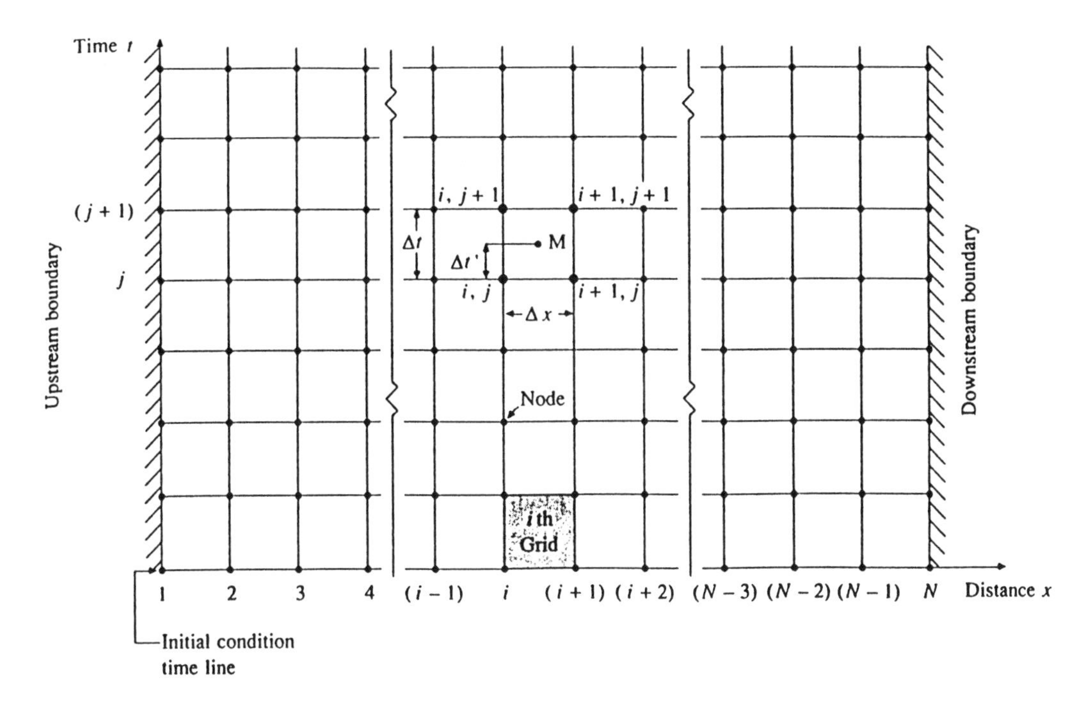

Figure 14-7: A space–time domain for solution of unsteady flow along a stream channel. The discharge and water surface elevation are computed at each point on the horizontal time lines as a function of the corresponding values of these variables on the previous time line. The horizontal axis represents distance along the channel (Source: Chow, Maidment, and Mays, 1988).

tion of continuity is combined with an algebraic equation for momentum to form an equation of motion for the system, and the result is converted from a partial differential equation to a set of algebraic equations by substituting for partial differential terms their finite difference equivalents. In groundwater flow, the momentum equation used is Darcy's Law:

$$Q = KA\ \partial h/\partial x \quad (14\text{-}4)$$

where Q is the discharge, K is the hydraulic conductivity (a measure of the ease of flow in the aquifer), A is the cross-sectional area, h is the height of the piezometric head surface, and x is the distance in the direction of flow.

Polygons

The third fundamental primitive in a GIS system is the polygon, a set of connected lines. Polygons are well known in hydrology as the boundaries of flow systems, such as watersheds and aquifers, and as the delineators of other spatially varying quantities, such as soil types. The spatial variation of hydrologic processes within polygons is normally captured by grid or TIN structures; so the polygon as a data structure does not normally form the basis of a hydrologic model itself, but rather is a boundary for a derived data structure such as a grid or TIN.

There are some cases, however, where polygon mappings of data can be directly used in hydrologic models. One case is the computation of probable maximum precipitation for a watershed, the precipitation that is the largest that is physically conceivable for the watershed location and area and duration of rainfall. This estimate of precipitation is used in studies of the consequences of catastrophic failures, such as the flooding of an area downstream of a dam as the result of a dam breach. As shown in Figure 14-8, the U.S. National Weather Service has developed a standardized design rainfall map having elliptical rainfall contours, or isohyets, of fixed shape and size. Each isohyet has a design rainfall depth associated with it whose magnitude depends on the watershed location and orientation with respect to the normal atmospheric moisture flow direction during extreme storms. This elliptical storm isohyetal map is laid over the watershed and rotated and moved until the average precipitation computed for the watershed is maximized. This would be an excellent procedure to be automated using GIS, because it is very time consuming to perform all the manipulations involved in this procedure by hand on paper maps, and the procedure involved has been standardized by the U.S. National Weather Service in reports

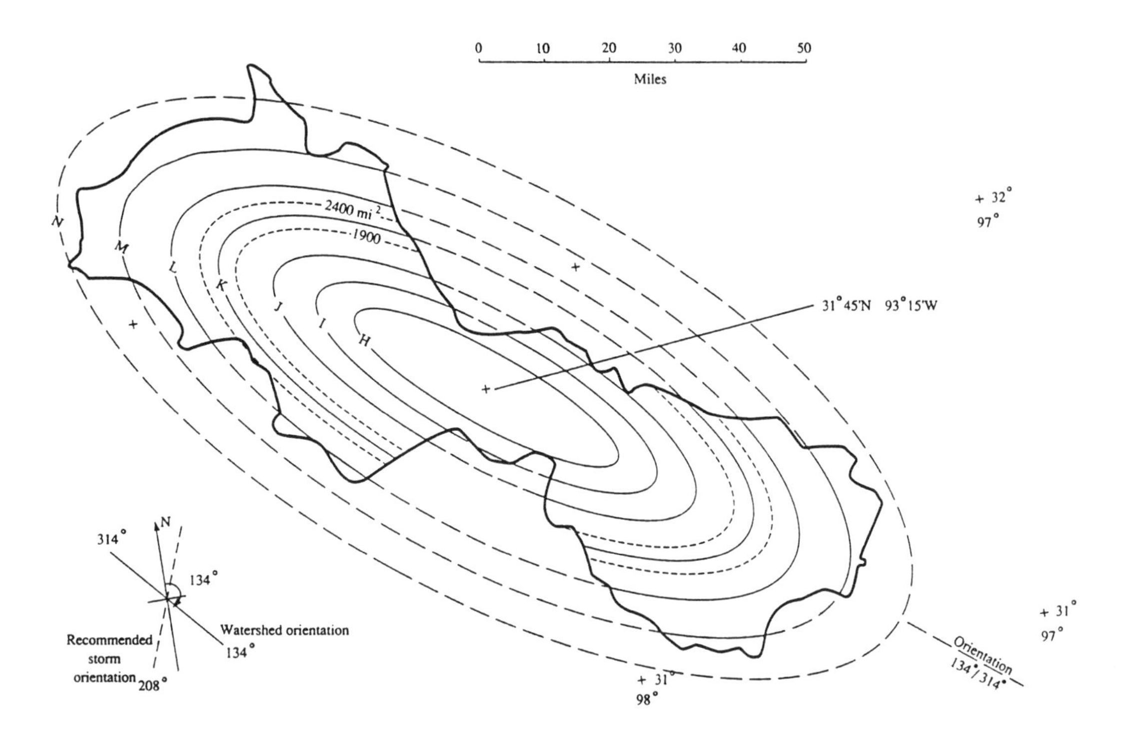

Figure 14-8: The probable maximum precipitation for Leon River near Belton, Texas, is calculated by overlaying a standard elliptical storm pattern developed by the U.S. National Weather Service onto the watershed in such a way as to maximize the average rainfall over the watershed (Source: Chow, Maidment, and Mays, 1988).

known as HMR-51 and HMR-52 (Schreiner and Reidel, 1978; Hanson, Schreiner, and Miller, 1982).

A second case where polygon mapping can be used directly in hydrologic modeling is in the mapping of snow cover, in which the current area of the snow pack can be remotely sensed and translated to become a direct input to a model of snow water balance. This is being done for snow melt forecasting in the western states of the United States.

Grids

Grids are built up in GIS by creating a rectangular mesh of points and joining them with lines, thus partitioning the domain into a rectangular pattern of subareas. In some systems, the term grid refers to the mesh points, while in others the subarea is called the grid and the lines and mesh points enclosing it a lattice. Grid structures are the basis of analysis in raster GIS systems, and they are a very good means of representing the topography of the land surface in vector GIS systems. The gridded land surface data used in such studies is called a digital elevation model.

In hydrologic modeling, grids are used as a finite difference representation of planar flow processes, that is, flow processes that are essentially horizontal, such as the plume migration shown in Figure 14-3. Finite difference schemes are called "mesh-centered" or "block-centered," depending on whether the dependent variables are computed at the nodal points or in the centers of the rectangles, respectively. Finite difference modeling is especially applicable to groundwater flow because the subsurface environment is sufficiently unknown that it can be reasonably approximated by a rectangular mesh of points.

Finite difference models for groundwater flow are complex, and their solution is not standardized. There are widely accepted models, some of which are listed in Table 14-1. The most logical way of integrating this type of hydrologic model with GIS is by interfacing the model as a GIS subsystem without altering the model in any way. The Intergraph Corporation is developing a groundwater analysis system called ERMA (Michael and Pearson, 1991) that incorporates the USGS ModFlow groundwater model and the AT123D contaminant transport model from the U.S. Department of Energy.

A considerable amount of study has been made of the hydrologic features of surface terrain using digital elevation models. The best known routines of this kind have been developed by Sue Jenson of the U.S. Geological Survey (Jenson and Domingue, 1988; Jenson, 1991). These procedures are based on a "pour-point" model in which water from each grid square flows onto the lowest of the eight adjacent grid squares (four on the coordinate axes and four on the diagonals). By tracking the simulated water motion from square to square a set of hypothetical flow paths is traced out; from these, watershed boundaries can be delineated from the set of squares that receive water from no adjacent squares. The stream network is delineated as the set of squares receiving water from at least a specified number of other squares, such as 100 or 200 other squares. This approach is useful for study of fairly large-scale rural watersheds, but is of limited utility in urban areas because imposed terrain of buildings and streets prevents water from flowing straight downhill, as the pour point model implies (see also Chapters 19 and 39).

The derivation of other hydrologic parameters such as the slopes of streams has also been studied using digital elevation models (Tarboton, Bras, and Rodriguez-Iturbe, 1991). The most useful parameters for surface water hydrology models that can be extracted from terrain models are the average land surface slope, and the length and slope of the streams, especially for the longest flow path to the outlet (which is used in computing the time of concentration of the watershed).

Triangulated irregular networks

A triangulated irregular network (TIN) is a triangular mesh drawn on a scattered set of (x,y) points. A TIN requires only a small percentage of the number of points that a DEM does in order to represent the surface terrain with equal accuracy, because the points can be placed strategically on the surface as is the case in land surveying. In hydrologic modeling, the triangular mesh is a familiar sight because it is a common basis for finite element solution of flow and transport problems. In finite element modeling, the dependent variables in the governing equations of motion, such as discharge and water surface elevation, are determined at each node in the triangular mesh, and their variation between nodal points is approximated by simple functions called basis functions (usually linear or quadratic functions), so that a surface of the dependent variable can be constructed over the (x,y) plane. By substituting the current values of the variables into the governing differential equations, a residual error is found for each equation at each nodal point that represents the degree to which that equation is not satisfied at that point. The values of the dependent variables are progressively adjusted until these residual errors are acceptably small, and thus the solution is determined.

Two-dimensional finite element modeling is widely applied in groundwater flow, but less so in surface water flow because the more rapid velocity of surface water flow creates numerical instabilities in the solution that are difficult to damp out. The linkage between the TIN data structure in GIS and the triangular mesh in finite element codes is an obvious way that finite element codes could be linked to GIS. Some widely used finite element codes are listed in Table 14-1.

TINs have also been used for watershed and stream network delineation, as shown in Figure 14-9 (Jones, Wright, and Maidment, 1990). Some automated routines have been developed by ESRI Germany called TIN cascading, by which a first approximation to surface water velocity and erosion rates can be derived from a TIN model of the land surface.

Networks

The network is the final derived data structure in GIS that is also used in hydrologic modeling. In GIS the network model is focused on traffic flow studies, and the routing routines available are fairly simple by hydrologic standards. Both ARC/INFO and Intergraph use Dykstra's algorithm (Larson and Odoni, 1981). Djokic and Maidment (1991) have studied how Dykstra's algorithm can be used for routing water in channel or pipe networks, in particular for flow in dendritic networks of stormwater pipes in a city. They showed that one can construct within GIS the function equivalent of the rational method in hydrology, which is a storm sewer design methodology in which the peak discharge Q at the inlet to a pipe is determined by the equation:

$$Q = CIA \qquad (14\text{-}5)$$

in which C is a runoff coefficient between 0 and 1, I is the precipitation intensity, and A is the drainage area. The

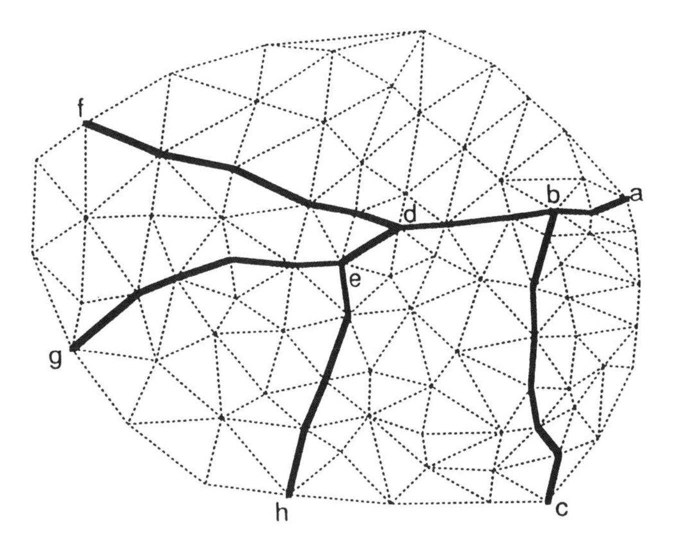

Figure 14-9: A triangulated irregular network model of a watershed and its stream system (Source: Jones, Wright, and Maidment, 1990).

runoff coefficient is a function of the land cover and land slope, and the rainfall intensity is a function of the time of concentration, or time of flow from the farthest point in the watershed to the pipe inlet. By using a polygon coverage of drainage areas for pipe inlets, a point coverage of inlets, and a line coverage of drainage pipes, Djokic and Maidment showed that the discharge Q can be progressively computed going downstream through the pipe network, and the adequacy of the pipe capacity determined. Similar computations could be performed to compute nonpoint source pollution from stormwater pipes.

Using a network model in association with the basic point, line, and polygon coverages in GIS diminishes the need to construct a digital terrain model of the land surface by DEM or TIN methods. But Dykstra's algorithm cannot be used to route unsteady flow so its utility in hydrology is limited. Nevertheless, the rational method is the simplest and most widely used design method in surface water hydrology, and its automation can be accomplished directly within GIS without having to invoke an external model, as is the case with finite difference and finite element codes.

Limits to effective coupling

The most critical limits are the difference in the data models and in the way relationships between variables are handled in GIS and in hydrologic models. The GIS data model is a spatial relational data model that is extremely efficient at processing vast quantities of data on individual layers of information in a uniform way over a large area of space. In hydrologic modeling, our focus is generally local rather than global—we may be concerned only with a small part of an aquifer or a watershed, or even a site only a few hundred square meters in area. But within this area, we want to build up a great deal of information and perform detailed analyses. The spatial relational data model is a good "soil" in which to build to give us a start on the digital representation we need, but we need more than what a GIS can provide to complete that representation.

It goes without saying that many hydrologic analyses are time varying, particularly for surface water flow (less so for groundwater). GIS really does not lend itself to time-varying studies because there is no explicit representation of time in the data structures. One cannot readily model the evolution through time of spatial variation in a phenomenon within GIS for this reason. But such variations are often needed in hydrology, for example, to look at tidal variation of flow and transport in an estuary. Until GIS has explicit time variation in its data structures, its role will largely be limited to an input data provider and an output display and mapping device.

GIS is beginning to include some capabilities to model spatial statistics such as kriging, and because this is an intrinsic geoprocessing technique, it is likely that GIS will become a prime environment for geostatistical studies in the future (see Chapters 40–47). It would be good to be able to model the kriging variogram explicitly in two space dimensions instead of as a correlation distance model whose parameters can be made directionally dependent as is now the case.

Beyond all these issues of data representation, there is another, perhaps more fundamental, limit that separates GIS and hydrologic modeling. In GIS the idea of a relationship is drawn from the classic relational database model of a relation being an association between two sets of data using a key item common to both. Thus, one can process and join vast sets of information on associated spatial features. But a database relation is really a fairly weak connection between two entities. I remember the first time I worked out what a relation meant, I thought, "This is all there is? They have built a successful technology on a measure of association as weak as this!"

When compared to the power inherent in the mathematical machinery commonly employed in hydrologic modeling, particularly when considering the solution of differential equations, a spatial relation does not seem to be a particularly useful device for capturing the algorithms we are accustomed to using in hydrology. A relation has no direction associated with it; it cannot be altered in degree—you cannot have a strong relation or a weak relation; you cannot write mathematical equations that act along a relation. And even if you could do these things, the spatial relational model just does not seem capable of handling the functional reality of the physical laws governing hydrologic processes in their native form. It is hard to see how the differential equations describing the conservation of mass, Newton's second law, the first law of thermodynamics, and the processes inherent in dispersion and transformation of constituents can be explicitly captured within GIS systems. What can be captured are snapshots of these processes at particular points in time, or time-averaged values of process variables over a long interval, such as a year, within which many of the intrinsic physical variations have been accumulated or averaged out, so we just do not worry about them. In this sense, the coupling of GIS and hydrologic models seems to be a logical direction, rather than assuming that one can program hydrologic equations within GIS, a utility that will always be limited by the data model and operators available in GIS.

NEW FRONTIERS

The most obvious accomplishment in linking GIS and hydrologic models is for GIS to provide a digital database representing the land surface environment that can be used to build up the input data for a hydrologic model without having to measure or planimeter those data from paper maps. Second, GIS can act as a display environment for the output from hydrologic models. Coupling GIS and hydrologic models thus will make the models more efficient and effective than they have been in the past.

A more interesting question is: What can be accomplished by a linked GIS–hydrologic model system that has simply been unattainable in the past? What new or better models could we create that would better represent hydrologic phenomena? How could we model phenomena that we have not been able to address previously? There appear to be several areas where these types of significant steps forward are possible.

Spatially distributed watershed properties

A study of the maps of soils and geology of a watershed often reveals that soils and geology are different in the lower areas of the watershed near the stream system than on the hillside areas higher up. Stream channels often lie in areas of alluvial deposits whose hydrologic properties are different from those of the thinner soils and more consolidated rocks on the hillside. In lumped modeling of watersheds, the relative proportion of these different geologies and soils can be accounted for, but the fact that one type is nearer the stream than another is not applicable. By building hydrologic models based on polygon coverages of soils and geology in GIS, instead of assuming spatially averaged properties, it seems possible to overcome this limitation even using lumped representations of the watershed.

Partial area flow

Most surface water hydrology models assume that streamflow is generated during rainfall by Hortonian overland flow, meaning that the soil becomes saturated by rainfall, which leads to overland flow to the stream, a process that grows in contributing area as rainfall proceeds until eventually it covers the whole watershed. Hortonian overland flow is applicable to watersheds with little water absorption capacity under very intense rainfall, but in most rainfalls on most watersheds, overland flow is not occurring at all, or if it is occurring, the area contributing to runoff is concentrated around the stream network. This concept of localized flow is called partial area flow, and although the concept has been known for about 25 years, it has not been incorporated into many surface water hydrology models because of a lack of ability to determine the size and location of the expanding and contracting contributing flow areas as rainfall increases and decreases. By using GIS coupled to models of soil saturation, it may be possible to develop more realistic models of streamflow generation than those assuming Hortonian overland flow. Recent requirements that urban runoff must be filtered to remove pollutants mean that storms of all sizes will be modeled, not just the very intense storms that generate floods.

Surface-water–groundwater interaction

Nearly all hydrologic models treat either surface water or groundwater, but not both together. For the surface water hydrologist, once water percolates down in the soil beyond the root zone of plants, it becomes a "loss" that may turn up in the stream as baseflow at a later time. The surface water hydrologist generally takes a bird's eye view of the watershed, viewing it from above, without a great deal of concern for what lies beneath the soil zone. For the groundwater hydrologist, any runoff that occurs is likewise a loss because it does not enter the soil or aquifer system. The groundwater hydrologist has a worm's eye view of the world, focused on a mental picture of a vertical cross section of layers of soil and geology, especially the water-bearing strata. The condition of the land surface is of secondary significance. For the groundwater hydrologist, a stream is a boundary condition, a point at which water leaves the system. The same comments apply to pollutants as to flow. If a pesticide or fertilizer is prevented from draining into groundwater, that is regarded as a victory for groundwater quality, whereas the reason that drainage did not occur might be because the pesticide or fertilizer was carried away by runoff to pollute surface water flow. Clearly, the surface water and groundwater views of the world are but parts of the same picture, useful for many purposes, but ultimately not completely satisfactory because they are not linked.

This lack of linkage is one reason that partial area flow models of streamflow have not been constructed: In this concept, the areas around the stream are being saturated by groundwater flow coming from below, rather than by rainfall coming from above, as Hortonian overland flow assumes. Moreover, the raising of the water table as rain percolates into the ground far from the stream has the effect of pushing out groundwater into the stream that may have been in the aquifer system for several months. So new water percolation far from the stream can, through transmission of pressure through the groundwater system, fairly quickly force older water resident near the stream out of the groundwater system and back into surface water again. These effects are accelerated if there are hillslopes near the stream as compared to a stream in a flat area.

It appears that to achieve a complete, coupled description of water movement in surface and groundwater, a series of layers of information is needed, for the surface and then for the soil and various subsurface strata. It is possible that these layers can be constructed in 2D but arranged vertically above one another using a terrain network that has a common set of (x,y) coordinates for all layers, but each layer has different descriptive attributes and a z attribute for vertical location at each (x,y) point. Some work has been done in Europe on this concept in a model called SHE (Système Hydrologique Européen), but it has yet to emerge from the research realm. By using GIS technology to represent the various required layers, coupled modeling of surface and groundwater flow may be more feasible, and thus a more complete surface-groundwater description could be achieved at a level capable of widespread application rather than just for research purposes.

Regional and global hydrology

Global climate modeling is now a well-accepted field, and the predictions of these models about global warming raise the concern of hydrologists as to the effect that such warming will have on hydrologic phenomena (particularly on droughts) and on sea water level because that will affect inundation of coastal lowlands. The GCMs are built on a grid model with huge grid squares, hundreds of miles on a side. Hydrology is characterized in GCMs with various abstractions of the soil water balance in order to estimate evaporation and surface heat balance, which are needed as boundary conditions of the atmospheric motion. The lack in GCMs of any characterization of watersheds, aquifers, rivers, and other hydrologic features means that it is difficult to connect climatologic and hydrologic modeling efforts.

A new digital database, called the Digital Chart of the World, has been produced by the Environmental Systems Research Institute and by the U.S. Defence Mapping Agency. This 1:1,000,000 representation of the world has rivers, lakes, wetlands, harbors, coastlines, and land surface contours, from which it may be possible to determine watershed boundaries for large basins automatically. It may thus be possible to speak realistically of large, regional, or even continental and global-scale hydrologic modeling using spatial units of watersheds that are hydrologically meaningful and that are large enough that overlaying GCM grid squares on them is a realistic exercise. One could verify the water balance computations being performed on these large basins by comparing the runoff computed from them with the streamflow recorded on gages in their lower reaches. A special study of this type is being launched in the United States on the Mississippi River basin, but similar efforts could be made elsewhere, though on a lesser degree of detail, using the Digital Chart of the World as a land surface representation.

A detailed study of the world water balance was made in the Soviet Union during the 1960s and 1970s that produced what is still the definitive data on this subject. There are plans being made to redo this study during the 1990s at the State Hydrological Institute in St. Petersburg, and GIS technology, especially coupled with the Digital Chart of the World, would be a very useful approach to employ in such a study.

Spatial patterns of droughts

It is remarkable that droughts have such a profound impact on society, as the latest California drought shows, yet so little is known about droughts. We know little about what causes them, how large or severe they will be, or when they will end. It is known that droughts are associated with particular patterns of atmospheric moisture flow that divert moisture from some areas for some period of time. But what causes these diversions seems lost in the endless turbulence of the atmosphere, coupled to the thermal motion of the oceans. It seems possible using GIS spatial analysis tools and some maps of atmospheric moisture flow, rainfall patterns, and ocean currents, perhaps coupled to GCM model computations, that some greater degree of insight into the circumstances surrounding droughts can be attained. Even if such insights were partial, and they took a great deal of time and effort to create, the benefit to society of some greater ability to anticipate the size and duration of droughts cannot be overestimated. So many water management decisions and other kinds of decisions could be improved by this information even if it was uncertain and incomplete.

RECOMMENDATIONS

It goes without saying that efforts to integrate GIS and hydrologic models should continue and that new frontiers should be pursued. The following recommendations are mainly addressed to those producing GIS information, as suggestions for additional items that would make that information useful for hydrologic purposes.

More remote sensing of land use

Although remote sensing has been possible for many years, data derived from this source are not directly used in many hydrologic models. The most active current use of remote sensing for direct measurement of water quantities is for snow hydrology, where aerial photography is used in conjunction with ground level measurements to assess the area and water content of snow packs.

Remote sensing has a significant potential to add to hydrologic modeling through quantification of land use. The USGS has a set of digital files called the LULC (Land Use Land Cover) files covering the United States with a 1:100,000 or 1:250,000 scale classification of land use according to level II of the Anderson classification system (Anderson et al., 1976; U.S. GeoData, 1990). For example, urban land use is a level I classification in this system, then at level II comes residential, commercial and services, industrial, and several other kinds of urban land use. This classification, when linked to soils and topography of the same area, provides the information needed in hydrologic modeling to determine how much rainfall will run off and how much will infiltrate during a storm.

It is likely that in detailed studies of urban hydrology, a further level of land use classification would be necessary, such as for different densities of residential land use, for example, because the percentage of impervious area, and hence the runoff and infiltration characteristics of the watershed, vary with density of development.

Better SCS curve numbers

In the United States, the most widely used single indicator of the hydrologic properties of a land surface is the Soil Conservation Service curve number, CN, a number which varies between 0 (rainfall produces no runoff) to 100 (all rainfall runs off). Most rural watersheds have CN values of about 50–70, while urban watersheds have CN in the 70–90 range. The table put out by the USDA Soil Conservation Service for determining CN was developed for use on soil conservation projects during the 1950s and is not adapted for use with remotely sensed assessments of land use. The SCS has done an excellent job of classifying all of the soils of the nation into four hydrologic soil groups, A through D, soil group A being porous sandy soils and group D being tighter clays. The curve number is determined from a look-up table based on soil group, land use, and topography. A new look-up table is needed using land use types classified by the Anderson system so that data from remote sensing can be more automatically applied to determine the curve number of a watershed.

Digital soil maps

Extensive digital databases of soil properties have been developed by the U.S. Dept. of Agriculture, such as the soils database maintained by Iowa State University, which contains about 30 characteristics of a very large number of soils. County soil maps have been developed for most of the counties in the U.S., which show the areal extent of these soil types. It does not appear, however, that there is a consistent national set of digital soil maps with attached characteristics that can be directly incorporated within GIS as can the USGS Digital Line Graph or Digital Elevation Models (see Chapter 38 for a more detailed discussion of current and future availability of digital soils data). The author is not aware of any centralized repository from which one can obtain digital coverages of soils. If such a repository does not exist, it needs to be created, because soil information is a critical factor in hydrologic computations, the most critical of all factors when infiltration is the focus of attention.

Line versus point representation of surface topography

There are really two types of coverage of the land surface in a GIS system: the coverage of cultural features, and the representation of the elevation of the land surface itself (sometimes called terrain modeling). Contour maps and DEM files specify surface elevation, but it is not clear how consistent are the measures derived from these two different sources of GIS data. In particular, one can derive stream networks from a grid or TIN representation of the land surface, but the stream network is also mapped in the USGS DLG files as a line coverage. How consistent is the stream network derived from the DEM data with that in the DLG files? How can one use existing line coverages of streams to guide automated watershed delineation routines in their partitioning of the land surface into watersheds (see Chapter 39)? It appears that some systematic study is needed to determine how best to use together the coverage and terrain model data sets that GIS offers.

Data structures for the space–time domain

Because most surface water hydrology flows are time varying and because much of the concern about groundwater flow deals with tracing out the motion of contaminants over time, it would be very helpful to have in GIS some explicit data structures that resemble the space–time domain depicted in Figure 14-5. With such data structures for point, line, and area primitives, it would be realistic to begin thinking about doing numerical modeling within GIS instead of keeping it in a separate code, as is now necessary. There is no doubt that computation is more efficient if it can rely on a single set of data structures instead of having data passed back and forth between different data structures.

3D GIS

The motion of water is truly a 3D phenomenon, and all attempts to approximate that motion in a space of lesser dimensions are just that, approximations. A true 3D modeling system could be utilized in hydrology because the equations of motion in 3D are well understood and because some experience in 3D modeling has been accumulated. The need for 3D GIS is especially acute if models of surface water and groundwater are to be linked up. Currently, surface water flow and groundwater flow are conceived as being quasihorizontal flows, while infiltration from the land surface to the water table is presented as a vertical flow. One might capture this combination with a series of layers vertically registered one beneath the other, but a more complete representation would be a true 3D representation environment where the point, line, and area primitives now present in GIS would be extended by the addition of a volume primitive.

CONCLUSIONS

As one who has been involved in hydrologic modeling for more than 20 years, I believe that GIS technology has the potential to place hydrologic modeling on a much firmer footing than has previously been possible. The waters of the Earth are so extensive, their motion is so complex, and so much about what happens in hydrology is determined by the flow environment through which the water passes, that GIS coupled with hydrologic modeling has a potential to help us do what we are doing now more quickly and efficiently, and to open up some new fields of study that have been previously inaccessible. It is probably true that the factor most limiting hydrologic modeling is not the ability to characterize hydrologic processes mathematically, or to solve the resulting equations, but rather the ability to specify the values of the model parameters representing the flow environment accurately. GIS will help overcome that limitation.

In linking GIS and hydrologic models, the most direct linkage is between hydrologic models of two-dimensional steady-state flow and transport in the horizontal plane. Many of these models exist for groundwater systems, and a lesser number for surface water systems. It is possible that for groundwater, analytical solutions to planar flow and transport could be incorporated within GIS systems, but the differential equations governing flow and transport in most real problems have to be solved numerically, usually by finite elements or finite differences, and those models probably need to be kept separate from the GIS system.

In linking GIS and lumped systems representations of watersheds and stream networks, an object-oriented data model seems to be a useful intermediate step between the spatial-relational model inherent in GIS and the data structures of the hydrologic model. Expert GIS with its linkage between object-oriented and spatial-relational data models may be a helpful tool in achieving this linkage.

Hydrologic phenomena are driven by rainfall and are thus always time dependent, even though by taking snapshots at particular points in time or by time averaging over long periods, a steady-state model can be created. To accomplish a complete linkage between GIS and hydrologic models would require GIS to have time-dependent data structures so that the evolution through time of the spatial distribution of hydrologic phenomena could be readily observed.

The concept of spatial analysis, as practiced by geographers and incorporated into GIS, has the goal of interpreting spatial data; that is, one is presented with a set of spatial features and associated descriptive data and one seeks to determine the patterns inherent in these data and by making intelligent queries of the data to define the optimal locations for activities. The concept of spatial analysis as practiced by hydrologists and incorporated into hydrologic models is that there are equations that govern the motion of water and its constituents through the spatial domain, and one uses these to infer what the flow and transport patterns will be in a particular circumstance with a model. These are two very different concepts of spatial analysis, but they are complementary, and if they can be brought together more closely through the integration of GIS and hydrologic modeling, both GIS and hydrology will be strengthened.

REFERENCES

Anderson, J.R., Hardy, E.E., Roach, J.T., and Witmer, R.E. (1976) A land use and land cover classification system for use with remote sensor data. *Geological Survey Professional Paper 964*, Washington, DC: U.S. Government Printing Office.

Arnold, J.G., Williams, J.R., Nicks, A.D., and Sammons, N.B. (1990) *SWRRB, A Basin Scale Simulation Model for Soil and Water Resources Management*, College Station, TX: Texas A&M University Press.

Chow, V.T., Maidment, D.R., and Mays, L.W. (1988) *Applied Hydrology*, New York: McGraw-Hill Publishing Company.

Djokic, D. (1991) Urban stormwater drainage assessment using an expert geographical information system. *PhD dissertation*, Austin, TX: University of Texas.

Djokic, D., and Maidment, D.R. (1991) Terrain analysis for urban stormwater modeling. *Hydrologic Processes*, 5(1): 115–124.

Freeze, R.A., and Cherry, J.A. (1979) *Groundwater*, Englewood Cliffs, NJ: Prentice Hall.

Hansen, E.M., Schreiner, L.C., and Miller, J.F. (1982) Application of probable maximum precipitation estimates—United States east of the 105th meridian. *NOAA Hydrometeorological Report No. 52*, Silver Spring, MD: U.S. National Weather Service.

Jenson, S.K. (1991) Applications of hydrologic information automatically extracted from digital elevation models. *Hydrologic Processes*, 5(1): 31–44.

Jenson, S.K., and Domingue, J.O. (1988) Extracting topographic structure from digital elevation data for geographic information system analysis. *Photogrammetric Engineering and Remote Sensing* 54(11): 1593–1600.

Jones, N.L., Wright, S.G., and Maidment, D.R. (1990) Watershed delineation with triangle-based terrain models. *Journal of Hydraulic Engineering*, 116(10): 1232–1251.

Larson, R.C., and Odoni, A.R. (1981) *Urban Operations Research*, Englewood Cliffs, NJ: Prentice Hall.

Le Blanc, D.R. (1984) Sewage plume in a sand and gravel aquifer, Cape Cod, MA. *USGS Water-Supply Paper 2218*, Washington, DC: U.S. Government Printing Office.

Michael, R.C., and Pearson, M.L. (1991) *The Environmental Resource Management and Analysis System*, Reston, VA: Intergraph Corporation, Reston VA, 11 pp.

Schreiner, L.C., and Reidel, J.T. (1978) Probable maximum precipitation estimates, United States East of the 105th meridian. *NOAA Hydrometeorological Report No. 51*, Silver Spring, MD: U.S. National Weather Service.

Tarboten, D.G., Bras, R.L., and Rodriguez-Iturbe, I. (1991) On the extraction of channel networks from digital elevation data. *Hydrologic Processes*, 5(1): 81–100.

Thomann, R.V., and Mueller, J.A. (1987) *Principles of Surface Water Quality Modeling and Control*, New York: Harper and Row.

U.S. GeoData (1990) Land use and land cover digital data from 1:250,000- and 1:100,000-scale maps. *Data Users Guide 4*, Reston, VA: National Mapping Program, U.S. Geological Survey.

GLOSSARY OF TERMS

Advection The motion of a contaminant being passively carried in a flow.

Continuity equation The equation of water flow that expresses conservation of mass (usually used for tracing the discharge through a system).

Darcy flux The discharge per unit area of porous medium in groundwater flow.

Deterministic model A model having no random components. A given input always produces the same output.

Dispersion The mixing of a contaminant as a result of molecular diffusion or mass mixing of water elements.

Distributed model A model in which the discharge, water surface elevation, or contaminant concentration is explicitly calculated as a function of location in the domain.

Finite difference model A numerical solution of the differential equations of motion on a rectangular grid.

Finite element model A numerical solution of the differential equations of motion on an irregularly spaced grid.

Groundwater flow Saturated flow below the water table.

Hydraulic conductivity A measure of the ease of flow in a porous medium.

Lumped model An abstract representation of spatial features in which properties are averaged over a watershed or stream segment.

Momentum equation The equation of water flow that relates the forces on a water body to the change in its momentum (usually used for determining the velocity of flow).

Nonpoint source pollution Pollution that is distributed over the land surface, such as drainage of agricultural chemicals.

Piezometric head The natural water surface elevation in a well drilled into an aquifer.

Steady flow Flow in which the discharge, water surface elevation, or contaminant concentration are not changing in time.

Stochastic model A model whose output is described by a probability distribution.

Unsteady flow Flow properties are time varying.

Variably saturated flow Flow that occurs between the land surface and the water table.

15

Integrated Use of a GIS and a Three-Dimensional, Finite-Element Model: San Gabriel Basin Groundwater Flow Analyses

JONATHAN HARRIS,
SUMANT GUPTA,
GREG WOODSIDE, AND
NEIL ZIEMBA

Although the use of GIS to support numerical modeling is becoming an almost commonplace, routine application of this technology, early efforts to integrate the two software environments faced the challenge of working with very different types of software not intended to communicate with each other. Much of the work described here occurred between 1987 and 1989, prior to the advances in GIS and computer technology of the past several years.

The motivation to pursue the development of such an interface was, in this case, the need to use numerical models to reconstruct historical conditions and generate potential future conditions of groundwater flow in the San Gabriel Basin. Four areas within this 170 square mile basin were placed on the National Priority List in 1984, following the discovery of volatile organic compound (VOC) contamination in groundwater. Early evaluations of conditions in these areas showed that the groundwater in the basin behaved as a single, dynamic system, in which stresses induced at one end could affect conditions many miles away (USEPA, 1986). Early on, therefore, it became evident that the entire basin would have to be evaluated as one.

This posed a logistical problem of considerable proportions. Large amounts of information were available, and substantially more information would become available as the Remedial Investigation/Feasibility Study (RI/FS) process moved forward. It was clear that a sophisticated database system was required, and the emerging GIS technology provided the means of housing this information within a spatial framework. At the same time, the modeling needs of the project required some early answers, and initial models of the basin using an industry-standard code (the USGS MODFLOW program) were developed in a standard fashion, outside the GIS (USEPA, 1986).

These initial modeling efforts by CH2M Hill, EPA's prime contractor for the San Gabriel RI/FS, relied on a fairly coarse grid designed to reproduce the regional behavior of the basin. As initial remedial planning efforts became more specific in their need to evaluate the effects of existing pumping patterns and of potential changes to these patterns, the need arose to understand the hydraulic response of the system on a finer scale. Approximately 80 percent of the water leaving this groundwater system is extracted from pumping wells. Consequently, changes in the magnitude and location of pumping can have enormous effects on the existing groundwater flow regime. Accurate spatial representation of pumping, both laterally and vertically, substantially enhances the ability of models to simulate the response of the system to changes in the pumping pattern. The need for better resolution in the simulations, coupled with the need to simulate contaminant transport, led to the search for an alternate modeling code.

A review of groundwater flow and contaminant transport codes culminated in the selection of CFEST (Coupled Fluid, Energy, and Solute Transport), a code developed by Battelle Memorial Institute to support the Department of Energy's search for a nuclear waste repository. CFEST was chosen primarily because it is well known, thoroughly tested, well documented, and features a finite-element solution technique that would enhance simulation of the effects of the geometry of the basin, faults, and specific well locations (Gupta et al., 1987).

The first use of GIS capabilities to assist modeling

efforts was in the conversion of the MODFLOW input files for CFEST. The GIS provided the capability to overlay grids and transfer information from the orthogonal, regular MODFLOW grid to a very irregular CFEST finite-element grid in an almost automated fashion. The ease and efficiency with which the MODFLOW/CFEST conversion was carried out underscored the potential value of continuing to use the GIS as a modeling platform.

Initial CFEST grids of the San Gabriel Basin, although based on the previous MODFLOW model, were finer in detail and required further calibration. Typically, calibration involves iterative simulation of historical conditions in which adjustments are made to input variables until the simulations reproduce historical, observed conditions with acceptable accuracy. This can be an arduous process requiring months of file editing and hundreds of repetitive model runs. The GIS, which already stored all the information needed to set up the model runs, also provided the capability of automating the process of modifying and generating input files. By developing menu-driven interfaces to perform the most repetitive steps, calibration was completed in perhaps half the time it would have taken if the GIS tools had not been available.

The value of the GIS in setting up and calibrating the San Gabriel model supported further development of a GIS/CFEST interface. Once calibration of the model was complete, ARC/INFO Arc Macro Language (AML) programs were written to support a variety of types of model-based evaluations. The role of ARC/INFO programming expanded to the display and analysis of CFEST output, rather than concentrating on automating the generation of CFEST input files. In time, throughout the process of setting up, calibrating, and using the model for a variety of evaluation purposes, it is estimated that savings of up to 60 percent have been realized over the time and expense associated with more traditional approaches to numerical modeling of large, complex systems.

The following paragraphs outline the general characteristics of the interface as developed so far. It should be noted that this interface has been developed through necessity, and has had to prove its worth every step of the way. There was no clearly thought out plan at the outset as to what the interface should look like and how it was to perform. There were no preconceptions nor large-scale plans for integrating diverse programming environments. Instead, efforts followed a day-by-day need to automate communication between programs. The resulting family of software provides an efficient means of modeling using GIS-based databases. By removing the time and expense associated with transferring information between different environments, modeling of natural systems for which data have been stored in GIS is potentially as straightforward as it would be if the GIS and modeling programs were one.

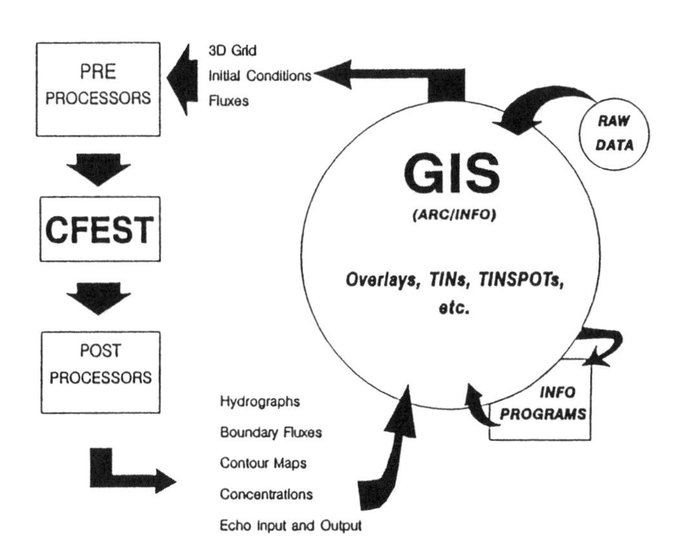

Figure 15-1: GIS/CFEST interface flow diagram. Raw data are stored and processed in the GIS, which is used to generate input to the numerical model. Model simulation results are postprocessed into a variety of tabular and graphical formats, and incorporated into the GIS, where they can be displayed in combination with other GIS information.

OVERVIEW OF GIS/CFEST INTERFACE

The general flow of information between the model and the database is shown in Figure 15-1. The GIS is used to generate files that are read by CFEST preprocessors. Once the model is run, the output can be brought back into the GIS as either raw information or as plots of processed information. This entire loop can be run repetitively using interfacing programs during both calibration of the model and subsequent simulations of scenarios.

CFEST

The CFEST code is actually a group of over twenty FORTRAN programs, of which only five perform the simulations. There are two file-generating, preprocessing programs, and most of the rest are postprocessing programs to display output.

Details of the CFEST model are best left to existing documentation (Gupta et al., 1987). Briefly, there are five main, computational programs that run consecutively. The first reads an input file containing the geometric and physical parameters of the problem and generates binary files read by subsequent programs. The second generates constant integration parameters to form the element matrix. The third assigns locations to a matrix coefficient system to estimate hydraulic heads and solute concentrations. These three programs together set up the basic parameters of a problem, and need be run only once after calibration. However, during calibration, these parameters typically undergo the most adjustment.

The fourth program reads a second input file containing time-dependent flux information. Specifically, this is where time steps, boundary conditions, infiltration rates, sources, and sinks are defined. The fifth and final program performs the actual simulation using the information set up by the previous programs.

As shown in Figure 15-2, there are essentially four steps in the setup and numerical simulation of a groundwater system. Initially, an adequate level of understanding of the physical system must be developed. This understanding is typically described as a conceptual model in which the physical parameters of the system are identified along with the components of the water budget, and if applicable, the magnitude and extent of contamination. The conceptual model is used as a basis for setting up the numerical model. When input requirements have been met, a series of simulations is begun to calibrate the model to best reproduce observed conditions.

For San Gabriel, the initial step of developing a conceptual model was accomplished over a period of several years (USEPA, 1989). This is, of course, an ongoing process that continues as additional data become available. The physical setup of the CFEST model of the basin was actually based to some extent on transferring information from the early MODFLOW model. However, substantial additional input information was required because the CFEST model covered a larger area on a finer scale and simulated contaminant transport in addition to groundwater flow. Following calibration, routine simulations have been run episodically to support a variety of investigations, but have typically also involved a series of repetitive tasks.

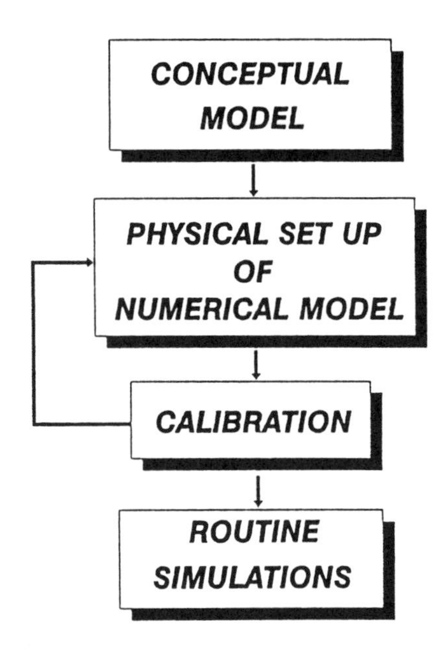

Figure 15-2: Modeling steps. Automating interface of data with a numerical model is efficient when repetition of similar steps is required.

GIS

The San Gabriel GIS database has evolved over a period of over 4 years. It was one of the first large-scale uses of GIS on EPA's Superfund program. It contains about 200 megabytes of data on water quality, well characteristics, land use, topography, geology, institutional and political boundaries, and interpretations of the extent of contamination (USEPA, 1989). In addition, numerous layers of CFEST model input requirements (grids, boundary conditions, elemental and nodal properties) and simulation results (heads, concentrations, flow vectors) are stored alongside recorded historical data.

The GIS is used on the project for more than data storage and model support. Data query and analysis capabilities are used to support a variety of investigations and evaluations. Automated procedures have been developed to summarize information in a variety of forms. For example, to revise and update interpretations of the extent and magnitude of contamination, macros have been written to query the water quality files for data on specific contaminants, at specific depths, from a specific time period, and output the information using predetermined symbology on maps that are used as a base on which interpretations are made. Data input procedures to upload information from PC data input stations are also automated.

The GIS played an important role in each of the four steps shown in Figure 15-2. GIS capabilities were extremely valuable in reviewing and evaluating the available information to form a conceptual model. As explained previously, the GIS was also a key to setting up the CFEST grid and transferring information from the MODFLOW model. In addition, the GIS was used to calculate nodal and elemental fluxes where spatial interpolations were required. But the third and fourth steps, which require the most iteration, are where the GIS proved most valuable.

Preprocessing procedures

Much of the work associated with setting up modeling input files involves assigning properties and fluxes to the nodes and elements of the grid. This can be particularly tedious in the case of an irregular, finite-element grid in which the elements are of unequal area and the nodes are not lined up in rows. The San Gabriel GIS includes automated procedures for assigning properties during the setup and calibration stage, and, to a somewhat lesser extent, for modifying fluxes in the process of running repetitive simulations.

The physical properties of an aquifer can be grouped and assigned as zones of constant value. A property like hydraulic conductivity, which may vary continuously over an area, can be defined in the model as a series of zones of contrasting value, instead of assigning different numbers to each node. Zoned parameters can greatly simplify the process of setting up and calibrating a model. Other parameters, like water table elevations, must be represented as continuously varying surfaces.

The spatial distribution of assigned values of a zoned parameter is further simplified in a GIS through the use of interactive query and calculation functions, and look-up tables. San Gabriel macros have been set up to revise most zoned variables with pull-down menus and interactive mouse-driven assignment procedures. In this manner the area occupied by zones of specific values can be easily manipulated, and the values assigned to each zone can be modified by editing a single value on a look-up table.

Parameters that are not zoned have been assigned using GIS surface modeling capabilities. Contoured or point data are used to generate continuous surfaces from which nodal values can be defined. Caution is advised, however, in the use of surface models based on widely scattered data or contours: The interpolation routine may generate a surface that deviates substantially from the data in areas with few data points. Assigning fluxes to simulate pumping wells and spreading basins can also be automated using GIS capabilities. In San Gabriel, over 400 pumping wells are assigned to individual nodes and elements.

Postprocessing procedures

CFEST postprocessors read files generated during simulation and translate the raw results into either tabular data summaries or plots, including contour maps and cross sections. Likewise, simulation results can be brought into the GIS as either data files or ARC/INFO coverages. Either may be preferable, depending on when in the modeling process a simulation is being performed (Figure 15-2). For example, during calibration it is valuable to compare simulated water levels as contour maps, which can be overlaid on existing contour maps of observed conditions. Later simulations may require more refined figures that can be generated from the raw data within the GIS.

Data files brought into ARC/INFO must be related to spatial attributes before they can be plotted on a map. This is easily accomplished if the original grid with node and element numbers is also available in the GIS. The data can be added to an existing file containing other node and element attributes, or, more simply, plotted by relating the new data file to an older one by node or element number.

CFEST postprocessors generate CALCOMP-format plot files that are converted into ARC/INFO coverages using the GENERATE routine. A family of translators has been developed that can pass on attribute data, delete undesired objects in the original plot, and reproduce color schemes. These translators provide an almost seamless interface between an operating system plotting environment and the GIS. Once these plots are brought in as coverages, they can be displayed along with other available information for comparison. In addition, the same types of spatial analyses can be performed on the simulated data as with information brought in with conventional methods.

Model calibration

The process of calibrating a model, following the initial set up of parameters, involves a repetitive cycle of (1) performing a simulation, (2) reviewing the results, and (3) modifying a parameter. It was during the calibration of the San Gabriel model that programs and macros were originally developed to perform the repetitive tasks of modifying input files and transferring data.

Calibration interface programs developed for San Gabriel fall into three categories: (1) ARC/INFO programs to modify parameters interactively and write files to the operating system; (2) CFEST FORTRAN preprocessors to read these files, integrate information, and prepare CFEST input; and (3) programs to translate CFEST output into ARC/INFO formats. The first of these types resulted in the greatest time savings by eliminating the need to edit very large files of input parameters manually. The second and third sets of programs form the actual interface and reduced the time required to move information into and out of various formats.

The result is a relatively seamless transition of information from the GIS into the model, and back into the GIS, as shown in Figure 15-1. The effort required to set up these interfaces is justified if, given a problem the size of the San Gabriel Basin, it is foreseen that sufficient repetition of tasks will ultimately result in significant time savings.

Comparative transport analyses

As with calibration, a variety of repetitive tasks were (and continue to be) required in using the model to support investigations, particularly feasibility studies comparing remedial options. The bulk of the effort in performing routine simulations is in the evaluation and display of simulated results, rather than in the modification of input files.

For example, a comparison of simulated areas of contamination is often required as a means of approximating the result of a remedial activity, versus the result of no action (e.g., USEPA, 1990). To compare simulated results of a no-action (or base case) scenario with the results of a simulation of a remedial scheme, two distributions of contaminants are overlaid. What is important in these overlays is the difference in area between contours of concentration, and the quantification of these differences is easily automated in ARC/INFO. In this manner, comparative evaluations can be performed repetitively to identify the most promising of a set of potential scenarios. The same evaluation, if performed outside the GIS, would require hours of visual inspection, or development of GIS-like software to perform a single function. In fact, for San Gabriel, these types of evaluations would not have been considered feasible had these tools not been available.

CONCLUSIONS

The problem posed in the undertaking of San Gabriel RI/FS activities was how to take advantage of both existing GIS and modeling software efficiently. Given the different structure of the two environments, transfer of information from one to another can be time consuming and expensive, potentially negating the benefits of using the two together. There are a number of ways of solving this problem. If the resources are available, pseudo-GIS capabilities could be built into the models to make available some of the spatial tools GIS provide to the modeling environment.

Alternatively, the modeling algorithms could be brought into the GIS environment as another tool available to the GIS user to manipulate information. The first approach would take advantage of the existing modeling software designed primarily to perform simulations, but that in many cases already includes a variety of data manipulation capabilities. The second would require reconstructing the architecture of the model to make it compatible with a GIS environment. The second option would also bring the model into direct contact with the multidimensional, structured data set provided by the GIS.

The approach described here represents an attempt to bring the GIS and model together by addressing the symptoms that make the transition between them painful. Instead of building a new, single environment, the two have been left apart, with all their capabilities left intact, but tied together with a network of programs that communicate between them. If developed sufficiently, this communication can make the merge seamless and transparent to the day-to-day user. Although less elegant than a true merge, it has been found, in this case, to work well enough to replace the immediate need to build a new environment.

In retrospect, it is not clear how this (or other) project(s) might have benefitted from a truly integrated GIS/CFEST system. Furthermore, a more ambitious and complete integration effort would never have been considered feasible, in terms of either time or money. The loose integration was cost effective because it provided a tangible savings in effort. It also preserved the strengths of the separate systems.

For example, like the individual systems, the San Gabriel integration was available for use in the short term (development time was less than time saved by its use). It was developed to be responsive to the needs of the project. It also retained the integrity of the individual systems: no testing, benchmarking, or validating of the codes was required. The reputation and recognition of both ARC/INFO and CFEST allowed project results to be much more defensible and generally acceptable than would have been the case if a new, single, integrated system was used. However, had such a system already been available, tested, documented, and distributed, the same or greater benefits would have been available without the burden of development.

REFERENCES

Gupta, S.K., Cole, C.R., Kincaid, C.T., and Monti, A.M. (1987) *Coupled Fluid, Energy, and Solute Transport Model*, Columbus, OH: Office of Nuclear Waste Isolation, Battelle Memorial Institute.

USEPA (1986) *Draft Supplemental Sampling Plan (SSP) Report, San Gabriel Basin, Los Angeles, California*, prepared for the U.S. Environmental Protection Agency, Region IX, by CH2M HILL.

USEPA (1989) *Draft Report of Remedial Investigations, San Gabriel Basin, Los Angeles County, California*, internal draft prepared for the U.S. Environmental Protection Agency, Region IX, by CH2M HILL (Public draft in preparation).

USEPA (1990) *Basinwide Technical Plan Report, Public Review Draft*, Volumes 1 and 2, prepared for the U.S. Environmental Protection Agency, Region IX, by CH2M HILL.

16

Modeling the Effects of Climate Change on Water Resources in the Gunnison River Basin, Colorado

LAUREN E. HAY,
WILLIAM A. BATTAGLIN,
RANDOLPH S. PARKER, AND
GEORGE H. LEAVESLEY

Changes in climate resulting from increasing concentrations of atmospheric carbon dioxide and other trace gases may alter various hydrologic characteristics in the drainage basins of the Rocky Mountains. These drainage basins provide a substantial water resource to users within, and outside, the Rocky Mountain region. Changes in hydrologic characteristics will affect components of the water balance, including snow pack accumulation and melt, evapotranspiration, streamflow, and recharge to subsurface storage. As part of the Gunnison River Climate Study, climatic and hydrologic processes are modeled to assess the effects of climat change or water resources.

The Gunnison River Basin (Figure 16-1) has a drainage area of 20,530 square kilometers, and elevations that range from 1,410 to 4,400 meters. The Gunnison River is a large contributor of water to the Colorado River system, providing more than 40 percent of the streamflow in the Colorado River at the Colorado–Utah state line (Ugland et al., 1990). The Gunnison River Basin is diverse and provides a challenge in defining the spatial distribution of various components of the water balance. Because the Gunnison River provides a substantial portion of water to a complex water-resource system, it is a useful basin to test the sensitivity of water resources to potential alterations in climate. A joint study to assess this sensitivity has been initiated between the U.S. Geological Survey and the U.S. Bureau of Reclamation.

The objectives of the Gunnison River Climate Study are to identify the sensitivity of the water resources within the basin to reasonable scenarios of climate change, and to develop techniques useful in assessing the sensitivity of water resources to changes in climate. To accomplish these objectives, the study has been divided into a number of work elements, two of which are the development of climate-change scenarios and watershed models. The objectives of the climate-change scenarios element are to describe the existing climate regime and develop alternative climate scenarios for use in sensitivity analyses of the water resource. The objective of the watershed-modeling element is to provide streamflow for current and hypothetical future climatic conditions. The watershed-modeling element has two major components—watershed or small basin modeling and streamflow/reservoir routing. Watershed modeling is conducted primarily on small, upland drainage basins. Watershed models will provide the input to downstream routing models that can be used for water resource management. The focus of this report will be the application of GIS to these elements.

MODELING WITH GIS—APPLICATIONS

GIS provides a link between data, researchers, and models. The ease of data access and the ability to develop flexible methods for the quantification of spatial variables over discrete areas makes GIS an integral tool to modelers working in the Gunnison River Basin. In general, computer programs are being written within GIS that expedite the transfer of information from the GIS to a model and automate the extraction of spatial data using a modeler-specified technique. The development of true interfaces between process models and the GIS is very programming intensive and is deemed beyond the scope of work for the Gunnison River Climate Study, although work is currently being done to develop a true interface to certain models (Leavesley and Stannard, 1990).

The Gunnison River Climate Study is using GIS to: (1) establish a common database for individuals working on

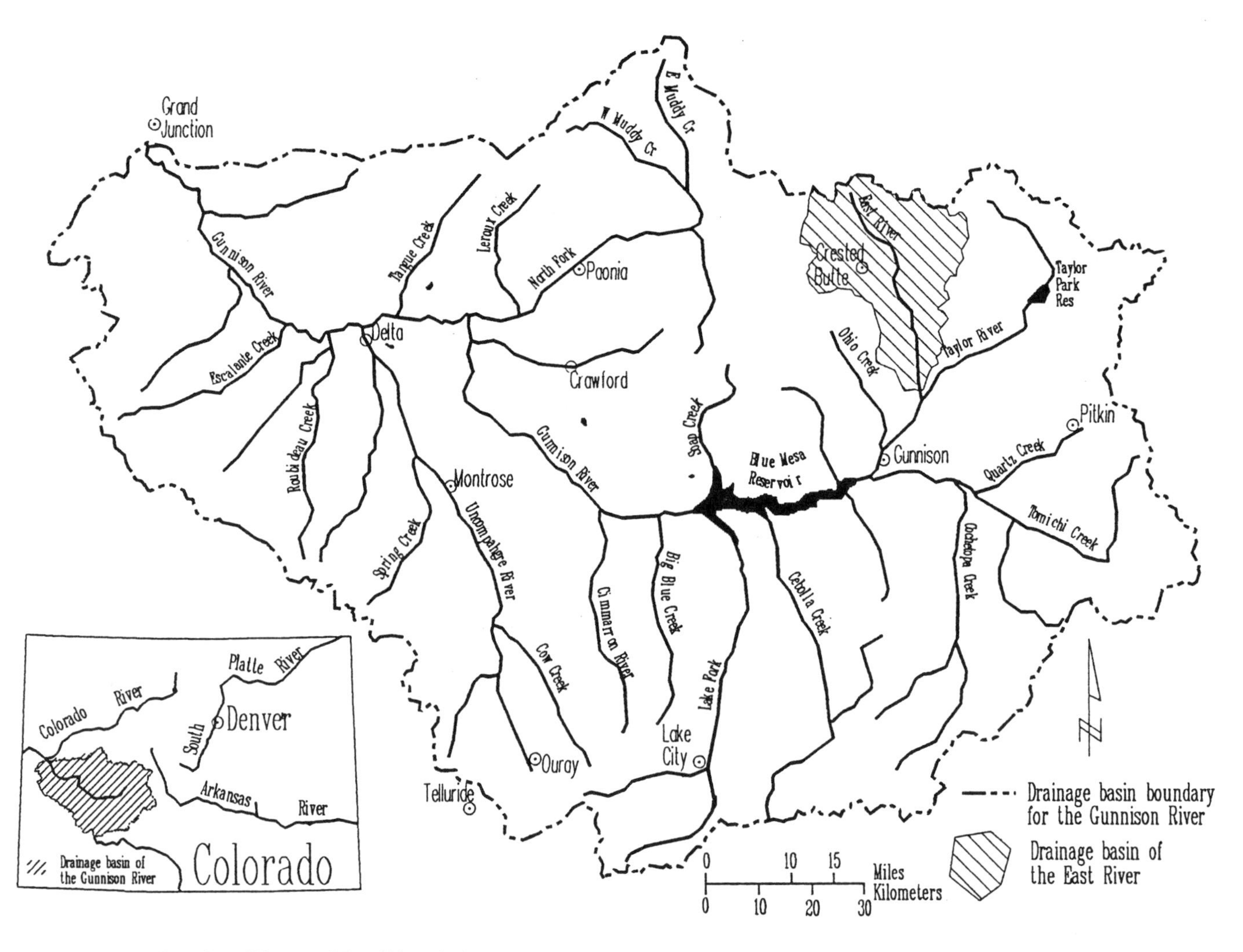

Figure 16-1: Gunnison River and East River drainages.

different aspects of the project, (2) develop methods for acquiring, generating, managing, and displaying spatial data required for modeling efforts, (3) execute algorithms that describe components of the water balance, (4) provide a means for verifying model results, (5) provide a means to investigate the effects of scale on model results, and (6) enhance the flow of information and ideas between project personnel with different specialties. Four example applications are described in this paper: (1) using GIS in conjunction with a precipitation runoff modeling system, (2) using GIS to estimate spatial distributions of potential evapotranspiration, (3) linking GIS to an orographic precipitation model, and (4) nesting atmospheric and hydrologic models to produce possible scenarios of climate change.

Precipitation runoff modeling system

The U.S. Geological Survey's Precipitation Runoff Modeling System (PRMS) (Leavesley et al., 1983) is a modular, distributed parameter, watershed modeling system that will be used to evaluate the effects of scenarios of climate change on watershed response in the Gunnison River Basin. PRMS will be used to model 20 upland watersheds such as the East River Basin (Figure 16-1) in which snow accumulation and melt are the dominant processes. The results of watershed modeling will provide input for a downstream flow routing model. GIS facilitates three major operations required of the PRMS modeler: basin characterization, parameter estimation, and model verification.

The distributed parameter capabilities of PRMS are enabled by partitioning a watershed into subareas that are assumed to be homogeneous in their hydrologic response, and are termed hydrologic response units (HRUs). Basin characterization by HRUs is done using the topographic variables—elevation, slope, and aspect, and the geographic variables—soil type, vegetation type, and precipitation distribution. Historically, the definition of HRUs was a labor-intensive, manual procedure that was likely to incorporate biases of the individual user. Leavesley and Stannard (1990) have developed a set of procedures for defining HRUs that utilize topographic data processing techniques developed by Jenson and Domingue (1988) and the overlaying functionality of the GIS to form HRUs in a more objective and repeatable manner.

Once HRUs are defined for a basin, the processes affecting the water budget within each HRU are modeled. These processes are driven by model inputs and are affected by parameters estimated from basin characteristics. The conceptualized hydrologic system used in PRMS is discussed in detail by Leavesley et al. (1983). The input data requirements for PRMS are daily values of air temperature, precipitation, and solar radiation. Some of the processes modeled by PRMS and the data required to model these processes are summarized in Table 16-1. Note that there is significant overlap in the data requirements for process modeling.

GIS technology also is being used to verify PRMS results. The work of Leavesley and Stannard (1990) demonstrates how simulated snow-covered areas from PRMS can be compared with remotely sensed data. The National Weather Service's National Remote Sensing Hydrology Center (Carroll and Allen, 1988) provides observed snow-covered areas from the Geostationary Operational Environmental Satellite (GOES) and the Advanced Very High Resolution Radiometery (AVHRR). Satellite data, in conjunction with streamflow data, enable the modeler to analyze measured and simulated snow-covered area and streamflow jointly, thus decreasing the uncertainty in defining sources of model error that occur when analysis is limited to the hydrograph alone (Leavesley and Stannard, 1990).

Estimates of potential evapotranspiration

Estimates of potential evapotranspiration (PET) are used by PRMS, in conjunction with vegetation-type and soil-type information, to calculate actual evapotranspiration (ET). In this example, PET is calculated for 10 km square grid cells (100 km^2 area), as a function of incident solar radiation and land-surface orientation, using a modification of the method described by Jensen and Haise (1963). Equation (16-1) shows that PET (in inches per day) is a function of R_s—the solar radiation expressed as inches of evaporation potential, and T_f—the average daily temperature in degrees Fahrenheit. The factor K is a function of elevation and is constant for each grid cell. C_m is a coefficient based on climate factors that can be varied by month.

$$\text{PET} = C_m\,(T_f - K)\,R_s \qquad (16\text{-}1)$$

The temperature (T_f) for each grid cell is defined by applying a lapse rate to values measured at a climate station to account for changes in elevation. Solar radiation (R_s) is the remaining unknown variable.

DRAD, the expected solar radiation received on a planar surface at the top of the atmosphere with a specified slope and aspect, is calculated using a modified routine from PRMS (Leavesley et al., 1983). DRAD is modified by factors that account for losses in incident solar energy due to atmospheric interference and cloud cover to produce R_s. The orientation of the plane describing the grid cell and the altitude of the grid cell control the calculated value of R_s.

A problem develops when trying to assign a single value for grid-cell (or HRU) elevation, slope, and aspect in topographically diverse regions. Figure 16-2 shows the elevation model described by approximately 400 30-arcsecond digital elevation model (DEM) point elevation values that fall within and around one mountainous 10 km grid cell on the south side of the Gunnison River Basin. The triangles, or facets, in Figure 16-2 represent the triangulated irregular network (TIN) that is formed by connecting the DEM points. Values for slope and aspect are known for each TIN facet. One can visualize from Figure

Table 16-1: Some of the processes modeled by PRMS and the basin characteristics data required to model these processes

Process	Basin characteristic data required
Interception	Vegetation type, vegetation density
Infiltration	Soil type, slope
Evapotranspiration	Vegetation type, vegetation density, soil type, slope, aspect, elevation
Snowmelt	Vegetation density, slope, aspect, elevation
Surface runoff	Vegetation type, vegetation density, slope, soil type, impervious area
Channel flow	Slope, stream length

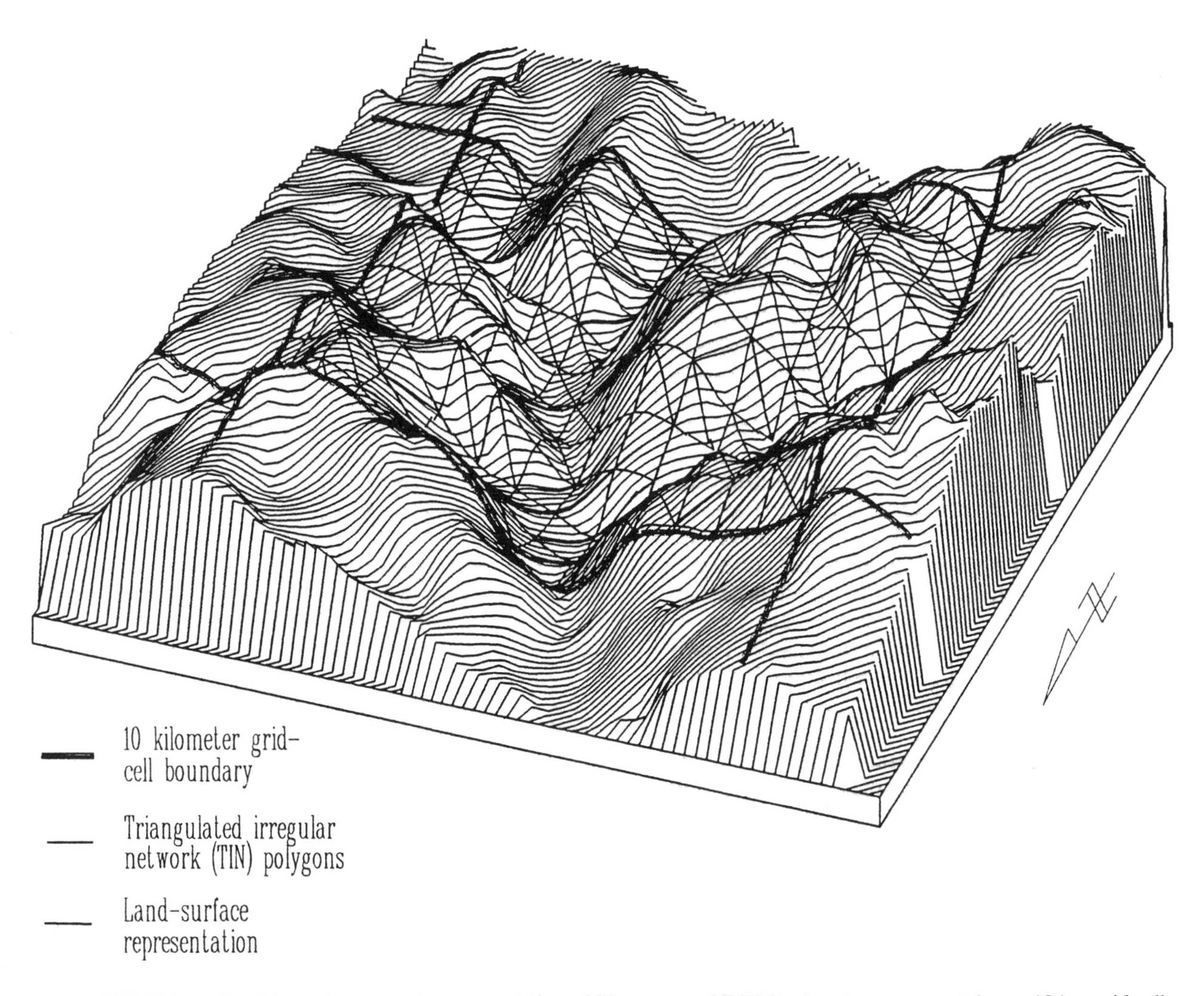

Figure 16-2: Triangulated irregular network representation of 30 arcsecond DEM points in one mountainous 10 km grid cell.

16-2 how poorly a single plane with one normal vector would describe the range of slope and aspect values occurring in this particular cell. The work of Wolock and Price (1989) and Jenson (1991) indicates that steeper slopes would be expected if finer-resolution DEM point elevation values (such as a 3 arcsecond DEM) were used.

In order to account for the slope and aspect within each grid cell better, the incident solar radiation for each grid cell was calculated by classifying the aspect values of the TIN facets in a grid cell into 9 classes—the 8 primary compass headings, and a horizontal surface. Then, the average slope of the TIN facets in each of the 9 classes is calculated. The grid cell is now described by 9 smaller planes, each having a unique normal vector. DRAD is then calculated as the area-weighted average of incident solar radiation on the 9 subplanes within a grid cell. Plate 16-1 shows the estimated areal distribution of PET in the Gunnison River Basin calculated for June 22, 1988. There is an inverse relation between PET and elevation in the basin.

Precipitation modeling

The spatial and temporal distribution of precipitation in mountainous regions is important in the application of distributed-parameter hydrologic models such as PRMS. Available data in mountainous basins tend to be sparse and restricted to the lower elevations. The Rhea–Colorado State University orographic precipitation model (Rhea, 1977; Branson, 1991) was modified for this study to produce estimates of precipitation over a range of spatial scales for the Gunnison River Basin. The precipitation model is steady state, multilayer, and two dimensional; one flow component is along the prevailing wind direction, and the other is vertical. The model requires as input twice-daily upper-air measurements (pressure, height, temperature, relative humidity, wind direction, and wind speed) and gridded elevation data. A different grid is used for each 10 degree change in wind direction. Upper-air data for each 50 millibar interval from 850 to 300 millibars are interpolated to the border of the elevation grid. The grid is selected by rounding the 700 millibar wind direction at the center of the study area to the nearest 10 degrees.

The orographic precipitation model follows the interaction of air layers with the underlying topography by allowing vertical displacement of the air column while keeping track of the resulting condensate or evaporation. As the layers flow across the region, part of the condensate precipitates. Evaporation of falling precipitation is accounted for as it falls through subsaturated layers, which decreases the amount of precipitation reaching the ground and moistens the subsaturated layers.

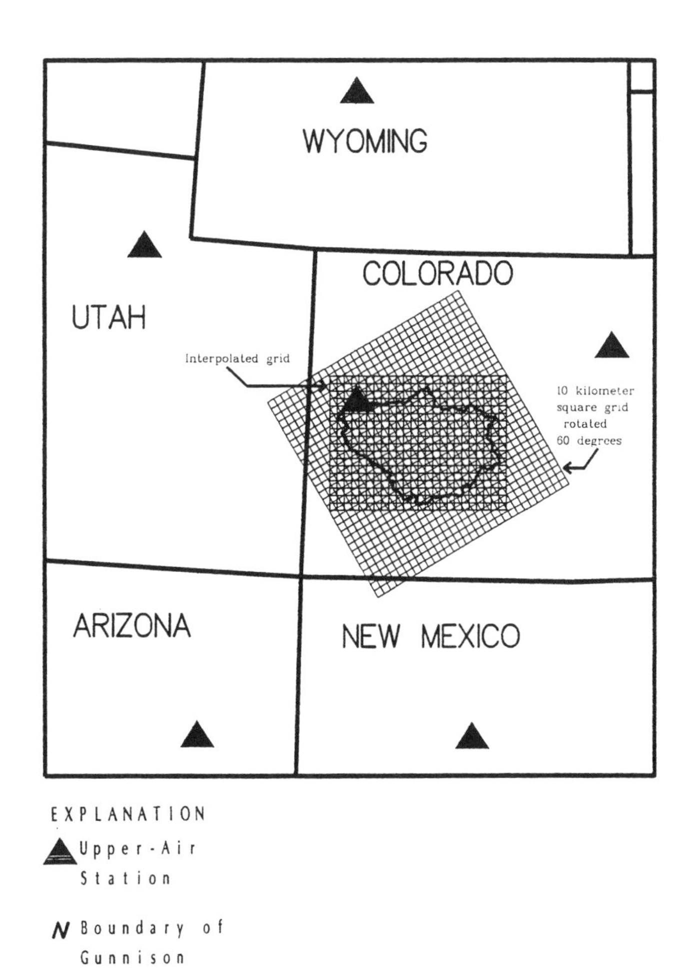

Figure 16-3: Upper air stations and rotated orographic precipitation model elevation grid.

One of the objectives of modification to the orographic precipitation model is to produce a model that is flexible and applicable to a wide range of geographic areas. This is achieved by linking the orographic precipitation model with a GIS. Modifications to the precipitation model also facilitate the transfer of model precipitation estimates to precipitation runoff models, which improves simulated streamflow in the Gunnison River Basin. GIS automates development of the elevation grids from DEM point elevation values for the orographic precipitation model, and enables the user to choose the method of elevation characterization within the grid cells. This modification eliminates what was the most labor-intensive step involved in applying the model to a new area (M. D. Branson, Colorado State University, Department of Atmospheric Science, personal communication, 1991).

Figure 16-3 shows the location of the upper air stations used to initialize the orographic precipitation model and an example of the elevation grid rotated 60 degrees. The rotated grid is used for the four wind directions corresponding to the four sides of the grid. Nine elevation grids are generated, one for every 10 degrees of rotation from 0 to 80 degrees. Mean, minimum, and maximum elevation values for each grid cell are calculated by overlaying each of the nine rotated elevation grids on DEM point elevation values. Precipitation estimates are made twice a day and interpolated to the inner grid (Figure 16-3).

Linking of GIS and an orographic precipitation model also makes it possible to examine the effects of the method of elevation characterization on precipitation estimation. Plate 16-2 shows 10 km elevation grids for the Gunnison River Basin characterized by maximum, mean, and minimum elevation within the grid cell, respectively. There is as much as 1600 m difference between maximum and minimum elevation within a single 10 km grid cell. This will have substantial effects on the results of a precipitation model that calculates air parcel moisture conditions during topographically induced ascents and descents.

Comparison of results using different methods of elevation characterization showed that minimum elevation did not produce enough relief to generate sufficient amounts of rainfall for the basin. Maximum elevation had too much relief, resulting in a premature depletion of moisture in the southwestern portion of the basin. Examination of these results indicated a shortcoming in the model's interpolation scheme to the inner grid. The model uses the inverse distance squared from the four closest rotated grid cells when interpolating to the inner grid or when interpolating to an observation station for comparisons of predicted versus observed precipitation at a point. Errors in the interpolation were evident and are most likely the result of failure to include elevation in the interpolation scheme.

Figure 16-4 shows an observation station, the distance

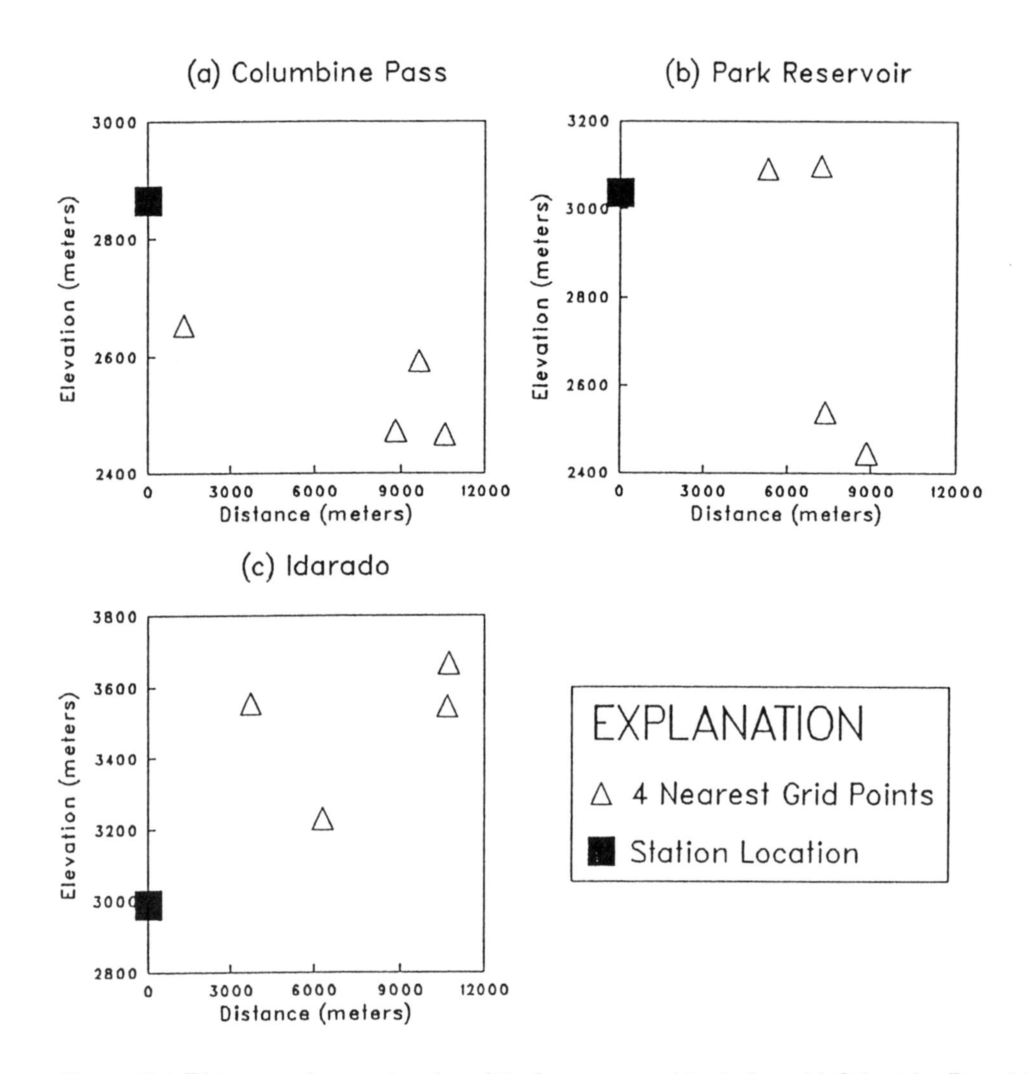

Figure 16-4: Distance and mean elevation of the four nearest grid cells from (a) Columbine Pass, (b) Park Reservoir, and (c) Idarado SNOTEL stations.

to, and mean elevation of, the four nearest grid cells used in the interpolation. Figure 16-5 shows the corresponding observed cumulative daily precipitation at these stations verses precipitation estimates using the maximum, mean, and minimum elevation grids for the 1987–1988 winter season. In Figure 16-4a, the SNOTEL (snow telemetry) station Columbine Pass is at a higher elevation than all the nearest four grid cells, resulting in an underestimation of precipitation using the minimum and mean elevation grids and an overestimation when using the maximum elevation grids (Figure 16-5a). In Figure 16-4b the SNOTEL station Park Reservoir is at an elevation that falls within the elevation range of the four nearest grid cells, resulting in close agreement between observations and precipitation estimated using mean elevation grids (Figure 16-5b). In Figure 16-4c the SNOTEL station Idarado is in a valley that is lower in elevation than the four nearest grid cells, resulting in close agreement between observations and precipitation estimated using minimum elevation grids (Figure 16-5c).

Linking of GIS and the orographic precipitation model makes it possible to assess the effects of topographic scale on precipitation estimates. A noticeable difference in detail can be seen between estimates of mean elevation for 10 and 5 km grids (Plate 16-3). It is thought that precipitation in mountainous terrain may vary over distances as small as 2 km (Rhea, 1977). Modifications to the model will allow the user to input grids finer than 5 km, but consideration of the effects of finer grids on model physics must first be examined. With a finer-resolution elevation grid, there will be more rises and descents of the modeled air parcels, which will have a direct effect on air parcel moisture conditions.

Comparison of precipitation estimates using 5 and 10

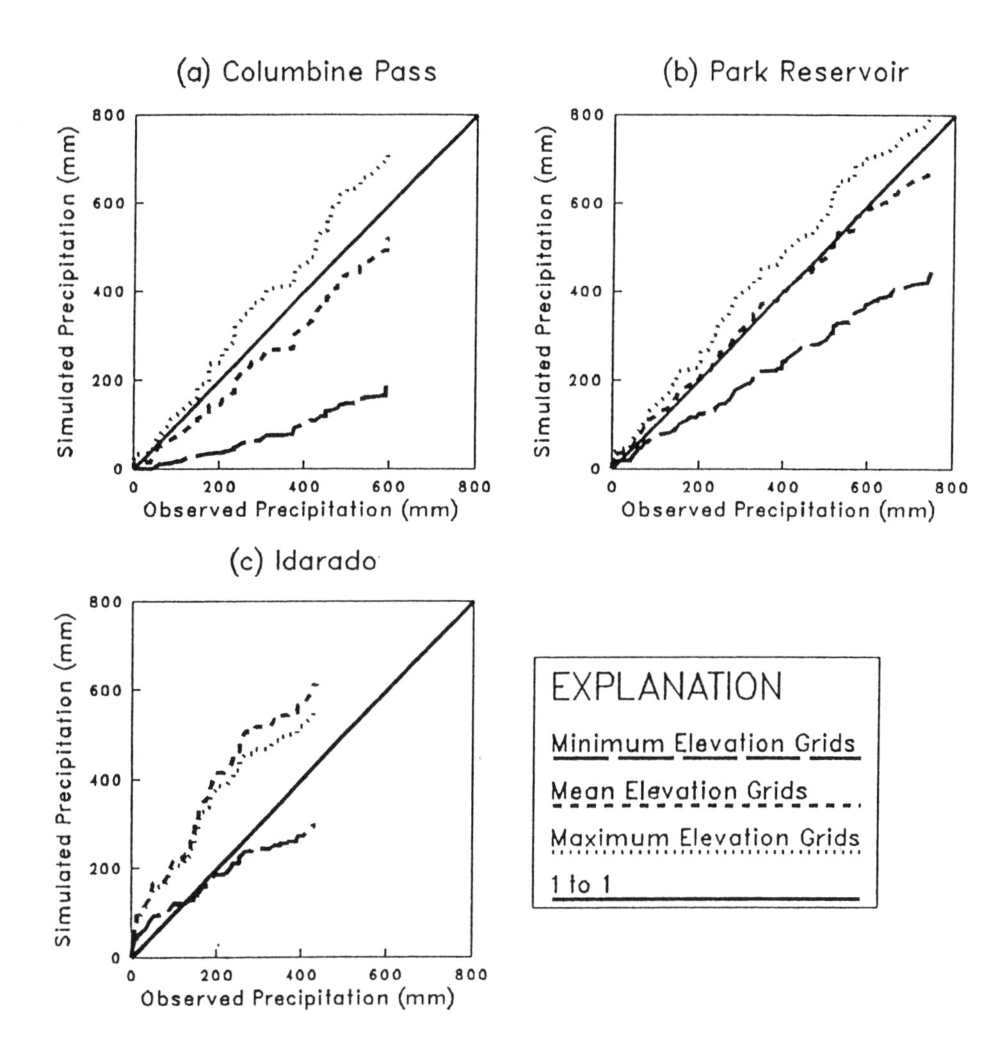

Figure 16-5: Observed cumulative daily precipitation at (a) Columbine Pass, (b) Park Reservoir, and (c) Idarado SNOTEL stations versus precipitation estimated using maximum, mean, and minimum elevation grids over the 1987–1988 winter season (October 1, 1987–April 30, 1988).

km elevation grids showed the 5 km grids producing more precipitation at the mountain peaks and less on the upwind mountain slopes. There are not enough reliable climate stations to determine which resolution grids produce the most realistic estimates of precipitation. Comparison with observation stations alone is not an adequate assessment of the model's accuracy due to error that is associated with point measurements at the observation stations and the error associated with the interpolation scheme used by the orographic precipitation model. Another method of verification is to use the 5 and 10 km precipitation estimates as input to a hydrologic model and compare the output hydrographs with hydrographs constructed using observed streamflow data. This application is discussed at the end of the following section.

Nesting of atmospheric and hydrologic models

Possible scenarios of the effects of climate change on watershed processes will be developed by nesting general circulation (GCM), GCM-mesoscale, precipitation, and watershed models. This approach attempts to account for potential changes in precipitation frequency, magnitude, and duration that are not accounted for in current approaches of modifying historic precipitation and temperature data. At the coarsest resolution, GCMs give full three-dimensional representation of the atmosphere, but fail to represent local processes. A coarse-resolution GCM can simulate the response of general circulation to global processes. A GCM-mesoscale model nested within a GCM can describe the effects of sub-GCM grid-scale forcing on atmospheric circulations (Giorgi and Mearns, 1991). Output from a finer-resolution GCM-mesoscale

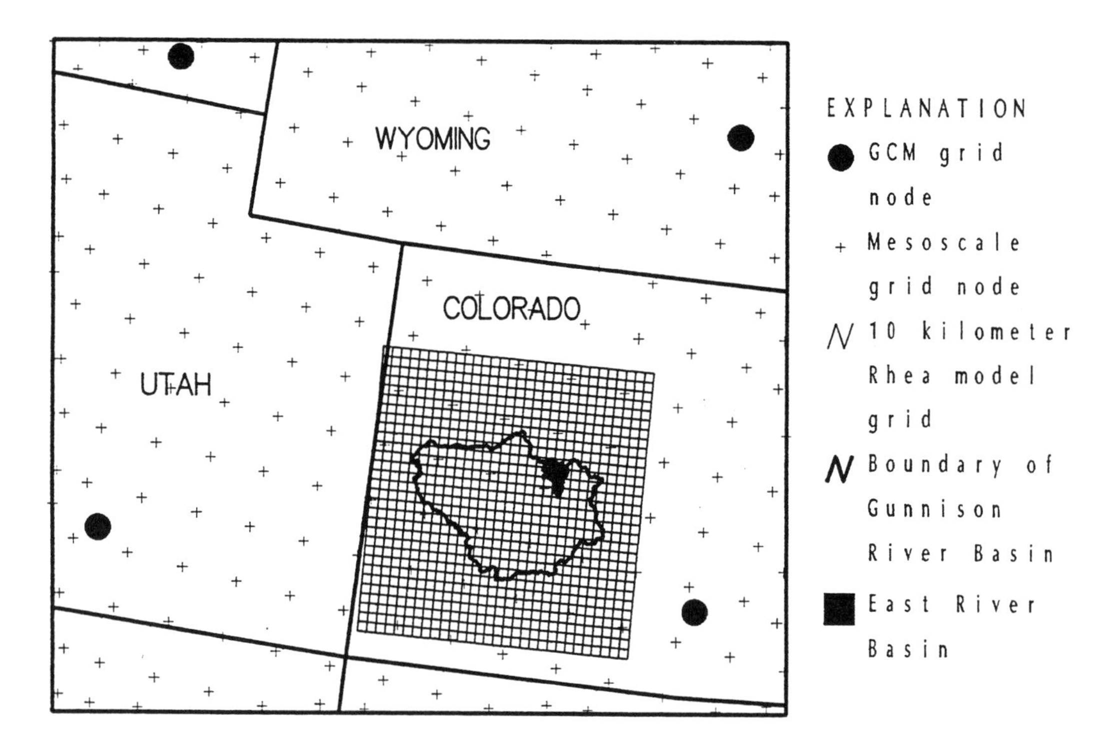

Figure 16-6: Grid spacing for general circulation model (GCM), mesoscale model and orographic precipitation model, and East River sub-basin.

model can be used to initialize the boundary conditions for the orographic precipitation model, which in turn will produce output that can be used in the watershed model. Figure 16-6 shows an example of the scales of the models to be nested. Shown are a GCM with a grid spacing of 4.5 by 7.5 degrees, a GCM-mesoscale model with a grid spacing of 60 km, the orographic precipitation model with a grid spacing of 10 km, and the East River subbasin within the Gunnison River basin, which contains seventeen HRUs.

Output from a GCM-mesoscale model nested within a GCM for current and double CO_2 conditions will be used to initialize the orographic model. Initial results show promise with linking GCM-mesoscale models with finer-resolution precipitation models. Output is currently available from an augmented version of the Pennsylvania State University/NCAR mesoscale model (version MM4) (Anthes et al., 1987), initialized with real data, for January 1988. The MM4 output was used to construct 50 millibar interval upper air output for input to the orographic precipitation model and results were similar to those produced using observed upper air data.

Output from the orographic model can then be used as input to PRMS to estimate basin response for current and future climate scenarios. Initial results are promising for linking the precipitation and watershed models. Precipitation estimates from the orographic precipitation model using both 10 and 5 km grids were used as input to seventeen HRUs, comprising the East River of the Gunnison River Basin (Figures 16-1 and 16-6). The resulting volume errors were on an average of less than 10% for the period 1981–1988. Actual volumes obtained using precipitation model estimates were comparable to those produced using observed data, but hydrograph shapes were better simulated using 10 and especially 5 km precipitation model estimates as input.

SUMMARY

The objectives of the Gunnison River Climate Study are being accomplished with the aid of GIS technology. GIS provides: a (1) common database for researchers; (2) a means for acquiring, generating, managing, and displaying spatial data; (3) a platform for executing algorithms that describe components of the water balance; (4) a means for verifying model results; (5) a means to investigate the effects of scale on model results; and (6) a means to enhance the flow of information and ideas between researchers with different specialties.

Four applications were described in this paper: (1) GIS used in conjunction with a precipitation runoff model; (2)

spatial distribution of potential evaporation estimated by linking the terrain modeling functionality of GIS with algorithms for determining incident solar radiation from PRMS; (3) elevation grids for an orographic precipitation model extracted from DEM point elevation values at several scales using three methods of elevation characterization; and (4) preliminary work on nesting models (GCM, GCM-mesoscale, orographic precipitation, and watershed) to assess the effects of climate change on basin response.

REFERENCES

Anthes, R.A., Hsie, E.Y., and Kuo, Y.H. (1987) Description of the Penn State/NCAR mesoscale model version 4 (MM4). *Technical Note NCAR/TN-282+STR*, Boulder, CO: National Center for Atmospheric Research, 66 pp.

Branson, M.D. (1991) An historical evaluation of a winter orographic precipitation model. *Master's Thesis*, Fort Collins, CO: Colorado State University, Department of Atmospheric Science.

Carroll, T.R., and Allen, M.W. (1988) *Airborne and Satellite Snow Cover Measurement—A User's Guide, Version 3.0.* Minneapolis, MN: National Weather Service, Office of Hydrology, 54 pp.

Giorgi, F., and Mearns, L.O. (1991) Approaches to the simulation of regional climate change: a review. *Reviews of Geophysics* 29(2): 191–216.

Jensen, M.E., and Haise, H.R. (1963) Estimating evapotranspiration from solar radiation. *Journal of the Irrigation and Drainage Division, American Society of Civil Engineers* 89(IR4): 15–41.

Jenson, S.K. (1991) Applications of hydrologic information automatically extracted from digital elevation models. *Hydrological Processes* 5: 31–44.

Jenson, S.K., and Domingue, J.O. (1988) Software tools to extract topographic structure from digital elevation data for geographic information system analysis. *Photogrammetric Engineering and Remote Sensing* 54(11): 1593–1600.

Leavesley, G.H., Lichty, R.W., Troutman, B.M., and Saindon, L.G. (1983) Precipitation-runoff modeling system—user's manual. *Water Resources Investigations Report 83-4238*, Washington, DC: U.S. Geological Survey, 207 pp.

Leavesley, G.H., and Stannard, L.G. (1990) Application of remotely sensed data in a distributed-parameter watershed model. *Proceedings of Workshop on Applications of Remote Sensing in Hydrology*, Saskatoon: National Hydrologic Research Centre, Environment Canada.

Rhea, J.O. (1977) Orographic precipitation model for hydrometeorological use. *PhD dissertation*, Fort Collins, CO: Colorado State University, Department of Atmospheric Science, 198 pp.

Ugland, R.C., Cochran, B.J., Hiner, M.M., Kretschman, R.G., Wilson, E.A., and Bennett, J.D. (1990) Water resources data for Colorado, water year 1990, Vol. 2, Colorado River Basin. *Water-Data Report CO-90-2*, Washington, DC: U.S. Geological Survey, 344 pp.

Wolock, D.M., and Price, C.V. (1989) Effects of resolution of digital elevation model data on a topographically based hydrologic model. *EOS, Transactions of the American Geophysical Union* 70(43): 1091.

17

Linkage of a GIS to a Distributed Rainfall-Runoff Model

XIAOGANG GAO,
SOROOSH SOROOSHIAN, AND
DAVID C. GOODRICH

Environmental modeling of natural phenomena is made difficult by characteristics such as spatial and temporal variability, dominating mechanisms that vary with environmental circumstances, uncertain factors, and scale effects. These effects are particularly inherent in distributed land surface, and near surface, rainfall-runoff modeling. GIS technologies offer powerful tools to deal with some of these problems. This chapter summarizes some experiences with the integration of GRASS (USACERL, 1988) and a distributed rainfall-runoff model. The system is simple to implement and is currently installed on a SUN workstation operating under the UNIX system.

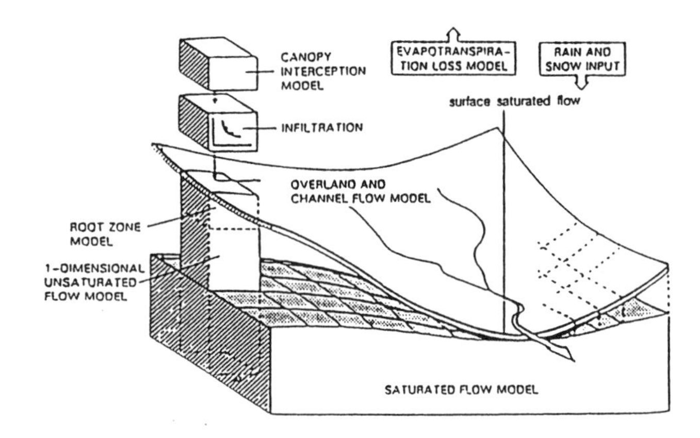

Figure 17-1: Description of the model structure.

MODEL DESCRIPTION

As part of a NASA-supported EOS (Earth Observation System) project (Kerr and Sorooshian, 1990), a physically based distributed model has been developed to study the dominant runoff production mechanisms under different environmental conditions and different spatial and temporal scales.

The model consists of several coupled modules, each describing a different hydrologic subprocess. Discretization of the horizontal land surface is done using a grid system of rectangular elements that correspond to the cells in the GIS. Vertical discretization of each soil column, with a cross section equal to the grid area, is done using layers to a user-specified depth. However, these are not stored as separate GIS layers due to memory constraints. The drainage network is superimposed along the edges of the grids (the river network becomes a series of straight line segments running along the grid edges). A schematic description of the model appears in Figure 17-1.

This model uses the finite difference method to solve a set of partial differential equations that characterize different hydrologic components. The following sequence of equations applies to the estimation of runoff from rainfall on an individual grid element and into a drainage network.

Rainfall input data are first processed through a "rainfall intensity module" and a "canopy interception module" to produce distributed "effective rainfall" on the soil surface. When the effective rainfall intensity is greater than the saturated hydraulic conductivity of the soil, a surface saturated zone in the soil is created. The infiltration rate is regulated by the Green–Ampt equation (Green and Ampt, 1911), and any infiltration excess results in overland flow. The 2D kinematic wave equation is used to simulate overland flow:

$$\frac{\partial d_0}{\partial t} + \frac{\partial}{\partial x}\left(\frac{\sqrt{s_0}}{n} d_0^{5/3} \cos\alpha\right) + \frac{\partial}{\partial y}\left(\frac{\sqrt{s_0}}{n} d_0^{5/3} \cos\beta\right) = R-I-E \tag{17-1}$$

where

$d_0(x,y,t)$ = depth of overland flow (output)
t = time
x,y = horizontal coordinates (geometric parameters from DEM)
s_0 = soil surface slope (soil parameter)
$(\cos\alpha, \cos\beta, \cos\gamma)$ = direction vector of overland flow along the slope (parameter)
n = Manning coefficient of the soil surface (surface friction parameter)
I = infiltration rate (input from Green–Ampt equation)
R = effective rainfall intensity (input)
E = potential evaporation rate (input).

In the surface saturated zone, subsurface lateral flow is generated by the force of gravitation. It flows parallel to the surface slope, and a wetting front that separates the surface saturated part from the unsaturated zone moves downward into the unsaturated zone. The 2D governing equation for this process is:

$$k_s\left(z_0 - z_w\right)\left(\frac{\partial^2 z_w}{\partial x^2} + \frac{\partial^2 z_w}{\partial y^2}\right) + k_s\left(\frac{\partial z_0}{\partial x} - \frac{\partial z_w}{\partial x}\right)\left(\frac{\partial z_w}{\partial x}\right) + k_s\left(\frac{\partial z_0}{\partial y} - \frac{\partial z_w}{\partial y}\right)\left(\frac{\partial z_w}{\partial y}\right) + \left(\theta_s - \theta_w\right)\frac{\partial z_w}{\partial t} = I - T \qquad (17\text{-}2)$$

where

$z_w(x,y,t)$ = wetting front elevation (output)
z_0 = elevation of the soil surface (geometric parameter from DEM)
θ_s = saturated soil moisture content (soil parameter)
θ_w = soil moisture content ahead of the wetting front (input state variable)
T = transpiration rate through the roots (vegetation parameter) (Dickinson et al., 1986)
k_s = soil saturated hydraulic conductivity (soil parameter).

Stream flow in the river network is routed using the St. Venant equations:

$$\frac{\partial By}{\partial t} + \frac{\partial Q}{\partial x} = q_l \qquad (17\text{-}3)$$

$$\frac{\partial Q}{\partial t} + \frac{\partial}{\partial x}\left(\frac{Q^2}{A}\right) + g\frac{\partial y}{\partial x} = gA\left(s_0 - s_f\right) - q_l\frac{Q}{A}$$

where

$y(x,t)$ = stream flow depth (output)
$Q(x,t)$ = stream flow discharge (output)
x = distance coordinate along the river from the upstream end
A = stream flow cross area (geometric parameter)
B = stream flow top width (geometric parameter)
q_l = total lateral inflow through a unit length of the bank [input from Eqs. (17-1), (17-2), and (17-5)]
g = gravitational acceleration (constant)
s_0 = river bed slope (geometric parameter)
s_f = friction slope (surface parameter)

If the effective rainfall intensity on a soil grid is small, then it becomes the flux boundary condition at the top of the soil column. The 1D Richard's equation is used to calculate the vertical moisture profile in the soil column:

$$C\frac{\partial \psi}{\partial t} = -\frac{\partial}{\partial z}\left[k\left(\frac{\partial \psi}{\partial z} + 1\right)\right] - T \qquad (17\text{-}4)$$

where

$\psi(z,t)$ = soil suction uniquely dependent on θ (Clapp and Hornberger, 1978) (output)
z = vertical coordinate with positive direction downward
$k(\theta)$ = unsaturated soil hydraulic conductivity (soil parameter)
$C(\theta)$ = soil specific moisture capacity (soil parameter).

Where the groundwater table is shallow, it plays an important role in runoff production. Groundwater flow is simulated using the 2D groundwater flow equation:

$$\frac{\partial}{\partial x}\left(k_s b\frac{\partial h}{\partial x}\right) + \frac{\partial}{\partial y}\left(k_s b\frac{\partial h}{\partial y}\right) - T + W = S_y\frac{\partial h}{\partial t} \qquad (17\text{-}5)$$

where

$h(x,y,t)$ = hydraulic head of groundwater (output)
b = groundwater depth [equal to $(h - h_b)$, where h_b bedrock elevation]
S_y = specific yield [equal to $(\theta s - \theta)$, from Eq. (17-4)]
W = groundwater recharge intensity [input from Eq. (17-4)]

The model is programmed according to the flow chart in Figure 17-2 while employing the equations presented. Further explanation of several computational points follows:

1. The calculation area is usually chosen to be a watershed. GRASS is used to analyze the digital elevation map (DEM) to create all watershed boundaries and river networks within a region map based on the algorithms of Band (1989).
2. The calculation grid size is usually larger than the DEM resolution. Using the GRASS function for changing map resolution, the watershed boundary and

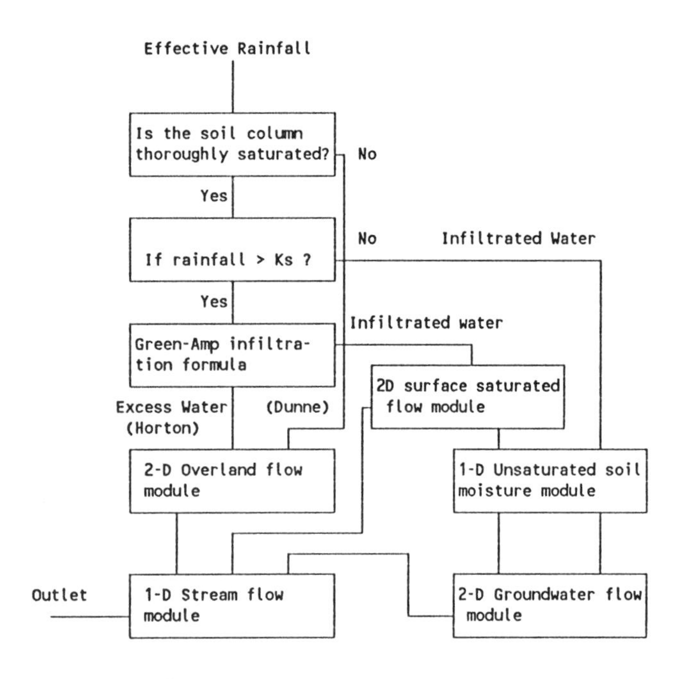

Figure 17-2: Flow chart of the model.

river network can be transformed to a series of straight line segments running along the grid sides so that they serve as boundaries for Equations (17-1), (17-2), and (17-5); the line segments corresponding to the river network collect the river inflows. While such a transformation changes the river length, the real river lengths calculated from the DEM are used when the river routing equations (17-3) are solved.

3. Geometric parameters such as s_0, $\cos\alpha$, $\cos\beta$, and $\cos\gamma$ (except those related to the river cross sections) can be calculated from the DEM using GRASS.
4. Soil parameters, vegetation parameters, and surface friction parameters can be determined at any grid element from soil, vegetation, and land use/cover maps using GRASS.
5. When numerically solving the equation set (17-1–17-5), some coupling is required to reflect the interactions between different processes. The coupling relationships in a soil column are as follows. When the wetting front of the surface saturated zone moves downward or as the groundwater table rises, the water from the saturated zones will fill the unsaturated region ahead of the front. The soil moisture content varies along the unsaturated soil moisture profile calculated by Equation (17-4). When the soil moisture profile is calculated, the depth of the unsaturated soil column is varied with the rising or falling of the groundwater table; therefore the lower boundary condition is set to $\psi = 0$ at the elevation of the groundwater table.
6. After the new soil moisture profile is created, a volume balance is conducted. The total volume difference between the successive profiles is defined as the groundwater recharge, W.

CASE STUDY

A simple example is presented here to illustrate how the integrated model-GIS system works: the hydrologic response of the Lucky Hills-104 semiarid watershed (4.4 ha) is simulated for a storm event that occurred on Aug. 1, 1974. The watershed is located in the USDA-ARS Walnut Gulch Experimental Watershed near Tombstone, Arizona; extensive rainfall and runoff data have been collected there on an ongoing basis as part of a long-term research program (Renard, 1970). A detailed 15 m x 15 m DEM database was obtained photogrammetrically using low-level aerial photography. Soil hydraulic information was obtained through field surveys at a number of points in the subwatershed (Goodrich, 1990). A planimetric illustration of the watershed and associated raingages is presented in Figure 17-3; a more detailed description of the watersheds can be obtained in Kincaid et al. (1966). In this watershed, the groundwater table is remote from the surface topography. Therefore, model components related to surfacegroundwater interaction are not exercised in this case study. Basins where groundwater interactions are significant will be examined in the future. The case study is used in the following section to illustrate linkage between the model and the GRASS GIS.

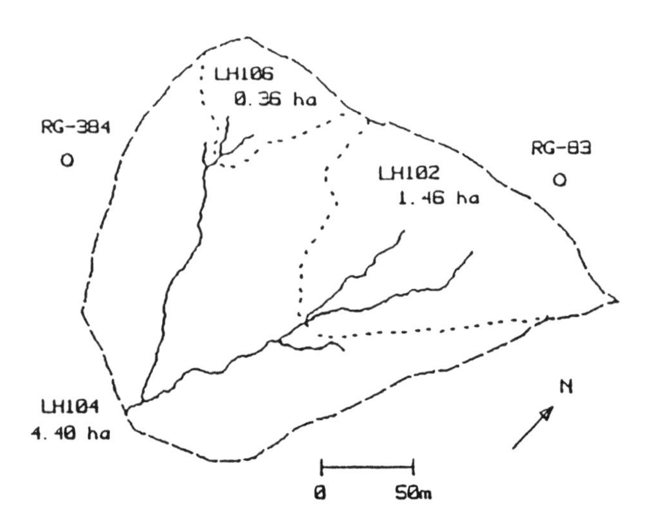

Figure 17-3: Lucky Hills-104 watershed.

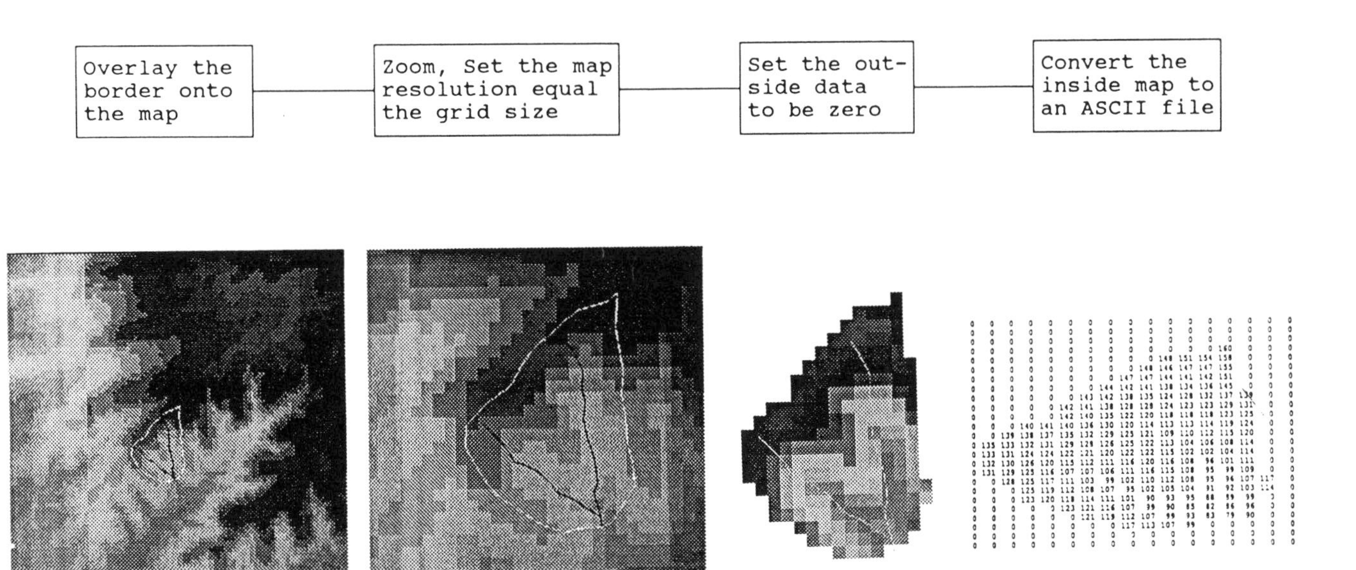

Figure 17-4: Watershed DEM data extraction.

THE MODEL-GIS SYSTEM

The model employs several idealized two-dimensional hydrological processes and their interactions to simulate a complex three-dimensional natural hydrological phenomenon. Almost all the input data, output results and parameters employed by the model are distributed on two-dimensional surfaces; many also vary with time. The primary linkage between the model and the GIS involves initial drainage basin characterization and handling and display of distributed model input and output. The use of a GIS greatly facilitates the data management and visualization tasks. In order to take full advantage of a GIS, some commonly-used functions were created: GRASS commands were used here. In addition, UNIX macros were formulated to link GRASS with the model to create an integrated system. When the model is run, GRASS is simultaneously initiated to provide the functions described in the following sections.

Ease of model setup and modification

A large amount of data describing the spatial distribution patterns of the input data and watershed parameters must be input into the model in the form of numerical matrices. The tedious nature of this work is especially apparent when unknown or uncertain parameters must be adjusted many times in order to calibrate the model. To simplify and facilitate this process, several GRASS-GIS commands are programmed to extract data required by the model from digital maps. The steps used in our illustrative example are:

1. Data retrieval: Given a watershed boundary and DEM data, extract the within-boundary data and transform it into a numerical matrix. The extraction of digital elevation data for Lucky Hills 104 is illustrated in Figure 17-4.
2. Data editing: Altering the map values (which represent certain categories) in a specified area, while keeping the remaining values in other areas unchanged. This useful macro can be realized through several GRASS procedures:
   a. Using the mouse, outline the boundary of a map area to be edited and record the boundary as a vector file.
   b. Convert the vector file to two special cell maps, map1: inside = assigned value, outside = 0; map2: inside = 0, outside = 1.
   c. Perform map operation: New map = (Original map) * map2 + map1.

Visualization of model processes

The use of GRASS-GIS in conjunction with other data visualization packages offers a simple way to monitor the modeling procedure; sequences of frames that depict the distributed dynamics of the modeled process can be easily constructed and viewed. Key physical variables and distributed physical fields can be readily displayed. This facilitates a better understanding of the hydrologic process and enables rapid error detection and correction. The basin outflow hydrographs at the outlet of Lucky Hills-104 and two nested subcatchments are illustrated in Figure 17-5. A schematic of the distributed overland flow depths (cm) at several times during the rainfall-runoff simulation event are illustrated in Figure 17-6.

Creation of maps at larger spatial scales

Map data and remotely sensed data are available at various different spatial resolutions. To combine data from different sources, the ability to reformulate model inputs and outputs at various scales is required. When rescaling from smaller to larger pixel sizes, issues regarding the loss of information by averaging must be addressed. For example, the brightness information in a 4 m × 4 m pixel may

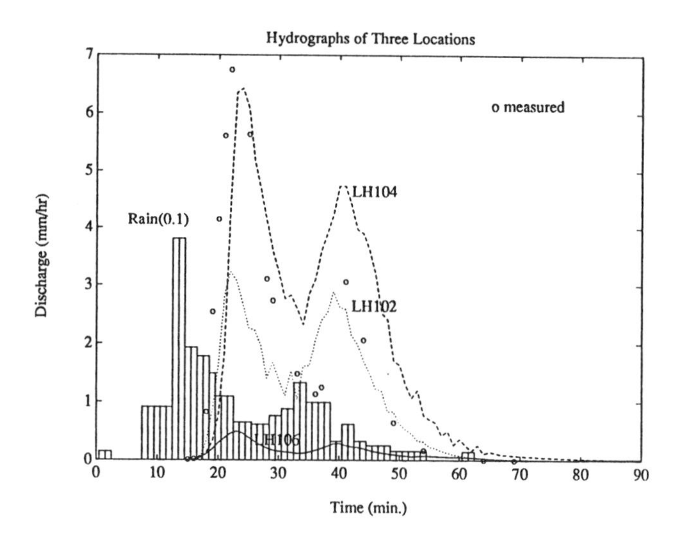

Figure 17-5: Simulated catchment and subcatchment hydrographs.

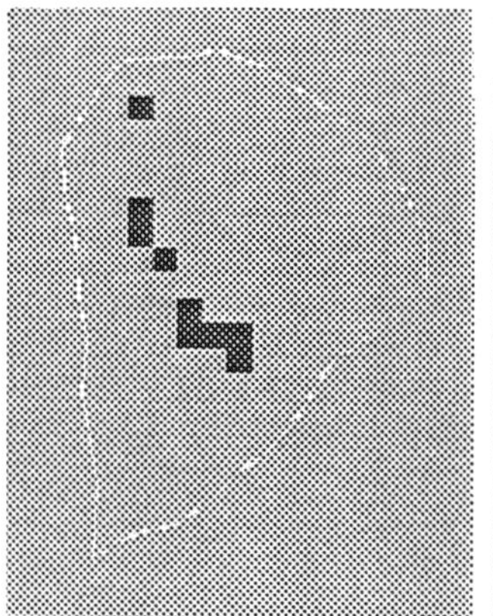
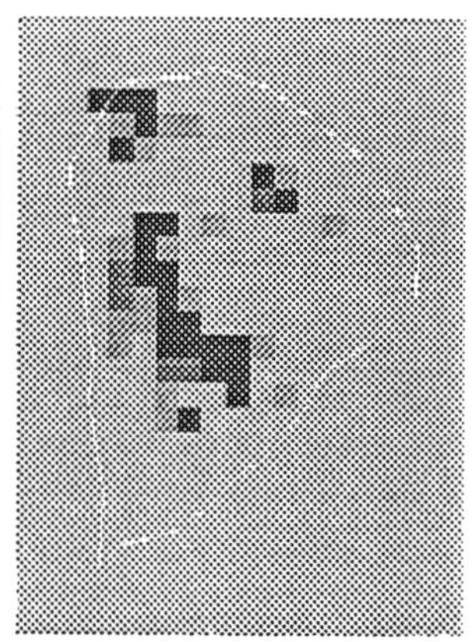
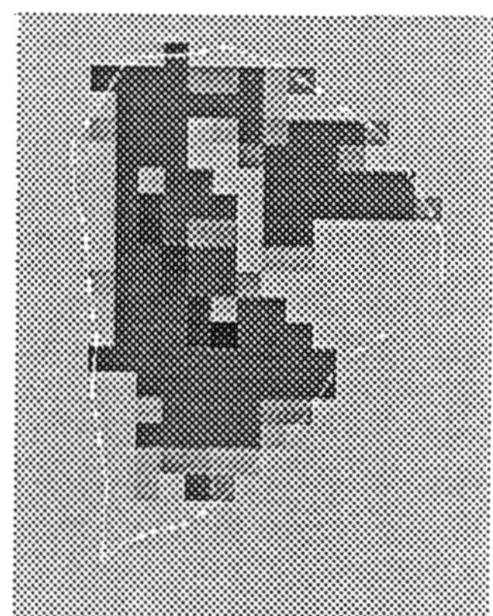
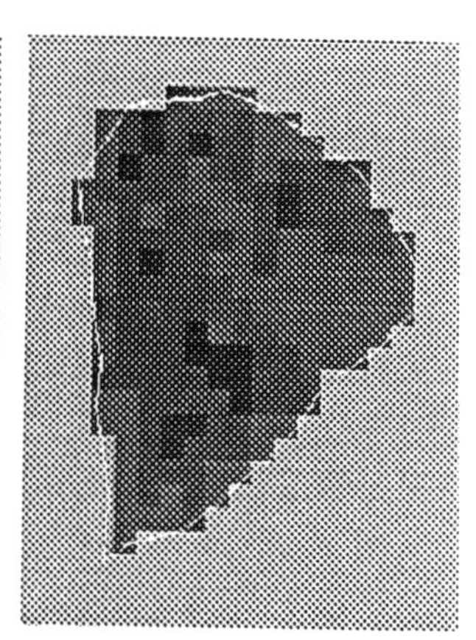

Figure 17-6: Variation of overland flow intensities.

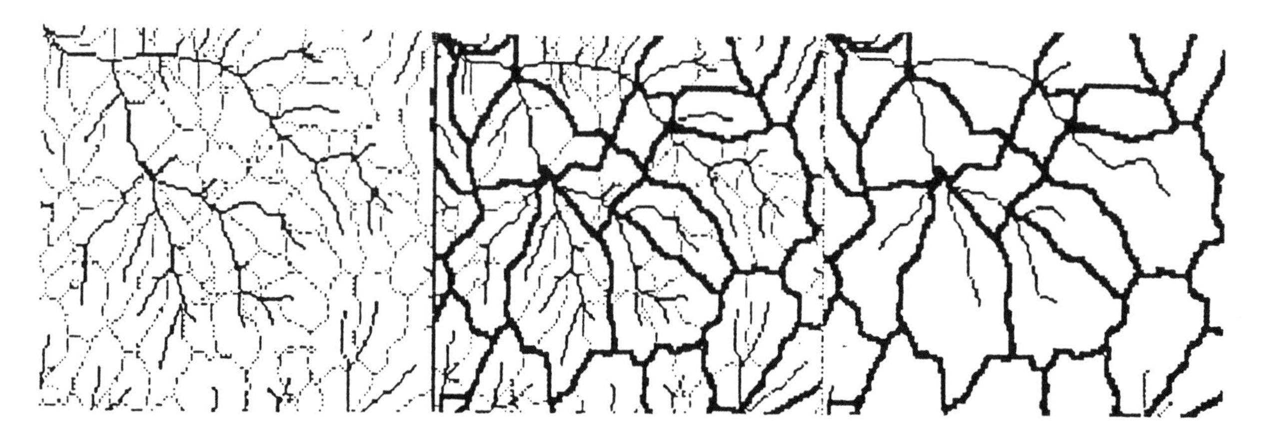

Figure 17-7: Combination of subcatchments via stream order reduction.

reflect some combination other than simple averaging of the information contained in the corresponding 16 1 m × 1 m pixels. Simple GRASS filters can be used to customize the method of grid averaging. In addition, GRASS can perform very complex user-defined map operations to generate composite maps. For example, Figure 17-7 illustrates the use of a GRASS macro to combine smaller subcatchments into larger subcatchments via a stream order reduction methodology. This facility is very useful for the study of issues related to scaling in hydrologic modeling.

CONCLUSIONS

Preliminary results indicate that the physically based distributed model described here can simulate the spatial variability of the rainfall-runoff phenomenon, although additional validation must be conducted. In the case study presented, parameters used by other models (Woolhiser et al., 1990) were applied without calibration. The simulated hydrograph at the Lucky Hills-104 watershed was found generally to match observed data. The model also provided a distributed dynamic description of the hydrologic response occurring in this watershed: processes such as soil moisture redistribution, runoff generation and diminution, stream flow variation, and so on were simulated. However, model accuracy is obviously limited by the requirement for large amounts of data of good quality. Validation of spatially distributed hydrologic response data (i.e., grid overland flow depth or average grid infiltration rates) is extremely difficult and costly and may not be realistic. If such data are not available, the model-GIS system discussed may be more properly used as an investigative tool to target additional data collection and identification of model components requiring improvement. Employed in this fashion, GIS technologies can provide powerful tools that can enhance the efficiency and effectiveness of numerical environmental models and simplify the investigation of the behavior of spatially distributed phenomena.

This paper has described some of our experience with the integration of GRASS (GIS) and a distributed rainfall-runoff model. The resulting system greatly facilitates visualization of the data, ease of model setup and analysis of model outputs. An important application of the system is the study of model-scale issues in surface hydrology. The experience acquired during our present research clearly illustrates the power and time-saving capabilities that GIS technologies can provide.

ACKNOWLEDGMENT

The authors would like to express their sincere appreciation to NASA (Contract # NAS5-33013) for financial assistance in the research and model development and to the USDA-ARS Southwest Watershed Research Center which provided financial support in development and maintenance of the long-term research facilities in Tucson and Tombstone, Arizona. The authors would also like to thank Dr. Vijay Gupta, Prof. Phil Guertin, and Dr. Robert MacArthur for valuable comments and suggestions used to improve this chapter.

REFERENCES

Band, L.E. (1989) A terrain-based watershed information system. *Hydrological Processes*, 3: 151–162.

Clapp, R.B., and Hornberger, G.M. (1978) Empirical equations for some soil hydraulic properties. *Water Resources Research* 14(4): 601–604.

Dickinson, R.E., Henderson-Sellers, A., Kennedy, P.J., and Wilson, M.F. (1986) *NCAR/TN-275+STR, NCAR Technical Note*, Boulder, CO: National Center for Atmospheric Research.

Goodrich, D.C. (1990) Geometric simplification of a distributed rainfall-runoff model over a range of basin scales. Unpublished PhD dissertation, University of Arizona, 361 pp.

Green, W.H., and Ampt, G.A. (1911) Studies on soil physics, 1. The flow of air and water through soils. *Journal of Agricultural Science* 4: 1–4.

Kerr, Y.H., and Sorooshian, S. (1990) Utilization of EOS data in quantifying the processes controlling the hydrologic cycle in arid/semi-arid regions. NASA proposal submitted in Response to A.O. No. OSSA-1/88, 107 pp.

Kincaid, D.R., Osborn, H.B., and Gardner, J.L. (1966) Use of unit-source watersheds for hydrologic investigations in the semi-arid southwest. *Water Resources Research* 2(3): 381–392.

Renard, K.G. (1970) The hydrology of semiarid rangeland watersheds. *Publication 41-162*, Washington, DC: US Department of Agriculture, Agricultural Research Service.

USACERL (1988) *GRASS 3.1 User Manual*, Champaign, IL: US Army Construction Engineering Research Laboratory.

Woolhiser, D.A., Smith, R.E., and Goodrich, D.C. (1990) KINEROS, a kinematic runoff and erosion model: documentation and user manual. *Publication ARS-77*, Washington, DC: U.S. Department of Agriculture, Agricultural Research Service, 130 pp.

18

Graphical Innovations in Surface Water Flow Analysis

D. R. RICHARDS,
N. L. JONES, AND
H. C. LIN

The U.S. Army Corps of Engineers is responsible for maintaining the coastal and inland waterways that form the backbone of waterborne commerce in the United States. In this context, maintaining waterways involves the construction and repair of locks, dams, levees, training structures, and navigation channel dredging. The Corps is currently maintaining safe navigation on these waterways in an environment where economic resources are scarce and environmental concerns are great. Quite often the waterway projects are very large and have the potential for failing or causing environmental problems if not engineered properly. Therefore, significant effort goes into understanding basic processes at each project.

The engineering process often requires the use of hydraulic models to determine the project environmental impacts. If the impacts are minimal, models are used to design the safest and most economical configuration. Unfortunately, many of the constraints on a project design are contradictory (e.g., flood protection versus wetland creation); so detailed analyses are required. As a result, the Corps has invested in the development of numerical models that can address these and other problems.

For the most part, these problems can be addressed by numerical solutions of the "shallow water equations," so called because they involve water depths that are very shallow in comparison to the water body's width or length. These equations adequately describe flows in rivers, lakes, estuaries, and wetlands, and other free-surface flows that are dominated by bottom friction. In general, the equations involve the solution of the continuity equation (conservation of mass) and the horizontal momentum equations (conservation of momentum). The form of the equations solved in this chapter is as follows:

$$\frac{\partial h}{\partial t} + \frac{\partial (uh)}{\partial x} + \frac{\partial (vh)}{\partial y} = 0$$

$$\frac{\partial u}{\partial t} + u\frac{\partial u}{\partial x} + v\frac{\partial u}{\partial y} + g\left(\frac{\partial h}{\partial x} + \frac{\partial a_0}{\partial x}\right) - \frac{e_{xx}}{\rho}\frac{\partial^2 u}{\partial x^2} - \frac{e_{xy}}{\rho}\frac{\partial^2 u}{\partial y^2} + \frac{gu}{C^2 h}\sqrt{u^2 + v^2} = 0$$

$$\frac{\partial v}{\partial t} + u\frac{\partial v}{\partial x} + v\frac{\partial v}{\partial y} + g\left(\frac{\partial h}{\partial y} + \frac{\partial a_0}{\partial y}\right) - \frac{e_{yx}}{\rho}\frac{\partial^2 v}{\partial x^2} - \frac{e_{yy}}{\rho}\frac{\partial^2 v}{\partial y^2} + \frac{gv}{C^2 h}\sqrt{u^2 + v^2} = 0$$

where

- u = horizontal flow velocity in the x direction
- t = time
- x = distance in the x direction (longitudinal)
- v = horizontal flow velocity in the y direction (lateral)
- y = distance in the y direction
- g = acceleration due to gravity
- h = water depth
- a_0 = elevation of the bottom
- ρ = fluid density
- e_{xx} = normal turbulent exchange coefficient in the x direction
- e_{xy} = tangential turbulent exchange coefficient in the x direction
- C = Chezy roughness coefficient (converted from Manning's n)
- e_{yx} = tangential turbulent exchange coefficient in the y direction
- e_{yy} = normal turbulent exchange coefficient in the y direction.

The solution of these equations provides 2D flow fields and water surface elevations that are sufficient for many applications. But there are times when a constituent or sediment transport analysis is needed that adds at least one additional equation to those shown. The transport equation solved is of the general form:

$$\frac{\partial(\rho c)}{\partial t} + \frac{\partial(\rho c u)}{\partial x} + \frac{\partial(\rho c v)}{\partial y} - \rho\sigma + \rho k c = 0$$

where

ρ = density of fluid
c = constituent concentration
u = instantaneous fluid velocity in the x direction
v = instantaneous fluid velocity in the y direction
σ = local sources of constituent mass
k = decay rate of constituent from the control volume.

When combined with flow fields from the shallow water equations, the transport equations can predict the fate of chemical constituents, sediment, and even such quantities as coliform bacteria, given that their existence in the water body can be described mathematically.

TABS-MD NUMERICAL MODELING SYSTEM

The analysis of 2D, shallow water flows has historically been the exclusive domain of a few universities, research labs, and highly specialized engineering firms. The reason for such exclusivity has been that the codes were not sufficiently generalized for a wide variety of shallow water problems to allow efficient usage by others (Richards, 1992). It is easy to understand the lack of generality when the impetus of each code was for solving a site-specific problem or conducting research that would ensure the completion of advanced degrees. More recently, however, some have sought to expand the base of applications and achieve more widespread acceptability (Lin et al., 1991).

The U.S. Army Corps of Engineers TABS-MD modeling system has been used in recent years to solve various problems in shallow water hydrodynamics and transport. The problems involved large-scale projects that were at first conducted by the U.S. Army Engineer Waterways Experiment Station in Vicksburg, Mississippi. After many such applications that included substantial verifications, the codes were turned over to Corps of Engineers field offices, other Federal agencies, and private consulting firms for general use. Such applications were largely successful, though the codes were too difficult to use on an infrequent basis; so efforts were made to improve user friendliness and productivity by developing a graphical user interface for setting up and running the codes on personal computers.

Requirements for interface design

Perhaps the most difficult and time-consuming problem in applying two-dimensional models has been constructing a topologically correct grid or mesh that accurately represents the bathymetric surface to be modeled. In the finite-element case, nodal points are digitized that represent the shoreline boundaries and important points on the interior of the water body. A mesh of elements is constructed from these points, and a connection table is then built that describes the relation between the nodes and elements. Previously, this task was done using simplistic software or manually by constructing connectivity and node coordinate tables. As long as meshes were small, this was practical if not very innovative. However, most significant waterways could not be satisfactorily described by such a small number of points. Therefore, an interface was needed to automate the mesh generation process and allow interactive mesh editing once the mesh was created.

A second problem in using numerical models is that mistakes in manual data entry cause errors that are difficult to trace due to the size of the input files. Input file sizes to describe medium to large geometries can easily be 0.5 to 1.0 megabytes in length. Data entry errors provide the largest loss in modeling productivity after the meshes are constructed. Automatic input file generation is a necessity to eliminate this source of error.

Displaying output from the models requires that vector and scalar representations of water elevations, current speeds, and constituent concentrations be viewed over a large 2D space through time. Since output data files can be quite large, anything but graphical analysis of the data is impractical.

The following specific requirements must be satisfied for a graphical user interface (GUI) to be successful in producing efficient flow modeling:

1. Easy/fast mesh creation and editing;
2. Easy/fast input file generation (boundary conditions and coefficients);
3. A minimum of manual character input to minimize typing errors;
4. Expert tools to determine adequacy of mesh, time step, residual error, etc.;
5. Graphical postprocessing tools to evaluate the results and write reports;
6. Database management tools to handle time-varying model data; and
7. Minimal training to achieve system proficiency.

Some of these requirements have caused modelers to evaluate GIS packages as a means of avoiding the substantial programming commitment required to develop model-specific graphics and database management software. Modern GIS software often provides some degree of macrolike programming that can be used to accomplish certain model input, output, and data handling tasks. However, GIS was primarily designed as a spatial database and not as a model pre- and post-processor.

Unfortunately, none of the more popular GIS packages sold today can be mastered with minimal training. Additionally, they require very large capacity hard disks and fast CPU speed. While these may be minor obstacles

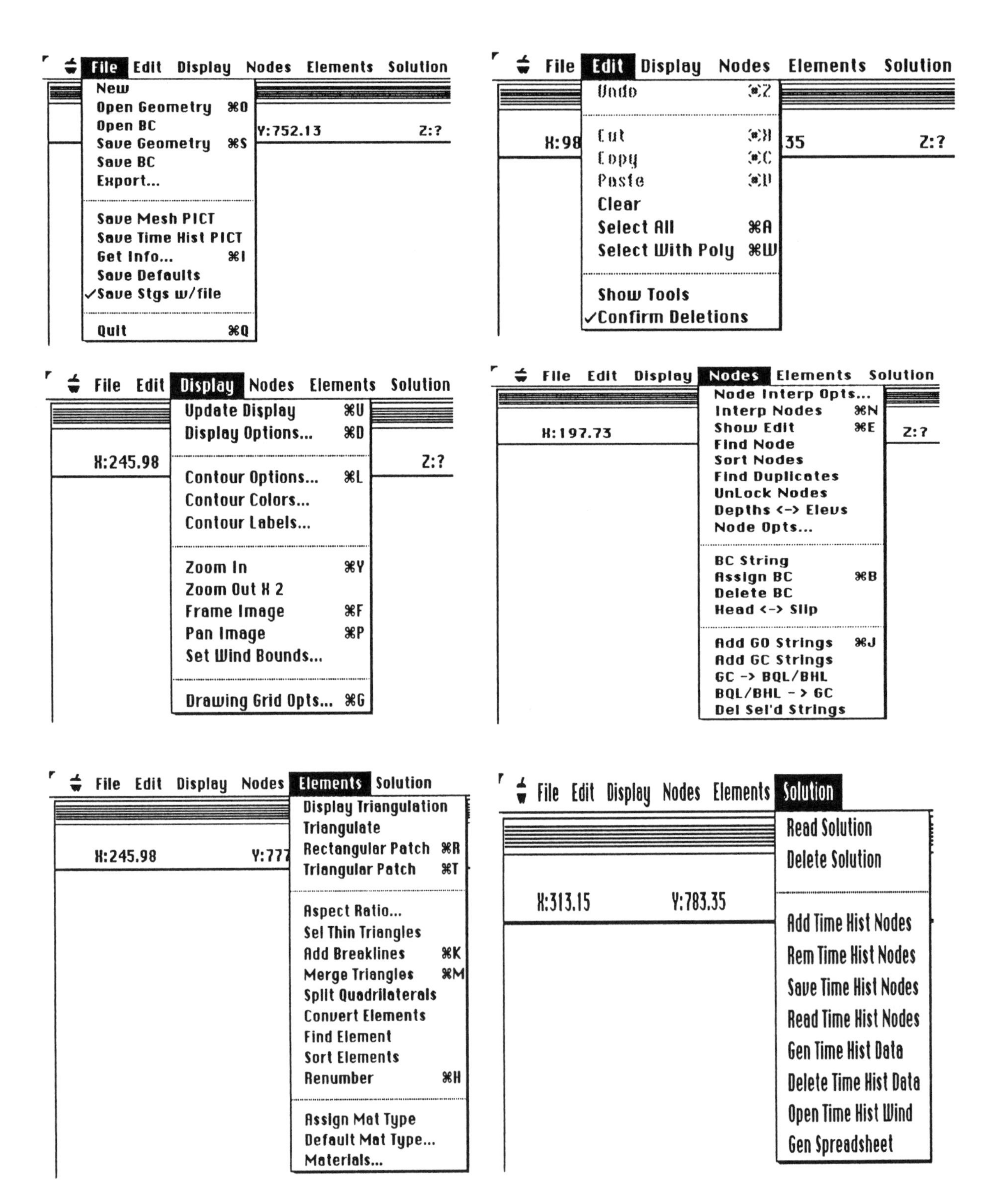

Figure 18-1: Menu items for the graphical user interface. Each line in the separate menus defines a tool that can be invoked by pulling down the cursor and releasing it over the selected tool.

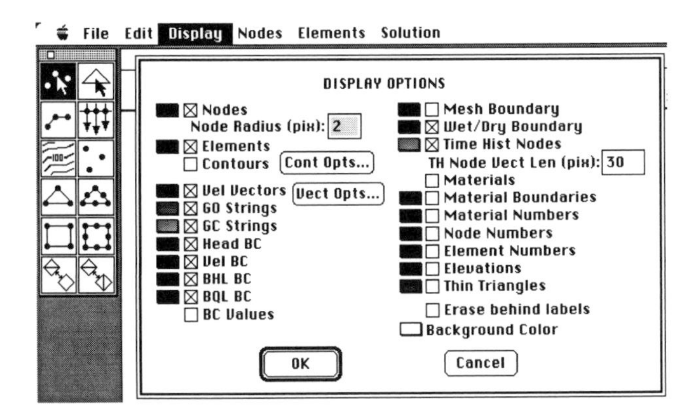

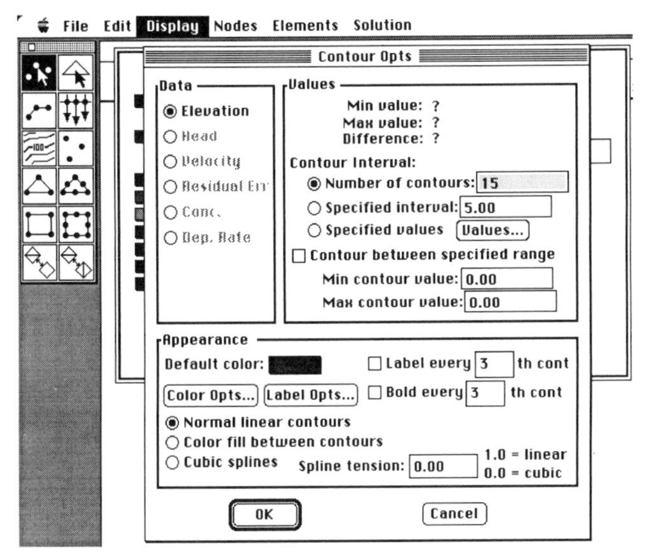

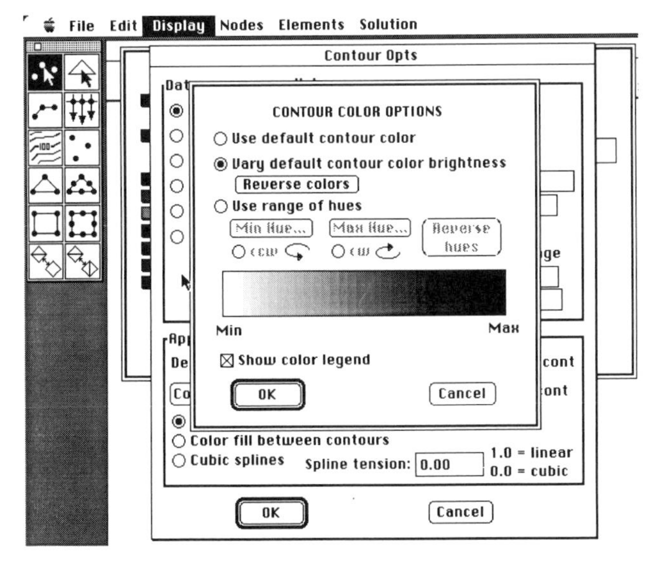

Figure 18-2: Display and contouring options. The top display defines the display form for each of the desired features. The middle display defines the types of contouring options desired for the quantities to be contoured (elevation, head, velocity, residual error, concentration, and sediment deposition rate).

for typical GIS users, they are insurmountable to many numerical modelers given the time and financial resources available for infrequent modeling applications. To address these requirements, a code was required that contained many of the attractions of GIS but was tailored in size and usefulness to numerical flow models.

Graphical user interface

The developed interface was based on triangular irregular networks (TINs). They were used successfully in digital terrain models (Jones, 1990), and their representation of the terrain corresponded with the TABS-MD finite-element models. The prototype system was developed on the Macintosh family of personal computers using the Macintosh graphical Toolbox, and significant effort was given to adhering to Macintosh style guides. The main menu bar and pull down menus are displayed in Figure 18-1. Figure 18-2 shows the various display and contouring options that are available. The menus contain a lengthy list of operations that can be performed in the creation and editing of finite-element meshes, boundary conditions, and the graphical display of model results. A users manual exists for the software but is rarely needed due to the intuitive nature of the tools and menus.

Modeling system capabilities

The Macintosh implementation of the TABS-MD system includes all of the software necessary to run flow and transport for both steady-state and unsteady applications. The codes have been used most commonly for general circulation studies but more recently have been used in near-field studies of training structures and studies of wetting and drying in wetlands. The amount of memory, available disk space, and time are the only constraints in determining the size of application that can be solved provided that the Macintosh has a math coprocessor. The software is not limited by operating system restrictions such as the 640K memory limitations of DOS, but the Macintosh cannot be efficiently used for long-term simulations with large meshes. Practically speaking, up to 2500-element meshes can be handled in steady-state simulations and up to 500-element meshes over a tidal cycle with 0.5 hour time steps. The graphics software could handle the analysis of much larger output files if the Macintosh were a faster machine. In some cases hydrodynamic results from large mainframes and supercomputers have been converted to Macintosh binary format to use the Macintosh and this software simply as an output display device.

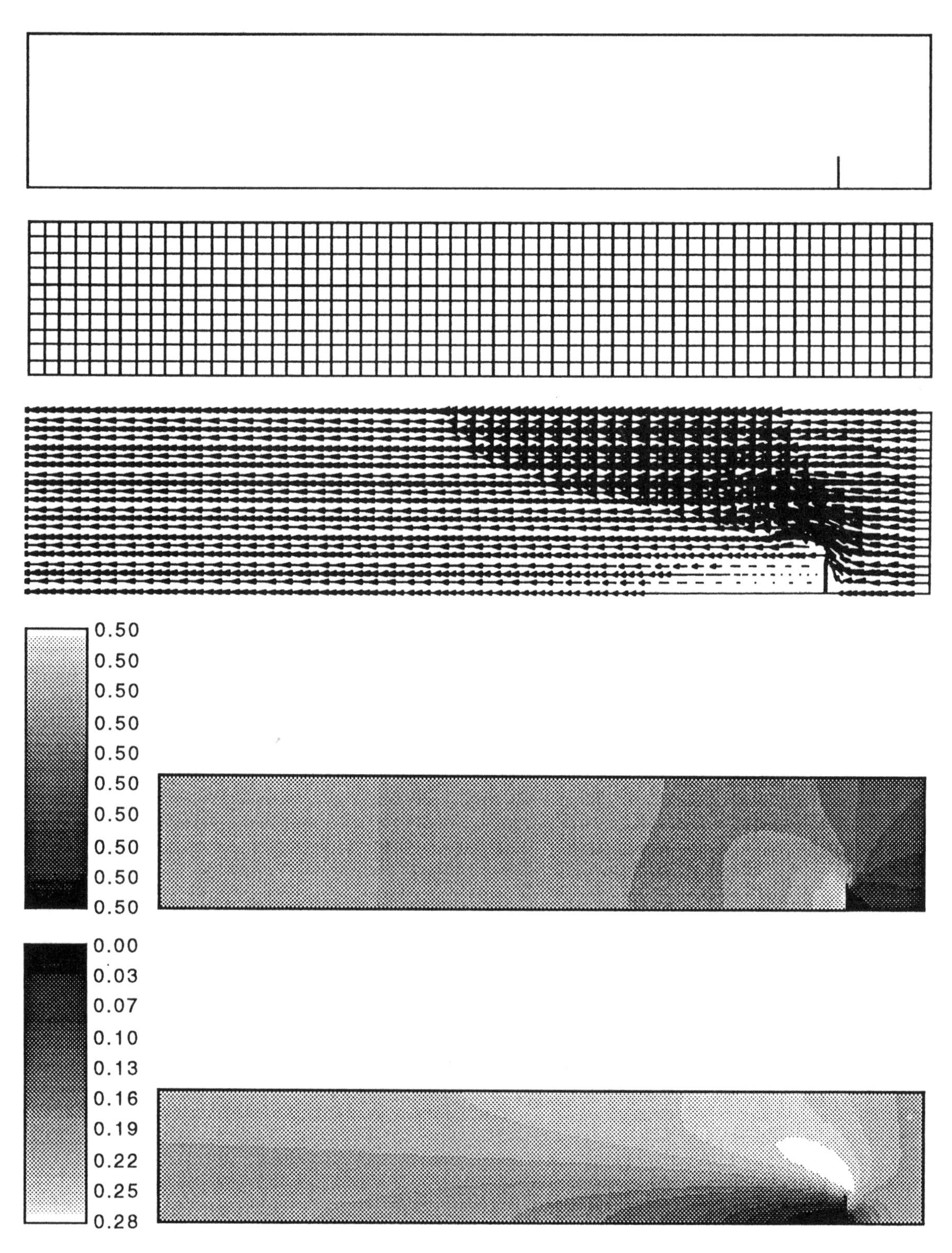

Figure 18-3: Solitary dike results from numerical flume. Flows enter the flat-bottomed rectangular flume from the right, then encounter a single dike that constricts the flow and causes a flow separation just downstream of the dike. The second figure defines the numerical mesh used for the calculation. The third figure is a zoomed-in view of the velocity vectors around the dike. The fourth figure shows a scalar representation of water surface elevation, with the higher elevations upstream of the dike. The bottom figure shows a scalar representation of current speed with the separation zone defined by the darkest color.

TYPICAL APPLICATIONS

Two recent hydrodynamic applications are presented that typify the class of problems that could easily be accomplished on personal computers. They were not chosen for their hydraulic peculiarities but rather to illustrate the range of applications that could be studied using the software.

The first application is a 2400-element flume study of flow around a solitary dike (Richards, 1992). The solitary dike problem was modeled to investigate the potential for using training structures to minimize maintenance dredging (Figure 18-3). These structures are commonly used to concentrate flows and increase the sediment carrying capacity of riverine and estuarine channels. This simulation employed steady-state (time-invariant) velocity input and exit water levels, which is sufficient for this problem. In actual river applications, there would likely be curved boundaries and multiple dikes considered at one time.

The second application is a dynamic (transient) tidal application of Noyo Harbor, California, that was used to evaluate, in a preliminary way, the effect of a proposed breakwater on tidal circulation (Figures 18-4 to 18-6). The Noyo Harbor mesh contains 126 elements and was run for a 12-hour tidal cycle with 30-minute time steps. This study is a typical modeling study that is included in the analysis of Corps construction projects. In such analyses, modelers determine differences between existing and plan hydrodynamics and provide these results to biologists, economists, and managers to determine project impacts. Quite often there is a need to fine tune project design to account for navigation or environmental impacts that were unforeseen before the modeling was completed.

The solitary dike application shown in Figure 18-3 includes displays of the external boundaries, the numerical mesh, a blowup of velocity vectors flowing around the dike, and shaded representations of water surface elevation and current speed. Although the figures are displayed here in gray scale, the data can be presented as monochrome contour lines or as color-shaded contour bands. Colors can be selected from a color wheel as wished to define a band or range of colors for displaying color-shaded contour plots. All of the figures can be saved as PICT files and printed on monochrome or color laser printers for use in reports.

The Noyo Harbor application in Figures 18-4 to 18-6 indicates how unsteady flow simulations are presented. Particular time steps in the simulation can be queried and quickly displayed over its 2D region. In addition, time history plots can be generated for water surface elevations or current velocities based on which nodes are selected on the screen. At no point is it necessary for the user to define the particular node number. All input and output is based on graphical interaction.

The bottom plot of Figure 18-6 is a depiction of resid-

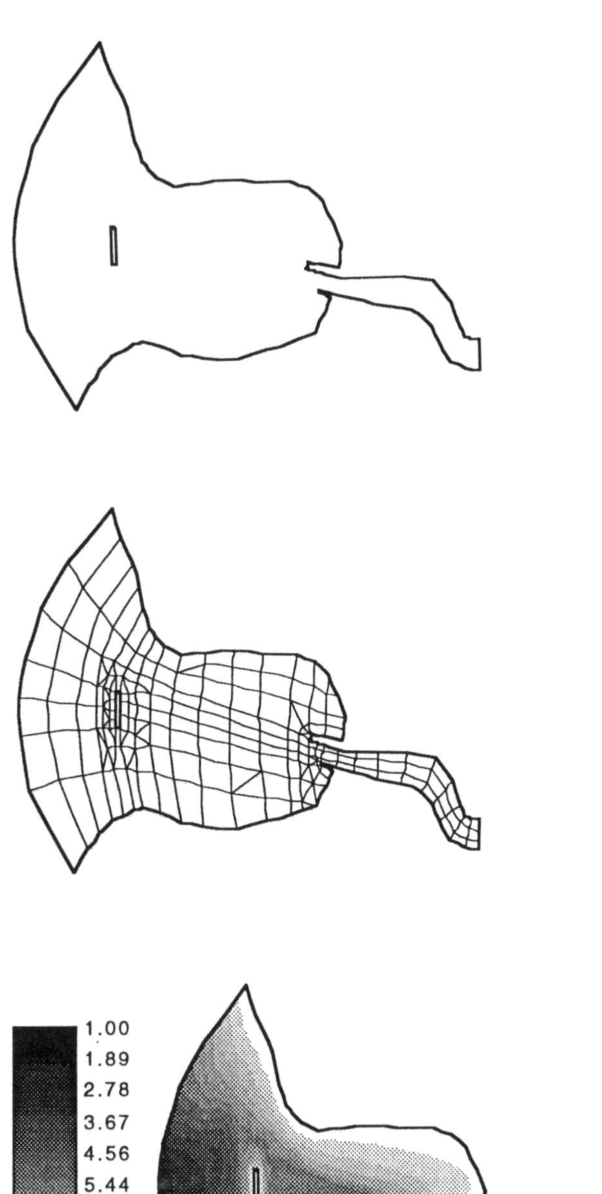

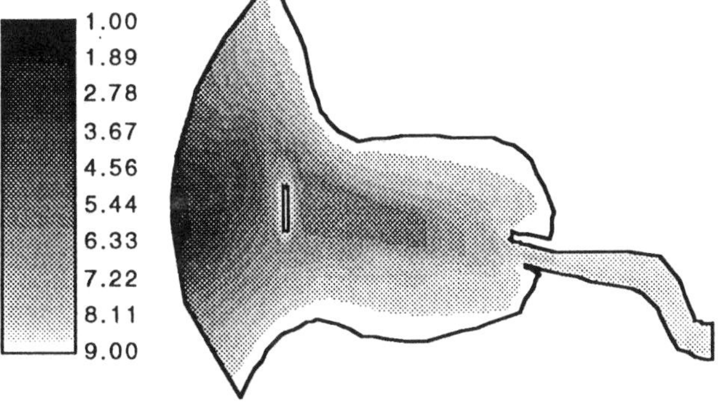

Figure 18-4: Noyo Harbor model boundaries, mesh, and bottom elevations. The top figure shows the Pacific Ocean boundary on the left, the shoreline boundary on the top and bottom, the Noyo River boundary on the right and the edges of the breakwater in the interior. The middle figure shows the numerical mesh with increased resolution around the breakwater. The bottom figure shows a contour map of bottom elevations (in feet).

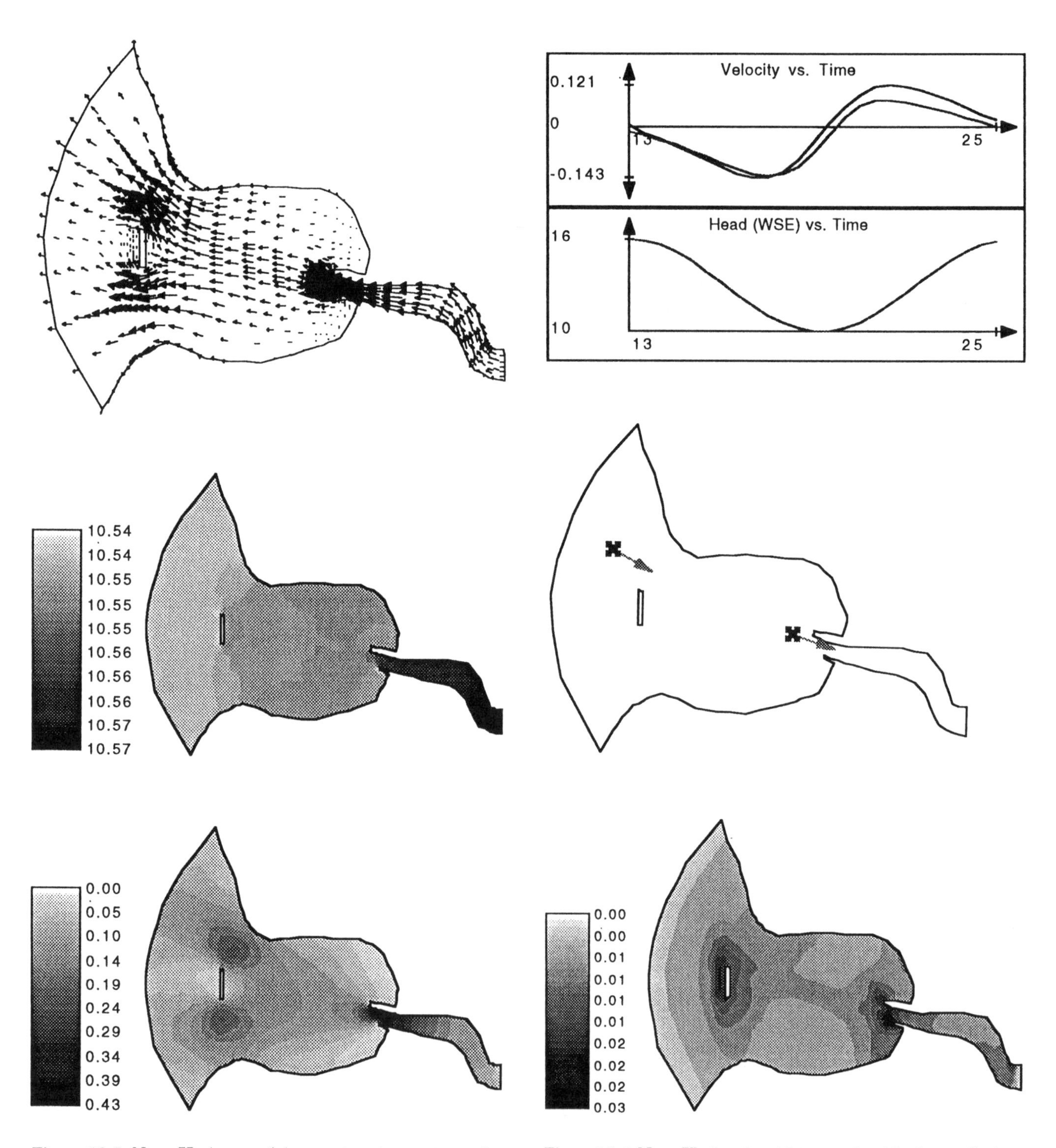

Figure 18-5: Noyo Harbor model current vector, water surface elevation, and current speed displays.

Figure 18-6: Noyo Harbor time history and residual error displays. The top two figures show how time histories of head and velocity are displayed at the two selected nodes; the flood angle is defined by dragging an arrow in the direction of flood. The bottom figure shows a map of residual error that indicates the hightest error near the breakwater and the entrance to the Noyo River. This display indicates that more resolution is needed in these regions where significant velocity gradients occur.

ual error in flow continuity at a particular time step. The residual error is an indication of the adequacy of the mesh resolution throughout the region. In this particular case, it appears that the breakwater and the jetty near the river entrance could use more resolution to ensure an even distribution of residual error throughout the region. Additionally, time histories of residual error can be plotted to show its response through time. This is a particularly important display of numerical adequacy that is not often found in 2D numerical models.

DEVELOPMENT PLANS

The current system can model most subcritical, unsteady, open-channel flow problems. By publication time, constituent transport will be modeled as well with concentrations (salinity, etc.) being displayed with the same tools currently used for velocity and water surface elevations. After the implementation of constituent transport is complete, we will port our existing sediment transport codes into this environment.

Perhaps the most important development plan is to convert the codes to X-Windows so that UNIX workstations can be used to solve much larger problems. They are typically 10 to 20 times faster than the fastest Macintosh, which is necessary for long-term simulations of sediment and constituent transport. This should be completed by the publication date of this volume.

ACKNOWLEDGMENTS

The work reported herein was supported by the Civil Works Program of the U.S. Army Corps of Engineers. Permission was granted by the Chief of Engineers to publish this information. The contents of this paper are not to be used for advertisement, publication, or promotional purposes. Citations of trade names do not constitute an official endorsement or approval of the use of such commercial products.

REFERENCES

Jones, N.L. (1990) Solid modeling of earth masses for applications in geotechnical engineering. *PhD Dissertation*, University of Texas at Austin.

Lin, H.C., Jones, N.L., and Richards, D.R. (1991) A microcomputer-based system for two-dimensional flow modeling. *ASCE National Conference on Hydraulic Engineering, Nashville, Tennessee.*

Richards, D.R. (1992) Numerical simulation of training structure influenced flows. *Master's Thesis*, University of Texas at Austin.

Thomas, W.A., and McAnally, W.H., Jr. (1990) *User's Manual for the Generalized Computer Program System: Open-Channel Flow and Sedimentation, TABS-2*, Vicksburg, MS: U.S. Army Engineer Waterways Experiment Station.

19

GIS and Land-Surface–Subsurface Process Modeling

IAN D. MOORE,
A. KEITH TURNER,
JOHN P. WILSON,
SUSAN K. JENSON, AND
LAWRENCE E. BAND

Generic, knowledge-based modeling techniques that are transportable across environments are rapidly becoming the basis of analytical methods used to examine and resolve complex resource and environmental issues. Models can incorporate descriptions of the key processes that modulate system performance or behavior with varying degrees of sophistication. A modeling framework provides a basis for evaluating options. Knowledge-based methods allow a variety of hypotheses to be explored and permit the calibration of physical system parameters from observations and prior knowledge. A generic framework is needed for the development of response strategies that can be applied to a variety of institutional arrangements, human settlement patterns, landscapes, and weather and climate input patterns.

Information systems and databases for addressing resource and environmental issues must be capable of integrating spatial information to representative and environmentally diverse landscapes. They also must be capable of integration with models and other analytical tools used for analysis. Geographic information systems that incorporate and spatially relate biophysical and socioeconomic data are an exciting new tool for manipulating and accessing these data.

There are many physical, biological, economic, and social models in use today. Richardson (1984) presents examples and describes the classification and characteristics of some of these "disciplinary-based" models. Some of the most interesting research being carried out today is aimed at developing integrated biophysical–socioeconomic models, databases and information systems to drive them, and decision support systems that permit their use by politicians, policy-makers, and managers who often lack intimate knowledge of the models being used.

This chapter is restricted to an examination of models of surface and subsurface processes operating at the "landscape" scale (i.e., the macro- and mesoscale). The domains dealt with by both subsurface and surface models and GIS technology are spatially distributed over a landscape; so the two technologies have a natural affinity. Moore and Gallant (1991) describe a variety of methods of classifying models. We will concentrate on computer-based, distributed parameter, semiempirical and physically based hydrological models. The hydrologic cycle provides the link between the physical, chemical, and biological processes within all the compartments of the earth (Eagleson, 1991). We will review how models of surface and subsurface processes can be interfaced with GIS to provide more robust and effective research and management tools for resource and environmental management. We attempt to address several key issues including: (1) problems of model and database structure; (2) the effects of scale and accuracy on GIS databases; (3) incompatibilities between the scales of many physically based models of land surface/subsurface processes, databases, and the intended end uses of model outputs; and (4) the specific problem of soil properties for environmental modeling, their spatial variability, and the constraints they place on land surface process modeling in particular.

MODELING AND GIS—APPLICATIONS AND PROBLEMS

The growing societal significance of non-point-source pollution, hazard identification, land resource assessment, and resource allocation problems at a variety of scales has fueled, to a degree, the development of spatially distributed surface/subsurface modeling technologies. These problems range from the development of site-specific management strategies to regional and national policy. GIS provide the technology to store and manipulate the spatial relational data demanded by many of these models.

GIS technology has been integrated with several surface/subsurface models in recent years. Several applications explore water quality in one or two dimensions at the catchment scale. The Simulator for Water Resources in Rural Basins (SWRRB) has been linked via a decision support system (Arnold and Sammons, 1989). The Finite Element Storm Hydrograph Model (FESHM) (Hession et al., 1987; Shanholtz et al., 1990), Areal Nonpoint Source Watershed Environment Response Simulation (ANSWERS) (Engel et al., 1991), and Agricultural Non-Point Source pollution (AGNPS) (Engel et al., 1991; Panuska et al., 1991) models have also been linked to a variety of GIS and image processing systems. These systems include the Geographical Resources Analysis Support System (GRASS), the Earth Resources Data Analysis Systems (ERDAS), and ARC/INFO, as well as several terrain analysis software packages. The next chapter illustrates how these linkages can be made. Specifically, it shows how a spatial decision support system can be interfaced with an agricultural non-point-source water quality model and the ease with which resource managers can examine alternative management options using this integrated technology. Methods for predicting areas at risk from catastrophic mass movement (i.e., landslides) in "young" landscapes and populated areas have combined GIS and modeling technologies in Europe, China, and other areas (Aste, 1988; Corominas et al., 1990; Noverraz, 1990; Carrara et al., 1991). GIS technology that incorporates physical models and spatial databases characterizing climate, soil nutrient status, and hydrology has provided valuable insights into the forcing functions that determine the distribution and abundance of flora and fauna in "natural" ecosystems (Verner et al., 1986; Turner and Gardner, 1990; Lees and Ritman, 1991; Mackey, 1991; Norton and Moore, 1991).

At a larger scale, the United States Environmental Protection Agency (USEPA) is using the DRASTIC index (Aller et al., 1985) as a screening tool to provide regional assessments of the vulnerability of ground water to pollution and for planning ground water monitoring programs. DRASTIC is a weighted numerical scheme that considers depth to ground water (D), net recharge rate (R), aquifer media (A), soil characteristics (S), topography (T), impact of the vadose zone (I) and hydraulic conductivity of the aquifer (C) and is well suited to GIS implementation. These assessments assist in determining resource allocation among the various regions for the USEPA's ground water pollution programs. There is also great potential for using integrated modeling and GIS technologies in national resource assessments such as the appraisal of soil, water and related resources on nonfederal land required under the Soil and Water Conservation Act of 1977 (USDA, 1989) and in implementing the conservation title of the Food Security Act of 1985.

Three-dimensional GIS-based systems are being developed to visualize and model three-dimensional space and handle the huge amounts of data that are generated in subsurface and Earth system studies. They have been used to study subsurface contamination at hazardous waste disposal sites, nuclear waste repositories, and petroleum spills (Fisher and Wales, 1990, 1991, 1992). The method can provide information where an adequate sampling density is too costly and allows conditions to be predicted in areas with no data and where nonintrusive survey techniques must be used (Fisher, 1991).

Although powerful, GIS-based environmental modeling approaches have characteristics and problems that may constrain their integration. These features are discussed next.

Mathematical models

Mathematical models integrate existing knowledge into a logical framework of rules and relationships. They represent one method of formalizing knowledge about how a system behaves (Moore and Gallant, 1991). Mathematical models range from simple empirical equations such as linear regression equations (functional models) to sets of complex differential equations derived from fundamental physics (physically based models).

There are at least two reasons for developing a mathematical model: (1) to assist in the understanding of the system that it is meant to represent, that is, as a tool for hypothesis testing; and (2) to provide a predictive tool for management (Grayson et al., 1992; Beven, 1989). These two purposes can be quite distinct. Hillel (1986, p. 42) identified four principles that should guide model development:

1. Parsimony: "A model should not be any more complex than it needs to be and should include only the smallest number of parameters whose values must be obtained from data";
2. Modesty: "A model should not pretend to do too much", "there is no such thing as THE model";
3. Accuracy: "We need not have our model depict a phenomenon much more accurately than our ability to measure it"; and
4. Testability: "A model must be testable" and we need to know "if it is valid or not, and what are the limits of its validity."

Many models of surface and subsurface flow processes arguably violate these guidelines. The quandary modelers

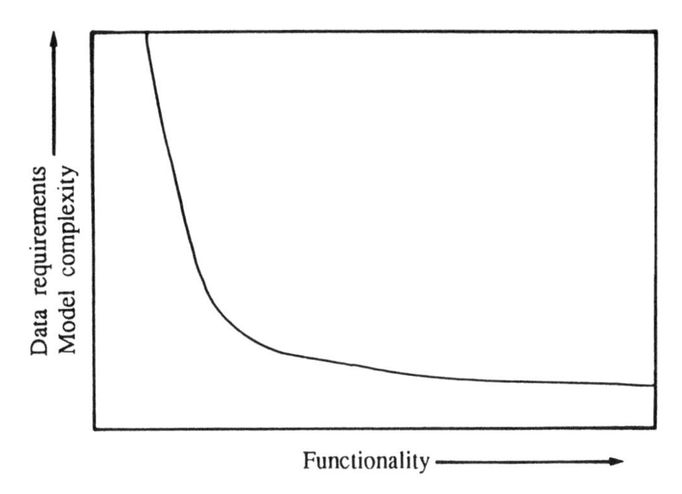

Figure 19-1: Data requirements and model sophistication versus model functionality.

often face was addressed by Denning (1990, pp. 497–498). He stated that "simplifying assumptions can introduce rather than resolve computational complexities, a possibility that looms larger for systems with many variables and for models with many simplifications" and that "it is easy for us to ignore these complexities by persuading ourselves that the model is real or that the simplifying assumptions are of no consequence." He later went on to say that "our drive to use scientific models to resolve world problems might not reflect the hubris of science. We need to ask ourselves whether some of the models of complex phenomena we seek to construct would gain us anything if we could find them." Figure 19-1 represents the tradeoff that inevitably must be made between model functionality or ease of use and data requirements or the sophistication of the process descriptions represented in the model. This tradeoff is an important consideration when attempting to interface surface/subsurface modeling techniques with GIS.

Some problems and limitations of distributed parameter, physically based mathematical models, and particularly hydrological models, were discussed by Klemes (1986), Beven (1989), Moore and Hutchinson (1991) and Grayson et al. (1992). In summary:

1. It is not clear that point-based equations adequately describe three-dimensional, spatially heterogeneous and time-varying systems in real landscapes;
2. Because of system nonlinearities, effective parameters at the grid scale may not be the same as spatially averaged values;
3. A priori estimation of model parameters is dependent on the initial method and scale of measurement;
4. Model structure can significantly influence parameter values; and
5. The ability to validate is dependent on the errors of both inputs and outputs, and requires different types of data than are traditionally available (Moore and Hutchinson, 1991).

Beven (1989, p. 157) concluded that "future developments in physically based (hydrological) modeling must take account of the need for a theory of the lumping of subgrid scale processes; for closer correspondence between equations and field processes; and for rigorous assessment of uncertainty in model predictions." Furthermore, "for a good mathematical model it is not enough to work well. It must work well for the right reasons. It must reflect, even if only in a simplified form, the essential features of the physical prototype" (Klemes, 1986, p. 178S).

Geographic information systems

Agencies adopting GIS should be aware that this technology creates a need to import data from outside the organization, often with little control by the agency. Even within an organization, one may find the need to integrate functions that were kept separate in the past. Many agencies have rushed to develop GIS with a view that it would be a panacea for providing solutions to their problems. Often the expectations of GIS technology are too high. Unfortunately, many GIS are simply being used as cartographic tools for displaying the basic input data in the form of colored maps. Many organizations are not realizing the full capabilities of this technology due to a lack of generic, knowledge-based (physically based) approaches for analyzing and interpreting data. These capabilities must be added by the user to versatile GIS such as ARC/INFO that provide a "toolbox" from which the user can build specific applications. Such approaches allow realistic answers to questions that might be asked, such as: "where are high erosion hazard zones located in a landscape?" Two other problems concerned with the implementation of GIS technology are that: (1) half-digested secondary data are often being input into them; and (2) data obtained at different scales are used with no consideration of scale effects. Primary data that modulate landscape processes and biological responses should be used as far as possible, and subjective data should be avoided. As ideas and concepts change and evolve, subjective data are often rendered useless.

Until recently GIS have been used most commonly as a cartographic tool to map essentially two-dimensional (2D) land surface phenomena such as land use, vegetation, or soil series boundaries. An extension of this traditional approach is required in soils and geological applications where representation of the vertical dimension (e.g., in characterizing soil horizons and geologic stratigraphy) is necessary and linkages to various data manipulation procedures require 3D data. Some of the most exciting developments in GIS are in the areas of topologically oriented 2D, 3D (space), and 4D (time) analyses and visualization (Gessler, 1991). These techniques have huge potential applications in the geological, soil, limnological, oceanic, and atmospheric sciences. These new GIS systems are perhaps best developed in the geological sciences, where they are called "geoscientific information systems" (GSISs) to differentiate the geologically oriented 3D and 4D systems from traditional 2D GIS products (Turner et al., 1990; Turner, 1991). In Chapter 21 Fisher illustrates how 3D GIS can be used to facilitate hazardous waste site investigations.

DATA AND MODEL STRUCTURES

Correct use of a model requires an understanding of both the biophysical system being studied and the operation of the model (Gersmehl et al., 1987a). Furthermore, a certain level of compatibility is required between the structure and spatial scale of the model and the database(s) needed to drive the model. Choosing a model is a "third-level" task. The first task consists of phrasing intelligent questions, and the second is to gather the appropriate data. All three levels must be linked (Gersmehl et al., 1987a). The problems and dangers in attempting to match spatial databases and quantitative models in land resource assessment are outlined by Burrough (1989).

There are at least two possible ways of representing topological data in a GIS: raster and vector representations (Burrough, 1986). These structures are described in detail in Chapters 3, 7 and 8. Raster structures are perhaps the simplest because entities are represented implicitly, whereas vector entities are explicitly stored in a linked database. Many low-cost GIS and large-scale environmental applications that use satellite data are based on a raster system. Vector systems usually incorporate topological structuring, which greatly increases their robustness and analytical capabilities. However, with vector systems much effort is expended in defining polygons for individual data layers and spatial operations (e.g., overlay, buffer) on two or more data layers.

Areal attribute data developed for either raster- or vector-based GIS may be stored in two ways: as a tag (classificatory) that describes each spatial unit as a whole based on the dominant attribute within it, or as a count that is an inventory of the frequency of occurrence of some phenomenon as a whole (Gersmehl et al., 1987a). Table 19-1 outlines the characteristics of the two methods. Gersmehl et al. (1987a) recommend that data collected by one method should not be used in models or to answer questions that require data of the other type.

Data problems and issues

Surface and subsurface models may require climatic/weather input data and model parameters derived from measurements of: (1) topography; (2) soil physical and chemical properties; (3) geology, including physical, chemical, and aquifer properties, and ground water characteristics; (4) land cover or land use; and (5) hydrography and water quality data. These data have vastly different temporal and spatial scales of variability and measurement. Often the scale and method of measurement are not directly compatible with the scale needed by a model. Furthermore, modeling applications require data georeferenced to a common system of grids, polygons, lines, or points. Often the lack of the necessary data to run the model or on which to estimate model parameters places the greatest constraints on the choice of model for a particular application.

Data required for biophysical modeling may be measured at a point (e.g., precipitation, soil properties, stream discharge) or over an area (e.g., remotely sensed land cover). Often point data are spatially interpolated before they are stored in GIS or used in surface and subsurface process modeling. Techniques commonly used to interpolate point data spatially include: (1) qualitative methods based on transfer by analogy, which is the dominant paradigm in soil survey; (2) local interpolation methods; (3) moving average methods; (4) kriging; and (5) partial thin plate (or Laplacian) smoothing splines (Moore and Hutchinson, 1991; McKenzie, 1991). Kriging has probably become the most widely used quantitative method of interpolating a variety of surface and subsurface data. Partial thin plate splines are formally related to kriging methods but offer some distinct advantages. The main advantage is that partial thin plate splines do not require the estimation of a spatial covariance or semivariogram function.

Different measurement and interpolation techniques characterize different biophysical phenomena occurring in landscapes, and these in turn raise different problems and issues as illustrated by the following discussion.

Climate/weather

Climate and weather variables are usually measured at a point. The gauging network is usually highly irregular, and the sampling density is nonuniform and often sparse. The records are often incomplete and of variable length. Precipitation is one of the most spatially and temporally variable types of climate data, and so we will use it as an example. Many hydrological models require hourly or even finer-resolution rainfall intensity data to drive them,

Table 19-1: Comparison of count and tag methods of data storage (reproduced from Gersmehl et al., 1987a)

Criterion	Count	Tag
Primary purpose	Regional inventory	Parcel classification
Question asked	How much of resource *R* is in area *X*?	What is the major resource in area *X*?
Typical answers	*X* has 50 ha cropland or 30 ha of *X* is woodland	*X* is a dense stand of eucalypt or *X* is mainly land capability class IV
Data scaling	Quantity measured (ratio data)	Category assignment (nominal data)
Data cell coding method	Each data cell is tabulated on the basis of the resource at the center (or some random point) within the cell	Each data cell is categorized on the basis of the areally (or economically) dominant resource within the cell
Method tries to maximize	Statistical validity in the aggregate	Descriptive accuracy in the particular
User is willing to sacrifice	Proper description of some individual tracts of land	Accurate tabulation of small or odd-shaped resource areas
What can (often does, even should) happen	The sample point "hits" a tiny woodlot in a big field and the entire cell is recorded as woodland	Narrow floodplain is "lost" from the data record because it is too small to dominate any data cells
Ideal procedure	Use enough sample points to make a valid count of even the least abundant resource	Use small enough cells to keep them homogeneous for even the smallest resource areas
Real-world compromise	Use sampling theory to limit error for major resources	Use cell size that can "capture" most major resources
Inevitable result	Point-count methods misclassify parcels	Area-tag methods miscalculate totals

but only daily data are commonly available from the National Weather Service. In addition, most models require some form of areal precipitation estimate rather than a point estimate. In recognition of these difficulties, hydrologists have derived several crude methods for estimating areal precipitation. They include Thiessen (i.e., Voronoi polygons), isohyetal, and reciprocal-distance-squared methods (Chow et al., 1988). None of these methods provides good estimates of the spatial distribution of temporally varying, fine time-resolution rainfall intensities on a catchment. Nor do they effectively handle the effects of moving storm cells. Recent developments in weather radar may overcome some of these problems, but it is not yet a fully developed technology.

Stochastic weather generators such as CLIGEN (Nicks, 1985; Nicks and Lane, 1989) and CLIMATE.BAS (Woolhiser et al., 1988) are being linked to crop growth, hydrological, erosion, and crop-productivity models such as SWRRB (Arnold et al., 1990), Water Erosion Prediction Project (WEPP) (Lane and Nearing, 1989), and Erosion Productivity Impact Calculator (EPIC) (Williams et al., 1983, 1984). They generate sequences of daily or hourly weather data (precipitation, maximum and minimum temperature, solar radiation, etc.) at a point as a function of the statistical moments of the historical data (e.g., monthly averages of daily means and standard deviations). The parameters of Nicks's model have been developed for nearly 200 stations in the United States.

Hutchinson (1989b) has recently developed monthly climatic surfaces for Australia (EStimation Of CLIMate—ESOCLIM) by interpolating irregularly spaced and variable record length historical climate data as a function of latitude, longitude, and elevation using thin plate smoothing splines. This method is particularly attractive because it allows the variance of the climate variable to be spatially varied. It also allows the mapping of the parameter values of the weather generator described. Running et al. (1987) and Hungerford et al. (1989) developed MTCLIM, a mountain microclimate simulator, that interpolates minimum and maximum daily air temperature, precipitation, humidity, vapor pressure deficit, and solar radiation over mountainous terrain from base station records by adjusting for elevation and aspect-related effects. MTCLIM provides information at a level of accuracy and precision necessary to operate forest ecosystem models over a growing season. ESOCLIM is being used to isolate climatic influences on forest structure (Mackey, 1991) and flora and fauna distribution and abundance. Both the ESOCLIM and MTCLIM techniques could be easily imbedded in the data analysis subsystem of a GIS.

Topography

Water and wind are the main erosive agents and transporting mechanisms for land surface processes. The distribution of water is also a major factor determining the distribution and abundance of flora and fauna in a landscape. Topography is a major determinant of these erosion and transport processes. Surface process models typically

Table 19-2: Primary topographic attributes that can be computed by terrain analysis from DEM data (adapted from Speight, 1974, 1980; Moore et al., 1991)

Attribute	Definition	Significance
Altitude	Elevation	Climate, vegetation, potential energy
Upslope height	Mean height of upslope area	Potential energy
Aspect	Slope azimuth	Solar insolation, evapotranspiration, flora and fauna distribution, and abundance
Slope	Gradient	Overland and subsurface flow velocity and runoff rate, precipitation, vegetation, geomorphology, soil water content, land capability class
Upslope slope	Mean slope of upslope area	Runoff velocity
Dispersal slope	Mean slope of dispersal area	Rate of soil drainage
Catchment slope*	Average slope over the catchment	Time of concentration
Upslope area	Catchment area above a short length of contour	Runoff volume, steady-state runoff rate
Dispersal area	Area downslope from a short length of contour	Soil drainage rate
Catchment area*	Area draining to catchment outlet	Runoff volume
Specific catchment area	Upslope area per unit width of contour	Runoff volume, steady-state runoff rate, soil characteristics, soil water content, geomorphology
Flow path length	Maximum distance of water flow to a point in the catchment	Erosion rates, sediment yield, time of concentration
Upslope length	Mean length of flow paths to a point in the catchment	Flow acceleration, erosion rates
Dispersal length	Distance from a point in the catchment to the outlet	Impedence of soil drainage
Catchment length*	Distance from highest point to outlet	Overland flow attenuation
Profile curvature	Slope profile curvature	Flow acceleration, erosion/deposition rate, geomorphology
Plan curvature	Contour curvature	Converging/diverging flow, soil water content, soil characteristics

* All attributes except these are defined at points within the catchment.

require topographic-based data defining catchment and subcatchment boundaries, drainage areas, land slopes, slope lengths, slope shape, channel slopes, aspects, channel networks, and cell or land unit connectivities that define how water moves through a landscape (Table 19-2). For example, approximately one third of the input parameters required by the AGNPS water quality model are terrain based (Young et al., 1989). Panuska et al. (1991) have shown how digital terrain modeling techniques can be interfaced with the AGNPS model to derive terrain-based model parameters automatically.

Topographic attributes can be divided into primary and secondary (or compound) attributes. Primary attributes are directly calculated from elevation data and include variables such as elevation, slope, and aspect. Compound attributes involve combinations of the primary topographic attributes and are physically based or empirically derived indices that characterize the spatial variability of specific processes occurring in the landscape. Table 19-2 lists the key primary topographic attributes and their significance in surface and subsurface process modeling. Moore et al. (1991) and Moore and Nieber (1989) describe a series of compound attributes used to characterize the spatial distribution of selected hydrological, geomorphological, and ecological processes. This approach is discussed in more detail later using two examples: the spatial distribution of soil water content and erosion potential in complex terrain.

These topographic attributes can be computed from a digital elevation model (DEM) using a variety of terrain analysis techniques and stored as a digital terrain model (DTM) as part of a GIS (Moore et al., 1991). It may be more efficient to include a terrain analysis technique within the analysis system of a GIS and compute terrain attributes from DEM data rather than store the large number of terrain attributes within the GIS. It is possible to compute the primary terrain attributes by either fitting a surface defined by some function $F(x,y,z)$ to the DEM and using elementary geometry to derive the attributes or by using finite difference techniques directly with the DEM. Methods that fit a surface sequentially to local regions of the DEM are called local interpolation methods, while those that fit a surface to the entire DEM are called global methods. Generally, global methods are computationally intractable except for small areas. Moore et al. (1991) present equations for calculating slope and aspect from surfaces fitted using local interpolation methods to a range of DEM structures. The equations are simple and easy to implement within a GIS and produce identical results to finite differencing methods, depending on the form of the finite differencing scheme used. The presentation of these relationships is beyond the scope of this paper.

The ideal structure for a DEM depends on the intended use of the data and how it might relate to the structure of a model. Figure 19-2 illustrates the three principal ways of structuring a DEM (Moore et al., 1991):

1. Grid-based methods may use a regularly spaced triangular, square, or rectangular mesh or a regular angular grid, such as the 3 arcsecond spacing used by the U.S. Defence Mapping Agency. The choice of grid-based method relates primarily to the scale of the area to be examined. The data can be stored in a variety of ways, but the most efficient is as *z* coordinates corresponding to sequential points along a profile with the starting point and grid spacing also specified.
2. Triangulated irregular networks (TINs) usually sample surface-specific points, such as peaks, ridges, and breaks in slope, and form an irregular network of points stored as a set of *x*, *y* and *z* coordinates together with pointers to their neighbors in the net (Peucker et al., 1978; Mark, 1975).
3. Vector or contour-based methods consist of digitized contour lines and are stored as digital line graphs (DLGs) in the form of *x*,*y* coordinate pairs along each contour line of specified elevation. These can be used to subdivide an area into irregular polygons bounded by adjacent contour lines and adjacent streamlines (Moore et al., 1988; Moore and Grayson, 1991). The method is based on the stream path analogy first proposed by Onstad and Brakensiek (1968).

The most widely used data structures consist of square grid networks because of their ease of computer implementation and computational efficiency (Collins and Moon, 1981). However, they do have several disadvantages (Moore et al., 1991). They cannot easily handle abrupt changes in elevation. The size of grid mesh affects the results obtained and the computational efficiency (Panuska et al., 1991). The computed upslope flow paths used in hydrological analyses tend to zigzag and therefore are somewhat unrealistic. This produces a lack of precision in the definition of specific catchment areas. Since regular grids must be adjusted to the roughest terrain, redundancy can be significant in sections with smooth terrain (Peucker et al., 1978). Triangulated irregular networks are more efficient and flexible in such circumstances. Olender (1980) compares triangulated irregular networks with regular grid structures. Grid-based DEMs generally provide the most efficient structures for estimating terrain attributes. Contour-based methods require an order of magnitude more data storage and do not provide any computational advantages. With TIN structures, there can be difficulties in determining the upslope connection of a facet. These can be overcome by visual inspection and

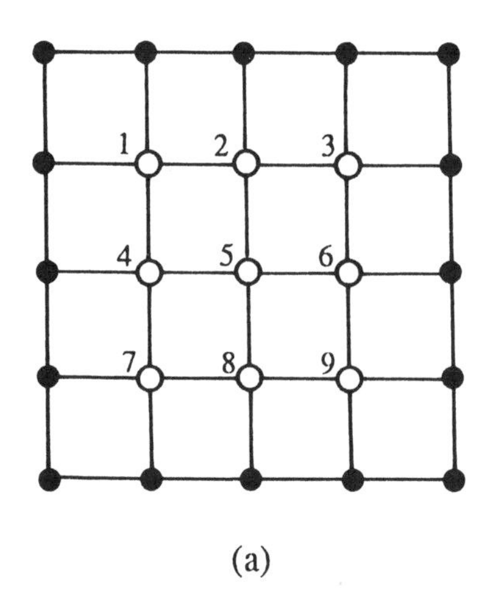

(a)

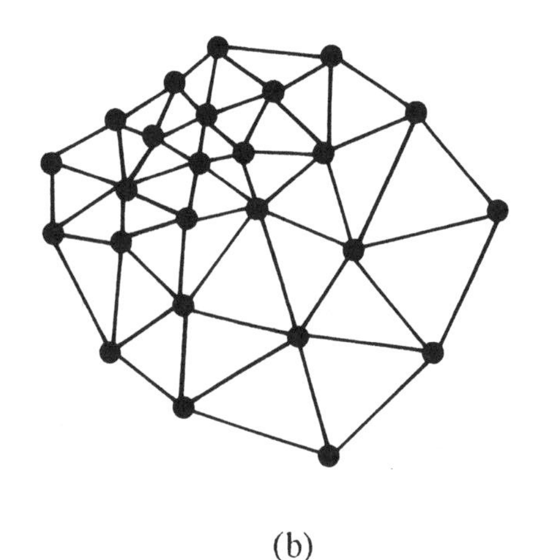

(b)

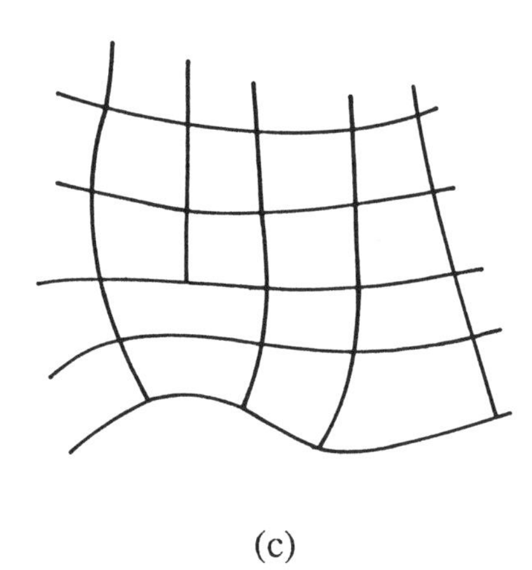

(c)

Figure 19-2: Methods of structuring an elevation data network: (a) square-grid network showing a moving 3 × 3 submatrix centered on node 5; (b) triangulated irregular network (TIN); and (c) vector- or contour-based network (adapted from Moore et al., 1991)

manual manipulation of the TIN network (Palacios-Velez and Cuevas-Renaud, 1989). The irregularity of the TIN makes the computation of attributes more difficult than for the grid-based methods.

Contour lines can be digitized automatically using the processes of raster scanning and vectorization (Leberl and Olson, 1982) or manually by using a flat-bed digitizer and software packages. Individual contours can be digitized and retained in contour form as DLGs or interpolated onto a regular grid or TIN. Similarly, spot heights can be digitized and analyzed as an irregular network or interpolated onto a regular grid. Hutchinson (1989a) describes a new efficient finite difference method of interpolating grid-DEMs from contour line data or scattered surface-specific point elevation data (ANUDEM). The method was used to develop the continent wide DEM for Australia that is described by Hutchinson in Chapter 39. The most common method of performing a contour-to-TIN transformation is via Delaunay triangulation (McLain, 1976).

Digital elevation data are available in several forms for selected areas in the United States. They can be obtained from the Earth Science Information Center (ESIC) of the U.S. Geological Survey (USGS). DEMs are available on a 30 m square grid for 7.5 minute quadrangle coverage, which corresponds to the 1:24,000 scale map series quadrangle. DEMs based on a 3 arcsecond spacing have been developed by the Defense Mapping Agency for 1 degree coverage, which is equivalent to the 1:250,000 scale map series quadrangle, and are also available from the ESIC. Digital contour lines are available as digital line graphs (DLGs) for certain 7.5 and 15 minute topographic quadrangles, the 1:100,000 scale quadrangle series, and 1:2,000,000 scale maps published by the USGS. A 30 arcsecond DEM is now available for the conterminous United States from the National Geophysical Data Center of the National Oceanic and Atmospheric Administration (NOAA). Stereo images available from the French Earth observation satellite SPOT can be used to produce orthophotos and DEMs in much the same way as conventional air photography. Satellite data have the advantage that they can be purchased in digital form and directly accessed by computers. However, they also requires special software for processing.

A new source of global elevation information will be provided by the U.S. Defense Mapping Agency's Digital Chart of the World (DCW) product. The DCW will contain fully structured and attributed vector data from the 1:1,000,000 Operational Navigation Charts (ONCs). It will be available on CDROM. For much of the world where DEMs are not available, the DCW hypsography and hydrography can be gridded to produce DEM data sets. The prototype DCW CDROM contains the vector data from the E-1 ONC map covering the United Kingdom. The hypsography and hydrography for a two-degree square area from 51 to 53 N and 0 to 2 W was gridded using ANUDEM (Hutchinson, 1989a) and compared to ETOPO5 (5 arcminute global DEM) and Defense Mapping Agency's three-arcsecond DEM for the area. Plate 19-1 illustrates the three DEMs and the DCW vector data. Major features are readily apparent in the DCW DEM, indicating that this will be a promising new source of topographic information.

Many published DEMs are derived from topographic maps; so their accuracy is dependent on the quality of the original source of the data and of the methods of interpolation. The most accurate DEMs produced by the U.S. Geological Survey are generated by linear interpolation of digitized contour maps. They have a maximum root mean square error of one-half contour interval and an absolute error no greater than two contour intervals in magnitude (USGS, 1987). The USGS DEMs are referenced to "true" elevations from published maps that include points on contour lines, benchmarks, or spot elevations (USGS, 1987). However, these "true" elevations also contain errors. All DEMs have inherent inaccuracies not only in their ability to represent a surface but also in their constituent data. It "behooves users to become aware of the nature and the types of errors" (Carter, 1989). The USGS DEM data are classified as Level 1, 2, or 3 in quality (USGS, 1987), with Level 3 data being the most accurate. The USGS does not currently produce Level 3 elevation data.

Soils

The U.S. Department of Agriculture Soil Conservation Service Soils-5 database provides information on the characteristics and interpretive properties of 13,788 soil series identified in the United States. These data are derived from the USDA-SCS County Soil Surveys that are mapped at scales ranging from 1:15,000 to 1:20,000, depending on the age of the survey. As an example application, the soil data required by the SWRRB model were derived from the Soils-5 database and are available on computer diskettes (Arnold et al., 1990). Lytle discusses the digital soils databases for the United States in more detail in Chapter 38.

The boundaries of mapping units in the County Soil Surveys are being digitized and stored in vector form using a variety of GIS software, but most commonly GRASS (U.S. Army, 1987), in many State Offices of the USDA-SCS. In Minnesota, Robert and Anderson (1987) have developed a Soil Survey Information System (SSIS) based on digitized County Soil Survey data. It has many of the features of a GIS. SSIS can retrieve, sort, display, highlight, and print soil survey information, as well as display and overlay other digitized data (such as vegetation cover) and develop and display interpretive maps. The system has been used to develop crop equivalent ratings, to prepare farm conservation plans and to help individual farmers and extension agents make better judgments on fertilizer and herbicide application rates using simple analytical models such as the Universal Soil Loss Equation (Wischmeier and Smith, 1978).

Specific soil property data within Soils-5 are presented as ranges. The values of some variables, particularly those describing the hydraulic properties of the soil, can vary by an order of magnitude. This limits the use of Soils-5 data to certain agro-forestry applications and macro-scale or basin-scale hydrological models (such as SWRRB) where there is a high degree of lumping of model parameter values and to large temporal scales probably exceeding one day. These data cannot be used directly, for example, in a model that might calculate infiltration, runoff, and subsurface water movement using physically based process descriptions such as the Richards equation (Richards, 1931) or even the greatly simplified Green–Ampt equation (Mein and Larson, 1973). In recognition of these major constraints, considerable effort has been directed towards developing methods of inferring air and water properties of soils using surrogates derived from soil morphological properties available from soil surveys (Rawls et al., 1982, 1983; McKeague et al., 1984; McKenzie and MacLeod, 1989; Williams et al., 1990; McKenzie et al., 1991). The most common surrogates used are soil texture, organic matter, soil structure, and bulk density. These methods have only been partly successful because the surrogates are usually measured at scales too coarse to capture the variability in the parameters used by the models (compare the variability of soil texture/soil series on a county soil map with the variability of infiltration and hydraulic conductivity that display great variability over distances of less than 10 m). Methods that also include landform descriptors derived from DEMs (e.g., Dikau, 1989) show potential for overcoming these problems (Moore et al., 1991; McKenzie, 1991).

In late 1991 the USDA-SCS National Soil Survey Laboratory in cooperation with the Blackland Research Center at Texas A&M University released the "Soil Pedons of the United States" database on CDROM. This includes laboratory data from the National Soil Survey Laboratory (16,000 pedons), soil description data (8,000 pedons), and mineralogy data for individual, geo-referenced pedons (Dyke, 1991). These data with DTMs and suitable spatial interpolation techniques may provide parametric methods of estimating specific soil properties for input into macro- and meso-scale models of land surface processes.

Geology

A major impediment to environmental modeling is the lack of coordinated mapping programs to produce geologic maps and digital databases at consistent scales. The lack of established standards is another problem, as is the inability of geologists working on adjoining map sheets to agree on feature locations and naming conventions. The basic areal unit for unconsolidated and surficial geologic materials can often be assumed to coincide with the surface drainage basin or catchment. For consolidated and bedrock material this is rarely true (Pfannkuch et al., 1987). Furthermore, most traditional 2D geological mapping is based on chronology (i.e., age) rather than lithology. Chronology often does not provide sufficient differentiation of physical and chemical geological properties needed in subsurface modeling.

The development of geologic conceptual models is the major process in subsurface characterization and poses unique demands on both data management and visualization (Turner, 1989). The procedure of using multiple working hypotheses is a fundamental tenet of geology. Given the sparseness of geological data, geoscientists must develop one or more "most probable" scenarios, or conceptual models, in order to expand measured data to create a continuous model of the entire subsurface. A variety of data types must be combined and synthesized. This requires a centralized database capable of handling a variety of data. The geoscientist using such a system desires to interact with the database in ways that retain the spatial relationships and allow 3D visualization.

During many geological investigations existing 2D geological maps and associated cross sections form an important data source. Two-dimensional mapping of surficial geology is also required for many types of biophysical modeling (such as water balance studies). Sometimes geology can be used as a surrogate when soils data are not available. Two-dimensional surficial and bedrock geology maps are commonly available at scales of 1:500,000 and 1:250,000, but more detailed maps are available for some local areas. Major sources of 1D data include borehole log records, well-pumping data, and water-depth data collected by the USGS and the petroleum industry. All these data sources must be incorporated into a 3D data structure for subsequent analysis and modeling activities. The methods for such incorporation are still under development. For example, environmental site characterization at uncontrolled hazardous waste sites typically involves the collection of data and development of a conceptual model of a site's geology, hydrogeology, and contaminant nature and extent. These investigations result in the generation of graphical and tabular data of different types and formats. Data integration, correlation, and synthesis are necessary for the development and testing of the required 3D computerized conceptual model(s). In addition, site data are often subjected to quality assurance and quality control (QA/QC) procedures. These range from prescribed adherence-related QC attributes, through sample collection and analysis techniques, to statistical verification techniques for identifying outliers and calculating parameter uncertainty. QA/QC procedures are often a prerequisite to data input into higher-level assessment and design activities, such as site characterization, feasibility studies, hazard assessments, and remedial design activities.

To accommodate the types of data flow and data uses described, the data management system must satisfy the following general data management requirements (Nasser and Turner, 1991):

1. Provide the capability to store, retrieve, manipulate, and view 1D, 2D, and 3D graphical objects in any user-defined viewplane in the 3D space;
2. Provide a direct link between the 1D, 2D, and 3D objects and their attributes through a relational database management system;
3. Accommodate the QA/QC procedures, including the level of confidence of the data being used and the RDBMS storage of data validation results;
4. Allow a user-defined, interactive graphical technique for the transition from the first dimension (borehole) into the second dimension (cross section) and then into the third dimension (3D solid block or 3D physical conceptual model);
5. Provide the tool to slice interactively through the created 3D object in any user-defined viewplane in 3D space;
6. Allow easy update of the created 3D model to take account of recently collected data; and
7. Provide application tools to support volumetric interpretive activities.

Typical application tools include spatial analysis, 3D statistical parameter estimation and interpolation (e.g., hydraulic conductivity interpolation by kriging analysis of measured hydraulic conductivity data), and integration of several 3D models to evaluate the possible relationship between these models (e.g., integration of a 3D contaminant plume model with a 3D hydrogeological model).

Land cover

Land cover (LC) is a critical component of most surface/subsurface models, whether they deal with plant growth, hydrology, erosion, or salinization. Land cover is a more useful variable than land use (LU) because land use is an economic concept and a given land use can be associated with a variety of land covers (Gersmehl et al., 1987). In addition, land cover is a dynamic quantity that often exhibits great seasonal variation. Some models, particularly crop-growth models, simulate vegetation canopy development, often via leaf area index (LAI), and so do not require land cover input data directly. Traditional sources of data include the USGS 1:100,000 (urban areas) and 1:250,000-scale (rural areas) LU/LC polygon maps and GIRAS digital data files. The USGS maps combine land use and land cover into a single classification that has four levels that range from very broad to detailed (Anderson et al., 1976). The major problem with these data is that they were collected for once-over cover of the conterminous United States and represent only a snapshot in time. The temporal incompatibilities associated with adjacent maps greatly limit their utility.

Land cover data are most commonly obtained from satellite imagery such as Landsat Thematic Mapper (TM) and Multi-Spectral Scanner (MSS) or SPOT1 and visual air photo interpretations. Satellite imagery is relatively inexpensive, can be rapidly processed and can provide data on the temporal variation in land cover. Landsat-5, for example, allows scenes of the same area to be obtained every 16 days, cloud cover permitting. The Advanced High Resolution Radiometer (AVHRR) satellites operated by the National Oceanic and Atmospheric Administration provide frequent large-area coverage with approximately 1 km resolution. A "greenness" measure derived from these data is frequently used to represent vegetation conditions. These data are composited at the EROS Data Center to provide digital biweekly United States greenness data sets. The availability of these kinds of data provides a basis for time series representations of land cover. Loveland describes these AVHRR derived data sets in more detail in Chapter 37.

Classification of land cover from satellite images is subject to misinterpretation because of problems with rectification and because spectral data cannot distinguish small features and can overlap in complex ways. The ambiguities in the interpretations can be reduced by using the terrain and soils data contained within a GIS to assist in the classification. Visual air photo interpretations are generally more accurate, particularly in urban areas, but are very expensive and time consuming to process (Gersmehl, 1987). Satellite data are provided in raster form, whereas data from air photo interpretations are usually provided in vector form.

Hydrography

Stream systems and channels are linear features and have a hierarchical organization based on the gravity flow of water. These features are best represented as vector data files. Discharge and water quality are measured at specific points along these linear features at different time scales. Mean daily discharge data are available through the USGS stream gauging database and databases maintained by some state agencies. Each stream gauge has a unique 8-digit number ordered in the downstream direction. These stream gauging records are of variable length and often contain missing records. Water quality is measured at a limited number of sites on a monthly or less frequent basis and provides only a snapshot in time. Furthermore, the water quality variables measured are inconsistent and because of the poor temporal cover are almost unusable for modeling purposes.

Many states also hold other hydrographic data, including information concerning lakes, wetlands, and coastal zones. Minnesota has MLMIS40 watershed boundary files, MLMIS40 water orientation files (all 40-acre parcels of land that adjoin or contain water), Systems for Water Information Management (SWIM) lake summary files (location, physical characteristics, water chemistry, use, diversion and other permits, development along shore line, public access, fishing resources), and River Kilometer Index (RKI) files (identifies all streams and ditches and records their lengths) (Brown et al., 1987). For modeling applications where runoff needs to be routed through a channel and lake network, data such as channel geometry, bed roughness and slope, channel stability conditions over short distances are required, but rarely available.

The "blue lines" drawn on topographic maps are a conservative representation of the stream network as they only represent permanently flowing or major intermittent or ephemeral streams (Mark, 1983). This is unfortunate because first-order channels are a major source of surface runoff in many environments. Rogers and Singh (1986) found that in Pennsylvania the number and distribution of first-order channels were inversely proportional to basin soil infiltration capacity and controlled surface runoff hydrograph shape, respectively. Not surprisingly, there has been an increasing effort directed towards the development of automatic, computer-based channel network models that use digital elevation data as their primary input. Hydrological, geomorphological, and biological analyses using GIS and remote sensing technology and computer-based cartographic systems can be more easily integrated with these methods than traditional manual methods, with a high degree of objectivity. Excellent reviews of channel network analysis are presented by Jarvis (1977), Smart (1978), Abrahams (1984), Mark (1988), and Tarboton et al. (1989).

Mark (1988) identified several algorithms for determining total accumulated drainage at a point for application to grid-based DEMs. Some are computationally impractical to implement from an operational and computational efficiency point of view. One of the first algorithms developed, Collins' (1975) method, required that the elevation data be sorted into ascending order. Although several efficient sorting algorithms are now available, the method is not practical for large elevation matrices compared to the alternative techniques to be described. A second and probably the most commonly used algorithm for determining drainage or contributing areas and stream networks was proposed by O'Callaghan and Mark (1984). After producing a depressionless DEM, they use the resulting drainage direction matrix, which gives the flow direction from a node to one of eight nearest neighbors (D8 algorithm—see Figure 19-2a, which shows the eight nearest neighbors to the central node 5), and a weighting matrix to determine a drainage accumulation matrix iteratively that represents the sum of the weights of all elements draining to that element. The element weights range from 0.0 (no runoff from the element) to 1.0 (entire element contributes to runoff). The drainage direction of each individual element is the flow direction from the element to one of its eight nearest neighbors based on the direction of steepest descent. If all the weights equal one, then the drainage accumulation matrix gives the total contributing area, in number of elements, for each element in the matrix. Streams are then defined for all elements with an accumulated drainage area above some specified threshold. A similar, but faster and more operationally viable, method of defining stream channels and drainage basins has been developed by Jenson and Domingue (1988) and is being widely used. Band (1986a, 1989) describes another variation of O'Callaghan and Mark's (1984) algorithm. Methods that use the D8 nearest-neighbor algorithm to assign drainage directions tend to produce parallel lines along preferred directions. Fairfield and Leymarie (1991) overcame this problem by introducing a quasirandom component to the D8 algorithm. Freeman (1991) and Quinn et al. (1991) recently developed algorithms that pass cell drainage to multiple downslope cells using a weighting factor calculated as a function of slope.

Band (1986b) describes another approach that uses Peucker and Douglas' (1975) algorithm for marking convex and concave upward points as ridge and stream points. For each element the highest of the neighboring eight elements is marked, and after two sequential sweeps (forward and backward) through the elevation matrix the unmarked elements identify the drainage courses. Similarly, identifying the lowest of the eight neighboring elements defines ridge lines that are approximations to the drainage divides. The drainage and ridge lines may not be continuous and may consist of multiple adjacent lines that must be interpolated and thinned to one-element-wide continuous lines. A connected channel network is developed using a steepest decent method that includes an elementary pit removal algorithm that often breaks down in flat terrain. In the final step each stream junction element is defined as a divide, thus anchoring the divide graph to the stream network, and the divide network is refined. Thus subcatchments are defined for each stream link. The method requires initial smoothing of the DEM but does have the advantage that no arbitrary threshold area needs to be specified (Tarboton et al., 1989).

Model structures

There are six model structures commonly found in surface/subsurface models, and most are based on topography: (1) lumped models that spatially integrate the entire area being modeled; (2) models based on identifiable hydrological response units (HRUs) consisting of so-called "homogeneous" subcatchments; (3) grid-based

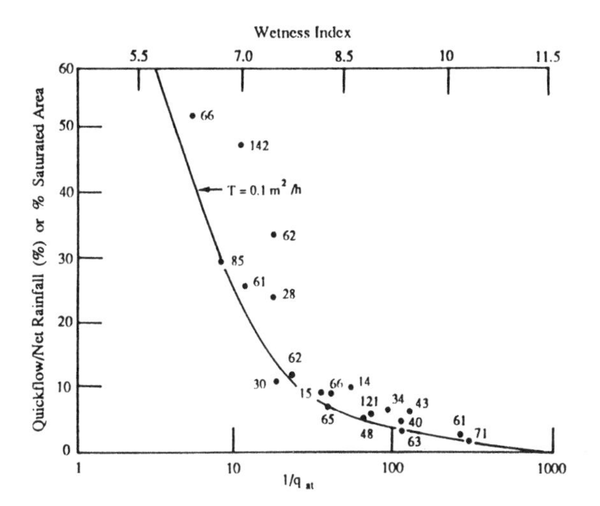

Figure 19-3: Percentage saturated source area as a function of wetness index [$\ln(A_s/\tan\beta)$] and average subsurface drainage rate per unit area, q_{at}, for a small catchment in south eastern Australia (adapted from Moore et al., 1986). The measured percentage of quickflow to net rainfall for individual storm events is plotted as the solid points and the net rainfall is shown for each storm. T is the catchment average transmissivity.

models; (4) TIN-based models; (5) contour-based models; and (6) 2D and 3D groundwater models. Types (2) and (3) to (5) could be viewed as extremes of the representative elemental area (REA) concept proposed by Wood et al. (1988) and are compatible with the DEM structures illustrated in Figure 19-2. The models presented as examples in this section can generally be classed as distributed-parameter, physically based models.

Lumped models

Lumped models are among the simplest forms of model. A major concern in using a lumped model is that it is probably not possible to obtain a single value of a spatially variable parameter that allows a model to predict the mean response of the modeled area (Moore and Gallant, 1991). Spatial integration to the catchment scale is difficult because of the typical nonlinear catchment response (Goodrich and Woolhiser, 1991). For example, Milly and Eagleson (1987) examined the effects of spatial variability of soil and vegetation on spatially and temporally averaged hydrologic fluxes. Only where there was little initial variability could an equivalent homogeneous soil reproduce the average behavior. Similarly, Addiscott and Wagenet (1985) warn that "gross over- or underestimates of solute and water movement may result from ignoring [parameter variations in the horizontal plane]" when applying leaching models.

If one process in a landscape dominates the response, it may be possible to use a simple lumped model to simulate the behavior if a quasidistributed representation of the effect of that process can be incorporated into the model. Moore et al. (1986) adopted this approach in their development of a lumped model of the dynamic hydrological response of a forested catchment in southeastern Australia. The response of the catchment was dominated by saturation overland flow. They used a distribution function of fraction of saturated source area versus mean base flow as the basic input to the model controlling the simulation of the effect on catchment runoff of the expansion and contraction of zones of surface saturation (Figure 19-3). This function was estimated using O'Loughlin's (1986) steady-state method of computing the wetness index [$\ln(A_s/\tan\beta)$, where A_s is the specific catchment area and β is the slope at a point] versus percent saturated area relationship (and in the process fitted the mean catchment transmissivity) from a DEM and historical rainfall-runoff data. This relationship, with the assumption of successive steady states in the time domain, provided the basic structure of a simple lumped-parameter model of forested catchment hydrological response.

HRU models

In hydrological modeling the traditional method of dealing with complex landscapes has been to subdivide them into "less complex" units or subcatchments. This approach finds its best expression in the Finite Element Storm Hydrograph Model (FESHM) (Ross et al., 1979; Hession et al., 1987). The model has two discretization schemes. The first divides a catchment into hydrological response units (HRU) determined by soil types and land use or land cover, as illustrated in Figure 19-4. This subdivision provides the basis for the computation of spatially distributed rainfall excess and infiltration. The second scheme divides the catchment into interconnected overland flow elements and channels based on topography and surface flow paths. Area-weighted HRU rainfall excess is then routed from successive elements to the catchment outlet. A limitation of this two-phase approach is the separation of the rainfall excess production mechanism from the overland flow routing mechanism. In practice interactions and feedbacks between the two mechanisms occur and sometimes can be significant (e.g., runoff-runon problem). The vast majority of hydrological and water quality models developed before the mid-1980s are based on some variation on this structure.

Over the past 5 years the Information Support Systems Laboratory at Virginia Tech in cooperation with the Virginia Division of Soil and Water Conservation have been pioneering the linkage of GIS and hydrological modeling and water quality modeling technology. For example, they have linked FESHM and other water quality models with both PC-ARC/INFO and VirGIS (Virginia Geographic

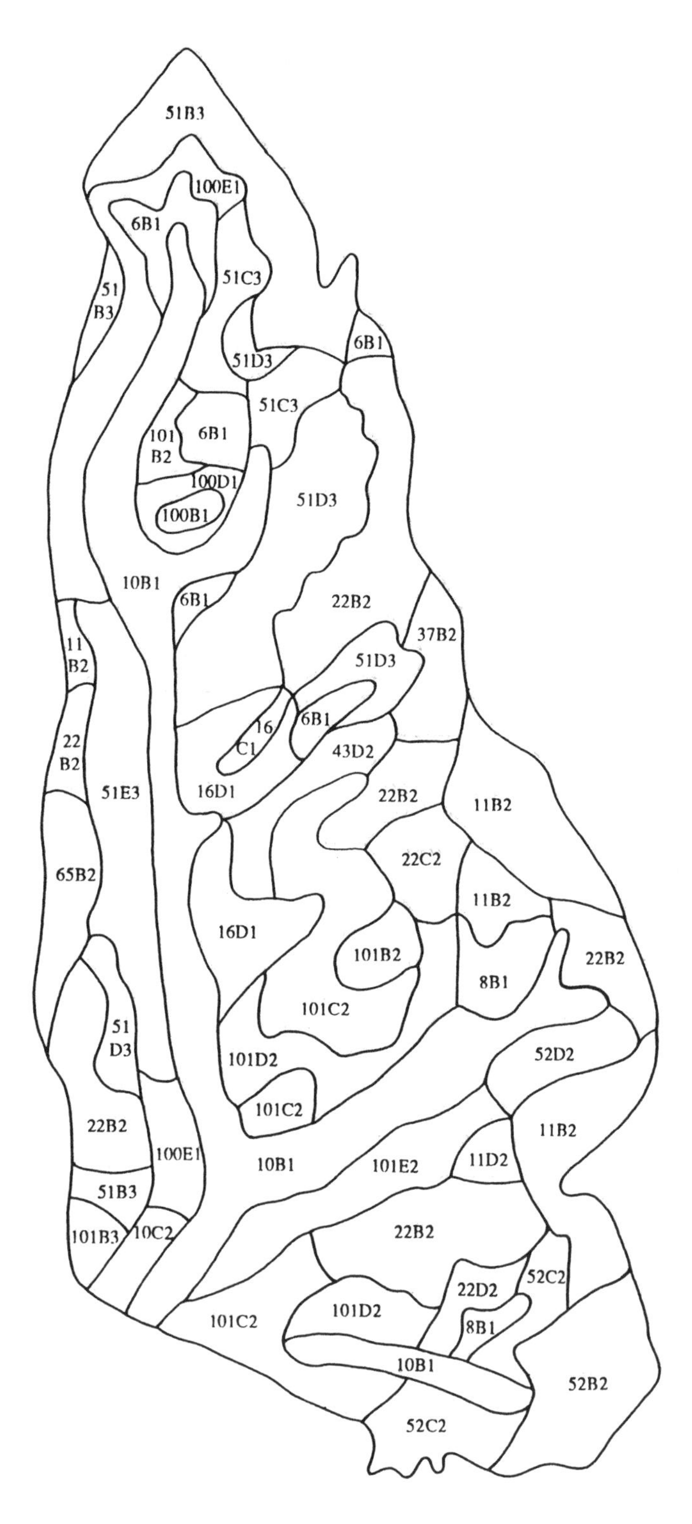

Figure 19-4: Subdivision of the Powells Creek catchment into hydrologic response units (HRU) on the basis of soil types (reproduced from Ross et al., 1982).

Information System). Also, they have built state-wide databases that these models can access (Hession et al., 1987; Shanholtz et al., 1990; Desai et al., 1990). In an evaluation of FESHM, Hession et al. (1987) concluded that "GIS technology did not provide for the entire range of management and manipulation of data desired for creating FESHM input files." A digital map-based hydrologic modeling system that can develop databases for topography, soils, land use, and point and radar-predicted rainfall using GIS has been developed by Johnson (1989). A variety of hydrologic models representing a range of processes with varying degrees of sophistication has been parameterized using this approach.

The following three model structures can be viewed as a scaling down and a merging of the basic HRU and routing element structure concepts. The apparent belief (or hope) is that successively smaller landscape units can be treated as internally homogeneous or uniform and that the spatial variability of the landscape is captured by the between unit variability. As the spatial scale of discretization decreases, the data and computational requirements of the models expand, almost exponentially. Hence the need for integrating fine resolution models with GIS.

Grid-based models

Grid or cellular approaches to subdividing the landscape provide the most common structures for modern dynamic, process-based hydrological models (Figure 19-2a). Some examples of models based on this structure include the ANSWERS (Beasley and Huggins, 1982), AGNPS (Young et al., 1989), and the very detailed Système Hydrologique Européen model (SHE) (Abbot et al., 1986). AGNPS is a semiempirical, peak flow surface water quality model and does not route the flow as such. SHE employs a two-dimensional form of the diffusion wave equations to partition flow. ANSWERS derives a weighting factor by partitioning a cell based on the line of maximum slope. A problem with these approaches is that the flow paths are not necessarily down the lines of greatest slope; so the resulting estimates of flow characteristics (e.g., flow depth and velocity) are difficult to interpret physically. ANSWERS uses a preprocessing program, ELEVAA, to generate slope steepness and aspect from grid-based DEMs for input into the model. Panuska et al. (1991) linked a grid-based terrain analysis method and grid-based DEMs with the AGNPS model via interface routines to generate the terrain-based parameter file. Terrain-based attributes represent about one-third of the input parameters required by the model. The use of grid structures for these models has a major benefit in that it allows pixel-based remotely sensed data, such as vegetation types and cover (i.e., canopy) characteristics, to be used to estimate the model parameters in each element or cell because of the inherent compatibility of the two structures.

Beven and Kirkby (1979) have developed a physically based, topographically driven flood forecasting model, TOPMODEL, that is beginning to be widely used. The model is based on the variable contributing theory of runoff generation and predicts the soil water deficit pattern and the saturated source area from a knowledge of topography and soil characteristics (Beven, 1986). Recent versions of TOPMODEL also consider Hortonian overland flow (Beven, 1986; Sivapalan et al., 1987). Beven and Wood (1983), Beven et al. (1984), Hornberger et al. (1985), Kirkby (1986), and Beven (1986) describe applications and developments of the model. Sivapalan et al. (1987) and Wood et al. (1988) used the model to examine hydrological similarity. They developed a series of dimensionless similarity parameters, one of which included a scaled soil-topographic parameter that was recently used to account for partial area runoff with the Geomorphic Instantaneous Unit Hydrograph (Sivapalan and Wood, 1990). An excellent review of the issues of similarity and scale in catchment storm response is presented by Wood et al. (1990). Band and Wood (1986, 1988) used grid-based DTMs to partition a catchment first into a set of stream link drainage areas and then extracted separate distribution functions of $\ln(A_s/\tan \beta)$, where A_s is the specific catchment area and β is the slope angle, for each subcatchment. The separate parameterization of each link catchment area revealed substantial differences in the $\ln(A_s/\tan \beta)$ distributions, and therefore in the hydrological response between typical exterior link and interior link catchments. The TOPMODEL approach is most commonly driven by a grid-based method of analysis but can easily be adapted to contour-based methods.

Several geomorphological models have been developed to simulate landscape evolution on geologic time scales. Landscapes are modeled as "open dissipative systems," where sediment transport is the dissipative process and the state equation for elevation is the sediment mass continuity equation, written in the form (Willgoose et al., 1989a,b):

$$\frac{\partial z(x, y, t)}{\partial t} = \text{sources} - \text{sinks} + \text{spatial coupling} \quad (19\text{-}1)$$

where $z(x,y,t)$ represents the land surface that is twice continuously differentiable in space (x,y) and once continuously differentiable in time, t (Smith and Bretherton, 1972). The spatial coupling is achieved via sediment transport that is usually expressed as a function of slope, β, and discharge, Q. This in turn is spatially linked via the water continuity equation (usually the overland flow equations). Examples of the processes represented in these models include: tectonic uplift (sources), bedrock weathering, rockfall, slow mass movement (creep), rain splash and overland flow erosion and sediment transport (Kirkby, 1971; Smith and Bretherton, 1972; Ahnert, 1987; Willgoose et al., 1989a,b). Models of the evolution of both 1D hillslopes (e.g., Kirkby, 1971; Smith and Bretherton, 1972; Ahnert, 1976) and 2D landscapes (Smith and Bretherton, 1972; Ahnert, 1987; Willgoose et al., 1989a,b, 1990, 1991) have shown that the shape of the final hillslope or landscape is dependent on the form of the governing sediment transport equation. Most models, such as Ahnert's (1987) SLOP 3D model, which simulates a wide range of landscape processes, do not specifically account for the linkages between the land system and the channel network. Hillslopes are equally dependent on the boundary conditions, and this is a deficiency when the hillslopes and channels do not interact. Willgoose et al. (1989a,b, 1990, 1991) recently developed a channel network and catchment evolution model that attempts to account for these interactions. However, in their model interactions are restricted to a parameterization of channel initiation that is prescribed and is not based on sediment continuity. Channels are fixed in position and cannot move laterally, nor can divides. This is an area that requires more research. All the 2D landscape evolution models developed to date are driven by grid-based DEMs.

TIN-based models

TINs (Figure 19-2b) have also been used for dynamic hydrological modeling (e.g., Palacios-Velez and Cuevas-Renaud, 1986, 1989; Silfer et al., 1987; Vieux et al., 1988; Maidment et al., 1989; Goodrich, 1990; Jones et al., 1990; Djokic and Maidment, 1991; Goodrich et al., 1991). Vieux et al. (1988) minimized the problems of using TINs for dynamic modeling by orienting the facets so that one of the three edges of a facet formed a streamline and there was only one outflowing edge. However, this required a priori knowledge of the topography, which was obtained from a contour map, thereby detracting from the advantages of the TIN method. For the more common case where the facets are not oriented so that the junction between facets forms a streamline, the modeling of the flow on the facets is not trivial because of the variable boundary conditions (i.e., either one or two inflowing and outflowing edges). A method of distributing the inflow and/or outflow for each facet to downslope facets under these conditions is described by Silfer et al. (1987). Maidment et al. (1989) solved the 2D St. Venant equations for flow on the facets using the TIN-based catchment discretization method of Jones et al. (1990). Djokic and Maidment (1991) linked a TIN surface terrain model with a GIS that defined stormwater intakes and drainage networks to model stormwater flow in an urban environment. Recently, Goodrich et al. (1991) have adapted a 2D kinematic cascade model to predict overland flow on a TIN representation of the land surface. Their method performs a within TIN facet finite element discretization based on a conformal rotation in the direction of steepest slope. This allows a 1D system of equations to be solved in a 2D domain. The TIN representation was an outgrowth of approaches proposed by Palacios-Velez and Cuevas-Renaud (1986, 1989).

Contour-based models

Contour- or isohypse-based methods of partitioning catchments (Figure 19-2c) and terrain analysis provide a natural way of structuring hydrological and water quality models because the partitioning is based on the hydraulics of fluid flow in the landscape. It is essentially a vector technique and was first proposed by Onstad and Brakensiek (1968), who termed it the stream path or stream tube analogy. Contour lines are assumed to be equipotential lines and pairs of orthogonals to the equipotential lines form the stream tube. An element is bounded by adjacent pairs of contour lines and stream lines (Figure 19-2c). With this form of partitioning only 1D flow occurs within each element, allowing water movement in a catchment to be represented by a series of coupled 1D equations. The equations can be solved using a 1D finite-differencing scheme.

Terrain Analysis Programs for the Environmental Sciences-Contour version–TAPES-C (Moore et al., 1988b; Moore and Grayson, 1991) and TOPOG (Dawes and Short, 1988) are examples of computer-based terrain analysis methods that use this approach. The applications oriented TOPOG is based on the concepts and methods first developed in TAPES-C, which can be considered the original research version. Both analysis packages have their origins in the work of O'Loughlin (1986).

Kozak (1968) and Onstad (1973) used this method to examine the distributed runoff behavior of small catchments. In each case the catchment partitioning was carried out by hand. The applications of Onstad and Brakensiek (1968) and Kozak (1968) involved small catchments with very simple topographies, while that of Onstad (1973) was for a simple hypothetical symmetric catchment. Tisdale et al. (1986) used this technique with an implicit 1D kinematic wave equation to examine distributed overland flow and achieved good accuracy when compared to steady-state solutions for flow depth and discharge over a hypothetical cone-shaped catchment. They found the method to be computationally faster than 2D finite element modeling methodologies, because it solves simpler model equations, while retaining an equivalent physical realism. Moore and Grayson (1991), Moore et al. (1990), and Grayson (1990) present two simple process-oriented hydrological models that simulate saturation overland flow and Hortonian overland flow using a contour-based network of elements and the kinematic wave equations for routing subsurface and overland flow between elements. In the solutions of the kinematic equations, the models use the slope and area of each element, the widths of the element on the upslope and downslope contours bounding the element, the flow path length across the element, and the connectivity of the elements calculated by the TAPES-C terrain analysis suite of programs (Moore et al., 1988b; Moore and Grayson, 1991).

Two- and three-dimensional ground water models

In most hydrogeological projects, the primary purpose of the subsurface characterization is to provide improved ground water flow and contaminant transport estimates. Finite element or finite difference numerical ground water models are the main methods used to simulate 2D and 3D ground water flow. There has been much debate concerning these two procedures (Faust and Mercer, 1980; Fetter, 1980). The finite difference method is widely used because it has a more easily understood intuitive basis, requires simpler data input, uses efficient numerical solution methods, and involves easier programming. The USGS MODFLOW model is the most widely used finite difference model (McDonald and Harbaugh, 1988). Finite difference has two major disadvantages: It may be inaccurate for some problems, and it cannot handle complex geometries in its relatively regular grids. The finite element method addresses these disadvantages, but at the price of more complex data entry requirements and more difficult programming (Poeter, 1983).

The ability of the finite element method to define complex geometries and its ability to evaluate internal flow heterogeneities more accurately has made it the logical choice for modeling applications that require more precision. Fortunately, some powerful 3D finite element ground water models have been developed recently. One example is the Coupled Fluid Energy and Solute Transport (CFEST) program developed at the Battelle Pacific Northwest Laboratories (Gupta et al., 1987). CFEST was designed to evaluate pressure or head, temperature, and solute concentrations for large, multilayered, natural hydrological systems using finite element methods. CFEST can simulate flow in a horizontal plane, in a vertical plane, or in a fully 3D region. Both steady-state and transient simulations are possible.

ROLE OF GIS IN SUBSURFACE MODELING

The heterogeneities of an aquifer must be known in order to simulate or predict the transport of contaminants in a ground water system, or the depletion of water resources within an aquifer. Researchers have focused on the inherent uncertainties associated with definitions of the subsurface, including the measurable properties and features at all scales of interest. To determine these properties exactly, every part of the region of interest would have to be tested. Stochastic modeling approaches have been used to solve the problem of subsurface uncertainties. A stochastic phenomenon or process is one that when observed under specified conditions does not always produce the same results. Stochastic methods are used extensively for reservoir characterization in the petroleum industry (Augedal et al., 1986; Haldorsen et al., 1987).

Many deterministic and/or stochastic geologic process simulation computer models have been developed for a number of geological environments. These models combine deterministic components, often using empirical formulae, with stochastic components to introduce a suitable level of complexity or uncertainty into the results. Measures of statistical or geometrical properties have demonstrated that these models replicate real systems. Use of these models can be considered a type of expert system. The expertise of many geologists is incorporated in the model formulation and the thought processes of experienced geologists are emulated to develop a conceptualization of subsurface conditions from limited data.

Domenico and Robbins (1985) and Domenico (1987) described methods, termed inverse plume analysis, that determined aquifer and contaminant source characteristics from the spatial distribution of contaminant concentration within a contaminant plume. Inverse plume analysis techniques determine the three orthogonal dispersivities, the center of mass of the contaminant plume, and the contaminant source strength and dimensions from contaminant concentration data and assumed aquifer characteristics. The original technique was restricted to isotropic and homogeneous aquifers, but Belcher (1988) has applied the method to heterogeneous aquifers.

For features such as bedding planes, the third dimension can be represented by surfaces that can be contoured or displayed as isometric views. Because the elevation of the surface is not an independent variable, these systems are best defined as "2.5D" systems and can only accept a single elevation value for any surface at any given location (Turner, 1989). Several important geologic structures that cause repetition of a single horizon at a given location, such as folds or faults, cannot be represented by these systems. In contrast, true 3D systems, containing three independent coordinate axes, can accept repeated occurrences of the same surface at any given location.

The ability to create and manipulate 3D images rapidly can materially assist understanding of the subsurface environment and aid in testing hypotheses for conceptual model development. For example, the geologic characterization of a given natural system is traditionally a complex task that requires the correlation and synthesis of a large amount of multidisciplinary data. The process usually involves manually developing a series of cross sections based on geologic correlations of borehole data. The representation of these cross sections and the resulting 3D conceptual model is often achieved through development of pseudo-3D fence diagrams that cannot be fully visualized by the often nonexpert decision maker. Further, development of new cross sections across the model in different directions in 3D space cannot be readily created based on the already developed cross sections or resulting fence diagram. Usually, the scientist must manually interpolate geologic data in space. This process becomes very tedious when the geologist is faced with updating and modifying the conceptual model to incorporate newly collected data or to test different hypotheses to conceptualize the system. The 3D spatial visualization capabilities of some existing GIS appear to offer promising tools to resolve the data management and 3D spatial visualization problems. However, the existing systems are immature, and none fully provides all the required data management and spatial capabilities. The prototype systems use distinctly different approaches:

1. Interactive volume modeling (IVM), developed by Dynamic Graphics Inc., which uses 3D contouring (Paradis and Belcher, 1990; Belcher and Paradis, 1991);
2. A prototype geoscience demonstration system developed by Intergraph Corporation based on nonuniform rational B-splines (NURBS) techniques currently used in their computer-aided design applications (Fisher and Wales, 1990, 1992); and
3. 3D component modeling developed by Lynx Geosystems for mine design and planning applications (Houlding, 1988; Houlding and Stokes, 1990).

Three-dimensional GIS alone cannot solve the hydrogeological analysis problems. The process of hydrogeological analysis can be considered in terms of four fundamental modules:

1. Subsurface characterization;
2. Three-dimensional GIS;
3. Statistical evaluation and sensitivity analysis; and
4. Ground water flow and contaminant transport modeling.

The 3D GIS must, therefore, be used in combination with the remaining modules that use many developed analytical tools. Figure 19-5 summarizes these concepts by showing the dominant information flows and cycles among these modules.

The process starts when the geologist investigator combines geological experience with limited field data to begin the subsurface characterization process (Figure 19-5). The subsurface characterization module contains a variety of analytical techniques, including geological process simulation models that may combine both stochastic and deterministic elements.

The information generated by the subsurface characterization module is linked directly to the 3D GIS module. This feedback loop is an important consideration in de-

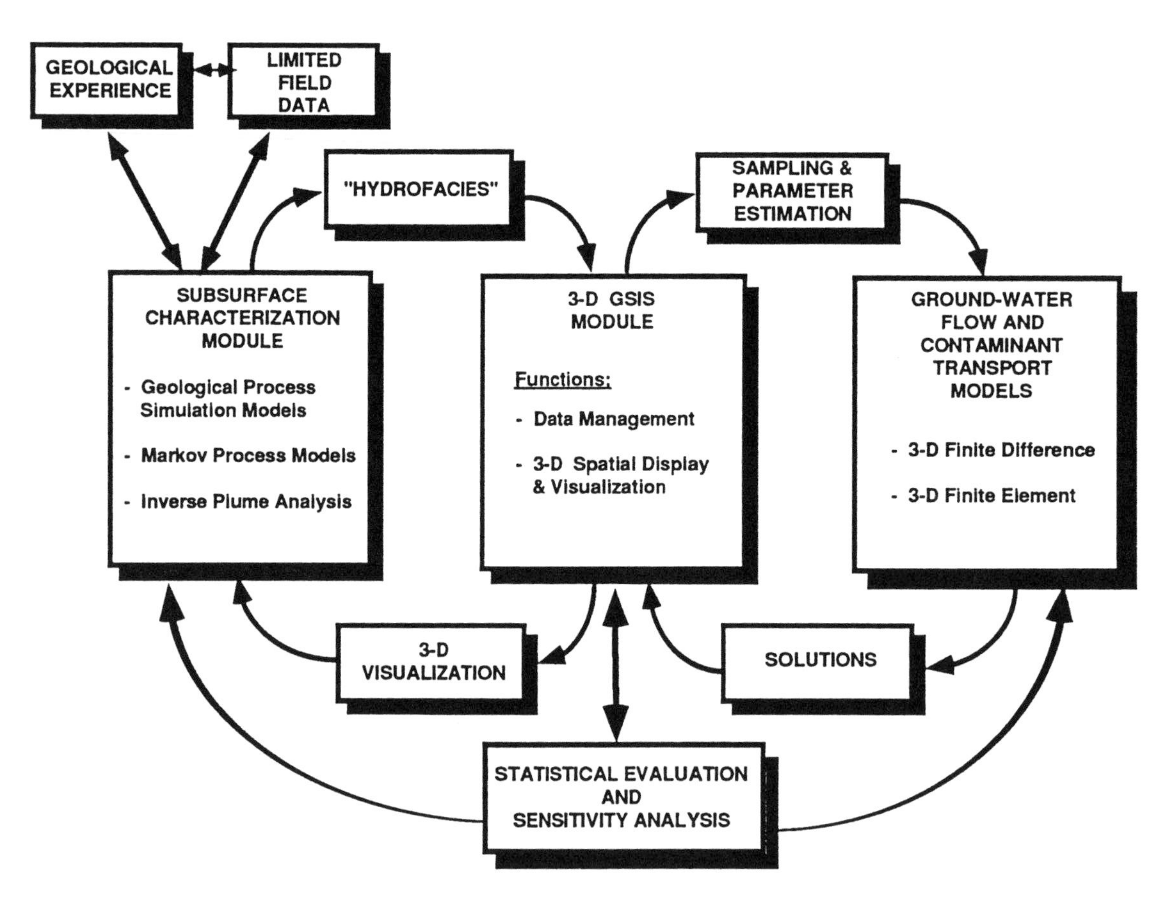

Figure 19-5: Information flow and the role of 3D GSIS for hydrogeology (after Turner, 1989).

fining appropriate interfaces between the GIS and the analytical tools within the subsurface characterization module. The interfaces must be designed to provide both data management and spatial visualization support. Several iterations are expected before the most probable subsurface conditions are defined. Sometimes a unique solution may not be achievable, and two, or more, alternative characterizations may be used.

Once a suitable subsurface characterization has been defined, the analysis continues with a second cycle. In this cycle 3D GIS is used to support appropriate ground water or contaminant transport models (Figure 19-5). This involves the creation of finite-difference or finite-element meshes by sampling from the database. The definition of an optimal mesh has been examined by Stam et al. (1989).

Figure 19-5 shows a strong linkage between the 3D GIS module and a module labeled "Statistical Evaluation and Sensitivity Analysis." This module contains methods for assessing the usefulness and reasonableness of the subsurface characterizations. The investigator can continue the analysis of the hydrogeological conditions only when the subsurface conditions are clearly defined and shown to be statistically acceptable. The spatial visualization capabilities of the 3D GIS are one way of making such reasonableness assessments. In addition, other more numerical approaches, including standard statistical screening methods and geostatistical techniques such as kriging (Olea, 1975; Clark, 1979; Lam, 1983), are usually required. Kriging assumes the data are time invariant. Another method, Kalman filtering, allows both time and spatial variation in the data (van Geer, 1987). These methods have been used for optimizing sampling networks, but appear to have special utility in analyzing seasonally varying contaminant data. The use of such techniques and the analytical methods contained within the other modules allows for a second level of information cycling and feedback (Figure 19-5). An important question that is often posed in groundwater contamination modeling studies concerns the sensitivity of the answers to variations or uncertainties in the input parameters. This sensitivity analysis requires the combined use of all the modules shown in Figure 19-5.

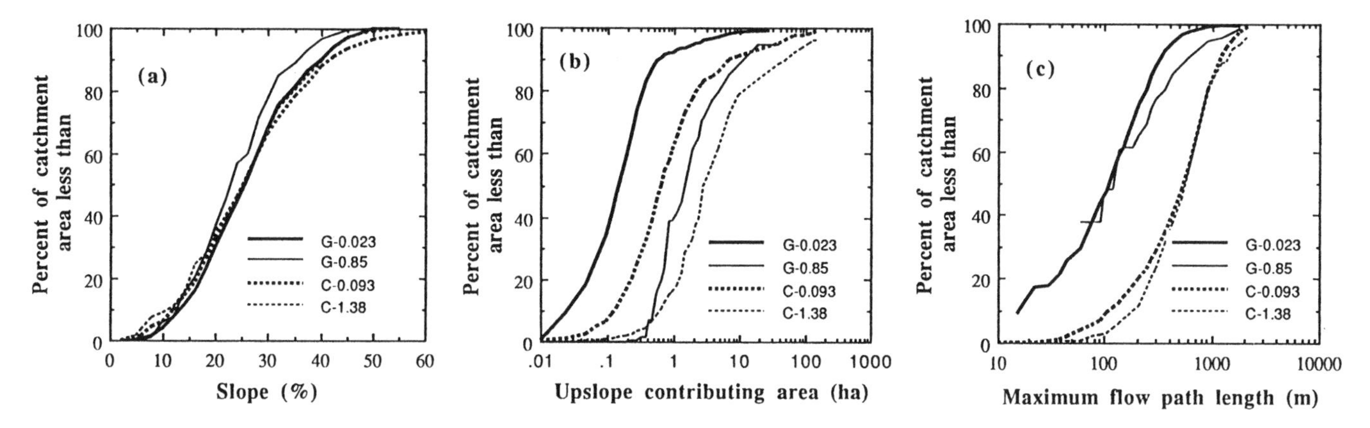

Figure 19-6: Examples of slope (a), upslope contributing area (b), and maximum flow path length (c) distributions computed from (G) grid- and (C) contour-based DEMs and terrain analysis methods for a range of average element sizes (numbers) (adapted from Panuska et al., 1991).

ROLE OF GIS IN SURFACE PROCESS MODELING

There are four major scale-related issues that influence the use of GIS in environmental modeling of surface processes: (1) GIS cell or polygon (element) size; (2) the method of analysis used to derive the attribute values (e.g., slope, aspect, etc.) or the method of measurement (soil physical and hydraulic characteristics); (3) merging data with different resolutions, accuracies, and structures; and (4) scale differences between model process representation and data available for model parameterization. Space does not allow in-depth discussion, but presentation of examples of the problems and (often ignored) difficulties in integrating GIS and surface process modeling may provide some perspective on the effects of scale.

Effects of scale and method of analysis

Panuska et al. (1991) examined the sensitivity of topographic attributes computed from different DEM structures for a range of element sizes on the 210 ha north fork of the Cottonwood Creek catchment in southwestern Montana. They considered grid- and contour-based DEMs and computed slope, upslope contributing area, and maximum flow path length using the TAPES-G and TAPES-C programs (Moore et al., 1991; Moore and Grayson, 1991), respectively. The average element sizes ranged from 0.023 to 1.38 ha. Representative cumulative frequency distributions of the three attributes are presented in Figures 19-6a–c. The numbers in the figures refer to the mean element size in hectares. More detailed results are presented in Panuska and Moore (1991). The computed topographic attribute distributions varied in sensitivity depending on the choice of DEM and terrain analysis method used and on element size. Overall, the slope distributions appeared to be the most stable (Figure 19-6a). However, at scales approaching the natural periodicities of the landscape or hillslope lengths (100–300 m), one might expect these relationships to break down. For the Cottonwood Creek catchment this occurs at grid sizes of about 130–140 m.

At a larger scale, the effect of DEM scale on slope for three different spatial resolutions for the conterminous United States and for the Elmira 1:250,000 scale map covering the Chemung area (40 to 43.3 N, 74 to 77.3 W) has been evaluated by Jenson (1991). The slope of each cell was calculated as the slope of a line connecting it with the cell of lowest elevation of its eight neighboring cells. The results are summarized in Figure 19-7. Increases in

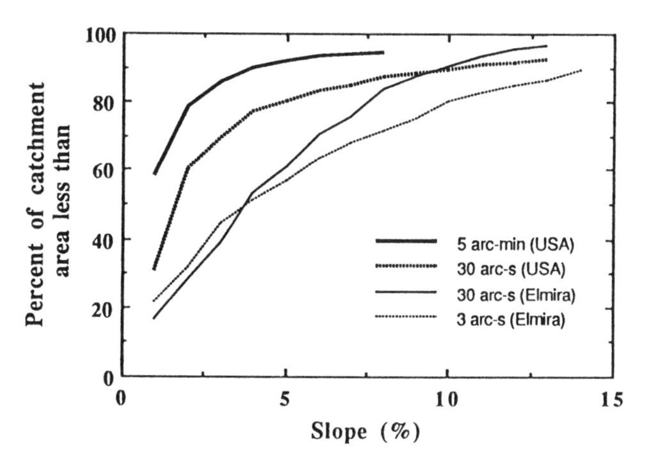

Figure 19-7: Comparison of slopes computed from DEMs of the conterminous US (thick lines) and Elmira 1:250,000 scale map sheet (thin lines) at various resolutions (adapted from Jenson, 1991).

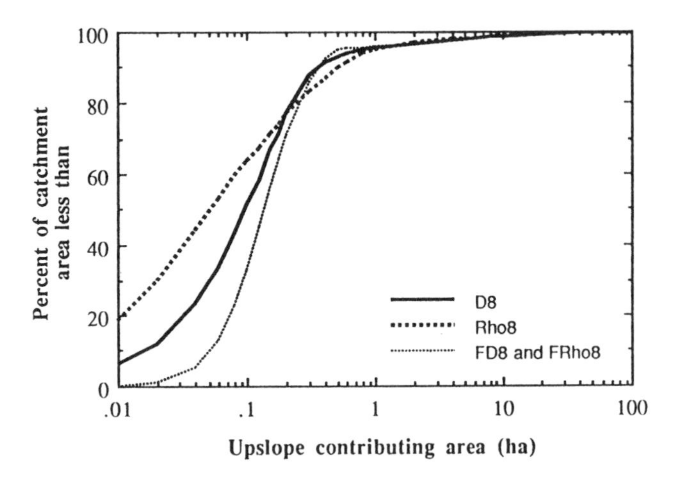

Figure 19-8: Upslope contributing area frequency distributions computed by the D8 (O'Callaghan and Mark, 1984; Jenson and Domingue, 1988), Rho8 (adapted from Fairfield and Leymarie, 1991), FRho8 and FD8 (adapted from Freeman, 1991, and Quinn et al., 1991) element connectivity algorithms from a 10 × 10m grid based DEM.

cell size produce lower slope values, which is consistent with the trends shown in Figure 19-6a at a finer scale.

Both the upslope contributing area and maximum flow path length distributions calculated by the two methods of analysis described depend on the computed connectivity of individual elements (i.e., the topology), but slope calculations were independent of this connectivity. The connectivity of the elements for the grid-based analysis depended highly on the element size, particularly in regions of high flow convergence or divergence (in valleys or ridges). Large element sizes could not account for the same degree of detail as could smaller element sizes. For example, the 90 m grid DEM predicted two flow paths along the edges of the valley in the upper part of the catchment, rather than one flow path along the valley floor (Panuska and Moore, 1991). The D8 algorithm developed by O'Callaghan and Mark (1984), and implemented by Jenson and Domingue (1988), was used to derive the element connectivity for the grid-based terrain analysis (TAPES-C) results presented here.

Figure 19-8 illustrates the impact that the form of the element connectivity algorithm can have on the distribution functions of attributes such as upslope contributing area computed from grid-based DEMs. Upslope contributing area is a basic attribute needed by most hydrological and water quality models. In Figure 19-8 a 0.01 ha square element (10 m grid-DEM) is used to estimate the element connectivities using three different approaches: (1) flow from an element to only one of eight nearest neighbors (Figure 19-2a), based on the direction of steepest descent—O'Callaghan and Mark (1984) and Jenson and Domingue's (1988) D8 algorithm; (2) Fairfield and Leymarie's (1991) modification of this algorithm to include a stochastic component when the D8 algorithm computes flow directions in N-E, S-E, S-W, and N-W directions—Rho8 algorithm; and (3) an adaption of the algorithms proposed by Freeman (1991) and Quinn et al. (1991) that allows flow to be distributed to multiple nearest-neighbor elements in upland areas above defined channels and the Rho8 or D8 algorithms below points of channel initiation—FRho8 and FD8 algorithms, respectively. The D8 and Rho8 algorithms cannot represent flow divergence or catchment spreading (Speight, 1974). In the FRho8 and FD8 algorithms the proportion of flow or upslope contributing area assigned to multiple downslope nearest neighbors is determined on a slope weighted basis. The FRho8 and FD8 algorithms require the specification of a critical upslope contributing area for channel initiation. It is our belief that the FRho8 algorithm is the better approximation of reality for grid-based methods of analysis. One advantage of contour-based methods of analysis (TAPES-C, Moore et al., 1988; Moore and Grayson, 1991; TOPOG, Dawes and Short, 1988) is that they explicitly represent both flow convergence and divergence.

Scale generalization

In many GIS applications there is an implicit assumption that the geographic areal units (raster cells or polygons) can be considered homogeneous in terms of attribute values. However, in many landscapes and at many scales of investigation this assumption is not valid. Jenson (1991) demonstrated that slopes calculated from raster elevation data vary significantly depending on cell size and data source, for example (Figure 19-7). If the topographic data from the three sources illustrated in Plate 19-1 were used for different parts of a catchment to build a GIS database for the entire catchment, the slopes would vary artificially within the database according to the variable topographic source data.

The challenge of merging data from different sources, and with different resolutions, accuracies, and data structures arises for most categories of data including topography, soils, and land cover. Land surface/subsurface models that incorporate multiple attributes as inputs add to this challenge because different parameters vary in different ways across landscapes (space). Loague (1988), for example, examined the impact of spatially variable rainfall and soil information on runoff predictions. Detailed hydraulic conductivity and rainfall data were aggregated to coarser resolutions, and subsequent impacts on hillslope runoff were assessed. The hydraulic conductivity data appeared to be more critical than rainfall information, although different runoff variables (peak rate, time to peak, runoff volumes, etc.) were also found to require different spatial scales.

The appropriate scale (areal unit) may vary with the type of landscape for many attributes as well. Panuska et al. (1991) noted that all DEMs are different representations of the actual land surface and that the accuracy with which a DEM represents the land surface depends on the accuracy and resolution of the original survey data. Hence, the smallest element size produces the most accurate representation of the terrain only if the dimensions of the element are greater than the horizontal resolution of the primary elevation data and if elevation differences between neighboring points are greater than the vertical resolution of the data. The adequacy of a DEM, therefore, depends on the characteristics of the terrain. For example, rolling terrain with moderate relief can be better represented by a larger element sized DEM than can a dissected catchment with sharp ridges, ravines, and abrupt changes in slope or elevation.

There are also the problems that arise because of scale differences between model process representation and the data available for model parameterization. Goodrich and Woolhiser (1991), in their recent review of catchment hydrology, stressed the need to continue experimental studies that explore spatial and temporal variability of catchment characteristics, states, and rainfall patterns. Consistent treatment of these processes and the validation of models requires that observational and model scales be commensurate. What little is known about the variability of water flow direction, solar insolation, infiltration, and a host of other physical processes, indicates a need to explore these phenomena over large numbers of events in a large variety of geographic settings.

Physically based, spatially distributed process representation

While the problems noted in the previous section mean that it may not be possible to move to a scale at which within-element variance is negligible, it is possible to store distributional information on important model parameters or indices (e.g., the attribute distributions presented in Figure 19-6a,c). It is useful to seek indices that scale system behavior from knowledge of areal mean or representative values over a given area. An example is the relationship between a mean wetness index and soil water content in a catchment. These parameters serve as similarity indices as they scale system behavior.

Many advances in the representation of surface and subsurface processes in the past 20 years have been in the characterization and mathematical description of processes at the plot scale. We can now successfully simulate point processes such as interception, infiltration, and evapotranspiration using equations derived from fundamental physics. The grey area occurs at the hillslope or catchment scale, what we will call the meso-scale. Since the first computer-based models were developed in the early 1960s, scientists have been attempting to use micro-scale process descriptions in meso-scale system models. Many meso-scale processes (e.g., hydrological processes) are highly nonlinear. Part of the drive behind the development of spatially distributed, physically based models (such as those described in the previous section) has been the belief that the sequential task of dividing a landscape into smaller units, elements, or grids that are assumed to be homogeneous, predicting the response of each element, and then aggregating the response (routing flow from element to element) will allow these nonlinearities and the spatially variable hydrological response to be predicted. Over the past 10 years evidence indicates that this approach has been only partially successful.

So why are meso-scale models important? Many solutions to environmental problems such as soil erosion and non-point-source pollution involve changes in management strategies of landscapes at the hillslope or catchment scale. In many parts of the world, resource agencies have been adopting the philosophy of "total catchment management." Physically based, distributed-parameter models (e.g., hydrological and water quality models) are being viewed as essential tools in developing and evaluating these land and water management strategies. But is this modeling approach the most effective method of fulfilling these needs, given the data and model constraints discussed above? An alternative approach that is more consistent with the available data, the management questions being posed (as opposed to the scientific rigor of the models) and the precision with which these question need (and can) be answered may be preferred.

We propose an index approach that is based on simplified representations of the underlying physics of the processes but includes the key factors that modulate system behavior (such as topography). The converse of this is that all other factors that are not explicitly accounted for by the index have low variance within the landscape. With this approach we sacrifice some physical sophistication to allow improved estimates of spatial patterns in landscapes (Moore et al., 1991). The method must be able to operate at different levels of sophistication depending on the availability of possible input data and the spatial resolution of those data. These simplified models are based on, and can operate with, what Nix (1981) termed "minimum data sets." Care must be taken in developing these techniques as simplifying assumptions can introduce rather than resolve computational complexities (Denning, 1990). The basic approach can be illustrated by examining two important hydrologic issues: (1) the distribution of soil water content in a landscape—a major factor determining the biophysical behavior of landscapes; and (2) the susceptibility of landscapes to erosion by water—an important issue in the recently enacted Food Security legislation in the United States.

Spatial distribution of soil water content

Hydrological prediction at the micro- and meso-scales is intimately dependent on the ability to characterize the spatial variability of soil water content. Soil water content and related soil properties exhibit extreme variability over distances of 1–100 m. This is supported by many reports in the literature (Brutsaert, 1986; Sharma et al., 1987; Webster, 1985; Yates and Warrick, 1987; Wilson and Luxmoore, 1988; Loague and Gander, 1990). Brutsaert (1986) posed an important question: How meaningful can a soil water index be at this scale? Several factors affect soil water content, including:

1. Soil characteristics: saturated hydraulic conductivity, thickness of the hydrologically active zone (A horizon), effective porosity, and the occurrence of deep percolation and preferential flow;
2. Topography: local slope (measure of hydraulic gradients), specific catchment area (measure of the potential maximum water flux), plan curvature (a measure of the rate of flow convergence and divergence), profile curvature (water concentrates where slopes flatten out), and aspect and topographic shading (which with slope influence the radiation regime and therefore evapotranspiration);
3. Vegetation: annual variation in surface cover and water use characteristics; and
4. Weather: net rainfall, net radiation and temperature.

In mountainous or hilly terrain soil, water distribution is controlled by vertical and horizontal water divergence and convergence, infiltration recharge, and evapotranspiration. The latter two terms are affected by solar insolation and vegetation canopy that vary strongly with exposure in semiarid areas. The divergence/convergence term is dependent on hillslope position. Most index approaches for predicting the spatial distribution of soil water can be expressed as:

$$\chi_i = \ln\left[\frac{T_e}{T\tan\beta}\frac{\sum_j \mu_j a_j}{b}\right]_i = \ln\left[\frac{1}{\tan\beta}\frac{\sum_j \mu_j a_j}{b}\right]_i + \left[\ln(T_e) - \ln(T_i)\right] \quad (19\text{-}2)$$

where χ is the wetness index, a_j is the area, b_i is the outflow width, μ_j is an area weighting coefficient, β_i is the slope, and T_i is the transmissivity (depth integrated hydraulic conductivity) of the ith or jth element and $\ln(T_e)$ is the areal average value of $\ln(T_i)$ (Moore and Hutchinson, 1991). This index was developed for predicting saturated source areas and depths to water tables. However, there is a conceptual jump required to apply these same relationships to the prediction of soil water spatial variability.

The most common form of Eq. (19-2) assumes that the weighting function, μ_j, is spatially uniform and equal to one (Beven and Kirkby, 1979; O'Loughlin, 1986; Moore et al., 1988b), so that Eq. (19-2) reduces to:

$$\chi_i = \ln\left[\frac{A_s T_e}{T\tan\beta}\right]_i = \ln\left[\frac{A_s}{\tan\beta}\right]_i + \left[\ln(T_e) - \ln(T_i)\right] \quad (19\text{-}3)$$

where A_s is the specific catchment area (catchment area draining across a unit width of contour). This index considers only topographic (first term) and soil factors (second term) and is derived from simple catchment drainage theory. In Eq. (19-3) A_s is a measure of the steady-state subsurface drainage flux, but assumes uniform infiltration over the entire catchment. The simplest method of accounting for vegetation evapotranspiration effects on the wetness index is to assume that

$$\mu_i = \mu = 1 - \frac{E}{P-R} \quad (19\text{-}4)$$

where E is the actual evapotranspiration, P is the precipitation, and R is the direct surface runoff plus deep drainage loss on a seasonal or annual basis. This approach assumes that evapotranspiration is spatially invariant. The lumped catchment term $E/(P-R)$ can be estimated directly from an analysis of long-term catchment precipitation and runoff records. For example, forested catchments along the south-east coast of Australia have an average value of $E/(P-R)$ of about 0.70–0.95. This index can only characterize soil water distribution in a landscape if evapotranspiration is conservative over the land area (to maintain stationary mean soil water content).

A GIS can be used to partition the region into areas of uniform evapotranspiration, E, such as hillslope facets. Band and Nemani (1990), Band (1991), and Band et al. (1991) took this approach, assuming that Eq. (19-3) controls the variability within hillslopes while canopy and microclimate variations dominate between hillslopes. An alternative approach is to recast Eq. (19-4) in the following form to account for spatially variable infiltration and evapotranspiration (Moore and Hutchinson, 1991):

$$\mu_i = 1 - f_i\frac{E_p}{(P_i - R_i)}\left[1 - \left(\frac{\theta_{fc}-\theta}{\theta_{fc}-\theta_{wp}}\right)_i^{C/E_p}\right] \quad \text{for } \theta_i \leq \theta_{fc,i}$$

$$\mu_i = 1 - f_i\frac{E_p}{(P_i - R_i)} \quad \text{for } \theta_i > \theta_{fc,i} \quad (19\text{-}5)$$

where E_p is the potential evapotranspiration on an unshaded horizontal surface, f_i is the ratio of the solar radiation on a sloping surface to that on a horizontal surface (0.5–1.2) and is a function of slope, β, aspect, η, and topographic shading, θ is the actual soil water content, θ_{fc} is the field-capacity soil water content, θ_{wp} is the wilting-point soil water content, and C is a constant. These equations would typically be applied on a seasonal or annual basis for examining the spatial distribution of "average" soil water content. The f_i term accounts for the effect of topography on E_p and assumes that E_p is directly proportional to the solar radiation received by the land surface. The term involving θ, θ_{fc} and θ_{wp} is simply an empirical expression for the κ factor, written in terms of the soil water content and the evaporative demand (Kristensen and Jensen, 1975), in the following equation relating actual evapotranspiration, E, to the potential evapotranspiration, E_p:

$$E = \kappa E_p \tag{19-6}$$

As a first approximation, the soil water content can be related to the wetness index via an empirical equation of the form:

$$\theta = b_1 \chi + b_2 \tag{19-7}$$

where b_1 and b_2 are fitted constants. Using the form of the weighting function given by Eq. (19-5), the wetness index, χ, appears on the left- and right-hand (via the μ term) side of Eq. (19-2) and must be solved iteratively. The solution technique is most efficient if it begins with the element of highest elevation and finishes with the element of lowest elevation at the catchment outlet. Plates 19-2a and 19-2b show the patterns of soil water distribution in a small catchment in Montana predicted using Eqs. (19-4) and (19-5), respectively. In both cases the values of the parameters were somewhat arbitrarily chosen to give a median soil water content of 0.27 $m^3 m^{-3}$.

Estimation of the spatial variation in soil water content using Eqs. (19-2) and (19-5) requires the estimation of local slope, aspect, flow paths (determines how the individual elements are connected), and topographic shading, which are all topographic attributes; field-capacity and wilting-point soil water contents and net rainfall (P–R), which are soil and weather attributes; and average catchment-wide potential evapotranspiration, which is a vegetation–weather attribute. Here we assume that the spatial scale of variation in E_p is much greater than that of the other variables, and so can logically be ignored. For large catchments this assumption may be invalid and so it may be desirable to use spatially distributed values of E_p, based on vegetation distribution. Equation (19-3) is a single equation that is an order of magnitude simpler than most distributed-parameter, physically based models, but accounts for the major factors affecting soil water content. However, the question remains as to how to estimate the distributed values of the soil attributes that are parameters in the equation.

Because of the difficulties associated with direct measurement or estimation of the spatial variability of soil properties, many people have used only topographic attributes to characterize soil water distribution. Moore et al. (1988a) found a strong correlation between $\ln(A_s/\tan\beta)_i$ and surface soil water distribution on a small fallow catchment near Wagga. They found a linear combination of aspect and $\ln(A_s/\tan\beta)_i$ to be the best predictor. Burt and Butcher (1986) found that plan curvature, $PLANC_i$, performed almost as well as $\ln(A_s/\tan\beta)_i$, but that the compound index $PLANC_i{*}\ln(A_s/\tan\beta)_i$ was best. Wood et al. (1990) demonstrate that the variation in the topographic variable, $\ln(A_s/\tan\beta)_i$, is far greater than the local variation in transmissivity, $\ln(T_i)$. The topographic variable alone is therefore a useful approximation. However, during interstorm periods the mean transmissivity, $\ln(T_e)$, plays a dominant role.

One rationale for only using topographic attributes to predict soil water content is that in many landscapes pedogenesis of the soil catena occurs in response to the way water moves through the landscape. Therefore, it can be hypothesized that the spatial distribution of topographic attributes that characterize these flow paths inherently captures the spatial variability of soil properties at the meso-scale as well. One of the most exciting new areas of research is the attempt to verify this hypothesis by examining the correlation between topographic attributes and soil properties. Parallel research is also underway aimed at exploring the correlation between topographic attributes and vegetation distribution in undisturbed landscapes. Climate, parent material, topography, vegetation, and other biotic agents are the dominant soil-forming process, but climate and parent material probably exert control at scales larger than we are considering here.

From the foregoing arguments we hypothesize that only net rainfall (P–R) and the spatial variability of topographic attributes need to be considered to provide adequate estimates of the spatial variability of soil water content. The other parameters can be spatially lumped, with little or no loss in predictive ability. For example, the variability in net rainfall has been ignored in the application of the index approach to field measurements (Burt and Butcher, 1986; Moore et al., 1988a), and this may be satisfactory for steady-state applications. Similarly, spatially distributed soil parameters have often been ignored because they are rarely available for applications in operational hydrology.

Care must be exercised in applying static indices like those described to predict the distribution of a dynamic process like soil water content that shows hysteretic effects, especially when threshold changes occur in the area of saturation. Because of fluctuating rainfall intensities and the short duration of rainfall events compared to the travel time of subsurface throughflow, the subsurface flow regime in a natural catchment rarely reaches steady state. Barling (1992) has observed that during storm events subsurface flow is only affected by a small proportion of the contributing area directly upslope. He has also observed that initial surface saturation does not necessarily occur at points with the largest $\ln(A_s/\tan\beta)$ values. Iida (1984) recognized this problem and proposed a hydrological method of estimating the effect of topography on saturated throughflow that involved the calculation of a subsurface flow travel time-specific area curve, $a_s(t)$, where $a_s(t_e)=A_s$ and t_e is the time to equilibrium. The time required for water to move from any point E to F along a flow line or stream tube is given by:

$$t_{EF} = \frac{\gamma}{k} \int_F^E \frac{ds}{\sin\beta \cos\beta} \tag{19-8}$$

where s is the horizontal component between E and F, k is the saturated hydraulic conductivity, and γ is the effective porosity. Iida assumed spatially uniform soil properties but did demonstrate the effects of slope angle, plan curvature, and profile curvature on the shape of the time–area curve. Barling (1992) has automated Iida's method and integrated it into Equation (19-2) to derive a spatially and temporally varying wetness index. This approach must still be validated.

Spatial distribution of erosion by water

A second example further illustrates the efficiency of the index approach. The physical potential for sheet and rill erosion in upland catchments can be evaluated by the product *RKLS*, a component of the universal soil loss equation (USLE), where R is a rainfall and runoff erosivity factor, K is a soil erodibility factor, and LS is the length–slope factor that accounts for the effects of topography on erosion (Wischmeier and Smith, 1978). The USLE is an empirical equation that includes terms representing the dominant factors affecting erosion by water. To predict erosion at a point in the landscape, the LS factor can be written as:

$$LS = (n+1)\left(\frac{A_s}{22.13}\right)^n \left(\frac{\sin\beta}{0.0896}\right)^m \tag{19-9}$$

where A_s is the specific catchment area, β is the local slope in degrees, $n=0.4$, and $m=1.3$. This equation was derived from unit stream power theory by Moore and Burch (1986). It is more amenable to landscapes with complex topographies than the original empirical equation because it explicitly accounts for flow convergence and divergence through the A_s term in the equation (Moore and Nieber, 1989). The erosion potential can be normalized using the concept of an allowable soil loss, E_a, to give the erodibility index, $RKLS/E_a$. This index can be mapped based on the topographic attributes of slope and specific catchment area, and soil and climatic attributes via K, E_a and R (Moore and Nieber, 1989). Each of these variables could be a primary data layer in a GIS. In the United States, highly erodible land is defined as land with an erodibility index greater than 8. This definition is used to identify land that can be bid into the Conservation Reserve Program (CRP) and taken out of production for 10 years. Plate 19-3 maps the erosion potential over two small catchments in southern Minnesota. It illustrates how a GIS with a simple knowledge-based analysis system can facilitate the land management decision-making process by efficiently identifying highly erodible land.

Vertessy et al. (1990) developed an erosion hazard index, derived in part as an abstraction of the physically based erosion model proposed by Rose et al. (1983a) and Rose (1985). The index is a useful diagnostic function that integrates, in a simple way, the effect of soil properties, vegetative cover, and runoff on the entrainment of soil particles by overland flow. The erosion hazard index, H, is given by:

$$H = P\eta \tag{19-10}$$

where

$$P = \text{stream power} = \rho g q \tan\beta \tag{19-11}$$
$$\eta = \text{entrainment efficiency} = \alpha_1 e^{-\alpha_2(1-C_r)}$$

and ρ is the density of water, g is the acceleration due to gravity, q is the overland flow discharge per unit width, β is the local slope, $1–C_r$ is the fraction of surface cover in contact with the ground, and α_1 and α_2 are constants that depend on soil type and surface management. The values of α_1 range from 0.2 to 0.7 for cultivated soils (Rose, 1985) and 0.05 to 0.06 for arid zone pasture (Rose et al., 1983b; Silburn, 1991). Increasingly sophisticated versions of the erosion hazard index can be obtained depending on the sophistication of the assumptions and model used to estimate the spatial variation of q. The simplest approach is to assume spatially uniform, steady-state rainfall excess over a catchment—the same assumption used to derive Eq. (19-9). With this assumption the erosion hazard index becomes:

$$H = \rho g A_s i \tan\beta\, \alpha_1 e^{-\alpha_2(1-C_r)} \tag{19-12}$$

where i is the rainfall excess rate. Vertessy et al. (1990) applied a more sophisticated version of the model that assumed runoff was generated from saturated source areas, which could be predicted as a function of topographic and soil attributes. Equation (19-12) is a function of topographic (A_s and β), soil (α_1 and α_2) and surface cover (C_r) attributes for a given rainfall excess regime, all of which could be data layers in a GIS.

SOIL PROPERTIES FOR SPATIALLY DISTRIBUTED ENVIRONMENTAL MODELING

Soil physical and chemical property data are essential for modeling surface and subsurface phenomena. The lack of these data at the required spatial resolution is probably the greatest impediment to the successful application of modeling and GIS technologies to analyzing resource and

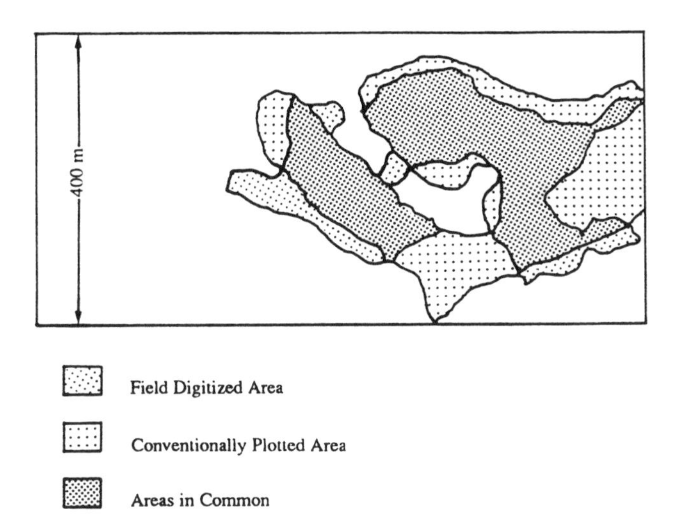

Figure 19-9: Comparison of boundaries between two soil series mapped using conventional soil survey techniques and by field digitizing using GPS (adapted from Long et al., 1991).

environmental problems. Until recently soil scientists have emphasized the vertical relationships of soil horizons and soil-forming processes rather than the horizontal relationships that characterize traditional soil survey (Buol et al., 1989). Soil spatial patterns have been captured and displayed as choropleth maps with discrete lines representing the boundaries between map units, which implies homogeneity within map units (Gessler, 1990). Two problems follow from this approach: (1) the lines drawn on the soil survey maps may not accurately depict the boundaries between map units [Figure 19-9, for example, shows the variations between map boundaries produced with traditional soil survey and global positioning system (GPS) technology within a 129 ha field in northern Montana]; and (2) the inferred homogeneities do not exist for many physical and chemical attributes needed for environmental modeling (see earlier discussion).

The prescribed standard accuracy for delineating soil boundaries in detailed soil surveys is 30 m (100 ft). However, significant errors can occur depending on the base map quality, the map reading skill and aptitude of the soil surveyor, and terrain conditions (Long et al., 1991). The global positioning system (Dixon, 1991) offers a relatively inexpensive, time-efficient, and accurate method of positioning on a base map, navigating to predetermined sites, and field digitizing soil boundaries compared to conventional positioning used in soil survey (particularly when surface features are difficult to observe from the ground), such as aerial photographs, and topographic maps (Long et al., 1991). In an evaluation of GPS errors, Long et al. (1991) concluded that positions measured by GPS receivers operating in autonomous mode (single receiver) are comparable with the accuracy of positions from 1:24,000 scale topographic maps. In differential mode (two receivers) the accuracy is within 5 m. GPS survey and digital terrain analysis technologies overlap and offer ways of reducing, or at least quantifying, the relational errors in point and vector land surface data that are used in environmental modeling and GIS.

From a soil science and hydrologic modeling perspective, there is considerable merit in using the soil horizon as the basic entity for modeling and quantifying soil properties in three-dimensional space rather than the map unit or soil series (Gessler, 1991). The hydrologically active zone generally corresponds to the A horizon, but varies greatly in thickness and in physical and chemical properties. It is possible to use soil series and pedon data as a basis for horizon-oriented, three-dimensional modeling and visualization of soil patterns. In one of the first applications of its kind, Gessler et al. (1989) used the horizon information within a GIS to analyze soil–vegetation–landuse patterns in southwest Wisconsin. To develop this horizon-based approach for modeling and quantifying soil properties in space requires the development of useful horizon classes by: (1) determining the distribution and arrangement of horizons in space; and (2) characterizing the chemical, physical, and biological properties of the horizon, for which recent work using fuzzy set theory shows potential (Powell et al., 1991; McBratney and DeGruitjter, 1992; McBratney, 1992).

The theory of similar media has been used to develop methods for micro-scaling of hydraulic properties and infiltration equations (Miller, 1980; Warrick, 1990). The variability of measured soil hydraulic data has been reduced by as much as 80% using these techniques. Tyler and Wheatcraft (1990) demonstrated that the power-law soil water retention function was a fractal process and could be analyzed using fractal scaling theory. An excellent review of scaling theory in soil physics is provided by Hillel and Elrick (1990).

In the past 10 years there have been many attempts to characterize the meso-scale spatial variability of measured soil attributes (Webster, 1985; Yates and Warrick, 1987; Loague and Gander, 1990). These attempts have concentrated on the characterization of patterns, rather than on the linking of pattern to process. Two techniques are commonly used: quantitative methods using kriging, which relate the spatial covariance function to the spatial separation of the data, and methods that relate soil properties to qualitative measures of landscape position such as toe-slopes and interfluves (an attempt to account for process). Both techniques require large databases, and their results are not transferable. Kriging techniques ignore pedogenesis, while methods based on landscape position have lacked a consistent quantitative framework.

Kriging has shown that spatial correlation lengths of soil properties, such as saturated hydraulic conductivity, are on the order of only tens of meters. Digital terrain modeling methods offer another way of stratifying measured soil properties based on the way the soil catena develops in response to water movement in the landscape (i.e., process) and quantifiable landscape units. Better correlations may possibly be obtained using kriging or partial splines if wetness index or plan curvature were included as independent variables. The incorporation of these variables via a parametric submodel of a partial thin plate spline is attractive. In this way, broad changes with position can be accounted for by a smooth dependence on the two spatial variables (x,y), and the parametric submodel can account for more local, process-based effects. This approach was successfully applied to the interpolation of temperature at the macro-scale by Hutchinson (1991).

A recent approach is to organize the land surface according to a formal geomorphological model of landform and interlandform relations (Speight, 1974; Weibel and DeLotto, 1988; Dikau, 1989; Lammers and Band, 1990; Gessler, 1990; Mackay et al., 1991). Instead of dealing with pixels, flow tubes, or triangular facets, there are advantages in using integrated systems of hillslopes, floodplains, and ridgelines. The geomorphological position of a site strongly influences soil and other important properties. For example, Dikau (1989) demonstrated how digital terrain analysis could be applied to quantitative relief form analysis to define basic relief units for geomorphological and pedological mapping. The main topographic attributes used to define these relief units were slope, plan curvature, and profile curvature (Figure 19-10). This approach provides a systematic and methodological basis for derivation of complex relief units. It may be possible to

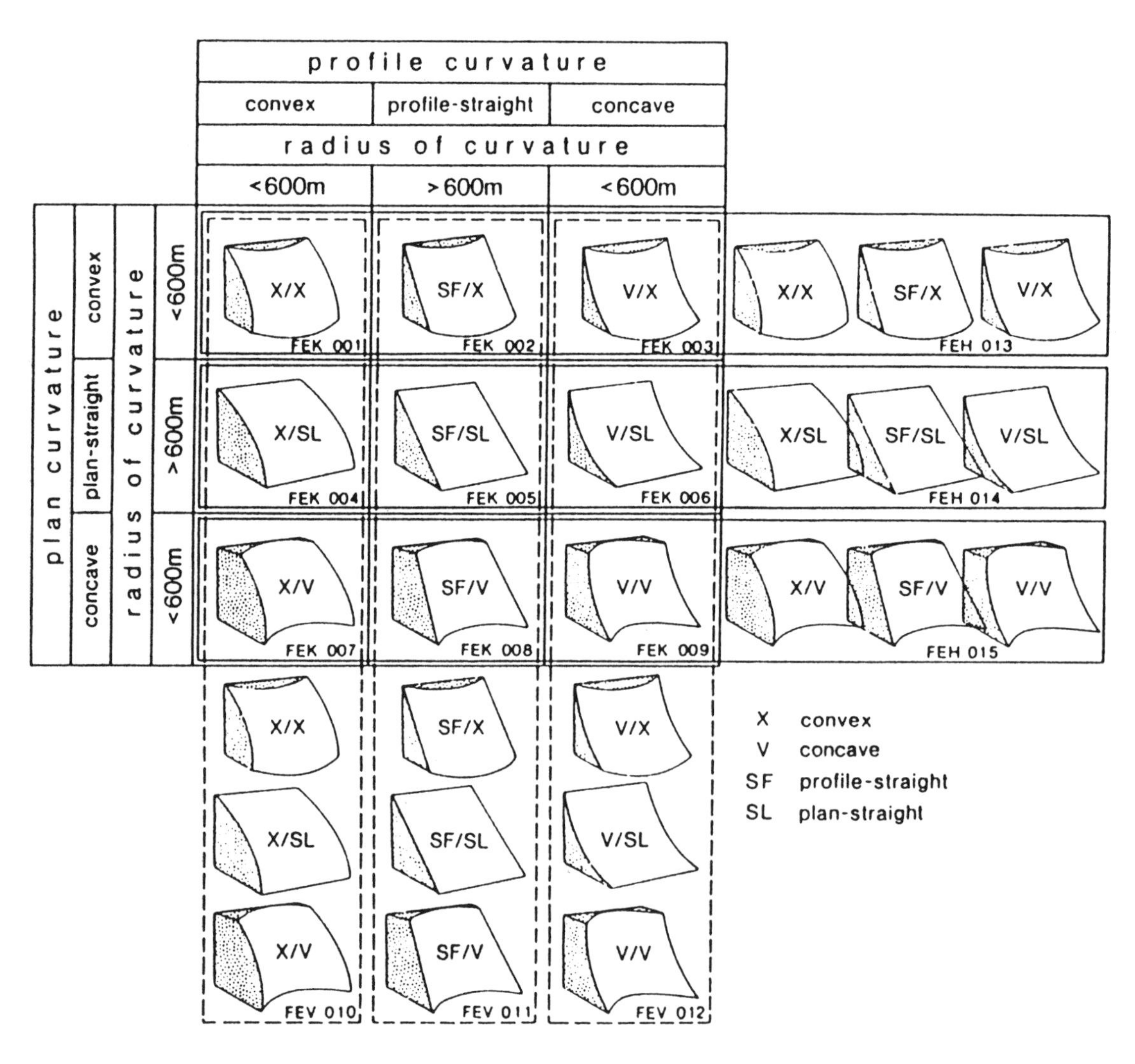

Figure 19-10: Classification of land form elements by plan and profile curvature for the determination of geomorphological relief units (reproduced from Dikau, 1989).

use these relief units to stratify the measured soil attributes and separate the micro- and meso-scale spatial variabilities.

Lammers and Band (1990) developed techniques for producing a set of landform files describing the morphometry, catchment position, and surface attributes of hillslopes and stream channels comprising a drainage basin. At this stage, these landforms can be said to comprise a "feature model" of the basin. Mackay (1990) and Mackay et al. (1991) extended the feature model by encapsulating the landform features derived from DTMs into a formal object model of the terrain, consisting of distinct object (landform) classes. Morphometric information on each landform object can be drawn from lower-level C programs, while a PROLOG model manipulates and addresses the objects and provides a graphical interface for object manipulation for the user. Mackay (1990) used this system to search for glacial features such as cirques or troughs based on local morphometry and global position. Mackey et al. (1991) used similar systems to structure and represent terrain knowledge for forested ecosystem applications.

RECOMMENDATIONS AND CONCLUSIONS

The overarching theme of this discussion has been integration: integration with respect to scientific disciplines, modeling, GIS technology, and data. The skills and resources necessary to build and operate effective models of environmental systems transcend individual disciplines. Today we need to be part hydrologist, geologist, soil scientist, geomorphologist, geographer, ecologist, biologist, atmospheric physicist, mathematician, statistician, and computer scientist. If we are concerned about how our models are used in decision-making and in the formulation of policy, then we should add sociologist and economist to this list. The recent literature is replete with references to multidisciplinary and transdisciplinary approaches, coupling of hydrologic and atmospheric models, and hybrid data structures. These integrative and innovative areas of research are, to us, the most promising and challenging areas requiring attention. New institutional and research structures that foster problem-oriented transdisciplinary research need to be encouraged. Often the traditional disciplinary division of academic institutions is an impediment to this type of activity. In particular, the biophysical science and spatial processing communities need to continue to join forces to provide the next generation of hybrid tools. For instance, elevation data can be generated from either map-based information or from stereo remotely sensed data. A hybridization of the two approaches could exploit both kinds of source materials.

Because the integration of modeling and GIS technology requires transdisciplinary skills, there is great potential for misuse and abuse of the technology. In addition, because of skill limitations, GIS technology is often solely used as a cartographic tool and so is not being fully utilized. There is thus a need for expanded training and education in the use of these tools. Training programs for users of these technologies are both lengthy and expensive. Training and education of a different sort is also required for the public at large. They need to understand what these systems can and cannot accomplish if they are to be accepted in various debates and forums, including legal challenges, which are so common in many environmental controversies. Furthermore, scientists have to make the effort to transfer their findings to the public in ways that can be easily understood.

A review of current GIS capabilities and their ability to support environmental modeling activities suggests that improved functionality is needed. The following issues are recommended for additional research and discussion:

1. *Emphasis on pattern as a function of process.* Models of land surface/subsurface phenomena that are linked with GIS should emphasize the identification of pattern as a function of process.
2. *Support for a broad spectrum of projects* dealing with land, water, and atmospheric environments at a variety of scales, including the site-specific, local, regional, and global scales.
3. *A variety of data interpolation algorithms are required.* GIS databases must incorporate the primary data that characterize the processes that modulate landscape and biological responses. The aggregation of data should be left to specific users since different users and applications may dictate different spatial scales of aggregation. Subjective and half-digested secondary data should be avoided as far as possible.
4. *Easy data input and output,* realizing that data may be derived from a very diverse suite of sources, and data may have to be exported for use by a variety of simulation models. This may be facilitated by some forms of expert systems.
5. *GIS should be supported by database management systems* that support two critical functions: (i) provide an "audit trail," or genealogy of the data; and (ii) provide methods of assessing data accuracy and uncertainty.
6. *Broadly accepted data standards.* Standards must be economically achievable, nonproprietary, and be enforceable if they are to be useful.
7. *Support interactive manipulation of model data.* Real-time data manipulation is desirable for the user to be able to add or delete certain data (e.g., an individual well, or some new information) and see how this addition or deletion affects the model.
8. *Provide ways for measuring model accuracy and/or defining the uncertainty in the results.* It is necessary to provide not only numerical values on data uncertainty,

but also to define the uncertainty associated with alternative interpretations. Monte Carlo techniques have been used in the past.

9. *Detailed field studies that explore the spatial and temporal variability of landscape processes must be continued.* These must be made available to the larger scientific community for continued model development and testing, especially the determination of proper model component process representation, the scales over which the processes are valid, and the integration of processes (Goodrich and Woolhiser, 1991).
10. *Improved soil survey and attribute mapping.* Current soils mapping techniques discard the pedon attribute data once a map unit line is drawn. Modern soil survey must use the pedon data as the primary information transfer mechanism (Burrough, 1986; Gessler, 1990). Also, soils attributes must be related to terrain, climate, vegetation/soil biology, parent material (which can be characterized in a GIS), and time.
11. *Improved spatial accuracy of relational data collected in the field.* This can be greatly aided by using global positioning system (GPS) technology.

New technology will only be adopted if it offers substantial time and financial savings over more traditional methods. Such time savings may be of great benefit in some environmental assessment and monitoring situations. Several issues need to be addressed if there is to be widespread adoption of GIS technology:

1. *Systems must become cheaper.* The environmental applications field is characterized by large numbers of smaller companies, and decentralized and relatively small groups of larger firms and Federal and state government agencies. These arrangements cannot support very expensive systems. However, they offer a large and diverse market for more modestly priced systems (e.g., the USDA-Soil Conservation Service's adoption of GRASS).
2. *There must be greater system integration.* For the reasons stated, environmental users cannot justify buying two or three systems to achieve the desired functionality; they require the functionality in a single system.
3. *Lower unit costs will yield an increased user base.* The anticipated reductions in the cost of acquiring GIS technology will encourage the acquisition of units into the smaller offices typical of the environmental consulting world. This will lead to much greater feedback, to the broad acceptance of technology, and to other benefits.
4. *There must be faster processing of 3D models.* Environmental projects often operate on short time frames, so rapid turnaround is critical.

ACKNOWLEDGMENTS

This study was funded in part by grant No. 90/82 from the Land and Water Resources Research and Development Corporation (Australia), grant No. 90/6334 from the Australia-USA Bilateral Science and Technology Program, grants SES-8912042 and SES-8912938 from the U.S. National Science Foundation, and by the Water Research Foundation of Australia. This paper is published as contribution No. J2713 of the Montana Agricultural Experiment Station. The authors thank Paul Gessler of the Division of Soils, CSIRO (Australia), Cliff Montagne, Steve Custer, and Jim Schmitt of Montana State University, and Mike Crane of the USGS–Rocky Mountain Mapping Center for their valuable comments and suggestions.

REFERENCES

Abbott, M.B., Bathurst, J.C., Cinge, J.A., O'Connell, P.E., and Hasmussen, J. (1986) An introduction to the European hydrological system-système hydrologique Européen, "SHE." 2: structure of a physically-based, distributed modeling system. *Journal of Hydrology* 87: 61–77.

Abrahams, A.D. (1984) Channel networks: a geomorphological perspective. *Water Resources Research* 20: 161–188.

Addiscott, T.M., and Wagenet, R.J. (1985) Concepts of solute leaching in soils: a review of modeling approaches. *Journal of Soil Science* 36: 411–424.

Ahnert, F. (1976) Brief description of a comprehensive three-dimensional model of landform development. *Zeitschrift fur Geomorphologie NF* 25: 29–49.

Ahnert, F. (1987) Uses of a theoretical model in the analysis of landform development. In Godard, A., and Rapp, A. (eds.) *Processes and Measurement of Erosion*, Paris: Editions du Centre National de la Recherche Scientifique, pp. 555–559.

Aller, L., et al. (1985) DRASTIC: a standard system for evaluating ground water pollution potential using hydrogeologic settings. *Rpt. No. EPA/600/2-85/018*, Ada, OK: U.S. Environmental Protection Agency, Robert S. Kerr Environmental Research Laboratory.

Anderson, J.R., Hardy, E.E., Roach, J.T., and Witmer, R.E. (1976) A land use and land cover classification system for use with remote sensor data. *Professional Paper No. 964*, Washington, DC: U.S. Geological Survey.

Arnold, J.G., and Sammons, N.B. (1989) Decision support system for selecting inputs to a basin scale model. *Water Resources Bulletin* 24.

Arnold, J.G., Williams, J.R., Nicks, A.D., and Sammons, N.B. (1990) *SWRRB: A Basin Scale Simulation Model for Soil and Water Resource Management*, College Station, TX: Texas A&M University Press.

Aste, J.P. (1988) Some reflections on the methodologies of landslide prevention in France. *Proceedings, International Workshop on Natural Disasters in European-Mediterranean Countries, Perogia, Italy, June 27–July 1, 1988*, 20 pp.

Augedal, H.O., Omre, H., and Stanley, K.O. (1986) SISABOSA—a program for stochastic modeling and evaluation of reservoir geology. *Proceedings, Reservoir Description and Simulation Conference, Institute for Energy Technology (IFE) Norway, Oslo, Norway, September 1986.*

Band, L.E. (1986a) Analysis and representation of drainage basin structure with digital elevation data. *Proceedings, Second International Conference on Spatial Data Handling, Seattle, Washington*, pp. 437–450.

Band, L.E. (1986b) Topographic partitioning of watersheds with digital elevation models. *Water Resources Research* 22: 15–24.

Band, L.E. (1989) A terrain-based watershed information system. *Hydrological Processes* 4: 151–162.

Band, L.E. (1991) Distributed parameterization of complex terrain. *Surveys in Geophysics* 12: 249–270.

Band, L.E., and Nemani, R. (1990) Spatial modeling of watershed ecosystem processes. *Proceedings, ISPRS Midterm Symposium, Victoria, British Columbia, Canada.*

Band, L.E., Peterson, D.L., Running, S.W., Coughlan, J., Lammers, R., Dungan, J., and Nemani, R. (1991) Ecosystem processes at the watershed level: basis for distributed simulation. *Ecological Modeling* 56: 171–196.

Band, L.E., and Wood, E.F. (1986) Computer graphics for distributed hydrologic modeling. *Transactions, American Geophysical Union* 67: 278.

Band, L.E., and Wood, E.F. (1988) Strategies for large scale, distributed hydrologic modeling. *Applied Mathematics and Computing* 27: 23–37.

Barling, R. (1992) Saturation zones and ephemeral gullies on arable land in eastern Australia. *Unpublished PhD Thesis*, Melbourne, Australia: Department of Civil and Agricultural Engineering, University of Melbourne.

Beasley, D.B., and Huggins, L.F. (1982) ANSWERS (Areal Nonpoint Source Watershed Environmental Response Simulation): user's manual. *EPA-905/9-82-001*, Chicago, IL: U.S. Environmental Protection Agency, 54 pp.

Belcher, R.C., and Paradis, A. (1991) A mapping approach to three-dimensional modeling. In Turner, A.K. (ed.) *Three-Dimensional Modeling With Geoscientific Information Systems*, Dordrecht, Netherlands: Kluwer Academic Publishers, pp. 107–122.

Belcher, W.R. (1988) Assessment of aquifer heterogeneities at the Hanford Nuclear Reservation, Washington, using inverse contaminant plume analysis. *Colorado School of Mines Master of Engineering Report No. ER-3594.*

Beven, K.J. (1986) Runoff production and flood frequency in catchments of order n: an alternative approach. In Gupta, V.K., Rodriguez-Iturbe, L., and Wood, E.F. (eds.) *Scale Problems in Hydrology*, Dordrecht: D. Reidel Publ. Co., pp. 107–131.

Beven, K. (1989) Changing ideas in hydrology—the case of physically-based models. *Journal of Hydrology* 105: 157–172.

Beven, K.J., and Kirkby, M.J. (1979) A physically-based variable contributing area model of basin hydrology. *Hydrological Sciences Bulletin* 24: 43–49.

Beven, K., Kirkby, M.J., Schofield, N., and Tagg, A.F. (1984) Testing a physically-based flood forecasting model (TOPMODEL) for three UK catchments. *Journal of Hydrology* 69: 119–143.

Beven, K., and Wood, E.F. (1983) Catchment geomorphology and the dynamics of runoff contributing areas. *Journal of Hydrology* 65: 139–158.

Brown, D.A., Anderson, K.L., and Gersmehl, P.J. (1987) Hydrographic data for a water resources GIS. In Brown, D.A., and Gersmehl, P.J. (eds.) *File Structure Design and Data Specifications for Water Resources Geographic Information Systems*, St. Paul, MN: Water Resources Research Center, University of Minnesota, pp. 12.1–12.16.

Brutsaert, W. (1986) Catchment-scale evaporation and the atmospheric boundary layer. *Water Resources Research* 22: 39S–45S.

Buol, S.W., Hole, F.D., and McCracken, R.J. (1989) *Soil Genesis and Classification*, 3rd ed., Ames, IA: Iowa State University Press.

Burrough, P.A. (1986) *Principles of Geographical Information Systems for Land Resource Assessment*, Oxford: Clarendon Press, 194 pp.

Burrough, P.A. (1989) Matching spatial databases and quantitative models in land resource assessment. *Soil Use and Management* 5: 3–8.

Burt, T.P., and Butcher, D.P. (1986) Development of topographic indices for use in semidistributed hillslope runoff models. In Slaymaker, O., and Balteanu, D. (eds.) *Geomorphology and Land Management*, Berlin: Gebruder Borntraeger, pp. 1–19.

Carrara, A., Cardinali, M., Detti, R., Guzzetti, F., Pasqui, V., and Reichenbach, P. (1991) GIS techniques and statistical models in evaluating landslide hazard. *Earth Surface Processes and Landforms* 16: 427–445.

Carter, J.R. (1989) Relative errors identified in USGS gridded DEMs. *Proceedings, Auto Carto 9: Ninth International Symposium on Computer Assisted Cartography, Baltimore, Maryland, April*, pp. 255–265.

Chow, V.T., Maidment, D.R., and Mays, L.W. (1988) *Applied Hydrology*, New York: McGraw-Hill 572 pp.

Clark, I. (1979) *Practical Geostatistics*, London: Applied Science Publishers, 129 pp.

Collins, S.H. (1975) Terrain parameters directly from a digital terrain model. *The Canadian Surveyor* 29: 507–

518.

Collins, S.H., and Moon, G.C. (1981) Algorithms for dense digital terrain models. *Photogrammetric Engineering and Remote Sensing* 47: 71–76.

Corominas, J., Esgleas, J., and Baeza, C. (1990) Risk mapping in the Pyrenees area: a case study. *Proceedings, International Conference on Water Resources in Mountainous Regions, Lausanne, Switzerland*, International Association for Scientific Hydrolology, pp. 425–428.

Dawes, W.R., and Short, D.L. (1988) *"TOPOG" Series Topographic Analysis and Catchment Drainage Modeling Package. User Manual—VAX/VMS Version*, Canberra: Australian Center for Catchment Hydrology, CSIRO Division of Water Resources, 74 pp.

Denning, P.J. (1990) Modeling reality. *American Scientist* 76: 495–498.

Desai, C.J., Shanholtz, V.O., and Flagg, J.M. (1990) Application of GIS technology in natural resources: system functionality. *ASAE Paper No. 90-3031*, St. Joseph, MI: American Society of Agricultural Engineers, 17 pp.

Dikau, R. (1989) The application of a digital relief model to landform analysis in geomorphology. In Raper, J. (ed.) *Three Dimensional Applications in Geographic Information Systems*, New York: Taylor and Francis, pp. 51–77.

Dixon, T.H. (1991) An introduction to the global positioning system and some geological applications. *Reviews of Geophysics* 29: 249–276.

Djokic, D., and Maidment, D.R. (1991) Terrain analysis for urban stormwater modeling. *Hydrological Processes* 5: 115–124.

Domenico, P.R. (1987) An analytical model for multidimensional transport of a decaying contaminant species. *Journal of Hydrology* 91: 49–58.

Domenico, P.R., and Robbins, G.A. (1985) A new method of contaminant plume analysis. *Ground Water* 23: 476–485.

Dyke, P.T. (1991) Temple, TX: Texas Agricultural Experiment Station, Texas A&M University (personal communication).

Eagleson, P.S. (1991) Hydrologic science: a distinct geoscience. *Reviews of Geophysics* 29: 237–248.

Engel, B.A., Srinivasan, R., and Rewerts, C.C. (1991) A GIS toolbox approach to hydrologic modeling. *Proceedings, GRASS 1991 User's Conference, Berkeley, California.*

Fairfield, J., and Leymarie, P. (1991) Drainage networks from grid digital elevation models. *Water Resources Research* 27: 709–717.

Faust, C.R., and Mercer, J.W. (1980) Groundwater modeling: numerical models. *Groundwater* 18: 395–409.

Fetter, C.W., Jr. (1980) *Applied Hydrogeology.* Columbus, OH: Bell and Howell, 488 pp.

Fisher, T.R. (1991) Concepts and approaches in multi-dimensional geologic modeling for environmental applications (in preparation).

Fisher, T.R., and Wales, R.Q. (1990) 3-D solid modeling of sandstone reservoirs using NURBS—a case study of Noonen Ranch Field, Denver Basin, Colorado. *Geobyte* 5: 39–41.

Fisher, T.R., and Wales, R.Q. (1991) 3-D visualization in areas of complex geologic modeling. *Proceedings, Houston Geotech '91.*

Fisher, T.R., and Wales, R.Q. (1992) Three dimensional solid modeling of geo-objects using non-uniform rational B-Splines (NURBS). In Turner, A.K. (ed.) *Three-Dimensional Modeling with Geoscientific Information Systems*, Dordrecht: Kluwer Academic Publishers (in press).

Freeman, T.G. (1991) Calculating catchment area with divergent flow based on a regular grid. *Computers and Geosciences* 17: 413–422.

Gersmehl, C.A. (1987) Land cover data for water resources GIS. In Brown, D.A., and Gersmehl, P.J. (eds.) *File Structure Design and Data Specifications for Water Resources Geographic Information Systems*, St. Paul, MN: Water Resources Research Center, University of Minnesota, pp. 7.1–7.39.

Gersmehl, P.J., Anderson, K.L., Greene, R.P., Dunning, N.P., Gersmehl, C.A., and Brown, D.A. (1987b) Hydrologic classification of land cover. In Brown, D.A., and Gersmehl, P.J. (eds.) *File Structure Design and Data Specifications for Water Resources Geographic Information Systems*, St. Paul, MN: Water Resources Research Center, University of Minnesota, pp. 6.1–6.30.

Gersmehl, P.J., Brown, D.A., and Anderson, K.L. (1987a) File structure and cell size considerations for water resources GIS. In Brown, D.A., and Gersmehl, P.J. (eds.) *File Structure Design and Data Specifications for Water Resources Geographic Information Systems*, St. Paul, MN: Water Resources Research Center, University of Minnesota, pp. 2.1–2.38.

Gessler, P.E. (1990) Geostatistical modeling of soil-landscape variability within a GIS framework. *Unpublished MSc Thesis*, Madison, WI: Soil Science Department, University of Wisconsin, 192 pp.

Gessler, P.E. (1991) Canberra: Division of Soils, CSIRO (personal communication).

Gessler, P.E., McSweeney, K., Kiefer, R.W., and Morrison, L.M. (1989) Analysis of contemporary and historical soil/vegetation/landuse patterns in southwest Wisconsin utilizing GIS and remote sensing technologies. *Technical Papers, 1989 ASPRS/ACSM Annual Convention, Baltimore, MD*, Vol. 4, pp. 85–92.

Goodrich, D.C. (1990) Geometric simplification of a distributed rainfall-runoff model over a range of basin scales. *PhD Dissertation*, Tucson, AZ: Department of Hydrology and Water Resources, University of Arizona, 361 pp.

Goodrich, D.C., and Woolhiser, D.A. (1991) Catchment hydrology. *Reviews of Geophysics, Supplement: U.S. Na-*

tional Report to International Union of Geodesy and Geophysics 1987–1990, pp. 202–209.

Goodrich, D.C., Woolhiser, D.A., and Keefer, T.O. (1991) Kinematic routing using finite elements on a triangular irregular network. *Water Resources Research* 27: 995–1003.

Grayson, R.B. (1990) Terrain-based hydrologic modeling for erosion studies. *Unpublished PhD Thesis*, Melbourne: Department of Civil and Agricultural Engineering, University of Melbourne, 375 pp.

Grayson, R.B., Moore, I.D., and McMahon, T.A. (1992) Physically-based hydrologic modeling—II. Is the concept realistic? *Water Resources Research* 26: 2659–2666.

Gupta, V.K., Cole, C.R., Kincaid, C.T., and MacDonald, C.J. (1987) Coupled fluid, energy, and solute transport (CFEST) model: formulation and user's manual. *ONWI Report 660*, Columbus, OH: Battelle Memorial Institute.

Haldorsen, H.H., Brand, P.J., and MacDonald, C.J. (1987) Review of the stochastic nature of reservoirs. *Presented at Seminar on the Mathematics of Oil Production, Robinson College, Cambridge University, July 1987.*

Hession, W.C., Shanholtz, V.O., Mostaghimi, S., and Dillaha, T.A. (1987) Extensive evaluation of the finite element storm hydrograph model. *ASAE paper No. 87-2570*, St. Joseph, MI: American Society of Agricultural Engineers, 34 pp.

Hillel, D. (1986) Modeling in soil physics: a critical review. *Future Developments in Soil Science Research (A Collection of Soil Science Society of America Golden Anniversary Contributions). Presented at the Annual Meeting, Soil Science Society of America, New Orleans, Louisiana*, pp. 35–42.

Hillel, D., and Elrick, D.E. (eds.) (1990) Scaling in soil physics: principles and application. *Special Publication No. 25*, Madison, WI: Soil Science Society of America, 122 pp.

Hornberger, G.M., Beven, K.J., Cosby, B.J., and Sappington, D.E. (1985) Shenandoah watershed study: calibration of a topography-based, variable contributing area hydrological model to a small forested catchment. *Water Resources Research* 21: 1841–1850.

Houlding, S.W. (1988) The evolution of 3D computer techniques in mining. *Engineering and Mining Journal* 189: 45–47.

Houlding, S.W. (1991) 3D geotechnical modeling in the analysis of acid mine drainage. *Proceedings, Second International Conference on the Abatement of Acid Mine Drainage, Montreal, Canada, September 1991.*

Houlding, S.W., and Stokes, M.A. (1990) Mining activity and resource scheduling with the use of 3-D component modeling. *Transactions of the Institute of Mining and Metallurgy* 99: A53–A59.

Hungerford, R.D., Nemani, R.R., Running, S.W., and Coughlan, J.C. (1989) MTCLIM: a mountain microclimate simulation model. *INT-414*, Washington, DC: U.S. Department of Agriculture, Forest Service, 52 pp.

Hutchinson, M.F. (1989a) A new procedure for gridding elevation and stream line data with automatic removal of spurious pits. *Journal of Hydrology* 106: 211–232.

Hutchinson, M.F. (1989b) A new objective method of spatial interpolation of meterological variables from irregular networks applied to the estimation of monthly mean solar radiation, temperature, precipitation and windrun. In Fitzpatrick, E.A., and Kalma, J.D. (eds.) *Need for Climatic and Hydrologic Data in Agriculture in Southeast Asia, Technical Memorandum 89/5*, Canberra: Division of Water Resources, CSIRO, pp. 95–104.

Hutchinson, M.F. (1991) The application of thin-plate smoothing splines to continent-wide data assimilation. In Jasper, J.D. (ed.) *Data Assimilation Systems, BMRC Research Report No. 27*, Melbourne: Bureau of Meteorology, pp. 104–113.

Iida, T. (1984) A hydrological method of estimation of the topographic effect on the saturated throughflow. *Transactions, Japanese Geomorphological Union* 5(1): 1–12.

Jarvis, R.S. (1977) Drainage network analysis. *Progress in Physical Geography* 1: 271–295.

Jenson, S.K. (1991) Application of hydrologic information automatically extracted from digital elevation models. *Hydrological Processes* 5: 31–44.

Jenson, S.K., and Domingue, J.O. (1988) Extracting topographic structure from digital elevation model data for geographic information system analysis. *Photogrammetric Engineering and Remote Sensing* 54: 1593–1600.

Johnson, L.E. (1989) MAPHYD—a digital map-based hydrologic modeling system. *Photogrammetric Engineering and Remote Sensing* 55: 911–917.

Jones, N.L., Wright, S.G., and Maidment, D.R. (1990) Watershed delineation with triangle-based terrain models. *Journal of Hydraulic Engineering* 116: 1232–1251.

Kirkby, M.J. (1971) Hillslope process-response models based on the continuity equation. *Special Publication, Institute of British Geographers* 3: 15–30.

Kirkby, M.J. (1986) A runoff simulation model based on hillslope topography. In Gupta, V.K., Rodriguez-Iturbe, L., and Wood, E.F. (eds.) *Scale Problems in Hydrology*, Dordrecht: D. Reidel Publishing Co., pp. 39–56.

Klemes, V. (1986) Dilettantism in hydrology: transition or destiny? *Water Resources Research* 22: 177S–188S.

Kozak, M. (1968) Determination of the runoff hydrograph on a deterministic basis using a digital computer. *International Association for Scientific Hydrology, Publication No. 80*, pp. 138–151.

Kristensen, K.J., and Jensen, S.E. (1975) A model for estimating actual evapotranspiration from potential evapotranspiration. *Nord. Hydrol.* 6: 170–188.

Lam, N.S. (1983) Spatial interpolation methods: a review. *The American Cartographer* 10: 129–149.

Lammers, R.B., and Band, L.E. (1990) Automating object representation of drainage basins. *Computers and Geosciences* 16: 787–810.

Lane, L., and Nearing, M. (eds.) (1989) USDA-water erosion prediction project: hillslope profile model documentation. *National Soil Erosion Research Laboratory Report No. 2*, West Lafayette, IN: U.S. Department of Agriculture, Agricultural Research Service.

Leberl, F.W., and Olson, W. (1982) Raster scanning for operational digitizing of graphical data. *Photogrammetric Engineering and Remote Sensing* 48: 615–627.

Lees, B.G., and Ritman, K. (1991) Decision-tree and rule-induction approach to integration of remotely sensed and GIS data in mapping vegetation in disturbed or hilly environments. *Environmental Management* 15: 823–831.

Loague, K. (1988) Impact of rainfall and soil hydraulic property information on runoff predictions at the hillslope scale. *Water Resources Research* 24: 1501–1510.

Loague, K., and Gander, G.A. (1990) R-5 revisited, 1: spatial variability of infiltration on a small rangeland catchment. *Water Resources Research* 26: 957–972.

Long, D.S., DeGloria, S.D., and Galbraith, J.M. (1991) Use of the global positioning system in soil survey. *Journal of Soil and Water Conservation* 46: 293–297.

Mackay, D.S. (1990) Knowledge based classification of higher order terrain objects on digital elevation models. *Unpublished MSc Thesis*, Toronto: Department of Geography, University of Toronto, 117 pp.

Mackay, D.S., Band, L.E., and Robinson, V.B. (1991) An object-oriented system for the organization and representation of terrain knowledge for forested ecosystems. *Proceedings, GIS/LIS '91*.

Mackey, B.G. (1991) The spatial extension of vegetation site data: a case study in the rainforests of the wet tropics of Queensland, Australia. *Unpublished PhD Thesis*, Canberra: Center for Resource and Environmental Studies, the Australian National University, 316 pp.

Maidment, D.R., Djokic, D., and Lawrence, K.G. (1989) Hydrologic modeling on a triangulated irregular network. *Transactions, American Geophysical Union* 70: 1091.

Mark, D.M. (1975) Computer analysis of topography: a comparison of terrain storage methods. *Geografiska Annaler* 57A: 179–188.

Mark, D.M. (1983) Automated detection of drainage networks from digital elevation models. *Proceedings, Auto-Carto 6, Automated Cartography: International Perspectives on Achievements and Challenges, Ottawa, Ontario*, pp. 288–298.

Mark, D.M. (1988) Network models in geomorphology. In Anderson, M.G. (ed.) *Modeling Geomorphological Systems*, Chichester: John Wiley, pp. 73–97.

McBratney, A.B. (1992) Some remarks on soil horizon classes. *Catena, Special Issue on Soil Horizon Classes* (in press).

McBratney, A.B., and DeGruijter, J.J. (1992) A continuum approach to soil classification and mapping: classification by modified fuzzy k-means with extragrades. *Journal of Soil Science* 43: 159–175.

McDonald, M.G., and Harbaugh, A.W. (1988) A modular three-dimensional finite difference ground-water flow model. *Techniques of Water Resources Investigations, Book 6*, Reston, VA: U.S. Geological Survey, 586 pp.

McKeague, J.A., Eilers, R.G., Thomasson, A.J., Reeve, M.J., Bouma, J., Grossman, R.B., Favrot, J.C., Renger, M., and Strebel, O. (1984) Tentative assessment of soil survey approaches to the characterization and interpretation of air-water properties of soils. *Geoderma* 34: 69–100.

McKenzie, N.J. (1991) Canberra: Division of Soils, CSIRO (personal communication).

McKenzie, N.J., and MacLeod, D.A. (1989) Relationships between soil morphology and soil properties relevant to irrigated and dryland agriculture. *Australian Journal of Soil Research* 27: 235–258.

McKenzie, N.J., Smettem, K.R.J., and Ringrose-Voase, A.J. (1991) Evaluation of methods for inferring air and water properties of soils from field morphology. *Australian Journal of Soil Research* 29: 587–602.

McLain, D.H. (1976) Two dimensional interpolation from random data. *The Computer Journal* 19: 178–181.

Mein, R.G., and Larson, C.L. (1973) Modeling infiltration during a steady rain. *Water Resources Research* 9: 384–394.

Miller, E.E. (1980) Similitude and scaling of soil-water phenomena. In Hillel, D. (ed.) *Applications of Soil Physics*, New York: Academic Press, pp. 300–318.

Milly, P.C.D., and Eagleson, P.S. (1987) Effect of spatial variability on average annual water balance. *Water Resources Research* 23: 2135–2143.

Moore, I.D., and Burch, G.J. (1986) Physical basis of the length-slope factor in the Universal Soil Loss Equation. *Soil Science Society of America, Journal* 50: 1294–1298.

Moore, I.D., Burch, G.J., and Mackenzie, D.H. (1988a) Topographic effects on the distribution of surface soil water and the location of ephemeral gullies. *Transactions, American Society of Agricultural Engineers* 31: 1098–1107.

Moore, I.D., and Gallant, J.C. (1991) Overview of hydrologic and water quality modeling. In Moore, I.D. (ed.) *Modeling the Fate of Chemicals in the Environment*, Canberra: Center for Resource and Environmental Studies, the Australian National University, pp. 1–8.

Moore, I.D., and Grayson, R.B. (1991) Terrain-based catchment partitioning and runoff prediction using vector elevation data. *Water Resources Research* 27:

1177–1191.

Moore, I.D., Grayson, R.B., and Ladson, A.R. (1991) Digital terrain modeling: a review of hydrological, geomorphological, and biological applications. *Hydrological Processes* 5: 3–30.

Moore, I.D., Grayson, R.B., and Wilson, J.P. (1990) Runoff modeling in complex three-dimensional terrain. *Proceedings, International Conference on Water Resources in Mountainous Regions, Lausanne, Switzerland*, International Association for Scientific Hydrology, pp. 591–598.

Moore, I.D., and Hutchinson, M.F. (1991) Spatial extension of hydrologic process modeling. *Proceedings, International Hydrology and Water Resources Symposium, Perth, 2–4 October*, Institute of Engineers of Australia, pp. 803–808.

Moore, I.D., Mackay, S.M., Wallbrink, P.J., Burch, G.J., and O'Loughlin, E.M. (1986) Hydrologic characteristics and modeling of a small forested catchment in southeastern New South Wales: prelogging condition. *Journal of Hydrology* 83: 307–335.

Moore, I.D., and Nieber, J.L. (1989) Landscape assessment of soil erosion and nonpoint source pollution. *Journal, Minnesota Academy of Sciences* 55: 18–25.

Moore, I.D., O'Loughlin, E.M., and Burch, G.J. (1988b) A contour-based topographic model for hydrological and ecological applications. *Earth Surface Processes and Landforms* 13: 305–320.

Nasser, K.H., and Turner, A.K. (1991) The development of computerized three-dimensional geohydrologic conceptual models using Geoscientific Information Systems. *Association of Engineering Geologists, Proceedings, 1991 Annual Meeting, Chicago.*

Nicks, A.D. (1985) Generation of climate data. *Proceedings of the Natural Resources Modeling Symposium*, Washington, DC: U.S. Department of Agriculture, Agricultural Research Service, pp. 297–300.

Nicks, A.D., and Lane, L.J. (1989) Weather generator. In Lane, L.J., and Nearing, M.A. (eds.) *USDA-Water Erosion Prediction Project: Hillslope Profile Model Documentation, National Soil Erosion Research Laboratory Report No. 2*, West Lafayette, IN: U.S. Department of Agriculture, Agricultural Research Service, pp. 2.1–2.19.

Nix, H.A. (1981) Simplified evaluation models based on specified minimum data sets: the CROPEVAL concept. In Berg, A. (ed.) *Application of Remote Sensing to Agricultural Production Forecasting*, Rotterdam: A.A. Balkema, pp. 151–171.

Norton, T.W., and Moore, I.D. (1991) Ecological applications of GIS and hydrological modeling: identification and management of Koala habitat in south east Australia. *Proceedings, International Hydrology and Water Resources Symposium*, Perth, 2–4 October, Institute of Engineers of Australia, pp. 647–648.

Noverraz, F. (1990) Essai de recensement cartographique des glissements de terrain et ecroulements rocheux sur le territoire Suisse. *Proceedings, International Conference on Water Resources in Mountainous Regions, Lausanne, Switzerland*, International Association for Scientific Hydrology, pp. 429–436.

O'Callaghan, J.F., and Mark, D.M. (1984) The extraction of drainage networks from digital elevation data. *Computer Vision, Graphics and Image Processing* 28: 323–344.

Olea, R.A. (1975) Optimum mapping techniques using regionalized variable theory. *Kansas Geological Survey Series on Spatial Analysis No. 2*, 137 pp.

Olender, H.A. (1980) Analysis of triangulated irregular network (TIN) terrain model for military applications. *Contract N00014-77-0698, Naval Analysis Program*, Arlington, VA: Office of Naval Research, 90 pp.

O'Loughlin, E.M. (1986) Prediction of surface saturation zones in natural catchments by topographic analysis. *Water Resources Research* 22: 794–804.

Onstad, C.A. (1973) Watershed flood routing using distributed parameters. In Schulz, E.F., Koezer, V.A., and Mahnood, K. (eds.) *Floods and Droughts, Proceedings, 2nd International Symposium in Hydrology, Sept. 11–13, 1972, Fort Collins, CO*, Fort Collins, CO: Water Resources Publications, pp. 418–428.

Onstad, C.A., and Brakensiek, D.L. (1968) Watershed simulation by stream path analogy. *Water Resources Research* 4: 965–971.

Palacios-Velez, O.L., and Cuevas-Renaud, B. (1986) Automated river-course, ridge and basin delineation from digital elevation data. *Journal of Hydrology* 86: 299–314.

Palacios-Velez, O.L., and Cuevas-Renaud, B. (1989) Transformation of TIN DEM data into a kinematic cascade. *Transactions, American Geophysical Union* 70: 1091.

Panuska, J.C., and Moore, I.D. (1991) Water quality modeling: terrain analysis and the Agricultural Non-Point Source Pollution (AGNPS) model. *Technical Report No. 132*, St. Paul, MN: Water Resources Research Center, University of Minnesota, 57 pp.

Panuska, J.C., Moore, I.D., and Kramer, L.A. (1991) Terrain analysis: integration into the agricultural nonpoint source (AGNPS) pollution model. *Journal of Soil and Water Conservation* 46: 59–64.

Paradis, A.R., and Belcher, R.C. (1990) Interactive volume modeling—a new product for 3-D mapping. *Geobyte* 5: 42–44.

Peucker, T.K., and Douglas, D.H. (1975) Detection of surface specific points by local parallel processing of discrete terrain elevation data. *Computer Graphics and Image Processing* 4: 375–387.

Peucker, T.K., Fowler, R.J., Little, J.J., and Mark, D.M. (1978) The triangulated irregular network. *Proceedings, Auto Carto III*, Falls Church, VA: American Congress on Surveying and Mapping, pp. 516–540.

Pfannkuch, H.O., Jones, P.M., and Guo, L. (1987) Groundwater systems analysis and monitoring. In Brown, D.A., and Gersmehl, P.J. (eds.) *File Structure Design and Data Specifications for Water Resources Geographic Information Systems*, St. Paul, MN: Water Resources Research Center, University of Minnesota, pp. 13.1–13.67.

Poeter, E. (1983) Computer codes for modeling in conjunctive ground water/surface water management programs. *Proceedings, NWWA Western Regional Conference on Ground Water Management, San Diego, California*.

Powell, B., McBratney, A.B., and MacLeod, D.A. (1991) Fuzzy classification of soil profiles and horizons from the Lockyer Valley, Queensland, Australia. *Geoderma* (in press).

Quinn, P., Beven, K., Chevallier, P., and Planchon, O. (1991) The prediction of hillslope flow paths for distributed hydrological modeling using digital terrain models. *Hydrological Processes* 5: 59–79.

Rawls, W.J., Brakensiek, D.L., and Saxton, K.E. (1982) Estimation of soil water properties. *Transactions, American Society of Agricultural Engineers* 25: 1316–1320, 1328.

Rawls, W.J., Brakensiek, D.L., and Soni, B. (1983) Agricultural management effects on soil water processes. Part I: Soil water retention and Green and Ampt infiltration parameters. *Transactions, American Society of Agricultural Engineers* 26: 1747–1752.

Richards, L.A. (1931) Capillary conduction of liquids in porous media. *Physics* 1: 318–333.

Richardson, J. (ed.) (1984) *Models of Reality: Shaping Thought and Action*, Mt. Airy, MD: Lomond Books, 328 pp.

Robert, P.C., and Anderson, J.L. (1987) A convenient soil survey information system (SSIS). *Applied Agricultural Research* 2: 252–259.

Rogers, W.F., and Singh, V.P. (1986) Some geomorphic relationships and hydrograph analysis. *Water Resources Research* 22: 777–784.

Rose, C.W. (1985) Developments in soil erosion and deposition models. *Advances in Soil Science* 2: 1–63.

Rose, C.W., Williams, J.R., Sander, G.C., and Barry, D.A. (1983a) A mathematical model of soil erosion and deposition processes. I. Theory for a plane land element. *Soil Science Society of America, Journal* 47: 991–995.

Rose, C.W., Williams, J.R., Sander, G.C., and Barry, D.A. (1983b) A mathematical model of soil erosion and deposition processes. II. Application to data from an arid-zone catchment. *Soil Science Society of America Journal* 47: 996–1000.

Ross, B.B., Contractor, D.N., and Shanholtz, V.O. (1979) A finite element model of overland and channel flow for assessing the hydrologic impact of landuse change. *Journal of Hydrology* 41: 1–30.

Ross, B.B., Wolfe, M.L., Shanholtz, V.O., Smolen, M.D., and Contractor, D.N. (1982) Model for simulating runoff and erosion in ungaged watersheds. *Bulletin No. 130*, Blacksburg, VA: Virginia Water Resources Research Center, Virginia Polytechnic Institute and State University, 72 pp.

Running, S.W., Nemani, R.R., and Hungerford, R.D. (1987) Extrapolation of synoptic meteorological data in mountainous terrain and its use for simulating forest evapotranspiration and photosynthesis. *Canadian Journal of Forest Research* 17: 472–483.

Sharma, M.L., Luxmore, R.L., DeAngelis, R., Ward, R.C., and Yeh, G.T. (1987) Subsurface water flow simulated for hillslopes with spatially dependent soil hydraulic characteristics. *Water Resources Research* 23: 1523–1530.

Shanholtz, V.O., Desai, C.J., Zhang, N., Kleene, J.W., Metz, C.D., and Flagg, J.M. (1990) Hydrologic/water quality modeling in a GIS environment. *ASAE Paper No. 90-3033*, St. Joseph, MI: American Soceity of Agricultural Engineers, 17 pp.

Silburn, M. (1991) Toowoomba: Agricultural Production Systems Research Unit, Queensland Department of Primary Industries (personal communication).

Silfer, A.T., Kinn, G.J., and Hassett, J.M. (1987) A geographic information system utilizing the triangulated irregular network as a basis for hydrologic modeling. *Proceedings, Auto-Carto 8, Baltimore, Maryland*, pp. 129–136.

Sivapalan, M., Beven, K., and Wood, E.F. (1987) On hydrologic similarity 2. A scaled model of storm runoff production. *Water Resources Research* 23: 2266–2278.

Sivapalan, M., and Wood, E.F. (1990) On hydrologic similarity 3. A dimensionless flood frequency model using a generalized geomorphic unit hydrograph and partial area runoff generation. *Water Resources Research* 26: 43–58.

Smart, J.S. (1978) The analysis of drainage network composition. *Earth Surface Processes and Landforms* 3: 129–170.

Smith, T.R., and Bretherton, F.P. (1972) Stability and the conservation of mass in drainage basin evolution. *Water Resources Research* 8: 1506–1529.

Speight, J.G. (1974) A parametric approach to landform regions. *Special Publication Institute of British Geographers* 7: 213–230.

Speight, J.G. (1980) The role of topography in controlling throughflow generation: a discussion. *Earth Surface Processes and Landforms* 5: 187–191.

Stam, J.M.T., Zijl, W., and Turner, A.K. (1989) Determination of hydraulic parameters from the reconstruction of alluvial stratigraphy. In Carlomagno, G.M., and Brebbia, C.A. (eds.) *Computers and Experiments in Fluid Flow: Proceedings, 4th International Conference on Computational Methods and Experimental Measurements, Capri, Italy*, New York: Springer-Verlag, pp.

383–392.

Tarboton, D.G., Bras, R.L., and Rodriguez-Iturbe, I. (1989) The analysis of river basins and channel networks using digital terrain data. *Technical Report No. 326*, Cambridge, MA: Ralph M. Parsons Laboratory, Department of Civil Engineering, Massachusetts Institute of Technology, 251 pp.

Tisdale, T.S., Hamrick, J.M., and Yu, S.L. (1986) Kinematic wave analysis of overland flow using topography fitted coordinates. *Transactions, American Geophysical Union* 67: 271.

Turner, A.K. (1989) The role of three-dimensional Geographic Information Systems in sub-surface characterization for hydrogeological applications. In Raper, J.F. (ed.) *Three Dimensional Applications in Geographic Information Systems*, London: Taylor and Francis, pp. 115–127.

Turner, A.K. (ed.) (1992) *Three-Dimensional Modeling With Geoscientific Information Systems*, Dordrecht: Kluwer Academic Publishers, 443 pp.

Turner, A.K., Downey, J.S., and Kolm, K.E. (1990) Potential applications of three-dimensional Geoscientific Information Systems (GSIS) for regional groundwater-flow system modeling, Yucca Mountain, Nevada. *Transactions, American Geophysical Union* 71: 1316.

Turner, A.K., and Kolm, K.E. (1991) Three-dimensional Geoscientific Information Systems for ground-water modeling. *Technical Papers, 1991 ACSM/ASPRS Annual Meeting*, Vol. 4, pp. 217–226.

Turner, M.G., and Gardner, R.H. (eds.) (1990) *Quantitative Methods in Landscape Ecology: The Analysis and Interpretation of Landscape Heterogeneity*, New York: Springer-Verlag, 536 pp.

Tyler, S.W., and Wheatcraft, S.W. (1990) Fractal processes in soil water retention. *Water Resources Research* 26: 1047–1054.

U.S. Army (1987) *GRASS Reference Manual*, Champaign, IL: Construction Engineering Research Laboratory, U.S. Army Corps of Engineers.

USDA (1989) *The Second RCA Appraisal: Soil, Water, and Related Resources on Nonfederal Land in the United States: Analysis of Conditions and Trends*, Washington, DC: U.S. Department of Agriculture, 280 pp.

van Geer, F.C. (1987) *Applications of Kalman Filtering in the Analysis and Design of Groundwater Monitoring Networks*, Delft: TNO Institute of Applied Geoscience, 130 pp.

Verner, J., Morrison, M.L., and Ralph, C.J. (1986) *Wildlife 2000: Modeling Habitat Relationships of Terrestrial Vertebrates*, Madison, WI: University of Wisconsin Press, 470 pp.

Vertessy, R.A., Wilson, C.J., Silburn, D.M., Connolly, R.D., and Ciesiolka, C.A. (1990) Predicting erosion hazard areas using digital terrain analysis. *Proceedings, IASH/AISH International Symposium on Research Needs and Applications to Reduce Erosion and Sedimentation in Tropical Steeplands, Suva, Fiji, 11–15 June, 1990.*

Vieux, B.E., Bralts, V.F., and Segerlind, L.J. (1988) Finite element analysis of hydrologic response areas using geographic information systems. *Modeling Agricultural, Forest, and Rangeland Hydrology, ASAE Publication 07-88*, St. Joseph, MI: American Society of Agricultural Engineers, pp. 437–446.

Warrick, A.W. (1990) Application of scaling to the characterization of spatial variability in soils. In Hillel, D., and Elrick, D.E. (eds.) *Scaling in Soil Physics: Principles and Application, Special Publication No. 25*, Madison, WI: Soil Science Society of America, pp. 39–51.

Webster, R. (1985) Quantitative spatial analysis of soil in the field. *Advances in Soil Science, Volume 3*, New York: Springer-Verlag.

Weibel, R., and DeLotto, J.S. (1988) Automated terrain classification for GIS modeling. *Proceedings, GIS/LIS '88, San Antonio, Texas*, pp. 618–627.

Willgoose, G., Bras, R.L., and Rodriguez-Iturbe, I. (1989a) A physically based channel network and catchment evolution model. *Technical Report No. 322*, Cambridge, MA: Ralph Parsons Laboratory, Department of Civil Engineering, Massachusetts Institute of Technology, 464 pp.

Willgoose, G., Bras, R.L., and Rodriguez-Iturbe, I. (1989b) Results from a new model of river basin evolution. *Earth Surface Processes and Landforms* 16: 237–254.

Willgoose, G., Bras, R.L., and Rodriguez-Iturbe, I. (1990) A model of river basin evolution. *Transactions, American Geophysical Union* 71: 1806–1807.

Willgoose, G., Bras, R.L., and Rodriguez-Iturbe, I. (1991) A coupled channel network growth and hillslope evolution model. 1. Theory. *Water Resources Research* 27: 1671–1684.

Williams, J.R., Jones, C.A., and Dyke, P.T. (1984) A modeling approach to determining the relationship between erosion and soil productivity. *Transactions, American Society of Agricultural Engineers* 27: 129–144.

Williams, J.R., Renard, K.G., and Dyke, P.T. (1983) EPIC: a new method for assessing erosion effects on soil productivity. *Journal of Soil and Water Conservation* 38: 381–384.

Williams, J., Ross, P., and Bristow, K. (1990) Prediction of Campbell water retention function from texture, structure and organic matter. *Proceedings, International Workshop on Indirect Methods for Estimating the Hydraulic Properties of Unsaturated Soils, University of California, Riverside, 11–13 October.*

Wilson, G.V., and Luxmoore, R.J. (1988) Infiltration, macroporosity, and mesoporosity distributions on two forested watersheds. *Soil Science Society of America, Journal* 52: 329–335.

Wischmeier, W.H., and Smith, D.D. (1978) Predicting

rainfall erosion losses, a guide to conservation planning. *Agriculture Handbook No. 537*, Washington, DC: U.S. Department of Agriculture, 58 pp.
Wood, E.F., Sivapalan, M., and Beven, K. (1990) Similarity and scale in catchment storm response. *Reviews of Geophysics* 28: 1–18.
Wood, E.F., Sivapalan, M., Beven, K., and Band, L. (1988) Effects of spatial variability and scale with implications to hydrologic modeling. *Journal of Hydrology* 102: 29–47.
Woolhiser, D.A., Hanson, C.L., and Richardson, C.W. (1988) Microcomputer program for daily weather simulation. *ARS-75*, Washington, DC: U.S. Department of Agriculture, Agricultural Research Service, 49 pp.
USGS (1987) *Digital Elevation Models: Data Users Guide.* Reston, VA: U.S. Geological Survey, National Mapping Division, 38 pp.
Yates, S.R., and Warrick, A.W. (1987) Estimating soil water content using cokriging. *Soil Science Society of America, Journal* 51: 23–30.
Young, R.A., Onstad, C.A., Bosch, D.D., and Anderson, W.P. (1989) AGNPS: a nonpoint source pollution model for evaluating agricultural watersheds. *Journal of Soil and Water Conservation* 44: 168–173.

20

A Spatial Decision Support System for Modeling and Managing Agricultural Non-Point-Source Pollution

BERNARD A. ENGEL,
RAGHAVAN SRINIVASAN, AND
CHRIS REWERTS

Environmental quality has become an international concern with interest continuing to grow. In the past, emphasis focused on point sources of pollution, but recently non-point-source pollution has been receiving increased attention. The cost of non-point-source pollution is estimated to be in the tens of billions of dollars annually in the U.S. alone (Committee on Conservation Needs and Opportunities, 1986). Identification of non-point-source pollution problem areas is often difficult because of the spatial, distributed nature of the processes involved. Once problem areas are identified, a variety of techniques can be used to minimize the impacts of agricultural and other activities on the environment. These practices are usually site specific, and thus general guidelines are not feasible. The cost and effectiveness of practices also vary significantly from site to site.

Evaluating alternative management practices through experiments at a site is generally not feasible. To identify non-point-source pollution problem areas within watersheds and to evaluate the effectiveness of hypothetical solutions, models and decision support systems are often used. However, the use of many of these tools has been limited because of the time, expertise, and cost in acquiring the model data, running the model, and interpreting model results. Thus the use of simple lumped parameter models such as the Universal Soil Loss Equation (USLE) (Wischmeier and Smith, 1978) to predict erosion due to water runoff is common despite the spatial nature of the erosion and runoff processes. The use of distributed parameter models for erosion modeling, such as the AGNPS (agricultural non-point-source pollution) model (Young et al., 1987), has been limited for these reasons despite the better information provided by these models. The integration of GIS with distributed parameter watershed models can eliminate many of the problems associated with the use of these models alone.

AGRICULTURAL NON-POINT-SOURCE POLLUTION PREDICTION MODELS

Research and modeling efforts involving agricultural non-point-source pollution have been concerned primarily with soil erosion. More recently, research has broadened its scope to include the movement of pesticides and nutrients in addition to soil movement. Numerous models have been developed and used for environmental management in the past 50 years. Most are lumped parameter models such as the USLE; however, distributed parameter models have become more prevalent in recent years. A lumped parameter approach uses some type of averaging technique to approximate characteristics of each parameter needed for computation in the model. Such approximations, regardless of how elaborate, introduce errors into the model, because they cannot account for the spatially varying factors within the watershed boundaries. Because of nonlinearity and threshold values, this can lead to significant error. The magnitude of error stemming from such approximations was demonstrated by Huggins et al. (1973).

The USLE

The USLE (Wischmeier and Smith, 1978) estimates annual sheet and rill erosion as affected by six factors: rainfall erosivity, soil erodibility, slope length, slope steepness, cover and management, and conservation practices. Values for each of the factors are determined by averaging or lumping factor values within the area for which an erosion estimate is desired. For example, if an area has several soil types with different soil erodibility factors, the soil erodibility factor used in the USLE would be an area-weighted average of the soil erodibility factors.

The USLE equation is:

$$A = RKLSCP \tag{20-1}$$

where

A = average soil loss for the time interval represented by factor R $[M L^{-2} T^{-1}]$,
R = combined erosivity of rainfall and runoff $[F T^{-2}]$,
K = soil erodibility factor $[M T L^{-2} F^{-1}]$,
L = slope length factor,
S = slope steepness factor,
C = cover-management factor,
P = supporting practices factor.

Wischmeier and Smith (1978) describe each of these factors and how to obtain their values.

AGNPS

Distributed parameter watershed models such as AGNPS are able to incorporate the influences of the spatially variable controlling parameters (e.g., topography, soils, land use, etc.) in a manner internal to computational algorithms. The primary advantage of a distributed parameter model is its potential for providing a more accurate simulation of the system being modeled. For watershed models, a second advantage of this approach is its ability to simulate conditions at all points within the watershed simultaneously. This allows simulation of processes that change both spatially and temporally throughout the watershed, such as erosion. Finally, this approach allows use of the relationships developed from "plot-size" studies to make predictions on a watershed scale. This can be important, since much of the non-point-source pollution research is conducted on "plot-sized" areas.

The AGNPS model (Young et al., 1987) has been developed to analyze non-point-source pollution in agricultural watersheds. It uses a distributed parameter approach to quantify a watershed by dividing the area into a grid of square cells as shown in Figure 20-1. Within this framework, runoff characteristics and transport processes of sediments and nutrients are simulated for each cell and routed to its outlet. This permits the runoff, erosion, and chemical movement at any point in the watershed to be examined. Thus, it is capable of identifying upland sources contributing to a potential problem and prioritizing those locations where remedial measures could be initiated to improve water quality. Runoff in AGNPS is predicted by applying the Soil Conservation Service (SCS) curve number runoff method to each cell. The curve number method is:

$$\mathrm{RO} = (P - 0.2S)^2 / (P + 0.8S) \tag{20-2}$$

where RO is the runoff volume in millimeters, P is the rainfall in millimeters, and S is a retention parameter in millimeters. The value of S is determined by:

$$S = 25400/\mathrm{CN} - 254 \tag{20-3}$$

where CN is the SCS curve number. Curve number values are calculated using the following information: land use, hydrologic soil groups, hydrologic condition, presence of conservation practices and antecedent soil moisture.

Erosion in AGNPS is predicted by a modified version of the USLE (Young et al., 1987) applied to each cell. The equation used is:

$$A = 2.24\, KLSCP\, \mathrm{EI}_{30}\, \mathrm{SSF} \tag{20-4}$$

where A, K, L, S, C, and P are the factors described in Equation (20-1); EI_{30} is the 30 minute energy intensity $[F T^{-2}]$; and SSF is a factor to adjust for the convex or concave nature of slopes within the cell.

Sediment routing is performed for five particle size classes: clay, silt, small aggregates, sand, and large aggregates. The approximate particle diameters for the upper limits of each size class are 0.002, 0.010, 0.036, 0.20, and 0.51 mm, respectively. Sediment is routed through the watershed as described by Young et al. (1987).

The nutrient movement components of AGNPS are adapted from CREAMS (Frere et al., 1980) and from a feedlot model (Young et al., 1982, 1989). Chemical transport calculations are divided into soluble- and sediment-adsorbed phases. Runoff, erosion, and nutrient movement within cells are routed to the watershed outlet. Because of space limitations and the complexity of the AGNPS model, the reader should consult the references provided for more details concerning the algorithms used within the AGNPS model. AGNPS can be used for watersheds up to 20,000 hectares in size with cell sizes of 0.4 to 16 hectares. In general, large cell sizes have been used because of the problems associated with distributed param-

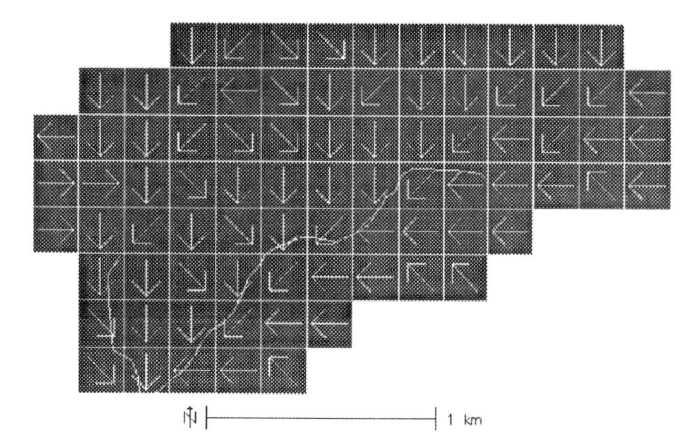

Figure 20-1: Grid cell and flow direction representation of a watershed.

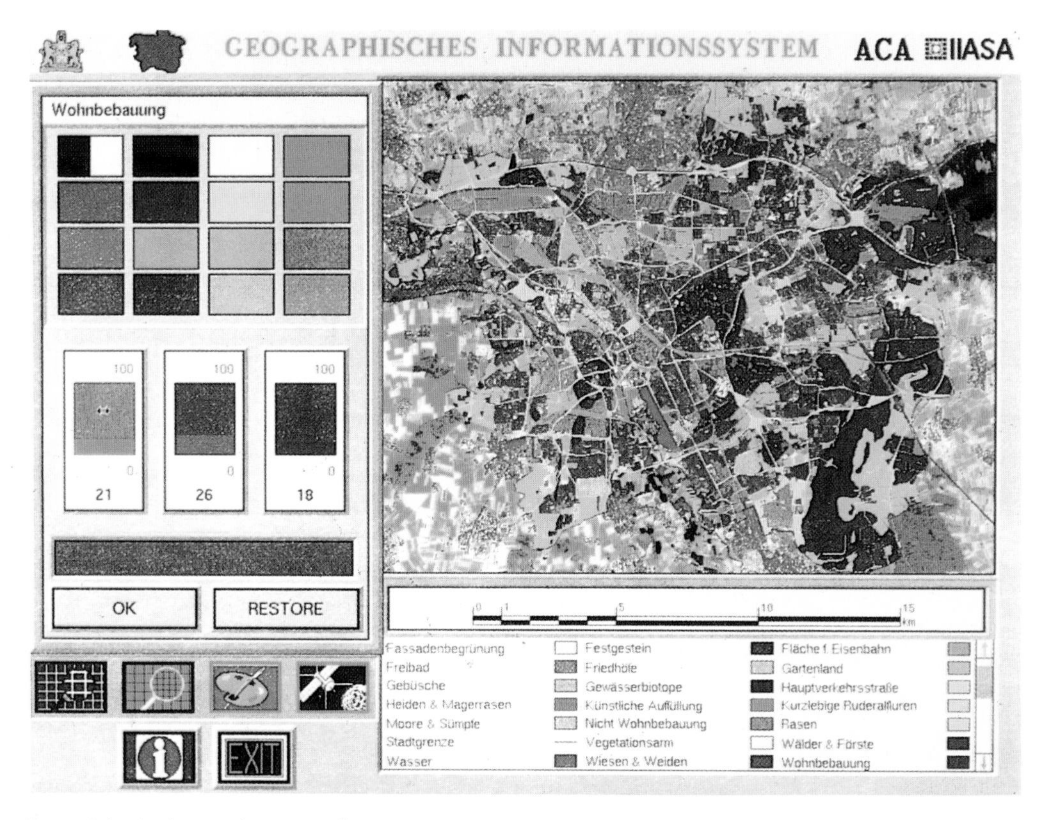

Plate 5-1: An interactive map editor.

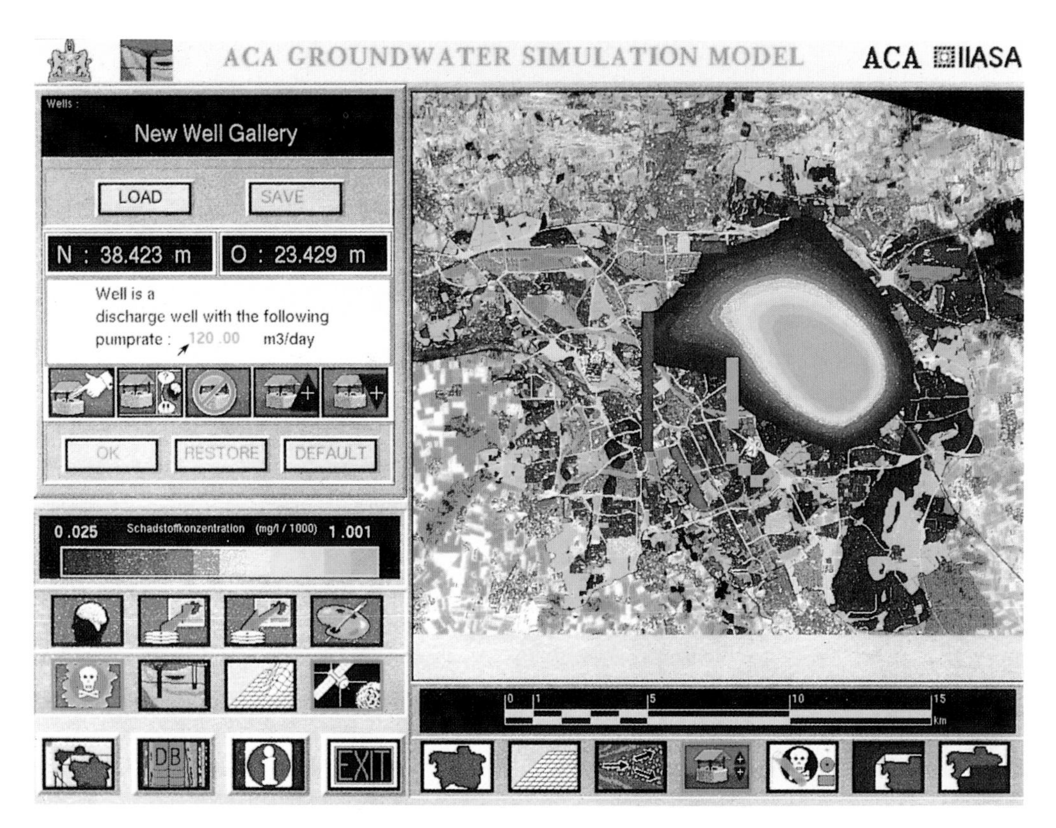

Plate 5-2: Groundwater model interface with integrated GIS functions.

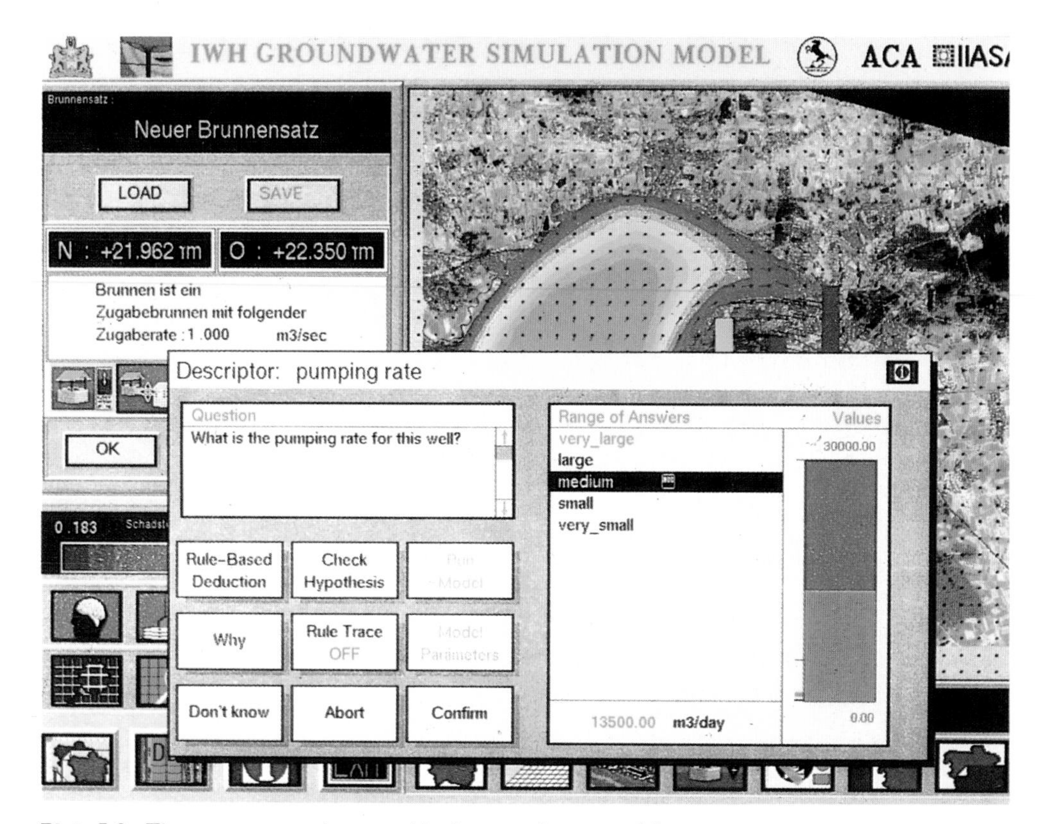

Plate 5-3: The expert system integrated in the groundwater model.

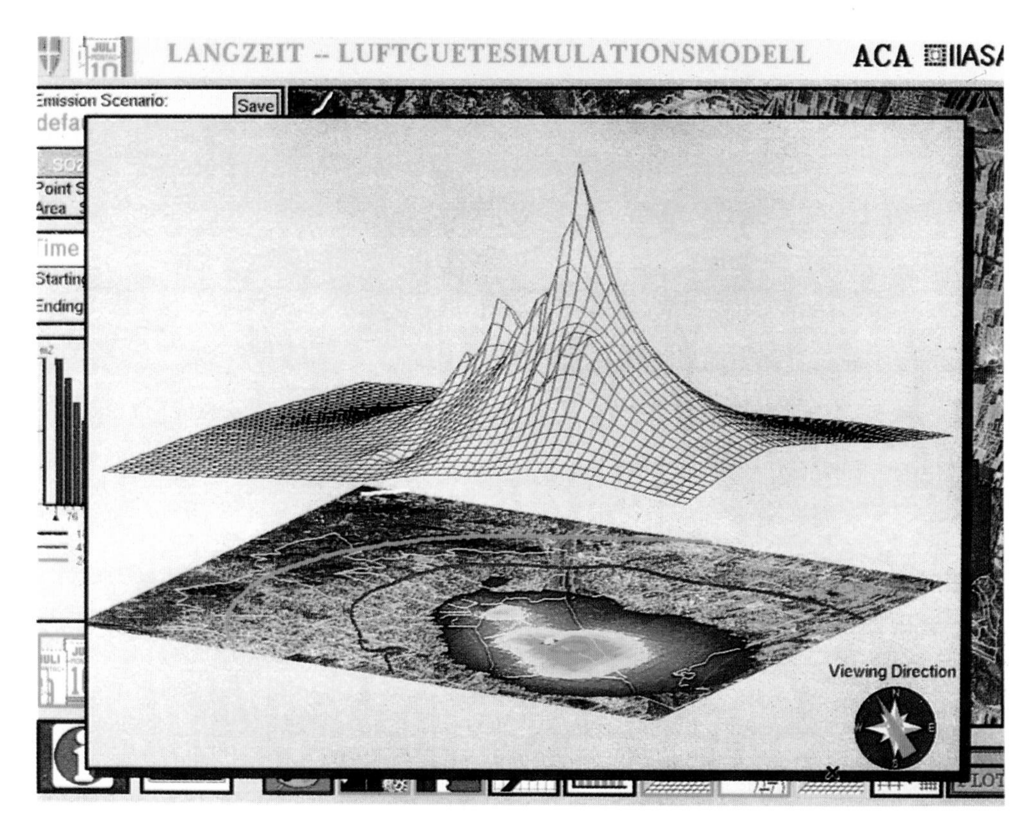

Plate 5-4: Air quality model with integrated GIS.

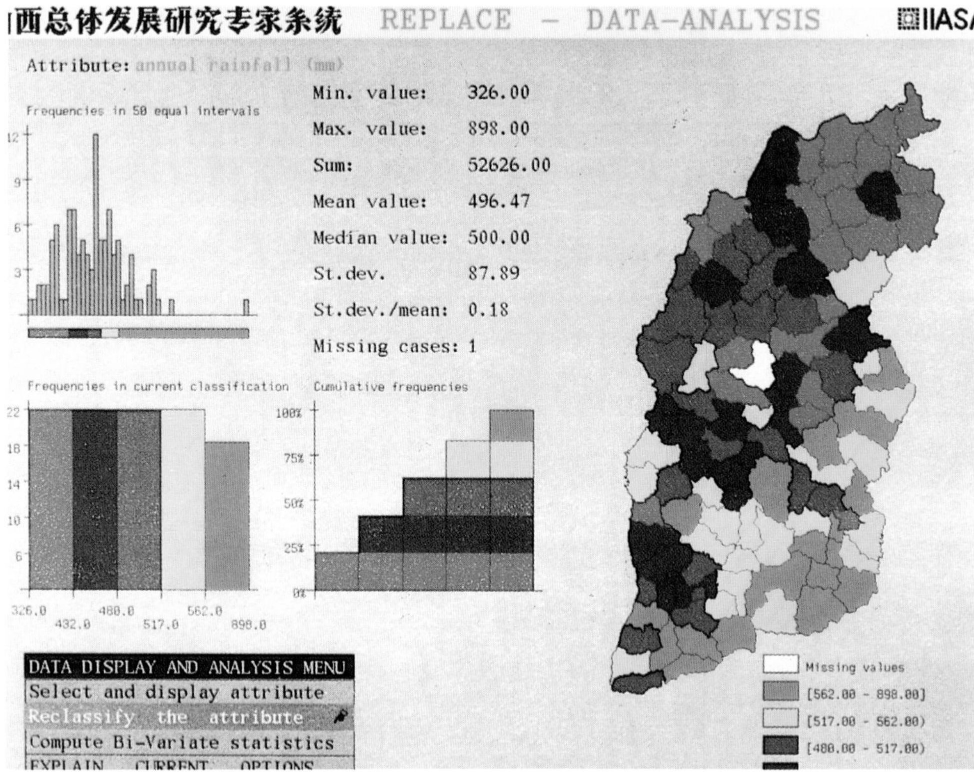

Plate 5-5: REPLACE, a site suitability analysis expert system.

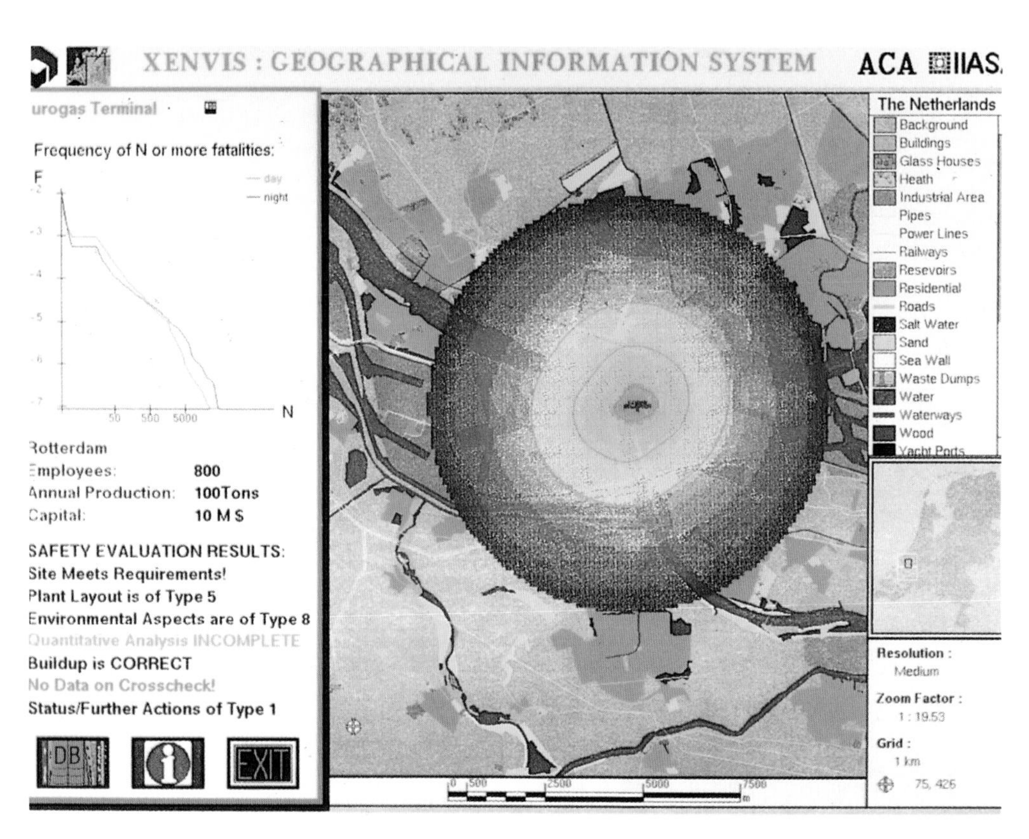

Plate 5-6: Risk contours in XENVIS.

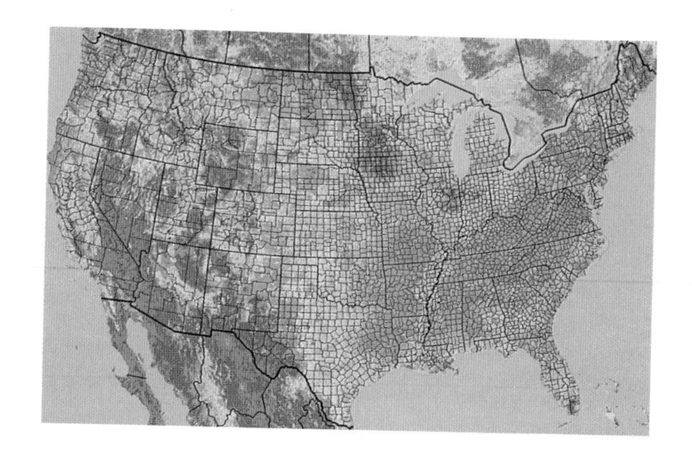

←

Plate 37-1: Conterminous United States composite Advanced Very High Resolution Radiometer normalized difference vegetation index data set for the May 1990 period.

Plate 37-2: Preliminary 1990 vegetation greenness classes derived form unsupervised classification of March October monthly Advanced Very High Resolution Radiometer normalized difference vegetation index composites.

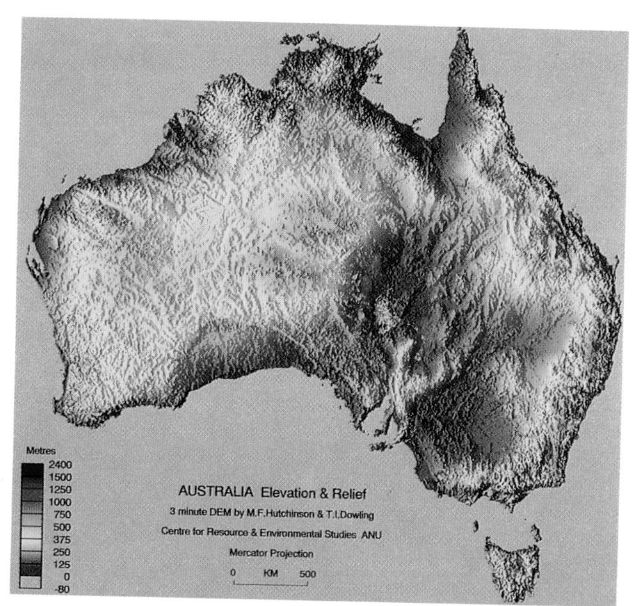

Plate 39-1: Shaded relief view of the 1/20th degree DEM of Australia, Mercator projection.

Plate 39-2: *(below, left and right)* Shaded relief views of the 1/10th degree and 1/40th degree DEMs of Tasmania.

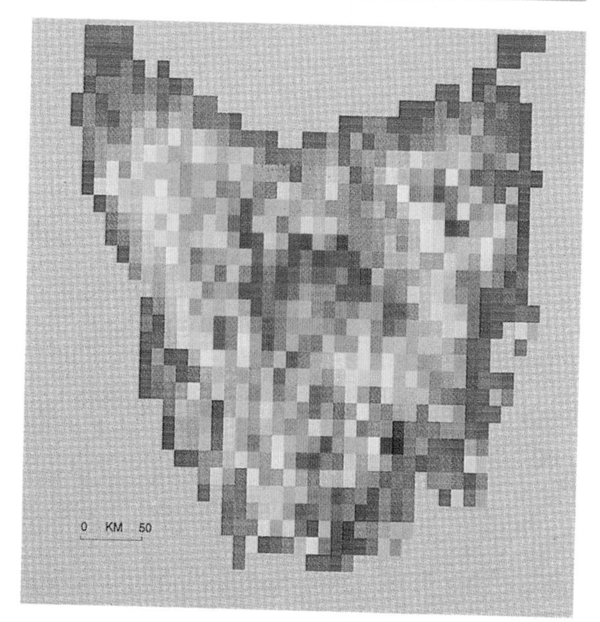

(a)

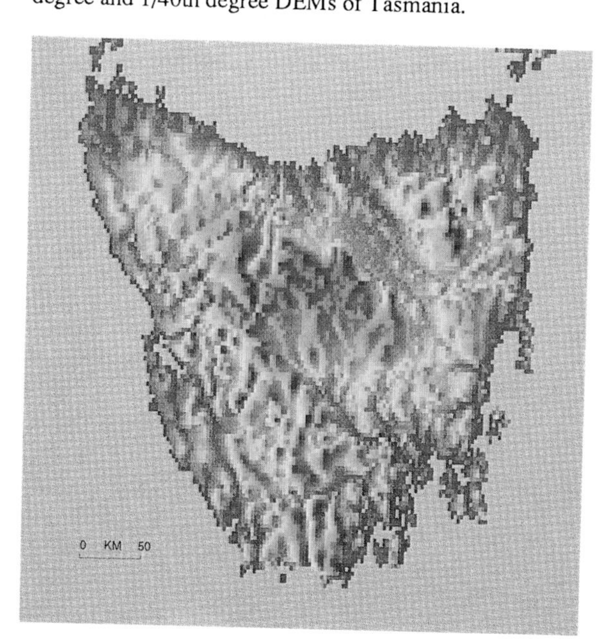

(b)

eter models as described previously. The model has been developed for personal computers under the DOS operating system, and with slight modifications it can also be run under the UNIX operating system. The model has been tested for runoff estimations with data from 20 different watersheds located in the north central United States (Frevert and Crowder, 1987; Setia et al., 1988; Wall et al., 1989). Parts of the model have also been tested for sediment yield estimates with data from experimental watersheds in Minnesota, Iowa, Nebraska, and Mississippi. In addition, the model is being used in several states to prioritize watersheds for potential severity of water quality problems, to pinpoint critical areas within a watershed contributing to pollution, and to evaluate effects of alternative management practices.

The AGNPS inputs required for each cell are shown in Table 20-1, and outputs by cell are shown in Table 20-2. As previously discussed, the problems with models such as AGNPS include the time, expertise, and cost of acquiring the model data, running the model, and interpreting model results. To help overcome these problems, AGNPS was integrated with a raster-based GIS tool.

Table 20-1: AGNPS cell input requirements

Cell number	Overland flow direction
Cell into which it drains	Soil texture
SCS curve number	Fertilization level
Average slope (%)	Fertilizer incorporation
Slope shape factor	Point source indicator
Average slope length	Gully source level
Average channel slope	Chemical oxygen demand factor
Manning's *n* for channel	Impoundment factor
USLE *K* factor	Channel indicator
USLE *C* factor	Channel side slope
USLE *P* factor	
Surface condition constant	

Table 20-2: AGNPS cell output

Runoff volume	Enrichment ratios by particle size
Delivery ratios by particle size	Chemical oxygen demand concentration
Peak runoff rate	Chemical oxygen demand mass
Sediment associated phosphorus mass	Sediment concentration
Sediment yield	Sediment particle size distribution
Soluble phosphorus mass	Soluble nitrogen mass
Upland erosion	Soluble nitrogen concentration
Soluble phosphorus concentration	Sediment associated nitrogen mass
Deposition	
Fraction of runoff generated	
Sediment generated	

INTEGRATING AGNPS WITH A RASTER-BASED GIS

A raster-based GIS tool was chosen since its data storage, retrieval, manipulation, analysis, and display characteristics could be used effectively with the AGNPS model, which uses a raster-based data representation and modeling approach as described in the previous section. The raster-based GIS tool GRASS (Geographical Resources Analysis Support System), a public domain software package developed by the U.S. Army (1987), was chosen for several reasons:

1. The availability of source code for GRASS allows software to be more easily integrated and allows customization;
2. Both AGNPS and GRASS are being used by the USDA Soil Conservation Service (SCS) and other local, state, and Federal organizations that could benefit from this integration; and
3. GRASS is capable of operating on a wide variety of computers.

A toolbox rationale was utilized in providing a collection of GIS programs to assist with the data development and analysis requirements of the model. This allows a modular development approach that offers several benefits. Many of the modules required for the integration of AGNPS with GRASS can be used alone or by other hydrologic, erosion, and chemical movement process models. Modules can be easily modified or replaced without having to rewrite the entire code. Multiple modules that accomplish the same types of operations can be provided, allowing the use of the best module for a particular application. The tools that have been developed for integrating AGNPS with GRASS can be categorized as either input or output tools. These tools were written in the "C" language and thus are directly compatible with existing GRASS functions (also written in "C") and thus are very portable. Although the tools have been developed on Sun workstations, they can be easily installed on other UNIX computers that are capable of running the GRASS GIS software.

Input tools

The input tools assist with preparation and extraction of data from the GIS database for use in the AGNPS model. For example, the watershed program within GRASS is used to delineate the watershed boundaries as well as determine slope percentages, slope lengths, and slope direction (used to indicate the primary direction water will flow through a cell) from elevation maps. Algorithms developed by Jenson and Domingue (1988) were adapted to eliminate problems with flow direction layers developed from elevation layers. Functions were added to GRASS to overlay flow direction arrows on GIS layers graphically, to assist with the visualization of assigned flow direction. A raster map editor was developed to allow individual cell flow direction values to be edited using the graphic display.

A tool to estimate SCS curve numbers for each cell within a watershed was developed (Srinivasan and Engel, 1991a). The tool requires input GIS layers describing land use, hydrologic soil groups, hydrologic condition, presence of conservation practices, and antecedent soil moisture. The SCS curve numbers are used in AGNPS to estimate runoff. Functions were developed to extract soils information from the Soils-5 database (Goran, 1983)—a national database providing hundreds of soil properties for each soil series—and create map layers of extracted data for soil properties of interest. A soil series layer is required as input to these functions. Each of these tools has many other potential applications in watershed modeling and other areas.

Using the tools described and other functions internal to the input tool, the 22 inputs required for each AGNPS cell (Table 20-1) are estimated from six GIS layers: soil series, elevation, land use, management practices, fertilizer nutrient inputs, and land preparation or type of farm machinery used. Once the data layers needed by AGNPS have been provided or derived, the input tool extracts the data from the GIS and builds the AGNPS input file (Srinivasan and Engel, 1991b). This approach was taken to provide additional flexibility and modularity. As either the GIS or the AGNPS model change, these functions will require minimal changes to allow the integrated GIS and model to continue to operate. A tighter integration would require more significant changes each time the GIS or the model changes.

Several inputs are required for the entire watershed including rainfall amount, rainfall energy, and the cell size. These data are input by the user through the input user interface to complete the AGNPS input file. Once the AGNPS input file has been built by the input tools, the model is run.

Output Tool Main Menu
Output Display Options
1. Watershed Summary including sediment (no graphics)
2. Soil Loss (graphics)
3. Nutrient Movement (graphics)
4. Feedlot Analysis (graphics)
5. Runoff Movement (graphics)
6. Analyze Different Scenarios
7. Save Output Maps
8. Exit to GRASS (to come back type 'return')
9. Quit
Enter the choice (1-9):

Figure 20-2: Initial screen of the output tool.

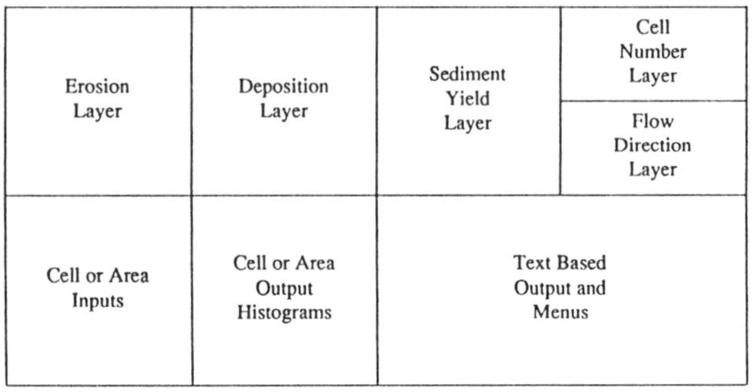

Output tools

After the AGNPS model is run, the output or visualization tool extracts the distributed parameter output data from the ASCII output file and builds GIS layers in GRASS for the parameters shown in Table 20-2 (Srinivasan and Engel, 1991c). The model summary for the watershed outlet is also extracted and placed in a text file for use with the visualization tool. Once data are extracted, the user is given a menu (Figure 20-2) with choices as will be described, to begin the decision-making process using the model results. This menu is located in the Text Based Output and Menus window shown in Figure 20-3.

1. *Watershed summary including sediment (no graphics)*: This option displays summary output data for the watershed in text form. This is the data form that has commonly been used in decision-making in past implementations of AGNPS since display of the spatial variability of the output layers was difficult without the assistance of a GIS.
2. *Soil loss (graphics)*: This option displays the GIS output layers related to erosion (erosion, deposition, and sediment yield) in the window locations as depicted in Figure 20-3. With this information, the user can quickly identify locations within the watershed that have erosion problems. The layout and the information displayed in the visualization portion of the tool have been selected based on interviews and discussions with potential users of the tool and with watershed modeling experts. When GIS output layers are displayed, the user is presented with a new menu, as described in Figure 20-4.
3. *Nutrient movement (graphics)*: This option is similar to

<table>
<tr><td rowspan="2">Erosion Layer</td><td rowspan="2">Deposition Layer</td><td rowspan="2">Sediment Yield Layer</td><td>Cell Number Layer</td></tr>
<tr><td>Flow Direction Layer</td></tr>
<tr><td>Cell or Area Inputs</td><td>Cell or Area Output Histograms</td><td colspan="2">Text Based Output and Menus</td></tr>
</table>

Figure 20-3: Output tool screen layout.

that for displaying erosion but displays results for nitrate, phosphorus, and COD in runoff and attached to sediment.

4. *Feedlot analysis (graphics)*: If point sources of nutrients are present within the watershed, this option displays their contributions to nutrient movement.
5. *Runoff (graphics)*: This option displays GIS layers showing runoff within each cell and runoff from upstream to downstream cells.
6. *Analyze different scenarios*: This option allows results of two watershed simulations to be viewed concurrently. This is useful in comparing various scenarios that have been simulated.
7. *Save output maps*: Since the user of these tools is likely to run many simulations of a watershed, the output GIS layers created are stored temporarily unless saved to the GIS mapset using this option.
8. *Exit to GRASS (to come back type "return")*: This option allows the user to exit to the GIS tool temporarily, and return later to the visualization tool.
9. *Quit:* This option exits the visualization tool.

When the user has selected one of the options (Figure 20-2) described that involves the display of graphical output data, a second menu with new options appears as shown in Figure 20-4. These options are described as follows:

1. *Zoom*: This allows the user to view a particular area of the watershed and is useful for examining potential problem areas.
2. *View a cell*: This option displays the inputs in textual form and outputs in histogram form of any cell in the lower left windows as shown in Figure 20-3. The user can select a cell using the mouse or can specify the cell number.
3. *View area inputs and outputs*: This option is similar to the previous option but displays average AGNPS inputs and outputs for a user-specified area. An area is defined using the mouse.
4. *Toggle between flow direction map/viewing area*: This allows the user to alternate viewing of a GIS layer showing the flow direction and a layer showing the location that has been zoomed to.
5. *Show the watershed summary output*: This option displays the textual output for the watershed outlet.
6. *Display range of output maps*: This option displays only the values of GIS maps that fall within user-specified ranges.
7. *Display user's choice of maps*: This option allows the display of other GIS layers that may be of interest. These map layers are displayed in an existing or a newly created window.
8. *Draw cumulative and frequency distribution statistics*: This option displays cumulative and frequency distribution graphs of the spatial data within GIS layers.
9. *Restore the initial screen*: This option restores the screen layout to that initially displayed.
10. *Exit to GRASS (to come back type "return")*: This option allows the user to exit to the GIS tool temporarily and return later to the visualization tool.
11. *Return to main menu*: This option exits from this section of the visualization tool and returns to the main visualization tool screen (Figure 20-2).

Cell input and output Display Options
1. Zoom
2. View a Cell
3. View area inputs and outputs
4. Toggle between flow direction map/viewing area
5. Show the watershed summary output
6. Display range of output maps
7. Display user's choice of maps
8. Draw cumulative and frequency distribution statistics
9. Restore the initial screen
10. Exit to GRASS (to come back type 'return')
11. Return to main menu
Enter the choice (1-11):

Figure 20-4: Options screen of the output tool for spatially distributed input and output..

Application of the integrated AGNPS/GIS system

The tools described previously have been applied to several watersheds, including the Animal Sciences Watershed located within the Indian Pine Natural Resources

Table 20-3: Simulated results for the Animal Sciences Watershed

	Rainfall (25.4 mm)(EI=17.4)		Rainfall (63.5 mm)(EI=52.0)	
Outputs	Cropped	Fallow	Cropped	Fallow
Erosion (kg)	11,600	29,300	40,900	92,300
N in runoff (kg)	56	19	205	89
P in sediment (kg)	19	41	48	93
Runoff (mm)	2	5	20	28

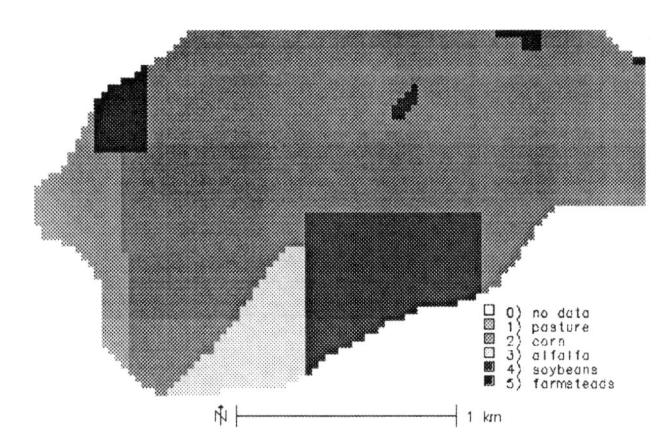

Figure 20-5: Land uses within the Animal Sciences Watershed for 1991.

Field Station near West Lafayette, Indiana. Figure 20-5 shows the land uses within the watershed during 1991. This watershed encompasses 329 hectares, has an average slope of approximately 2.6%, and is characterized by row cropped agricultural land uses on soils that are silty clay loams. This watershed is representative of much of the Midwestern U.S. agricultural region. The watershed is gauged at its outlet. Since the installation of the gauging station, the weather has been abnormally dry, and there have been no events that produced significant runoff from the watershed. Despite the lack of precipitation and associated runoff, the integrated AGNPS/GIS system may still be used to simulate watershed conditions. For example, watershed simulation results for the Animal Sciences Watershed outlet are presented in Table 20-3 for two rainfall events during a year. Various scenarios and rainfall events such as those presented in Table 20-3 are easily simulated using the tools described. Using the visualization tools, problem areas can be quickly identified, analysis focused, and alternative management and land use scenarios examined. Preliminary results suggest that these tools can reduce the decision-making time by orders of magnitude compared to the use of AGNPS without GIS.

SUMMARY

The AGNPS model was integrated with the GRASS GIS tool to develop a decision support tool to assist with management of runoff, erosion, and nutrient movement in agricultural watersheds. The integrated system assists with development of AGNPS input from GIS layers, running the model, and interpretation of the spatially varying results. The system is currently being evaluated for several watersheds in the Indian Pine Natural Resources Field Station located near Purdue University. Preliminary results suggest that the integrated GIS/AGNPS model significantly reduces the time required to obtain the data needed by AGNPS, simplifies operation of AGNPS, but most important, allows the identification of problem areas very quickly. Once problem areas are identified, land use, management, and structural practices can be proposed to reduce the problem and their effectiveness simulated using the decision support tool.

RECOMMENDATIONS

The development of the integrated AGNPS and GRASS system was facilitated by the similar raster data representation used in both of these tools. The availability of source code for both AGNPS and GRASS was vital to their integration and to the development of supporting tools. Work continues on the integrated system to provide additional tools to assist with model input and to provide suggestions based on model results.

Based on the work presented here, there are several limitations that must be overcome to facilitate the integration of environmental models and GIS.

1. Digital spatial databases containing data sets that support environmental analyses are often limited in their geographic coverage and content. Availability of such data sets would increase the interest in integrating GIS and models.
2. Both GIS and models should be modular. In addition, the integrated system should be as modular as possible.
3. Availability of source code and/or libraries for both GIS tools and models is essential for integration such as the one described in this chapter.
4. More research is needed to determine how best to integrate lumped parameter models with GIS.
5. Access to systems analysis tools including expert systems capabilities within GIS would be useful to many model and GIS integration efforts.

REFERENCES

Committee on Conservation Needs and Opportunities (1986) *Soil Conservation: Assessing the National Resource Inventory*, Vol. 1, Washington, DC: National Academy Press, 114 pp.

Frere, M.H., Ross, J.D., and Lane, L.J. (1980) The nutrient submodel. In *CREAMS, A Field Scale Model for Chemicals, Runoff, and Erosion from Agricultural Management Systems*, Washington, DC: U.S. Department of Agriculture, Conservation Research Report 26, pp. 65–85.

Frevert, K., and Crowder, B.M. (1987) Analysis of agricultural nonpoint pollution control options in the St. Albans Bay watershed. *Staff Report No. AGES870423*, Washington, DC: Natural Resource Economics Division, Economic Research Service, U.S. Department of Agriculture.

Goran, W.D. (1983) An interactive soils information system users manual. *Technical Report N-163*, Champaign, IL: U.S. Army Construction Engineering Research Laboratory.

Huggins, L.F., Burney, J.R., Kundu, P.S., and Monke, E.J. (1973) Simulation of the hydrology of ungaged watersheds. *Technical Report 38*, West Lafayette, IN: Water Resources Research Center, Purdue University, 70 pp.

Jenson, S.K., and Domingue, J.O. (1988) Extracting topographic structure from digital elevation data for geographic information system analysis. *Photogrammetric Engineering and Remote Sensing* 54: 1593–1600.

Setia, P.P., Magleby, R.S., and Carvey, D.G. (1988) Illinois Rural Clean Water Project: an economic analysis. *Staff Report No. AGES880617*, Washington, DC: Resources and Technology Division, Economic Research Service, U.S. Department of Agriculture.

Srinivasan, R., and Engel, B.A. (1991a) GIS estimation of runoff using the CN technique. *ASAE Paper No. 91-7044*, St. Joseph, MI: American Society of Agricultural Engineers.

Srinivasan, R., and Engel, B.A. (1991b) A knowledge based approach to extract input data from GIS. *ASAE Paper No. 91-7045*, St. Joseph, MI: American Society of Agricultural Engineers.

Srinivasan, R., and Engel, B.A. (1991c) GIS: a tool for visualization and analysis. *ASAE Paper No. 91-7574*, St. Joseph, MI: American Society of Agricultural Engineers.

U.S. Army (1987) *GRASS-GIS Software and Reference Manual*, Champaign, IL: U.S. Army Corps of Engineers, Construction Engineering Research Laboratory.

Wall, D.B., McGuire, S.A., and Magner, J.A. (1989) *Water Quality Monitoring and Assessment in the Garvin Brook Rural Clean Water Project Area*, St. Paul, MN: Minnesota Pollution Control Agency, Division of Water Quality.

Wischmeier, W.H., and Smith, D.D. (1978) Predicting rainfall losses—a guide to conservation planning. *Agricultural Handbook No. 537*, Washington, DC: U.S. Department of Agriculture, 58 pp.

Young, R.A., Onstad, C.A., Bosch, D.D., and Anderson, W.P. (1987) AGNPS, Agricultural Non-Point-Source Pollution Model: a watershed analysis tool. *Conservation Research Report 35*, Washington, DC: U.S. Department of Agriculture.

Young, R.A., Onstad, C.A., Bosch, D.D., and Anderson, W.P. (1989) AGNPS: a nonpoint-source pollution model for evaluating agricultural watersheds. *Journal of Soil and Water Conservation* 44(2): 168–173.

Young, R.A., Otterby, M.A., and Roos, A. (1982) A technique for evaluating feedlot pollution potential. *Journal of Soil and Water Conservation* 37(1): 21–23.

21

Use of 3D Geographic Information Systems in Hazardous Waste Site Investigations

THOMAS R. FISHER

Three-dimensional geographic information systems (3D GIS), which have their roots in both multidimensional geologic modeling and two-dimensional geographic information systems (2D GIS), are bringing a powerful new set of tools to the realm of environmental modeling, even though they are an immature and emergent branch of the science. In time, 3D GIS may become a discipline separate, but complementary to, the more customary 2D GIS (though many would desire that the two be tightly integrated). The subject of 3D GIS is of increasing interest among geoscientists and has spawned a number of conferences and texts devoted entirely to the subject (e.g., Raper, 1989a; Turner, 1991; Pflug and Harbaugh, 1991). Such systems are causing a revolution in the way geoscientists think and in the way they perceive data (Kelk, 1991). Because most 3D GIS focus on the geosciences and, in particular, subsurface geology, Turner (1991) has suggested that these systems are more properly called 3D geoscientific information systems (GSIS) to distinguish them from their 2D counterparts, which focus more on geography.

Certainly, traditional 2D GIS have been shown to be effective tools for environmental management. Harris et al. (Chapter 15), Atkinson et al. (1990), Morganstein et al. (1990), and Estes et al. (1987), for instance, have shown the value of GIS for data management, mapping, spatial analysis, monitoring, and some forms of modeling aimed at improving management of hazardous waste sites. However, data from hazardous waste sites are typically multidimensional (at least 3D) and have both spatial and time-dependent qualities. Most of the interest is in the subsurface, where the objectives are locating buried materials; determining the presence of contaminant plumes, their three-dimensional distribution, direction, and rate of movement; and characterizing hydrogeologic conditions (Foster et al., 1987). This is where 2D GIS are less suitable because they provide only a limited set of functions to describe the complex "geo-objects" encountered in the subsurface. Additionally, 2D systems stress absolute position over interaction between objects.

3D GIS provide a third dimension by accommodating subsurface data to afford solutions for efficient visualization, modeling, and interpretation of multiple geologic or other attributes in their true (i.e., x, y, z; z being independent of x and y) 3D spatial relationships (Van Driel, 1989). These characteristics make 3D GIS suitable vehicles for modeling geological frameworks for use in studying geologic and hydrologic processes and supplying parameters for sophisticated 3D groundwater flow models.

Three studies involving a range of available approaches to 3D GIS, including an optimized form of "voxels" known as geo-cellular models, three-dimensional gridding and isosurfacing, and complex mathematical functions called nonuniform rational *b* splines (NURBS), illustrate the present capabilities of emerging commercial systems. All of these systems emphasize the geometric aspects of the geology. However, some show promise for use in more dynamic modeling approaches. Affordable, fully functional, three-dimensional GIS products have been slow in coming to the market, initially due to lack of perceived demand and costs (Turner, 1990). However, rising interest, a crossing of the critical performance/cost threshold, and new uses for these products are forcing development of new systems with greater functionality.

Because the field of 3D GIS is new and immature, the functionality that users would like to have in these packages is far from complete. Some areas of importance for future research and development include: improved data management capabilities (a major missing component of most existing 3D GIS, and a reason many still consider these systems simply 3D modeling tools, and not GIS at all); additional functionality for visualizing and navigating a geologically correct model; scaling and visualizing differing levels of detail as scale changes; robust approaches for creation of heterogeneous property distributions; modeling and displaying time-dependent phenomena; "fuzzy-data" handling (e.g., a sandstone body with no true "zero" edge, wherein the sandstone is gradational to another lithology); and finally, integration with appropriate analytic and numerical modeling applications.

THE NEED FOR 3D GIS

Successful interpretation of the hydrogeologic setting of a hazardous waste site (HWS) requires data detailing three-dimensional distribution of the natural surficial soils and manmade fill material, geologic strata and structure, and groundwater conditions both locally and regionally (Foster et al., 1987). These requirements, coupled with an augmentation in sophistication of analytical techniques, a corresponding multiplication of tight standards for cleanup, and an increasing level of detail, has increased the volume and diversity of data to be processed, even for a relatively limited number of boreholes or samples, to the level that computer methods are necessary to analyze all the relationships between variables (Duplessis, 1990).

Manual methods of geological interpretation (e.g., contour maps, cross sections, fence diagrams, and isometric surfaces) are no longer adequate to meet the needs of today's hazardous waste site investigations. Most of the investigator's data are multidimensional and will vary continuously in time and real three-dimensional space. Visualizing combinations or relationships between more than a few variables at a time is nearly impossible using manual 2D methods. Introducing new data into a previous interpretation becomes a laborious and time-consuming effort that may result in rebuilding the entire model (Fisher and Wales, 1991b).

Traditional 2D mapping techniques may also introduce error through the premature averaging of parameters. For instance, in contouring geologic properties or other attributes (e.g., porosity, contaminant distributions), significant averaging can occur (Jones and Leonard, 1990). This may produce a map or model inconsistent with reality, a severe problem when estimating the cost of removing a volume of contaminated soil from a site. Studies by Denver and Phillips (1990) suggest that averaging extreme values during contouring and prior to multiplication of independent variables may yield inaccurate volumes of greater than 20 percent. Further inconsistencies may arise because contouring algorithms do not take into account several variables at once (Jones, 1988). Mapping porosity independently of rock facies and groundwater contaminants independently of porosity and piezometric surfaces may result in putting porosity where there are no porous lithofacies at all, and no contaminants where, really, contaminants are occurring.

The need for more detail in modeling groundwater movements at hazardous waste sites and especially high-level nuclear waste disposal grounds also points up the limitations of both 2D and "2.5"D methods ("2.5"D methods make reference to "apparent" 3D displays, e.g., an isometric projection of a 2D mathematical surface, where the third dimension, z, is dependent upon x and y). Sophisticated 3D groundwater simulation models, which have been demonstrated (Turner, 1989) to be capable of efficiently and accurately calculating hydrodynamic flow

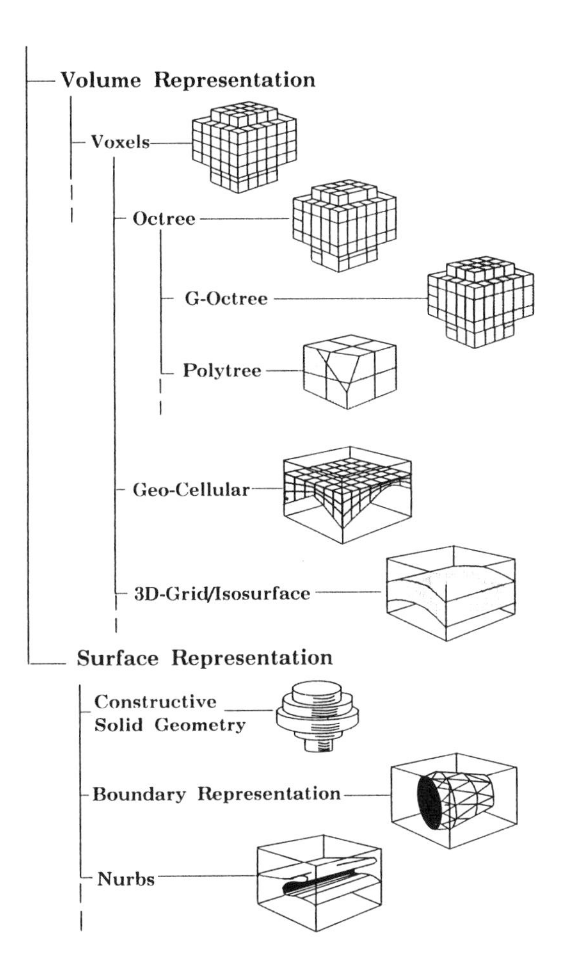

Figure 21-1: Three-dimensional spatial representation methods (modified after Fried and Leonard, 1990).

characteristics of a site are highly sensitive to parameter selection. These models work only if suitably accurate three-dimensional characteristics of the modeled geological materials can be supplied. Traditional calibration methods may fail to identify problems, and 2D methods cannot supply suitable parameters. Using true 3D GIS products appears to be the best hope of solving both spatial visualization and data management problems facing geoscientists involved in hazardous waste site investigations (Turner, 1989).

3D GIS with continuous volumetric data structures and appropriate analytical functions have the advantage in that they do not prematurely average data, and at the same time, they give the geoscientist tools to integrate the mass of data being derived from HWS characterization efforts. Methods are further capable of providing information where an adequate sampling density is too costly and allow conditions to be predicted in areas with no data or where nonintrusive survey techniques must be used (Fisher, 1991). In many ways, these systems have begun to meet requirements outlined by Brian Kelk of the British Geological Survey. These are:

Computer graphics systems which allow interactive creation of spatial models (3D or greater) of the physical nature of parts of the Earth's crust (normally, that portion very near the surface) including the geometry of the major lithologic or stratigraphic units, the variation of their internal composition, their displacement or distortion both in space and time, by faults or other tectonic forces, and the flow of fluids through them.

The ultimate goal of such systems should be to:

- Allow the geoscientist complete freedom of interpretation (i.e., the system imposes no constraints on the users' manipulation and analysis of the data);
- Combine disparate data types (e.g., scalar, vector, raster, text); and
- Allow creation of any model of geologic reality.

"GEO-OBJECTS" AND 3D GIS

As 3D GIS are described here, an understanding of the nature of the geologic phenomena with which we are dealing is in order. Though this may be, to some, a "restatement of the obvious," it is as good to remember that we have rarely "visualized" our data in the manner that 3D modeling allows. Raper (1989b) defined those "distinctive geological features or conditions, located within the subsurface, and having measurable spatial limits or boundaries in three dimensions" as *"geo-objects."* These he further subdivided into sampling-limited and definition-limited objects, which are further described here. The inherent characteristics of geo-objects are the most important factor in the choice of approach to 3D modeling. Geo-objects are, at once, multidimensional, heterogeneous, and spatially and temporally dynamic (i.e., over the span of time, geo-objects may vary in their characteristics, geometry, and location in space). In the process of 3D modeling, an important qualitative difference is seen between the identification of geo-objects that are believed to have a discrete spatial identity (e.g., a perched aquifer, or fault block) and those that vary in identity in space, but that can be visualized by choosing threshold parameter values for inspection (Raper, 1989b).

Objects with discrete spatial identity are classed as the sampling-limited type; the more samples taken, the better defined the object becomes. Secondary or *indicative* data may help to define the object further. The 3D spatial identity of definition-limited objects becomes a matter of searching a population for the boundaries of the defining conditions. We, therefore, have an abstraction of the real world, wherein the geo-object itself is defined by the sampling or selection of parameters established by a data model (Raper, 1989b). This leads to nonunique solutions and the need to handle fuzzy data (because no true "zero" edge exists to define the envelope of the object). Changing the definition (i.e., the defining parameters) changes the object's shape and perhaps characteristics, and this change may not occur in a linear fashion.

CURRENT APPROACHES TO 3D GIS

Current techniques for 3D GIS rely heavily on the power of computer graphics capability and stress geometry over interaction between geo-objects, but do attempt to provide solutions for creation, visualization, and analysis of complex geo-objects, spatial, and entity-data relationships found in the real world (Fisher and Wales, 1991b). Most of these volume modeling or rendering methods rely on a semitransparent depiction of both surface and internal features and have no manual equivalents (Van Driel, 1989). Solid volume modeling techniques range from 3D representations using relatively simple polygon meshes and piecewise linear interpolation (Mallet, 1991), complex 3D gridding and isosurface techniques (Belcher and Paradis, 1991), and voxels (Samet, 1990a,b; Kavouras, 1987), to curved surface and solid modeling based on advanced mathematical functions such as splines (Fisher and Wales, 1991a,b). Most of these methods may be categorized as either volume or surface representations (Figure 21-1) and are elaborated on only briefly here. More detailed discussions may be found in papers by Jones and Leonard (1990) and Fried and Leonard (1990). Appropriate functions and conceptual design of 3D GIS utilizing these techniques are discussed by Raper (1989a,b, 1991).

Volume representations

Many commercial 3D GIS on the market today use some variation of voxels ("volumetric pixel elements") for volumetric rendering. Examples of voxels are shown in Figure 21-1. These are octrees and their variations, polytrees (polyhedral trees), and g- (grey-scale) octrees; geo-cellular models; 3D grids and isosurfaces. Voxels may be thought of as 3D pixels, and, because they bear similarities to 2D raster GIS applications (Fried and Leonard, 1990), they are often called "raster" systems. Voxels are, in fact, the regional quadtree extended to represent three-dimensional binary data. The resulting structure is called a region octree or simply an octree (Samet, 1990a,b).

Octrees use a structured organization of space analogous to quadtree "adaptive gridding," where space is recursively divided, at the edge of the modeled object, into octants down to some required resolution (Fried and Leonard, 1990). But, because the world is not made of tiny cubes, object boundaries thus produced are not smooth. Smoothness can be added by further subdividing the modeled volume into successively smaller elements; however, this does not increase the accuracy of the model. Polytrees allow more complex geometry (edges, vertices, etc.) within the smallest nodes to represent bounding surfaces more accurately, but some roughness remains in the model. A chief advantage of voxels and their variations is that they allow easy distribution of heterogeneous attributes throughout the 3D volume. Their major disadvantage is that they require large amounts of data space and processing.

More complex voxel techniques (e.g., 3D grids and isosurface methods) utilize the value at each of the 3D voxel's eight nodes to describe a dense 3D orthogonal grid. Isosurfaces ("3D contours") are then calculated for the grid. The result is an onion peel model, which may then be graphically manipulated to view the interior of the model. Geo-cellular modeling, still another variation on voxels, permits complexity and discontinuities to be built into the model by use of gridded surfaces to control voxel cell geometry and distribution (Denver and Phillips, 1990).

Surface modeling techniques

Surface modeling methods describe solids through various boundary representation techniques. Some of these are illustrated in the lower half of Figure 21-1. Boundaries may be constructed from simple geometric elements (primitives), facets (e.g., triangular polygons), quadrilateral meshes, or function-based representations (e.g., splines). Constructive solid geometry (CSG) builds complex objects using an aggregation of simpler forms (i.e., primitives: cubes, cylinders, cones, etc.). Simple boundary representation (B-Rep) describes surfaces and solids through use of facets or polygons. The French-built system described by Mallet (1991) is an example, and uses a polygon mesh and piecewise linear interpolation to describe the surface. The disadvantage of this method is that the shape cannot easily be manipulated interactively; a change at one point may affect the entire model.

Functional representations such as nonuniform rational *b* splines (NURBS) find their greatest use in surface representation. This technique describes surfaces using mathematical formulae, instead of a large array of small elements. NURBS and certain other forms of splines can describe all complex surfaces and provide a single uniform and precise mathematical form capable of representing the common analytical shapes, primitive quadrics, free-form curves, and surfaces necessary for geologic modeling (Fisher and Wales, 1991a). It has also been shown in theory by Herring (1991) that splines can be used for heterogeneous volume representations, but, as of yet, no commercial application uses this approach. Transformations and analyses based on this representation are rapid and efficient, however, as a representation it enforces a continuity of curvature between known points; this is an acceptable assumption for many, but not all, geoscientific applications (Raper and Kelk, 1991).

CASE STUDIES

Modeling DNAPL distributions

Radian Corporation undertook to evaluate the extent of contamination of dense nonaqueous phase liquids (DNAPL), for example, perchloroethylenes (PCEs) and trichloroethylenes (TCEs), at an Operable Unit (OU) of a major military base in the southwestern U.S. This area was contaminated by aircraft washing operations, and over the years DNAPLs seeped into soils and eventually groundwater via the highly porous and permeable glacial outwash, lake, and alluvial fan deposits that underlie the site. DNAPL compounds are present as a separate phase within the aquifer system and present an extremely difficult situation to model.

DNAPL may be present in multiple phases. Product vapors may exist in the vadose zone, light product may be floating on the water table, and dissolved product exists in the saturated zone. The fluid tends to sink through the water column until it encounters a zone of lower permeability such as a clay-confining layer or bedrock surface (Preslo and Stoner, 1991). Once the DNAPL is in the saturated zone, it flows with gravity along the bedrock or confining clay layers, enters fractures, or collects within depressions on the confining surface. It continues to disperse by advection into groundwater at the fringes of the concentration. Analyzing this behavior is extremely important when characterizing and cleaning up the subsurface; DNAPL can, under the right circumstances, flow with gravity in a direction opposite that of groundwater. Both monitoring and extraction wells must be designed to accommodate this anomalous occurrence (Sara, 1991).

Stratamodel Inc.'s Stratigraphic Geo-Cellular Modeling (SGM) product was employed to model the subsurface at the OU. The objectives were to define the relationships of porosity and clay-confining layers, the subsurface topography, and the distribution of DNAPLs in relationship to these factors. Information supplied by monitoring wells and sample bores provided the base data set. Other information available included the known regional and local groundwater gradients, rainfall measurements, and a detailed picture of the surface geology, although these were not incorporated into the models discussed. One of the resultant models is shown in Plate 21-1, which displays the 3D model as a series of interlocking "fence diagrams." These were "sliced" from the true 3D model after its generation and show distribution of porosity and well control for the OU.

SGM is a true voxel product and uses discrete cells to model the volume of interest. The method has been optimized to permit the geoscientist considerable control of model properties. The basic geologic model is built up as a stratigraphic sequence that follows the geologic history of the volume being modeled. Imported 2D grids represent the various geologic horizons and discontinuities and are used to place constraints on the interpolation routines. The geoscientist has control of the density of cell layers in the vertical sense within each stratigraphic unit so defined. This permits a degree of control on the heterogeneity of any one layer, based on well bore data. Each cell may be encoded with one or more attributes of interest. The value of a cell is then represented by color. Because any cell may be encoded with multiple attributes, any attribute's distribution may be controlled by the presence, absence, or value of another attribute. This feature allows attribute-dependent interpolation (Denver and Phillips, 1990). This same functionality also permits semicomplex 3D queries of data (e.g., "show areas that have *x* percent or greater of the contaminant, *y* percent porosity and that lie in the saturated zone"). The results of a query may then be visualized by displaying only those cells that meet the specified criteria. A more complete model of porosity with all cells displayed is shown in Plate 21-2.

Lawrence Livermore site remediation investigations

Aircraft washing operations some 40 years ago at a former Naval base on what is now the site of Lawrence Livermore National Laboratory (LLNL) in California permitted perchloroethylenes (PCEs) and trichloroethylenes (TCEs) to enter the local groundwater system via a series of naturally occurring fluvial and alluvial channel deposits meandering throughout the region of the Livermore Valley (Mahoney, 1991). These chemicals are considered carcinogenic, and under EPA regulations the contaminated soils and groundwater must be removed or otherwise remediated to acceptable levels. LLNL geologists Qualhein and Daley (see Qualhein and Daley, 1990) utilized Dynamic Graphics Inc.'s 3D grid and isosurface-based modeling system, IVM, to determine the extent of the contaminant plume, pathways of transport, volume of soils, and groundwater affected and predict future contaminant plume movement. Additionally, the system was used to direct placement of further sampling wells and monitor effectiveness of corrective efforts (Qualhein and Daley, 1990).

One model resulting from Qualhein and Daley's study is shown in Plate 21-3. This model reveals the concentration of PCEs in an approximate one-mile square (1.45 km square) area beneath the former naval base, which is shown by the aerial photo of the area that was scanned into the system, overlaid, and accurately registered with the model. The extent of the plume is shown by graphically stripping away isosurfaces or the "onion peels" to some predefined level of concentration. This model, then, is a 3D representation of a definition-limited geo-object.

Since 1983, over 300 wells have been drilled in order to determine the extent of the PCE plume. Analyses of core samples were used to create the three-dimensional model of the plume shown in Plate 21-3. Dynamic Graphics' IVM product uses the voxel technique of 3D gridding and isosurfacing. A three-dimensional extension of the minimum tension algorithm is used to calculate a true three-dimensional grid from scattered (point) data of a physical property. The grid then represents the modeled distribution of the input property throughout a defined volume. Once the three-dimensional grid is calculated, the user has considerable latitude in how the data are viewed. Intervals may be specified over which isosurfaces are computed within the grid. These surfaces may then be shaded to reveal the internal geometry of the geo-object. Graphic manipulation of the grids permits "slicing and dicing" the model to view its internal structure at various orthogonals on the grid. The user has the ability to use bounding surfaces (imported as 2D grids) to influence the internal geometry of generated models. An important feature of the IVM system is that the 3D grid permits data generated in finite-element flow and dispersion simulations to be imported into the model and simultaneously displayed with the isosurfaces. This is shown by a theoretical simulation for the LLNL site in Plate 21-4. Here, vectors representing flow directions, with color coding indicating velocity of flow, are superimposed on a dispersing contaminant plume.

The Yucca Mountain project

Yucca Mountain is a mesa of faulted, Tertiary, ash-flow tuff, about 70 miles (112 km) northwest of Las Vegas, Nevada, and has been proposed by the U.S. Department of Energy as the nation's first repository for commercial high-level nuclear waste (Borns et al., 1990). The site is in the Basin and Range Structural Province of the western U.S. and is part of the Southwest Nevada Volcanic Field (SWNVF). The work of site characterization, which began in 1989, is expected to take 7 years and demands a one-of-a-kind engineering design. It also presents the geological sciences with an unprecedented challenge: predicting geologic and hydraulic events at the site for 10,000 years to come (Thompson and Frishman, 1989).

Of primary interest are the relationships between groundwater systems, fluid flow, and tectonic features of the region (Fisher and Wales, 1991b). Distribution of rock and engineering properties (e.g., mineralogy, porosity, permeability, water saturation, thermal properties) of the rock slated to contain the waste is also of major importance (Nimick and Williams, 1984; Ortiz et al., 1985). The presence of certain clays (e.g., zeolite) and minerals may inhibit migration of radionucleides, while the presence of water may threaten integrity of the containment site.

Because extensive drilling of test bores could compromise the integrity of the site, noninvasive techniques must be used, as much as possible, to gather geologic information. However, Thompson and Frishman (1989) have pointed up the inadequacies of available geophysical testing technologies for the delineation of subsurface structure and stratigraphy to the degree of detail required. The problem is further compounded by the geologic complexity of the site, which assures that a simplistic idealized groundwater flow pattern is not justified; actual flow parameters within the heterogenous system must be understood. This quickly leads us to what Turner (1989) so aptly called the "parameter crisis" and the inability to provide needed parameters to the sophisticated 3D groundwater flow models. Multidimensional models, such as those described here, are expected to become important tools in determining the suitability of the site and providing the framework for modeling geologic processes operative in the area, as well as predicting geologic events that may compromise the integrity of the site in the future (Fisher and Wales, 1991b).

To begin to create a picture of the complexity of the Yucca Mountain site and supply parameters to the flow equations at the physical sites of interest, two 3D GIS methods were used. 3D grid and isosurface techniques of the previously discussed Dynamic Graphics' IVM product were combined with models created using an experimental prototype NURBS-based system, based on Intergraph's object-oriented EMS product, to delineate the known fault systems and lithology distributions further. The complexity of the situation and the inability, at that time (Dynamic Graphics has since implemented a prototype fault handling capability in its IVM product), of the 3D grid and isosurfacing system to handle discontinuities and faults adequately, and the NURBS system's inability to display heterogeneous attributes within the modeled volume, dictated an integrated approach.

The test data set provided consisted of a 3 arcsecond, 1:250,000 scale digital elevation model (DEM) for the area (approximately 11 by 6 miles or 18 x 10 km), geological and engineering information from 15 wells, and a map of the 31 main faults present in the area of interest showing the fault traces at sea level (0 m) and as projected to an elevation of 6000 ft (1828 m) above MSL. Additional information including interpreted geologic cross sections and maps was obtained from an earlier interpretation of the Yucca Mountain area by Nimick and Williams (1984), Ortiz et al. (1985), and Scott and Bonk (1984).

A structural model of the Yucca Mountain site was built using the NURBS-based system because of its surface handling capability. The NURBS system stores descriptions of surfaces, in this case, faults and topographic surfaces, as mathematical functions (i.e., piecewise polynomials). A thorough and detailed discourse on splines is beyond the scope of this chapter; however, an overview of NURBS may be found in Tiller (1983); for a general discourse on splines, see Bartels et al. (1987). A more detailed description of NURBS and their use in geologic modeling is provided by Fisher and Wales (1991a,b).

The result of the NURBS modeling at Yucca Mountain is shown in Plate 21-5. This model combines the DEM, which has been converted into a spline surface, and the faults represented as vertical spline surfaces. More complexity could be added to this model by joining it with spline surfaces representing the various geologic horizons. Because the NURBS prototype could not generate distributions of attributes, such as lithology, it could not render the volume of the zeolite-bearing zones with respect to the faults and piezometric surfaces, based on scattered point data provided by the wells.

An attempt to model the lithology distribution based on point data and using Dynamic Graphics' IVM product's 3D grid and isosurface capabilities proved more successful, as is shown in Plate 21-6. Shown is a completed lithology model of Yucca Mountain using a multizone approach. Geologic horizons input as gridded surfaces were used to constrain data interpolations. The model shows the combined DEM and the subsurface model, which has been sliced along orthogonals of the 3D grid to display the interior of the block. By graphically "lifting off" successive layers of the model, variations in lithology were revealed. These were sufficient, in some cases, to explain observed anomalies in geochemical samples. The obvious lack of ability to input discontinuities such as faults in the modeled volume hampers its usefulness for supplying parameters to 3D groundwater models. The model would appear considerably different if faulting could have been included in the display and the configuration of each fault block taken into account.

A combination of two models, a structural model generated by the NURBS system, and a 3D lithology model produced by 3D grid and isosurfacing techniques, revealed much about the Yucca Mountain site but left many unanswered questions. Had these systems been available as one truly integrated package, much more could have been accomplished, such as gaining an understanding of the interaction and contribution of faults, rock porosity, and permeability on control of regional and local groundwater flow patterns.

FUTURE DIRECTIONS FOR RESEARCH AND DEVELOPMENT

Certainly, there is already a large measure of functionality available in 3D GIS that cannot be ignored and should be useful at many levels of practical application for characterizing and modeling hazardous waste sites. The potential value of 3D GIS in the modeling and managing of hazardous waste sites and especially for supplying parameters for sophisticated 3D numerical models such as groundwater flow and contaminant transport is obvious. However, the field of 3D GIS is immature, and its full potential is yet to be realized. Therefore, the list of required and desirable new functionality for 3D is quite long, and much research and development will be necessary before it is realized. Space does not permit addressing this topic to the extent desired; however, the following gives a relative idea of the task ahead. The reader is also referred to works by Turner (1991), Pflug and Harbaugh (1991), Raper (1989a), and Raper and Kelk (1991) for additional views on the topic.

Data management

Data management must be addressed by existing and future 3D GIS. Currently, no 3D GIS, to this author's knowledge, provides database management capability to any extent. This is a critical missing component. Systems must provide capabilities for easy entry of many disparate data types, as is available in more robust 2D counterparts. This must include scalar, vector, raster, and textual data. Tools must also be developed for data encoding, validation, structuring, restructuring, and transformation. Capabilities must also be provided for creating and altering multiple data models, as well as functions for data tracking and audit trails, to record, as the model is developed, all changes and manipulations of the data.

Visualization

User interface and visual interactivity with the modeling process is a must. This should be done from the user's viewpoint and handled in the user's terminology. More tools for easy navigation of the model as it is constructed and as completed must be developed. Perhaps answers lie in virtual reality systems. Visualization tools must provide more than a capability to simply "render an artistic concept." The system must be capable of producing a geologically correct model without the introduction of graphic artifacts, or the introduction of steps (e.g., in dynamic modeling) that are mechanically or geologically impossible. This will necessarily touch upon the underlying modeling assumptions and data interpolation techniques. The ability to visualize more than one or two variables at once within the same model is highly desirable. This will, no doubt, provide a challenge for researchers for some time to come.

Model validation

Model validation could perhaps be termed the single most critical area of 3D GIS. Many systems employ accepted and "mathematically proved" interpolation or geostatistical techniques. However, in untrained hands, the models produced, while colorful, neat, and plausible, may be absolutely wrong; nothing near physical reality; or reflect physical processes acting in an area. Too often modelers begin with a numerical representation of a physical process and then attempt to force reality to fit the model rather than fit the model to the domain it represents. They fail to begin with "what's there" and work toward a model. 3D GIS must be proved by actual field investigations and measurements. Now that we can see our data in supposedly true spatial relationship, we must ask ourselves the hard question: "Are we ready to accept the model produced by the 3D GIS as a true representation of reality?"

Distribution of attributes

Most of the 3D GIS available today can manipulate, at most, only a few variables at a time; few provide attribute-dependent interpolation. None provides the ability to control distribution of attributes according to some physical process nor take into account the many variables interacting in that process. This is an area critical to both development of realistic geologic framework models and the sophisticated numerical simulators to which the models supply parameters. Additionally, research needs to be done on fuzzy data handling. When dealing with definition-limited geo-objects or even sampling-limited objects, we rarely have a sharp contact or "zero" edge that defines the limits of the object. Instead, we have a gradational change from one lithology to another, or one fluid property to another. This functionally is a must to deal with multiphase flow models and processes acting at contaminated sites.

Spatial relationships

Modeling 3D spatial relationships and creating and maintaining that topology is an area for future research and one possible key to integrating existing 2D GIS with 3D GIS. A truly useful and powerful system would be one that would allow the user to formulate a query about some surface facility and simultaneously recover 3D graphics or text information about the subsurface beneath or near that facility. Additionally, a 3D GIS must properly portray and reference all geo-objects with spatial accuracy. The user should, for instance, be able to visualize spatial relationships readily, for example, a sandstone body and the enclosing shales, along with the body's shape, geometry, and internal properties and obtain a correct read-out of their spatial coordinates.

Internal structures and trends

One of the main purposes of a 3D system is to enhance visualization and analysis of the internal structure of geologic features beyond that of traditional 2D tools. A 3D GIS system should, in conjunction with distribution of attributes, display and model trends, the character of these trends or properties, and the surfaces interrupting and bounding certain properties within the larger geo-object. It follows that all subcomponents be capable of being assigned characteristics and properties of their own (e.g., a fault remains impermeable to fluid movement until fluid pressures in the surrounding carrier beds reach a level exceeding certain capillary pressures).

Scaling and hierarchies

With development of the capabilities to model trends and structures internal to a geo-object, the problem of scaling and hierarchies comes into play. A 3D GIS must be capable of working at multiple scales, that is, differing levels of detail within the same model. Functionality needs to be developed that will permit producing a model at the micro scale, but allow one to "zoom" out by filtering data for a regional or field scale view, yet maintaining correctness and accuracy of the model. In other words, we wish to produce detail at all scales.

Time-dependent data

Little research has been focused on handling time-dependent data in 3D GIS, yet it is again an obvious area needing to be addressed if 3D GIS is to move from purely static modeling to dynamic modeling. Data structures for both 3D and 4D situations must be developed. These must support true time-dependent phenomena (e.g., fluid movement through rock, salt movements, volume changes, folding, and thrusting). A great leap forward would be to provide the capability to: (1) model and display fluid movements in three and four dimensions through a "static" model; and (2) model and display volume and geometric changes in rock bodies in three and four dimensions through a continuum of time (versus stepwise "snapshots" in time).

CONCLUSIONS

Traditional two-dimensional geographic information systems are well suited for environmental management applications involving mainly surface information, administrative, and analytic data. However, complex subsurface geologic data derived from hazardous waste sites cannot be adequately manipulated, modeled, and visualized using the limited functions provided by traditional 2D systems. 3D GIS are necessary to accommodate subsurface data, provide continuous data structures and appropriate functionality to integrate data derived from site investigations, and afford solutions for efficient visualization, modeling, and interpretation of multiple geologic or other attributes. The state of this technology is still immature and must be integrated with its 2D counterpart and appropriate numerical, analytic modeling, and simulation applications in order to provide scientists with a robust analytical capability. The current technology does, however, provide a start towards a vehicle for development of geologic frameworks convenient for the study of geologic and hydrologic processes. And, while useful, much research and development remains before the full potential of 3D GIS is realized.

ACKNOWLEDGMENTS

I wish to thank my employer, Radian Corporation of Austin, Texas, for supporting development of this paper. Acknowledgment and thanks also go to Stratamodel Inc., Houston, Texas; Dynamic Graphics Inc., Alameda, California; and Intergraph Corporation, Huntsville, Alabama, for their help in preparing models and providing graphics discussed here. I also thank A.K. Turner and K.E. Kolm of the Institute of Ground Water Research and Education, Colorado School of Mines for their continued encouragement of my efforts in 3D modeling and GIS.

DISCLAIMER

Stratamodel and Stratamodel SGM are registered trademarks of Stratamodel Inc., Houston, Texas; IVM is a trademark of Dynamic Graphics Inc., Alameda, California; I/EMS is a registered trademark of Intergraph Corporation, Huntsville, Alabama. Mention herein of certain software products and their manufacturer is for example only and is not an endorsement by Radian Corporation of any one product or manufacturer, nor of the suitability of any one product for a specific task.

REFERENCES

Atkinson, S.F., Hunter, B.A., and Marr, P.G. (1990) Spatial analysis of ground water pollution. *Conference Proceedings, NCGA GIS '90, Houston*, pp. 30–38.

Bartels, R.H., Beatty, J.C., and Barsky, A. (1987) *An Introduction to Splines for Use in Computer Graphics and Geometric Modeling*, Los Altos, CA: Morgan Kaufmann.

Belcher, R.C., and Paradis, A.R. (1991) A mapping approach to three-dimensional modeling. In Turner, A.K. (ed.) *Three-Dimensional Modeling With Geo-scientific Information Systems*, NATO ASI Series C: Mathematical and Physical Sciences, Vol. 354, Dordrecht: Kluwer Academic Publishers.

Borns, D.J., Sass, J.H., and Schweikert, R.A. (1990) Proposed study of the Basin and Range from Death Valley to Yucca Flat. *EOS Transactions of the AGU* 71(31): 1012–1013.

Denver, L.F., and Phillips, D.C. (1990) Stratigraphic geocellular modeling. *Geobyte* 5(1): 45–47.

Duplessis, C. (1990) Contaminated site assessment: the microcomputer edge. *Hazardous Materials Management Magazine*, January/February: 28.

Estes, J.E., McGwire, K.C., Fletcher, G.A., and Foresman, T.W. (1987) Coordinating hazardous waste management activities using geographical information systems. *International Journal of Geographical Information Systems* 1(4): 359–377.

Fisher, T.R. (1991) Concepts and approaches in multi-dimensional geologic modeling. *Die Geowissenschaft*, Essen.

Fisher, T.R., and Wales, R.Q. (1991a) Three-dimensional solid modeling of geo-objects using non-uniform rational B-splines. In Turner, A.K. (ed.) *Three-Dimensional Modeling With Geo-Scientific Information Systems*, NATO ASI Series C: Mathematical and Physical Sciences, Vol. 354, Dordrecht: Kluwer Academic Publishers.

Fisher, T.R., and Wales, R.Q. (1991b) Rational splines and multi-dimensional geologic modeling. In Pflug, R., and Harbaugh, J.W. (eds.) *Three Dimensional Computer Graphics in Modeling Geologic Structures and Simulating Processes*, Lecture Notes in Earth Sciences, Vol. 41, Heidelberg: Springer-Verlag.

Foster, A.R., Veatch, M.D., and Baird, S.L. (1987) Hazardous waste geophysics. *The Leading Edge* 6(8): 8–13.

Fried, C.C., and Leonard, J.E. (1990) Petroleum 3-D models come in many flavors. *Geobyte* 5(1): 27–30.

Herring, J.R. (1991) Using spline functions to represent distributed attributes. *Technical Papers Auto-Carto 10*, Bethesda, MD: ACSM/ASPRS, Vol. 6, pp. 46–58.

Jones, T.A. (1988) Modeling geology in three dimensions. *Geobyte* 3(1): 14–20.

Jones, T.A., and Leonard, J.E. (1990) Why 3-D modeling? *Geobyte* 5(1): 25–26.

Kavouras, M. (1987) A spatial information system for the geosciences. *PhD Dissertation*, Fredericton: University of New Brunswick.

Kelk, B. (1991) Personal communication.

Mahoney, D.P. (1991) Mapping toxic spills. *Computer Graphics World* 14(1): 89–90.

Mallet, J-L. (1991) gOcad: a computer aided design program for geological applications. In Turner, A.K. (ed.) *Three Dimensional Applications With Geo-Scientific Information Systems*, NATO ASI Series C: Mathematical and Physical Sciences, Vol. 354, Dordrecht: Kluwer Academic Publishers.

Morganstein, K.A., Witt, P.V., and Flood, D. (1990) Active site discovery using a geographic information system. *Proceedings 11th National Conference, HMCRI, Superfund*, pp. 35–41.

Nimick, F.B., and Williams, R.L. (1984) A three-dimensional geologic model of Yucca Mountain, Southern Nevada. *SAND83-2593*, Sandia National Laboratories, pp. 1–68.

Ortiz, T.S., Williams, R.L., Nimick, F.B., Whittet, B.C., and South, D.L. (1985) A three-dimensional model of reference thermal/mechanical and hydrologic stratigraphy at Yucca Mountain, Southern Nevada. *SAND84-1076*, Nevada Nuclear Waste Storage Investigations Project, Sandia National Laboratories, pp. 1–72.

Pflug, R.D., and Harbaugh, J.W. (eds.) (1991) *Three Dimensional Computer Graphics in Modeling Geologic Structures and Simulating Processes*, Lecture Notes in Earth Sciences, Vol. 41, Heidelberg: Springer-Verlag.

Preslo, L.M., and Stoner, D.W. (1991) The overall philosophy and purpose of site investigations. In Neilsen, D.M. (ed.) *Practical Handbook of Ground-Water Monitoring*, Chelsea, MI: Lewis Publishers, pp. 69–95.

Qualhein, B.J., and Daley, P.F. (1990) Environmental restoration: dynamic planning and innovation. *Report UCRL-52000-90-4*, Lawrence Livermore National Laboratory, pp. 1–25.

Raper, J.F. (ed.) (1989a) *Three Dimensional Applications in Geographic Information Systems*, London: Taylor and Francis.

Raper, J.F. (1989b) The 3-dimensional geo-scientific mapping and modeling system: a conceptual design. In Raper, J.F. (ed.) *Three Dimensional Applications in Geographic Information Systems*, London: Taylor and Francis.

Raper, J.F., and Kelk, B. (1991) Three dimensional geographical information systems. In Maguire, D.J., Goodchild, M.F., and Rhind, D.W. (eds.) *Geographical Information Systems: Principles and Applications*, Harlow: Longman, pp. 299–317.

Samet, H. (1990a) *The Design and Analysis of Spatial Data Structures*, New York: Addison-Wesley.

Samet, H. (1990b) *Applications of Spatial Data Structures*, New York: Addison-Wesley.

Sara, M.N. (1991) Ground-water system design. In Neilsen, D.M. (ed.) *Practical Handbook of Ground-Water Monitoring*, Chelsea, MI: Lewis Publishers, pp. 17–68.

Scott, R.B., and Bonk, J. (1984) Preliminary geologic map of Yucca Mountain, Nye County, Nevada with geologic sections. *Open File Report 84-494*, Washington, DC: U.S. Geological Survey.

Thompson, J.F., and Frishman, S. (1989) The challenge of Yucca Mountain. *Civil Engineering* 59(4): 44–46.

Tiller, W. (1983) Rational B-splines for curve and surface representation. *IEEE Computer Graphics and Applications* 3(6): 61–69.

Turner, A.K. (1989) The role of three-dimensional geographic information systems in subsurface characterization for hydrogeological applications. In Raper, J.F. (ed.) *Three Dimensional Applications in Geographic In-*

formation Systems, London: Taylor and Francis, pp. 115–127.

Turner, A.K. (1990) Three dimensional GIS. *Geobyte* 5(1): 31–32.

Turner, A.K. (ed.) (1991) *Three Dimensional Modeling With Geoscientific Information Systems*, Nato ASI Series C: Mathematical and Physical Sciences, Vol. 354, Dordrecht: Kluwer Academic Publishers.

Van Driel, J.N. (1989) Three dimensional display of geological data. In Raper, J.F. (ed.) *Three-Dimensional Applications in Geographical Information Systems*, London: Taylor and Francis, pp. 1–9.

22

Spatial Models of Ecological Systems and Processes: The Role of GIS

CAROLYN T. HUNSAKER,
ROBERT A. NISBET,
DAVID C. L. LAM,
JOAN A. BROWDER,
WILLIAM L. BAKER,
MONICA G. TURNER, AND
DANIEL B. BOTKIN

This overview provides a foundation for discussion of methods and techniques to integrate spatial ecological models with the technology of geographic information systems (GIS). The discussion of models is limited to those that simulate ecological dynamics in two-dimensional (horizontal or vertical) or three-dimensional space. We briefly review historical modeling approaches for terrestrial, freshwater, and marine ecosystems that have some spatial component or can be extended into space. Examples of current spatial models are discussed for these ecosystems, and the use of distributional mosaic models for the simulation of broad-scale landscape disturbances such as fire and pests is examined. We exclude models of organisms in patches, animal dispersion models, and insect- and disease-dispersion models that consider space without location (coordinates) or configuration of subareas. This distinction is important for the application of GIS techniques to spatial ecological models (Baker, 1989). Future needs and directions that will enable ecological models to capitalize on the capabilities of GIS are discussed with regard to database management; remotely sensed data; landscape-level models; model-base management systems; data and model sensitivity, errors, and uncertainties; and environmental risk assessment.

Ecological processes operate at a variety of scales in space and time. Historically ecologists focused on changes in time and single sites or small geographic areas; however, during the past two decades ecological modelers have worked on incorporating spatial pattern into models and applying models to large geographic areas (Okubo, 1975; Levin, 1976). Currently much of this work is being championed by the new discipline of landscape ecology, which tends to combine work from many disciplines including geography, physics, biology, and ecology. A better basic understanding of the spatial aspects of ecosystem processes is needed. As Hunsaker et al. (1990) point out, effective long-term management and protection of valuable natural resources require a better understanding of how the scale of the environmental hazard affects ecological processes and over what scales the effects should be monitored and examined.

Not all ecological topics require models that are explicitly spatial; progress has been made, as we will review, with models that have no explicit spatial interactions. In deciding whether spatial aspects are necessary in a model, one must consider two conceptual questions: (1) What are the spatial units and their interactions, and (2) how do the relative sizes and locations of these spatial units affect such state variables as biomass and population size and such ecosystem functions as energy flow, biological productivity, and population change?

Several issues are involved. First, what are the "natural" units of a landscape? Sometimes the answer is obvious to the human observer, but other times the spatial units can be determined only after detailed analyses of plant and animal distributions. Sometimes there may be no correct single unit since units differ for different organisms. Once spatial units are defined, we next must consider how the state of one type of unit can affect the future state of another. The null hypothesis is that the future state of a landscape unit is independent of adjacent units. When substantive spatial interactions occur, this is not true.

We know of cases in which there are spatial effects of one unit on another. Sprugel (1985), for example, showed that upwind trees protect other trees from wind and frost damage. It follows that the removal of an exposed, upwind forest plot can alter the future condition of a previously interior plot. Woodby (1991) found that future forest composition was affected by seed tree availability in neighboring locations. Such results suggest that spatial models are needed to address questions related to forest growth and regeneration. However, where we are confident that there are no substantive spatial influences, spatial components to models may not be necessary.

The development of models with spatial processes is one way to examine landscape effects. Another is to analyze existing landscape patterns. A third is to develop an understanding of processes, such as seed dispersal and wind effects on forests, that create spatial patterns. Although most of this chapter emphasizes methods and approaches for the integration of ecological models with GIS, questions about spatial units and their interactions underlie the discussion.

HISTORICAL MODELING APPROACHES

In the past, ecological modeling has had a temporal emphasis, that is, projections of state changes at one location over time. Different approaches were necessary for terrestrial, aquatic, and marine models. Terrestrial plant models did not consider organisms moving over the ground, whereas animal models did. Aquatic models had to consider that organisms were embedded in a medium (water) that moved; only rarely was movement of the terrestrial medium (the soil) an important factor in terrestrial models (e.g., models of organisms on eroding or disintegrating landscapes). Also considered in aquatic models was the notion that animals and plants in oceans and lakes are always immersed in some form of chemical solution. As a result, aquatic models often incorporated chemical and toxicity submodels. It was necessary in aquatic models to integrate time and spatial processes (two dimensional and three dimensional) for as many as ten species or more and hundreds of model grid cells. Many aquatic models are linked to water circulation models. These models are good examples of the linkage of process models and GIS.

The simplicity of the basic relationships represented in models often leads to incomplete perceptions of nature. Most models are based on a series of differential or difference equations, Markov chains, or Leslie matrices, without spatial considerations and with explicit steady-state assumptions. Early ecologists viewed spatial heterogeneity as a "problem," whereas more recently heterogeneity is viewed as an important and tractable source of spatial variation in biophysical processes (Kolasa and Pickett, 1991; Levin, 1974). Also recently, some ecologists have questioned this steady-state or "balance of nature" concept; they claim that nature is better characterized by dynamic change with large fluctuations in magnitude of state variables (Egerton, 1973; Botkin and Sobel, 1975; Botkin, 1990). In order to set the stage for the discussion of the role of GIS in spatial ecological modeling, we briefly discuss the historical background of ecological modeling, but the reader should understand that we cannot possibly do the topic justice in a few pages. A tremendous amount of ecological modeling has taken place since the late 1960s, and to aid the reader we refer to published reviews of models. Spatial statistics is not reviewed here, but the reader is referred to Patil et al. (1971), Cliff and Ord (1973), Pielou (1977), Sokal (1979), Turner et al. (1989a), S.J. Turner et al. (1991), Cressie (1991) and the spatial statistics section of this volume.

Terrestrial plant models

There are a large variety of highly developed nonspatial models of grasslands, shrublands, and forests (e.g., Lauenroth et al., 1983), as will be covered in the following sections:

Dynamical systems models

"Dynamical systems models" refer here to those models that use both discrete and continuous mathematics. Solutions to a partial differential equation in continuous space have been used to model changes in age–size distribution of patches of mussel beds in Washington (Paine and Levin, 1981). This approach might be scaled up to the landscape level and incorporated with GIS. However, it may be more practical to determine states of landscape subareas using discrete mathematics.

Difference equation models

All difference equation landscape models use discrete state spaces (Baker, 1989). The most common forms of difference equation models use stochastic Markov models or the deterministic Leslie matrix models. In the Leslie matrix model, the matrix contains values that are rates of change, not probabilities. This approach uses explicit birth functions for modeling biological populations, and modifications have been used to model populations with nonstationary transition probabilities and spatial effects (see Rees and Wilson, 1977). One forest succession model was linked with a highly modified Leslie matrix to predict moose population changes in relation to changes in forest composition and structure through time (Botkin and Levitan, 1977).

The Markov chain models calculate changes in state by matrix multiplication, where the column vector representing the population's age classes is multiplied by a transition matrix containing birth and death change probabilities. Markov chains have been used to model succession (Horn, 1975) and vegetation-type changes on small plots (Lough et al., 1987), and on large areas (Lippe et al., 1985). However, the assumption in Markov chains that the transition probabilities are stationary (do not change over time) is questionable. Markov chains cannot easily accommodate higher-order effects (interactions in time and space) and have difficulty in accounting for spatial heterogeneity (Baker, 1989). Transition probabilities between ecological states can be estimated from time-series remotely sensed data, and these estimates can be used in Markov models. For example, Hall et al. (1991) examined disturbance frequencies in wilderness and managed areas in northern Minnesota boreal forests in this manner. The steady-state solution of the model (e.g., percentage of the landscape in different states) was determined by the frequency of disturbance.

Some of the more complex ecosystem models were developed by teams of scientists during the 1960s and 1970s as part of the International Biological Program. The grassland ecosystem model, ELM, for example, has submodels for water, temperature, nutrients, producers, plant phenology, decomposers, insect consumers, and mammalian consumers that change on daily time steps through the growing season (Innis, 1978). The model has been used to analyze the effects of variation in weather, herbivory, and nutrients at several grassland sites. Grassland succession models, in contrast, typically have one trophic level and biomass changing at seasonal or yearly time steps (Bledsoe and Van Dyne, 1971). Forest photosynthesis models usually consider only one trophic level with model flows of carbon during daily time steps (Sollins et al., 1974), but more complex forest ecosystem models have also been developed (Swartzman, 1979). Forest succession models usually consider only one trophic level and population levels changing at yearly time steps (Botkin et al., 1970; Horn, 1975).

A group of stochastic difference equation models (the JABOWA family of models) has been used to model forest growth in many parts of the world (Botkin et al., 1972; Shugart and West, 1977; Prentice, 1986). The JABOWA-type forest growth models use difference equations to model diameter increment of individual trees (the states). The models use explicit, stochastic birth and death functions. Death (mortality), for example, is driven by a random function, affected by tree age and stress. Competition among trees for light on a small plot (1/100 to 1/10 ha) is modeled as a function of total leaf weight (proportional to leaf area index). Several simplifying assumptions eliminate the need for large amounts of data on growth changes over time and environmental conditions in the forest canopy. Extrinsic variables (temperature and moisture) control soil water balance and tree growth. The use of the JABOWA-type model as a landscape submodel will be discussed here under mosaic landscape models, and Nisbet and Botkin present a case study in Chapter 23.

The development of terrestrial plant models led to finer temporal resolution, increasing complexity of process modeling, multispecies models, and increasing attention to the effects of changes in exogenous variables, like climate. Simultaneously, a surge of interest in spatial processes in ecology, such as "gap dynamics" in forest canopies on a landscape scale, has set the stage for the expansion of spatial modeling to be discussed later.

During the past 20 years, much has been learned from models about the dynamics of single species and multispecies vegetation associations. However, vegetation response in space has only recently begun to be analyzed. One lesson ecologists have learned is that more is often not better. Analyses of some complex terrestrial ecological models are constrained by large data and computational requirements. Extension of these models to the landscape level greatly magnifies these constraints. One solution to the problem is to simplify the models; differences distinguished by high-resolution nonspatial models may have far less significance at the landscape level. Another solution is offered by recent developments in computer technology that greatly increase model processing speeds. Soon desktop workstations will be as fast as today's supercomputers, and computational speed will be far less of a hindrance on the development of complex spatial ecological models.

Freshwater models

Ecological models for lakes and streams often consider not only the plants and animals within several trophic levels but also the uptake of chemical nutrients and the regeneration of detrital materials. Early aquatic models were empirical steady-state models of nutrient input and primary production (Vollenweider, 1965). The concept of mass balance (i.e., accounting for the loss and gain of the biological and chemical variables by considering different physical, chemical, and biological processes) was widely used in aquatic models. In the simplest case, nutrient loading (chemical) balances algal production (biological) for a given lake's retention time (physical). Over the past two decades, most aquatic models have evolved into complex ecosystem models emphasizing interactions among all chemical, physical, and biological components. Some models emphasize only the biological component (Jorgensen, 1978). Ecological models usually include a spatial component, whereas biological models may not.

Several state-of-the-art reviews on aquatic ecological models are available (e.g., Straskraba and Gnauck, 1985; Sonzogni et al., 1986; Booty and Lam, 1989). We discuss two types of aquatic models, fish yield models and nutrient/plankton models, that were developed without the use of GIS.

Fish yield models

Fish yield models are mostly empirical ones (e.g., Matuszek, 1978; Lee and Jones, 1982) that were designed for a whole lake, and they use a seasonal or yearly time step. Typical inputs are lake retention time, nutrient load, and total dissolved solid concentration. The output includes fish yield, population abundance, and commercial catch. Some detailed simulation models were also developed, using such factors as food consumption, metabolism, swimming speed, ingestion, and excretion (Stewart et al., 1983).

Nutrient/plankton models

Nutrient/plankton models were developed to answer questions about eutrophication (excess algal overgrowth resulting from rich nutrient conditions). Model equations were a set of differential equations with state variables covering, in some cases, 4 to 6 phytoplankton species, 3 to 6 zooplankton species (in 3 or 4 life stages), 1 to 2 benthic animals, and 1 to 4 fish groups (in 3 or 4 life stages) (DiToro et al., 1981; Park et al., 1975). These were developed for a variety of one-, two-, and three-dimensional representations of the lake or water body (Lam et al., 1982). The time step was hourly, daily, or seasonal. In the 1970s, nutrient/plankton models were at their peak of development. Hundreds of horizontal grid cells with several vertical layers were used, sometimes directly adapted from ocean/lake hydrodynamic models, which provided transport mechanisms between cells (Lam and Halfon, 1978). These spatially complex models preceded the advent of GIS. The dynamic presentation of the two- and three-dimensional results of so many variables would pose serious challenges even to current GIS.

There are two major lessons learned from the past two decades of experience in developing aquatic models. One lesson is that a spatially or ecologically complex model was not necessarily better than a simpler one in terms of predictive capability (Lam and Halfon, 1978). The choice of model complexity depends on the questions being asked and the availability of knowledge and data. The other lesson is that an ecosystem model that includes biological, chemical, and physical components can provide broader answers than a simple biological model. Lam et al. (1983), for example, used an ecosystem model to simulate the effect of plankton production/decay that resulted from changes in water level, nutrient loading, and weather (climate).

Marine models

Ecological models developed for marine systems include ecosystem models, migration models, spatial stochastic models, regression-based models relating abundance to variable environmental factors, colonization models, and multispecies models. Although we focus on ecosystem models applied to fisheries, we also present some discussion of spatial models applied to fisheries.

Ecosystem models

Multitrophic-compartment marine ecosystem models have been developed to evaluate the effect of fishing on fish stocks. Development has been along three pathways: static energy or materials budgets (Polovina, 1984; Christensen and Pauly, 1991); dynamic simulation models (Andersen and Ursin, 1977; Browder, 1983); and dynamic spatial simulation models (Laevastu and Favorite, 1981; Longhurst, 1978). Marine ecosystem models have had limited acceptance in fisheries research, possibly because traditional fishery biologists distrust complex models with large numbers of state variables, coefficients, and assumptions. The advancement and refinement of ecological modeling techniques for marine systems have been slow because of lack of support. Static models are better represented than dynamic models in the marine fishery literature (i.e., Cohen et al., 1982; Christensen and Pauly, 1991). Only a few dynamic spatial marine ecosystem models exist. Until recently, the computer core and run time required for the execution of models of this type have hampered their development.

Marine ecosystem models that emphasize primarily plankton also have been developed (Kremer and Nixon, 1977) but have not emphasized spatial heterogeneity.

Other spatial models in marine systems

The earliest spatial models in fisheries science were simple conceptual models that attempted to explain seasonal fish movements on the basis of seasonal changes in temperature. More recently numerical models of fish migration have been developed, such as a stochastic compartmental model of the migration of juvenile salmon (Lee, 1991) and a stochastic compartmental model for the migration of marine shrimp (Grant et al., 1991). Migration models also have been developed for marine mammals—for instance, to describe the movements of the northern fur seal (French et al., 1989) and the sperm whale (Botkin et al., 1978). These models move animals through space, although they do not examine the effect of spatial heterogeneity or organization of spatial subunits on migration patterns or condition of stocks.

Spatial models have recently been developed to test the validity of common procedures for estimating stock abundance indices in a spatially heterogeneous marine environment. Along parallel lines, remote sensing and statistical analysis techniques are being used to examine fish and marine mammal distributions in relation to oceanographic and bathymetric features (e.g., Podesta et al., in review). Therefore, there are many needs for spatial information in fisheries apart from those required for ecosystem modeling.

Summary

In their review article, Sklar and Costanza (1991) make the following observations. Animal population models in ecology are similar to the demographic models in geography in that both deal with birth, death, and diffusion. Ecologists, however, have spent more time on temporal density-dependent interactions. Plant ecologists have long recognized the correlation between abundance and spatial heterogeneity. Although plant ecologists use demographic principles to model spatial dynamics, they have a tendency to emphasize community-level rather than population-level interactions and have attempted to deal more explicitly with spatial–time processes than animal ecologists. In the past, ecological modelers have rarely combined geometric, statistical, or mechanistic approaches with the incorporation of spatial characteristics or processes into a model.

CURRENT STATE OF SPATIAL MODELING

We review the current state of spatial models for ecological processes by ecosystem type—terrestrial, freshwater aquatic systems, and marine systems—and by general types. This approach produces a fairly dramatic contrast among disciplines. To date much more attention has been given to spatial relationships and the use of GIS in the first two ecosystems than in marine systems. In Chapter 25 Johnston discusses the use of GIS and spatial models from the viewpoint of population, community, and landscape ecology. The complexity of many modern concerns such as climate change and cumulative effects from one or more hazards begs for more holistic ecological models, and the discipline of landscape ecology and research on "integrated modeling" are moving ecological models in that direction (see Chapters 26–29).

Terrestrial models

Modeling for terrestrial ecosystems usually involves simulation over time of landscape-level ecological attributes related to the location and configuration of area subunits or cells. The classification of terrestrial spatial models follows Baker (1989). The discipline of landscape ecology (Forman and Godron, 1986) attempts to describe dynamics and interactions of biological resources in a spatial context and thus provides the "where" aspect of reality to add to the "when" and "how" aspects of temporal mechanistic modeling. Although not restricted to terrestrial systems, most of the modeling work in landscape ecology has occurred in terrestrial ecosystems; thus our discussion of landscape ecology is included in this section. Baker (1989) and Sklar and Costanza (1991) provide reviews of landscape models.

A very promising approach to landscape-model development appears to be the incorporation of nonspatial distributional models (sometimes called process-based models) into spatial landscape models with GIS (Baker, 1989). A distributional model is one in which the distribution of values of a variable (e.g., distribution of land area among landscape elements) is modeled without regard to location. Evaluation of distributional models requires determination of the states of such variables as vegetation type, species population density and size, and elevation. In the past some have tried to estimate the change in state variables over time by Markov analysis (Shugart et al., 1973; Horn, 1975; Marsden, 1983) or by one or a series of differential or difference equations. Interactions in space may use continuous or discrete mathematics, but a description of discrete state space may be easier in landscape ecology than a description of continuous space.

True spatial models use the location and configuration of landscape subareas to project change in the landscape. These models may be raster based (grid cell maps) or vector based (polygon maps defined by coordinates). Most spatial landscape maps are mosaics of subareas (either cells or polygons). Each cell or polygon is characterized by a single value per layer (single discrete state) or a distribution of values even though the values may be taken from a continuous state space (e.g., topographic map). Spatial models with multiple layers are also possible in which each layer is composed of cells whose states are characterized by variables mapped in two-dimensional space. Mosaic spatial models can be classified as whole mosaic models or distributional mosaic models.

Whole mosaic models

Cells in whole mosaic models have only one value per data layer rather than a distribution of values. Browder et al. (1985) used this approach to simulate the spatial configuration of land and water in a disintegrating marsh. They first used a probabilistic spatial model to develop a knowledge base that was then used by an expert system to select clustering coefficients best approximating the observed pattern of land and water in specific Louisiana marsh sites. The appropriate clustering coefficients were subsequently employed to model the entire sequence of land disintegration in the GIS raster representation of each specific marsh. The project objective was to determine how the length of the ecologically important shoreline (land–water interface) changed over time as a function of stage of disintegration and degree of clustering or spatial dependence in disintegrating coastal marshes. Influences of neighboring cells have also been incorporated into a whole landscape model of land-use changes (Turner, 1987).

Another type of a whole mosaic model is a cartographic model in which values for environmental or ecological variables are accorded to each cell in a GIS data layer (Tomlin, 1991). Each variable is represented by a data layer in the GIS and treated as a component in an equation that simulates some ecological response. For example, the locations of present and potential sage grouse strutting grounds (leks) in Pine Valley, Utah, were simulated in a cartographic model that incorporated multivariate analyses of existing lek sites and mapped variables representing elevation, temperature, water sources, and vegetation type (Nisbet and Reed, 1983). A Bayesian model can also be used with data layers in a GIS to make spatially explicit projections. Milne et al. (1989) used this approach to predict winter deer habitat independently at each of 22,750 grid cells based on 12 variables stored in a GIS. A grid-cell-based model was used to explore the effects of vole and deer herbivory in slowing the invasion by trees of a utility right-of-way (Hyman et al., 1991). This model used a unique approach in which data layers with different spatial resolutions were used to simulate differently scaled processes. This approach also increased computational efficiency and decreased computer storage requirements. Johnston gives additional examples of habitat suitability modeling in Chapter 25.

The final type of whole landscape model is based upon a "cellular automaton," an array of cells whose dynamics are determined only by interactions with neighbors (Wolfram, 1984). This approach has been used to model hypothetical plant populations competing in a landscape subject to fires (Green, 1989; Green et al., 1990).

Distributional mosaic models

In these models, sometimes referred to as process-based landscape models (Sklar and Costanza, 1991), each cell (or group of cells) contains a distributional landscape submodel. Several models have used a JABOWA-type model as a submodel in landscape-level analyses. The biggest problem with this approach is the difficulty in scaling up from the plot level to the landscape level.

One approach to scaling up from cells to the landscape is to simulate growth for a small number of stand types characterized by different climate and soil conditions, and extend the results geographically for each stand type to all areas of similar type (Shugart and Noble, 1981; Pearlstine et al., 1985; Dale and Gardner, 1987; Pastor and Post, 1985). A JABOWA-type model (BRIND), for example, was used to simulate forest response in 16 conditions of elevation and fire probabilities, from which landscape-level inferences of fire probability were made, although this was not an explicitly spatial model (Shugart and Noble, 1981). In another application, bottomland forest growth was projected by a version of the JABOWA model (FORFLOW) for various flood elevations of the Santee River in South Carolina (Pearlstine et al., 1985). Results from simulations before river diversion and from two diversion options were noted for changes in vegetation type and mapped into a GIS for display.

The other approach to scaling up is to simulate state changes (growth of individual trees) in a grid of plots in a landscape. This requires extensive computing resources and characterization of climate, soil, and initial forests for all plots. Smith and Urban (1988) used this approach with ZELIG, a JABOWA derivative that incorporates spatial interactive effects on the establishment, growth, and death of trees in a grid of 900 plots in an east Tennessee forest. In another application, which considered interplot dynamics in a riparian corridor, a version of the FORFLOW model was used to predict rates of seed dispersal by gravity, mammals, and birds (Hanson et al., 1990).

An interesting mosaic landscape model (CELSS) was constructed of differential equation submodels that determined exchanges of water, salts, and suspended sediments among 1 km cells in a Louisiana coastal marsh (Costanza et al., 1990). Analysis of landscape-level flows showed that past and future climate, ecological variables (e.g., primary productivity), and past activities (e.g., canal construction) significantly affect landscape evolution. A related model (Boumans and Sklar, 1990) used fixed subareas (polygons) rather than grid cells to model succession in a degrading Louisiana wetland. In another wetland application a model called ISLAND was developed using difference equation (Markov chain) submodels to simulate the response of a barrier island landscape to changes in rates of sand deposition, seaward erosion, and loss of a grass species (Rastetter, 1991). In an ecosystem-oriented application, a grid-cell model of nitrogen dynamics in the Walker Branch Watershed in Oak Ridge, Tennessee, was constructed (Bartell and Brenkert, 1991). Difference equations were used to model 50-year changes in the distribution of nitrogen among soil, litter, understory vegetation, and overstory vegetation pools.

One of the most successful groups of models that has been linked to a GIS simulates effects of fire on chaparral shrublands and coniferous forests (Kessell and Cattelino, 1978; Kessell, 1979; Kessell, 1990). Data are input from the GIS and manipulated by gradient analysis submodels to estimate vegetation and fuel data for each grid cell. These models have been extended to predict the effects on mammals of changes in landscape structure (Kessell et al., 1984). Kessell's models are important in the development of landscape modeling applications because they incorporate environmental and vegetation data with explicit modeling of landscape-scale natural disturbances in a form useful for resource managers (Baker, 1989).

Distributional mosaic models have been used extensively in the simulation of broad-scale landscape disturbances (e.g., fire, pests, pathogens, exotic species, and windthrow). Prediction of disturbance spread is important in terms of both a practical standpoint (e.g., concerning how far and how fast a pest species might spread) and a more general understanding of the dynamics of the vegetation mosaic in disturbed landscapes (Turner and Dale, 1991). Some models use a simple probabilistic relationship to determine whether a disturbance will spread spatially from one site to another (Turner et al., 1989b). Probabilistic models of fire spread, which require no knowledge of physical mechanisms, have been used to explore the conditions under which a fire spreading through a simple, homogeneous, two-dimensional space can become "critical"—spread from one end of the space to the other, for instance (Albinet et al., 1986; Ohtsuki and Keyes, 1986; Hirabayashi and Kasahara, 1987; von Niessen and Blumen, 1988). To apply a probabilistic fire model to an actual landscape, predictions from empirical models of fire spread can be used to develop probabilities of spread, perhaps based on thresholds of wind speed and fuel moisture (Turner et al, in press; Turner et al., 1992), for various fuel categories across a landscape. Mechanistic detail and temporal dynamics have been incorporated in a probabilistic fire spread model developed by Baker et al. (1991) in which a set of data layers stored in a GIS are linked with a disturbance-spread model that incorporates the effects of temporally varying weather conditions on fire size.

Key strengths of the probabilistic modeling approach are its generality and the ability to produce testable hypotheses (Turner and Dale, 1991). The approach can be applied to numerous types of disturbances that spread spatially across heterogeneous landscapes. Even when an initial landscape pattern is constant, replicated simulations of a stochastic model can be used to project both the magnitude and range of potential disturbance effects. However, the simulations must often be interpreted and evaluated at the scale of the whole landscape rather than on a site-by-site basis. Because the model is probabilistic, it is not designed to predict whether a particular hectare or plot will be disturbed; rather, the objective of the model is to predict disturbance patterns over a large area. Physically based models that will predict the actual location and other properties of individual disturbances are available, however (e.g., Rothermel, 1972; Kessell, 1979).

Another distributional mosaic model links a forest growth model and a Leslie matrix insect-dynamics model and applies the coupled model across interacting patches that vary in elevation and other site-specific environmental conditions (Dale et al., 1990). This model couples biotic dynamics that occur at vastly different temporal scales, emphasizes the importance of fine-scale dynamics in generating broad-scale patterns (Turner and Dale, 1991), and demonstrates the potential of making simple and stochastic difference equation models spatially explicit.

Models of other processes, such as foraging dynamics, can also be distributed across a landscape and linked with a GIS. A spatially explicit winter foraging model, for example, was developed to explore interactions between the abundance and spatial distribution of resources and ungulate densities in Yellowstone National Park (Turner et al., 1990; Wu et al., 1991). During the simulation, ungulates search the landscape and move on the basis of the distribution and arrangement of resources represented in a GIS, and foraging is simulated by using a differential equation. Forage intake, ungulate energy balance, and ungulate survival are projected under alternative resource distributions and winter severity.

Freshwater aquatic models

Aquatic models use similar methodologies as described here for spatial model development. Although some models use the Markov process and stochastic treatment of random (heterogeneous) spatial patterns, the majority of models fall into the mass balance category by using partial differential equations in time and space that are solved by finite difference and finite element techniques (Gray, 1986). It is not possible to describe all types of aquatic models in this discussion. Some ecological models are of significant importance in advancing ecological theories. Other models have practical importance, including those on benthic and groundwater biology (MacQuarrie et al., 1990). The potential to make use of GIS technology is greater when the spatial component in these models is well developed. The following are two examples of aquatic spatial models that have been linked with GIS.

Biological/toxics models

The issues of toxic waste, chemical pollution, and human health have gained much attention. Two types of models have been developed: toxicity models and contaminant pathway and fate models. In toxicity models (Minns et al., 1991), threshold values were used to establish survival rates for several taxa of aquatic biota (fishes, algae, zooplankton, zoobenthos, invertebrate, amphibians, wetland birds, etc.). Toxicity is commonly measured by the concentration of the pollutant (metals, organic and inorganic chemicals, acidity, and radionuclides, etc.) needed to achieve a 50% mortality rate of a species (e.g., rainbow trout over a given period, such as 4 days). This empirical indicator is known as the LC50 and is widely used in toxicity models. One interesting use of the model is to back-calculate the loading of the toxicant needed to achieve a safe guideline or water quality objective for all the lakes over a large region. For example, critical values of sulfate deposition load were derived from watershed acidification models (Lam et al., 1989) for a number of species (Minns et al., 1991) in more than 2000 lakes in eastern Canada.

In contaminant pathways and fate models, chemical equilibria are usually assumed among the media, which for a lake or sea normally include water, suspended solids, and biota. The toxicant often goes through various media before accumulating in the higher trophic animals such as fish and humans. Thus, it is not possible to run the model for one particular medium by itself and expect the computed concentration to be at equilibrium with the concentration in other media. The model equations and equilibrium conditions have to be solved for all media simultaneously. For example, Thomann and Connolly (1984) adopted this approach for computing the dynamic equilibrium concentration of polychlorinated benzenes in water, plankton, and fish. A more comprehensive approach for contaminant pathways and fate models is proposed by Brewers et al. (1991). As in the case of nutrient/plankton models, GIS techniques can be used to present the spatial patterns of the computed results and to input detailed spatial information (e.g., satellite data) for model enhancement.

Habitat models

GIS technologies have been applied to simulate the home range and habitat of water-based plants (algae) and animals (fish, waterfowl, turtles, etc.). Data on temporal and spatial distribution of the animal can be collected by various methods, including radar transmitters and remote sensors. Such habitat data as depth, light, and temperature are converted from point measurement to interval classes by statistical clustering and contouring. GIS overlay analyses and modeling can then be performed to relate the biological data with the habitat variable (Painter, 1991).

Marine models

Dynamic spatial ecosystem models

Spatial models are needed in marine resource studies because the marine environment is spatially heterogeneous, biomass varies as a result of the heterogeneity in environmental variables, and fish and marine mammals move from one environment to another. Pola (1985) stated that "Migrations are an important feature of marine ecosystem dynamics because they cause temporal changes in the spatial distribution of biomass and influence prey availability and vulnerability to predation, accessibility to fishing gear, and exposure to environmental conditions." Fish and marine mammals migrate to feeding grounds, to spawning grounds, or to areas of improved environmental conditions. At least four types of periodic movements can be observed in marine ecosystems: (1) seasonal long-distance migrations, usually along a latitudinal gradient, between summer feeding grounds and winter spawning grounds; (2) seasonal onshore–offshore migrations; (3) ontogenetic onshore–offshore movements in which the different life stages move in different directions; and (4) daily excursions between resting and feeding grounds. Marine mammals also make long-distance migrations in connection with their feeding and reproduction. In some species, separate populations occupy adjacent habitats such as coastal nearshore and offshore waters.

Despite the importance of fish and marine mammal migrations to energy flow in ecosystems, few marine ecosystem models are constructed to take into account spatial variation and movements of animals and material between systems. For the most part, ecosystem modelers have avoided consideration of the effects of spatial heterogeneity by defining system boundaries so that the flows across boundaries are minimal. This means the averaging of state variables and conditions across the entire area, which could lead to erroneous estimates of environmental effects and prey–predator interactions.

Spatial marine ecosystem models are needed to (1) estimate biomass from results of resource surveys, (2) quantify prey–predator and other interspecies interactions, (3) estimate effects of variable environmental factors on recruitment and stock size, and (4) estimate effects of fishing. Laevastu et al. (1982) emphasized the critical importance of installing the appropriate spatial and temporal resolution in ecosystem models. We discuss three types of spatial models for marine ecosystems: multitrophic-compartment models that compute change in the magnitude and distribution of fish biomass over time; a dynamic simulation model for carbon and nitrogen flow; and a simple spatial biomass model.

Multitrophic-compartment models

Laevastu and associates developed DYNUMES, a gridded model that computes change in the magnitude and distribution of fish biomass over time in the eastern Bering Sea, and PROBUB, a simplified version of DYNUMES that computes equilibrium biomasses for larger spatial units (polygons rather than grid cells) and covers the Gulf of Alaska and the Bering Sea (Laevastu and Larkins, 1981).

Initial inputs of the DYNUMES model are depth, sea–land table, surface and bottom temperatures, and nature of the bottom. Covering an area of 1,500,000 km^2, DYNUMES has a spatial resolution of 63.5 km to a cell side. Overlaid grids represent different depth layers (surface and bottom). There is a biomass submodel for each point in a cell.

The DYNUMES model is "process oriented" with rate variables that determine state variables. The emphasis is on fish. Standing crops of phyto- and zooplankton are simulated with a harmonic formula to represent seasonal variability. Consumption of fish by marine mammals and birds is included in the model, based on their respective estimated biomasses, which vary seasonally but not as part of the simulation. Growth of fish biomasses, fishery yields, mortalities, consumptions (predation), and migrations are computed at each time step. Feeding is food-density dependent. Distributions affect predation and the availability of food. Growth influences migration. Temperature has an impact on both growth and migration.

In simulating migration, biomass is moved from one grid point to another according to season or specified events (i.e., changes in currents, temperature anomalies, etc.) (Pola, 1985). Finite difference equations such as those employed in estimating diffusion (Laevastu and Larkins, 1981) are used to approximate migrations. There are two migration components: a predetermined, or fixed, component based on season; and a variable component based on temperature change with time. The four-dimensional DYNUMES model can simulate the distribution of any species or species group in space and time.

Such models as DYNUMES ideally should be coupled to hydrodynamical–numerical models, but because of the large computer core memory and run time required for both model types, the environmental models have to be run separately and their output fields stored at locations from which they are read into ecosystem models in the desired time steps (Laevastu and Favorite, 1981).

In PROBUB (Prognostic Bulk Biomass Model), the area modeled is separated into nine regions, each of which is treated as internally homogeneous (Laevastu and Larkins, 1981). The regions roughly conform to fishery statistical areas that have distinct differences in biomass density and landings. Separate, linked models are created and initialized for each region. In the initialization of the model, an estimate is made of the biomass of a top predator. Then biomasses of other species under equilibrium conditions are determined iteratively, while the original input of biomass remains fixed (Laevastu and Favorite, 1981). Results of this initialization stage of a dynamic simulation model correspond to a static model or budget. Once equilibrium biomasses are computed, the model can be executed to simulate long-term and cyclic changes in the ecosystem caused by the fishery and other factors, such as climate change. In PROBUB, migrations across boundaries are computed empirically.

GEMBASE (General Ecosystem Model for Bristol Channel and Severn Estuary) is another example of a dynamic ecosystem simulation model (Longhurst, 1978; Longhurst and Radford, 1978). It simulates the carbon and nitrogen flow between ecological state variables and seven geographic regions. The model contains about 150 equations and 225 parameter values. It uses hydrodynamical models to transfer materials between adjacent geographic regions. NORFISK, a biomass model with six spatial subcomponents, has recently been developed for the North Sea (Bax et al., 1984).

Proposed marine ecosystem models

As stated earlier, few marine ecosystem models have been constructed to take into account spatial variation and movements of animals and materials. We propose three systems in which the application of spatial marine ecological models is especially relevant: coral reefs, estuaries, and migrations of large pelagic fish in the Atlantic.

Coral reefs are one spatial component of shallow-water tropical marine systems. The fish biomass of coral reefs is supported not only by the primary production of the reef but also by the primary production in nearby seagrass beds and mangroves. The carrying capacity of a marine area may be determined by the mosaic of reef, seagrass, and mangrove area rather than by reef area alone or by the total area of all three habitats. Initially a comparative biomass budget could be prepared for several locations having different mosaics of reef, seagrass, and mangrove area. Results of such studies would (1) provide help in educating resource managers and the public about the importance of seagrass and mangrove areas, (2) guide the effective placement of artificial reefs, and (3) provide a perspective for selecting "source" areas to establish as preserves to protect reef fish species from overfishing.

The estuary is the most spatially heterogeneous part of the marine environment. Physical habitat usually consists of coastal wetlands, shallow tidal creeks, submerged shoreline (the littoral zone), seagrass beds, mud bottom, and open water, all of which are important to the production of estuarine-dependent fishery species. The chemical variation of an estuary is best characterized by salinity, in which a gradient is established from its freshwater source to its outlet to the sea. Both physical characteristics and salinity determine suitability of habitat (Browder and Moore, 1981). Freshwater flow determines the relationship between salinity contours and physical habitats. A continuing challenge for marine ecologists is for each estuary to (1) demonstrate the relationship between freshwater runoff and fishery production for an estuary and (2) answer the question "How much water does the estuary need to best maintain fishery production?" These are complex tasks that can best be addressed with ecological landscape models.

Both large-scale and small-scale spatial variability are characteristic of the ocean. On the large scale, the equatorial convergence zone and regions of upwelling, continental margins receiving terrestrial nutrients are much more productive than the ocean at large. Even within such regions, boundaries between water masses, denoted by discontinuities in temperature or salinity, are much more productive than adjacent areas.

The spatial heterogeneity of the ocean is an important factor in the ecology of tunas and other large oceanic pelagic fish, which migrate long distances between spawning grounds in the tropics and subtropics and feeding grounds in northern temperate waters. The tunas appear to have strong temperature preferences. Furthermore, studies suggest that they are attracted to temperature discontinuities, or fronts, where prey are thought to concentrate. Persistent and recurrent fronts and upwelling areas appear to correspond to topographic features of the ocean floor—the shelf break, islands, and seamounts, for example. Fronts also occur on the boundaries of major current systems like the Gulf Stream. Although upwellings and fronts recur in the same general area, they are not always in the exact location but may be displaced by several kilometers seasonally or annually. Year-to-year variation in the timing and rate of seasonal temperature change could affect migration patterns and the availability of prey.

Year-to-year variations in the location, strength, and persistence of frontal and upwelling systems that stimulate primary productivity and concentrate prey also could affect food availability for tunas. Dynamic spatial bioenergetic–migration models could systematically test the potential effects of extreme feeding conditions in various parts of tuna range on bioenergetics and recruitment. The model also could test the potential effect of variable migration patterns on annual abundance indices.

Stock assessment problems caused by the uneven spatial distribution of fish have led to the development of other types of spatial models in fisheries research. For instance, Farber (1990) developed the stochastic spatial simulation model BILSIM to examine how the aggregation of fish along environmental attractants affected the reliability of relative abundance estimates based on catch rate (catch per unit effort). Edwards and Kleiber (1989) used a spatial model to investigate how the nonrandom distribution of dolphin in the Pacific Ocean affected estimates of relative abundance based on dolphin sightings by observers on tuna vessels. Results of both studies focus attention on the need to (1) improve the general understanding of small-scale spatial and temporal distributions of large pelagic fish and marine mammals and (2) develop data stratification methods that are more robust to the effects of small-scale randomness. Such stratification might require detailed spatial data on ocean surface and bottom features.

The future in marine modeling

The future of spatial ecological models in the marine system is strongly tied to the future of ecosystem models, other spatial models, and other approaches to spatial analysis. The spatial marine models by Laevastu and associates were pioneering efforts. Recently applied in Norway and in the Black Sea, these ecosystem models provide a rich source of useful information on spatial approaches to marine modeling. The work of Pauly, Christensen, and others (e.g., Pauly et al., 1987) and the symposium publication edited by Christensen and Pauly (1991) are generating new interest in marine ecosystem modeling as well as strengthening the conceptual basis of model development. Although the models are neither spatial nor dynamic, they strengthen the foundation of marine ecosystem modeling in general by improving the estimation and comparison of critical ecosystem parameters between areas and species. Furthermore, this volume represents an encouraging interest in ecosystem modeling by those who have a strong foundation in traditional fishery biology. The increased power of personal computers and the recent development of workstations with large memory capacity and fast processing speed will rapidly increase the progress of ecosystem modeling in the marine environment. Geographic information and image analysis systems are other technological advances that will improve the opportunities for marine ecosystem modeling by making these models easier to develop.

FUTURE NEEDS AND DIRECTIONS

GIS is not necessary for all ecological modeling. Indeed, some ecological studies are mainly concerned with temporal aspects of ecological processes. Only those models that have a spatial component or that have the potential to be incorporated into a spatial modeling framework should be linked with a GIS. For this discussion, a spatial model can be loosely defined as one that has either one or more state variables that are a function of space or can be related to other space-dependent variables. The three factors that primarily have limited the use of GIS by ecological modelers are the difficulty and expertise required to link models with GIS, the time required to develop the necessary databases, and the cost of software and hardware. These are not necessarily unique to ecological modeling, although some of the databases may not be of interest to other scientists. However, many novel and challenging research concepts will shape future spatial model development using GIS as well as ideas for GIS enhancement to meet the research needs.

Landscape-level models

Many mechanistic ecological models exist, but new tools are needed to extend these models to the landscape level and to use GIS technologies. Landscape models have the potential of: (1) mapping the flows of energy, materials, and information; (2) designating source, sink, and receptor areas; (3) predicting succession in two- and three-dimensional space; (4) determining cumulative thresholds for anthropogenic substances; and (5) addressing questions of scale (Reiners, 1988). Models could be reduced into modules that can be related explicitly to location or configuration of subareas on the landscape (Baker, 1989; Sklar and Costanza, 1991). However, to manage these modules a modular system requires an interface (with

BASIC, FORTRAN, "C," etc.) within the computer system or GIS or, even better, a macro language facility (Lam and Swayne, 1991). Such an improvement is needed particularly when temporal models are generalized from one location to many locations with spatial interactions and feedbacks. The problem of integrating models with different temporal and spatial scales is not trivial and requires special methodologies (Turner and Gardner, 1990).

Sensitivity, errors, and uncertainties

When site models are generalized to regional models, it is usually necessary to interpolate data, and hence the data uncertainty increases. The majority of so-called map data are generated by contouring methods based on a finite number of observations. Propagation of uncertainties from input and boundary conditions based on map data will pose new problems for spatial models. In particular, data generated from GIS are often given as ranges or class intervals, not point values. Indeed, some spatial models now use intervals and classes as the measure for state variables (Camara et al., 1990). Sensitivity analysis and model calibration must consider pattern statistics and new techniques to display and compare computed and observed results (Turner et al., 1989a; Cai, 1991). These should be nested as a library of functions within GIS systems. GIS vendors should consider more sophisticated contouring methods with optimization constraints and error estimation using kriging and semivariogram methods to enable modelers to track down the GIS-generated interpolation errors. Future GIS will probably require fast, large-memory computers with parallel processors.

Database management

Spatial models require a large set of mapped landscape data. An effective database management system (Cowen, 1988) is needed to pass two- and three-dimensional data between models and GIS. Currently many GIS can archive and retrieve various types of data including cartographic (map), satellite (image), and digitized (raster and vector) data from different sources. A user-friendly (for modelers) interface is required to access and integrate these data. Standardization of data format by the GIS industry will facilitate easy transfer of data but must incorporate data integrity control (e.g., in computer networking). In particular, ecological modeling requires a GIS that can store time-varying data and include/exclude data according to specified regions or time periods. Common GIS usage includes applying overlay delineation, contouring and plotting, calculating energy budget and transport gradients between grid cells, and computing influence or probability zones. A demand exists for high-resolution maps (oceans, lakes, forests, and land use) and detailed zoom-in powers.

Remotely sensed data

Traditionally ecologists have focused on the manipulation of satellite and radar images in GIS for terrestrial vegetation data and phytoplankton chlorophyll-density data; more research is needed in the design of fully interactive image analysis software and in the area of optical modeling (e.g., relating optical spectral density to chlorophyll concentration and the presence or absence of organisms). The use of radar tracking of organisms is unique in the field of ecological habitat and home range modeling. Future model development with these and other GIS technologies through the inclusion of more landscape information will broaden the scope of application in habitat management.

Model-base management system

A GIS is often used as only a display device for the presentation of model results. However, graphical presentation is a fundamental and mundane function for GIS. On the other extreme, some GIS offer a modeling capability themselves and make the models GIS subroutines. One has to learn a new modeling protocol within the GIS and follow special procedures and database structures to program an ecological model. Although this GIS-driven modeling capability is an excellent tool for overlaying map variables and deriving relationships among them, it is strictly limited to the methodologies defined by the GIS and not suitable for complex models.

Therefore, what is needed is something between the extremes of just a display device and a GIS-specific modeling protocol. Expert systems with GIS capability (Lam and Swayne, 1991) and expansion of GIS capability with artificial intelligence (Coulson et al., 1987) are integrative solutions. Such systems provide an open architecture (e.g., complete with such modeling languages as DYNAMO) and could allow the user to decide which models should be resident within or outside the GIS. (DYNAMO is a continuous-time simulation model, useful for modeling flows of material between compartments.) Interfaces for GIS data to be transferred to the models and vice versa through a complete library of GIS functions are essential. An example is the GIS system called GRASS (USA-CERL, 1988), which consists of a set of some 300 GIS programs and functions for GIS operations that can be called directly from external simulation programs, making the GIS and all its capabilities effectively available as subroutines for environmental models. GRASS's map functions, for example, have been used in a spatial model of natural disturbances in landscapes (Baker et al., 1991), using a driving program written in the popular simulation language SIMSCRIPT II.5 (CACI, 1987). Unlike DYNAMO, SIMSCRIPT has an event-related language that permits analysis of discrete time models. However, most simulation languages can make "system calls" to external routines or use external libraries of functions and can thus directly use GRASS capabilities. This capability for open, direct, command-line access to a library of GIS functions provides a unique and powerful link between environmental models and GIS capabilities.

Expert systems provide a simple query-and-answer interface so that choices of data, models, or model parameters can be made according to a set of expert rules (Lam and Swayne, 1991; Lam et al., 1989; Crowe and Mutch, 1992). An example is the RAISON system (Lam and Swayne, 1991), which is also programmable and can integrate external components of database, spreadsheet, graphics, statistics, simulation models, and expert systems. In Chapter 24 Lam presents a case study of a fish species richness model that uses GIS and expert systems. Coulson et al. (1991) discuss intelligent GIS that contain an artificial intelligence component capable of using qualitative information. An intelligent GIS is a powerful tool because it can blend methodologies for representation, analysis, and interpretation of quantitative, spatially referenced data with heuristic knowledge of experts. Open architecture systems are rapidly emerging and will probably form an integral part of future GIS.

Environmental risk assessment

Ecological risk assessment provides a framework to enable (1) a quantitative basis for balancing and comparing risks associated with environmental hazards and (2) a systematic means of improving the estimation and understanding of those risks. Graham et al. (1991) show that the spatial heterogeneity of the landscape is especially important for regional ecological risk assessment and identify GIS as an important tool for regional assessments (Hunsaker et al., 1990). Society needs a quantitative and systematic way to estimate and compare the impacts of environmental problems at various spatial scales, and future ecological models must provide probabilistic numerical and graphical estimates of change for issues as varied as marine fish migration, forest production and succession, non-point-source pollution, fire and pest movement, climate change, and identification of sensitive and unique ecological areas. Those models that consider the dynamic interactions of species populations with extrinsic variables such as sea levels and climate (Botkin et al., 1991) will be most useful if there is a built-in GIS component.

CONCLUSIONS AND RECOMMENDATIONS

Ecological modeling differs greatly among the subdisciplines of ecology, but it is developing in the same general direction—integration of temporal and spatial effects. Most advances in the past concerned temporal effects, and considerable progress has been made. The next logical step is to consider spatial phenomena. As stated, the null hypothesis in spatial analysis is that the future state of a unit on a landscape is independent of adjacent units. This assumption is often implicit in many ecological models, but the degree to which it is true remains unclear. Most demonstrations of spatial aspects in ecological models just display independent effects as if they were integrated in space, but there is no spatial interaction between cells or subareas. True integration of spatial aspects in ecological models is more an ecological problem than a display problem. Sklar and Costanza (1991) ask ecological modelers to give explicit and concentrated attention to the modeling of larger geographic areas (regional and global), and GIS technology will be important in making the transition from local models to larger areas. The next step for ecologists is to test the null hypothesis to determine at what scale it is true or false, and this test is particularly important when considering issues at regional and global scales.

From a pragmatic viewpoint models are developed for two purposes: to better understand ecological processes; and to help make resource-management and policy decisions. Often these two purposes correspond to the classification of model approaches into stochastic (probability-based) and process-based models, although these are not mutually exclusive. Grant (1988) reminds ecologists of two important points to keep in mind as ecological models are improved and GIS technology is better incorporated:

> One is that the quality of a model does not depend on how realistic it is, but rather on how well it performs in relation to the purpose for which it was built. The other is that the level of resolution included in a model must fit the resolution required by the problem at hand. Modelers must learn to compromise between the known complexity of ecosystems and the need to address a problem with limited data and time.

In light of the limited use of GIS by ecologists during the past 10 years, GIS technology offers a relatively untapped potential for ecological modeling of spatial processes. One future direction is to design ecological models to accommodate spatial analysis by using GIS modules as subroutines for data storage, manipulation, and display. Another approach is to code future ecological models in a GIS modeling language that can take full advantage of GIS spatial analysis and display.

Most GIS systems are still relatively expensive. A system of software and workstation hardware can cost $100,000 or more; however, some GIS are available for less than $1000. These include both workstation and PC versions such as the GRASS system developed by the U.S. Army Corps of Engineers; the pMAP cartographic modeling package, available from Spatial Data Systems; RAISON, available from Canada's National Water Research Institute; and IDRISI, developed by Clark University. Widespread use of GIS in ecological studies could be encouraged greatly by the availability of less expensive GIS software.

Future integration of GIS and ecological modeling will require close cooperation between modelers and GIS vendors. Vendors of GIS should be encouraged to develop: (1) open software architectures that provide access by model subroutines to GIS modules, such as in the GRASS GIS; (2) comprehensive modeling languages (e.g., the RAISON Programming Language) that permit complex spatial analyses, such as calculation of gradients between cells and movement of materials based on gradients and probabilities; (3) the ability to store, analyze, and display time-varying data by pixel or polygon; and (4) libraries of commonly used data, such as regional and national topography, temperature, and rainfall. Such close cooperation between developers of ecological models and GIS is necessary to accomplish the huge task of modeling the current state of regional and global ecological phenomena. The products of such cooperation will provide the tools to assess the effects on ecological phenomena and conditions from regional and global changes.

ACKNOWLEDGMENTS

We thank D.L. DeAngelis of Oak Ridge National Laboratory and M.A. Harwell of the Rosenstiel School of Marine and Atmospheric Sciences, University of Miami, for their helpful manuscript reviews. S.A. Levin, Cornell University, also provided helpful comments. Work was performed at Oak Ridge National Laboratory, managed by Martin Marietta Energy Systems, Inc., under contract DE-AC05-84OR21400 with the U.S. Department of Energy, Publication No. 3902, Environmental Sciences Division.

REFERENCES

Albinet, G., Searby, G., and Stauffer, D. (1986) Fire propagation in a 2-D random medium. *Journal de Physique* 47: 1–7.

Andersen, K.P., and Ursin, E. (1977) A multispecies extension to the Beverton and Holt theory of fishing with accounts of phosphorus circulation and primary production. *Danmarks Fiskeri- og Havundersgelser* NS7: 319–435.

Baker, W.L. (1989) A review of models of landscape change. *Landscape Ecology* 2: 111–133.

Baker, W.L., Egbert, S.L., and Frazier, G.F. (1991) A spatial model for studying the effects of climatic change on the structure of landscapes subject to large disturbances. *Ecological Modelling* 56: 109–125.

Bartell, S.M., and Brenkert, A.L. (1991) A spatial–temporal model of nitrogen dynamics in a deciduous forest watershed. In Turner, M.G., and Gardner, R.H. (eds.) *Quantitative Methods in Landscape Ecology*, New York: Springer-Verlag, pp. 379–398.

Bax, N., Sunnana, K., Godo, O.R., and Dragesund, O. (1984) NORFISK—an ecosystem simulation model for studies of the fish stocks off the coast of Norway. Council Meeting, International Council for Exploration of the Sea, *Report 1984/B:25*, 19 pp.

Bledsoe, L.J., and Van Dyne, G.M. (1971) A compartment model simulation of secondary succession. In Patten, B.C. (ed.) *Systems Analysis and Simulation in Ecology*, New York: Academic Press, pp. 479–511.

Booty, W.G., and Lam, D.C.L. (1989) Freshwater ecosystem water quality modeling. In Davies, A.M. (ed.) *Modeling Marine Systems*, Boca Raton: CRC Press, Inc., pp. 387–432.

Botkin, D.B. (1990) *Discordant Harmonies: A New Ecology for the Twenty-First Century*, Oxford: Oxford University Press, 241 pp.

Botkin, D.B., Janak, J.F., and Wallis, J.R. (1970) Rationale, limitations, and assumptions of a northeast forest growth simulator. *IBM Journal of Research and Development* 16: 101–116.

Botkin, D.B., Janak, J.F., and Wallis, J.R. (1972) Some ecological consequences of a computer model of forest growth. *Journal of Ecology* 60: 849–872.

Botkin, D.B., and Levitan, R.E. (1977) Wolves, moose, and trees: an age-specific trophic-level model of Isle Royale National Park. *IBM Research Report RC 6834*, Yorktown, NY: IBM Research Center.

Botkin, D.B., Schimel, D.S., Wu, L.S., and Little, W.S. (1978) Some comments on the density dependent factors in sperm whale populations. *Annual Proceedings of the International Whaling Commission*, pp. 83–88.

Botkin, D.B., and Sobel, M.J. (1975) Stability in time-varying ecosystems. *American Naturalist* 109: 625–646.

Botkin, D.B., Woodby, D.A., and Nisbet, R.A. (1991) Kirtland's warbler habitats: a possible early indicator of climate warming. *Biological Conservation* 56: 63–78.

Boumans, M.J.R., and Sklar, F.H. (1990) A polygon-based spatial (PBS) model for simulating landscape change. *Landscape Ecology* 4(2/3): 83–97.

Brewers, J.M., Blankton, J.O., Davies, A.M., Gurbutt, P.A., Hofmann, E.E., Jamart, B.M., Lam, D.C.L., Takahashi, M., and Verboom, G.K. (1991) A conceptual model of contaminant transport in coastal marine systems. *AMBIO* 21: 166–169.

Browder, J.A. (1983) A simulation model of a near-shore marine ecosystem of the north-central Gulf of Mexico. In Turgeon, K.W. (ed.) *Marine Ecosystem Modeling*, Washington, DC: U.S. Department of Commerce, NOAA/National Environmental Satellite, Data, and Information Service, pp. 179–222.

Browder, J.A., Bartley, H.A., and Davis, K.S. (1985) A probabilistic model of the relationship between marsh-land–water interface and marsh disintegration. *Ecological Modeling* 29: 245–260.

Browder, J.A., and Moore, D. (1981) A new approach to determining the quantitative relationship between fishery production and the flow of fresh water to estuaries. In Cross, R., and Williams, D. (eds.) *Proceedings, National Symposium on Freshwater Inflow to Estuaries, Vol. 1, FWS/OBS-81/04*, Washington, DC: U.S. Fish and Wildlife Service, Office of Biological Services, pp. 403–430.

CACI (1987) *SIMSCRIPT II.5 Programming Language*, La Jolla, CA: CACI Products Co.

Cai, Y. (1991) GLE—landscape ecological analysis software using the GIS GRASS. *MA Thesis*, University of Wyoming, Laramie, WY, 137 pp.

Camara, A.S., Ferreira, F.C., Loucks, D.P., and Seixas, M.J. (1990) Multidimensional simulation applied to water resources management. *Water Resources Research* 26: 1877–1886.

Christensen, V., and Pauly, D. (1991) Trophic models of aquatic ecosystems. *ICLARM Conference Proceedings No. 26.*

Cliff, A.D., and Ord, J.K. (1973) *Spatial Autocorrelation*, New York: Pion Press.

Cohen, E.B., Grosslein, M.D., Sissenwine, M.P., Steimle, F., and Wright, W.R. (1982) In Mercer, M.C. (ed.) *Multispecies Approaches to Fisheries Management Advice*, Ottawa: Canadian Special Publication in Fisheries and Aquatic Sciences, No. 59.

Costanza, R., Sklar, F.H., and White, M.L. (1990) Modeling coastal landscape dynamics. *Bioscience* 40: 91–107.

Coulson, R.N., Lovelady, C.N., Flamm, R.O., Spardling, S.L., and Saunders, M.C. (1991) Intelligent geographic information systems for natural resource management. In Turner, M.G., and Gardner, R.H. (eds.), *Quantitative Methods in Landscape Ecology*, New York: Springer-Verlag.

Cowen, D.J. (1988) GIS versus CAD versus DBMS: what are the differences? *Photogrammetric Engineering and Remote Sensing* 54: 1551–1555.

Cressie, N. (1991) *Statistics for Spatial Data*, New York: John Wiley.

Crowe, A.S., and Mutch, J.P. (1991) EXPRES: an expert system for assessing the fate of pesticides in the subsurface. *Environmental Monitoring and Assessment* 23: 20–38.

Dale, V.H., and Gardner, R.H. (1987) Assessing regional impacts of growth declines using a forest succession model. *Journal of Environmenal Management* 24: 83–93.

Dale, V.H., Gardner, R.H., DeAngelis, D.L., Eagar, C.C., and Webb, J.W. (1990) Broad-scale effects of the balsam wooly aphid on the southern Appalachian spruce-fir forests. *Canadian Journal of Forest Research* (submitted).

DiToro, D.M., Fitzpatrick, J.J., and Thomann, R.V. (1981) Documentation for water quality analysis simulation program (WASP) and model verification program (MVP). *EPA-600/3-81-044*, Washington, DC: U.S. Environmental Protection Agency.

Edwards, E.F., and Kleiber, P.M. (1989) Effects of nonrandomness on line transect estimates of dolphin school abundance. *Fisheries Bulletin* 87: 859–876.

Egerton, F.N. (1973) Changing concepts of the balance of nature. *Quarterly Review of Biology* 48: 322–350.

Farber, M.I. (1990) Evaluating statistical bias in using catch-rate indices from the U.S. recreational billfish fishery for estimating abundance by the use of a simulation model. *PhD Dissertation*, University of Miami, Coral Gables, FL, 231 pp.

Forman, R.T.T., and Godron, M. (1986) *Landscape Ecology*, New York: John Wiley & Sons.

French, D.P., Reed, M., Calambokidis, J., and Cubbage, J.C. (1989) Assimilation model of seasonal migration and daily movements of the northern fur seal. *Ecological Modelling* 48: 193–219.

Graham, R.L., Turner, M.G., and Dale, V.H. (1991) Increasing CO_2 and climate change: effects on forest. *Bioscience* 40: 575–587.

Grant, W.E. (1988) Models for conservation and wildlife management. *Ecological Modelling* 41: 325–326.

Grant, W.E., Matis, J.H., and Miller, T.H. (1991) A stochastic compartmental model for migration of marine shrimp. *Ecological Modelling* 54: 1–15.

Gray, W.G. (1986) *Physics-Based Modeling of Lakes, Reservoirs, and Impoundments*, New York: ASCE Monographs.

Green, D.G. (1989) Simulated effects of fire, dispersal and spatial pattern on competition within forest mosaics. *Vegetatio* 82: 139–153.

Green, D.G., Reichelt, R.E., van der Laan, J., and MacDonald, B.W. (1990) A generic approach to landscape modeling. *Mathematics and Computers in Simulation* 32: 237–242.

Hall, F.G., Botkin, D.B., Strebel, D.E., Woods, K.D., and Goetz, S.J. (1991) Large-scale patterns of forest succession as determined by remote sensing. *Ecology* 72: 628–640.

Hanson, J.S., Malanson, G.P., and Armstrong, M.P. (1990) Landscape fragmentation and dispersal in a model of riparian forest dynamics. *Ecological Modelling* 49: 277–296.

Hirabayashi, F., and Kasahara, Y. (1987) A fire-spread

simulation model developed as an extension of a dynamic percolation process model. *Simulation* 49: 254–261.

Horn, H.S. (1975) Markovian processes of forest succession. In Cody, M.L., and Diamond, J.M. (eds.) *Ecology and Evolution of Communities*, Cambridge, MA: Belknap Press, pp. 196–211.

Hunsaker, C.T., Graham, R.L., Barnthouse, L.W., Gardner, R.H., O'Neill, R.V., and Suter II, G.W. (1990) Assessing ecological risk on a regional scale. *Environmental Management* 14: 325–332.

Hyman, J.B., McAninch, J.B., and DeAngelis, D.L. (1991) An individual-based simulation model of herbivory in a heterogenous landscape. In Turner, M.G., and Gardner, R.H. (eds.) *Quantitative Methods in Landscape Ecology*, New York: Springer-Verlag, pp. 443–475.

Innis, G.S. (ed.) (1978) *Grassland Simulation Model*, Ecological Studies, Vol. 26, New York: Springer-Verlag, 298 pp.

Jones, R.A., and Lee, G.F. (1982) Recent advances in assessing impact of phosphorus loads on eutrophication-related water quality. *Water Research* 16: 503–515.

Jorgensen, S.E. (1978) A model of fish growth. *Ecological Modelling* 4: 303–313.

Kessell, S.R. (1979) *Gradient Modeling: Resource and Fire Management*, New York: Springer-Verlag.

Kessell, S.R. (1990) An Australian geographical information and modeling system for natural area management. *International Journal of Geographical Information Systems* 4(3): 333–362.

Kessell, S.R., and Cattelino, P.J. (1978) Evaluation of a fire behavior information integration system for southern California chaparral wildlands. *Environmental Management* 2: 135–159.

Kessell, S.R., Good, R.B., and Hopkins, A.J.M. (1984) Implementation of two new resource management information systems in Australia. *Environmental Management* 8: 251–270.

Kolasa, J., and Pickett, S.T.A. (eds.) (1991) *Ecological Heterogeneity*, New York: Springer-Verlag, 332 pp.

Kremer, J.N., and Nixon, S.W. (1977) *A Coastal Marine Ecosystem*, New York: Springer-Verlag, 217 pp.

Laevastu, T., and Favorite, F. (1981) Holistic simulation models of shelf-seas ecosystems. In Longhurst, A.R. (ed.) *Analysis of Marine Ecosystems*, New York: Academic Press, pp. 701–727.

Laevastu, T., Favorite, F., and Larkins, H.A. (1982) Resource assessment and evaluation of the dynamics of the fishery resources in the northeastern Pacific with numerical ecosystem models. In Mercer, M.C. (ed.) *Multispecies Approaches to Fisheries Management Advice*, Ottawa: Canadian Special Publication in Fisheries and Aquatic Sciences, No. 59, pp. 70–81.

Laevastu, T., and Larkins, H.A. (1981) *Marine Fisheries Ecosystem*, Farnham, England: Fishing News Books Ltd., 162 pp.

Lam, D.C.L., and Halfon, E. (1978) Model of primary production, including circulation influences in Lake Superior. *Applied Mathematics and Modeling* 2: 30–40.

Lam, D.C.L., Schertzer, W.M., and Fraser, A.S. (1982) Mass balance models of phosphorus in sediments and water. *Hydrobiologia* 91/92: 217–226.

Lam, D.C.L., Schertzer, W.M., and Fraser, A.S. (1983) *Simulation of Lake Erie Water Quality Response to Loading and Weather Variations*, Ottawa: Inland Water Directorate Scientific Series, No. 134.

Lam, D.C.L., and Swayne, D.A. (1991) Integrating database, spreadsheet, graphics, GIS, statistics, simulation models and expert systems: experiences with the RAISON system on microcomputers. *NATO ASI Series* G26: 429–459.

Lam, D.C.L., Wong, I., Swayne, D.A., and Storey, J. (1989) Watershed acidification models using the knowledge-based approach. *Ecological Modelling* 47: 131–152.

Lauenroth, W.K., Skogerboe, G.V., and Flug, M. (eds.) (1983) *Analysis of Ecological Systems: State-of-the-Art in Ecological Modeling*, Amsterdam: Elsevier.

Lee, D.C. (1991) A stochastic, compartmental model of the migration of juvenile anadromous salmonids in the Columbia River Basin. *Ecological Modelling* 54: 227–245.

Levin, S.A. (1974) Dispersion and population interactions. *The American Naturalist* 108(960): 207–228.

Levin, S.A. (1976) Population dynamic models in heterogeneous environments. *Annual Review of Ecological Systems* 7: 287–310.

Lippe, E., DeSmidt, J.T., and Glenn-Lewin, D.C. (1985) Markov models and succession: a test for a heathland in the Netherlands. *Journal of Ecology* 73: 775–791.

Longhurst, A.R. (1978) Ecological models in estuarine management. *Ocean Management* 4: 287–302.

Longhurst, A.R., and Radford, P.J. (1978) *GEMBASE 1*, Plymouth: Institute for Marine Environmental Research.

Lough, T.J., Wilson, J.B., Mark, A.F., and Evans, A.C. (1987) Succession in a New Zealand alpine cushion community: a Markovian model. *Vegetatio* 71: 129–138.

MacQuarrie, K.T.B., Sudicky, E.A., and Frind, E.O. (1990) Simulation of biodegradable organic contaminants in groundwater: 1. Numerical formulation in principal directions. *Water Resources Research* 26: 207–222.

Marsden, M.A. (1983) Modeling the effect of wildfire frequency on forest structure and succession in the northern Rocky Mountains. *Journal of Environmental Management* 16: 45–62.

Matuszek, J.E. (1978) Empirical predictions of fish yields from large North American lakes. *Transactions, American Fisheries Society* 107: 385–394.

Milne, B.T., Johnston, K.M., and Forman, R.T.T. (1989) Scale-dependent proximity of wildlife habitat in a spatially-neutral Bayesian model. *Landscape Ecology* 2:

101–110.
Minns, C.K., Moore, J.E., Schindler, D.W., and Jones, M.L. (1990) Assessing the potential extent of damage to inland lakes in Eastern Canada due to acidic deposition. III. Predicted impacts on species richness in seven groups of aquatic biota. *Canadian Journal of Fisheries and Aquatic Sciences* 47: 821–830.
Nisbet, S.B., and Reed, K.L. (1983) A spatial model of sage grouse habitat quality. In Lauenroth, W.K., Skogerboe, G.V., and Flug, M. (eds.) *Analysis of Ecological Systems: State-of-the-Art in Ecological Modeling*, Amsterdam: Elsevier, pp. 267–276.
Ohtsuki, T., and Keyes, T. (1986) Biased percolation: forest fires with wind. *Journal of Physics A* 19: L281–L287.
Okubo, A. (1975) *Ecology and Diffusion*, Tokyo: Tsukiji.
Paine, R.T., and Levin, S.A. (1981) Intertidal landscapes: disturbance and the dynamics of pattern. *Ecological Monographs* 51: 145–178.
Painter, S. (1991) Establishing habitat goals and response in an area of concern using a geographic system. *Proceedings, Wetlands of the Great Lakes Symposium*, Berne, NY: Association of Wetland Managers (in press).
Park, R.A., Scavia, D., and Clesceri, L.S. (1975) CLEANER, the Lake George model. In Russel, C.S. (ed.) *Ecological Modeling in a Management Context*, Washington, DC: Resources for the Future, pp. 49–81.
Pastor, J., and Post, W.M. (1985) Development of a linked forest productivity/soil process model. *ORNL/TM-9519*, Oak Ridge, TN: Oak Ridge National Laboratory, 162 pp.
Patil, G.P., Pielou, E.C., and Waters, W.E. (eds.) (1971) *Spatial Patterns and Statistical Distributions*, University Park: Pennsylvania State University Press.
Pauly, D., Soriano, M., and Palomares, M.L. (1987) Improved construction, parameterization and interpretation of steady-state ecosystem models. *Kuwait Bulletin of Marine Science* No. 11.
Pearlstine, L., McKellar, M., and Kitchens, W. (1985) Modeling the impacts of a river diversion on bottomland forest communities in the Santee River floodplain, South Carolina. *Ecological Modelling* 29: 283–302.
Pielou, E. (1977) *Mathematical Ecology*, New York: Wiley.
Podesta, G.P., Browder, J.A., and Hoey, J.J. (in review) Swordfish catch rates and thermal fronts on U.S. longline grounds in the western North Atlantic. Miami, FL: University of Miami Rosenstiel School of Marine and Atmospheric Science.
Pola, N.B. (1985) Numerical simulation of fish migrations in the eastern Bering Sea. *Ecological Modelling* 29: 327–351.
Polovina, J.J. (1984) Model of a coral reef ecosystem. 1. The ECOPATH model and its application to French Frigate Shoals. *Coral Reefs* 3: 1–11.
Prentice, I.C. (1986) The design of a forest succession model. In Fanta, J. (ed.) *Forest Dynamics Research in Western and Central Europe*, Wageningen: Pudue.
Rastetter, E.B. (1991) A spatially explicit model of vegetation-habitat interactions on barrier islands. In Turner, M.G., and Gardner, R.H. (eds.) *Quantitative Methods in Landscape Ecology*, New York: Springer-Verlag, pp. 353–378.
Rees, P., and Wilson, A.G. (1977) *Spatial Population Analysis*, London: Edward Arnold.
Reiners, W.A. (1988) Achievements and challenges in forest energetics. In Pomeroy, L.R., and Alberts, J.L. (eds.) *Concepts of Ecosystem Ecology*, New York: Springer-Verlag, pp. 75–114.
Rothermel, R.C. (1972) A mathematical model for predicting fire spread in wildland fuels. *USDA Forest Service Research Paper INT-137*, Ogden, UT: Intermountain Research Station.
Shugart, H.H., Crow, T.R., and Hett, J.M. (1973) Forest succession models: a rationale and methodology for modeling forest succession over large regions. *Forest Science* 19: 203–212.
Shugart, H.H., and Noble, I.R. (1981) A computer model of succession and fire response of the high-altitude Eucalyptus forest of the Brindabella Range, Australian Capital Territory. *Australian Journal of Ecology* 6: 149–164.
Shugart, H.H., and West, D.C. (1977) Development and application of an Appalachian deciduous forest succession model. *Journal of Environmental Management* 5: 161–179.
Sklar, F.H., and Costanza, R. (1991) The development of dynamic spatial models for landscape ecology: a review and prognosis. In Turner, M.G., and Gardner, R.H. (eds.) *Quantitative Methods in Landscape Ecology: The Analysis and Interpretation of Landscape Heterogeneity*, New York: Springer-Verlag.
Smith, T.M., and Urban, D.L. (1988) Scale and resolution of forest structural pattern. *Vegetatio* 74: 143–150.
Sokal, R.R. (1979) Testing statistical significance of geographical variation patterns. *Systematic Zoologist* 28: 227–232.
Sollins, P., Waring, R.H., and Cole, D.W. (1974) A systematic framework for modeling and studying the physiology of a coniferous forest ecosystem. In Waring, R.H., and Edmonds, R.L. (eds.) *Integrated Research in the Coniferous Forest Biome*, Coniferous Forest Biome Bulletin 5, Seattle: University of Washington, pp. 7–20.
Sonzogni, W.C., Canale, R.P., Lam, D.C.L., Lick, W., MacKay, D., Minns, C.K., Richardson, W.L., Scavia, D., Smith, V., and Strachan, W.M.J. (1986) *Uses, Abuses and Future of Great Lakes Modeling*, Report to the Great Lakes Science Advisory Board, Windsor: IJC Publications.
Sprugel, D.G. (1985) Changes in biomass components

through stand development in wave-regenerated balsam fir forests. *Canadian Journal of Forest Research* 15: 269–278.

Stewart, D.J., Weinberger, D., Rottiers, D.V., and Edsall, T.A. (1983) An energetics model for lake trout, Salvelinus namaycush; application to the Lake Michigan population. *Canadian Journal of Fisheries and Aquatic Science* 40: 681–698.

Straskraba, M., and Gnauck, A.H. (1985) *Freshwater Ecosystems Modeling and Simulation*, Amsterdam: Elsevier Science Publishers.

Swartzman, G.L. (1979) Simulation modeling of material and energy flow through an ecosystem: methods and documentation. *Ecological Modelling* 7: 55–81.

Thomann, R.V., and Connolly, J.P. (1984) An age-dependent model of PCB in a Lake Michigan food chain. *EPA-600/S3-84-026*, Washington, DC: U.S. Environmental Protection Agency.

Tomlin, C.D. (1991) *Geographic Information Systems and Cartographic Modelling*, Englewood Cliffs, NJ: Prentice-Hall, 249 pp.

Turner, M.G. (1987) Spatial simulation of landscape changes in Georgia: a comparison of 3 transition models. *Landscape Ecology* 1: 29–36.

Turner, M.G., Costanza, R., and Sklar, F.H. (1989a) Methods to evaluate the performance of spatial simulation models. *Ecological Modelling* 48: 1–18.

Turner, M.G., and Dale, V.H. (1991) Modeling landscape disturbance. In Turner, M.G., and Gardner, R.H. (eds.) *Quantitative Methods in Landscape Ecology*, New York: Springer-Verlag, pp. 323–351.

Turner, M.G., and Gardner, R.H. (1990) Future directions in quantitative landscape ecology. In Turner, M.G., and Gardner, R.H. (eds.) *Quantitative Methods in Landscape Ecology*, New York: Springer-Verlag, pp. 519–525.

Turner, M.G., Gardner, R.H., Dale, V.H., and O'Neill, R.V. (1989b) Predicting the spread of disturbance across heterogeneous landscapes. *Oikos* 55: 121–129.

Turner, M.G., and Romme, W.H. (in press) Landscape dynamics in crown fire ecosystems. In Laven, R.D., and Omi, P.N. (eds.) *Pattern and Process in Crown Fire Ecosystems*, Princeton: Princeton University Press.

Turner, M.G., Romme, W.H., and Gardner, R.H. (1992) Relationships between spatial heterogeneity and large-scale fire on subalpine plateaus in Yellowstone National Park. *Unpublished Manuscript.*

Turner, S.J., O'Neill, R.V., Conley, W., Conley, M.R., and Humphries, H.C. (1991) Pattern and scale: statistics for landscape ecology. In Turner, M.G., and Gardner, R.H. (eds.) *Quantitative Methods in Landscape Ecology*, New York: Springer-Verlag, pp. 17–49.

Turner, M.G., Wu, Y., Romme, W.H., and Wallace, L.L. (in press) A landscape simulation model of winter foraging by large ungulates. *Ecological Modelling.*

USA-CERL (1988) *GRASS 3.0 User's Manual*, Champaign, IL: U.S. Army Corps of Engineers Construction Engineering Research Laboratory.

Vollenweider, R.A. (1965) Calculation models of photosynthesis depth curves and some implications regarding day rate estimation in primary productivity measurements. *Mem. Ist. Ital. Idrobiol. Suppl.* 18: 452–457.

Von Niessen, W., and Blumen, A. (1988) Dynamic simulation of forest fires. *Canadian Journal of Forest Research* 18: 805–812.

Wolfram, S. (1984) Cellular automata as models of complexity. *Nature* 311: 419–424.

Woodby, D.A. (1991) An ecosystem model of forest tree persistence. *PhD Dissertation* (in preparation), University of California, Santa Barbara, CA.

Wu, Y., Turner, M.G., Romme, W.H., and Wallace, L.L. (1991) A landscape simulation model of winter foraging by large ungulates. *Ecological Society of America Bulletin.*

23

Integrating a Forest Growth Model with a Geographic Information System

ROBERT A. NISBET AND
DANIEL B. BOTKIN

A promising path of development for landscape models of ecological processes is to combine process-level models with GIS (Jarvis, 1989; Hunsaker et al., Chapter 22). The JABOWA forest growth model and its derivatives have been successful in simulating forest growth in many areas of the world during the past 20 years, and this model would seem to be a good starting point for such an integration. During the past 5 years, several attempts have been made to achieve this integration, but GIS has been used only as an output display device, not as an integral part of a dynamic simulation. That is, the JABOWA model projects forest growth for a number of independent plots with a variety of site conditions; the GIS is used to display data for each plot in its proper geographic location. However, there has been only very limited linkage of the state of one plot with the states of surrounding or more distant plots on a landscape.

This case study will discuss the JABOWA model briefly, review past progress to extend the model to determine spatial effects, and make some suggestions for the direction of future research in the development of landscape models of ecological processes.

The JABOWA model

The JABOWA model was developed to project forest growth for Northeastern United States forests (Botkin et al., 1972, 1973), and it has been extended to many other parts of the world (Shugart and West, 1977; Shugart et al., 1981; Pastor and Post, 1986; Solomon, 1986; Prentice, 1987; Kienast and Kuhn, 1989). The latest version of the model is described in detail elsewhere (Botkin, 1992). A few key features of the model are given here. The model is a simplified representation of forest growth as affected by available light, temperature, soil moisture, and available nitrogen in the soil (a measure of soil fertility). Light, soil moisture, and soil nitrogen factors are each expressed by functions representing stress-tolerant, stress-intermediate, and stress-intolerant effects. The temperature function is expressed as a parabola varying between a maximum of 1 to 0 at minimum and maximum values for growing degree days (a measure of the yearly heat accumulation above 40°F) (Figure 23-1). Soil moisture effects are expressed in terms of a dryness factor and a wetness factor. Effects of the dryness factor (SD) vary from 1 at a zero evaporative demand index to 0 at some higher evaporative demand index, depending on species-specific parameters (Figure 23-2). The wetness function reflects the negative effects on tree growth of lack of oxygen at the ends of roots. As a first approximation of this effect, the site wetness is expressed simply as the reciprocal of the water table depth (Botkin and Levitan, 1977):

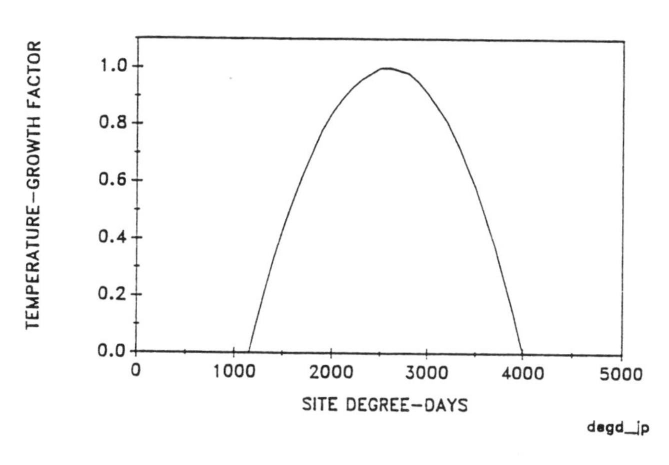

Figure 23-1: Temperature–growth factor by degree-days for jack pine (Pinus banksiana).

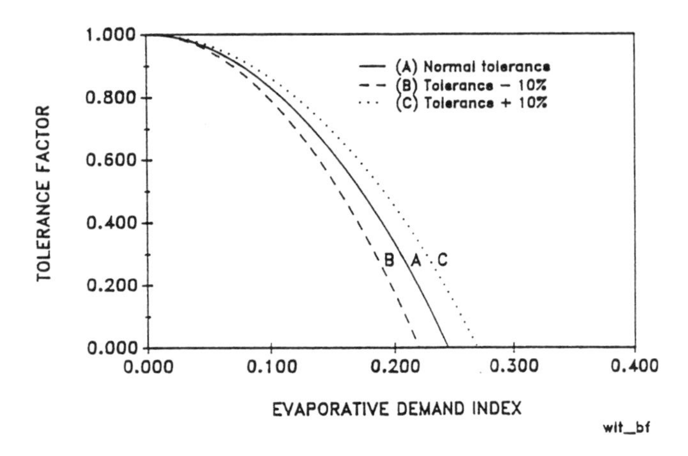

Figure 23-2: The dryness factor for balsam fir (Abies balsamifera) showing the effect of an error of plus and minus 10 percent in the estimation of this parameter on the response function (from Botkin, 1992).

$$WeF_i = \max(0,1\text{-}DTMIN_i/DT) \qquad (23\text{-}1)$$

where $DTMIN_i$ is the minimum distance to the water table tolerable for species i and DT is the depth to the water table. The functions for light, temperature, soil wetness, soil dryness, and soil nitrogen are expressed as factors ranging from 0 to 1.0 and are multiplied together to reduce the maximum growth rate ($GROWTH_{max}$) for a species under the climate conditions to project actual tree growth for the year:

$$\text{Actual growth} = GROWTH_{max}*AL*TMP*SW*SD*SN \qquad (23\text{-}2)$$

where $GROWTH_{max}$ is the maximum growth rate under optimal conditions, AL is the available light factor, TMP is the temperature factor, and SW, SD, and SN represent the soil wetness, soil dryness and soil nitrogen factors.

Plot size is chosen so a large tree can shade the entire plot in all locations at least once during the year. Shade-intolerant species have relatively low growth rates under the low light conditions of the understory, but they increase in growth more rapidly than shade-tolerant species as light intensity increases. This relationship is expressed in the available light factor and functions to restrict growth of understory trees competitively.

Regeneration is determined as a random function of the maximum number of trees reaching the sapling stage in a given year. Mortality is expressed in the model as age-related mortality and stress-related mortality.

JABOWA AND SPATIAL MODELING

GIS as a display device

The work of Pearlstine et al. (1985) represents an early attempt to combine a JABOWA-type model and a GIS. The growth of a bottomland hardwood forest in South Carolina was modeled in under several flooding scenarios related to river diversion (Pearlstine et al., 1985). The coastal drainage of the Santee River has been radically modified during the past 45 years by diversion for hydroelectric power generation (Figure 23-3). The extensive shoaling problems in Charleston harbor caused by the diversion led to a rediversion in 1968 of 80 percent of the river flow back through the Santee River, which in turn led to local flooding and decline of a bottomland hardwoods forest. A forest growth model was required to model effects of this rediversion and project effects on forest growth of a modified rediversion design.

A descendant of the JABOWA model (FORET; Shugart and West, 1977) was modified to account for effects of flood height and duration and used to project bottomland hardwoods forest growth with full rediversion (control) and a partial rediversion during the months of April through July each year. The model was run for sites at 15 cm elevation intervals. After each run, elevations were recorded at which a change in forest type occurred. These changes were used as guides in redrawing forest-type maps for comparison of each rediversion option, and the new forest-type maps were then mapped into a GIS for display (Figure 23-4). This study was important because it was the first demonstration of the integration of a JABOWA-type model with GIS.

A new version of JABOWA (JABOWA-II) was integrated with a GIS for display of model projections of forest growth under normal climate and global warming conditions (Van Voris et al., 1989). The model was run for 200 years for sites located at 100 m intervals between 150 and 1500 m elevation using the Hubbard Brook Experimental Forest data. Results for the normal and global warming simulations were mapped onto a topographic surface in 30 year time steps. The GIS display of total biomass at year 90 under normal conditions indicated that the highest total forest biomass levels occur at lower elevations, while under global warming conditions, the total forest biomass is increased at higher elevations and is greatly reduced at lower elevations.

Pearlstine et al. (1985) and Van Voris et al. (1989) used the GIS only as an output display of model results. There

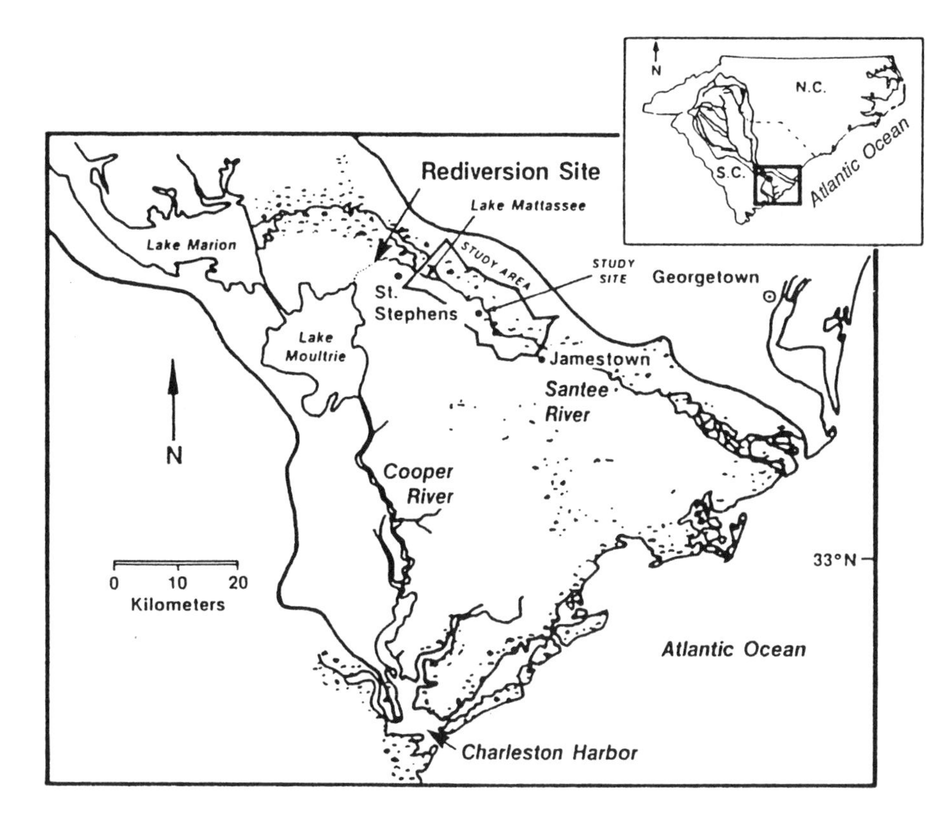

Figure 23-3: Location map of the Santee River floodplain and the rediversion site (from Pearlstine et al., 1985).

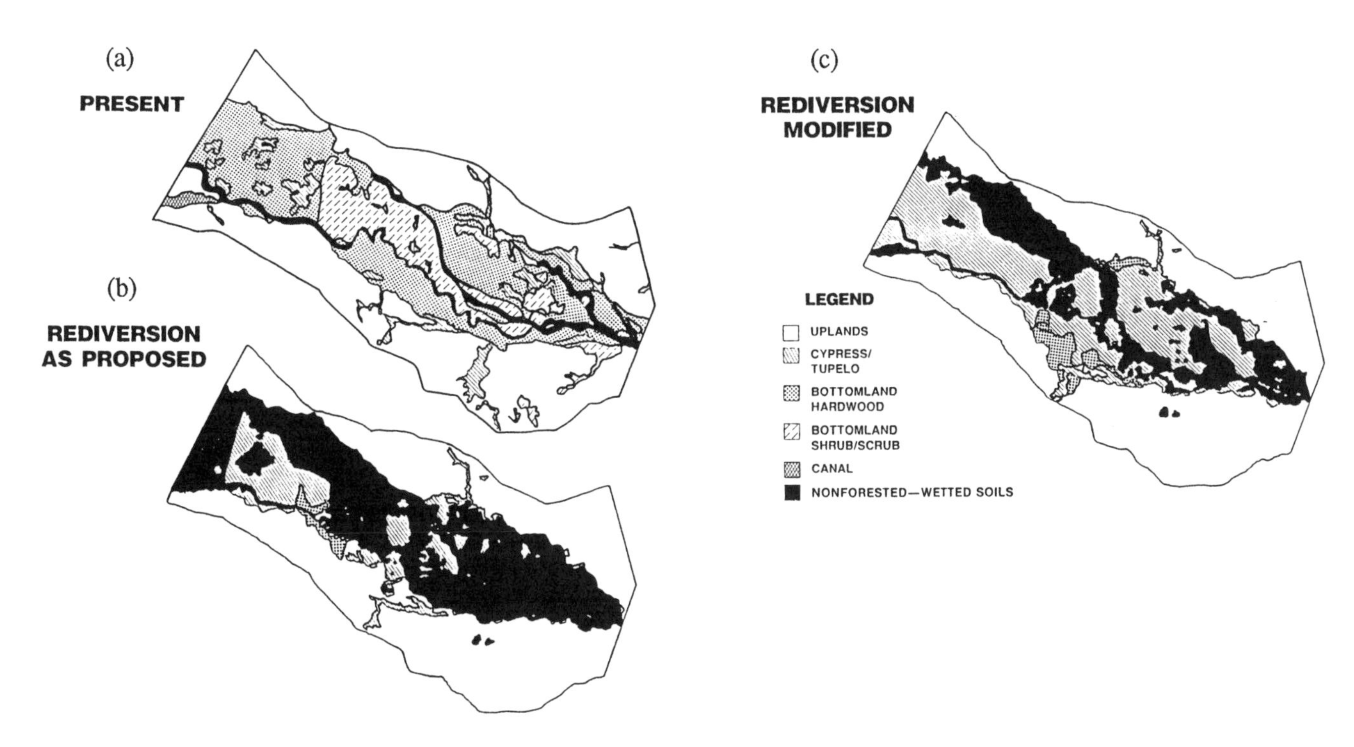

Figure 23-4: Area extent of habitat types in 1985 (a) and rediversion options: proposed plan (b) and modified plan (c) (from Pearlstine et al., 1985).

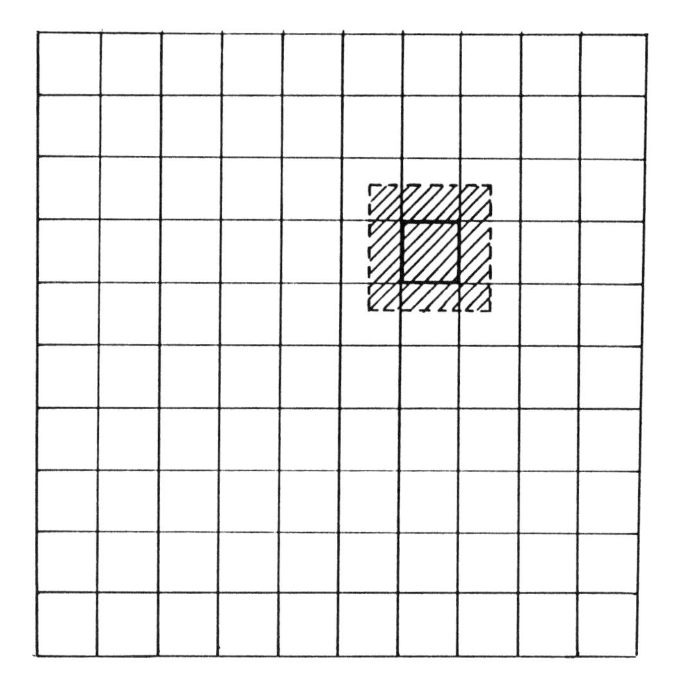

Figure 23-5: Schematic representation of a 0.04 ha zone of influence around a 0.01 ha grid cell. Leaf area and stand biomass aggregated for the area corresponding to the shaded area influenced tree growth for the grid cell (from Smith and Urban, 1988).

were no mechanisms to link change in the state of one plot to the states of surrounding plots. Use of process models such as JABOWA in landscape analysis requires "scaling up" of the analysis from a small plot basis (small area with a large geographic scale) to a much larger area with a smaller geographic scale (landscape). Scaling up to the landscape level raises several questions: (1) Does a whole forest respond to controlling variables the same way as a single tree? (2) Does scale affect the magnitude or direction of the forest response? (3) Does the state of one part of the landscape affect (feedback to) the states of other parts? (4) Are there any controlling variables that operate on the landscape level that do not operate at the plot level?

Spatial interactions

Recently, several studies using descendants of the JABOWA model have attempted to analyze landscape level effects in forests. While none of these studies used GIS, the spatial analysis techniques used are readily adaptable to GIS analysis. In the first study, a descendant of JABOWA was modified to keep track of biomass and leaf area half a grid cell distant from each simulated 10 × 10 m plot within a 30 × 30 grid of plots (Smith and Urban, 1988). This method simulated a roving window of influence that moved over the landscape and affected forest regeneration, growth, and mortality processes in the model (Figure 23-5). Principal components analysis was performed on the projections from the analysis of the 900 plots to distinguish important axes of spatial variation.

Another study using JABOWA-II (the latest version of JABOWA) considered the extinction probabilities of trees as a function of site heterogeneity and the proximity of seed trees in the surrounding landscape of 100 plots (10 × 10 grid) (Woodby, 1991). The model simulated growth on all 100 plots for 1000 years. Spatial interaction was represented by seed transport from surrounding plots to each of the 100 plots. The number of species persisting on the plot (species richness) was found to be dependent on patch size (number of contiguous plots), spatial variation in soil characteristics, and occurrence of infrequent severe disturbance (Figure 23-6). Only eight species persisted on a 100 plot grid (1 ha) under uniform control soil conditions and no added disturbance. However, when both moisture stress and available nitrogen were allowed to vary at random between 50 and 150 percent of control conditions, 55 species persisted to year 1000 on the 1 ha area. When severe canopy disturbances were permitted (by removing trees in the mortality routine), the number of species persisting on the 1 ha area nearly doubled to about a hundred. These results imply that the design of nature preserves should consider the inclusion of spatial variation in soils (and other site factors) and disturbance regimes, such as thinning or controlled burning. This approach could be extended in space to determine the size of a nature preserve that optimizes species persistence within other considerations, such as cost and land availability.

FUTURE DIRECTIONS

The insights provided by previous spatial applications of the JABOWA-type model suggest that a suite of forest growth models could be developed in the short term and applied to a variety of site conditions in major forest types on a landscape scale. Results could be composed into an atlas of forest response to projected normal climate and global warming conditions. One model would assume the null hypothesis: There are no spatial interactions. Additional models would assume simple spatial interactions (similar to the studies of Smith and Urban, 1988, and Woodby, 1991), and complex interactions could consider: (1) transport of materials by wind, water, and biological vectors; (2) windshed effects; for example, trees on the edge of a mountain slope protect interior trees from wind damage (Sprugel, 1985); and (3) pollution plume effects.

Desirable longer-term developments could include development of a complex modeling language to process

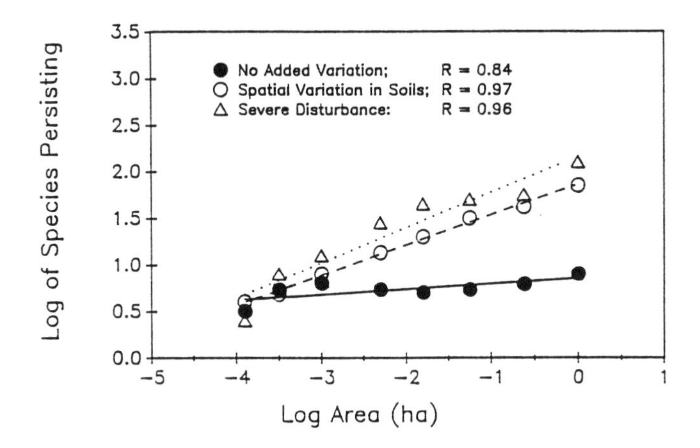

Figure 23-6: The number of species persisting as a function of area after 1000 years for three scenarios of spatial variability: spatial variation in soil characteristics, severe disturbances, and control (no added spatial variability).

Alternatively, GIS vendors could provide for system calls to model subroutines from within the GIS environment. Improved 2D and high-resolution 3D display capabilities would greatly enhance the effectiveness of modeling activities with GIS. Lam (Chapter 24) provides an excellent example of combining an ecological model with GIS and an expert system.

The JABOWA forest growth model and its descendants have proven to be valuable tools in analyzing forest growth. Further integration of the model with rapidly developing GIS technology presents an opportunity to increase the realism and extend the scale of forest growth projections. In summary, a GIS should be used for the source of spatial data to drive model processes; it should provide complex spatial analysis capabilities (available either by system calls by the model program, or calls to model subroutines); and it should provide display and storage capabilities of output data at various model time steps. The result of such integration can be truly a landscape-level analysis of forest growth that considers effects of spatial pattern and interactions, site heterogeneity, and diversity of disturbance regimes. Only by developing such capabilities can we analyze realistically the complex spatial relationships in forests.

REFERENCES

Botkin, D.B. (1993) *Forest Dynamics: An Ecological Model*, Oxford University Press.

Botkin, D.B., Janak, J.F., and Wallis, J.R. (1972) Rationale, limitations, and assumptions of a northeastern forest growth simulator. *IBM Journal of Research and Development* 16: 101–116.

Botkin, D.B., Janak, J.F., and Wallis, J.R. (1973) Some ecological consequences of a computer model of forest growth. *Journal of Ecology* 60: 849–872.

Botkin, D.B., and Levitan, R.E. (1977) Wolves, moose, and trees: an age-specific trophic-level model of Isle Royale National Park. *IBM Research Report RC 6834*, Yorktown, NY: IBM Research Center.

Jarvis, P.J. (1989) Atmospheric carbon dioxide and forests. *Philosophical Transactions, Royal Society of London* B324: 369–392.

Kienast, K., and Kuhn, N. (1989) Simulating forest succession along ecological gradients in southern Central Europe. *Vegetatio* 79: 7–20.

Pastor, J., and Post, W.M. (1985) Development of a linked forest productivity-soil process model. *ORNL/TM-9519*, Oak Ridge, TN: Oak Ridge National Laboratory, 162 pp.

Pearlstine, L., McKellar, H., and Kitchens, W. (1985) Modeling the impacts of a river diversion on bottomland forest communities in the Santee River floodplain, South Carolina. *Ecological Modelling* 29: 283–302.

Prentice, I.C. (1987) Description and simulation of tree-layer composition and size distributions in a primaeval Picea-Pinus forest. *Vegetatio* 69: 147–156.

Shugart, H.H., Mortlock, A.T., Hopkins, M.S., and Burgess, I.P. (1981) The development of a successional model for a subtropical rainforest and its application to assess the effects of timber harvest at Wiangarree State Forest, New South Wales. *Journal of Environmental Management* 11: 243–265.

Shugart, H.H., and West, D.C. (1977) Development of an Appalachian deciduous forest model and its application to assessment of the impact of the chestnut blight. *Journal of Environmental Management* 5: 161–179.

Smith, T.M., and Urban, D.L. (1988) Scale and resolution of forest structural pattern. *Vegetatio* 74: 143–150.

Solomon, A.M. (1986) Transient response of forests to CO_2-induced climate change: simulation modeling experiments in Eastern North America. *Oecologia* 68: 567–579.

Sprugel, D.G. (1985) Changes in biomass components through stand development in wave-regenerated balsam fir forests. *Canadian Journal of Forest Research* 15: 269–278.

Van Voris, P., Botkin, D.B., Nisbet, R., Woodby, D., Bergengren, J., Demarais, T., Millard, D., Thomas, John and Thomas, James (1989) TERRA-Vision: terrestrial environmental resource risk assessment—computer visioning system. *Exploratory Research Project Report*, Richland, WA: Batelle Pacific Northwest Laboratory, 21 pp.

Woodby, D.A. (1991) An ecosystem model of forest tree persistence. *Ph.D. dissertation*, Santa Barbara, CA: University of California.

24

Combining Ecological Modeling, GIS and Expert Systems: A Case Study of Regional Fish Species Richness Model

DAVID C. L. LAM

Ecological models for lakes and streams consider not only the plants and animals within several trophic levels, but also the physical, chemical, and biological processes. In a traditional approach, a comprehensive model involving all possible processes that are considered significant is implemented. However, it often requires an unnecessarily enormous amount of data. In another approach, models are developed for specific sites. The models are simplified and incorporate only those processes that are needed. These, however, are not generally applicable for all situations. In this paper, we present a third approach in which technologies such as expert systems and GIS are used to select appropriate models according to the classification of available data and geographical regimes. We will illustrate this new approach by using the lake acidification problem in Eastern Canada as a case study.

Acid rain is a major environmental problem in Eastern Canada. The main source of acidic deposition stems from industrial emissions of sulfur dioxide transported over long distances. Much of the acidic precipitation does not fall onto the lakes directly, but contacts land surfaces, follows hydrological pathways, and undergoes chemical changes including neutralization processes, before entering surface waters. Often, lakes and streams may have natural sources of organic acids from marshes and bogs. High acidity leads to ecological stresses, of which a common indicator is fish species richness, that is, the percentage of potential fish species present.

Several models have been developed to predict the effects of acidic deposition on the acid neutralization capacity (ANC) of lakes based on some key processes such as soil buffering (Thompson, 1982), weathering (Schnoor et al., 1986) and organic acidity (Lam et al., 1989a). We will explain the integration and role of GIS, along with expert system technology, in data aggregation and model selection over a large region. The objective is to estimate the damage to the fish species richness for the lakes in Eastern Canada, divided into subregions through GIS overlay analysis of deposition and geochemistry patterns.

GIS AND EXPERT SYSTEMS

The method that we propose is based on the RAISON (Regional Analysis by Intelligent Systems ON a microcomputer) software package (Lam and Swayne, 1991). The RAISON system (Figure 24-1) is a fully integrated database, spreadsheet, and graphic interpretive system with GIS and expert system capabilities for PC/AT microcomputers. While these subcomponents form a complete system, they can be used as a toolkit that links internal programming functions with external files from popular software. The program is written in the "C" language, and the package offers the RAISON Programming Language (RPL) which is similar to the "BASIC" language for more advanced computations such as modeling. The system is fully integrated through basic manipulation such as transfer of data, text, map, and graphics within the system, as well as seamless operation between these steps and modeling. A good example is the case where a model written in RPL solicits input from a map via the GIS and displays output on the map.

Expert or rule-based systems are computer programs that solve problems that normally call for expert knowledge and judgment. They rely on a database of knowledge, that is, a knowledge base, about a particular subject and skills at solving the problems. The main logical component of an expert system is the inference engine, which contains the procedures for reasoning and for making conclusions using the knowledge base. A rule-based system is a simple expert system that uses a series of rules structured in a logical sequence (known as IF–THEN–ELSE rules) that permits a correct solution to a question to be selected from a series of possible outcomes.

In the RAISON system, a logical inferencing subsystem using rule-based knowledge has been implemented (Figure 24-1). The expert rules can be easily entered using spreadsheet techniques (Lam and Swayne, 1991) and can then be executed using information from the database (Figure 24-1). In this paper, stations were first grouped by

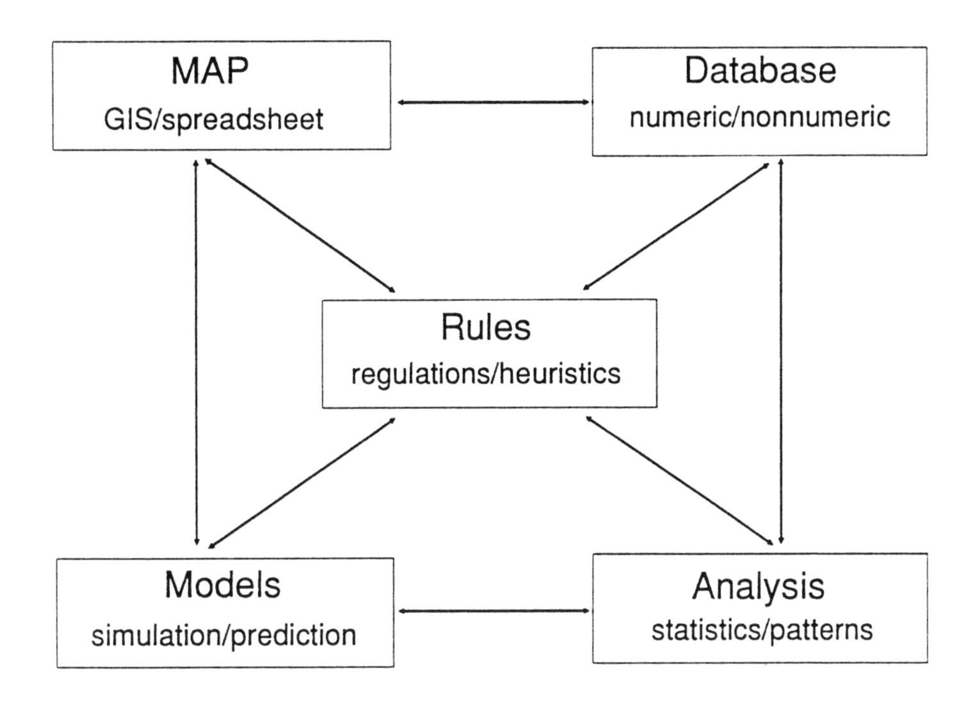

Figure 24-1: RAISON paradigm: a fully integrated system with five primary components linked to each other.

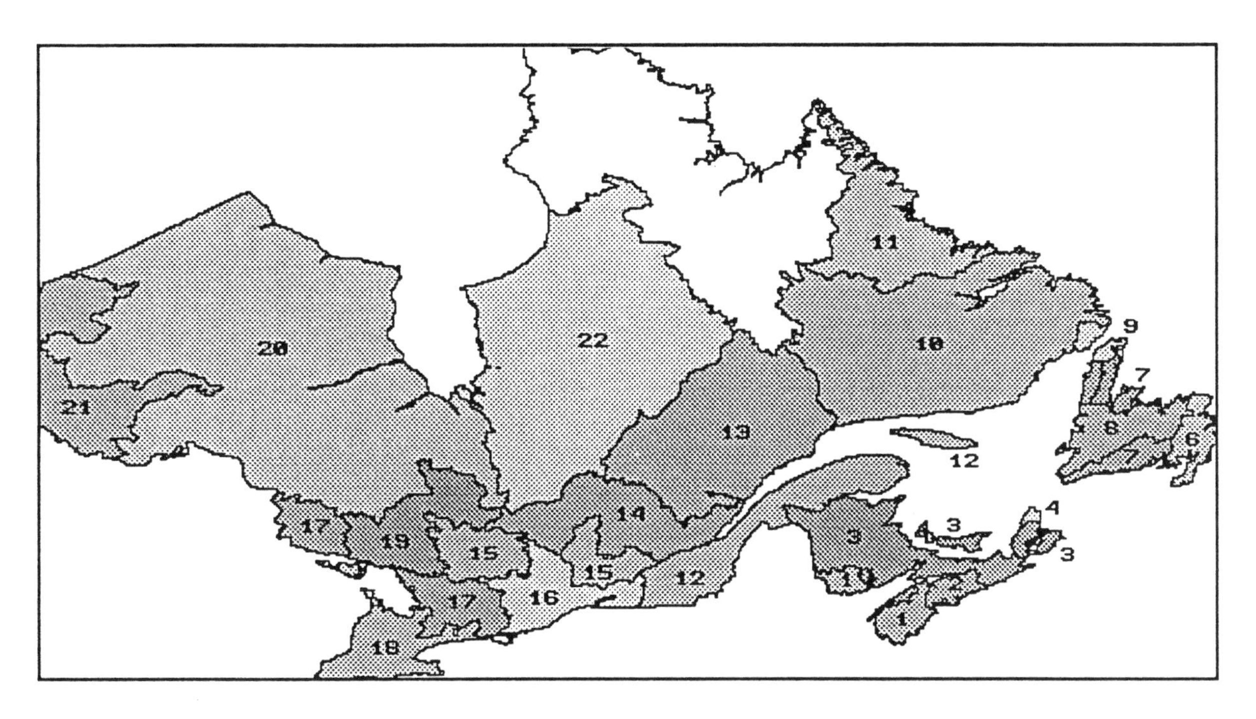

Figure 24-2: The 22 subregional watersheds for Eastern Canada: 1. South Nova Scotia and New Brunswick, 2. Mid-eastern Nova Scotia, 3. North Nova Scotia and New Brunswick, 4. Cape Breton Highlands, 5. Avalon, 6. East Newfoundland, 7. North and South Newfoundland, 8. Central and West Newfoundland, 9. Belle Isle, 10. East Quebec and South Labrador, 11. North Labrador, 12. St. Lawrence South shore, 13. Saguenay, 14. Laurentide, 15. Southwest Quebec, 16. Ottawa Valley, 17. Central Ontario, 18. South Ontario, 19. Sudbury–Noranda, 20. North Ontario, 21. Northwest Ontario, and 22. North Quebec.

subregions. Model selection for each station was then made according to expert rules. From these optimized results, regional distributions of lake pH were predicted for several sulfur-loading scenarios, and the fish species richness under these scenarios was derived using maps from GIS overlay analysis.

REGIONAL AGGREGATION OF LAKE DATA

In the past decade, 20,629 data records on lake chemistry from 8,505 stations contained in 325 tertiary watersheds in Eastern Canada have been assembled. Over 2,000 stations were found to have sufficient data for modeling purposes and were used in this study (i.e., satisfying ionic balance requirement). In order to obtain geographic units amenable to presentation of spatial variations, we identified 22 subregions (Figure 24-2), each of which is an aggregation of tertiary watersheds. The aggregation or grouping of watersheds is based on the criterion that it must minimize intragroup variance and maximize intergroup variance using specific conductivity as an indicator. Subjective knowledge of local variability in geology and sulfate deposition was used in defining the boundaries of the subregions. The final boundaries of the subregions were decided by a group of experts through several iterations of statistical calculations and GIS overlay analysis in the RAISON system.

MODELS AND EXPERT SYSTEMS

We consider two inorganic acidity models, the CDR model which is based on the cation denudation rate (Thompson, 1982), and the TD model, which is based on weathering rate (Schnoor et al., 1986), and two organic acidity models, the BO model, which uses a monoprotic formulation (Oliver et al., 1983) and the LTH model, which uses a triprotic formulation (Lam et al., 1989a). These models can be combined (e.g., CDR-BO) to calculate the ANC or alkalinity of the lake at steady-state equilibrium with a given acidic deposition (Lam et al., 1992). Since each of the six possible combined models (Figure 24-3) emphasized only one or two major processes, it may not be applicable for lakes where these processes are less important. As an example, Figure 24-3 shows the computed median ANC values for those subregions with observed ANC medians less than 0.1 milliequivalents/liter (meq/L). Selecting the best model is difficult, since no one model is clearly superior over all cases.

Lam et al. (1988) proposed that instead of choosing only one of the several models for all hydrogeochemical regimes, the appropriate model should be chosen for a given type of regime and available data. A simple set of expert rules was used to delineate model selection. For example, organic acid models are used to account for natural acidity, if water color > 30 Hazen units and/or

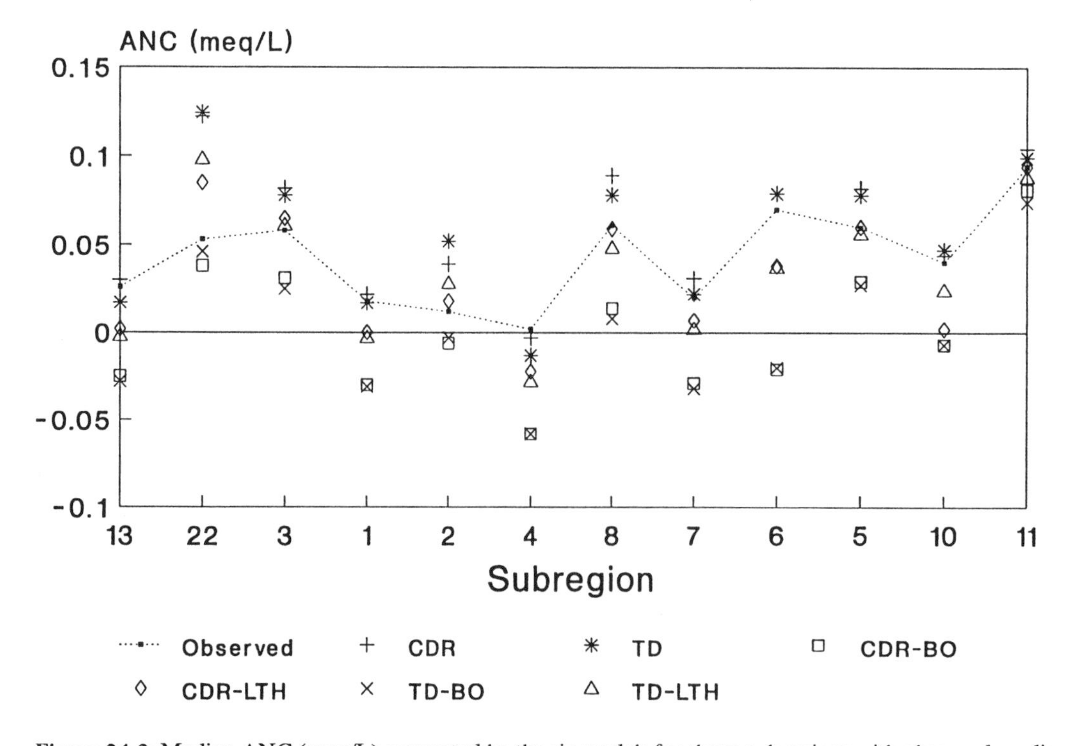

Figure 24-3: Median ANC (meq/L) computed by the six models for those subregions with observed median ANC less than 0.1 meq/L.

dissolved organic carbon > 4 mg/L. These rules were described in detail in Lam et al. (1992) and were applied to every station in each subregion. In this way, the most appropriate of the six models was chosen for each station. Comparison with observed data from 396 stations in South Quebec showed that the expert system model results are superior to the individual model results (Lam et al., 1989b). For Eastern Canada, similar conclusions were made on the results of lake ANC for various subregions (Figure 24-4), with a mean relative error of about 33%, which is well within the variability of the observed data.

FISH SPECIES RICHNESS MODEL

Most biological damage models use damage transfer functions expressed in terms of pH, not ANC. The pH can be computed from the ANC, based on an inverse hyperbolic sine function (Lam et al., 1992). For each subregion, the observed pH was used to calibrate the function based on the ANC and organic anion present, producing a unique regional pH–ANC relationship. Presence/absence data for fish species are available in Canada and useful for developing fish species richness models. For example, Matuszek and Beggs (1988) found that based on lake data of Ontario, fish species numbers increased with lake area according to an empirical equation:

$$\log(\text{species number}) = 0.52 + 0.22 \log(\text{lake area}) \quad (24\text{-}1)$$

After filtering out the lake area effects, they then obtained a linear relationship between pH and the mean residual number of species (i.e., reduction in the number of species). This relationship was assumed to hold for lakes in Eastern Canada. The fish damage (i.e., the percentage reduction in the number of fish species) due to different acid loads can be calculated by computing the ANC for a given load with the expert model, obtaining the pH from

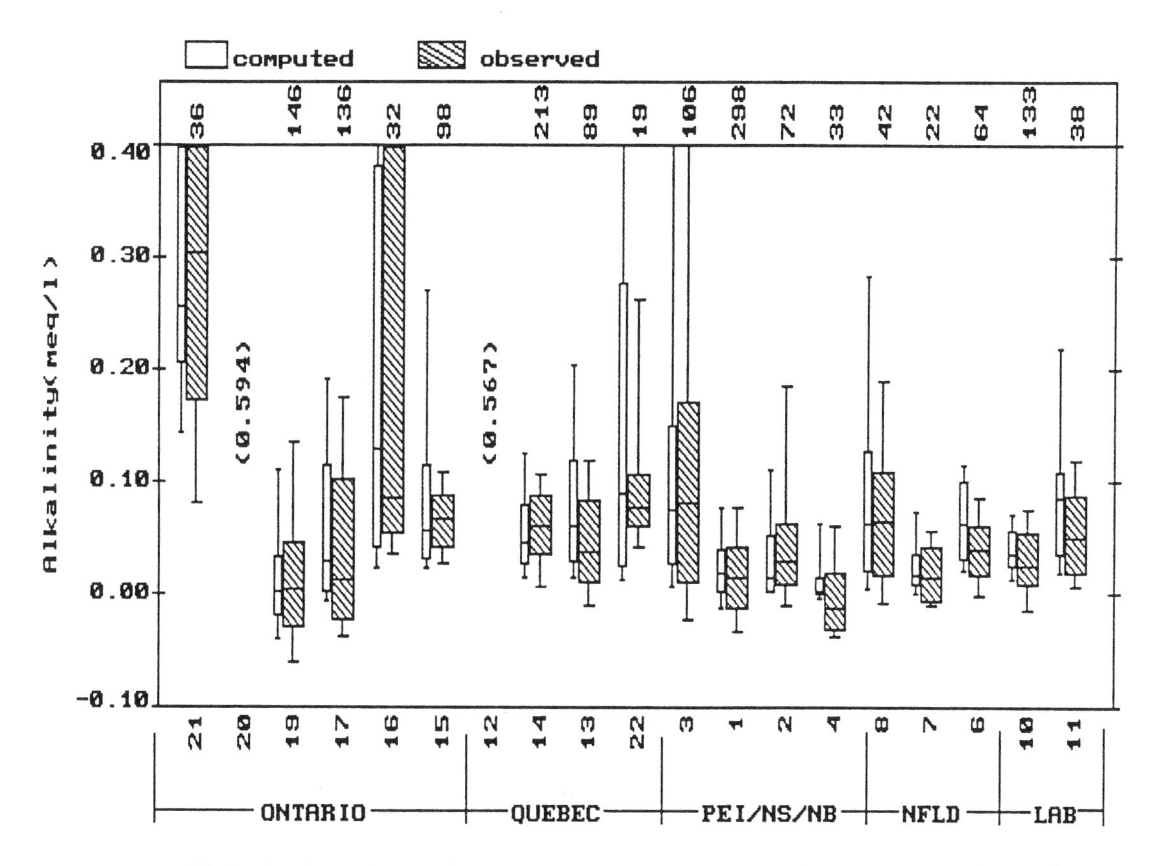

Figure 24-4: ANC distributions (with median denoted by the bar in the box which contains the 25th to 75th percentiles) observed and computed by expert model for selected subregions. The subregion numbers are given on the *x* axis and grouped by provinces: Ontario, Quebec, Prince Edward Island (PEI), Nova Scotia (NS), New Brunswick (NB), Newfoundland (NFLD), and Labrador (LAB). The number of stations used in each subregion is shown at the top of the figure (for subregions 20 and 21, the observed median ANC value is given in parenthesis as it exceeds the scale of the *y* axis).

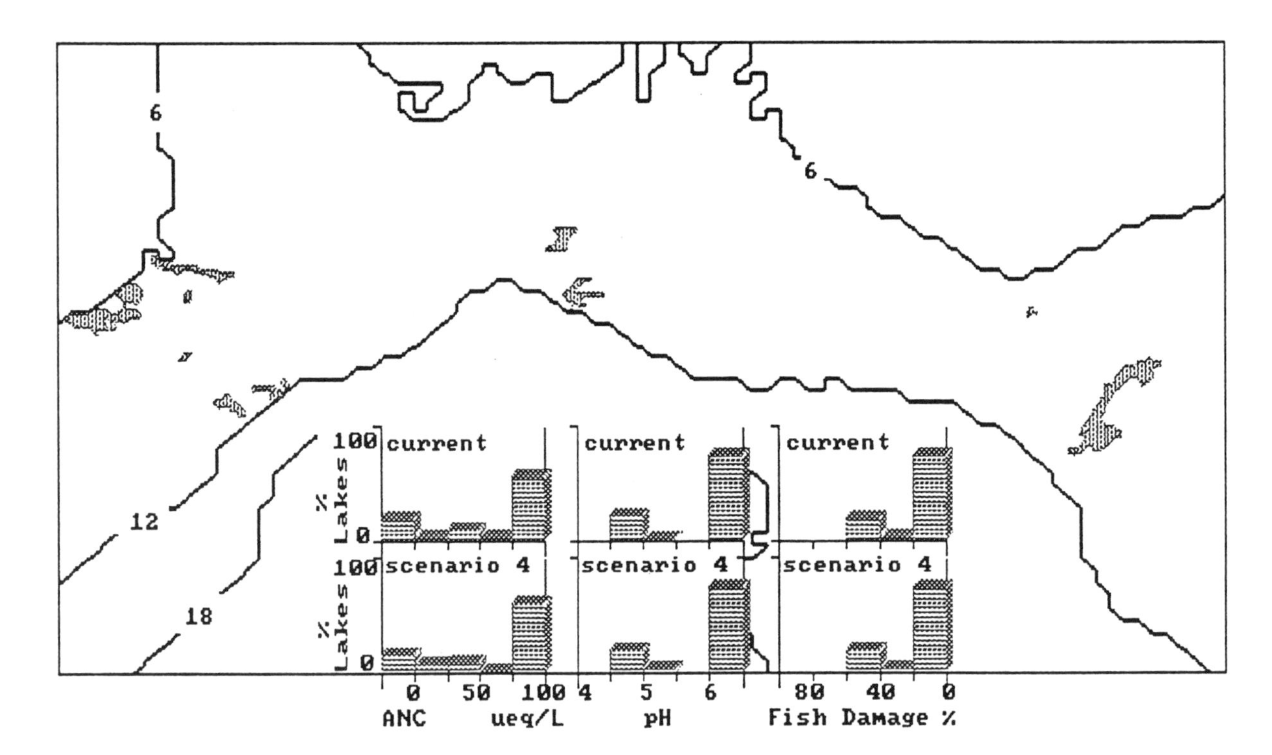

Figure 24-5: Percentage of lakes predicted to have a class of ANC, pH, and fish damage for current and future scenarios for low deposition and low sensitivity areas, with SO_4 load contours (kg/ha/y).

the regional pH–ANC characteristics and then estimating the damage by the empirical pH–species relationship.

Four scenarios of wet sulfate deposition were originally considered. However, for this chapter, we use only two scenarios. Scenario 1, or the base case, consists of the observed average deposition between 1982 and 1986. Scenario 4 represents the deposition load for Canadian and U.S. sulfate reduction programs as currently scheduled. In order to present results for these deposition scenarios, it is necessary to classify areas according to deposition levels. Thus, the deposition is deemed high if it is over 18 kg/ha/y, medium if it is between 12 and 18 kg/ha/y, low if it is between 6 and 12 kg/ha/y, and at background level if it is less than 6 kg/ha/y.

To illustrate the use of GIS techniques in the RAISON system further, we use the map of sulfate deposition as well as a map of the so-called terrestrial sensitivity (Environment Canada, 1988). The terrestrial sensitivity is classified into three levels (high, medium, and low) based on soil and bedrock properties (mineralogy, texture, and thickness). According to this classification scheme (Environment Canada, 1988), terrain ranked with a high sensitivity is more susceptible to acid rain than those ranked as low sensitivity.

As an example, we use the GIS feature in RAISON to generate two new special regions. In the first special region, the overlap areas with low deposition and low terrestrial sensitivity were defined. These areas are represented by the shaded areas in Figure 24-5, shown with the observed wet sulfate deposition contours. This map uses the same scale and covers the geographic area (not shown in Figure 24-5) as Figure 24-1.

The second special region is defined by high deposition and high terrestrial sensitivity (shaded areas in Figure 24-6). These two special regions represent two extreme cases. In Figure 24-5, the results were for little or no change in deposition, and, since the land was well buffered, no noticeable changes in ANC, pH, and fish species were predicted. In Figure 24-6, the results were for high deposition with substantial changes between Scenarios 1 and 4, and for poorly buffered lands. The percentage of lakes with low ANC values (e.g., negative values) were predicted to decrease from Scenario 1 to Scenario 4, showing a similar transformation of low pH lakes to high pH and from more damaged lakes to less damaged.

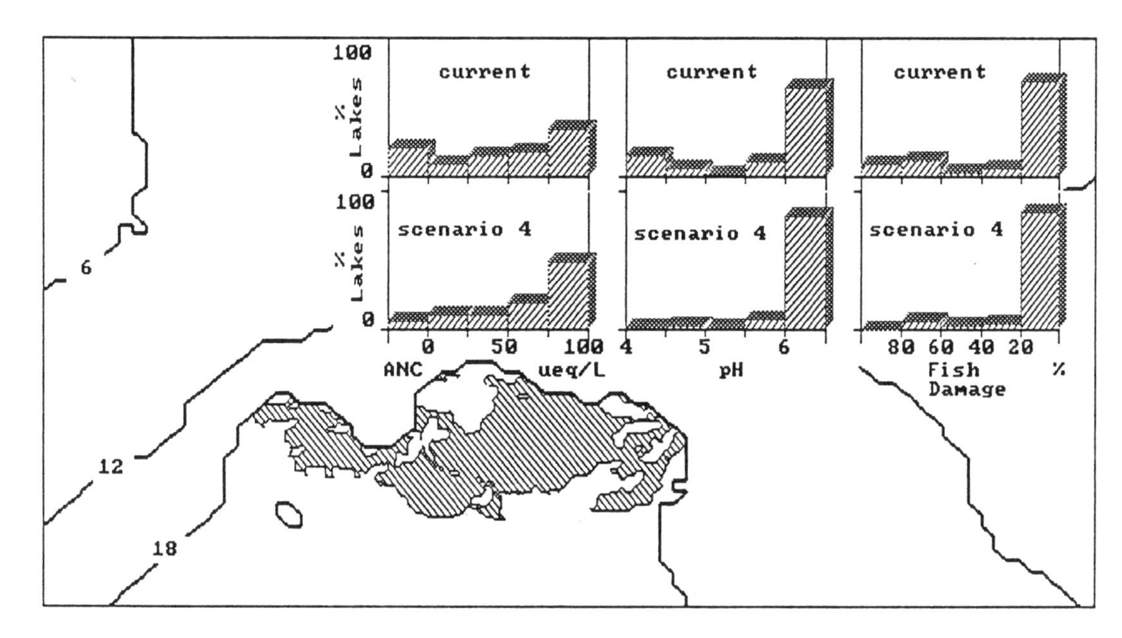

Figure 24-6: Percentage of lakes predicted to have a class of ANC, pH, and fish damage for current and future scenarios for high deposition and high sensitivity areas, with SO4 load contours (kg/ha/y).

CONCLUSIONS

This case study shows how the use of GIS in lake acidification modeling can be greatly enhanced with an intelligent interface, not only for manipulating the database but also for aggregation of areas with similar characteristics and for chemical ionic balance diagnostics. A fully integrated system with programming capability such as RAISON offered a viable approach (Hunsaker et al., Chapter 22) to combining GIS, models, and expert systems.

ACKNOWLEDGMENTS

The author thanks D.A. Swayne, I. Wong, J. Storey, and J.P. Kerby for development work and D.S. Jeffries, J. Gunn, C.K. Minns, and A.S. Fraser for data and advice.

REFERENCES

Environment Canada (1988) Acid rain: a national sensitivity assessment. *Environmental Fact Sheet 88-1*, Ottawa: Inland Waters and Lands Directorate.

Lam, D.C.L., Fraser, A.S., Swayne, D.A., and Storey, J. (1988) Regional analysis of watershed acidification using the expert systems approach. *Environmental Software* 3: 127–134.

Lam, D.C.L., Bobba, A.G., Bourbonniere, R.A., Howell, G.D., and Thompson, M.E. (1989a) Modeling organic and inorganic acidity in two Nova Scotia rivers. *Water, Air, and Soil Pollution* 46: 277–288.

Lam, D.C.L., Swayne, D.A., Fraser, A.S., and Storey, J. (1989b) Watershed acidification using the knowledge-based approach. *Ecological Modeling* 47: 131–152.

Lam, D.C.L., and Swayne, D.A. (1991) Integrating database, spreadsheet, graphics, GIS, statistics, simulation models and expert systems: experiences with the RAISON system on microcomputers. *NATO ASI Series, Vol. G 26*, Heidelberg: Springer-Verlag Publ., pp. 429–459.

Lam, D.C.L., Wong, I., Swayne, D.A., and Storey, J. (1992) A knowledge-based approach to regional acidification modeling. *Journal of Environmental Assessment and Monitoring* 23: 83–97.

Matuszek, J.E., and Beggs, G.L. (1988) Fish species richness in relation to lake area, pH and other abiotic factors in Ontario lakes. *Canadian Journal of Fisheries and Aquatic Science* 45: 1931–1941.

Oliver, B.G., Thurman, E.M., and Malcolm, R.L. (1983) The contribution of humic substances to the acidity of colored natural waters. *Geochimica et Cosmochimica Acta* 47: 2031–2035.

Schnoor, J.L., Lee, S., Nikolaidis, N.P., and Nair, D.R. (1986) Lake resources at risk to acidic deposition in the eastern United States. *Water, Air, and Soil Pollution* 31: 1091–1101.

Thompson, M.E. (1982) The cation denudation rate as a quantitative index of sensitivity of Eastern Canadian rivers to atmospheric precipitation. *Water, Air, and Soil Pollution* 18: 215–226.

25

Introduction to Quantitative Methods and Modeling in Community, Population, and Landscape Ecology

CAROL A. JOHNSTON

Ecologists have traditionally focused on organisms. Population ecologists have studied groups of organisms of the same species, while community ecologists have studied assemblages of organisms of different species. In either discipline, the primary emphasis has been on interactions that affect organism numbers, and changes in those numbers over time. Spatial interactions, such as the distribution of organisms relative to each other and/or to their physical environment, were generally overlooked or oversimplified.

With the development of landscape ecology in the 1980s, however, ecologists began to focus more on the effect of location. Landscape ecology is defined as the study of the structure, function, and change of interacting ecological entities in a heterogeneous land area composed of groups of organisms (including humans) interacting with their nonliving environment (Forman and Godron, 1986). Since spatial distribution is inherent to the discipline, landscape ecologists have widely embraced GIS as an important analysis and modeling tool. Innovative GIS modeling techniques are also being used in the more traditional ecological disciplines (Johnston, 1991).

The distinction between landscape ecology and population or community ecology is largely a matter of extent: Landscape ecologists study populations and communities, but over larger areas than do population or community ecologists. Landscape ecologists also pay attention to the effect of location, which is generally ignored by scientists in more traditional ecological disciplines. Since GIS pay attention to location by necessity, most studies of populations or communities that employ a GIS could be considered to be landscape ecology studies. Therefore, this chapter is divided into sections on GIS for the analysis of populations and communities, each beginning with the types of research questions asked by more traditional ecologists and ending with those addressed by landscape ecologists.

MODELING POPULATIONS WITH GIS

Population ecology is the study of the growth, demographics, and interactions of groups of individuals of the same species (Wilson and Bossert, 1971). Population growth at a particular location is affected by both intrinsic (e.g., birth rate, death rate, survivorship) and extrinsic factors (e.g., physical environment, interactions with other species). Spatial limits are implicit in population ecology: A population is a count of organisms that exist within a unit area, or an area with defined boundaries (e.g., a country, a wildlife refuge).

Most population models seek to answer the basic question "How many organisms are there now, and how many will there be in the future?" When this question cannot be answered directly, other population questions come into play: "Where are they (or where are they likely to be)?"; "What factors influence how many there are and where they are?"; "How do populations influence each other?"; "How do populations influence their environment?" GIS modeling has been applied to all of these questions.

How many are there?

Populations are rarely mapped, because: (1) individual organisms are too small to be detected by most remote sensing techniques, (2) ground-based mapping of individual organisms is time-consuming and expensive, and (3) with the exception of long-lived plants (e.g., trees), most populations are so dynamic that maps of their distribution would become rapidly outdated.

While the resolution of remote sensors is generally too coarse to detect the individuals in a population, remote sensing has been used to map vegetation disturbances caused by certain populations, such as insect pests. Empirical relationships between population numbers and disturbance effects can be developed and used to estimate populations based on remotely sensed information alone.

This approach was used by Broschart et al. (1989) to predict beaver colony density based on vegetation alteration caused by beaver dam building. By regressing field counts of beaver colonies against GIS-derived data from air photo maps of beaver pond vegetation, they developed empirical models for predicting population based on parameters derivable from air photo interpretation. This type of empirical model can then be used to estimate populations in areas lacking field data.

Where are they?

Range maps provide information about the geographic distribution of populations, although they provide no information about population numbers. Range maps can be overlaid in a GIS to determine the spatial coincidence of several species, which provides a measure of biodiversity (Davis et al., 1990). Although range maps are traditionally drawn as polygons subjectively circumscribing the locations of field observations, GIS techniques can be used to improve range map accuracy by recording actual observations (McGranaghan, 1989). This also provides the potential for quantitative analysis of population distributions, which can be used to assess the significance of changes in species ranges (Morain and Harrington, 1990).

Location of organisms is dynamic, due to both organism invasions/extinctions and the movement of mobile organisms. Field monitoring methods such as radiotelemetry have long been used by wildlife ecologists to record the movement of animals, and this information can be combined with GIS-derived environmental data to determine habitat use at a scale and location appropriate to the organism (Young et al. 1987). Field sightings and remote sensing can be useful in detecting change in organism location, such as the spread of exotic invaders (e.g., gypsy moth, zebra mussel). The predicted rate of spread based on biological mobility can be compared with monitored rates of spread to determine if other modes of transport, such as human transportation networks, are accelerating their dispersal.

At the landscape level, the theory of island biogeography has applied principles of population demographics to account for how migrating species coexist or go extinct (MacArthur and Wilson, 1967). More recently, population ecologists have begun to build simulation models that incorporate dispersal of organisms with models of population demographics (Fahrig, 1991). A simulation experiment with these models concluded that to ensure survival in the presence of a disturbance, it was most important for a species to have a high dispersal rate and a high disperser survival rate. When the models were applied to a simulated patchy landscape, the most important determinants of mean local population size were the fraction of organisms dispersing from the patches and the probability that they would detect new patches (Fahrig, 1991). The linkage of such models with real landscapes in a GIS would facilitate the building and testing of new population models that incorporate organism movements.

What factors influence how many there are and where they are?

Populations vary in their sensitivity to environmental conditions. Under conditions where unlimited growth may occur, populations increase exponentially, approaching infinity over time. In situations where the environment is limiting, however, population growth is predicted by the Verhulst growth equation:

$$dN/dt = rN\,[(K-N)/K] \tag{25-1}$$

where the growth of a population (dN) increases to an upper asymptote known as the carrying capacity of the environment (K), as a function of its own density (N) and intrinsic rate of increase (r) (Pearl and Reed, 1920).

Table 25-1: Life history strategies of populations (MacArthur and Wilson, 1967; Pianka, 1970)

PARAMETER	r SELECTION	K SELECTION
Population controlled by:	Maximal intrinsic rate of natural increase (r_{max})	Carrying capacity of the environment (K)
Population size:	Variable in time Density independent Nonequilibrium Below carrying capacity Frequent recolonization	Fairly constant in time Density dependent Equilibrium Near carrying capacity Recolonization unnecessary
Environmental conditions:	Variable, unpredictable climate Frequent disturbance Abundant resources	Constant climate Little disturbance Limited resources

Growth curves for populations limited by their environment are logistic, in which growth starts slowly, accelerates rapidly, then decelerates and plateaus at a more or less constant rate (Wilson and Bossert, 1971).

Populations fall between two extremes of life history strategies, depending on whether they are environmentally or growth limited (Table 25-1). Populations whose growth is controlled by their intrinsic rate of natural increase (r_{max}) are known as *r* strategists. Their high reproduction rate allows *r* strategists (e.g., insects, annual plants) to colonize ecological vacuums rapidly, such as areas where disturbance or catastrophic mortality have reduced pre-existing populations. Populations whose growth is limited by environmental carrying capacity are known as *K* strategists. *K* strategists (e.g., perennial plants, terrestrial invertebrates) have a much lower intrinsic rate of increase, and a population size more or less at equilibrium near the carrying capacity of the environment.

The use of GIS to analyze and model populations varies according to their life history. With *K*-strategists, it is possible to model populations cartographically (Tomlin, 1990) based on mapped environmental variables that are related to their survival, such as energy (light, food), water, oxygen (for aquatic organisms), nutrients, and physical habitat (rooting medium, space). For example, vegetation and road databases could be analyzed to determine coincidence of food, cover, and isolation requirements for an herbivore such as moose or caribou, yielding a habitat suitability map (Figure 25-1a). GIS modeling of habitat suitability has been used with many *K*-strategist species: wild turkeys (Donovan et al., 1987), golden-cheeked warblers (Shaw and Atkinson, 1988), wood storks (Hodgson et al., 1988), deer (Stenback et al., 1987; Tomlin et al., 1987), the gopher tortoise (Mead et al., 1988), and the California condor (Scepan et al., 1987). Cartographic models (i.e., whole mosaic models) are further discussed by Hunsaker et al. in Chapter 22.

When the influence of environmental factors on populations is not known, a GIS can be used to evaluate the spatial coincidence of populations relative to their environment. For example, animal locations monitored by radiotelemetry have been used in a GIS to analyze the land cover characteristics of the home ranges of bats (Cook et al., 1988) and spotted owls (Young et al., 1987). Not only does this type of analysis provide relationships that can be used in models that infer population numbers based on habitat characteristics, but it also provides information on the spatial extent and overlap of home ranges that can be used to analyze interactions among individuals in a population. Population/environment relationships can also be analyzed using historical field sightings of animals (Agee et al., 1989). Empirical relationships between organisms and their environment can yield information essential to the forecasting of environmental impacts on populations.

Since *r* strategists are not limited by environmental carrying capacity, it is more difficult to predict their populations by static habitat variables (Morse, 1986). GIS can be used, however, in combination with remote sensing to display changes in the distribution of *r* strategist species that damage vegetation, such as gypsy moths (Rowland, 1986) and oak wilt (Ware and Maggio, 1990). Some *r* strategist species are associated with mappable disturbances, such as fire or flooding, which could be used to predict the probability of their occurrence (Figure 25-1b).

A more sophisticated approach to predicting the occurrence of *r* strategist species is to combine GIS with expert systems, computer systems that advise on or help solve real-world problems that would normally require a human expert's interpretation. Three types of rules are

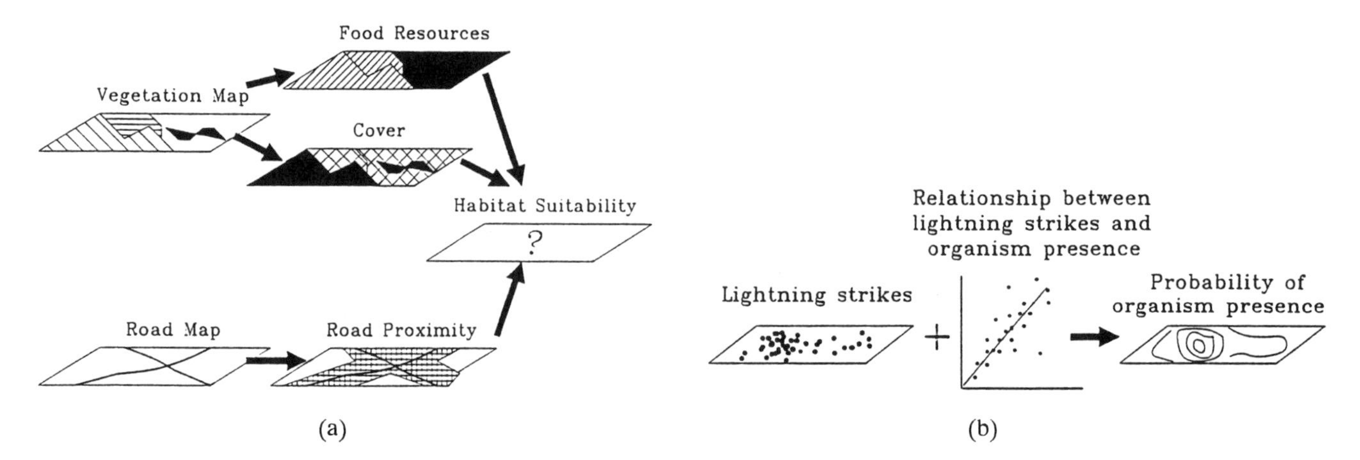

Figure 25-1: (a) Cartographic modeling of habitat suitability for a *K*-strategist population. The distributions of environmental variables known to affect organism abundance are used to predict the location of suitable habitat. (b) Probabilistic GIS modeling of an *r*-strategist population. An empirical relationship is developed between the population and a disturbance that affects it, and used to plot a probability surface of organism distribution.

developed and programmed into the expert system: (1) database rules to evaluate numerical information, (2) map rules to evaluate mapped categorical variables, and (3) heuristic rules to evaluate the knowledge of domain experts. Coulson and co-workers (1991) have used this approach to develop an "intelligent GIS" for predicting the hazard of southern pine beetle outbreaks, based on environmental factors (i.e., forest composition) and disturbance probabilities (i.e., stress potential, lightning strikes). Once an expert system has been developed for a species, it can be used not only for scientific research, but also to predict hazards and provide decision support for management (White, 1986; Johnston, 1987). Lam (Chapter 24) presents a case study using an expert system and GIS for modeling regional fish species richness.

Although most of the previous examples have dealt with small spatial areas, it is possible with the use of simulation models to bridge the gap between organism-scale phenomena up to landscape and even continental scales. For example, the FORET forest-stand model simulates the birth, growth, and death of individual trees based on deterministic, intrinsic stand variables (shading, crowding) and stochastic environmental variables (heat sums, temperature extremes, soil moisture) (Shugart and West, 1980). Pastor and Post (1988) used a version of FORET to simulate the changing distribution of forest species in eastern North America in response to projected climatic change under a doubling of current CO_2 concentration. Their model assumed that species migrated to new sites to the extent that temperature and soil water availability were optimal and that growth was limited by temperature, water, nitrogen, or light, whichever was most restrictive under the altered climate. Forest biomass was predicted by species for twenty climatic stations across eastern North America, and interpolated and displayed using an ERDAS GIS (Pastor and Johnston, in press). Nisbet and Botkin (Chapter 23) present a case study on integrating a similar forest growth model with GIS.

The key to extrapolating organism-scale models successfully to landscapes is to focus on ecological properties that transcend spatial and temporal scales. In the example of the forest growth models (Pastor and Johnston, in press; Nisbet and Botkin, Chapter 23), the same model was applicable at different spatial scales because the cumulative growth of individual trees could be summed over a larger area to estimate biomass accrual of an entire forest. Another approach is to link separate models that operate within narrow spatial and temporal limits into a coupled model that can address ecological processes across a variety of scales (e.g., Turner and Dale, 1991).

How do populations influence each other?

Populations do not exist in a biological vacuum; they interact with each other in ways that may be beneficial, detrimental, or neutral with regard to their population growth. Interactive biological models with some spatial characteristics include competitive interaction models, patch dynamics, interference and exploitation, and niche space (Sklar and Costanza, 1991). Other interactive models are improved by adding a spatial element, such as the widely used Lotka–Volterra equations describing predator–prey interactions:

$$dN_{prey}/dt = rN_{prey} - \alpha_{prey} N_{pred} N_{prey} \tag{25-2}$$

$$dN_{pred}/dt = \alpha_{pred} N_{prey} - mN_{prey} \tag{25-3}$$

where prey densities (N_{prey}) are controlled by an intrinsic rate of increase r, predator densities (N_{pred}), capture efficiency ($\alpha_{prey}/\alpha_{pred}$), and predator mortality m (Lotka, 1925; Volterra, 1926). Nachman (1987) designed a more stable predator–prey model by incorporating spatial realism and synergistic feedbacks such as environmental patchiness, dispersal characteristics, behavioral responses, and demographic stochasticity. Other computer simulation studies have increased our knowledge of the effect of dispersal on predator–prey interactions (Hastings, 1977; Caswell, 1978; Turner and Dale, 1991). Spatial models of this type are conducive to use with a GIS, particularly when the spatially distributed variables are in the form of maps.

How do populations influence their environment?

Habitat characteristics are often thought of as invariant, but both plants and animals are capable of altering their environment to varying degrees. Plants do so by taking up moisture and returning litter to the soil, animals do so by foraging and physical habitat alteration (e.g., mixing soils by burrowing). This alteration can create feedbacks that can further alter population–environment interactions. Although such feedback mechanisms are built into many ecosystem-level simulation models (Pearlstine et al., 1985; Pastor and Post, 1988; Burke et al., 1990), there are fewer examples of population models that predict population effects on environment (e.g., Hyman et al., 1991).

Activities by animal populations can alter environments at the landscape level. Johnston and Naiman (1990b) used GIS analysis to study long-term landscape alteration by beaver (Castor canadensis). By changing the flow of water in the landscape, beaver impoundments convert terrestrial to aquatic systems, increasing landscape heterogeneity by creating a spatial mosaic of aquatic and semiaquatic patches in an otherwise forested matrix. Maps of beaver ponds prepared from six dates of historical aerial photography taken between 1940 (when the population was low due to overtrapping) and 1986 (when the population was high) were analyzed with a raster-based

GIS, and overlay techniques were used to quantify hydrologic and vegetation changes between map dates. The results indicated that not only did beaver influence a large proportion of landscape area (13%), but that the rate of landscape alteration over the first 20 years of recolonization was much higher than the rate of population increase, indicating that beaver colonies were moving to new habitat sites for reasons other than population growth (Johnston and Naiman, 1990b).

The GIS analysis of the landscape-level environmental changes caused by beaver pond creation yielded transition probabilities among four major impoundment types: (1) valley bottoms not currently in ponds, (2) flooded ponds with little or no vegetation, (3) seasonally flooded impoundments with emergent macrophytes, and (4) moist meadows dominated by graminoids. These transition probabilities were derived for each decade of beaver occupancy (1940s, 1950s, 1960s, 1970s), and assembled into Markov matrices to form simulation models of change in areas of different cover types over time (see following section on "Modeling Communities with GIS"). In addition, the techniques of linear algebra were used to discern major pathways of change. All matrices have a set of mathematical properties known as eigenvalues, each of which is associated with a unique eigenvector satisfying the equation:

$$\mathbf{A}\mu = \lambda\mu \qquad (25\text{-}4)$$

where $\mathbf{A}$ is the matrix of transition probabilities and λ is a particular eigenvalue associated with the eigenvector μ. In a Markov matrix, the eigenvector associated with this eigenvalue will give the stable acreage distributions of landscape elements for that set of transition probabilities. Plotting the stable acreage fraction of flooded land versus moist meadow (impoundment types 2 and 4 above) for each of the four decades analyzed revealed that the moist class rose and fell in inverse relation to that of the flooded class, and that the variation in acreage among the two classes was almost perfectly correlated ($r = -0.912$). Thus, the extremely complicated patterns seen from the maps were reduced by GIS and algebraic analysis to a simple description of a landscape whose principal dynamic is an alternation of flooded wetlands with moist meadows as ponds are occupied, abandoned, and reoccupied by beaver (Pastor and Johnston, in press).

MODELING COMMUNITIES WITH GIS

Community ecology is the study of assemblages of different species in relation to each other and their environment (Whittaker, 1970). The classical representation of plant communities is as discrete patches distinguished by sharp boundaries between species asssemblages. Most vegetation maps are drawn this way, with areas of contiguous vegetation deemed to be homogeneous lumped into a single polygon. Although sharp boundaries can exist in nature, the more common scenario is a gradual change in species composition across environmental gradients. While both plants and animals form communities, GIS studies of animal communities are rare.

GIS provides ecologists with a new tool for analyzing the influence of the environment on plant community distribution. Although the traditional techniques of collecting ecological data at a few points and extrapolating to larger areas have worked well within homogeneous areas, the use of a GIS to develop community–environment interactions has the advantages that it: (1) allows researchers to compare variables across the entire data surface, rather than just a few points, and (2) reduces the bias introduced by subjective sample point selection. If statistically valid relationships can be developed between plant communities and environmental variables, they can be used to predict community distributions from environmental conditions alone (static conditions), or community response to environmental change (dynamic conditions). This type of plant community modeling has been done with remotely sensed data for both terrestrial (Davis and Goetz, 1990) and aquatic systems (Welch and Remillard, 1988).

Community and landscape ecologists are currently interested in the role of ecotones, zones of transition between adjacent ecological systems, on biodiversity and the flux of materials and energy in the landscape (Johnston and Bonde, 1989). GIS techniques can be used to extract information about ecotones from a larger database, the ecotones can be classified by their topology to identify the adjacency of different community types, and statistical analysis can be used to determine if plant communities are contiguous to each other in greater proportion than would be expected from their overall abundance in the landscape as a whole (Pastor and Broschart, 1990). For example, the data from a GIS analysis of ecotones derived from a Landsat Thematic Mapper image for the forested Horsehead Lake area of northern Minnesota revealed vegetation relationships that were not evident from the original map nor its summary statistics: Water bodies shared more border with lowland shrub areas than with any other cover type, even though shrubs had the lowest cumulative ecotone length of any upland vegetation type (Johnston and Bonde, 1989). This indicates a preferential association between water and lowland deciduous shrubs that may influence fluxes of materials between the two plant communities. In a similar analysis, Pastor and Broschart (1990) found that conifer bogs in a pristine northern Michigan forest were usually surrounded by hemlock and dissociated with hardwood and mixed hemlock–hardwood stands, which they attributed to the influence of fire. They hypothesized that the bogs served as refugia for hemlock seeds during fire episodes, after which the hemlocks regenerated into the surrounding upland.

Temporal change is an integral part of community and

landscape functioning: Pioneer plant communities are succeeded by secondary plant communities, landscape patchiness is altered by disturbance, and ecotone locations are affected by climate change. GIS provide a means of quantifying temporal change using such sources as historical maps, air photos, and remotely sensed images (Iverson and Risser, 1987; Johnston and Naiman, 1990a; Hall et al., 1991). Changes in the spatial distribution of land classes can be summarized by overlaying maps of different dates and analyzing their spatial coincidence. Changes fron one land class to another can be mathematically described as probabilities that a given pixel will remain in the same state or be converted to another state. Mathematically, the expected change in landscape properties can be summarized by a series of transition probabilities from one state to another over a specified unit of time:

$$p_{ij\tau} = n_{ij} / \Sigma_j n_{ij} \qquad (25\text{-}5)$$

where $p_{ij\tau}$ is the probability that state i has changed to state j during time interval t for any given pixel and n_{ij} is the number of such transitions across all pixels in a landscape of n states (Anderson and Goodman, 1957). When assembled in a matrix and used to generate a temporal series, known as a Markov chain, these transition probabilities form a simulation model of changes in areas of different cover types over time:

$$\mathbf{n}_{t+1} = \mathbf{A}\mathbf{n}_t \qquad (25\text{-}6)$$

where $\mathbf{n}$ is a column vector of the acreage distribution of cover types in the landscape at time t or $t+1$ and $\mathbf{A}$ is the matrix of transition probabilities.

Markov models have been assembled to simulate succession of forests (Waggonner and Stephens, 1970; Shugart et al., 1973; Johnson and Sharpe, 1976; Hall et al., 1991), heathland (Jeffers, 1988), and wetland vegetation associated with beaver ponds (Pastor and Johnston, in press). While early Markov models were parameterized using succession rates measured or observed in the field, the development of remote sensing and GIS techniques has enabled researchers to calculate less biased transition probabilities from the full extent of the landscape (Hall et al., 1991; Pastor and Johnston, 1992). See Hunsaker et al. (Chapter 22) for an additional discussion of difference equation models.

CONCLUSIONS

The power of GIS for any ecological research lies in its ability to analyze spatially distributed data, whether the GIS is used alone or linked with more sophisticated models. Ecological models are built on known relationships between organisms and their environment, and GIS can be instrumental in deriving those relationships. GIS can be used to interpolate data, analyze topological relationships, evaluate the spatial coincidence of ecological variables, and detect temporal change. Other scientific tools, such as air photo interpretation, remote sensing, field measurements, laboratory experiments, and multivariate or spatial statistics can be integrated with GIS to enhance the modeling. The derived information can then serve as input to a GIS-driven model, or be exported to an external model. Once modeling is completed, a GIS can be used to display results or produce hard copy. Continued development of GIS technology, as well as the integration of ecological science and technology, is essential to solving complex ecological problems at population, community, and landscape scales.

REFERENCES

Agee, J.K., Stitt, S.C.F., Nyquist, M., and Root, R. (1989) Geographic analysis of historical grizzly bear sightings in the North Cascades. *Photogrammetric Engineering and Remote Sensing* 55: 1637–1642.

Anderson, T.W., and Goodman, L.A. (1957) Statistical inference about Markov chains. *Annals of Mathematical Statistics* 28: 89–110.

Broschart, M.R., Johnston, C.A., and Naiman, R.J. (1989) Predicting beaver colony density in boreal landscapes. *Journal of Wildlife Management* 53: 929–934.

Burke, I.C., Schimel, D.S., Yonker, C.M., Parton, W.J., Joyce, L.A., and Lauenroth, W.K. (1990) Regional modeling of grassland biogeochemistry using GIS. *Landscape Ecology* 4: 45–54.

Caswell, H. (1978) Predator mediated coexistence: a nonequilibrium model. *American Naturalist* 112: 127–154.

Cook, E.A., Gardner, J.E., Gardner, J.D., and Hofmann, J.E. (1988) Analyzing the home range and habitat selection of the Indiana bat using radio telemetry and GIS techniques (abstract). *Ecological Society of America Bulletin* 69(2): 105.

Coulson, R.N., Lovelady, C.N., Flamm, R.O., Spradling, S.L., and Saunders, M.C. (1991) Intelligent geographic information systems for natural resource management. In Turner, M.G., and Gardner, R.H. (eds.) *Quantitative Methods in Landscape Ecology*, New York: Springer-Verlag, pp. 153–172.

Davis, F.W., and Goetz, S.J. (1990) Modeling vegetation pattern using digital terrain data. *Landscape Ecology* 4: 69–80.

Davis, F.W., Stoms, D.M., Estes, J.E., Scepan, J., and Scott, J.M. (1990) An information systems approach to the preservation of biological diversity. *International Journal of Geographical Information Systems* 4: 55–78.

Donovan, M.L., Rabe, D.L., and Olson, C.E., Jr. (1987) Use of geographic information systems to develop habitat suitability models. *Wildlife Society Bulletin* 15: 574–

579.

Fahrig, L. (1991) Simulation methods for developing general landscape-level hypotheses of single-species dynamics. In Turner, M.G., and Gardner, R.H. (eds.) *Quantitative Methods in Landscape Ecology*, New York: Springer-Verlag, pp. 415–442.

Forman, R.T.T., and Godron M. (1986) *Landscape Ecology*, New York: John Wiley & Sons.

Hall, F.G., Botkin, D.B., Strebel, D.E., Woods, K.D., and Goetz, S.J. (1991) Large-scale patterns of forest succession as determined by remote sensing. *Ecology* 72: 628–640.

Hastings, A. (1977) Spatial heterogeneity and the stability of predator-prey systems. *Theoretical Population Biology* 12: 37–48.

Hodgson, M.E., Jensen, J.R., Mackey, H.E., Jr., and Coulter, M.C. (1988) Monitoring wood stork foraging habitat using remote sensing and geographic information systems. *Photogrammetric Engineering and Remote Sensing* 54: 1601–1607.

Hyman, J.B., McAninch, J.B., and DeAngelis, D.L. (1991) An individual-based simulation model of herbivory in a heterogenous landscape. In Turner, M.G., and Gardner, R.H. (eds.) *Quantitative Methods in Landscape Ecology*, New York: Springer-Verlag, pp. 443–475.

Iverson, L.R., and Risser, P.G. (1987) Analyzing long-term changes in vegetation with geographic information systems and remotely sensed data. *Advances in Space Research* 7: 183–194.

Jeffers, J.N.R. (1988) *Practitioner's Handbook on the Modelling of Dynamic Change in Ecosystems*, New York: John Wiley & Sons.

Johnson, W.C., and Sharpe, D.M. (1976) An analysis of forest dynamics in the northern Georgia Piedmont. *Forest Science* 22: 307–322.

Johnston, C.A. (1991) GIS technology in ecological research. In *Encyclopedia of Earth System Science, Vol. 2*, Orlando: Academic Press, pp. 329–346.

Johnston, C.A., and Bonde, J.P. (1989) Quantitative analysis of ecotones using a geographic information system. *Photogrammetric Engineering and Remote Sensing* 55: 1643–1647.

Johnston, C.A., and Naiman, R.J. (1990a) The use of a geographic information system to analyze long-term landscape alteration by beaver. *Landscape Ecology* 4: 5–19.

Johnston, C.A., and Naiman, R.J. (1990b) Aquatic patch creation in relation to beaver population trends. *Ecology* 71: 1617–1621.

Johnston, K.M. (1987) Natural resource modeling in the geographic information system environment. *Photogrammetric Engineering and Remote Sensing* 53: 1405–1410.

Lotka, A.J. (1925) *Elements of Physical Biology*, Baltimore: Williams and Wilkins.

MacArthur, R.H., and Wilson, E.O. (1967) *The Theory of Island Biogeography*, Princeton: Princeton University Press.

McGranaghan, M. (1989) Incorporating bio-localities in a GIS. *GIS/LIS'89 Proceedings*, Bethesda: American Society for Photogrammetry and Remote Sensing, pp. 814–823.

Mead, R.A., Cockerham, L.S., and Robinson, C.M. (1988) Mapping gopher tortoise habitat on the Ocala National Forest using a GIS. *GIS/LIS'88 Proceedings*, Falls Church: American Society for Photogrammetry and Remote Sensing, pp. 395–400.

Morain, S.A., and Harrington, M.W. (1990) A GIS for predicting environmentally induced changes in the flora of the Rocky Mountains. *GIS/LIS'90 Proceedings*, Bethesda: American Society for Photogrammetry and Remote Sensing, pp. 803–813.

Morse, B. (1986) Forecasting forest pest hazard with a geographic information system. *Proceedings of Geographic Information Systems Workshop*, Falls Church: American Society for Photogrammetry and Remote Sensing, pp. 255–262.

Nachman, G. (1987) Systems analysis of acarine predator–prey interactions: I. A stochastic simulation of spatial processes. *Journal of Animal Ecology* 56: 247–265.

Pastor, J., and Broschart, M.R. (1990) The spatial pattern of a northern conifer–hardwood landscape. *Landscape Ecology* 4: 55–68.

Pastor, J., and Johnston, C.A. (1992) Using simulation models and geographic information systems to integrate ecosystem and landscape ecology. *New Perspectives for Watershed Management*, New York: Springer-Verlag, pp. 324–346.

Pastor, J., and Post, W.M. (1988) Response of northern forests to CO_2-induced climate change. *Nature* 334: 55–58.

Pearl, R., and Reed, L.J. (1920) On the rate of growth of the population of the United States since 1790 and its mathematical representation. *Proceedings of the National Academy of Sciences* 6: 275–288.

Pearlstine, L., McKellar, H., and Kitchens, W. (1985) Modeling the impacts of a river diversion on bottomland forest communities in the Santee River Floodplain, South Carolina. *Ecological Modelling* 29: 283–302.

Pianka, E.R. (1970) On *r*- and *K*-selection. *American Naturalist* 104: 592–597.

Rowland, E. (1986) Use of a GIS to display establishment and spread of gypsy moth infestation. *Proceedings of Geographic Information Systems Workshop*, Falls Church: American Society for Photogrammetry and Remote Sensing, pp. 249–254.

Scepan, J., Davis, F., and Blum, L.L. (1987) A geographic information system for managing California condor habitat. *GIS '87, Proceedings*, Falls Church: American Society for Photogrammetry and Remote Sensing, pp. 476–486.

Shaw, D.M., and Atkinson, S.F. (1988) GIS applications for golden-cheeked warbler habitat description. *GIS/LIS'88 Proceedings*, Falls Church: American Society for Photogrammetry and Remote Sensing, pp. 401–406.

Shugart, H.H. Jr., and West, D.C. (1980) Forest succession models. *BioScience* 30: 308–313.

Shugart, H.H. Jr., Crow, T.R., and Hett, J.M. (1973) Forest succession models: a rationale and methodology for modeling forest succession over large regions. *Forest Science* 19: 203–212.

Sklar, F.H., and Costanza, R. (1991) The development of dynamic spatial models for landscape ecology: a review and prognosis. In Turner, M.G., and Gardner, R.H. (eds.) *Quantitative Methods in Landscape Ecology*, New York: Springer-Verlag, pp. 239–288.

Stenback, J.M., Travlos, C.B., Barrett, R.G., and Congalton, R.G. (1987) Application of remotely sensed digital data and a GIS in evaluating deer habitat suitability on the Tehama Deer Winter Range. *GIS '87, Proceedings*, Falls Church: American Society for Photogrammetry and Remote Sensing, pp. 440–445.

Tomlin, C.D. (1990) *Geographic Information Systems and Cartographic Modeling*, Englewood Cliffs, NJ: Prentice Hall.

Tomlin, C.D., Berwick, S.H., and Tomlin, S.M. (1987) The use of computer graphics in deer habitat evaluation. In Ripple, W.J. (ed.) *Geographic Information Systems for Resource Management: A Compendium*, Falls Church: American Society for Photogrammetry and Remote Sensing, pp. 212–218.

Turner, M.G., and Dale, V.H. (1991) Modeling landscape disturbance. In Turner, M.G., and Gardner, R.H. (eds.) *Quantitative Methods in Landscape Ecology*, New York: Springer-Verlag, pp. 323–351.

Volterra, V. (1926) Varizioni e fluttuazioni del numero d'individui in specie animali conviventi. *Mem. R. Acad. Naz. dei Lincei (ser. 6)* 2: 31–113.

Waggoner, P.E., and Stephens, G.R. (1970) Transition probabilities for a forest. *Nature* 225: 1160–1161.

Ware, C., and Maggio, R.C. (1990) Integration of GIS and remote sensing for monitoring forest tree diseases in central Texas. *GIS/LIS'90 Proceedings*, Bethesda: American Society for Photogrammetry and Remote Sensing, pp. 724–732.

Welch, R., and Remillard, M.M. (1988) Remote sensing and geographic information system techniques for aquatic resource evaluation. *Photogrammetric Engineering and Remote Sensing* 54: 177–185.

White, W. (1986) Modeling forest pest impacts—aided by a geographic information system in a decision support system framework. *Proceedings of Geographic Information Systems Workshop*, Falls Church: American Society for Photogrammetry and Remote Sensing, pp. 238–248.

Whittaker, R.H. (1970) *Communities and Ecosystems*, London: MacMillan.

Wilson, E.O., and Bossert, W.H. (1971) *A Primer of Population Biology*, Sunderland: Sinauer Associates.

Young, T.N., Eby, J.R., Allen, H.L., Hewitt, M.J., III, and Dixon, K.R. (1987) Wildlife habitat analysis using Landsat and radiotelemetry in a GIS with application to spotted owl preference for old growth. *GIS '87, Proceedings*, Falls Church: American Society for Photogrammetry and Remote Sensing, pp. 595–600.

26

Spatial Interactive Models of Atmosphere–Ecosystem Coupling

DAVID S. SCHIMEL AND
INGRID C. BURKE

The simulation of the atmosphere and terrestrial ecosystems as parts of a coupled system is currently an area of active research. This work is motivated by several concerns. First, vegetation and soils control attributes of the land surface that are significant to climate and weather. These attributes include albedo, sensible versus latent heat partitioning, and surface roughness. Second, vegetation controls aspects of the hydrological cycle, most significantly through evapotranspiration (latent heat flux) but also via interception and stemflow. Vegetation control of evapotranspiration leads to vegetation having significant control over runoff as a fraction of rainfall, influencing fluvial processes (Vorosmarty et al., 1989). Finally, the vegetation and soils of terrestrial ecosystems are major sources and sinks of atmospheric trace gases. Ecosystems thus influence the atmospheric burden of 'greenhouse' gases (CO_2, CH_4, N_2O), the oxidant status of the atmosphere (via CO, NO_x, hydrocarbons), and the pH of aerosols and rainfall (via NH_4) (Mooney et al., 1987).

Modeling of these interactions has progressed rapidly, to the point where global analyses of many of the couplings have been conducted, and detailed regional analyses completed and tested (reviewed in Schimel et al., 1991). Modeling of land–atmosphere coupling whether through biophysics or chemistry forces modelers to consider landscape heterogeneity. This is because first the land surface only influences the atmosphere when land processes are integrated over fairly large areas (Dickinson, in press) and second, because the arrangement of landscape types may in itself influence the atmosphere through differential energy exchange (Pielke et al., 1991). Current models of biophysical exchange and biogeochemistry have developed from nonspatial models, designed for site-specific applications. Recent efforts to couple these models to atmospheric processes have resulted in substantial effort, ranging from basic studies of the spatial controls over basic processes (e.g., Schimel et al., 1991; Matson and Vitousek, 1987) to considerable effort in the development of GIS databases and spatial analyses (Burke et al., 1990; Schimel, Davis, and Kittel, in press).

In this chapter, three models are discussed. These are used to address different aspects of land–atmosphere coupling, presenting first the basic science of the models and second the modeling and data analytical challenges involved in using these models in a spatial context. The theme in all cases will be the relationship between the processes included in the models and the types of spatial analysis required to use them for regional to global analyses.

THE CLIMATE–VEGETATION INTERACTION

Recent analyses have made it clear that the land cover influences climate in midcontinental areas (e.g., Sato et al., 1989). Vegetation influences albedo, surface roughness, and energy exchange between the land surface and the atmosphere. Recently, several models have been developed to simulate these interactions, for example, the biosphere–atmosphere transfer scheme (BATS) as described by Dickinson et al. (1986), and the simple biosphere model (SiB) as discussed by Sellers et al. (1986). It has been demonstrated that these interactions significantly affect the simulation of regional climate in a global model (Lean and Warrilow, 1987; Shukla et al., 1990). This suggests that simulations of soil moisture and other climate change and drought-related parameters should be done using a land surface package coupled to a GCM.

Several models simulate the land surface's interactions with the atmosphere via water and energy exchange. These include the BATS model of Dickinson et al. (1986), the land–ecosystem atmosphere–feedback (LEAF) model of Lee et al. (Chapter 26), and the SiB model of Sellers et al.

(1986; and see Chapter 27). The SiB model will be described as an example.

In SiB, the world's land cover types are classified by a simplified description of vegetation cover that is prescribed prior to simulation so that vegetation is not interactive with climate. Each vegetation type (bare ground, shrub dominated, grassland, savannah, forest) is described by a list of optical, physical, and biophysical properties, which must be described spatially. The model calculates the transfer of radiation through the canopy using a simple approximation known as the "two stream approximation." The model simulates the exchange of sensible and latent heat with an electrical resistance analogy that is dependent upon canopy resistance to gas (water) exchange. This is the "big leaf" approach to integrating leaf-level stomatal resistance to gas exchange to large areas. Increasingly, theoretical and field studies support the use of simplified canopy models for application at large spatial scales. In recent versions of SiB, the well-known dependence of stomatal resistance on photosynthetic rate is incorporated, coupling the carbon and hydrological calculations in the model.

SiB requires spatial information on land surface properties, which are not normally treated as time-varying parameters. It also requires time-varying information on the state of the atmosphere, updated frequently (order 1–15 minutes). The atmospheric data are derived from the coupling of SiB to an atmospheric general circulation model, or in 1D applications, from station data. The requirement for relatively static and extremely dynamic spatial data is common in spatial models of land–atmosphere interactions. Data for models like SiB are not normally managed using GIS software, and current GIS do not well accommodate the above combination of data types. Also, GIS–model links are not now efficient enough to run in the GCM context. Clearly, as SiB-type models become more sophisticated, they will require more detailed and extensive land surface data, and these data may have a time dimension. As the data needs increase, so will the incentive to couple these models to data systems capable of managing, rescaling, and displaying information.

It is worth noting again that models of the SiB class are used to improve the simulation of climate over short time scales. That is, while long-term changes in vegetation can have long-term impacts on climate, the sort of interactions simulated by SiB influence climate on diurnal and subseasonal time scales. Longer-term changes are simulated by prescribing changes in the vegetation description—changes that are imposed by the modeler rather than calculated by the model. Biosphere–atmosphere interaction models designed to improve simulation of the hydrological cycle and energy balance typically operate with high temporal resolution, matched to the atmosphere, and low spatial resolution, matched to the scale of large-scale atmospheric phenomena and models. This combination of high temporal resolution and low spatial resolution is different from most ecologically derived models, as will be described.

HYDROLOGY–ECOSYSTEM MODELS

The dynamics of growing vegetation and ecosystem water budgets are inextricably linked. This is because evapotranspiration is a necessary result of photosynthesis in most plants, and at the same time a major control over water budgets. Thus, models for the calculation of ecosystem or watershed water budgets must consider evapotranspiration, and models of ecosystem carbon gain must simulate water use. Note that these models are usually forced by climate information, and in general, do not feed back to the climate simulation.

A widely used ecosystem–hydrology model is the forest–BGC (BGC, biogeochemical cycling) model (for example, see Running and Coughlan, 1988). This model is also described elsewhere (Nemani et al., Chapter 28); so the details will not be repeated, except to point out its critical links to spatial information and to highlight the specific requirements of this type of model.

Forest–BGC uses the leaf area index (LAI) as the most important parameter controlling carbon gain and water loss. The forest canopy is treated as a homogeneous volume with depth proportional to LAI. Other components include wood elements, soil water-holding capacity in the rooting zone, and snowpack. The model is driven by daily meterological variables, maximum and minimum air temperature, rainfall, and humidity. Note that these are considerably less resolved in time (daily) than the climate information exchanged with Sib in a climate modeling context. Equations for hydrological balance and net carbon balance (photosynthesis, respiration) are solved daily. Carbon storage terms (allocation to roots, leaves, etc., growth respiration and decomposition) are solved annually.

While the number of variables required for the core forest–BGC simulation are minimal (that is, LAI and soil water-holding capacity plus climate in its normal mode of application), the model requires considerable spatial detail in these variables. Most applications of forest–BGC have been in mountainous terrain. Such applications require additional topographic information for computation of meteorological variables. In mountainous terrain, weather changes dramatically over short distances because of relief. Weather station data do not provide the resolution to describe such variations. Thus, forest–BGC is normally used with a microclimate simulator (MT-CLIM) that calculates expected changes in climate according to elevation and topography, that is, the effects of adiabatic lapse rate and of slope and aspect on the radiation balance. Application of forest–BGC with MT-CLIM requires data from a high-resolution (30 m) digital elevation model (DEM). This information is obviously more detailed than the data input to SiB, which is designed to operate within global climate model (GCM) cells with a spatial resolution of more than 100 km. In addition to computing microclimate, the DEM-derived information can be used to route the runoff component of the hydrological calculations so that realistic runoff can be included in the water budget. Such applications of forest–BGC with a DEM are ongoing and promise to offer the possibility not only of watershed budgeting, but also of validation against hydrographs.

In summary, models of the forest–BGC class require low temporal resolution of climate data, but high resolution in space. They require, in addition, resolution of topography and soils on scales matched to microclimate variability. These are considerably different requirements than those imposed by coupled simulations within GCMs and are more closely matched to the experience of ecologists of the data needs for modeling ecosystem and watershed dynamics.

BIOGEOCHEMISTRY

The previous two sections dealt with models whereby land and atmospheric processes are coupled by the exchange of energy and water (biophysical processes). Land and atmospheric processes are also influenced by the exchange of trace gases. This links terrestrial processes to the atmosphere at a variety of scales from regional, for reactive chemistry (O_3 cycle), to global, for the long-lived greenhouse gases. Trace gas exchange is largely controlled by ecosystem carbon and nitrogen cycles, and the foundation for any predictive model of emissions must begin with simulation of these processes. The Century model is briefly described with a discussion of its application to the carbon cycle.

In the Century formulation (see Parton et al., 1987) net primary production is determined by water availability (rainfall plus storage), subject to the constraint that the plant must maintain critical C:element (N,P,S) ratios (note: "C" stands for carbon, "N" for nitrogen, "P" for phosphorus, and "S" for sulfur). Thus C:N ratios (for the whole plant) must not exceed about 100, and similar limits (analogous to Redfield ratios) are set for the other elements. Control by light restricts growth as light (i.e., radiation) interception nears 100 percent. While different schemes are used for production in other models, all embody the same concept: joint regulation by climate and a suite of limiting soil-derived nutrients with light competition acting as a constraint. This concept is quite different from the assumptions underlying many biophysical treatments.

Soil properties influence primary production through nutrient availability and water-holding capacity. The ability of a soil to hold water is related to its structure, texture, and organic matter content. Soil water-holding capacity is most strongly controlled by texture, or grain size distribution. Fine-grained soils hold more water, and hence under a common climate, may have higher NPP (net primary production). In dry environments, this pattern may reverse as coarse soils, which allow deeper infiltration, may have lower evaporation and higher transpiration than fine soils.

Nutrients are derived from three sources: mineralization of soil organic matter, the atmosphere, and weathering of soil minerals. Mineralization of nutrients occurs as organic matter is decomposed by microorganisms. A huge literature exists on this subject, covering its biochemistry, microbiology, the role of larger organisms (protozoa, microarthropods), and control by the abiotic environment. In its essentials, decomposition is controlled by substrate organic chemistry, nutrient supply, and environmental variables regulating microbial activity rates. Substrate organic chemistry is generally summarized by "lignin" content. Lignin is a mix of polymers resistant to microbial decomposition. As lignin content increases, decomposition rate decreases. Decomposer organisms require nutrients for growth, some of which may be derived from the substrate. However, most plant substrates are low in nutrients relative to microbial tissue, with C:N ratios of 40–100 in plant material compared with 6–12 for most heterotrophs. Thus, nutrients are required from the environment, most of which are derived from the turnover of soil organic matter (C:N of about 10) and of the microbial population itself. Plant material with high lignin or low nutrient content will constitute a sink for nutrients, slowing down nutrient turnover and reducing plant nutrient availability, with low-lignin or high-N material behaving in the opposite fashion. In practice, the organic chemistry of plant litter is often summarized by the lignin:N ratio.

Soil organic matter (SOM) is central to nutrient cycling and organic matter storage. Much of the world's organic carbon is stored in soils, and this organic matter reserve is also a major reserve of plant nutrients. Although alternate formulations for decomposition and soil organic matter formation exist, all agree that plant debris is progressively decomposed to less labile and more nutrient rich forms. As the more decomposed forms are reached, decomposition rate slows, but nutrient release per unit carbon respired or metabolized increases. Nutrient mineralization may be dominated by intermediate forms in the decomposition sequence, neither fresh litter nor stabilized humus.

Stabilization of humus appears to be governed largely by the overall decomposition rate (controlled by temperature and moisture) and by the formation of organo-mineral compounds. Organic matter stabilization decreases as mean annual temperature increases, with low soil carbon levels and high mineralization rates in some tropical ecosystems providing an extreme example. Formation of organo-mineral compounds is enhanced by increasing surface activity of soil particles, and so increases with increasing clay content. Different types of clay minerals also differentially stabilize organic matter. In general, nutrient turnover (nutrient mineralized/unit organic nutrient) decreases with increasing clay content, but absolute mineralization rate may increase, because of larger pool sizes.

The major greenhouse species under direct human influence is, of course, CO_2. CO_2 is produced from fossil fuel consumption and by organic matter decomposition, and consumed by photosynthesis. Exchange with the carbonate cycle in rock and water significantly modulates the cycle, with oceanic uptake of CO_2 being a major sink over years to centuries and the rock cycle dominating over geological time. Considerable uncertainty exists over the sinks for industrial CO_2, as well as over biological sources, for example, land conversion to agriculture and effects of climatic change. Considerable uncertainty also exists as to how the global carbon cycle will function in the future if atmospheric CO_2 concentration and climate vary from the present. Century has been used to probe biological and biophysical controls over ecosystem carbon storage at time scales of years to centuries.

The changing global environment will change the way in which the terrestrial carbon cycle functions. Enhanced CO_2 will lead to increased primary production under certain conditions. This may be a transient, as some evidence shows that plants may acclimate to increased CO_2 and show a gradually decreasing response. This acclimation should be larger in perennial than annual plants.

The effects of CO_2 may be attenuated by constraints from other limiting factors. Nutrients or water may limit CO_2 uptake at levels only slightly greater than under current CO_2 concentrations. The interaction of CO_2 fertilization with other limiting factors requires far more study in a range of ecosystems with varying limiting factors. CO_2 fertilization has the potential to alter nitrogen and water-use efficiencies by allowing increased enzyme efficiencies (photosynthetic enzymes contain large amounts of N) or by increasing water-use efficiency. Under most circumstances, these changes in water- or N-use efficiencies will result in production of plant tissue with reduced content of N and other nutrients. This is because CO_2 fertilization does not enhance nutrient availability through any known mechanism and, with increased efficiency, more biomass must be produced on the same amount of nutrients. The production of plant tissue with higher C:element ratios will increase microbial uptake of nutrient when that tissue is decomposed, competing with plants and reducing nutrient availability. This feedback will tend to reduce the effects of CO_2 fertilization on primary production, homeostatically. The consequences of increasing CO_2 on carbon storage have been evaluated using Century, and the behavior described here was found to apply; specifically, feedback through plant–microbial competition for nutrients limited the effects of CO_2-induced increases in resource use efficiencies (see Schimel et al., 1991).

If greenhouse gas emissions lead to increased temperatures, this will also affect the carbon cycle. Increased temperatures, other factors being equal, will accelerate decomposition and cause loss of stored carbon, increasing the atmospheric inputs of CO_2. In general, decomposition is quite sensitive to temperature. In a recent simulation exercise, it was demonstrated that across a range of grasslands, increased temperature with no CO_2 fertilization effect resulted in carbon losses despite enhanced plant production due to CO_2 enrichment (Schimel et al., 1991; Burke et al., 1991). While this result is only for one ecosystem type, other models exhibit similar behavior for other ecosystem types, suggesting it may be a fairly general result. If this is true, then CO_2 fertilization responses may not keep pace with respiration losses resulting from increases in temperature.

In subsequent analyses that included the direct effects of CO_2, it was shown that, despite minimal increase in net primary productivity, as discussed, CO_2-caused changes in litter chemistry resulted in slower, decreased decomposition rates. Two lines of cause and effect occurred with opposing consequences. Increased temperatures caused higher decomposition rates, resulting in soil carbon loss but higher primary production. Increased CO_2 caused higher plant resource use efficiency, leading to lower quality litter, slower decomposition, and lower N availability. The net effect was that the temperature and CO_2 effects were opposed and resulted in the global change scenario having minimal effects on state variable amounts. This type of analysis is very dependent upon the quantitative strength of relationships and can only be pursued with modeling.

Century has also been adapted to simulate nitrogen trace gases, especially N_2O, and is being modified to simulate methane production in anaerobic environments, and methane oxidation in aerobic soils. These fluxes are simulated in a two-step process in which, first, substrate pools are generated (i.e., NO_3, acetate) and then the effects of temperature, oxygen status, pH, and other microbial controls are considered. This approach allows changes in fluxes to be considered, both as direct consequences of the changing abiotic environment and via carbon/nitrogen cycle feedbacks.

Geographic applications of Century have defined the land surface as a series of cells with homogeneous properties (Burke et al., 1991). These cells may be a prescribed grid or irregular polygons. Each cell is associated with tabular data describing salient ecosystem properties. Other data layers contain model output. The choice of properties described for each cell is determined by the parameter structure of the model. In recent regional simulations with CENTURY, an irregular grid determined by an overlay of a climate parameter (potential evapotranspiration/precipitation; PET/P) on a map of soil texture was used. A common climate time series and soil texture were used within each cell.

Aggregation and interpolation are common problems in developing spatial databases. Climate data come from point measurements or climate model cell centroids, and so must be interpolated to produce surfaces. However, when cells are defined, these surfaces must be segmented. When multiple data layers are interpolated separately to produce surfaces and then overlaid, artificially small cells ("splinters") can emerge of a size not warranted given the resolution of the original data. CENTURY uses monthly climate data; an annual precipitation index was used rather than choosing cell boundaries based on monthly climate information, which would have required overlaying a dozen climate maps for each parameter. On the other hand, problems with large cells occur when mapped variables are nonlinearly related to responses. Then, considerable detail must be preserved or nonlinear weighted averages calculated within cells to avoid error in the final estimates. For example, soil texture is nonlinearly related to soil carbon storage. Fine-textured soils have disproportionately high carbon storage, and so fine-textured soils must be mapped with care. Soil texture may be averaged within a polygon, but the weighting function for the average must weight the fine-textured soils more heavily than the mean for the polygon (Burke et al., 1990).

Remote sensing plays a critical role in spatial applications of ecosystem models. The model of Fung et al. (1987) allowed calculation of a critical flow in terrestrial biogeochemistry but required the empirical calculation of biome-specific conversion efficiencies (carbon fixed/light intercepted). Their approach suggests an interesting inversion strategy for the CENTURY class of model. While ecosystem models have not conventionally been parameterized in terms of light interception (in contrast to biophysical models), light interception is related to a number of key ecosystem variables. Research findings have suggested that the instantaneous normalized difference vegetation index (NDVI) is an estimate of phytosynthetic capacity, or chlorophyll density. Subsequent research has broadly supported this conclusion.

Chlorophyll density and photosynthetic capacity are strongly related to canopy N, and their time series (rather than integral) may indicate N uptake and dynamics in the canopy. While the NDVI has often been interpreted as leaf area index (LAI, leaf area/ground area) or biomass, chlorophyll density and by correlation N density and photosynthetic capacity may be more useful interpretations. That NDVI predicts LAI or biomass in some cases may reflect the constrained ratios of N per unit mass or LAI within a study region or biome.

Can light interception be used as an input to models of ecosystems? Promising results have been obtained. Running et al. (1989) have demonstrated a successful application of such a technique for a model of photosynthesis and transpiration. Data from the first ISLSCP field experiment (FIFE) study in Kansas grasslands showed strong relationships between light interception and canopy N mass and N allocation.

Long-term success of satellite data-driven models will depend next on the development of algorithms to predict partitioning of C and N below ground for a given above-ground time series of light interception. Current models assume that allocation above and below ground will reflect the relative importance of above-ground (light) and below-ground resources (water, nutrients) as factors limiting production. One simple model assumes that the partial derivatives of production with water, N uptake, and light should be equal. Well-constrained solutions to such an algorithm will be difficult if only light interception (NDVI) is available as an input. Such a treatment could be better used in a model where several environmental parameters were used as inputs, with soil moisture or rainfall being most useful, as in a hybrid model including geographic and remotely sensed inputs.

CONCLUSIONS

Current models address land–atmosphere–hydrology interactions at a wide range of spatial and temporal levels of resolution. Models intended for climate calculations over time scales of the diurnal and seasonal cycles have more difficult data requirements than ecosystem models intended to calculate carbon storage over decades. Current GIS technology does not support any of these approaches very well, having limited abilities to couple to dynamic simulations, inadequate data analytical capabilities for interpolation/extrapolation, and poor links to satellite data. Scientific GIS for modeling will have to address these issues.

REFERENCES

Burke, I. C., Schimel, D. S., Yonkers, C. M., Parton, W. J., Joyce, L. A., and Lauenroth, W. K. (1990) Regional modeling of grassland biogeochemistry using GIS. *Landscape Ecology* 4(1): 45–54.

Burke, I.C., Kittel, T.G.F., Lauenroth, W.K., Snook, P., Yonker, C.M., and Parton, W.J. (1991) Regional analysis of the central Great Plains. *Bioscience* 41(10): 685–692.

Dickinson, R.E. (1991) Global change and terrestrial hydrology—a review. *Tellus* 43AB: 176–181.

Dickinson, R.E., Henderson-Sellers, A., Kennedy, P.J., and Wilson, M.F. (1986) Biosphere–atmosphere transfer scheme (BATS) for the NCAR community climate model. *NCAR Technical Note NCAR/TN-275+STR*, 69 pp.

Fung, I.U., Tucker, C.J., and Prentice, K.C. (1987) Application of AVHRR vegetation index to study atmosphere-biosphere exchange of CO_2. *Journal of*

Geophysical Research 92: 2999-3016.

Lean, J., and Warrilow, D.A. (1989) Simulation of the regional climate impact of Amazon deforestation. *Nature* 342: 411–413.

Matson, P.A., and Vitousek, P.M. (1987) Cross-system comparisons of soil nitrogen transformations and nitrous oxide flux in tropical forest ecosystems. *Global Biogeochemical Cycles* 1: 163-170.

Mooney, H.A., Vitousek, P.M., and Matson, P.A. (1987) Exchange of materials between terrestrial ecosystems and the atmosphere. *Science* 238: 926-932.

Parton, W.J., Schimel, D.S., Cole, C.V., and Ojima, D.S. (1987) Analysis of factors controlling soil organic levels in Great Plains grasslands. *Soil Science Society of America* 51: 1173–1179.

Pielke, R.A., Dalu, G.A., Snook, J.S., Lee, T.J., and Kittel, T.G.F. (1991) Nonlinear influence of mesoscale land use on weather and climate. *Journal of Climate* 4: 1053–1069.

Running, S.W., and Coughlan, J.C. (1988) A general model of forest ecosystem processes for regional applications I: hydrologic balance, canopy gas exchange and primary production processes. *Ecological Modeling* 42: 125–154.

Running, S.W., Nemani, R.R., Peterson, D.L., Band, L.E., Potts, D.F., Pierce, L.L., and Spanner, M.A. (1989) Mapping regional forest evapotranspiration and photosynthesis by coupling satellite data with ecosystem simulation. *Ecology* 70: 1090-1101.

Sato, N., Sellers, P.J., Randall, D.A., Schneider, E.K., Shukla, J., and Kinter, III, J.L. (1989) Effects of implementing the Simple Biosphere Model in a General Circulation Model. *Journal of Atmospheric Science* 46(18): 2757–2782.

Schimel, D.S., Davis, F.E., and Kittel, T.G.F. (in press) Spatial information for extrapolation of canopy processes: examples from FIFE. In Ehleringer, J.R., and Field, C.B. (eds.) *Scaling Physiological Processes: Leaf to Globe*.

Schimel, D.S., Kittel, T.G.F., and Parton, W.J., (1991) Terrestrial biogeochemical cycles: global interactions with the atmosphere and hydrology. *Tellus* 43AB: 188–203.

Sellers, P.J., Mintz, Y., Sud, Y.C., and Dalcher, A. (1986) A Simple Biosphere Model (SiB) for use within General Circulation Models. *Journal of Atmospheric Science* 43(6): 505–531.

Shukla, J., Nobre, C., and Sellers, P. (1990) Amazon deforestation and climate change. *Science* 247: 1322–1325.

Vorosmarty, C.J., Moore II, B., Grace, A.L., Gildea, M.P., Melillo, J.M., Peterson, B.J., Rasletter, E.B., and Steudler, P.A. (1989) Continental-scale models of water balance and fluvial transport: an application to South America. *Global Biogeochemical Cycles* 3: 241–265.

27

A Brief Review of the Simple Biosphere Model

YONGKANG XUE AND
PIERS J. SELLERS

The terrestrial surface occupies about 30 percent of the Earth's surface and interacts with the overlying atmosphere through radiative, hydrologic, and aerodynamic processes; and trace gas exchange. After more than a decade of sustained work in global climate change studies, many scientists have reached a consensus that the variations in the land surface characteristics can have a significant impact on climate. The tools that meteorologists use to investigate such interactions include general circulation models (GCMs).

Before the early 1980s the surface properties that modulate the exchange between the land and atmosphere in GCM were prescribed as boundary conditions, such as albedo, surface roughness, and the ratio of the sensible heat and latent heat fluxes. The bucket model was used to represent the surface hydrology. Using such a basic surface model, Charney et al. (1977), Shukla and Mintz (1982), and Sud and Smith (1985) showed that changing the land surface albedo, the available soil moisture, and land surface roughness could produce significant changes in the large-scale atmosphere circulation and precipitation. More realistic models, such as SiB, are designed to simulate these effects not only qualitatively, but also quantitatively. In SiB the albedo is one of the morphological parameters, surface roughness is one of the physical parameters, and soil moisture is predicted by the model.

The human effects on global climate change are another important motivation to develop more sophisticated surface models for studying biosphere–atmosphere interactions. According to Skoupy (1987) and Lanly's (1982) studies, over the past 30 years the plant cover has been significantly degraded across the African continent. Tropical forests have been destroyed at the rate of 1.3 million hectares per year. The savanna, which occupies a vast area in West and Central Africa and the large part of the eastern horn, also has been experiencing degradation. In the early 1980s, only 35 percent of the former area of productive savanna was left. In the Amazon basin, the deforestation process has also progressed rapidly. From 1975 to 1989 the average deforestation rate in the Brazilian Amazon was about 20000 km^2 per year (Nobre et al., 1991). If deforestation were to continue at this rate, most of the Amazonian tropical forests would disappear in less than 100 years. The assessment of the effects of desertification in sub-Saharan and of the deforestation in the Amazon region on climate could contribute constructively to the debates on economic development and policy-making.

Over the past two decades considerable progress has been made in understanding the micrometeorology of vegetated surfaces. The resulting models, based on small-scale studies, have been extended to much larger scales. In 1984, Dickinson (1984) developed a biosphere model for GCM studies. It was later called the biosphere–atmosphere transfer scheme (BATS, Dickinson et al., 1986) and implemented into the NCAR community climate model. In 1986, Sellers et al. (1986) developed the simple biosphere model (SiB) for GCM studies. This is a biophysically based model of land–atmosphere interactions. It attempts to simulate realistically the controlling biophysical processes. SiB was designed for use as the land surface process model in the GCMs, to provide fluxes of radiation, momentum, sensible heat, and latent heat. At present there is no description of trace gases in this model.

A BRIEF REVIEW OF SiB'S DEVELOPMENT

There are three soil layers and two vegetation layers in SiB. The model has eight prognostic variables: soil wetness in three soil layers; temperature at the canopy, surface, and deep soil layer; and water stored on the canopy and ground. The force–restore method is used to predict the time variation of the soil temperatures. In the three-layer soil model, water movement is described by a finite-difference approximation to the diffusion equations. The governing equations for the interception water storage are based on energy conservation equations.

SiB consists of three major parts: the calculation of albedo, the aerodynamic resistances, and the surface resistances, which determine the radiative transfer at surface, momentum flux from the surface, and surface energy partition into sensible heat and latent heat fluxes, respectively.

The schemes for albedo and stomatal resistances in SiB are described in more detail in Sellers (1985). The two-stream radiation method, as described by Meador and Weaver (1980) and Dickinson (1983), has been used to model radiative transfer in plant canopies; Sellers (1985) obtained the analytic solutions for the equation set for the radiative fluxes, and hence the albedo and absorptance of the vegetated surfaces. This procedure is carried out for five radiation components: the direct and diffuse fluxes for visible and near-infrared radiation, and thermal infrared (diffuse only) radiation. An efficient method to calculate the albedo has been developed by Xue et al. (1991). For a specific vegetation type, the albedo is a function of canopy state, the solar zenith angle, and snow cover. Since the variation of the albedo with these variables is quite regular, a quadratic fit to the results produced by the two-stream method was used to summarize the diurnal variation of albedo.

The resistances to the transfer of water vapor from the canopy, ground cover, and upper soil layer to the adjacent exterior air include the canopy resistance r_c, ground cover resistance r_g, and soil surface resistance r_{soil}. An empirical equation based on the results of Camillo and Gurney (1986) was used to obtain r_{soil}. The parameterization for the stomatal resistance in SiB was based on the work of Jarvis (1976), which was for an individual leaf. Three stress terms were included in this scheme to describe the dependence on the atmospheric temperature, the leaf water potential, and the vapor pressure deficit. An analytic solution for the bulk stomatal resistance was introduced in Sellers (1985). An empirical equation between the soil moisture and the adjustment factor of the stomatal resistances was developed (Xue et al., 1991), which is very simple and required fewer parameters than more comprehensive models, while producing similar results.

There are three aerodynamic resistances in SiB: the resistance between the soil surface and the canopy air space, r_d, the resistance between all of the canopy leaves and the canopy air space, r_b, and the resistance between the canopy air space and the reference height, r_a. The eddy diffusion concept was used to calculate all three resistances. The eddy fluxes were assumed to be constant below and above the canopy; within the canopy, the K theory was applied. A triangular distribution of leaf area density was later implemented in the SiB to improve the simulation for the tall canopies. Since extrapolation of the log-linear profile has been shown to yield underestimates of the turbulent transfer coefficient close to the top of the plant canopies, this profile was assumed valid only above a certain transition height. Between this height and the top of the canopies, an empirical adjustment to the profile was made to compensate for the model's inability to reproduce the effects of higher-order transfer processes directly. The results from this model were comparable to those obtained with a second-order closure scheme. The theory was introduced in Sellers et al. (1986), and the detailed presentation can been found at the appendices of Sellers et al. (1989). The original parameterization of the resistance between the reference height and top of the canopy was based upon the equations of Paulson (1970) and Businger et al. (1971) in SiB. Since this was too time consuming to apply directly in a GCM, a linear relationship between Richardson number and aerodynamic resistance was developed (Xue et al., 1991). Those linear equations were able to reproduce the previous results satisfactorily.

A number of numerical experiments have been carried out to test SiB in an off-line version, where the meteorological forcing was provided by observations (Sellers and Dorman, 1987; Sellers et al., 1989). The micrometeorological and biophysical measurements from surface experiments conducted over arable crops in West Germany and the United States, a forested site in the United Kingdom, and a tropical rain forest in the Amazon were used to test the operation of SiB. The predicted partitioning of the sensible and latent heat fluxes from the surface was close to the observations, and the various subcomponents of the model appeared to operate realistically. It was estimated that the model will generate uncertainties of the order of 7% in the calculated net radiation, and about 15% in the calculated evapotranspiration rate.

MODEL INPUT PARAMETERS AND THE COMPUTATIONAL FLOW

Input parameters

There are twelve vegetation types in the SiB model. These include tall vegetation, short vegetation, arable crops, and desert (Dorman and Sellers, 1989). For GCM applications, it is necessary to have a simplified classification of the world's vegetation and a vegetation map, which shows the distribution of the vegetation types in the world. The main sources of data for the distribution of the world vegetation types were the physiognomic classification of Kuchler (1983), which recognized 32 natural vegetation communities, and the land use database of Matthews (1984, 1985); the latter was used to outline areas that had at least 50% cultivation by area. The 32 Kuchler surface types plus the cultivated land use cover were grouped into 12 major classes. The vegetation conditions present in the Gobi, Arabian, and Middle East deserts were described as bare soil in the vegetation map instead of the sparse grass cover defined by Kuchler (1983).

An input parameter set for each of the 12 SiB vegetation types was created based on a variety of sources (Dorman and Sellers, 1989; Willmott and Klink, 1986). Different types are distinguished by the assigned values of their parameters. Each vegetation type consists of three types of land surface data sets: morphological parameters, such as vegetation cover and leaf area index; physiological parameters, such as green leaf fraction; and physical parameters, such as surface albedo and roughness length. In most cases, the physiological parameters specified for each biome are taken from reports of the field measurements in the literature.

Many parameters in SiB are invariant with season. However, monthly values of leaf area index and green leaf fraction are prescribed. The treatment of the agricultural zones is different from that for other vegetation types: the vegetation cover, leaf area index, greenness, leaf orientation, and root length are varied according to growing season, which varies with both latitude and date.

The SiB has more than 40 parameters (Sellers et al., 1986). In the simplified version there are just 23 parameters (Xue et al., 1990), which can be divided into three categories: morphological parameters, physiological parameters, and physical parameters, and will be described here in detail.

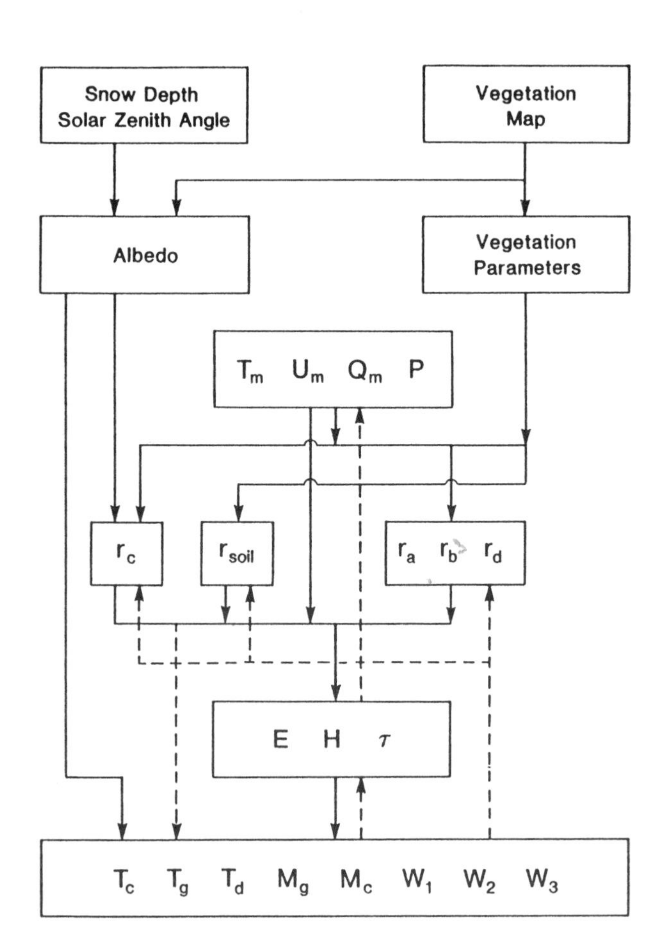

Figure 27-1: Sequence of calculations of the simplified SiB model. Symbols are defined in the text.

There are eight morphological parameters in this model. Five are for vegetation: (1) fractional area covered by the canopy, V_c; (2) canopy leaf and stem area density, L_c; (3) height of the canopy top, Z_2; (4) leaf angle distribution, O_c; and (5) rooting depth, Z_d. Three are for soil layers: the depths of three soil layers, D_1, D_2, and D_3. Among the eight parameters, V_c and L_c are key parameters, and are used in almost every part of this model. Z_2, O_c, and Z_d are used for the calculations of aerodynamic resistances, stomatal resistances, and transpiration, respectively.

In this study, physical parameters were divided into three subgroups. They are (1) surface albedo, (2) the parameters for the aerodynamic resistances: roughness length, Z_0, displacement height, d, the transfer coefficients between surface to canopy, C_d, and canopy to air space, C_b, and (3) soil property parameters: soil pore space, θ_s, soil layer slope, α, soil moisture potential at saturation, ψ_s, hydraulic conductivity of saturated soil, K_s, and an empirical constant, B, which is used to describe variation of soil hydraulic properties with changes in moisture content.

There are five sets of physiological parameters: the green leaf fraction, light-dependent stomatal resistance coefficients, and coefficients for temperature, water vapor deficit, and soil moisture adjustment. All of these are used for the stomatal resistance calculation. Greenness is also used for the radiative transfer calculation.

Computational flow

The main computation flow is shown in Figure 27-1. For simplicity, some secondary connections have been eliminated in this figure.

In a GCM coupled with SiB, a vegetation map has to be read in first and parameters assigned to each grid square based on vegetation type and month. These parameters plus the solar zenith angle and snow cover determine the surface albedo and the surface radiation budget.

With knowledge of the surface state and the atmospheric condition at reference height, including temperature T_m, wind field U_m, and humidity Q_m, and precipitation P, the surface resistances, r_c and r_{soil}, and aerodynamic resistances, r_a, r_b, and r_c can be calculated. These resistances are used to obtain the sensible heat H, the evapotranspiration rate E, and momentum flux τ using an interaction procedure. Subsequently, the atmospheric variables at the reference height are also updated.

Finally, the eight prognostic variables are updated. The updated variables will affect the resistances and fluxes calculation in the next time step. The model will provide the GCM with the land–atmosphere fluxes of latent heat, sensible heat, momentum, and radiation.

SiB APPLICATIONS

A great number of experiments have been carried out using the SiB model, which include the Sahel desertification and reforestation experiments, Amazon deforestation experiments, U.S. 1988 drought studies, and FIFE experiments. FIFE stands for the First ISLSCP Field Experiment, where ISLSCP is the International Satellite Land Surface Climatology Program. All of these studies showed that a more realistic surface model is very important for the land–atmosphere interaction studies.

The SiB model was implemented in a GCM by Sato et al. (1989). This GCM was developed in the Center for Ocean–Land–Atmosphere Interactions (COLA) (Kinter et al., 1988). Compared with the same GCM using a bucket model, the coupled SiB–GCM was elaborated to simulate the diurnal variation of temperature and humidity and produced a more realistic partitioning of energy at the land surface. Generally, SiB–GCM produced more sensible heat flux, less latent heat flux over vegetated land, and rainfall was reduced over the continents.

Nobre et al. (1991) used the coupled numerical model of the global atmosphere and biosphere (COLA GCM) to assess the effects of Amazonian deforestation on the regional climate. For deforestation runs, the model results showed mean surface temperature increases of about 2.5°C. The evaporation, precipitation, and runoff were reduced by 30, 25, and 20 percent, respectively. Their studies suggested that it might be difficult for the tropical forests to reestablish following massive deforestation.

Since the late 1960s, the Sahel has suffered severe droughts. Research has been carried out to study the mechanism for African drought (Charney et al., 1977; Walker and Rowntree, 1977; Laval and Picon, 1986). In those studies, just one or two parameters, such as albedo and initial soil moisture, were artificially changed. Because the prescribed changes in the surface condition were extreme, the results from those studies are controversial (Folland et al., 1991).

Using the COLA GCM with a simplified version of SiB, Xue and Shukla (1991) conducted a number of numerical experiments to test the effects of desertification on the Sahelian climate. In contrast to previous studies, this work tried to simulate the effect of desertification realistically. In the experiment, the Sahara desert was extended to the south. The vegetation types in this region were changed to shrubs with bare soil. The GCM was run for 90 days with and without anomalous surface conditions.

After the surface conditions were changed, the rainfall was reduced in the desertification area compared to the control run. The rainfall anomaly patterns for the model simulations are similar to those in the African dry years. Figure 27-2 shows the accumulative rainfall, evaporation, and runoff from both control run and the anomaly run. All these results are the averages over the prescribed desertification area. After desertification, the rainfall was reduced by 132 mm over the next 90 days. However, despite the reduced rainfall, the surface runoff increased by about 49 mm. This is because less rainfall was intercepted by leaves and a large portion of the precipitation reaches the ground directly. Meanwhile, the soil layer had less capacity to hold water in the desert area. The evaporation was reduced by about 60 mm during that period. This amount of reduction is not large enough to compensate for the deficit caused by the reduced rainfall and the enhanced runoff. The total soil water content was reduced by 121 mm in the desertification area.

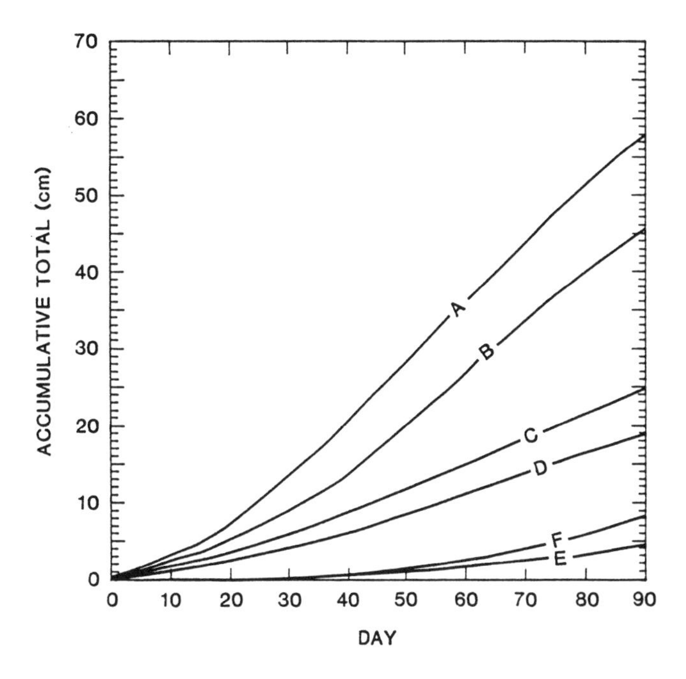

Figure 27-2: Result of 90-day simulation for Sahel desertification run and control run, COLA GCM experiment. A: rainfall of control run; B: rainfall of desertification run; C: evaporation of control run; D: evaporation of desertification run; E: runoff of control run; F: runoff of desertification run.

CONCLUDING REMARKS

SiB uses a sophisticated approach to incorporate biophysical processes into the GCMs through land surface energy balance and water content balance. The application of SiB on the land–surface–atmosphere interaction studies, especially on the Sahel drought studies, shows that this model is able to produce more realistic results than the simple ones.

There are 23 input parameters in the simplified SiB. The model results depend on these input data. Our sensitivity studies suggested that some of these parameters, such as albedo, soil moisture, and the parameters for aerodynamic resistance, play a critical role in the land–atmosphere interaction.

Improved global land surface data sets with seasonal components are vital to the further development of the biosphere models and enhanced simulation of land surface effects. More reliable information on land surface characteristics from geographical information systems (GIS) would benefit environmental modeling.

REFERENCES

Businger, J.A., Wyngaard, J.C., Zumi, Y.I., and Bradley, E.G. (1971) Flux-profile relationships in the atmosphere surface layer. *Journal of Atmospheric Science* 28: 181–189.

Camillo, P.J., and Gurney, R.J. (1986) A resistance parameter for bare-soil evaporation models. *Soil Science* 2: 95–105.

Charney, J.G., Quirk, W.K., Chow, S.-H., and Kornfield, J. (1977) A comparative study of the effects of albedo change on drought in semi-arid regions. *Journal of Atmospheric Science* 34: 1366–1385.

Dickinson, R.E. (1983) Land surface processes and climate—surface albedos and energy balance. *Advances in Geophysics* 25: 305–353.

Dickinson, R.E. (1984) Modeling evapotranspiration for three-dimensional global climate models. In Hansen, J.E., and Takahasi, T. (eds.) *Climate Processes and Climate Sensitivity,* Washington, DC: American Geophysical Union, pp. 58–72.

Dickinson, R.E., Henderson-Sellers, A., Kennedy, P.J., and Wilson, M.F. (1986) Biosphereatmosphere transfer scheme (BATS) for the NCAR community climate model. *NCAR/TN-275+STR*, Boulder, CO: National Center for Atmospheric Research.

Dorman, J.L., and Sellers, P. (1989) A global climatology of albedo, roughness length and stomatal resistance for atmospheric general circulation models as represented by the Simple Biosphere Model (SiB). *Journal of Applied Meteorology* 28: 833–855.

Folland, C., Owen, J., Ward, M.N., and Colman, A. (1991) Prediction of seasonal rainfall in the Sahel region using empirical and dynamical methods. *Journal of Forecasting* 1: 21–56.

Jarvis, P.G. (1976) The interpretation of the variations in leaf water potential and stomatal conductance found in canopies in the field. *Philosophical Transactions*, Royal Society of London B273: 593–610.

Kinter III, J.L., Shukla, J., Marx, L., and Schneider, E.K. (1988) A simulation of the winter and summer circulations with the NMC global spectral model. *Journal of Atmospheric Science* 45: 2486–2522.

Kuchler, A.W. (1983) World map of natural vegetation. *Goode's World Atlas*, 16th Edition, New York: Rand McNally, pp. 16–17.

Lanly, J.P. (1982) Tropical forest resources. *FAO Forestry Paper 30,* Rome: Food and Agriculture Organization, 106 pp.

Laval, K., and Picon, L. (1986) Effect of a change of the surface albedo of the Sahel on climate. *Journal of Atmospheric Science* 43: 2418–2429.

Matthews, E. (1984) Prescription of land-surface boundary conditions in GISS GCM II: a simple method based on high-resolution vegetation data bases. *NASA Technical Memorandum 86096*, Washington, DC: National Aeronautics and Space Administration, 20 pp.

Matthews, E. (1985) Atlas of archived vegetation, land-use and seasonal albedo data sets. *NASA Technical Memorandum 86199*, Washington, DC: National Aeronautics and Space Administration, 53 pp.

Meador, W.E., and Weavor, W.R. (1980) Two-stream approximations to radiative transfer in planetary atmospheres: a unified description of existing methods and a new improvement. *Journal of Atmospheric Science* 37: 630–643.

Nobre, C.A., Sellers, P., and Shukla, J. (1991) Amazon deforestation and regional climate change. *Journal of Climate* 4(10): 957–988.

Paulson, C.A. (1970) Mathematical representation of wind speed and temperature profiles in the unstable atmospheric surface layer. *Journal of Applied Meteorology* 9: 857–861.

Sato, N., Sellers, P.J., Randall, D.A., Schneider, E.K., Shukla, J., Kinter III, J.L., Hou, Y.-T., and Albertazzi, E. (1989) Effects of implementing the simple biosphere model in a general circulation model. *Journal of Atmospheric Science* 46: 2757–2782.

Sellers, P.J. (1985) Canopy reflectance, photosynthesis and transpiration. *International Journal of Remote Sensing* 6: 1335–1372.

Sellers, P.J., and Dorman, J.L. (1987) Testing the simple biosphere model (SiB) using point micrometeorological and biophysical data. *Journal of Applied Meteorology* 26: 622–651.

Sellers, P.J., Mintz, Y., Sud, Y.C., and Dalcher, A. (1986) A simple biosphere model (SIB) for use within general circulation models. *Journal of Atmospheric Science* 43: 505–531.

Sellers, P.J., Shuttleworth, W.J., Dorman, J.L., Dalcher, A., and Roberts, J. (1989) Calibrating the simple biosphere model for Amazonian tropical forest using field and remote sensing data. Part I: Average calibration with field data. *Applied Meteorology* 28: 727–759.

Shukla, J., and Mintz, Y. (1982) Influence of land-surface evapotranspiration on the Earth's climate. *Science* 215: 1498–1501.

Skoupy, J. (1987) Desertification in Africa. *Agricultural Meteorology Program, Proceedings of Regular Training Seminar on Drought and Desertification in Africa, Addis*

Ababa, Geneva: WMO, pp. 33–45.

Sud, Y.C., and Smith, W.E. (1985) The influence of roughness of deserts on the July circulation—a numerical study. *Boundary Layer Meteorology* 33: 1–35.

Walker, J., and Rowntree, P.R. (1977) The effect of soil moisture on circulation and rainfall in a tropical model. *Quarterly Journal, Royal Meteorological Society* 103: 29–46.

Willmott, C.J., and Klink, K. (1986) A representation of the terrestrial biosphere for use in global climate studies. *ISLSCP: Proceedings of an international conference held in Rome, Italy, 2–6 December 1985*, Rome, Italy: European Space Agency, pp. 109–112.

Xue, Y., Sellers, P.J., Kinter III, J.L., and Shukla, J. (1991) A simplified biosphere model for global climate studies. *Journal of Climate* 4: 345–364.

Xue, Y., and Shukla, J. (1991) A study of the mechanism and impact of biosphere feedback on the African climate. *Tenth Conference on Biometeorology and Aerobiology and the Special Session on Hydrometeorology, September 10–13, 1991, Salt Lake City, Utah*, Boston, MA: American Meteorology Society.

28

Regional Hydroecological Simulation System: An Illustration of the Integration of Ecosystem Models in a GIS

RAMAKRISHNA NEMANI,
STEVEN W. RUNNING,
LAWRENCE E. BAND, AND
DAVID L. PETERSON

The integration of simulation models in a GIS is a new concept in environmental modeling. GIS provide a valuable tool in bringing disparate data sources together in an organized fashion. In addition, the spatial data handling abilities in a GIS allow modelers to explore such diverse questions as global carbon cycling or groundwater contamination in a small watershed. Unfortunately, many recent efforts at integrating models in a GIS involve only overlaying various data layers to parameterize the simulation models. While this simplistic approach may be appropriate in certain cases, coupling process models involves more than data overlays. Since process models are inherently complex and nonlinear, the way in which continuous geographic information is aggregated has far-reaching effects on model output.

For example, most calculations of the flux of energy and mass from terrestrial ecosystems have been executed at two different spatial scales. First, global-scale computer models for climate (Manabe and Weatherald, 1980) or for the global carbon cycle (Emmanuel et al., 1984) treat the entire land surface as a single entity or as discrete units of approximately 1 degree latitude by 1 degree longitude. Considerable work has been done at the other spatial extreme—single plant to stand level process models of ecosystems (Parton et al., 1987; Bonan, 1991). None of the existing GIS packages can adequately support modeling at both scales. The problem here is twofold. First, current GIS packages do not support any reasonable logic for polygon definition except simple overlays. This could potentially lead to large variances in computed parameter fields, resulting in erroneous output when used in conjunction with a nonlinear process model. Second, many of the current generation of process models are highly scale dependent in terms of both parameter requirements and process representation. Implementing such models in a GIS without prior experimentation is not appropriate. We must note that choosing an appropriate scale is hard, as it depends on the process being modeled, the inherent variation in the subscale phenomenon, data availability, and computational limitations. Therefore, there is a need for an integrated modeling system designed in such a way that it is scale flexible, and optimized for parameter and computational requirements at each of the desired scales.

In this chapter, we present such a system, RHESSys (Regional Hydroecological Simulation System), in which a stand-level forest ecosystem model is used to calculate ecosystem flux rates at different spatial scales ranging from a single hillslope to a 100×50 km region.

REGIONAL HYDROECOLOGICAL SIMULATION SYSTEM (RHESSys)

RHESSys is a data and simulation system designed to compute and map carbon, water, and nutrient fluxes at various spatial scales, see Figure 28-1 (Running et al., 1989; Band et al., 1991a,b). The design criteria for RHESSys are (1) to provide easy interface of climatological and hydroecological models with remote sensing, climate, topographic, and soils data, (2) to allow variable representation of climate and surface parameterizations (e.g., hillslopes, subcatchments, catchments) for model simulations. Three simulation models in the system, in combination, perform the computation of ecosystem flux rates across the landscape: FOREST–BGC, an ecosystem simulation processor; TOPMODEL, a simulation model for computing hillslope-level hydrologic processes; and MT-CLIM, a topoclimatological model for extrapolating

REGIONAL HYDRO-ECOLOGICAL SIMULATION SYSTEM (RHESSys)

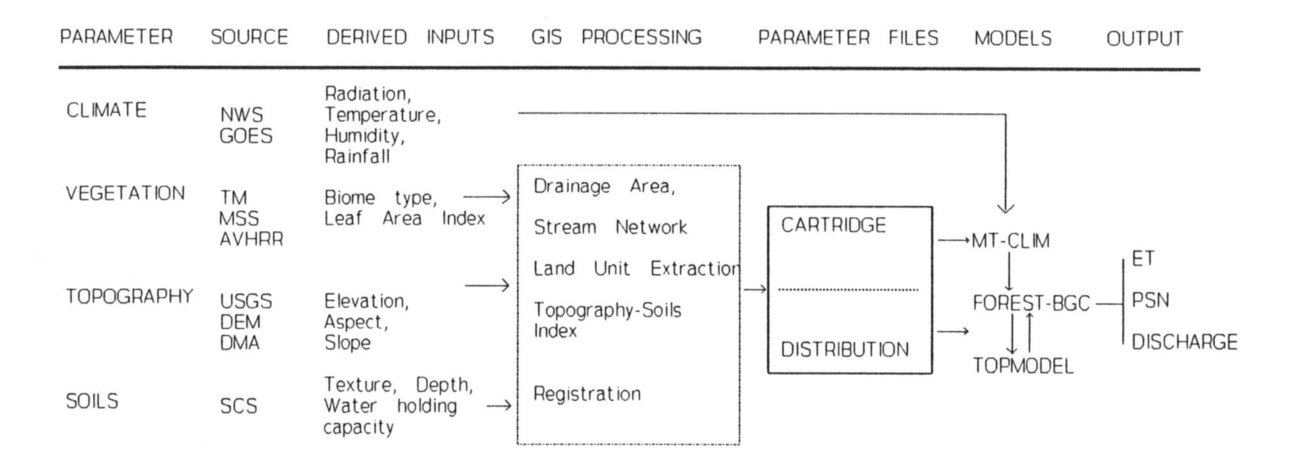

NWS - National Weather Service
GOES - Geostationary Operational Environ. Satellite
TM - Landsat/Thematic Mapper
MSS - Landsat/Multispectral Scanner
AVHRR -NOAA/Advanced Very High Resolution Radiometer
USGS - United States Geological Survey
DEM - Digital Elevation Model
DMA - Defence Mapping Agency
SCS - Soil Conservation Survey

CARTRIDGE - Land unit parameterization
DISTRIBUTION -Within Land unit Parameterization

MT-CLIM -Mountain Microclimate Simulator
FOREST-BGC -Forest Ecosystem Simulator
TOPMODEL -Hydrologic Routing Simulator

ET - Evapotranspiration
PSN -Net Photosynthesis
Discharge -Stream Discharge

nemani/NTSG

Figure 28-1: An organizational diagram of RHESSys showing the sources of raw climatic and biophysical data, derived data, and GIS processing needed to produce the parameter files, and the simulations required for generating the evapotranspiration maps shown in Figure 28-3. Note that geographic information processing is required only once per target area. Once parameter files (cartridge and distribution) are generated at various levels of landscape aggregation, simulations can be performed automatically at any given scale by inserting an appropriate set of cartridge/distribution files.

meteorological data across the landscape. The system includes a complement of digital terrain analytical methods for implementation of a formal object model of a given landscape. Brief descriptions of the key components of RHESSys will be given.

Ecosystem models

The forest ecosystem model, FOREST–BGC

FOREST–BGC (bio–geo–chemical cycles) is a process-level ecosystem simulation model that calculates the cycling of carbon, water, and nitrogen through forest ecosystems (Figure 28-2; Running and Coughlan, 1988). The model requires daily data for standard meteorological conditions, maximum–minimum temperatures, dew point, incident shortwave radiation and precipitation, and definition of key site and vegetation variables. The model calculates canopy interception and evaporation, transpiration, soil outflow of water; photosynthesis, growth, and maintenance respiration, allocation, litterfall, and decomposition of carbon; and deposition, uptake, litterfall, and mineralization of nitrogen. The model has a mixed time resolution, with hydrologic, photosynthetic, and maintenance respiration processes computed daily, and the other carbon and all nitrogen processes computed yearly. As a brief summary of the model's function, daily precipitation is routed to snowpack or soil dependent on air temperature, a distinction that greatly influences the timing of future hydrologic activity. A canopy interception fraction, effectively a capacitance defined by LAI, is

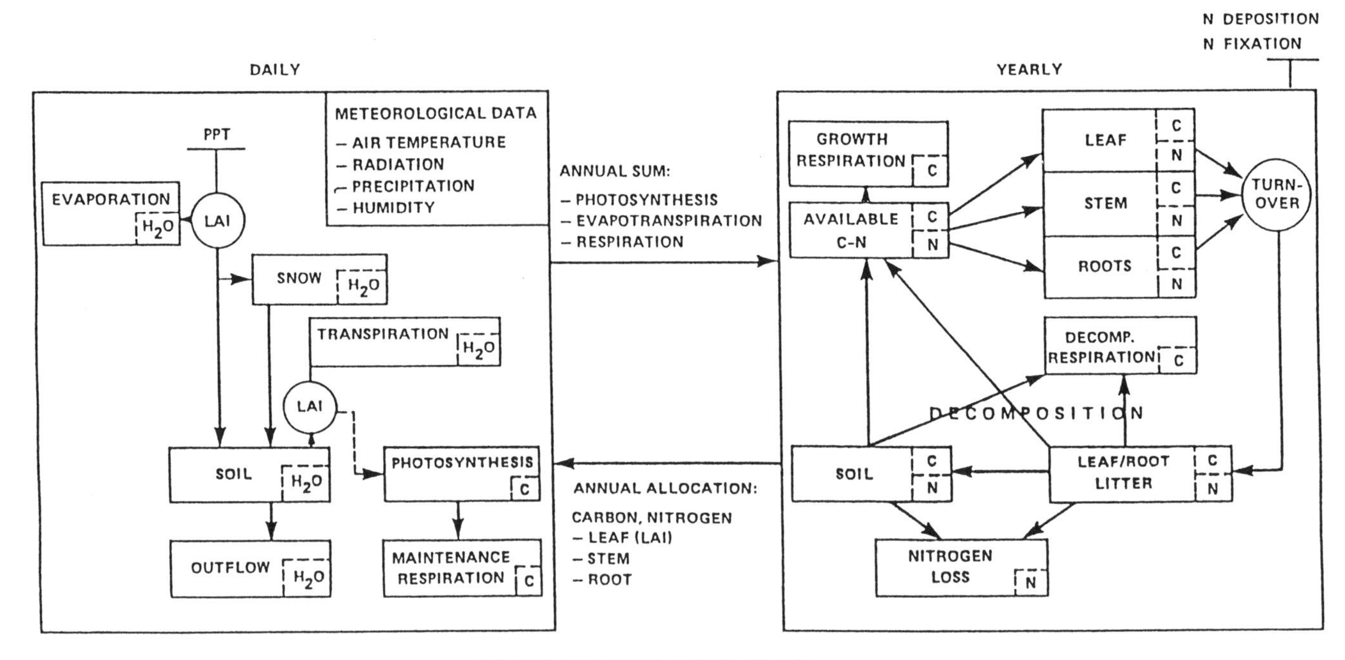

Figure 28-2: Compartment flow diagram for FOREST–BGC, a simulation model of carbon, water, and nutrient cycling processes for forest ecosystems. The two submodels integrate daily and yearly time steps and require daily meteorological data. FOREST–BGC was designed to accept remotely sensed observations, particularly of leaf area index (Running and Coughlan, 1988).

subtracted and evaporated with a Penman equation, and the remaining water goes to a soil compartment. Modeling canopy interception is a critical hydrologic partitioning point because this process determines whether incoming precipitation remains in the system for a few hours and evaporates, or is stored in the soil where it may be transpired many weeks later or routed to streamflow. Transpiration is calculated with a Penman–Monteith equation.

The canopy conductance term is a complex function of air temperature, vapor pressure deficit, incident radiation, and leaf water potential. Air temperatures below 0°C reduce canopy conductance to cuticular values, and are an important determinant of the active physiological season. Canopy conductance is linearly reduced to a default cuticular value when either average daily vapor pressure deficit exceeds 1.6 kPa or predawn leaf water potential, estimated from soil water availability, decreases below –1.65 MPa. Aerodynamic resistance is assumed as 5.0 s/m in the Penman–Monteith equation, modified slightly by LAI to define surface roughness, because wind-speed data are rarely available.

Canopy photosynthesis is calculated by multiplying a CO_2 diffusion gradient by a mesophyll CO_2 conductance and the canopy water vapor conductance, both of which are controlled by radiation and temperature. The light response surface for mesophyll conductance is asymptotic. A function of –0.5/LAI is used for Beer's law attenuation of incident radiation to produce canopy average radiation. An inverse parabolic response surface defines high- and low-temperature compensation points respectively. Net photosynthesis (PSN) is computed by subtracting a maintenance respiration term, calculated as an exponential function of air temperature with a $Q_{10} = 2.3$ for all stem and root biomass, and then subtracting 35% of the remaining gross photosynthesis for growth respiration.

Hillslope hydrology, TOPMODEL

The soil hydrology in FOREST–BGC is based on simple bucket-type logic, where the size of the bucket is set by the water holding capacity of the site. Consequently, it cannot account for horizontal redistribution of soil water, which

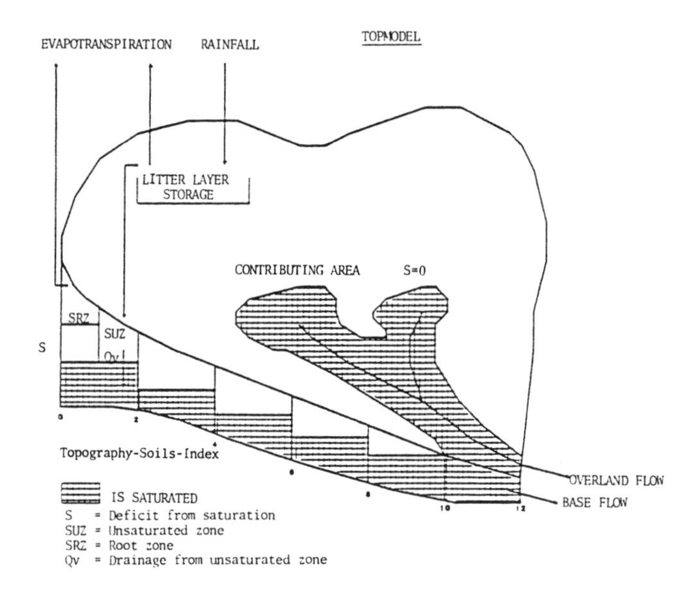

Figure 28-3: Conceptual diagram showing hillslope level hydrologic processes represented in TOPMODEL (after Hornberger et al., 1985).

often dominates soil hydrologic processes in steep mountainous terrain. To remedy this problem, the simple bucket-type logic was replaced by a topography-based hydrologic model (TOPMODEL) (Beven and Kirkby, 1979; Band et al., 1991b), which explicitly treats the lateral flow of soil water based on topography and soil properties (Figure 28-3).

TOPMODEL was originally designed by Beven and Kirkby (1979) to simulate baseflow and overland flow generation in moderate to steep topography. The model is based on principles of hillslope hydrology, and handles variations in soil water content and soil water flux as functions of the surface topography and distribution of soils. TOPMODEL is considered to be "quasi-distributed" as it uses a topography–soils similarity index (TSI) to stratify zones of differing soil water dynamics. Any two points with the same similarity index are then considered hydrologically identical so that simulation is only necessary over the index distribution, and not explicitly over the full terrain.

In the context of RHESSys, TOPMODEL serves two important roles: (1) It allows better estimates of stream discharge in terms of timing as well as quantity, and (2) it allows better representation of soil moisture patterns across the landscape that have strong influence on rates of evapotranspiration and photosynthesis.

Microclimate extrapolation, MT-CLIM

Even if proper formulae are used, and accurate definitions of vegetation and soils are available, regional ET estimates require accurate meteorological drivers at daily time resolution or greater. Although many government agencies collect meteorological data for specific purposes, only the NOAA Environmental Data Service reports data that are consistent and widely available. Because air temperatures, vapor pressure deficit, incoming shortwave radiation, and precipitation are required every day for simulation, standard data sources are necessary. However, multiple problems have been encountered in acquiring these data for regional simulations. First, the "standard" daily data collected by the U.S. National Weather Service and cooperating stations are maximum–minimum temperature and precipitation. The density of primary stations recording humidity and solar radiation in any form is less than 1/100,000 km^2 throughout the western United States. Second, standard meteorological data are not available for diverse mountain sites, as most regular weather stations are in populated valleys. Although one can monitor a few mountain sites with research instrumentation, routine monitoring of entire regions is not feasible. Satellite observations can partially solve this problem, but the spatial resolution is very coarse, 8x8 km for GOES, and cloud-cover problems restrict dependable daily observations. Third, meteorological data specific to ecological processes, such as a photosynthetically active definition of day length, are never available from routine sources (Figure 28-4a).

To solve these problems, we developed MT-CLIM, a model that allows us to build a daily meteorological map of our study region based on a few direct NWS measurements extrapolated by climatological principles to the surrounding terrain (Running et al., 1987). The centerpiece of MT-CLIM is a submodel that calculates daily potential incoming shortwave radiation for any elevation, slope, aspect, and latitude based on the original algorithm of Garnier and Ohmura (1968). The observed maximum–minimum diurnal temperature amplitude is then used to calculate an atmospheric transmissivity (which incorporates cloud cover), a surprisingly robust logic developed by Bristow and Campbell (1984) that eliminates the need for using the inaccurate visual cloud-cover estimates provided by the NWS (Figure 28-4a). Elevational adiabatic lapse rates are used to correct temperatures, and the differential radiation loading to different slopes, attenuated by LAI, also adjusts final estimates of slope surface temperatures (Figure 28-4b). If humidity data are not available, night minimum temperature substitutes for dew point, then allowing estimation of daytime vapor pressure deficits (Figure 28-4a). Mountain precipitation is adjusted by the ratio of annual isohyets of the base station relative to the study site.

Our validations of this model on three north–south mountain slope pairs in Montana found these simple climatological calculations to produce meteorological data of sufficient accuracy for simulation of seasonal

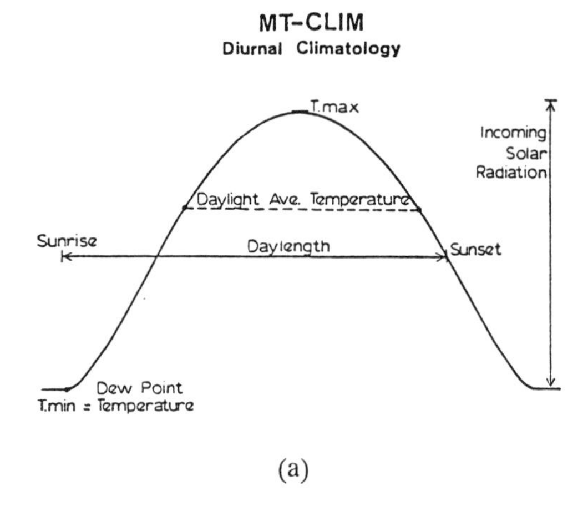

(a)

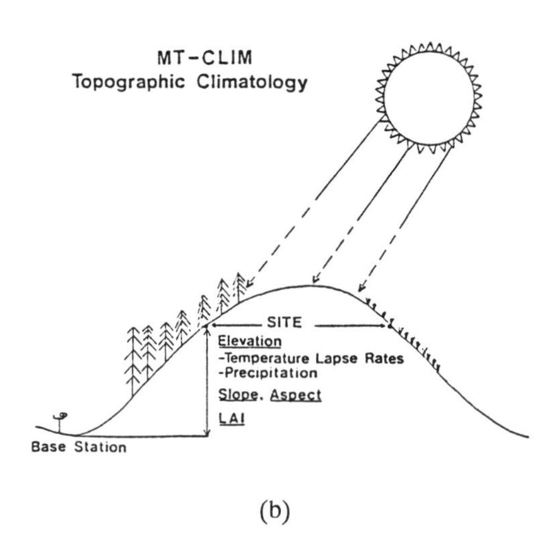

(b)

Figure 28-4: Conceptual diagrams showing the climatological principles used in MT-CLIM to derive site specific meteorological data from base station data: (a) diurnal climatology and (b) topoclimatology.

evapotranspiration and photosynthesis of our forests (Running et al., 1987).

Field data

Leaf area index observations from remote sensing

Critical to our logic for regional implementation is a simple definition of the hydrologically significant vegetation properties. If the basic biome type (forest, grassland, crop, etc.) can be defined and general physiological controls assumed for a biome, then the most important general vegetation parameter needed is LAI. For example, LAI is used as a principal independent variable in FOREST–BGC, for calculating many processes: interception of rainfall and radiation, transpiration, and photosynthesis.

A practical reason for choosing LAI as the key vegetation parameter is our ability to estimate LAI by satellite. Nemani and Running (1989) observed a strong curvilinear relation between LAI and NDVI derived from NOAA AVHRR satellite for 53 conifer stands in Montana. Similar results were attained using the higher spatial resolution TM sensor, where the NIR/red ratio of Channels 4/3 was correlated with measured LAI of coniferous forests, $R^2 = 0.91$ (Peterson et al., 1987). Other investigators have also found good correlations between LAI and NIR and red reflectances of vegetation (Asrar et al., 1984).

Soils data

Of all the data required for running RHESSys, soil information is the most difficult to obtain. Soil conditions vary on the order of a few meters in mountainous terrain. In most cases, available soil information is either too coarse or it does not provide the required information. For example, most soil surveys do not include water holding capacity or depth. Currently, we adjust the coarse resolution soil maps in two ways: (1) We set soil properties on talus and rock outcrop areas, delineated from remotely sensed observations, to low water holding capacity (WHC) and high transmissivity values; (2) depth of soil is increased from shallow at ridge tops to deep soils at the valley bottoms using the topography data. We are working on an object-oriented soil inferencing system in which soil properties required for RHESSys (Figure 28-1) would be inferred using remote sensing, geological maps, and topographic position of a given object on the landscape. The procedure involves generating rules of soil–climate–topography–vegetation associations from soil experts, and then automating the procedure for entire landscapes.

Digital terrain analysis

Landscape aggregation

The dominant physical processes of energy and mass exchange, transpiration, and photosynthesis are characterized by strong nonlinearities. The manner in which continuous geographic information fields are aggregated into discrete landscape units is found to have significant impact on both the parameterization and simulation results (Band et al., 1991a). Therefore, the primary criteria for landscape aggregation are: (1) to minimize within-unit variance so the model behaves linearly; and (2) to maximize between-unit variance of parameter fields so as to minimize redundant model runs.

Table 28-1: An example of a cartridge file showing various parameter entries of site specific information for each landscape unit as defined by the digital terrain analysis

Unit #	Aspect	Elevation	Slope	LAI	SWC (m)	Area (ha)	Mean TSI	# of intervals
1	177.20	1600.50	30.48	10.6	0.10	129.0	5.70	10
2	25.05	1941.00	30.64	4.9	0.12	25.0	5.52	6

In our study sites of western Montana, the strong influence of topographically controlled microclimate on surface ecologic processes results in a geographic pattern of forest cover that closely follows the pattern of hillslope facets. Consequently, we chose topographic partitions to act as a template to organize all other parameter fields. The methods of hillslope extraction over a range of scales from digital terrain data use a formal geomorphic model of watershed structure (Band, 1989). The scale flexibility is obtained by computing the drainage area upslope of each pixel in a DEM. By setting a minimum threshold drainage area required to support a drainage channel, the scale of the stream network may be defined. A high threshold would produce fewer partitions and vice versa (see Plate 28-1b, which compares 5 and 170 partitions).

INFORMATION PROCESSING AND MODEL PARAMETERIZATION

The preparation of the landscape template and estimation of the model parameters for each landscape unit involve a combination of digital terrain analysis, remote sensing, and geographic information processing. A significant portion of the computation and analysis is devoted to the processing and combination of primary data files (vegetation, soils, and topography). Briefly, remote sensing and soils data are registered to topography. While deriving LAI from remotely sensed data, topographic data are used to correct the radiances for variable illumination. Topographic information is also used to adjust the coarse-resolution soils information as described earlier. Finally, the soil and topographic information is combined to produce a Topography–Soils Index (TSI), defined as $\ln(aT_e)/T_i\tan B$. While a (drainage area upslope of each pixel) and B (slope of a given pixel) are derived from digital terrain analysis, T_e (mean transmissivity of soil for the landscape unit) and T_i (transmissivity of a given pixel) are derived from the soil maps. The products of these steps are two landscape parameterization files: (1) cartridge, and (2) distribution. The cartridge file holds mean parameter values for each landscape unit (e.g., hillslope): slope, aspect, elevation, LAI, soil water holding capacity (WHC), etc. (Table 28-1). On the other hand, the distribution file holds the frequency distribution of vegetation and soil parameters within each landscape unit/hillslope (Table 28-2). These frequency distributions are generated based on the topography–soils index. Each entry in the cartridge file represents a landscape unit as outlined by the digital terrain analysis. Once the cartridge/distribution files are generated at various landscape partitions, the current cartridge/distribution files can be pulled out and another set of cartridge/distribution files inserted to model another representation of the landscape.

The cartridge file describing site-specific characteristics, along with meteorological data collected at a base station, are used in MT-CLIM to generate site specific climatological data for each entry in the cartridge file. If weather data are available from more than one site in the study area, an inverse distance function (based on the distance between the centroid of the landscape unit and the weather stations) is used to scale the influence weather conditions at each station on a given landscape unit.

Once the seasonal site-specific meteorological data are generated, the hydroecological simulation runs begin at the lowest TSI interval and proceed to the highest TSI interval for each of entry in the cartridge file. FOREST–BGC is initialized at the beginning of each run with generic physiological parameters for conifer forests. At the end of each day, after performing the simulations for all the intervals, a new mean soil water deficit is calculated for the landscape unit. In this fashion the simulations continue for a whole year for each unit and are repeated for all the entries in the cartridge file.

RHESSys allows several output options. For example, variables like snow pack, soil water, discharge, transpiration, evaporation, and photosynthesis can be output for each entry in the cartridge file and/or for each TSI interval for selected days or for a whole year. In addition, image products showing spatial variation around the landscape in soil moisture snowpack, evapotranspiration, and photosynthesis can be also generated on any given day.

Currently, RHESSys runs under Unix environments on SUN and IBM RS/6000 workstations. While the simulation models are in FORTRAN, digital terrain analysis is done in C. Graphic output and visualization of spatial patterns are accomplished by using Precision Visuals' WAVE package.

Table 28-2: An example of a distribution file showing the frequency distributions of vegetation and soil properties for each landscape unit calculated based on topography–soils index (TSI)

Unit #	TSI Interval	Area	LAI	Hyd. Cond (cm/sec)
1	2.5	1.17	7.0	0.93
1	3.5	8.55	11.3	0.91
1	4.5	23.94	11.5	0.90
1	5.5	49.41	10.4	0.90
1	6.5	32.31	10.5	0.90
1	7.5	8.28	10.7	0.89
1	8.5	2.97	11.5	0.90
1	9.5	1.35	10.4	0.92
1	10.5	1.26	9.3	0.90
1	11.5	0.09	7.0	0.90
2	3.5	3.06	4.8	0.73
2	4.5	7.47	4.3	0.73
2	5.5	5.31	4.7	0.72
2	6.5	5.58	6.0	0.70
2	7.5	2.61	5.4	0.70
2	8.5	1.08	4.4	0.72

SCALING ECOSYSTEM PROCESSES

We have recently completed the implementation of RHESSys over the Seeley–Swan valley (100 by 50 km) of northwestern Montana. In this area climatic influences on vegetation, mainly coniferous forests, are pronounced, because two mountain ranges (Mission mountains on the left, Swan mountains on the right, Plate 28-1a) are separated by a valley, creating strong precipitation and temperature gradients. In general, precipitation varies from 40 cm at the southwest corner to about 200 cm at the northeast corner in Plate 28-1a, resulting in large differences in leaf area index. The computed LAIs, about 2–16, seem to be within the range of measured LAIs in this area (Nemani and Running, 1989). Extensive clear-cutting of forests over the past few decades is evident from the low LAI observations computed in the valley.

Plate 28-1b shows evapotranspiration in the Seeley–Swan valley calculated by RHESSys at two levels of spatial aggregation. First, ET is calculated by aggregating the entire study area into 5 partitions using a high threshold value (drainage area) for extracting the stream network. Then, the threshold value is reduced, resulting in 170 partitions for the same area. Though the absolute magnitudes of ET (approximately 50–70 cm) computed at each partition level are similar, there is considerably more spatial detail at 170 partitions. The telescoping capability of RHESSys is illustrated with ET calculations performed for Soup Creek. Soup Creek, a 13 km^2 watershed, cannot be distinguished in the case of 5 partitions. But at 170 partitions Soup Creek is represented by two partitions. When high-resolution information is required, in cases like watershed management, RHESSys can perform simulations for Soup Creek over a range of spatial aggregations (2–66 partitions). For example, Plate 28-1b shows an ET map of Soup Creek computed for each TSI interval of 22 partitions. Again as we move down into a single watershed, considerable spatial variation in ET is evident. At this scale, knobs on hillslopes and ridge tops show lower ET rates relative to valley bottoms and stream pixels.

It is clear from Plate 28-1b that the importance of factors controlling ET rates changes as we telescope from a region down to a single watershed. At regional scale (5 partitions) climatic factors, precipitation, and temperature exert strong influence on ET rates. At the other extreme of a single watershed, microclimate along with lateral redistribution of soil water due to microtopographic differences strongly influences ET patterns. The ability to represent the controlling factors at each scale correctly has important implications. For example, a mesoscale climate modeler may find the spatial patterns of ET computed at 170 partitions appropriate for describing the boundary conditions. On the other hand, a forest manager interested in increasing water yield of a catchment would like to see spatial patterns of ET at catchment scale.

GIS/MODELING

The integration of GIS tools with environmental modeling has its own pitfalls. It may, in some cases, be appropriate to run a model with parameters derived by overlaying various data layers in a GIS (i.e., linear regression models of stand growth or soil erosion). However, in integrating a process model in a GIS, care must be taken as to the model assumptions, variance in parameter fields, and process representation at different scales. For example, in RHESSys a tree/stand-level model (FOREST–BGC) is used at various scales only after careful thought is given to model requirements and process representation details as the scale increased from stand to region (climate extrapolation, definition of vegetation through LAI, etc.).

Many of the commercially available GIS packages are too rigid in their data representation formats, and often the user does not have access to the specific data formats. This creates ample problems for modelers trying to integrate their models in a GIS. Modelers often spend more than 90% their efforts trying to gather data. When a GIS package does not allow easy interfacing of their models with various data sets, they resort to writing their own software. The inevitable result is that models are difficult to adapt to new data sources, and merge with seemingly complementary modeling efforts.

The effectiveness of regional ecosystem modeling depends on both the quality of models used and the effectiveness of the information processing techniques used to integrate various data sources. It is apparent from our experience with RHESSys that ecosystem modelers still have to spend a considerable amount of time designing appropriate information processing techniques. None of the commercially available GIS packages can accomplish half the tasks involved in RHESSys. Clearly there is a need for separating modeling concerns from data representation details. Towards that goal, we have initiated an object-oriented modeling system called REIS (Regional Ecosystem Information System). REIS is being designed to allow modelers to concentrate on ecosystem modeling rather than data modeling through two key concepts: representation-independent modeling schemes and data context sensitivity. Additionally, REIS will also provide visualization and statistical tools under the same modeling environment.

ACKNOWLEDGMENTS

Funding for this research has been provided by grants from NASA (NAGW-952) and NSF (BSR-8817965). We would like to acknowledge Lars Pierce, Richard Lammers, and Ray Ford for their contributions.

REFERENCES

Asrar, G., Fuchs, M., Kanemasu, E.T., and Hatfield, J.L. (1984) Estimating absorbed photosynthetic radiation and leaf area index from spectral reflectance in wheat. *Agronomy Journal* 76: 300–306.

Band, L.E. (1989) Spatial aggregation of complex terrain. *Geographical Analysis* 21: 279–293.

Band, L.E., Patterson, P., Nemani, R., and Running, S.W. (1991b) Forest ecosystem processes at the watershed scale: incorporating hillslope hydrology. *Agricultural and Forest Meteorology* (in press).

Band, L.E., Peterson, D.L., Running, S.W., Coughlan, J.C., Lammers, R., Dungan, J., and Nemani, R. (1991a) Forest ecosystem processes at the watershed scale: basis for distributed simulation. *Ecological Modeling* 56: 171–196.

Beven, K.J., and Kirkby, M.J. (1979) A physically based, variable contributing area model of basin hydrology. *Hydrological Sciences Bulletin* 24(1): 43–69.

Bonan, G.B. (1991) Atmosphere–biosphere exchange of carbon dioxide in boreal forests. *Journal of Geophysical Research* 96: 7301–7312.

Bristow, K.L., and Campbell, G.S. (1984) On the relationship between incoming solar radiation and daily maximum and minimum temperature. *Agricultural and Forest Meteorology* 31: 159–166.

Emmanuel, W.R., Killough, G.G., Post, W.M., and Shugart, H.H. (1984) Modeling terrestrial ecosystems in the global carbon cycle with shifts in carbon storage capacity by land-use change. *Ecology* 65: 970–983.

Garnier, B.J., and Ohmura, A. (1968) A method of calculating the direct shortwave radiation income of slopes. *Journal of Applied Meteorology* 7(5): 796–800.

Hornberger, G.M., Beven, K.J., Cosby, B.J., and Sappington, D.E. (1985) Shenandoah watershed study: calibration of a topographically based, variable contributing area hydrologic model to a small forested catchment. *Water Resources Research* 21: 1841–1850.

Manabe, S., and Wetherald, R.T. (1980) On the distribution of climate change resulting from an increase in CO_2 content of the atmosphere. *Journal of Atmospheric Science* 37: 99–118.

Nemani, R.R., and Running, S.W. (1989) Testing a theoretical climate–soil–leaf area hydrologic equilibrium of forests using satellite data and ecosystem simulation. *Agricultural and Forest Meteorology* 44: 245–260.

Parton, W.J., Schimel, D.S., Cole, C.V., and Ojima, D.S. (1987) Analysis of factors controlling soil organic matter levels in Great Plains grasslands. *Soil Science Society of America Journal* 51: 1173–1179.

Peterson, D.L., Spanner, M.A., Running, S.W., and Teuber, K.B. (1987) Relationship of Thematic Mapper Simulator data to leaf area index of temperate coniferous forests. *Remote Sensing of Environment* 22: 323–341.

Running, S.W., and Coughlan, J.C. (1988) A general model of forest ecosystem processes for regional applications. I. Hydrologic balance, canopy gas exchange and primary production processes. *Ecological Modeling* 42: 125–154.

Running, S.W., Nemani, R.R., and Hungerford, R.D. (1987) Extrapolation of synoptic meteorological data in mountainous terrain, and its use for simulating forest evapotranspiration and photo-synthesis. *Canadian Journal of Forest Research* 17: 472–483.

Running, S.W., Nemani, R.R., Peterson, D.L., Band, L.E., Potts, D.F., Pierce, L.L., and Spanner, M.A. (1989) Mapping regional forest evapotranspiration and photosynthesis by coupling satellite data with ecosystem simulation. *Ecology* 70: 1090–1101.

29

Simulation of Rangeland Production: Future Applications in Systems Ecology

JON D. HANSON AND
BARRY B. BAKER

Rangelands are relatively low-productivity agricultural systems that comprise about 40 percent of the Earth's land surface. The host of plant species populating native range provide food and habitat for domestic livestock and many species of wildlife. Complex interactions between system components hinder our understanding of the structure and function of rangelands. National legislation, during the past several years, has induced USDA's Agricultural Research Service to develop several systems-level models (Figure 29-2). Clean water legislation led to the development of ACTMO (agricultural chemical transport model) in 1975, and subsequently CREAMS (chemicals, runoff and erosion from agricultural management systems) in 1979 (Knisel, 1980). In 1984, EPIC (erosion/productivity impact calculator) was completed (Williams et al., 1983) in response to the 1977 Soil and Water Conservation Act. The ALMANAC model is an enhancement of EPIC and was developed to analyze regional phenomena. In 1985, ARS began work on WEPP (Water Erosion Prediction Project) in an effort to respond to several conservation laws. In 1978, a range modeling workshop was held to begin planning the development of a general grassland model; the model was officially started in 1981 and given the name SPUR (simulation of production and utilization of rangelands).

The integration of models representing complex ecological systems is very difficult. Models of varying resolution (point models to basin scale) describing plant growth and development have been produced in the past decade. These models range in complexity from regression models to extremely complex process models (such as the SPUR model). Some models are based on a metaphor that considers the plant to be some physical or biochemical system, such as an enzymatic system. Other models were developed using a systems modeling approach, which makes them particularly useful in describing and synthesizing the conceptual structure of the system. These models are often nonlinear and usually incorporate maximum reduction techniques. Different approaches used to structure plant growth models within any particular system include linear versus nonlinear, empirical versus mechanistic, and deterministic versus stochastic (Hanson et al., 1985). Through the use of these different approaches, models of a selected ecosystem emphasize those factors that most limit plant growth.

Plant growth models generally assume that the processes determining plant production (photosynthesis, respiration, carbon allocation, etc.) do not differ appreciably between species of various ecosystems. Available water and nitrogen usually limit plant growth on arid and semiarid grasslands (Reuss and Innis, 1977; Lauenroth, 1979). Soil moisture depends not only on the amount of precipitation, but also on plant species composition and production, soil water holding characteristics, and rainfall distribution (Duncan and Woodmansee, 1975). Water stress reduces plant growth, but little mechanistic information is available that describes the detailed physiological recovery of postdrought plants.

In addition to the obvious effects of nutrients and water availability, models for rangeland ecosystems must consider the effect of large domestic herbivores, various wildlife species, and insects. These grazers not only alter plant community structure by direct consumption of plants and plant parts, but they also trample vegetation and recycle nutrients. Arid and semiarid grassland models must also deal with landscape spatial heterogeneity. Unless stocking rates are high or various management techniques are employed, cattle do not graze pastures uniformly. Rather, they tend to graze selectively and leave large areas of the pasture effectively ungrazed. Models developed to simulate grasslands under grazing must consider the patchy grazing behavior of cattle. Parton and Risser (1980) found the ELM model (Innis, 1978) would not work correctly without considering spatial distribution of the vegetation.

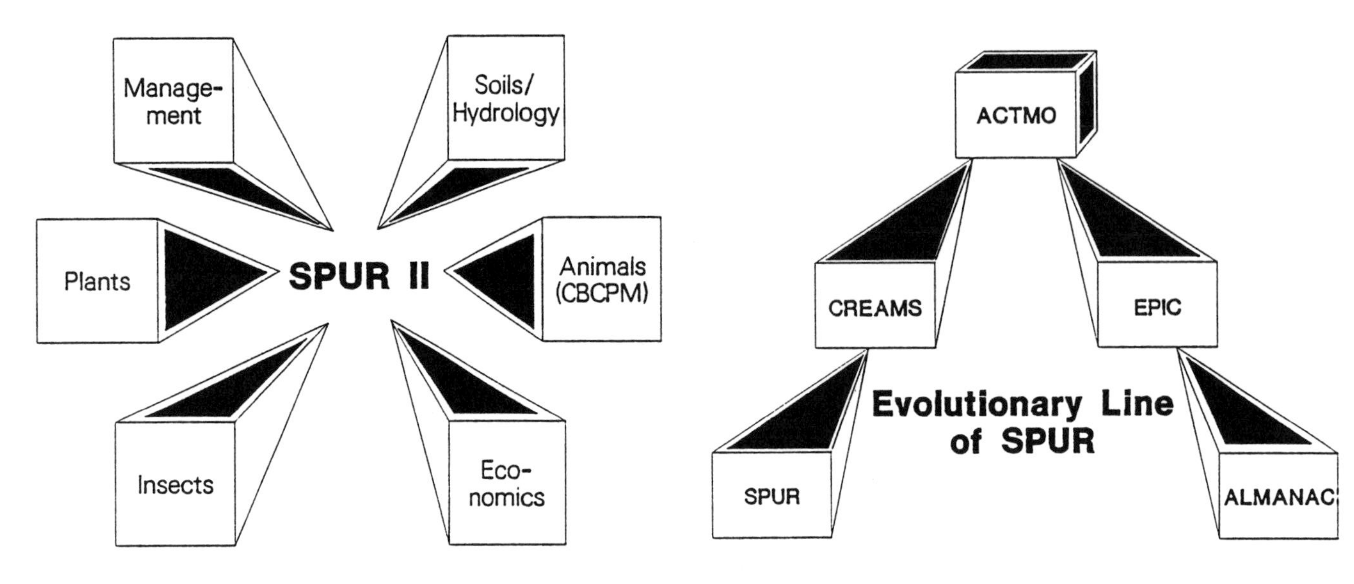

Figure 29-1: Components of the SPUR model.

Figure 29-2: Historical lines of the SPUR model.

The irregular grazing patterns found in pastures may be induced by the variable distribution of nutrients throughout the grassland. Schimel et al. (1985) reported that carbon, nitrogen, phosphorus, and clay increased from the top to the bottom of a slope; the lower slope position seemed to have a larger soil nitrogen pool that appeared to have a slower turnover rate than those of corresponding upslope soils. Other obvious factors that need to be considered in modeling grazing behavior on grasslands include fences, salt blocks, water tanks or holes, plant species composition, and type of grazer.

MODEL DESCRIPTION

SPUR is a process-oriented, grassland model designed to simulate various edaphic, biotic, and economic scenarios of the rangeland ecosystem (Figures 29-1 and 29-2) (Hanson et al., 1985). The field-scale version of SPUR can simulate the growth of up to seven plant species or species groups growing on up to nine different rangeland sites within a grazing unit. SPUR simulates herbage removal by cattle and other herbivores from rangelands under moderate grazing rates and predicts steer weight (and more recently, cow/calf dynamics) over a normal growing season. The user has the option of outputting several hundred variables. These include snowmelt, soil–water potential, carbon and nitrogen content of various vegetative components, cattle production, and forage consumption by domestic herbivores and wildlife species.

The SPUR model is driven by daily inputs of precipitation, maximum and minimum temperatures, solar radiation, and daily wind run. These variables are derived either from existing weather records or from use of a stochastic weather generator. The soils/hydrology component calculates upland surface runoff volumes, peakflow, snowmelt, upland sediment yield, and channel streamflow and sediment yield. Soil–water tensions, used in the model to control various aspects of plant growth, are generated using a soil–water balance equation. Surface runoff is estimated by the Soil Conservation Service curve number procedure and soil loss is computed by the modified Universal Soil Loss Equation. The snowmelt routine employs an empirical relationship between air temperature and energy flux of the snowpack.

Evaporation and transpiration are calculated using the Ritchie (1972) equation. Potential evaporation (E_o) is computed as a function of daily solar radiation, mean air temperature, and albedo. Potential evaporation at the soil surface (E_{so}) is computed as a function of leaf area index (LAI) of the field and the amount of litter that covers the soil surface. Actual soil evaporation (E_s) is computed in two stages, based on the soil moisture status of the upper layer. Stage 1 evaporation is limited by energy, meaning the soil profile is saturated and the process is dependent on the amount of solar radiation reaching the soil surface. Stage 2 evaporation is dependent on the movement of water to the soil surface and the amount of water in the soil profile; it is probably the more common of the two mechanisms in operation in the arid and semiarid regions of the western U.S. Potential plant transpiration (E_{po}) is calculated as a function of E_s and LAI (when LAI < 3.0) or as a function of E_o and E_s (LAI > 3.0). If soil water is limited, plant transpiration (E_p) is computed as a function

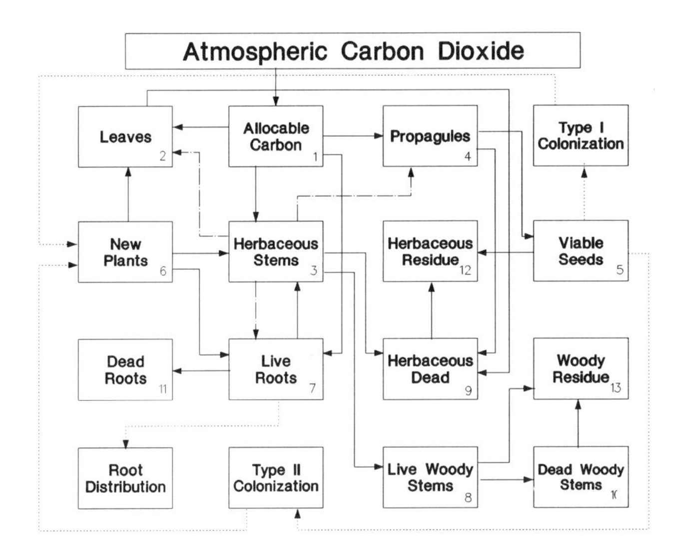

Figure 29-3: Compartmentalized view of the SPUR plant growth component.

of E_{po}, soil water in the root zone, and the field capacity. Solar radiation and the soil-surface area covered by above-ground biomass govern water loss from the soil profile.

The water balance equation used in SPUR is given in the following (Renard et al. 1987). Water balance for a single time step at a single point on a watershed is:

$$SW = SW_0 + P - (Q + ET + PL + QR)$$

where:

SW	=	current soil water content
SW_0	=	previous soil water content
P	=	cumulative precipitation
Q	=	cumulative amount of surface runoff
ET	=	cumulative amount of evapotranspiration
PL	=	cumulative amount of percolation loss to deep groundwater storage
QR	=	cumulative amount of return flow

For evaporation from the soil to occur, water must infiltrate through the soil surface and into the soil profile. Any soil water that is not lost to percolation, to root uptake, or to evaporation remains in the interstitial area between soil particles and is termed "storage." The process of infiltration reduces the amount of water available for overland flow component of runoff (Q), and increases the amount of water for evapotranspiration (ET) and for deep percolation (PL). Once water has percolated through the root zone of the soil, it is essentially lost to this hypothetical single point on the watershed. Such percolated water may return to the surface down the slope as return flow to another point on the watershed or as flow into a channel. Otherwise, this water remains in an aquifer, but is not available for transpiration nor for evaporation. If precipitation occurs while air temperatures are below freezing, that moisture will not be available for any of the hydrologic processes mentioned, as it will be tied up in the snow pack. The consequence is a reduction in the effective precipitation for a given time step. This water may show up at a later time step once it melts and it will then be available for infiltration and for runoff.

In the plant component, carbon and nitrogen are cycled through several compartments, including standing green, standing dead biomass, live root biomass, dead root biomass, seed biomass, litter, and soil organic matter (Figure 29-3). Allocable carbon enters the system via photosynthesis (Hanson, 1991). The new photosynthate is partitioned, dependent on present environmental conditions and plant growth stage, into leaves, herbaceous stems, propagules, woody stems, and roots. Translocation from herbaceous stems to leaves, propagules, and live roots is also considered in the conception of the model. All plant parts are susceptible to mortality, and biomass is eventually pooled in the herbaceous residue (above-ground soft tissue), woody residue (above-ground woody tissue), and dead roots (below ground). New plants can enter the system via the colonization model (Hanson and Olson, 1987). Two types of colonization are considered: Type I is for plants that flower late in the growth cycle, and Type II is for plants that flower early in the growing season. Finally, the plant model distributes roots through the soil profile. Roots are distributed according to the environmental fitness of the layer. Environmental fitness is determined as a function of the current temperature, water content, calcium content, aluminum content, and soil pH.

Soil inorganic nitrogen is also simulated (Hanson et al., 1988). The model simulates competition between plant species and the impact of grazing on vegetation. Required initial conditions include the initial biomass content for each compartment and parameters that characterize the species to be simulated (Hanson et al. 1988).

Recently, CBCPM (Colorado Beef Cattle Production Model), a second-generation beef-cattle production model that was a modification of the Texas A&M Beef Model (Sanders and Cartwright, 1979), was linked with the SPUR model. CBCPM is a herd-wide, life cycle simulation model and operates at the level of the individual animal. The biological routines of CBCPM simulate animal growth, fertility, pregnancy, calving, death, and demand for nutrients. Currently, fourteen genetic traits related to growth, milk, fertility, body composition, and survival can be studied. Intake of grazed forage is calculated by FORAGE, a deterministic model that interfaces CBCPM and SPUR (Baker et al., 1992). The model is driven by weight from the animal growth curve, animal demand for forage, and the quantity and quality of forage available for each time step of the simulation. FORAGE determines the intake of grazed forage by simulating the rate of intake and grazing time of each animal in the time step.

The economics component of SPUR uses biological and physical relationships as a basis for evaluation. Most of the information needed for the economics component is not part of SPUR, but must be derived external to the model. The analysis in SPUR is a simple application of benefit–cost analysis. These include gross returns or benefits, costs and net returns, or benefits alone.

APPLICATION

SPUR has been subjected to numerous validation tests and investigative research. As examples, results from studies describing the abiotic and biotic components will be discussed here.

Abiotic components

Richardson et al. (1987) successfully validated the weather generator. Renard et al. (1983) tested the hydrology module by conducting a 17-year (1965–1981) simulation for two Arizona watersheds and obtained high correlations between observed and predicted runoff (r=0.878 and 0.941). Springer et al. (1984) used the hydrology component of SPUR to predict the hydrological response of three Idaho watersheds over a 5-year period: the explained variance ranged from 63 to 85 percent. They concluded that the model did a good job of simulating the timing and amount of monthly runoff, but was a poor predictor of erosion and sedimentation. The snow accumulation and snowmelt components were tested in a watershed in southwest Idaho by Cooley et al. (1983). Using parameters derived from the 1980 snow season, they obtained a correlation of 0.91 between observed data from 1970–1972 and 1977 and model predictions.

Biotic components

Skiles et al. (1983) simulated the growth of the two dominant grasses in the shortgrass steppe of Colorado and concluded that the SPUR plant growth module adequately reproduced the biomass production of the grasses and matched the dynamics of the growing season. In a test of the plant–animal interface, Hanson et al. (1988) showed that the SPUR model correctly predicted domestic animal weight gains as a function of stocking rate for a Colorado grassland, which is also an indirect validation of the plant module. SPUR has been successfully used to predict animal gains and plant biomass production on pastures in West Virginia (Stout et al., 1990). It has been used to provide simulated forage for a modified heifer module (Field, 1987), and to estimate biomass for a grazing behavior model (Baker et al., 1992).

SPUR has also been used in an extensive study employing general circulation models (GCMs) to evaluate how climate change could potentially affect those regions of the United States that have a high dependency on grazing cattle (Hanson and Baker, 1992). Doubling of CO_2 alone caused only slight increases of plant production, but when CO_2 doubling was coupled with changes in precipitation and temperature, plant production increased significantly. The most limiting factors on rangeland are moisture and available nitrogen. Thus, as expected, moisture accounted for the greatest increases in plant production. However, wet–dry cycles have tremendous consequences on rangeland production. Normal, periodic drought is as important an issue as human-induced global climate change when considering agriculture in arid and semiarid lands.

Six key points were inferred from these simulation experiments. First, the climate-change scenarios tended to predict an increased length of growing season. Grasslands rely mostly on spring moisture while summers are generally very dry. The GCMs produced results that accentuated this pattern, and consequently the growing season was longer (by some 30 days). Subsequently, production increased, particularly in the spring and fall. Second, animal production decreased for the California and Southern Great Plains climate-change scenarios. Reduced animal production occurred coincident with increased plant production because of decreases in soil inorganic nitrogen and concomitant decreases in forage quality. The availability of nitrogen in rangeland systems is a controlling factor and is closely related to the decomposition of soil organic nitrogen. Third, livestock production tended to shift northward because of adverse conditions (primarily high temperatures) in the southern regions. Fourth, climatic variance between years was slightly greater for the climate-change scenarios than for the nominal run. If the variation for yearly production does indeed increase, then uncertainty regarding plant growth increases, thereby resulting in uncertainty in management decisions. Fifth, management will be able to compensate for climate change. Management of livestock will ultimately guide animal performance. In bad years the manager must either sell livestock or feed them. If vegetation production is more variable, then stocking rates must be decreased to reduce risk and to ensure good animal vigor. And finally, the more intense the management, the more the cost to the operator. Thus, even though the livestock may perform at similar levels, in the end, beef production costs may increase.

FUTURE DIRECTIONS FOR MODELING RANGELAND

Computer technology seems to be in an exponential growth phase, which makes predicting future developments in coupled models very difficult. As machines become faster and more powerful, models that are computationally impossible today may be feasible in the very near future. For example, many of the projects presented in this volume would have been impossible 5 years ago. Although we may not be able to forecast precisely future developments in ecosystem modeling, we are confident that models such as coupled terrestrial–atmospheric models and simulation models combining GIS technology will be important research tools in the future of ecological studies.

Linking models, spatial information systems, and behavior systems

We believe, however, that another set of modeling technologies will become important, especially in modeling ecosystem structure, function, and behavior. These techniques incorporate the use of behavior algorithms. A conceptual model that would link mathematical simulation models, spatial information systems, and system behavior algorithms to simulate the behavior and physiological processes of an ecosystem has been developed (Figure 29-4). Some of these links have already been established (see Chapters 13, 26, and 28).

The first link (Figure 29-4, Link I) is the connection of mathematical simulation models to spatial information systems. Information about physiological processes for individuals and populations is simulated by the mathematical models and geographically referenced via the spatial information system. Spatially referenced information is processed and passed back to the mathematical model.

Next, Link II in the conceptual model of Figure 29-4 represents the system behavior algorithms. These describe the behavior of the individuals and populations that are simulated in the system. There are several types of algorithms to choose from for this task: linear programs, expert systems, L systems, fractal theory, and genetic algorithms. Which algorithm to choose depends on the question being addressed by the modeling effort. This connection also has been established (Langton, 1989; Goel et al., 1991).

Finally, the link between behavior algorithms and spatial information systems has not yet been completed (Figure 29-4, Link III). Geographically referenced information is passed to behavior algorithms where the behavior of the system is determined by spatial proximity and spatially referenced resources. This link could integrate behavior and function between hierarchical scales. Behavior algorithms in this link determine how individual and population behaviors and physiological processes interact with and influence population/community process and behavior.

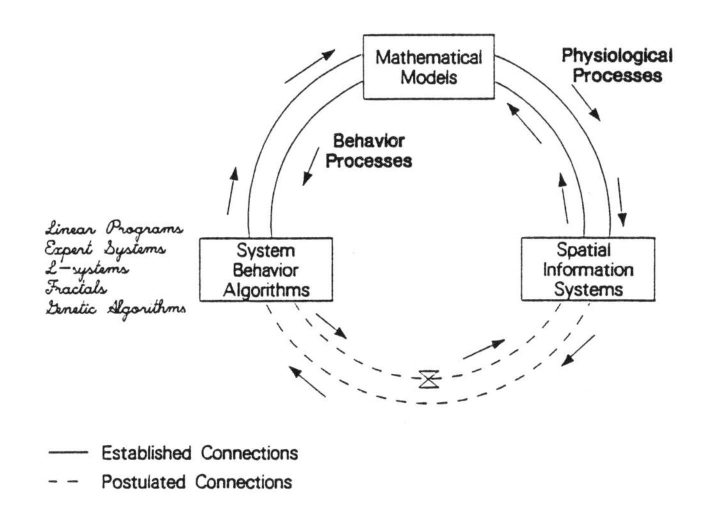

Figure 29-4: Conceptual model for linking physical and behavior processes of an ecostystem.

Expert systems and artificial intelligence

The application of artificial intelligence techniques, particularly expert system development and artificial life, can enhance the performance of simulation models and the interpretation of spatial data. Expert systems are computer programs that use extensive, high-quality human knowledge about some narrow problem area to answer very specific questions (Waterman, 1986). In the past few years, expert systems have been applied to the biological sciences (Olson and Hanson, 1988). Areas of application include natural resource management (Coulson et al., 1987), farming (Lemmon, 1986), and ecological modeling (Olson and Hanson, 1987).

Expert systems can be used in conjunction with simulation models in at least two ways. First, the technology can be used to parameterize components of large-scale models, such as the SPUR plant growth model, thereby enhancing the usefulness of the simulation. Second, an expert system can serve as an intelligent linkage between several different components of coupled models, for example, linear programs and simulation models. Expert systems technology enables the computerization of necessary expertise to interpret the output of one program in light of another. Such linkage could form a decision-support system capable of examining problems at various temporal and spatial scales.

Model parameterization

Expert system technology can be used to enhance the utility of large-scale, dynamic ecosystem models. In particular, these models are often encumbered with many parameters whose values are difficult to estimate because of a lack of data or uncertainty about the meaning of the parameters. Thus, the process of parameterization is a time-consuming, repetitive function normally performed by the model developer. At the same time, it requires genuine expertise in terms of model function and biological behavior.

Parameterization of systems models is not a trivial process, even for the "expert" (i.e., the model developer). Thus, a "portable expert," one that travels with the program and assists with parameterization, is needed. The parameterization of the SPUR plant growth component is difficult and requires substantial expertise. Thus, a knowledge-based system (ESTIS, Expert System Technology Interfaced with SPUR) is being developed that produces parameter sets for SPUR that allow accurate predictions of plant dynamics. The initial parameter values are determined by combining user-supplied information with general knowledge contained within the system. Then, in an iterative fashion, SPUR is run, model results are compared with field data, and appropriate adjustments are expertly made to the parameter set.

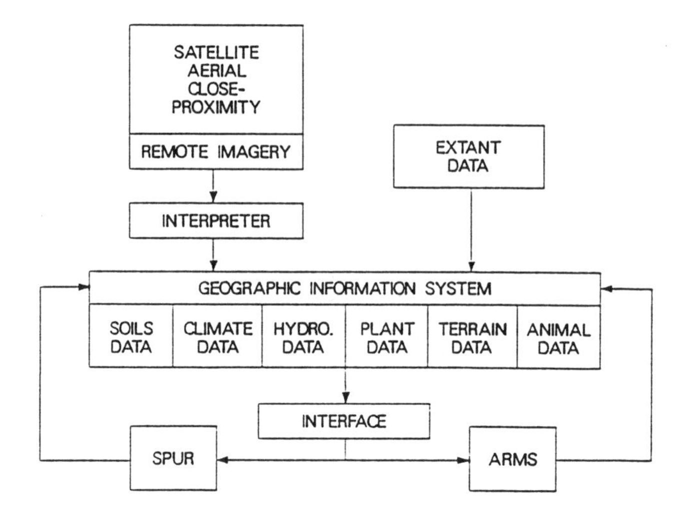

Figure 29-5: Diagram of a geographically referenced information system.

Spatial information technology

Mathematical models such as SPUR have assisted researchers in synthesizing information about ecological systems, but such models can demand data that are expensive to obtain or that are often unavailable. Development of spatial information technology has provided us with the ability to deal rapidly and effectively with large amounts of complex data (Anderson and Hanson, 1991) (Figure 29-5). Integrating highly sophisticated tools, such as SPUR (simulation modeling), GRASS (spatial analysis), and ERDAS (image processing) into a process-oriented, spatial assessment package involves information from several sources. These include spatial technology, and system modeling will allow us to simulate system processes within a space–time framework. "Spatial modeling" will result in a more realistic portrayal of system processes as they vary from location to location and from time to time. The result will be a product that spatially emulates system processes and provides a spatial connectivity through which adjacent systems can be linked and examined.

Under some circumstances, remote sensing techniques can help estimate rate processes such as evapotranspiration (Jackson et al., 1983) and primary production (Asrar et al., 1985). State variables such as cover (Graetz and Gentle, 1982) and green leaf area (Pearson et al., 1976) may also be estimated by this means. The information provided by remote sensing techniques can be used to parameterize models and verify their predictions (Pech et al., 1986). The simultaneous use of simulation modeling and remote sensing techniques is useful for describing ecological processes. When used together, these tools can increase our understanding of the spatial variability of ecosystems. Remotely sensed data can also be combined with simulation models to predict the accumulation of plant biomass, as surrogate variables within the models, or as correlative information for monitoring entire ecoregions.

The desired outcome of our work is to develop and validate process-oriented models that can accept spatial information, along with necessary in situ data, as input. Specific research needs include: (1) determining the dependability of remotely sensed data and simulation results in measuring biomass production and accumulation, (2) evaluating the relationship between spectral transformations derived from satellite imagery and key variables simulated by an ecological model, and (3) demonstrating the utility of integrating spatial data and modeling processes for regional monitoring of ecological processes. Vegetation communities form a natural mosaic of different sized patches. The identification, quantification, or description of these communities differs depending on the scale used to observe them. Thus, an investigator may be able to distinguish between two different plant communities on the ground (be it native range or cropland), but these vegetation types may not be distinguishable using remotely sensed data from satellites. Improved methods for evaluating regional and even global processes through the combined use of simulation models and spatial data

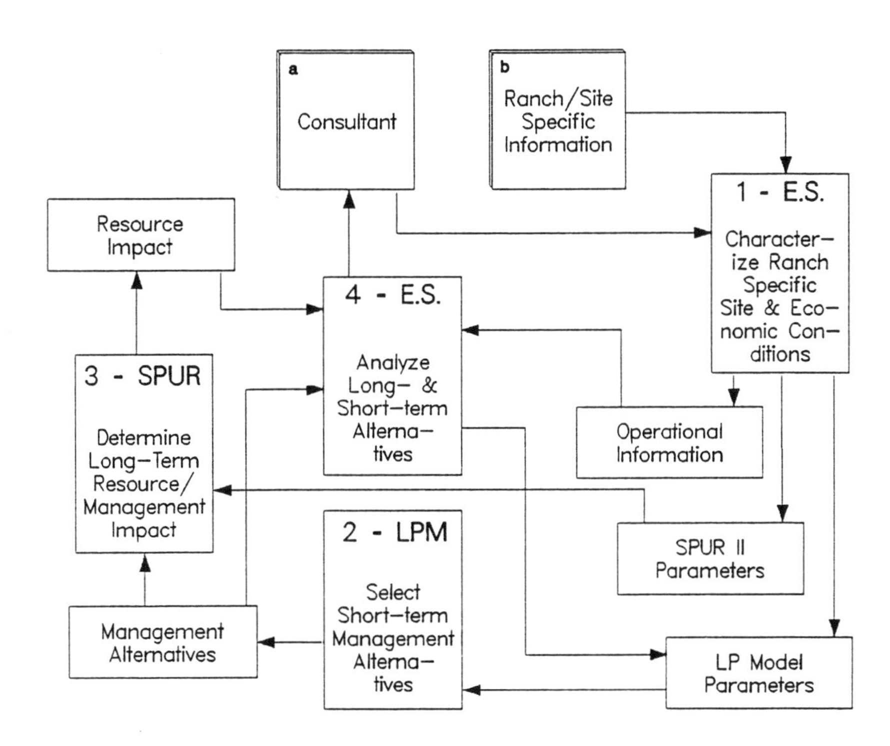

Figure 29-6: Diagram of the proposed ARMS decision support system.

are conceivable. With these tools, we will be able to monitor and forecast the effect of climatic and edaphic changes on vegetation at regional scales. Such research will help us take significant strides in understanding functional relationships between high- and low-resolution spatial data, determining the spatial resilience of complex ecological models, and determining the feasibility of using spatial data in conjunction with model output for monitoring regional phenomena.

Decision-support systems

Included in our research program is the development of a decision-support system that integrates short-term economic optimization techniques and the long-term predictive power of simulation models with the management expertise of ranchers. ARMS (Agricultural Resource Management System) is a decision-support system conceived for this purpose (Baker and Hanson, 1991) (Figure 29-6). The combination of a linear programming model and a simulation model operating at different temporal scales could aid ranchers in making decisions that are predicted to be financially feasible yet consistent with long-term resource conservation goals. ARMS will allow ranchers, other land managers, and consultants to evaluate the overall effects of each alternative available to them. Short-term effects include the economic consequences of different management options, for example, the costs and expected returns from feeding supplements or leasing additional pasture. Longer-term evaluation, over 5 years or more, brings in the environmental as well as economic results of different management decisions. Options to be evaluated include different grazing systems, supplemental feed crops, placing some marginal croplands in the Conservation Reserve Program, cross-fencing, developing additional water supplies, and revegetating depleted ranges. In addition to anticipated economic returns or costs, ARMS would predict the effects of such options on forage species composition and production, soil erosion, water runoff and other impacts. Finally, the system would provide the rancher with the information that will enable him to compete better for borrowed funds and investor participation.

RECOMMENDATIONS

To accomplish the tasks we have discussed in this chapter, systems ecology applied to rangelands must result in:

1. Development of a systematic representation of the

ecological assumptions useful for developing simulation models. Concurrently, a logic-based, computerized representation of the simulation should be constructed that will epitomize and test the ecological theory contained within the model.
2. Development of a strong foundation of modeling expertise and capability, including knowledge-based systems and GIS so that various kinds of spatial data can be used to help parameterize and execute models and validate their output. This will allow extrapolation of point data on soils, climate, and plants and the results from experimental plots to larger scales, that is, pastures, ranches, grazing allotments, landscapes, and regions.
3. Integration of ranch-scale linear programming models with simulation models. This approach may require the use of artificial intelligence technology. These decision-support systems will integrate short-term economic optimization and long-term predictive power with the expertise of ranchers.

These efforts will enhance our understanding of the theoretical aspects of rangeland ecology as well as provide practical tools for improving our ability to manage our priceless natural resources.

REFERENCES

Anderson, G.L., and Hanson, J.D. (1991) A geographically referenced information delivery system. In Hanson, J.D., Shaffer, M.J., Ball, D.A., and Cole, C.V. (eds.) *Sustainable Agriculture for the Great Plains, Symposium Proceedings*, Washington, DC: US Department of Agriculture, Agricultural Research Service, pp. 97–106.

Asrar, G.E., Kanemasu, E.T., Jackson, R.D., and Pinter, P.J. Jr. (1985) Estimation of total aboveground phytomass production using remotely sensed data. *Remote Sensing of Environment* 17: 211–220.

Baker, B.B., Bourdon, R.M., and Hanson, J.D. (1992) FORAGE: a simulation model of grazing behavior in beef cattle. *Ecological Modelling* (in press).

Baker, B.B., and Hanson, J.D. (1991) A management tool for rangeland systems. In Hanson, J.D., Shaffer, M.J., Ball, M.A., and Cole, C.V. (eds.) *Sustainable Agriculture for the Great Plains, Symposium Proceedings*, Washington, DC: US Department of Agriculture, Agricultural Research Service, pp. 107–114.

Cooley, K.R., Springer, E.P., and Huber, A.L. (1983) SPUR hydrology component: snowmelt. In Wight, J.R. (ed.) *SPUR—Simulation of Production and Utilization of Rangelands: A Rangeland Model for Management and Research*, Washington, DC: U.S. Department of Agriculture, Miscellaneous Publication No. 1431.

Coulson, R.N., Folse, L.J., and Loh, D.K. (1987) Artificial intelligence and natural resource management. *Science* 237: 262–267.

Duncan, D.A., and Woodmansee, R.G. (1975) Forecasting forage yield from precipitation in California's annual rangeland. *Journal of Range Management* 28: 327–329.

Field, L.B. (1987) Simulation of beef-heifer production on rangeland. *Unpublished thesis*, Fort Collins, CO: Department of Animal Science, Colorado State University.

Goel, N.S., Knox, L.B., and Norman, J.M. (1991) From artificial life to real life: computer simulation of plant growth. *International Journal of General Systems* 18: 291–319.

Graetz, R.D., and Gentle, R.R. (1982) The relationships between reflectance in the Landsat wavebands and the composition of an Australian semiarid shrub rangeland. *Photogrammetric Engineering and Remote Sensing* 48(11): 1721–1730.

Hanson, J.D. (1991) Analytical solution of the rectangular hyperbola for estimating daily net photosynthesis. *Ecological Modelling* 58: 209–216.

Hanson, J.D., and Baker, B.B. (1992) The potential effects of global climate change on rangeland livestock production for the United States. *Research and Development Report*, Washington, DC: U.S. Environmental Protection Agency, Office of Policy, Planning and Evaluaton (in press).

Hanson, J.D., and Olson, R.L. (1987) The simulation of plant biography. *Proceedings of Symposium: Seed and Seedbed Ecology of Rangeland Plants*, Washington, DC: U.S. Department of Agriculture, Agricultural Research Service, Miscellaneous Publications.

Hanson, J.D., Parton, W.J., and Innis, G.S. (1985) Plant growth and production of grassland ecosystems: a comparision of modelling approaches. *Ecological Modelling* 29: 131–144.

Hanson, J.D., Skiles, J.W., and Parton, W.J. (1988) A multispecies model for rangeland plant communities. *Ecological Modelling* 44: 89–123.

Innis, G.S. (ed.) (1978) *Grassland Simulation Model*, New York: Springer-Verlag, 298 pp.

Jackson, R.D., Hatfield, J.L., Reginato, R.J., Idso, S.B., and Pinter, P.J. (1983) Estimation of daily evapotransporation from one time of day measurements. *Agricultural Water Management* 7: 351–362.

Knisel, W.G. (ed.) (1980) *CREAMS: A Field-Scale Model for Chemicals, Runoff, and Erosion from Agricultural Management Systems*, Washington, DC: U.S. Department of Agriculture, Conservation Research Report No. 26, 640 pp.

Langton, C.G. (ed.) (1989) *Artificial Life. A Proceedings Volume in the Santa Fe Institute Studies in the Science of Complexity*, Redwood City, CA: Addison-Wesley, 655 pp.

Lauenroth, W.K. (1979) Grassland primary production:

North American grasslands. In French, N. (ed.) *Perspectives in Grassland Ecology*, New York: Springer, pp. 3–21.

Lemmon, H. (1986) Comax: an expert system for cotton crop management. *Science* 233: 29–33.

Olson, R.L., and Hanson, J.D. (1987) A knowledge-based deduction system for curve-shape analysis. *AI Applications in Natural Resource Management* 1(2): 3–11.

Olson, R.L., and Hanson, J.D. (1988) AI tools and techniques: a biological perspective. *AI Applications in Natural Resource Management* 2(1): 31–38.

Parton, W.J., and Risser, P.G. (1980) Impact of management practices on the tallgrass prairie. *Oecologia (Berlin)* 46: 223–234.

Pearson, R.L., Tucker, C.J., and Miller, L.D. (1976) Spectral mapping of shortgrass prairie biomass. *Photogrammetric Engineering and Remote Sensing* 42: 317–323.

Pech, R.P., Graetz, R.D., and Davis, A.W. (1986) Reflectance modeling and the derivation of vegetation indices for an Australian semiarid shrubland. *International Journal of Remote Sensing* 7(3): 389–403.

Renard, K.G., Shirley, E.D., Williams, J.R., and Nicks, A.D. (1983) SPUR hydrology component. Upland phases. In Wight, J.R. (ed.) *SPUR—Simulation of Production and Utilization of Rangelands: A Rangeland Model for Management and Research*, Washington, DC: U.S. Department of Agriculture, Miscellaneous Publication No. 1431, pp. 17–44.

Renard, K.G., Shirley, E.D., Williams, J.R., and Nicks, A.D. (1987) Hydrology component. Upland phases. In Wight, J.R., and Skiles, J.W. (eds.) *SPUR—Simulation of Production and Utilization of Rangelands: Documentation and User Guide*, Washington, DC: U.S. Department of Agriculture, Agricultural Research Service, pp. 17–30.

Reuss, J.O., and Innis, G.S. (1977) A grassland nitrogen flow simulation model. *Ecology* 58: 379–388.

Richardson, C.W., Hanson, C.L., and Huber, A.L. (1987) Climate generator. In Wight, J.R., and Skiles, J.W. (eds.) *SPUR—Simulation of Production and Utilization of Rangelands: Documentation and User Guide*, Washington, DC: U.S. Department of Agriculture, Agricultural Research Service, pp. 3–16.

Ritchie, J.T. (1972) Model for predicting evaporation from a row crop with incomplete cover. *Water Resources Research* 8: 1204–1213.

Sanders, J.O., and Cartwright, T.C. (1979) A general cattle production systems model. I. Description of the model. *Agricultural Systems* 4: 217–227.

Schimel, D., Stillwell, M.A., and Woodmansee, R.G. (1985) Biogeochemistry of C, N and P in a soil catena of the shortgrass steppe. *Ecology* 66: 276–282.

Skiles, J.W., Hanson, J.D., and Parton, W.J. (1983) Simulation of above- and below-ground nitrogen and carbon dynamics of Bouteloua gracilis and Agropyron smithii. In Lauenroth, W.K., et al. (eds.) *Analysis of Ecological Systems: State-of-the-Art in Ecological Modelling*, Elsevier Scientific Pub. Co., pp. 467–473.

Springer, E.P., Johnson, C.W., Cooley, K.R., and Robertson, D.C. (1984) Testing the SPUR hydrology component on rangeland watersheds in southwest Idaho. *Transactions, American Society of Agricultural Engineers* 27: 1040–1046, 1054.

Stout, W.L., Vona-Davis, L.C., Skiles, J.W., Shaffer, J.A., Jung, G.A., and Reid, R.L. (1990) Use of the SPUR model for predicting animal gains and biomass on eastern hill land pastures. *Agricultural Systems* 34: 169–178.

Waterman, D.A. (1986) *A Guide to Expert Systems*, Reading, MA: Addison-Wesley.

Williams, J.R., Dyke, P.T., and Jones, C.A. (1983) EPIC—a model for assessing the effects of erosion on soil productivity. In Lauenroth, W.K., et al. (eds.) *Analysis of Ecological Systems: State-of-the-Art in Ecological Modelling*, Amsterdam: Elsevier, pp. 553–572.

IV

From Modeling to Policy

MICHAEL F. GOODCHILD

Models of environmental processes may help us to understand the operation of the Earth system, but they are useful to society as a whole only if they can play a role in the development and formulation of policy. But moving models from the laboratory and ivory tower into the pragmatic and politicized world of policy formulation is no simple task. Models must be integrated, since it is not adequate to know the impact of a policy on only one aspect of the environment. Results must be displayed in a form that is understandable by the decision-maker, who will likely be more easily influenced by pictures and maps than by words or numbers. Models must be spatially disaggregated, to allow policy to be evaluated at a detailed spatial scale using accurate spatial data. Software to evaluate "what-if" scenarios must be well designed and easy to use, to avoid dysfunctional separation of the roles of analyst and decision-maker. And finally, software must allow the results of modeling to be presented in a form suitable for policy development, for example, by aggregating to political or administrative units—counties, states, or nations.

All of these objectives are at odds to some degree with those of normal science—publication in refereed journals, repeatability, and predictive accuracy. Too often science and policy development appear to operate as separate worlds, each within its own communities of interest. Models may be essential to policy, but as King and Kraemer argue in Chapter 34, the content of models may not matter much in practice. Without the "smoking gun"—the effects of DDT accumulation in the food chain, or the ozone hole—the scientific validity of models may have very little relevance to policy formulation.

In this context, it is perhaps not surprising to find that the computer is a popular platform for modeling. Computer results are impersonal, and perhaps speak louder as a consequence. Computers make it easier to integrate models, to disaggregate them to greater levels of spatial detail, and to reaggregate results over politically relevant areas. The GIS function of map display can be used to make striking products that carry far more weight than tables of numbers, and the interactive nature of much current GIS allows the decision-maker to work with complex models in a comfortable, reassuring environment that does not demand a great depth of scientific understanding.

So is GIS, and computerization more generally, the essential bridge between science and policy that will make science more relevant and policy development more scientifically honest? Certainly the environmental policy agencies have invested heavily in GIS and other forms of computerization in recent years, and continue to do so. The chapters in this section explore these issues in depth, beginning with an introduction by Hewitt to the concept of risk. The overview of risk and hazard modeling by Rejeski (Chapter 30) discusses the use of GIS in the assessment of risk as an interaction between three cultures—science, policy formulation, and the public at large. New tools such as GIS change the dynamic tensions between the three groups, but not the basic conflict of objectives.

In Chapter 31, Wadge, Wislocki, and Pearson examine the use of GIS to assess natural hazards, with an example of landslide instability. They compare two approaches to GIS use—a weighting method based on the locations of past events, and a simple mathematical model of the hazard process. In Chapter 32, von Braun looks at the use of GIS at two Superfund sites, as a tool for communicating risk to the public, and for evaluating remedial scenarios. As always, there are dangers of credibility and miscommunication, and von Braun discusses the role of uncertainty in this context. In the final chapter on risk and hazard modeling, Parrish et al. (Chapter 33) describe an exercise in risk assessment by Region 6 of the U.S. Environmental Protection Agency, and illustrate the inevitable interplay of subjectivity and science in assessing environmental vulnerability.

The final chapter of this section, by King and Kraemer (Chapter 34), presents the perspective of two social scientists who many years ago asked the fundamental question: "Why do policy makers like models?" Although much of their work has been in the area of economic modeling and forecasting, where problems are at least as severe as in environmental modeling, they are able to offer many useful insights into the political context. Their conclusions are both encouraging and discouraging. On the one hand, it seems that if "numbers beat no numbers every time," and a picture is worth a thousand words, then the display of model results in map form using GIS can indeed move environmental modeling and policy formulation closer together. On the other hand, if it does indeed matter little to the policy-maker where the numbers (or map) came from, then GIS will create even greater potential for abuse.

Risk and Hazard Modeling

MASON J. HEWITT III

Life is a risky business. How many times did mother remind us to calculate the risks of crossing the street without looking or the risk of playing with sharp sticks? Even in adulthood we are reminded daily of the risks associated with smoking or not using seat belts. Risk is a part of life, and our identification and calculation of risk is virtually second nature. It has been suggested that humans unconsciously perceive, calculate, and map risk constantly (Figure IVA-1). We seem to carry with us an "*n*-dimensional" map through which we perceive risks as functions of threat. Perhaps this is a remnant of our hunter psychology drawing upon our primitive memory. I once observed a group of hikers from the eastern U.S. on their first outing in the Southwest. I was with them when we made contact with an unseen, but very alarmed, rattlesnake. There was immediate recognition of the threat being communicated by the snake to the person, despite the fact that none of them had ever heard a rattlesnake before and none could see the threat.

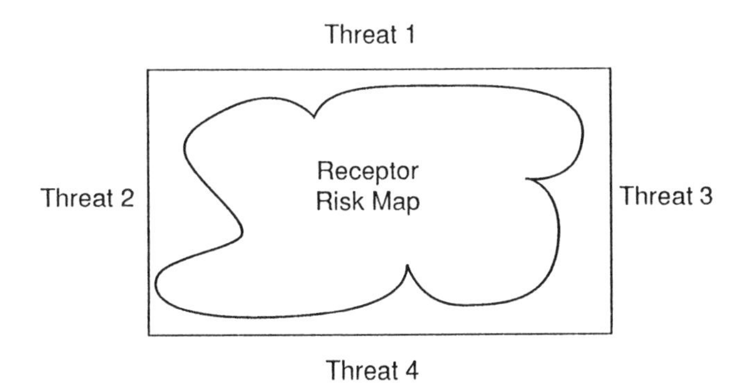

Figure IVA-1: Schematic representation of how humans continuously assess and "map" risks as a function of threats in their everyday life.

Unfortunately, we often have to deal with such unseen risks; those that do not communicate their presence as well as a rattlesnake are not readily recognized from past experience, and may not produce noticeable effects for years. For example, the risk of building a house on the edge of a cliff may be perceived and recognized empirically or it may be deliberately ignored. Yet other risks associated with industrial development are not well perceived. We hear of many ills that seem to be symptoms or indicators of risks that have gone undetected. The endangered species list and the ozone "hole" are examples of the results of threats that were undetected until an indicator of a systems failure became evident.

The challenge before modern decision-makers is to operate with a new paradigm having a greater focus on long-term strategies for risk communications and risk reduction. This paradigm will require the definition of systems and means for their management by which we will be able to detect, assess, respond to, and communicate risk. One necessary component of risk assessment is the coupling of models with information systems, like GIS, in order to mimic the risk mapping that occurs in our minds.

The chapters in this section are a start towards the design of the new paradigm. The authors have sought solutions for assessing risk to individuals, communities, and ecosystems by beginning to use models to provide order and structure to their data. Results of such efforts have been useful to bound policy questions, supplement dialogue between organizations, or improve understanding of environmental processes. These authors have used GIS to map the location and proximity of risk to selected receptors; they have recorded the results of their efforts to communicate risks, and they have identified some of the issues yet to be addressed.

The challenge to the reader is to join in the efforts to design, improve, and implement the elements of a new risk assessment paradigm.

30

GIS and Risk: A Three-Culture Problem

DAVID REJESKI

For the past 20 years, discussions about environmental risks and their assessment have been punctuated by debates between members of various subcultures in our society. The character of these debates has been both relativistic and moralistic, often resulting in battles over who is right about risks and who is to be believed.

A few examples may suffice to illustrate this point. One of the more recent attempts to define the various subcultures and the differences in their perspectives has focused on the types of "rationality" that guide risk assessment: the "technical rationality" of the scientists, or the "cultural rationality" displayed by the public, which draws on traditional peer groups and shared social perceptions rather than expert opinion (Krimsky and Plough, 1988). This bifurcated view of rationality built upon long, and often acrimonious, debates about "the public" and the mechanisms underlying public risk perception. Was the public: (1) an atomized network of individuals whose risk perception depended on different patterns of cognition and perceptual biases (cognitive science view); (2) a group of utility maximizing, informed agents carefully weighing risks against benefits (economic view); or (3) part of a bounded social group whose shared belief system affected the perception and acceptability of risks (anthropological view) (Douglas, 1985)? Despite years of research, conferences, and publications, those involved with risk assessment have yet to reach a consensus on many of these issues.

As the GIS community begins to wander into this multicultural landscape, it is important to understand the nature of the tribal wars, the history of intellectual treaty making, the position of the conceptual fault lines, and the new frontiers that remain open to exploration. Those using GIS to model and estimate risks are pioneers in this landscape, caught between the subcultures of science, policy, and the public (Figure 30-1). Navigation in and between these various cultural spaces presents some unique problems and challenges.

For instance, the manager of risks has a very different need for GIS-based mapping than the risk assessor. The goal of the manager is not simply to navigate in physical space (Where are the risks?), but to navigate in decision space (What can be done to effectively reduce these risks?). Members of the public affected by the distributions of environmental risks have yet another navigational question (Where are these risks within the perceptual space of individual and social concern? Or, in other words, "Should I/we worry?").

Members of these three cultures also face different hurdles to effective, rational action. The scientists attempting to estimate risk burdens are confronted with uncertainties of knowledge. This uncertainty comes in

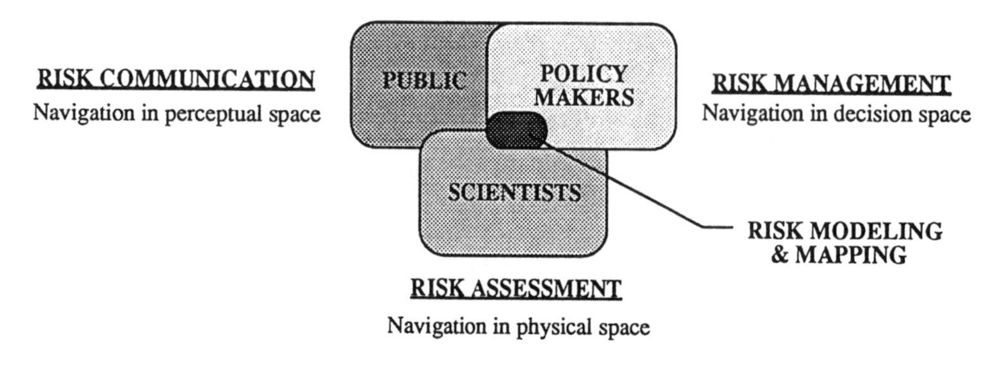

Figure 30-1: The three cultures of GIS and risk.

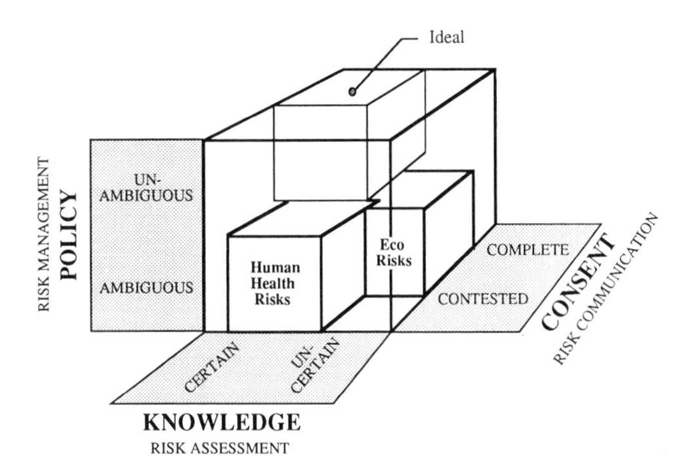

Figure 30-2: Three dimensions of risk assessment: policy, knowledge, and consent.

many forms and may include a lack of understanding about the primary, secondary, and tertiary effects of particular risks, or questions about how certain risk-induced changes should be estimated and valued (Russell, 1990). Though uncertainty imposes significant limitations on risk estimates, it can often be resolved by obtaining additional information or constructing more sophisticated models.

Risk managers and policy-makers live in a complex world of choices where ambiguity dominates. Ambiguity cannot be resolved simply by gathering more information or by thinking smarter; it can only be resolved through shared agreements about what is important and unimportant in our society (the differences between uncertainty and ambiguity are discussed by Feldman, 1989). Even if we had accurate assessments of the health risks of a given subpopulation from radon exposure, ambiguity remains concerning the possible policy interventions—education and outreach, mandatory testing, building retrofit subsidies, changes in building codes, etc. Ambiguity is an intrinsic, structural component of most policy problems and it separates risk management from risk assessment.

Finally, the risk problem is one of both knowledge and consent (Douglas and Wildavsky, 1982). From the point of view of the public, social consent and a trust in institutions is crucial in defining what risks will be accepted, and by whom. Though reducing ambiguity and uncertainty may help in achieving consent, ultimately, a more consultative and participatory system may have to be developed to facilitate consensus building among the public, policy-makers, and scientists (Nelkin and Pollack, 1979; Johnson and Covello, 1989).

In short, the current dilemma of risk assessment can be characterized by high policy ambiguity, considerable uncertainty, and a lack of consensus over the proper courses of action (see Figure 30-2). Though the degree of these problems may vary between areas of human health and ecological risk, they are not likely to disappear in the near future. This situation provides some unique challenges and opportunities for those interested in risk modeling with geographic information systems. Opportunists in this new landscape see the possibility of using the GIS technology to "regionalize" the risk analysis process—moving it from its traditional focus on site-specific problems to a true macro-scale planning and policy tool. This will open up a new arena of potential uses (see Figure 30-3).

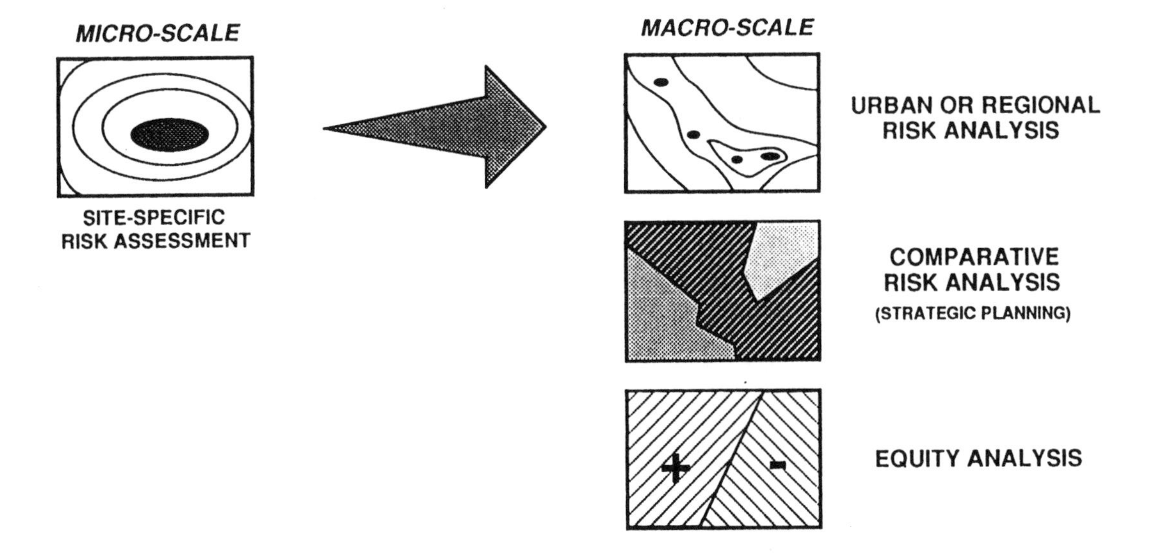

Figure 30-3: Potential uses for a macro-scale risk analysis process.

These new uses will also generate new challenges, and one is reminded of the saying that "The chief cause of problems is solutions" (Severeid, 1979). GIS cannot solve many of the mentioned problems inherent in the risk analysis process; in fact, it may add additional problems particular to the technology and its utilization. It can, however, shift our perspective on the risk problem, both conceptually and spatially, and stimulate long-needed debates on contentious or avoided issues.

Of the many issues that must be addressed, four appear particularly urgent and will be used to structure the remainder of this chapter:

- Believability—are the models and supporting data properly chosen?
- Honesty—have uncertainties inherent in the analysis been conveyed?
- Decision utility—does the analysis provide a clear basis for action?
- Clarity—are the maps understandable and sensitive to perceptual differences of the intended audience?

These are core issues, which cut across the various subcultures of risk and subsume other issues of model structure, data, and the GIS technology itself.

BELIEVABILITY

Many of the disputes concerning believability revolve around questions of model appropriateness. The primary challenge is to choose models and supporting data that provide reasonable representations of the systems, and system interactions, that we are interested in analyzing.

Over the past 20 years, the risk assessment community has developed a variety of established and accepted models for the estimation of both human health and ecological risks. Though significant progress has been made, there are many issues that continue to affect the believability of these models, such as the difficulty in extrapolating from high to low doses, the problem of comparing endpoints between species (animal/human), and general problems with data availability.

Despite these shortcomings, the ability of GIS to integrate spatial variables into risk assessment models is a major incentive for methodological experimentation with models, and has attracted both GIS and risk assessment enthusiasts. Initial barriers to model development have been data related. Many of the risk assessment procedures are site specific and data intensive, resulting in significant costs for spatial data acquisition and preparation if procedures are to be transferred into GIS environments and applied to larger geographic areas.

In examining the data–model fit and the underlying believability issue, it is important to understand the limitations of the data, the function of the models, and exactly where informational certainty gives way to etiological guesswork. These issues can be explored using the risk process diagram and model taxonomy illustrated in Figure 30-4.

Most of the support data for risk assessment focuses on the description of sources, the estimation of releases of specific agents, and the toxicological or ecotoxicological characteristics of these agents. Large uncertainties occur when we attempt to model exposures and outcomes, requiring leaps of scientific faith to be made across the cause-and-effect chain. Often, the relationship between source emissions and exposure is poorly defined. In addition, the correlation between exposure dose (as defined by ambient levels), body dose (the amount inhaled or ingested), and target dose (the amount reaching a sensitive organ) may be affected by numerous variables that are poorly understood and difficult to model.

Despite caveats, we can use this general risk process model to explore modeling approaches with GIS and discuss their strengths and weaknesses. To avoid confusion, we will continue to use the word "risk" in association with these approaches, although in many cases we are actually talking about hazard, not risk modeling (hazards can be defined as threats to humans and what they value as opposed to risks, which express the conditional probabilities of experiencing harm; Hohensemser, Kates, and Slovic, 1985).

Spatial coexistence

The simplest type of GIS-based risk analysis attempts to answer the question: "Do specific objects or events coexist in space and/or time?" For instance, do large numbers of people live near hazardous waste sites or industrial facilities emitting high levels of toxics? Most input data come from inventories of chemical releases (for instance, the Toxic Release Inventory) and are combined with demographic data to "estimate" population exposure (USEPA, 1990a).

Though the focus is normally on the identification of spatial coexistence, a lack of coexistence may also be of interest, especially for the characterization of ecological risks. For instance, recent attempts to use GIS to assess the protection afforded biodiversity have focused on identifying "gaps" between endangered species and the habitats crucial to their existence (Scott et al., 1990).

Spatial coexistence models draw largely from the general class of empirical models in which a relationship between various attributes has been observed, but the exact mechanism is not clearly understood. These models generally serve as techniques for risk "screening" rather than for orthodox risk assessment and can be classified into the following subgroups (Wheeler, 1988).

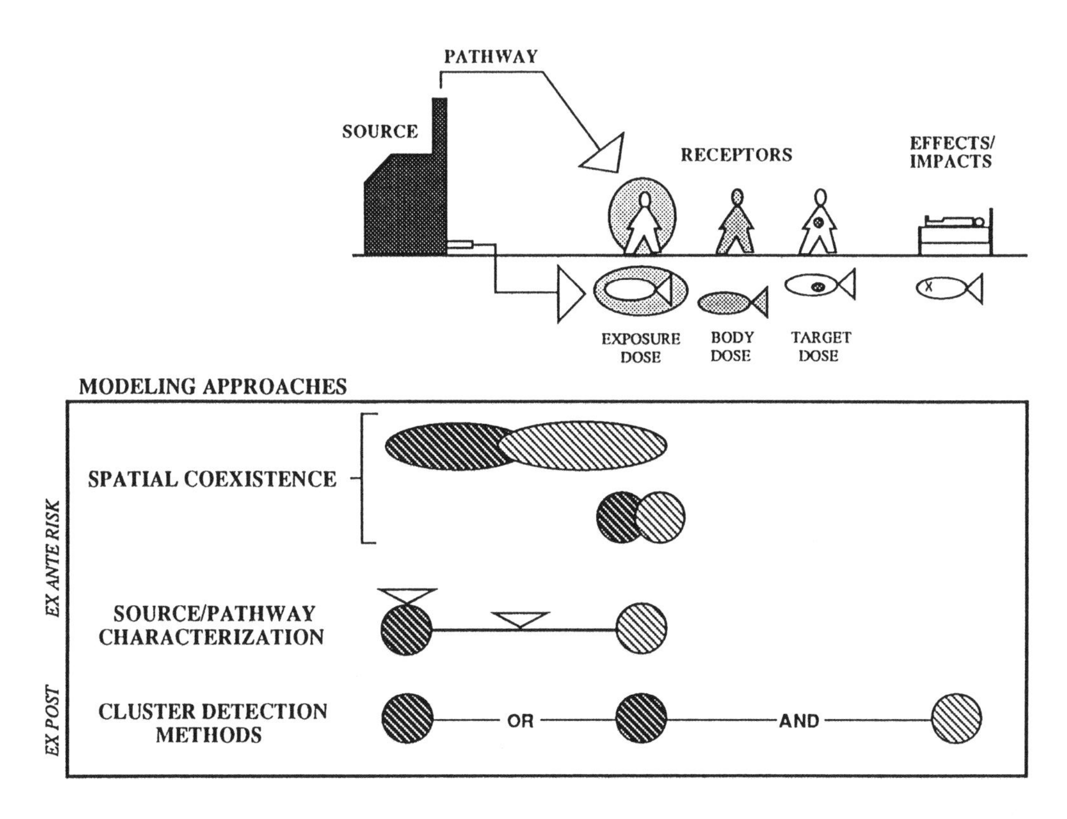

Figure 30-4: The risk process, and a taxonomy of modeling approaches.

1. Binary (nominal) models

A majority of the GIS-based risk modeling currently being done uses binary models. In such applications, the GIS user looks for yes/no answers to queries and assumes that the coexistence of certain parameters implies high risks (for instance, areas with high ambient levels and areas with high population density). Modifications of this approach involve looking for crossed thresholds or proximities. Though simplistic, binary models implemented in GIS environments may be extremely helpful in identifying risk "hot spots" or generating hypotheses concerning risk distributions and affected populations over large geographical areas.

2. Weighted (ordinal) models

Moving beyond the mere presence or absence of parameters allows more information to be contained in the model. In weighted models, it is assumed that particular substances, or the characteristics of the receptor(s), may play a disproportionate role in risk and need to be accounted for in the model. Such weightings may be based on the relative toxicity of released chemicals, or the vulnerability/sensitivity of receptor populations or indicator species. Though intuitively attractive, a consistent quantitative basis for weighting is often missing. In such cases, ranking becomes a discretionary act, incorporating the subjective biases of the modeler or modeling institution. The goal is not to eschew weighting, but to recognize the need for open peer review of ranking schemes and "debiasing" of the modeling approach (Fischoff, 1982). Specific care must be taken with the use of weighting schemes in GIS environments, since weighting errors can propagate in multilayered map analysis.

3. Quantitative (interval/ratio) models

In quantitative models, GIS users assign absolute numbers to various parameters or thematic map layers. For instance, combining population density with ambient concentrations for a specific chemical (micrograms/m^3) and the unit risk factor for that chemical (risk/microgram/m^3) allows a map to be produced showing the risk per unit of population. Because quantitative models are based on continuous mathematical functions, it may be easier to model uncertainty simultaneously.

As with most modeling, these approaches represent a simplification of reality and need to be used with caution. Spatially random data distributions can appear interesting when mapped and lead to false conclusions concerning the distribution of risk. In addition, coexistence does not prove causality, and these types of GIS analyses can be easily misinterpreted or misused to support "guilt by association" arguments. Spatial coexistence analysis tells us something about potential risk, but leaves critical parts of the risk model unaddressed.

Source/pathway characterization

One way to improve estimates of risk significantly is through the characterization of the exposure pathway affecting the transport and fate of risk-inducing agents. While not directly measuring exposure levels, this method acknowledges the fact that certain physical and environmental variables can have significant impacts on exposure. Adding information about these variables to the risk model can increase its predictive accuracy.

The simplest approach is to add more spatial data to the GIS model to describe the transport medium and its effect on the distribution or dispersion of risk agents empirically. For instance, one could add maps describing soil infiltration rates and depth-to-groundwater to a GIS model estimating drinking water risks. It is also possible to use weighting schemes described earlier to refine exposure estimates. The maps in Plate 30-1 show the result of overlaying a windrose-derived weighting on an omnidirectional air dispersion pattern to estimate risks associated with tropospheric ozone.

Though this general approach normally focuses on exposure pathways, risk sources can also be characterized, especially when data from monitoring programs are missing, or data collection proves too expensive. For instance, information on heating fuel consumption, heating fuel type, furnace efficiencies, and emission factors can be combined in a GIS environment to estimate and map the type and quantity of emissions associated with residential, commercial, and industrial heating.

It is also possible to link/integrate more sophisticated deterministic models with GIS to characterize risk situations better. The earliest systems to accomplish this were developed for emergency response situations, and combined air dispersion models with a chemical database and quasi-GIS to map areas and populations affected by chemical releases and spills. Examples are: ALOHA (Areal locations of Hazardous Atmospheres) from the U.S. Department of Commerce NOAA Hazardous Materials Response Branch; and CAMEO (Computer-Aided Management of Emergency Operations) from the U.S. Environmental Protection Agency. Another area where linking/integrating has occurred is for the mapping of natural hazards resulting from geological and geomorphological processes (landslides, rock slides, sink holes, etc.) (Pearson, Wadge, and Wislocki, 1991).

Though there exist a wide variety of readily available and robust physical process models for characterizing potential media pathways and risk sources, those using GIS for risk analyses have failed to take full advantage of these tools. This can be partly explained by the lack of large-scale spatial databases needed to support the models, but insecurity on the part of risk analysts with sophisticated models from other disciplinary fields also plays a role. In addition, very little has been done to explore whether such external models can be "internalized" in the GIS environment using a refined cartographic modeling language found in such raster-based programs as pMAP, MapBox, or IDRISI.

Cluster detection methods

This last category of models is extremely well developed, but has unfortunately received little serious attention from GIS users. Cluster detection refers to a variety of approaches pioneered and used primarily by epidemiologists to study the occurrence of disease (Raubertas, 1988). They are designed to answer the question: "Do we find significant clusters of disease in a specific geographical area that we would not have expected to see?"

Because the focus is on validating health outcomes rather than estimating probabilities, these approaches provide strong potential evidence implicating environmental stressors in disease (if statistically valid links to proximal sources can be established). Two approaches, which might be of particular value in raster-based GIS systems, will be described here.

Cell occupancy: This method, commonly called EMM after its originators (Edere, Myers, and Mantel, 1964), makes use of an equidistant grid placed over geo-referenced data. A time frame is chosen and then cell counts are taken to determine whether more disease events than expected occurred within the defined spatio-temporal regions.

Space–time regression: This method determines the correlation between the disease events in space and the disease events in time. The goal is to determine whether the interevent distances in space have a similar distribution as the interevent distances in time. Space–time regression is particularly useful in detecting "hot spots" caused by general exposure to pollutants, as opposed to true clinical disease excesses that can occur downwind or downstream of pollution sources.

The science of epidemiology provides a rich history and depository of models suitable for implementation in GIS environments. Geographers have begun to contribute to this knowledge base by developing new approaches to assess spatial and temporal variations (Thomas, 1990). Recent attempts to identify and study disease patterns have used the capabilities of GIS to support the query and statistical analysis of multiple data sets on population, disease incidence, and exposure factors (Gatrell and

Dunn, 1990; Bashor, Turri, and Pickering, 1989).

Though we can clearly benefit from more predictive, multivariate models for risk analysis, the key question often becomes how many variables. There are significant tradeoffs between the predictive accuracy of models and their associated costs. The acquisition, preparation, and management of spatial databases for GIS analysis are extremely expensive; so great care must be taken in the choice of decision-relevant data sets. There are no simple solutions to this problem, but a willingness on the part of the risk analysts and GIS support people to negotiate about modeling tradeoffs is a prerequisite to rational model choice.

A recent survey conducted by the U.S. Environmental Protection Agency indicated that the risk models currently being implemented in GIS environments are very basic and focus on screening risks based on rather broad criteria such as absence/presence or proximity (Kapuscinski and Rejeski, 1990). There is a clear need to integrate empirical and deterministic modeling approaches and draw on the wide variety of epidemiological modeling techniques. The level of modeling currently being used is probably reflective of a number of interrelated factors that need to be addressed:

- Lack of the necessary spatial data to support more sophisticated models;
- Insecurity over methodological approaches such as weighting;
- Difficulty in constructing certain types of complex models in the GIS environment or linking existing external models with GIS;
- Lack of communication and knowledge/technology transfer between the risk and GIS communities.

HONESTY

Modeling honesty involves a truthful representation of model limitations and uncertainties. Though it is true that most decision-makers would like high levels of analytical certainty, they are usually acutely aware of the complexities and uncertainties underlying certain analytical problems and sensitive to attempts to disguise uncertainty or simplify complexity. Uncertainties need to be dealt with in an open manner without reverting to simplistic reductionism.

In the GIS world, uncertainty has been greeted by a conspiracy of silence, even among highly credible practitioners. Methods for classifying errors in spatial databases and dealing with error propagation in GIS environments have only recently begun to be discussed and developed, and certainly have not received the level of intellectual interest or financial support awarded other GIS topics (Veregin, 1989; Berry, 1987). In the risk world, when quantification of uncertainties is presented at all, what is quantified is normally some derivative of simple experimental sampling error. This is often only a trivial portion of the total overall uncertainty (Mendez, Hattis, and Ashford, 1980).

One of the fundamental changes that must take place is the recognition that uncertainty is valuable information. The maps shown in Figure 30-5 illustrate this point. The generation of an additional uncertainty surface allows the decision-maker to divide the area of concern further into three subcategories, each representing a different strategy with differing policy and resource implications (see Figure 30-6).

Though this is a simple example, the willingness of both risk modelers and GIS users to engage in this type of exercise has been extremely limited. This has been partly due to deficiencies of the GIS technology, which make it difficult, if not impossible, to trace errors quickly through an analysis. In an ideal world, we would have both the methodologies and supporting technologies to deal quickly with the following types of uncertainty relevant to risk analysis.

Spatial uncertainty

By making the spatial characteristics of risk parameters explicit, GIS introduces an additional element of uncertainty into analyses. The most common form of locational error that impacts risk modeling involves point data: for instance, the location of monitoring stations, outfall pipes, hazardous waste sites, and emissions sources. Often the required parameter data exist, but cannot be linked to accurate lat–long data for modeling and mapping. Though this problem poses significant barriers to analysis, the development of locational data standards and the increased use of global positioning systems will reduce its significance in the future.

The second form of spatial uncertainty common to risk modeling occurs when data are aggregated to a spatial scale that is too coarse to support the required analysis. Demographic and medical data are often delivered at an aggregated level making it difficult to explore cause-and-effect relationships spatially.

Finally, whether we are analyzing human health or ecological risk, there are virtually no hard boundaries. Often, we are dealing with significant transition zones (between air sheds or different population groups) and/or actual dynamic boundaries (coastal and tidal phenomena or population mobility). Cartographic representations need to capture this uncertainty through the use of "fuzzy boundaries." Scientists involved in risk analysis need to provide information to GIS technicians about the nature of boundaries and their transition characteristics (for instance, develop a boundary characterization inventory). In addition, major improvements are needed in GIS tech-

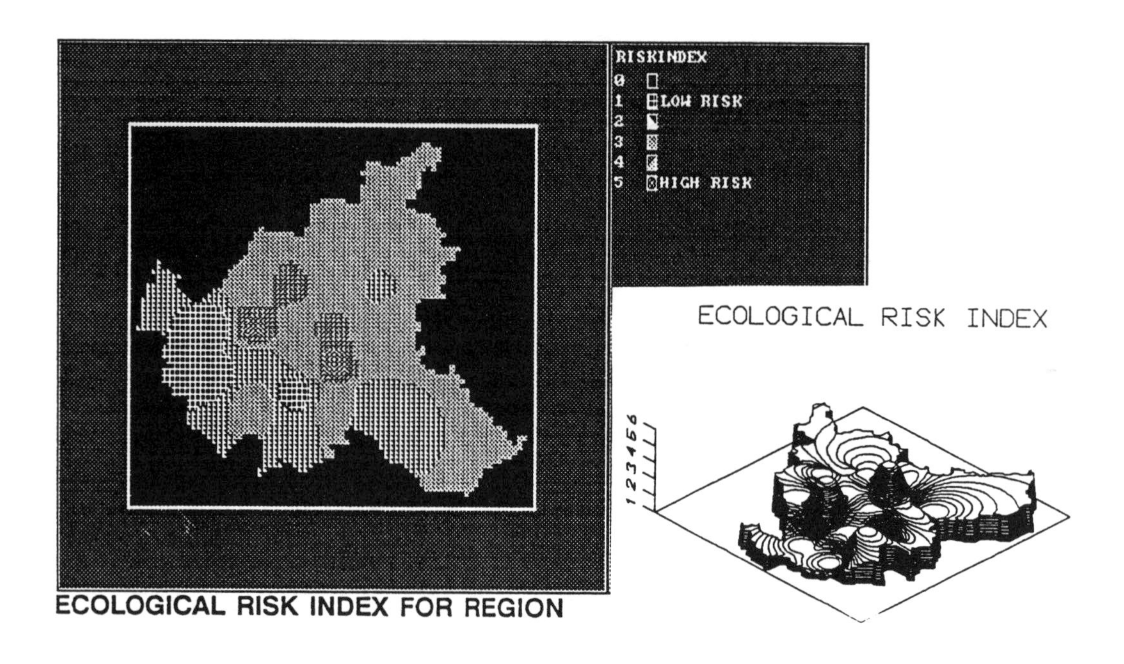

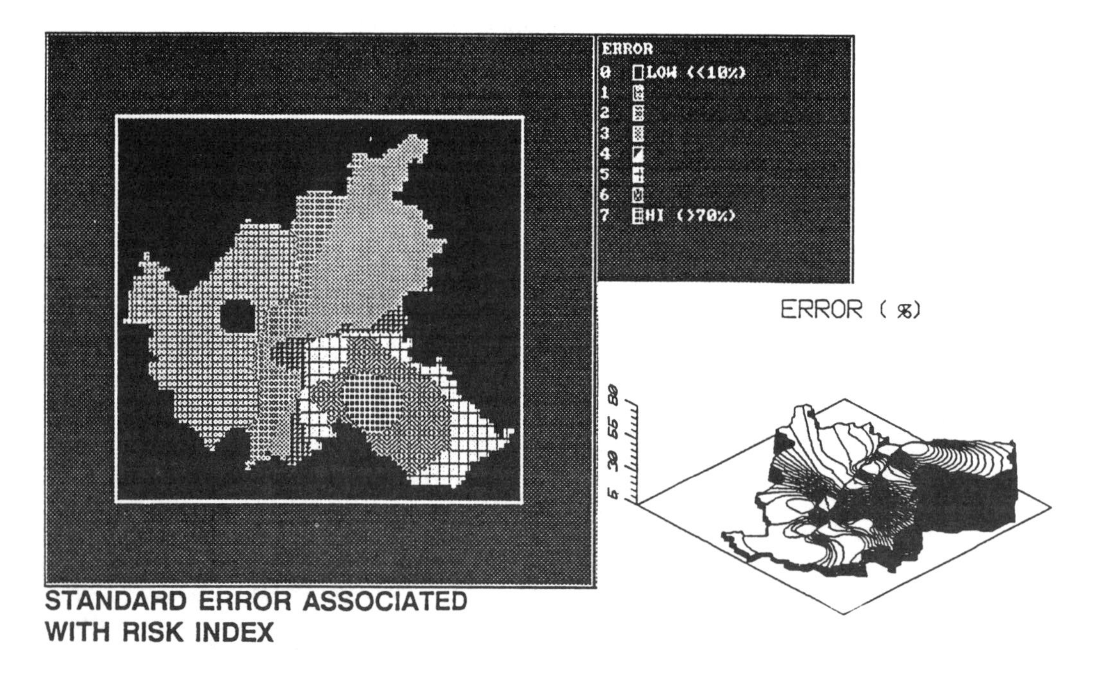

Figure 30-5: Addition of an uncertainty surface to a measure of ecological risk.

nology so that spatial imprecision can be mapped and displayed visually (Leung and Leung, 1991).

Linguistic uncertainty

The ambiguity of our language provides us with both creative potential and with classificatory problems, as we try to attach words and concepts to numbers. It is important to keep in mind that many policy decisions are made not on the basis of exact quantitative data, but on a qualitative description of those data (this temperature change is above normal; this level of erosion is severe; this risk is high, low, acceptable, tolerable, or below regulatory concern). In the area of risk assessment, many debates are waged on the linguistic battlefield (witness the problems the EPA is having in defining "acceptable" risk). There have even been accusations of "linguistic detoxification" occurring when agencies provide benign labels to environmental risks perceived as significant by the public.

Research has shown that there is often a surprisingly low level of agreement, even among experts, about the quantitative ranges associated with various linguistic descriptors (Robinson and Wong, 1987). Since risks (especially ecological risks) are often classified using an ordinal scale with descriptors, greater attention must be paid to the mismatches and misrepresentations that can occur. Recent advances in combining GIS with fuzzy logic and linguistic exploratory analysis may help to address problems with linguistic uncertainty. This would also allow us to represent different cognitive maps of the same risk situation (see Figure 30-6).

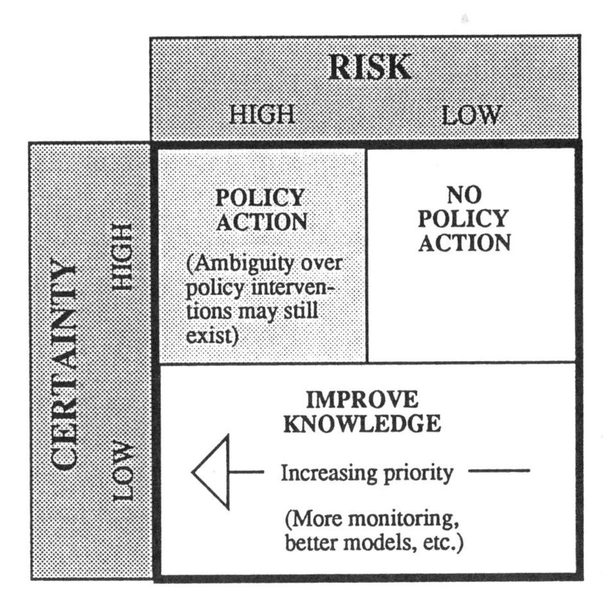

Figure 30-6: A measure of uncertainty increases the strategies open to the policy maker.

Model uncertainty

Given the twisted path between anthropogenic pressures and system responses, a wide variety of uncertainties can be introduced through the risk models presented in the previous section. These errors arise from the form or incompleteness of the mathematical functions used to represent the causal processes producing damage, and include:

- The damage pathways analyzed do not represent all important modes of damage;
- Inappropriate spatial aggregation or needless disaggregation in the system model (for instance, when inhomogeneous data are grouped together for analysis);
- Inappropriate attribution of cause, and/or neglect of an important causal determinant;
- Inappropriate definition of system boundaries (Hattis, 1989).

Parameter uncertainty

Many of the parameter inputs into health and ecological risk models are plagued with large uncertainties that need to be articulated and captured. Parameter uncertainties arise from possible errors in the expected values of specific parameters in the causal models, or probability density functions used to describe the estimated errors in those parameters. These include:

- Unknown uncertainty or unwillingness on the part of experts to articulate the error associated with parameter estimates;
- General misestimation (usually underestimation) of uncertainties of individual parameters (toxicities, exposure factors, vulnerability indexes, etc.);
- Misperceptions concerning parameter interrelationships (e.g., normal versus lognormal, unimodal versus multimodal).

Quantifying parameter uncertainties becomes crucial in GIS analyses because statistical error propagation during cartographic modeling can result in the production of meaningless maps. The use of verbal uncertainty indicators alone often prohibits the effective tracking of errors across modeling stages. More robust methods that allow the input of variance with parameters must be integrated into the GIS technology.

Though there are currently no standardized procedures for handling uncertainties in GIS-based risk models, a number of approaches appear promising, especially for dealing with parameter errors. The first is the use of Monte Carlo simulation techniques in which the ranges of variables and their associated probability distributions

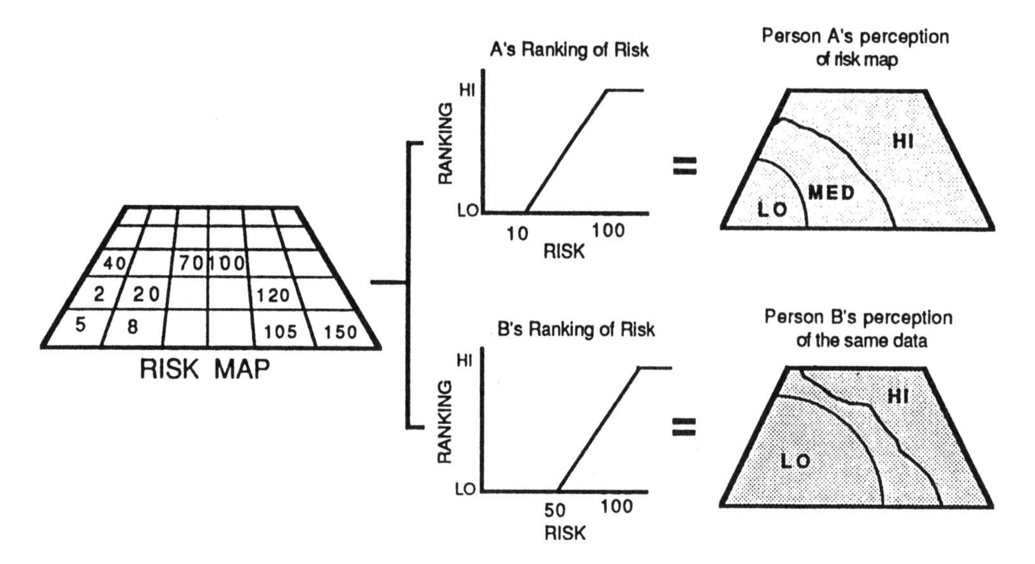

Figure 30-7: The effects of interpersonal variation on the mapping of risk.

can be input into a GIS system and carried through all model stages (Charlton, Openshaw, and Carver, 1989). A simpler approach would involve allowing one or two sensitive variables in a GIS/risk model to take extreme values and define the endpoints of a risk scenario space (see Figure 30-7).

In general, a greater use of probabilistic approaches for GIS-based risk analysis is required. This may have additional benefits beyond the study of errors. It has been pointed out that many emerging environmental problems, such as climate change, have a probabilistic character rather than a deterministic or manifest one; this is also the case for some existing problem areas such as the eutrophication of lakes (Svedin and Heurling, 1988).

Honesty in GIS-based risk modeling will require major changes in the perceptions and attitudes of both risk assessors and GIS technicians. If, and when, this renaissance occurs, methods for the handling of error must be institutionalized, and improvements in the GIS technology must follow. The GIS vendor community must begin to develop another generation of software (or linked packages) that will allow error to be input and calculated as a normal GIS operation. As a general point, we must carefully assess the marginal economic utility of increased investments in improving data quality, versus the development and increased use of uncertainty measures.

DECISION UTILITY

There can be no doubt that increasing the believability and honesty of GIS products will increase their utility for decision-makers. However, developing a close fit between map analysis and policy goals is of such critical importance that it requires a considerable level of effort and foresight on the part of the analysts and map makers. When maps fail to have an impact, it often has little to due with model nuances or data gaps, but more with a lack of fit between cartography and decision reality.

The presentation of human health and ecological risks must be a tightly woven logical argument for action. In this regard, there is a considerable difference between methodological experimentation and the production of analysis for decision-making. If GIS-based risk work is to have a significant impact in the future, the technology will have to help bridge the gap between the two worlds of risk assessment and risk management. It will have to help answer the question: "Now that I have identified the risks in space and time, what should I do about them?" To date, very little work has been done to answer this question or to provide decision-makers with a more articulated and multifaceted view of risks. This has little to do with any inherent limitations imposed by the GIS technology, nor is it necessarily due to a lack of appropriate data. Simply, few risk managers are aware that such map-based analyses are possible, and risk modelers have had little incentive to produce them.

Consider, however, supplementing a GIS/risk analysis with the following types of spatially and decision-relevant information:

Cost of remediation: GIS is beginning to be used to examine the spatial distribution of risks around toxic sources such as Superfund sites or smelters. The ability of some GISs to calculate volumes would allow soil removal

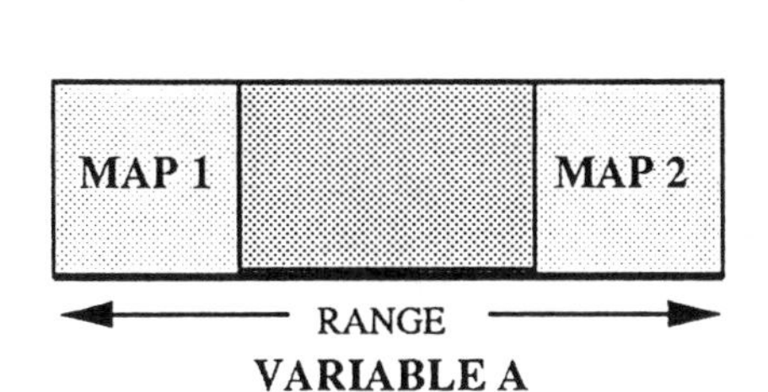

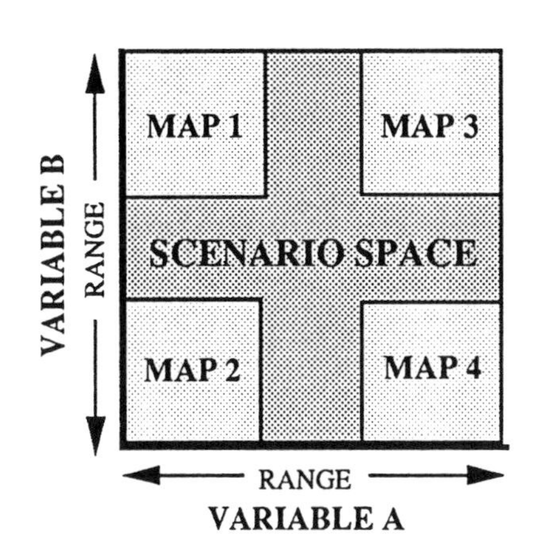

Figure 30-8: Sensitivity of a GIS/risk model to one and two input variables.

costs to be estimated. The costs to transport contaminated soil or waste material could also be estimated and optimized using network models, providing risk managers with a more complete set of strategies for risk reduction (von Linden and von Braun, 1986).

Risk reduction strategies: In order to reduce risks in any substantial way, we need to broaden our kit of tools for environmental protection beyond traditional enforcement and regulation techniques to include the provision of information, market incentives, pollution prevention, and cooperation with other organizations (USEPA, 1990b). Once risks have been mapped using GIS, it may be possible to match risks to risk reduction strategies and to delineate spatially whre money needs to be invested, and in what strategies. We could then geographically target investments that could affect multiple risk problems (for instance, public outreach for radon and indoor air risks).

Risk reduction budgets: The reduction of risks is as much a financial question as a scientific one, requiring a careful consideration of the risk reduction potential achieved for each marginal dollar invested. In this regard, decision-makers would benefit from maps displaying financial information associated with the reduction of risks. This could involve displaying funding for the reduction of different risks by region, state, or county, or a more articulated mapping showing investments by strategy, that is, how much is invested in enforcement versus education and outreach.

Equity: The idea of environmental justice, based on an equitable distribution of risks, is emerging as a critical theme of the 1990s. To date, the disputes concerning equity have focused mainly on the issue of toxic waste sites and associated disproportionate ethnic and racial risk burdens (United Church of Christ, 1990; Citizens Fund, 1990). GIS will become the tool of choice for exploring equity issues, in moving beyond the calculation of risk based on abstract probabilities for statistically average people, and exploring the question: risk to whom (or to what species). The release of the 1990 census data will help support such inquiries, but new methodological approaches will have to be developed for GIS environments to look at the multidimensional question of equity (geographic, social, cumulative, and intergenerational) (Kasperson and Dow, 1991). In mapping the risks faced in actual neighborhoods and communities, GIS will force policy makers finally to focus on the quality of life of real people in real places (Landy, Roberts, and Thomas, 1990).

Future risk: To date there has been virtually no GIS/risk work focused on the anticipation of future risks and their prevention. The simplest type of approach would be to hold today's risks constant and vary the exposed population based on demographic projections (including population growth, mobility, and changing characteristics of vulnerable subgroups). One could add to this analysis by varying toxic loadings based on differing emissions and regulatory scenarios. More sophisticated models could be built that examine future risk based on changes in industry/service mix, agricultural intensity, energy use, transportation density, etc. As scientists, we all remember Bacon's credo that "knowledge is power." Unfortunately, we have forgotten Comte's observation that "foreknowledge is power."

CLARITY

Over the past 5 years, risk analysis has increasingly found its way outside scientific and regulatory institutions and into the public realm—into briefings, into public meetings, and often into the popular press and media. This is the result of two interrelated trends. The first is a general decline in the public's willingness to defer to public institutions over important decisions concerning risks (Laird, 1989). The second is a fundamental change in our perception of the public and their potential role in the management of risk. Unlike earlier periods, when public opinion concerning risk was largely ignored, or seen as requiring

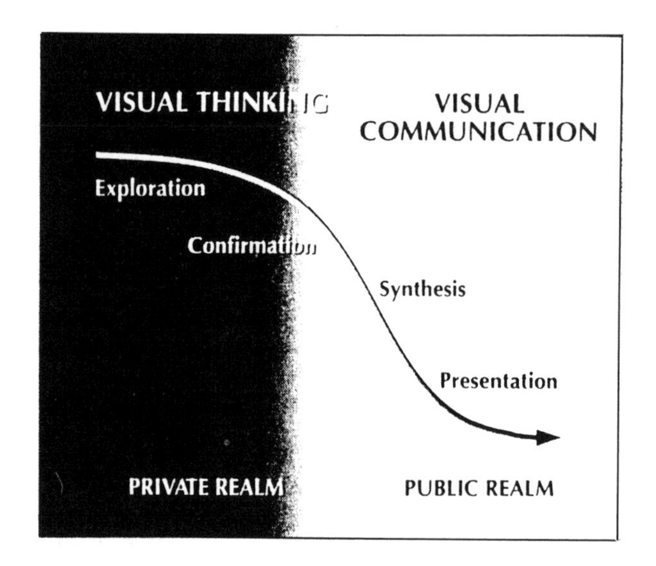

Figure 30-9: Fundamental shift of focus from visual thinking in the private realm to visual communication in the public realm.

correction, a dialogue with the public has become a basic tenet governing most institutional strategies for risk management. The first cardinal rule of risk communication developed by the Environmental Protection Agency states: "Accept and involve the public as a legitimate partner" (USEPA, 1988).

The intuitive appeal of maps for visual communication will thrust GIS technicians into the public limelight, pulling along the risk modelers who have traditionally been guaranteed a high degree of anonymity. Increasingly, modelers will be forced to explain, defend, and illustrate their assumptions in public forums where maps will play an ever-increasing communicative role. To meet this challenge, GIS/risk modeling projects must be integrated into an overall risk communications strategy that considers the following points:

Concerns: GIS analyses should directly address the concerns and sensitivities of the intended audience, be they policy-makers or the public. Public concerns may not be immediately obvious, so that time and effort may have to be expended to identify and frame the proper issues. Based on a growing body of research, there appears to be significant variability in perceptions of environmental risk related to ethnicity, race, socioeconomic status, and gender (Vaughan and Nordenstam, 1991). For instance, the polluting of coastal areas provokes a great deal of concern across all groups in the United States, but a significantly greater percentage of whites that nonwhites report great personal concern about this issue (Gallup, 1989). Ethnic minorities, however, have consistently shown a greater concern than other social groups for the possible risks associated with the disposal of radioactive and hazardous waste (Pilisuk and Acredolo, 1988; Bullard, 1990). If GIS-generated risk maps are to be used in public settings, or by the media, it is important to understand the differential sensitivities of the potential audience and the possible equity issues that might be raised by such maps. If only the risk concerns of the map makers and modelers guide the content of the maps, a significant potential for misunderstanding or misrepresentation exists (Monmonier and Johnson, 1990).

Presentation: Maps are powerful visual tools that can help risk communicators get their message across. They are, however, not without limitations. Since mapped messages can be manipulated, a considerable amount of care must be given to the design of cartographic presentations (Monmonier, 1991). The maps used to present risk situations to policy-makers or the public may need to be significantly different from those developed and used during the risk estimation and analysis process. As we move out of the world of risk assessment, we fundamentally shift our focus from visual thinking in the private realm to visual communication in the public realm (see Figure 30-9) (DiBiase, 1990).

Maps developed for risk presentation must be built with attention to graphic design, the avoidance of complex jargon, the inclusion of required caveats, and a focus on providing a comprehensible and relevant message. In this regard, we may want to go beyond printer and plotter outputs. Paper maps may not be the best way to communicate risks to public audiences. The use of slide shows, videos, and optical disk techniques need to be explored more fully. Given the fact that most health and ecological risks have both spatial and temporal aspects, a much greater use of animation will be required in the future (MacEachren and DiBiase, 1992). Various methods of presenting risks should be pretested on a small sample of the intended audience to match presentation with preference and ensure comprehension.

Context/involvement: Messages concerning risk should be put into an appropriate substantive and temporal context. Defining the context should include a look at risks versus benefits, risks versus the costs of risk reduction, and/or risks versus the cost of substitutes. If past data are available, it is desirable to develop a mapped time series showing how risks have changed over time (especially in response to earlier policy initiatives, regulations, changes in sectoral activities, or demographic shifts). It is important to link the risk message to a plan for risk management and use GIS to estimate future risks based on various policy assumptions and organizational actions.

Informed judgment is an important principle affecting public acceptance of risk. GIS studies should be used in a process that invites, rather than precludes, public participation, and provides affected parties with sufficient information concerning the nature of the risks and their options for risk reduction.

Credibility: It is important to keep in mind that the best risk analyses will have little impact on public opinion if the credibility of the organization or presenter is questionable. In the public's eye, there is a high correlation between credible risk assessment and management and their trust in institutions. Unfortunately, the public's trust in experts from regulatory institutions is often low, far below the trust allocated to independent scientists or physicians. This problem is exacerbated by the fact that physician experts have often taken different views on risks than their engineering counterparts, further confusing the public and policy debate (Tarr and Jacobson, 1989). In cases where organizational credibility is lacking, or expert opinions severely diverge, GIS-based risk projects should be given to intermediary institutions perceived as having greater objectivity. If studies are likely to raise serious equity and policy questions, the appropriate people need to be ready to respond competently and compassionately. In cases where the risk issue is contentious, institutional perceptions will overpower any concerns about model structure and map design.

CONCLUSIONS/RECOMMENDATIONS

We have come full circle and explored, in varying degrees of detail, the three subcultures of the risk world and four core issues that cut through these worlds. The most fundamental recommendation one can make follows directly from our exploration—the need to form bridges between these three subcultures.

Organization: From an organizational standpoint, we need to align roles, responsibilities, and relationships to address the wider task of analyzing risks, communicating them to the affected parties, and impacting crucial policies. The functional segregation of modelers from mappers, and from communicators and decision-makers can no longer be supported. Interdisciplinary teams will be needed with the necessary analytical and intellectual diversity to support comprehensive risk analysis as a macro-level planning and policy tool. In addition, the transfer of technologies and know-how between organizations working on risk problems and issues needs to receive greater attention (possibly through the organization of cross-cutting conferences, newsletters, and electronic communications networks).

Policies: We need a more wide-ranging debate on the impact of uncertainty on GIS/risk analysis, including the development of polices to deal with error problems. Policies and guidance for risk communication need to be expanded to include map products, their design, and their use with different audiences.

Technology: Though GIS technology has developed rapidly during the past 5 years, a number of limitations still exist that affect its utility for risk analysis. Specifically, we need tools to track and display uncertainty. In addition, techniques that allow the analysis and display of spatio-temporal relationships and changes need further development.

Methods: As was pointed out, there is a significant amount of methodological groping occurring in the area of GIS/risk analysis. The development of methods, both for analyzing and displaying risks, needs to be accelerated and coordinated. Promising approaches need to be peer reviewed and possibly standardized for wider dissemination and use (dissemination could take place through a macro library). The development of intra- or interagency task groups or modeling forums should be considered as a mechanism for the systematic management of model and guidance development (USEPA, 1989).

Data: If GIS-based risk analysis is to become an organizational goal, sufficient funds will have to be allocated for the acquisition and preparation of supporting spatial data at appropriate levels of resolution. Spatial data acquisition plans of agencies need to reflect a growing emphasis on risk, as well as the need to coordinate interagency data exchange (for instance, working through the Federal Geographic Data Committee). Specific emphasis needs to be placed on quality control assurance of crucial data sets and the avoidance of redundant digitization efforts.

REFERENCES

Bashor, B., Turri, P., and Pickering, K. (1989) *The Memphis Study*, Memphis, TN: Environmental Epidemiology, Tennessee Department of Health and Environment.

Berry, J.K. (1987) Computer-assisted map analysis: potential and pitfalls. *Photogrammetric Engineering and Remote Sensing* 53(10): 1405–1410.

Bullard, R. (1990) *Dumping in Dixie: Race, Class, and Environmental Quality*, Boulder, CO: Westview Press.

Charlton, M.E., Openshaw, S., and Carver, S.J. (1989) Monte Carlo simulation of error propagation in GIS. Paper presented at the 6th European Colloquium of Theoretical and Quantitative Geography, Chantilly, France.

Citizens Fund (1990) *Poisons in Our Neighborhoods: Toxic Pollution in the United States*, Washington, DC: Citizens Fund.

DiBiase, D. (1990) Visualization in the earth sciences. *Earth and Mineral Sciences*, University Park, PA: Penn State University, College of Earth and Mineral Sciences.

Douglas, M. (1985) *Risk Acceptability According to the Social Sciences*, New York: Russell Sage Foundation.

Douglas, M., and Wildavsky, A. (1982) *Risk and Culture: An Essay on the Selection of Technical and Environmental Dangers*, Berkeley, CA: University of California Press.

Ederer, F., Myers, M.H., and Mantel, N. (1964) A statistical problem in space and time: do leukemia cases come in clusters? *Biometrics* 20: 626–638.

Feldman, M. (1989) *Order Without Design: Information Production and Policy Making*, Stanford, CA: Stanford University Press.

Fischhoff, B. (1982) Debiasing. In Kahneman, D., et al. (eds.), *Judgment under Uncertainty: Heuristics and Biases*, Cambridge: Cambridge University Press.

Gallup Organization (1989) General environmental concern. *Gallup Report*, 285 pp.

Gatrell, A., and Dunn, C. (1990) GIS in epidemiological research: analyzing cancer of the larynx in North West England. *Proceedings, EGIS '90*, Utrecht: EGIS Foundation.

Hattis, D. (1989) Scientific uncertainties and how they affect risk communication. In Covello, V., et al. (eds.), *Effective Risk Communication: The Role and Responsibility of Government and Nongovernment Organizations*, NY: Plenum Press.

Hohensemser, C., Kates, R., and Slovic, R. (1985) A causal taxonomy. In Kates, R., et al., *Perilous Progress: Managing the Hazards of Technology*, Boulder, CO: Westview Press.

Johnson, B., and Covello, V. (eds.) (1989) *The Social and Cultural Construction of Risk: Essays in the Perception and Selection of Risks*, Dordrecht, Holland: Reidel.

Kapuscinski, J., and Rejeski, D. (1990) *Risk Modeling with Geographic Information Systems: Approaches and Issues*, Washington, DC: U.S. Environmental Protection Agency, Office of Information Resources Management, National GIS Program.

Kasperson, R., and Dow, K. (1991) Developmental and geographical equity in global environmental change: a framework for analysis. *Evaluation Review* (in press).

Krimsky, S., and Plough, A. (1988) *Environmental Hazards: Communicating Risks as a Social Process*, Dover, MA: Auburn House.

Laird, F. (1989) The decline of deference: the political context of risk communication. *Risk Analysis* 9: 543–550.

Landy, M., Roberts, M., and Thomas, S. (1990) *The Environmental Protection Agency: Asking the Wrong Questions*, New York: Oxford University Press.

Leung, Y., and Leung, K.S. (1991) Analysis and display of imprecision in raster-based information systems. *Unpublished paper*, Hong Kong: Department of Geography, Chinese University of Hong Kong.

MacEachren, A., and DiBiase, D. (1992) Animated maps of aggregate data: conceptual and practical problems. *Cartography and Geographic Information Systems* (in press).

Mendez, W., Hattis, D., and Ashford, N.A. (1980) Discussion and critique for the carcinogenicity assessment group's report on population risk due to atmospheric exposure to benzene. Report to the Office of Air Quality Planning and Standards of the U.S. Environmental Protection Agency. *Publication No. CPA-8-1*, Cambridge, MA: MIT Center for Policy Alternatives.

Monmonier, M. (1991) *How to Lie with Maps*, Chicago, IL: University of Chicago Press.

Monmonier, M., and Johnson, B. (1990) *Design Guide for Environmental Maps*, Trenton, NJ: New Jersey Department of Environmental Protection, Division of Science and Research.

Nelkin, D., and Pollak, M. (1979) Public participation in technological decisions: reality or grand illusion. *Technology Review*, August/September.

Pearson, E., Wadge, G., and Wislocki, A. (1991) Mapping natural hazards with spatial modelling systems. *Proceedings, EGIS '91*, Utrecht: EGIS Foundation.

Pilisuk, M., and Acredolo, C. (1988) Fear of technological hazards: one concern or many? *Social Behavior* 3: 17–24.

Raubertas, R. (1988) Spatial and temporal analysis of disease occurrences for detection of clustering. *Biometrics* 44: 1121–1129.

Robinson, V.B., and Wong, R. (1987) Acquiring approximate representations of some spatial relations. Paper for the Eighth International Symposium on Computer-Assisted Cartography, Baltimore, Maryland.

Russell, M. (1990) Evaluating ecological impacts: a conceptual framework. In Grodzinski, W., Cowling, E.B., and Breymeyer, A. (eds.) *Ecological Risks: Perspectives from Poland and the United States*, Washington, DC: National Academy Press.

Scott, M., et al. (1990) *GAP Analysis: Protecting Biodiversity Using Geographic Information Systems*. Handbook prepared for a workshop held at the University of Idaho, October 29–31.

Severeid, E. (1979) *Town and Country*, May.

Svedin, U., and Heurling, B. (eds.) (1988) Swedish perspectives on human response to global change. *Report 88:3*, Stockholm: Swedish Council for Planning and Coordination of Research.

Tarr, J., and Jacobson, C. (1989) Environmental risk in historical perspective. In Johnson, B., and Covello, V. (eds.), *The Social and Cultural Construction of Risk*, Dordrecht, Holland: D. Reidel Publishing Co.

Thomas, R.W. (ed.) (1990) *Spatial Epidemiology*, London: Pion, Ltd.

United Church of Christ (1990) *Toxic Wastes and Race*, Washington, DC: United Church of Christ.

USEPA (1988) *Seven Cardinal Rules of Risk Communications*, Washington, DC: U.S. Environmental Protection Agency.

USEPA (1989) Resolution on the use of mathematical models by EPA for regulatory assessment and decision-making. *Science Advisory Board, EPA-SAB-EEC-89-012*, Washington, DC: U.S. Environmental Protection Agency.

USEPA (1990a) *Toxics in the Community: The Toxics Release Inventory Annual Report*, Washington, DC:

U.S. Environmental Protection Agency.

USEPA (1990b) Report of the Strategic Options Subcommittee, Relative Risk Project. Appendix C of *Report by the Science Advisory Board, SAB-EC-90-021C*, Washington, DC: U.S. Environmental Protection Agency.

Vaughan, E., and Nordenstam, B. (1991) The perception of environmental risks among ethnically diverse groups in the United States. *Journal of Cross-Cultural Psychology*, March, pp. 2–11.

Veregin, H. (1989) A taxonomy of error in spatial databases. *Technical Paper 89-12*, Santa Barbara, CA: National Center for Geographic Information and Analysis, University of California.

von Linden, I., and von Braun, M. (1986) The use of geographic information systems as an interdisciplinary tool in smelter site assessment. *Proceedings, Management of Uncontrolled Hazardous Waste Sites—Superfund '86 Conference, Washington, DC, Dec 1–3.*

Wheeler, D. (1988) Model building with geographic information systems. *Proceedings, GIS/LIS'88, San Antonio, Texas*, Falls Church, VA: ASPRS/ACSM, Vol. 2, pp. 580–585.

31

Spatial Analysis in GIS for Natural Hazard Assessment

G. WADGE,
A. P. WISLOCKI, AND
E. J. PEARSON

Maps have been the principal medium for the summary and representation of hazards posed by natural processes (e.g., Varnes et al., 1984; Crandell et al., 1984). Such maps typically present hazards as high, medium, and low ordinal zones, as values of probability of occurrence for individual cells, or as a dimensionless ratio summarizing a physical state.

GIS has an important role to play in hazard assessment because of the following advantages over traditional methods:

1. Spatial modeling and map creation can be done on the same computer;
2. A variety of models can be created and displayed to reflect different hazard scenarios and in forms other than the traditional map; and
3. The implications of hazard in terms of risk and planning can be made understandable to planners.

The role of spatial analysis for hazard assessment is outlined in the first part of the chapter in terms of deductive and inductive techniques and illustrated with examples from the field of mass movements with a much wider relevance. A case example of landslide hazard assessment in Western Cyprus is examined in this manner, and the implications of the requirements of spatial analysis in terms of system design and operation are discussed.

MODELING NATURAL HAZARDS USING GIS

Natural hazards are the result of a subset of processes that affect the surface of the Earth that are perceived as threats to societies. Ideally, we should be able to identify the process responsible for the hazard, devise an appropriate mathematical representation of the process, and use this to provide a computer simulation of the hazard over the area of interest. In practice this deterministic approach may not be readily applied because:

1. The hazard may be created by a combination of coupled processes;
2. The process(es) may be too poorly understood to be represented by a physical model;
3. Simulation may be too time consuming; and
4. Information on physical variables may not be available.

An alternative way to model natural hazards is to use spatial and/or temporal characteristics of previous hazard events to infer the ambient hazardous states of a number of environmental variables and forecast the liability to hazard in the future. The problems with an empirical approach are that:

1. The variables available for measurement may not have the same values as those obtaining during the hazard event;
2. Choice of what environmental variables to use is largely guesswork, and the model may only explain a fraction of the hazard variance; and
3. Cross-tabulation ignores potentially useful information in the spatial autocorrelation of the data.

Both of these modeling approaches can be implemented within a GIS. The empirical or inductive approach is one that fits most closely with the analytical capabilities of a vector GIS and has been most employed in hazard analysis. The deterministic or deductive approach is more usually associated with simulation modeling in standalone computer programs rather than as a component of a GIS.

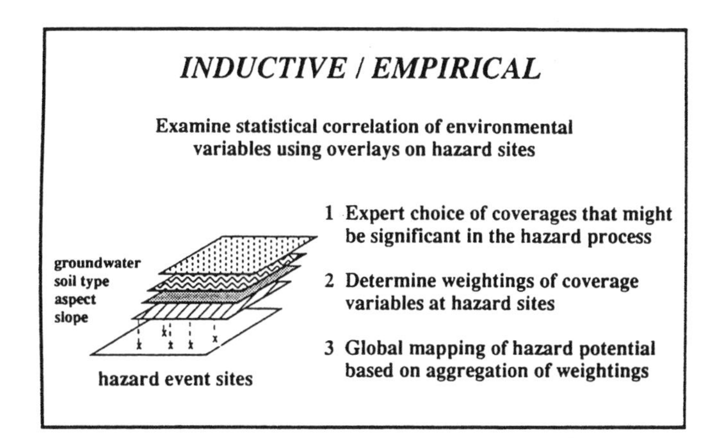

Figure 31-1: A schematic overview of the process of inductive modeling in GIS applied to natural hazards.

Inductive methods

Figure 31-1 illustrates the principles of the inductive approach to GIS modeling of natural hazards. Spatial locations of past hazard events are identified by field survey or other information sources, and these locations are overlaid with a number of environmental variable coverages, typically using point-in-polygon operations. The choice of variables used should be based on an expert judgment of the most relevant factors. Values of the variables at the hazard sites are analyzed using some form of multivariate statistical technique to derive global functions, which are then applied to produce a mapping of hazard potential. This application may be via a raster or a polygonal segmentation of the map. Some of the most common variables used in inductive GIS analysis of landslide potential are listed in Table 31-1. Rarely are all the original variables employed in the final model, and often after ranking their importance only a subset is retained.

Several statistical methods have been used to derive global functions for hazard potential mapping in a GIS. At its simplest this approach involves calculating the number of point hazard events per unit area for each polygon and for each category of independent variable (e.g., Gupta and Joshi, 1990). Formal statistical methods used to date fall into three groups:

1. *Linear regression*. This is appropriate if the independent variables have values that can be expressed as ratio-scaled numbers. Preparatory to multivariate regression the point data are cross-tabulated with each variable to form a multidimensional table. Usually, the GIS coverages comprise data compressed into a few categories (e.g., Jibson and Keefer, 1989).
2. *Logistic regression*. Although this technique also operates on the sort of multidimensional table used in linear regression, here the dependent variable is of the nominal type (e.g., a landslide is either present or absent), and some of the independent variables are of the interval/ratio type (e.g., slope angle) and perhaps some of nominal (e.g., soil type) or ordinal (e.g., stratigraphic position) types. This technique produces a mapping of the probability of hazard occurrence and is generally iterative with no exact solution (e.g., Bernknopf et al., 1988).
3. *Discriminant analysis*. A classification method that maximizes the separation between groups of data using functions of the original data values. The application of Carrara et al. (1991) uses a polygonal segmentation of the land surface into stable and unstable classes. The method used a priori classification, on the basis of field mapping, of the presence or absence of previous landslides, and stepwise discriminant analysis of the independent variables created a list of discriminant functions for each of these.

These statistical methods are often not available in the existing commercial GIS, and recourse to external software is needed. The choice of environmental variables to analyze is equivalent to making the a priori hypothesis that those variables are responsible for the hazard process. This is rarely justified explicitly. In some situations the hazard to be modeled may be very poorly correlated with available variables; this is often the case with subsidence hazard, where there is negligible subsurface information. Here reliance must be placed on the spatial pattern of subsidence alone as the indicator of future events, perhaps with the help of some reasonable assumptions, such as increased susceptibility along previous event alignments.

Table 31-1: Commonly used GIS variables for inductive analysis of landslides

Slope angle	Slope height
Bedrock lithology	Soil thickness
Distance to faults	Presence of construction
Slope aspect	Groundwater height
Soil type	Land use
Distance to earthquake	

Deductive methods

These methods assume there is sufficient knowledge of the physical process of hazard to be able to state the governing equations and that the information on the variables in these equations will be available in the GIS database. Figure 31-2 shows an example of a simple deductive method of modeling rockfall runout using an energy equation (Scheiddeger, 1975) that is discussed in detail later. Examples of past hazard events are usually not used in deductive modeling except to verify results.

Wadge (1988) has discussed the use of some of these deductive methods in GIS for modeling gravity flow and slope instability hazards. Simple stability equations such as the infinite slope equation can be implemented in GIS (Brass et al., 1991). Determination of the kinetic aspects of hazard requires neighborhood and connectivity operations or distance transforms. An accurate digital elevation model (DEM) of the land surface is of fundamental importance to model the path of fluid flow (e.g., Jenson and Domingue, 1988) such as water, mudflow, debris avalanche, or lava. An extreme implementation of this type of modeling uses the numerical solutions to appropriate laws governing mass and energy transport such as the Navier–Stokes equations. These are usually performed by finite difference approximations to partial differential equations and are currently too computationally expensive to be run within GIS. Future technical improvements will remove this obstacle.

Issues of scale

The nature of the model will depend on the scale of the mapping within the GIS. If the size of the smallest element (e.g., raster) in the GIS is greater than the smallest map dimension of the hazardous process, then a deterministic model cannot be created. Earlier hazard events can be treated as points by inductive methods. Many hazardous processes have minimum map dimensions in the range 10–100 m (e.g., translational landslides, subsidence collapse and debris flows) and cannot be adequately represented at scales smaller than 1:100,000.

Van Westen (1989) describes the GIS methods that are appropriate for hazard analysis at different scales. At regional scales (greater than 1:100,000) deductive methods are inappropriate. He advocates the use of "terrain mapping units," in which the surface is segmented into units sharing common values of variables such as landform type, lithology, and soil. Remotely sensed data are a useful source of such regional information. At medium (1:50,000–1:25,000) to large (1:10,000–1:5,000) scales, deductive methods using DEMs are appropriate because the topography can be represented to sufficient accuracy. Also, data sampled at field stations can best be associated with engineering geological mapping by renumbering within the GIS at these scales. Just as the applicability of deductive methods increases at larger scales, those inductive techniques that treat former hazard events as points become less valid as the finite dimensions of the hazard process become important. Careful thought needs to be given to the appropriate scale at which to model hazard within the GIS.

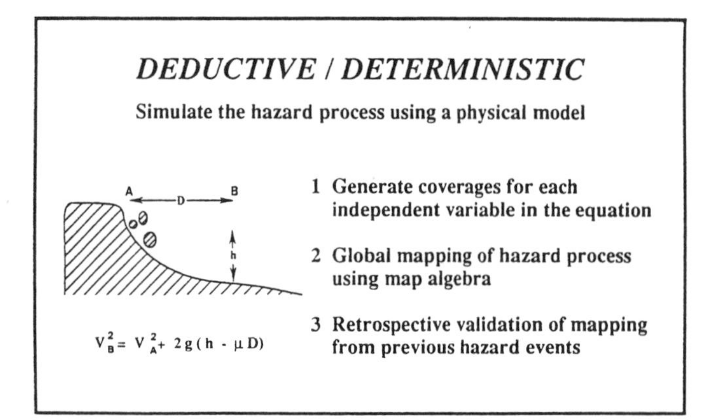

Figure 31-2: A schematic overview of the process of deductive modeling in GIS applied to natural hazards.

Probabilistic aspects

We have already discussed the logistic regression technique that predicts the probability of occurrence of a binary event. There are two other aspects of probability that commonly have an impact on GIS analysis of hazard: spatial uncertainty of variables, and the use of the magnitude–frequency data on past hazard events.

Spatial uncertainty. This is a problem common to many fields of spatial analysis. There may be the need to produce a map of a physical variable based on the limited point sampling of its values in the field. If we assume no uncertainty, then an interpolation of a surface honoring the values of the data points may be appropriate. More realistically, we will know that the field values are uncertain but can characterize this variance. One common technique in this situation is to apply Monte Carlo simulation sampling from probability distribution functions derived from the data. This may be constrained within discrete regions if we have additional knowledge (e.g., boundaries based on field observations). A considerable number of statistical techniques have been applied to spatial uncertainty, for example, in geotechnics (Rethati, 1988).

Magnitude–frequency analysis. Many natural phenomena display a characteristic relationship between the frequency and magnitude of their occurrence. In particular, many hazardous events obey a relationship in which there are rare, large-magnitude events and frequent, small-magnitude events (e.g., an exponential one in the case of earthquakes). Typically, there is a need to calculate a recurrence interval for a particular magnitude of hazard event. Often historical records are too short for us to be certain that the largest possible events have been experienced, and hence there is the need for extending this calculation using extreme value, or Gumbel analysis (e.g., storm rainfall). Many hazard events only occur when a certain threshold of a temporal variable is exceeded (e.g., landslides being triggered only by earthquakes giving a minimum ground acceleration).

LANDSLIDE HAZARD IN WESTERN CYPRUS

Nature of the problem

Immediately west of the Troodos Mountains of Cyprus is an area that suffers from chronic surface instability. This area was studied by engineering geologists from the British Geological Survey and the Cyprus Geological Survey (Northmore et al., 1986), who produced engineering geology maps at a scale of 1:10,000. The technical content of these maps makes them largely unreadable by planners. Our task is to use the information content of these maps and the accompanying technical report, together with topographic data, to produce within a GIS simple hazard maps that can be easily understood, making best use of the complex information already assembled. Many studies of hazard analysis using GIS assume a "clean sheet" approach in which the required data are generated from primary sources and any previous interpretative mapping is ignored. This approach is neither practicable nor desirable in this case. Instabilities in the Phiti study area are largely a result of the local geology where weak, fissured, montmorillonite-rich clay rocks underlay highly permeable, chalky limestones. Earthquakes and winter rainstorms are the destabilizing agents. Additionally, water passing through the limestone aquifer drains from springs onto the clays promoting localized high pore pressures and resultant failures. The mapping revealed both deep and shallow rotational and translational landslides, debris flows, and rockfalls. A crude tripartite division of the movements based on age was possible: (1) old "areal" deposits composed of numerous deposits whose individual boundaries are no longer distinguishable, (2) "dormant" individual deposits whose boundaries can be mapped and that represent single movement events, and (3) "active" deposits whose geomorphological appearance suggests recent movement.

Cartographic versus GIS representations

A consequence of using thematic maps of interpreted data drawn by traditional cartographic techniques is that considerable editing and reinterpretation is introduced. The formality of the vector topology data model (created within ARC/INFO on a SUN) imposes a greater rigor on the logic of how a landslide is represented than is required on the original maps. "Open" polygons, gradations in symbolic representation, and implicit spatial relationships are some of the problems encountered (Figure 31-3). In part these imprecisions reflect the interpretative uncertainties in the mind of the field mapper, but they may equally hide inadequate mapping, vague thinking, and errors. Had the original mappers known their work was to be used within a vector GIS, then a more helpful scheme of representation could have been devised. Such a scheme would benefit from the hierarchical representation of features derived from the same movement event, such as a rotated block polygon nested within a landslide polygon whose boundary is partly coincident with the source scarp (see Figure 31-3). Here the class landslide contains subclasses of scarp and deposit, the latter comprising landslide deposit and an intact rotated block deposit.

Role of GIS

The variety of past hazards at Phiti (landslides, debris flows, and rockfalls) suggests multiple processes at work. Hence no one deductive or inductive method will adequately model the potential hazard. Additionally, the scale of mapping (1:10,000) means that inductive methods driven by statistical analysis of point hazard events in polygons will not work because the source scarps (linear) and landslide deposits (polygons) are finite features and

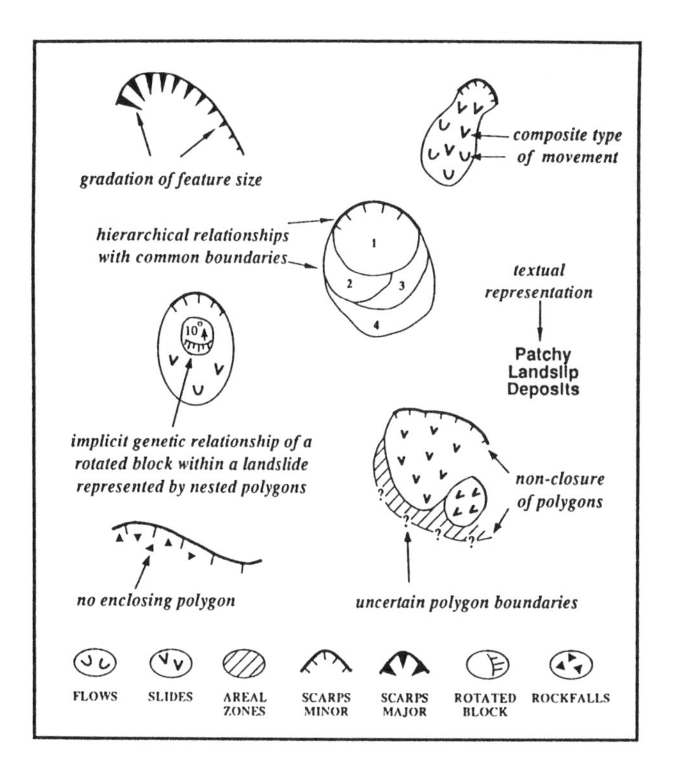

Figure 31-3: Graphical summary of the types of problems faced in converting a traditional cartographic map of engineering geology for western Cyprus to a vector GIS representation.

highly correlated with other map elements. There are three principal roles of the GIS used for this problem:

1. Modeling those hazard processes that are amenable to deductive techniques;
2. Applying inductive techniques to test hazard-related hypotheses, posed explicitly and implicitly, by the engineering geological study; and
3. Integrating the results of (1) and (2) to provide simple maps of hazard potential intelligible to nonspecialists.

In the following we discuss briefly examples of how (1) and (2) are performed.

Rockfall hazard

Rocks falling from cliffs is a hazard at Phiti as evidenced by mapped scree deposits. Simple deductive models of such rockfall can be devised. For example, in Figure 31-2 the velocity at B (V_B) increases as potential energy is lost in its fall from A, overcoming rolling/bouncing friction along its trajectory. To produce a mapping of areas at hazard requires the runout paths of all potential rockfalls to be generated. Once the source of the rockfall is identified, this can be achieved by repeatedly propagating V_B into the neighboring downslope cell as the value of V_A until the velocity falls to zero. The GIS steps required are:

1. *Identify sources.* Two rock types (limestone and sandstone) are known to form the source escarpments where slopes are in excess of about 45 degrees. Overlays of the lithology coverage with a slope coverage threshold at 45 degrees define the source areas for rockfall.
2. *Map downslope runout.* This is best performed in a raster GIS; van Dijke and van Westen (1990) describe one implementation in the ILWIS GIS, and the MAPBOX software of Tomlin's (1990) map algebra also permits a similar model. The first operation is to create a steepest-descent mapping from the source cells using a neighborhood operation on the DEM. Next a height-difference map layer (equivalent to h in Figure 31-2) is created for the steepest-descent cells. If the friction coefficient (μ) is known to vary with surface type, then a mapping of this value can be used. In the case of bouncing, it may be appropriate to sample the coefficients of restitution and the rebound angle as stochastic variables (Paronuzzi, 1989). As D and g are constants in the equation (Figure 31-2) V_B can be calculated using values of h, μ, and V_A from the relevant neighbor using a connectivity operator. When V_B falls to zero, the edge of the hazard zone is defined.

Hypothesis testing by induction

The map and report of Northmore et al. (1986) proposed a number of hypotheses relating to hazard processes at Phiti. These hypotheses are a valuable source of knowledge. Most of them can be tested using the spatial analytical capabilities of the proprietary GIS products. Such testing helps justify the choice of variables in a complete mapping of hazard because, in general, the tests only involve one or two variables at a time, and the effects of spatial autocorrelation can be better appreciated. Some examples of these hypotheses are:

- Failure planes of major rotational landslides occur at the junction of the clay-rich rocks and the limestones.
- The commonest (most recent) landslide mechanism is the reactivation of old slides.
- Translational landslides in clay-rich rocks are close to ambient limiting equilibrium.

Let us look at the last of these in detail. The first GIS task is to identify translational landslides in clay lithology. Unfortunately, although the report discussed translational landslides, no explicit distinction was made on the map between rotational and translational types. In translational slides the failure surface is planar and subparallel to the ground surface; rotational slides have the failure surface concave upwards and involve obvious rotation, particularly coherent components (rotated blocks) of the resultant deposit. We assumed that all landslides without nested rotated blocks (see Figure 31-3) were translational and selected only those that overlay clay lithology. Ambient limiting equilibrium means that an approximate balance between shear strength of the material and shear stress at the failure plane currently exists. This balance is traditionally expressed by the infinite-slope equation for surface-parallel planes of failure. Northmore et al. (1986) used a simplified version of this equation in which cohesion was assumed to be zero and the pore–water pressure affected the whole failed mass (i.e., the phreatic surface reached the ground surface). Cast in terms of slope angle, this equation is:

Limiting slope angle = arctan (1 – density of water/density of soil) × effective residual angle of shear

Using averages of soil bulk densities and shearing resistance angles produced limiting slope angles of 8.5 degrees for one of the clay lithologies that compared with values from Abney-leveled profiles across translational landslides measured in the field in the range 8.5–10 degrees. The GIS can test the slope angles for all areas of the translational landslides, not just selected profiles. Figure 31-4 summarizes these results. The main mode is slightly higher than the field range, but the distribution is wider

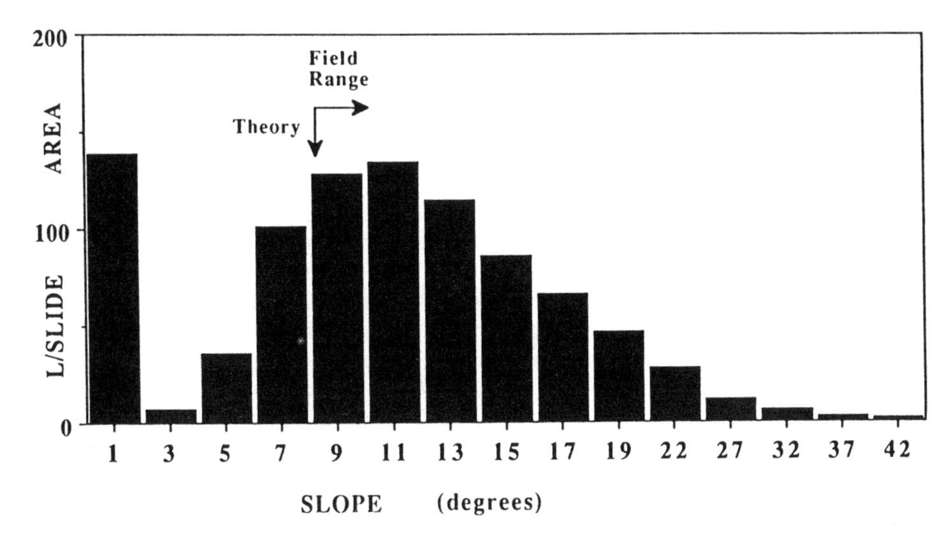

Figure 31-4: Histogram of the areal frequency of slopes found on the population of translational landslides underlain by the Kannaviou Clay within the Phiti area of western Cyprus determined by GIS operations. The "Theory" and "Field Range" arrows indicate values discussed in the text.

than indicated by the hypothesis. Several reasons can be invoked to explain this:

1. We should regard the theoretical limiting slope angle of 8.5 degrees as a minimum because cohesion was ignored and the soil was assumed to be saturated.
2. If we have included any rotational landslides in our analysis, then we might expect some high slope angles.
3. The presence of a distinct mode at 0–2 degrees may indicate damming of the slide deposits behind topographic obstructions.
4. The DEM used to calculate the slopes was derived from 1:5,000 scale contours in the TIN module of ARC/INFO that uses the Delaunay triangulation algorithm.

A well-known defect of this algorithm is that, if the spacing between crenellate contours is greater than the amplitude of the crenellations, then benches of flat triangles are formed around the contours and the intervening areas are artificially steepened (Robinson et al., 1989). This would have the effect of producing a mode at 0–2 degrees and contribute to the higher end of the distribution (Figure 31-4). We cannot pursue these arguments further here, except to note that in testing the original hypothesis a number of connected hypotheses can be brought to bear on the problem.

IMPLICATIONS FOR SYSTEM DESIGN

It is clear that any list of GIS functionality required to model even a restricted set of natural hazards would be very long. In any GIS used to assess more than one hazard, a variety of spatial analytical functions will be required, not all of which will be available in the host GIS. The assessors of natural hazard are usually earth scientists, and the users of the assessments are usually planners. As Marker and McCall (1990) argue there is a need to make earth scientists aware of the forms in which the planners can best understand and use the information present in a hazard/risk map. We see the design requirements of a GIS for natural hazard assessment to be: comprehensive spatial analytical functionality; interfaces to external software; and user-friendly systems that can be easily interrogated and manipulated by the interested parties. The application of expert system technology would assist in meeting these specifications and would perform the following tasks:

1. Provide application-oriented perspectives (both earth scientist and planning) to the user interface, hiding the GIS interface.
2. Manage a suite of spatial modeling routines including inductive and deductive methods utilizing both in-built GIS functionality (e.g., AML in ARC/INFO) and external spatial and statistical software via a software bridge between the GIS, expert system, and external programs.
3. Use an object-oriented hierarchy of spatial knowledge about hazard conditions that could be manipulated using rules that permit hypotheses to be invoked and tested.

ACKNOWLEDGMENT

This work was undertaken within a project funded by the ESRC/NERC Joint Programme for Geographical Information Handling. NUTIS is supported by NERC contract F60/G6/12.

REFERENCES

Bernknopf, R.L., Cambell, R.H., Brookshire, D.S., and Shapiro, C.D. (1988) A probabilistic approach to landslide hazard mapping in Cincinnati, Ohio, with applications for economic evaluation. *Bulletin of the Association of Engineering Geologists* 25(1): 39–56.

Brass, A., Wadge, G., and Reading, A.J. (1991) Designing a GIS for the prediction of landsliding potential in the West Indies. In Cosgrove, J., and Jones, M. (eds.) *Neotectonics and Resources*, London: Belhaven Press, pp. 220–230.

Carrara, A., Cardinali, M., Detti, R., Guzzetti, F., Pasqui, V., and Reichenbach, P. (1991) GIS techniques and statistical models in evaluating hazard. *Earth Surface Processes and Landforms* 16: 427–445.

Crandell, D.R., Booth, B., Kazumadinata, K., Shimozuru, K., Walker, G.P.L., and Westercamp, D. (1990) *Source-Book for Volcanic Hazards Zonation*, Paris: UNESCO, Natural Hazards Vol. 4, 97 pp.

Gupta, R.P., and Joshi, B.C. (1990) Landslide hazard zoning using the GIS approach—a case study from the Ramganga catchment, Himalayas. *Engineering Geology* 28: 119–131.

Jenson, S.K., and Domingue, J.O. (1988) Extracting topographic structure from digital elevation data for GIS analysis. *Photogrammetric Engineering and Remote Sensing* 54(11): 1593–1600.

Jibson, R., and Keefer, D. (1989) Statistical analysis of factors affecting landslide distribution in the New Madrid seismic zone, Tennessee and Kentucky. *Engineering Geology* 27: 509–542.

Marker, B.R., and McCall, G.J.H. (1990) Applied earth-science mapping: the planners' requirement. *Engineering Geology* 29: 403–411.

Northmore, K., Charalambous, M., Hobbs, P.R.N., and Petrides, G. (1986) Engineering geology of the Kannaviou, 'Melange' and Mamonia complex formations—Phiti/Statos area, SW Cyprus. *E.G. & R.P.R.G. Report, EGARP-KW/86/4*, Keyworth: British Geological Survey, 205 pp.

Paronuzzi, P. (1989) Probabilistic approach for design optimization of rockfall protective barriers. *Quarterly Journal of Engineering Geology* 22: 175–183.

Rethati, L. (1988) Probabilistic solutions in geotechnics. *Developments in Geotechnical Engineering* 46: 451.

Robinson, G.R., Pearson, E.J., and Settle, J.J. (1989) The use of digital elevation models in GIS applications. *Proceedings, AGI Conference "GIS—a corporate resource," Birmingham, 11–12 October*, pp. D.3.1–D.3.5.

Scheiddeger, A.E. (1975) *Physical Aspects of Natural Catastrophies*, Amsterdam: Elsevier.

Tomlin, C.D. (1990) *Geographic Information Systems and Cartographic modeling*, Englewood Cliffs, NJ: Prentice-Hall, 249 pp.

van Dijke, J.J., and van Westen, C.J. (1990) Rockfall hazard: a geomorphological application of neighbourhood analysis with ILWIS. *ITC Journal* 1: 40–44.

van Westen, C.J. (1989) ITC–UNESCO project on GIS for mountain hazard analysis. *Simposio Suramericano De Deslizamientos, 7–10 August, Paipa, Colombia*, pp. 214–224.

Varnes, D.J., and The International Association of Engineering Geology Commission on Landslides and other Mass Movements on Slopes (1984) *Landslide hazard zonation—a review of principles and practice*, Paris: UNESCO, Natural Hazards Vol. 3, 64 pp.

Wadge, G. (1988) The potential of GIS modeling of gravity flows and slope instabilities. *International Journal of Geographical Information Systems* 2(2): 143–152.

32

The Use of GIS in Assessing Exposure and Remedial Alternatives at Superfund Sites

MARGRIT VON BRAUN

Hazardous waste site cleanups typically involve analyzing contaminants in numerous environmental media, including the air, groundwater, surface water, and soil. Decisions about remedial strategies are largely driven by the actual or predicted risk to the target organism or "receptor" associated with the environmental exposure. Media-specific (e.g., air or groundwater) contaminant fate and transport models are used to describe the environmental pathways of contaminants to receptors. GIS are particularly useful for integrating the modeling results in time and space for assessing exposure and risk and in assisting remedial decision-making.

The Comprehensive Environmental Response, Compensation, and Liability Act of 1980 (CERCLA or "Superfund"), as amended by the Superfund Amendments and Reauthorization Act of 1986 (SARA), established a national program for response to release of hazardous substances into the environment. The overall approach for determining appropriate remedial actions at Superfund sites is provided by a Remedial Investigation/Feasibility Study (RI/FS). This process is fundamentally driven by an assessment of exposures and related risks to human health and the environment. Continuous communication between the three subcultures of scientists, policy-makers, and the public, as described in the overview chapter of this section (Rejeski, Chapter 30) is imperative. Because issues of public health and welfare are involved, this communication must occur in an open, public forum. GIS can provide an evolving site-specific database and offer a common format within which these groups interact.

This chapter discusses two hazardous waste sites illustrating the flexibility of GIS applications. At the Bunker Hill Superfund site, extensive data on human exposure to pollutants were available. The GIS was used primarily to communicate risk and facilitate remedial activities. However, at the Tucson site, few exposure data were available. The GIS was used primarily to develop estimates of historical exposure based on the results of extensive external models. At both sites the term "GIS" is used to represent a combination of software, including ARC/INFO (ESRI, 1985), pMAP (Spatial Information Systems, 1986), SAS (SAS Institute, 1986), dBASE (Ashton–Tate Corp., 1990), Lotus 1-2-3 (Lotus Development Corp., 1989), Golden Graphics (Golden Software Inc., 1987), and AutoCAD (Autodesk Inc., 1987) that access a common spatially related database. A variety of external, media-specific environmental models (e.g., of pollutant transport in groundwater) were linked to the GIS in a common format. Many of these models had already been employed in site-related activities. Their uncertainty and rigor is documented and quantifiable. Therefore, it was considered advantageous to couple the results of these models to the GIS rather than to use the GIS itself for multimedia modeling.

CASE STUDY #1: BUNKER HILL SITE

Site background

The Bunker Hill Superfund site encompasses a 21 square mile area surrounding a primary lead/zinc smelting complex in Northern Idaho and contains 5000 persons living in five towns (Woodward Clyde Consultants and TerraGraphics, 1986; Dames and Moore, 1988) (see Figure 32-1). A long history of contaminant releases from mining and smelting activities led to excess blood levels in over 99% of preschool children in Smelterville, Idaho, in 1974 (Yankel et al., 1977; Wegner, 1976; Jacobs Engineering Group and TerraGraphics, 1988). Although smelter operations shut down in 1981, community-wide testing in 1983 revealed that 25% of the preschool children continued to demonstrate blood lead levels exceeding Centers for Disease Control (CDC) criteria (CDC, 1986).

Subsequent studies linked this excess absorption to contaminated soil and dust exposures in the community.

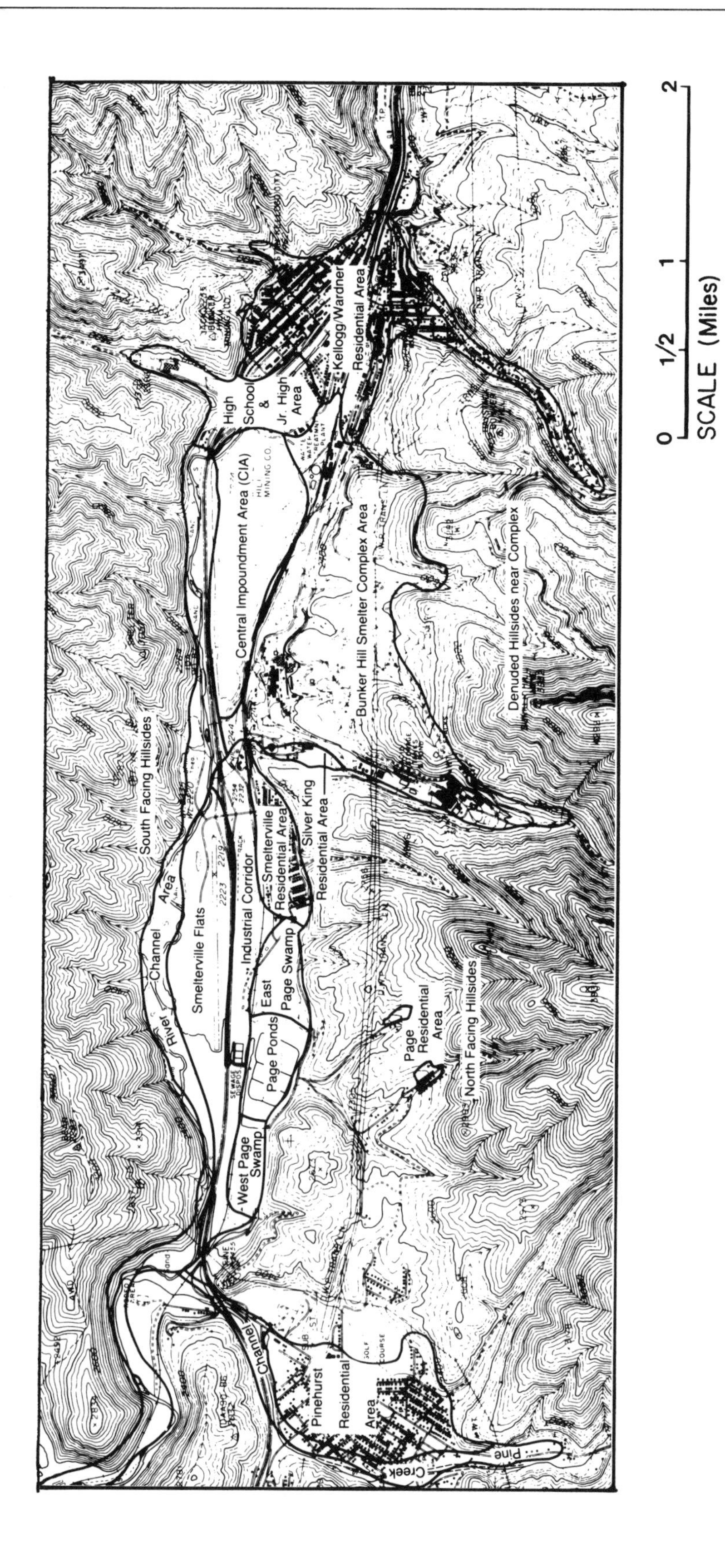

Figure 32-1: Study area and GIS base map for the 21-square mile Bunker Hill Superfund site, Idaho.

As a result of these studies, the Bunker Hill Site was placed on the NPL in 1983. In 1985 a large multiphase RI/FS commenced and several expedited response activities have been undertaken. Initial remediation activities have focused on ranking and excavating private residential yards and undeveloped properties.

Database development

In implementing the project structure on this complex site, the USEPA recognized the need for effective information management and decided to employ a GIS-based strategy. That system has been described by von Lindern and von Braun (1986). This system has been used extensively in populated areas to integrate health, population, and property-related databases for risk assessment, inventory, and notification purposes.

The Overall Base Map encompasses the site as a 3 × 7 mile rectangle centered on the smelter complex (see Figure 32-1). The populated portions of the study area are maintained as a series of subunit base maps representing each town. Attribute files are maintained in ASCII format indexed to the Property ID Number. Three types of attribute files were developed for this effort: (1) county tax records providing ownership details and legal descriptions of the property (Shoshone County Tax Assessor, 1989); (2) health census data containing population, health, and risk factors such as blood lead levels, yard conditions, number of children, existence of smokers (Panhandle Health District, 1988); and (3) sampling regimes results of soil metals surveys (TerraGraphics, 1986a,b). A base map of each property was generated showing the Property ID Number. Figure 32-2 shows the map for the City of Smelterville. The combination of base maps and attribute files provides a complete and comprehensive summary of the information available for each property on the site.

GIS analysis

GIS analyses served five basic functions in the home yard soil removal project. Each of those functions is briefly discussed:

1. *Data verification and owner/resident notification.* The relational database aspects of the GIS were exploited to prepare individual property summaries for 1500 homes containing the ownership, childhood census, sample results, and risk indices obtained for that property. These were used to notify homeowners and residents of the data collected, provide interpretation of sampling results, and communicate the risk involved.
2. *Public display and risk communication.* For public presentation, it was necessary to use maps and displays that contained no identifiable individual results. GIS techniques were used to prepare nonconfidential maps for risk communication purposes in public meetings. Figure 32-3 shows the results for Smelterville. Neighborhoods (such as "Smelterville 1st Addition") were defined, and summary statistics were developed. A subchronic hazard ranking (SHR), defined as soil lead concentration in ppm divided by 1000, was developed to describe to residents how their soils compared to proposed national criteria.
3. *Providing master lists and maps for project managers.* Confidential property summary sheets containing all data for each property were prepared for select project personnel. The reclassification functions of GIS were used to develop a number of maps for confidential project use. These maps were similar to Figure 32-2, except sample concentration and health survey attributes were substituted for lot ID numbers. These maps included such items as (a) top-inch soil metal levels, (b) litter metal levels, (c) soil and litter lead hazard indices (color coded), (d) children's blood lead levels, and (e) sample status (e.g., whether sites had been sampled, owners contacted, etc.). This series of summary sheets and maps allowed project managers to access and evaluate individual data quickly when dealing with residents and parents.
4. *Ranking properties for remediation.* Project resources were available to remediate about 100 homes each construction season. GIS techniques were used to help select which properties should be remediated. Preschool children and pregnant women are those groups at greatest health risk from lead absorption. Reclassification functions substituting health census data were used to develop maps of homes where young children or pregnant women resided. Using overlay functions, these maps of "high-risk residences" were overlaid with the soil lead concentration map; resulting properties were ranked by lead concentration. Various cleanup scenarios were selected to maximize targeting high-risk properties while minimizing costs associated with equipment and crew logistics. Cost effectiveness of various strategies could be evaluated by reclassifying mapped data to derive estimated dollar values, calculating the volume of soil to be excavated and associated costs. A similar approach is described by von Lindern and von Braun (1988).
5. *Tracking remedial progress.* Records of contacts with residents and of their permissions for remediation and for continued monitoring were tracked as attribute files in the GIS. This aids in record-keeping, providing progress maps, and logistical assistance in future remediations. In addition to optimizing remediation of nearly 1800 home yards, unoccupied or undeveloped properties are also being evaluated using GIS spatial analysis. The database is being expanded to include information such as access to utilities and roads, flood-

Figure 32-2: City of Smelterville residential property map showing individual property ID numbers used to index attribute files.

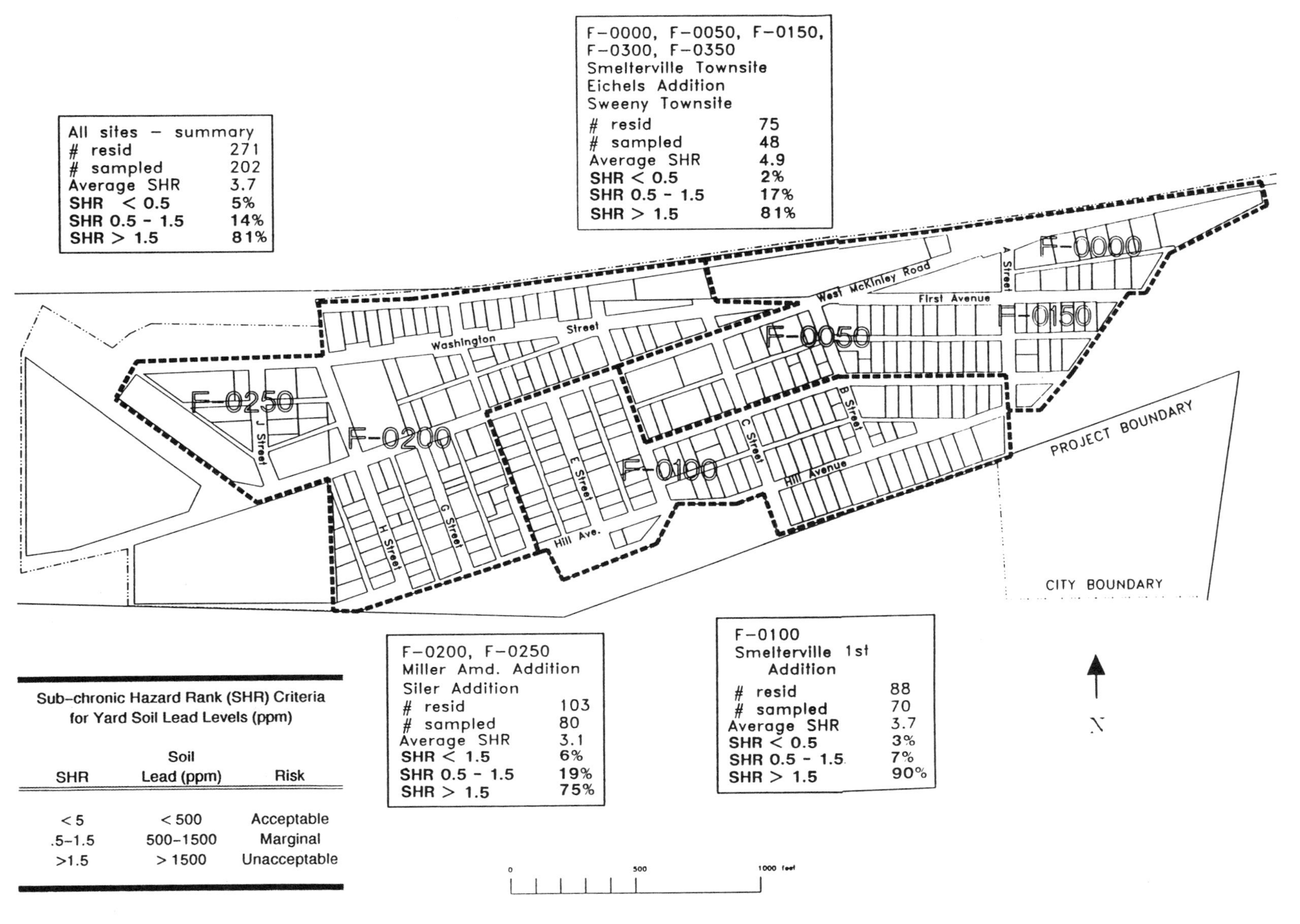

Sub-chronic Hazard Rank (SHR) Criteria for Yard Soil Lead Levels (ppm)

SHR	Soil Lead (ppm)	Risk
< 5	< 500	Acceptable
.5–1.5	500–1500	Marginal
>1.5	> 1500	Unacceptable

Figure 32-3: Nonconfidential map produced to communicate risk due to soil lead levels in Smelterville.

plains, zoning, and soil types, so that each property's potential for development can be used as a criterion for remedial selection.

CASE HISTORY #2: TUCSON AIRPORT AREA SITE

Site background

At the Tucson Airport Area (TAA) site in Arizona, over 25 years of improper waste water and industrial solvent disposal practices resulted in a large plume of contaminated groundwater underlying residential areas. Several public drinking water wells were contaminated with trichloroethylene (TCE) and subsequently closed by health authorities. However, these wells were not closed until 1981, nearly three decades after initial contamination reports. During this period, water containing varying degrees of contaminants was extracted from the aquifer and consumed by the growing community. Despite indications of adverse effects associated with past exposures (Caldwell, 1986a,b; Goldberg, 1989), government officials cited the lack of adequate historical exposure estimates as precluding quantitative assessment of risks to human health (ATSDR, 1988).

Both current data from the Superfund RI/FS and historic information collected from local government, industry, and research institutions were developed and analyzed. Results from a series of external models were integrated using the GIS to infer past exposures. As shown in Figure 32-4, computerized base maps of (1) residential locations, (2) plume locations, (3) well locations, and (4) water system layouts (and others) were created. These maps resulted from specific models of (1) the population's residential locations over a 30-year period, (2) pollutant transport in the groundwater, and (3–4) the city of Tucson's own water distribution system, respectively. These base maps were combined and manipulated to create new maps that became part of the site database. In this way, the data layers could be analyzed simultaneously to evaluate their interactions. The flowchart depicted in Figure 32-5 illustrates the overall model used to estimate exposures to contaminants from the polluted water supply. As illustrated, the intersections of numerous raw data planes are identified in a spatial and temporal "overlay and reduction" technique to arrive at the final combination of individual and population-based contaminant exposures.

The basic elements of the analyses included:

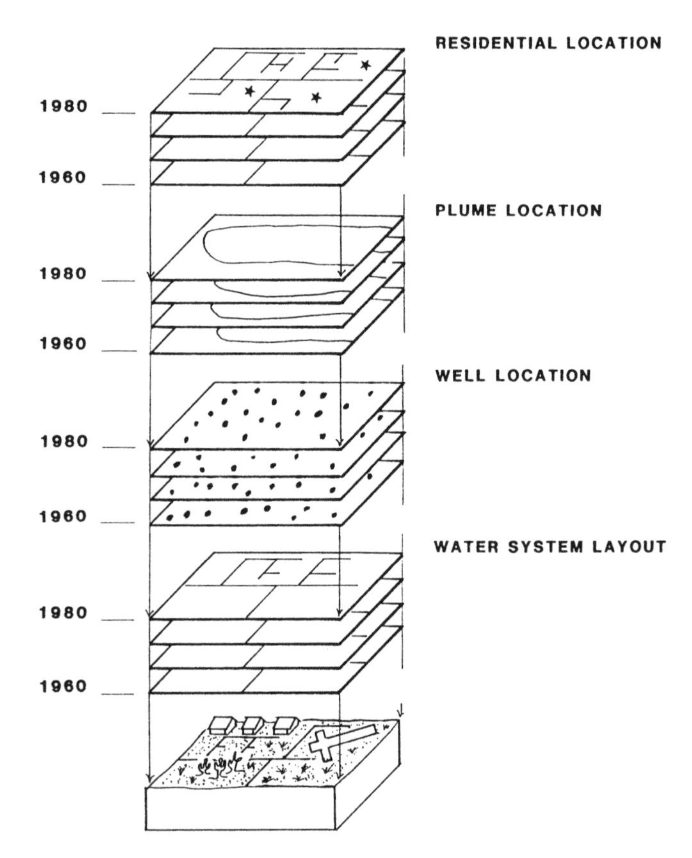

Figure 32-4: Illustration of Tucson Airport area Superfund site spatial database with temporal components.

1. *Historical plume location* (i.e., "where was the plume when?"). Groundwater flow and contaminant transport models were developed to predict future exposures and determine source contributions as part of the site Superfund effort (CH2M Hill, 1987). The results of these efforts were used in conjunction with GIS cartographic analyses to project the plume back in time to 1950 in 5-year intervals. Figure 32-6 shows the 1975 "snapshot" of plume location under two different hydraulic conductivity (K) scenarios to incorporate some of the uncertainty associated with the model.
2. *Contaminant production by area wells* (i.e., "how were wells affected?"). Using data from the City of Tucson (1989), the uptake of polluted waters via municipal production wells and development of contaminant levels in wells was accomplished. This analysis used the well production, depth, and construction data in conjunction with the historical plume location (from Part 1). Spatial and temporal variation was significant, as different wells pumped different amounts at different depths throughout the 30-year period.
3. *Distribution of contaminated drinking water* (i.e., "where did the contaminated water go?"). Historical water service areas (WSA) were developed by cartographically combining the contaminant production

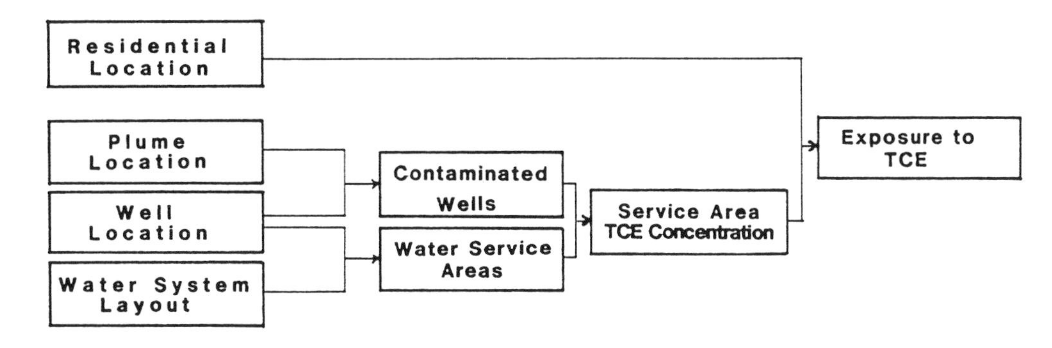

Figure 32-5: Flowchart of cartographic model used to assess exposure to trichloroethylene (TCE).

(from Part 2) with community demand and system hydraulics.

4. *Residential exposure* (i.e., "who drank it?"). The analyses from Part 3 resulted in annual contaminant concentration estimates for each specific geographic area. These results were cross-indexed with maps of demographic characteristics to develop detailed exposure profiles. Once the conceptual model was applied for a given year, an iterative cartographic modeling procedure was applied to obtain profiles for each year. These results were then combined to provide estimates of the degree and extent of contaminated water service over time. Stratified exposure estimates for the area population (see Table 32-1) and exposure profiles for individual residents were developed and presented for geographic areas, target populations, and individuals. This analysis is provided in greater detail by von Braun (1989).

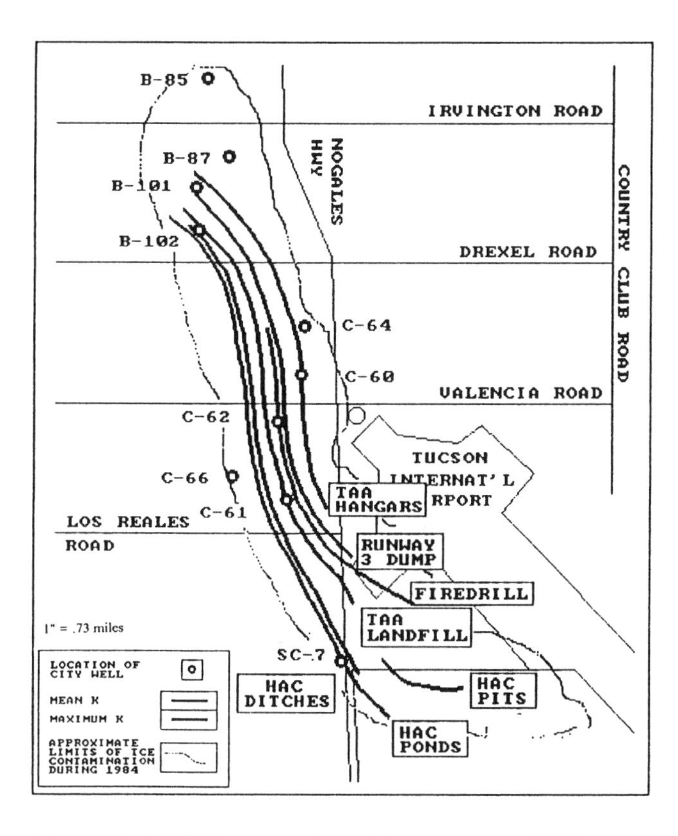

Figure 32-6: 1975 "snapshot" of trichloroethylene (TCE) plume location using mean and maximum values of hydraulic conductivity (*K*).

DISCUSSION AND RECOMMENDATIONS

The four issues of believability, honesty, decision utility, and clarity discussed in the overview chapter of this section (Rejeski, Chapter 30) were encountered at both of these sites.

At the Bunker Hill site, the availability of measured blood lead levels negated the need for developing rigorous external exposure estimates. Instead, GIS analyses were most useful in identifying at-risk populations, optimizing remedial scenarios, communicating risk, and tracking project information. Several instances of conveying believability and honesty with GIS were encountered at this site. At one point in the project, the use of geostatistical techniques to estimate home yard lead levels was explored, to avoid the expense of sampling each individual yard. Such analyses are routine at Superfund sites, and often utilize GIS. It was determined early on that from a risk communication viewpoint this would not work at the Bunker Hill site. No amount of statistical assurances could convey to a homeowner that his or her yard would not be remediated, while the neighbor's would be. The difficulty with "drawing the line" in such a controversial situation was mitigated by, in fact, sampling every yard. However, because of privacy and property value

Table 32-1: Estimate of population exposed to TCE (1955–1980) (fictional data for demonstration only)

Year	Class 1 5–24 ppb.	Class 2 25–99 ppb.	Class 3 100–249 ppb.	Class 4 ≥ 250 ppb.	Population exposed	Total population	Percent exposed
1955	0	0	0	1000	1000		
1956	0	0	0	1000	1000		
1957	0	0	0	1000	1000		
1958	0	0	0	1000	1000		
1959	0	249	0	1000	1249		
1960	604	249	0	1000	1853	12382	15
1961	670	518	0	1268	2456	13253	19
1962	736	787	0	1535	3059	14124	22
1963	798	1036	0	1782	3616	14930	24
1964	864	1305	0	2050	4219	15801	27
1965	925	0	0	0	925	16607	6
1966	2868	0	0	0	2868	17478	16
1967	2941	2790	0	2093	7824	20574	38
1968	1657	1416	2958	2309	8340	21816	38
1969	3157	0	3138	2540	8835	23145	38
1970	4147	2517	0	2756	9420	24422	39
1971	2653	4132	2972	0	9758	25628	38
1972	4109	6918	8391	0	19418	26869	72
1973	5944	3398	9439	0	18781	28196	67
1974	5088	11344	5164	0	21596	29432	73
1975	4360	10272	8336	0	22968	29952	77
1976	3308	9568	5952	0	18828	30420	62
1977	856	7228	3720	0	11804	30604	39
1978	12668	0	8780	0	21448	30784	70
1979	3392	6788	2300	0	12480	30932	40
1980	2236	1408	7752	0	11396	31088	37

issues, homeowners were not allowed to know each other's results. GIS techniques were used to provide the summary statistics for neighborhoods for public display while allowing the use of individual data points within the project and maintaining confidentiality.

One of the more difficult problems encountered in integrating various pathway models into a multimedia format is assessing and conveying uncertainty. These efforts have two fundamental drawbacks. The results can be no stronger than the weakest link in the chain, and uncertainties in one link can be perpetuated or magnified through the system. On the other hand, these integrated systems allow for more sophisticated sensitivity analyses because variable effects on one medium can be assessed in outcome estimates on other media. For example, for the TAA site, assumptions about hydraulic conductivities could be evaluated directly in terms of their effect on household exposures many years later.

At the TAA site, the GIS was used to create static representations or "snapshots" in space and time by incorporating the results of specific external models. Although some GIS users favor the development of dynamic and inclusive multimedia models within GIS with 3D capabilities and enhanced graphics, such models run the risk of implying a false certainty and becoming "black box" applications. Being able to portray the growth of the plume and the transport of contaminants in the Tucson water supply dynamically over the 30-year period might be helpful for visualizing the situation; however, it seems that it would imply a false accuracy and precision to the viewers. Exposure and risk analyses are often deliberately "simple" just to prevent such misconceptions. The routines employed within GIS applications are often more flexible but less rigorous than those in dedicated external models. The uncertainty of most external models is more easily quantified. The responsibility of quantifying uncertainties in the external models must fall to and remain with the scientist user. As a result, more reliable and believable results can likely be obtained through use of outside dedicated models employing GIS techniques to integrate the results and inputs required. How to convey this uncertainty in GIS systems is an area requiring investigation and research in the coming years.

REFERENCES

Ashton–Tate Corp. (1990) *dBase IV. Programming with dBase IV*, Torrance, CA: Ashton Tate Corp.

ATSDR (1988) *Health Assessment for the Tucson International Airport Site, Tucson, Arizona*, Atlanta, GA: Agency for Toxic Substances and Disease Registry.

Autodesk, Inc. (1987) *AutoCad. Release 9*, Sausalito, CA: Autodesk, Inc.

Caldwell, G. (1986a) *Inter-Office Memorandum: June 20, 1986*, Committee Recommendations for Follow-up Studies in Southside Tucson Area of Contaminated Drinking Water—Final Report. To Lloyd F. Novick, Arizona Department of Health Services, Phoenix, Arizona.

Caldwell, G. (1986b) *Mortality Rates on Tucson's Southside*, Phoenix, AZ: Arizona Department of Health Services.

Centers for Disease Control (1986) *Kellogg Revisited—1983 Childhood Blood Lead and Environmental Status Report.*

CH2M Hill (1987) *Draft Assessment of the Relative Contribution to Groundwater Contamination from Potential Sources in the Tucson Airport Area, Tucson, Arizona*, prepared for the U.S. Environmental Protection Agency Region IX, San Francisco, California, Contract No. 68-01-7251.

City of Tucson (1989) *Tucson Water Files, 1950–1989.*

Dames and Moore (1988) *Bunker Hill RI/FS: Data Evaluation Report.*

ESRI (1985) *ARC-INFO, Version 3.0*, Redlands, CA: Environmental Systems Research Institute.

Goldberg, S.J. (1989) Human cardiac teratogenesis of TCE. *Progress Report to the Arizona Disease Control Research Commission.*

Golden Software, Inc. (1987) *Golden Graphics. Version 3.0*, Golden, CO: Golden Software, Inc.

Jacobs Engineering Group, Inc., and TerraGraphics (1988) *Final Draft Endangerment Assessment Protocol for the Bunker Hill Superfund Site*, EPA Contract No. 68-01-7531.

Lotus Development Corp. (1989) *Lotus 1-2-3. Version 2.2*, Cambridge, MA: Lotus Development Corporation.

Panhandle Health District (1988) *Summary Report 1988 Lead Health Screening Program*, Silverton, ID.

SAS Institute, Inc. (1986) *SAS, Statistical Analysis System*, Cary, NC: SAS Institute.

Shoshone County Tax Assessor (1989) *1988 Update of Property Owner Files*, Wallace, ID.

Spatial Information Systems (1986) *pMAP. The Professional Map Analysis Package*, Omaha, NE: Spatial Information Systems.

TerraGraphics (1986a) *1986 Residential Soil Survey Status Report.*

TerraGraphics (1986b) *Bunker Hill Site RI/FS, Soils Characterization Report.*

von Braun, M.C. (1989) Use of a geographic information system for assessing exposure to contaminants released from an uncontrolled hazardous waste Site. *PhD Dissertation*, Washington State University.

von Lindern, I.H., and von Braun, M.C. (1986) The use of geographic information systems as an interdisciplinary tool in smelter site remediations. *Proceedings, National Conference on Management of Uncontrolled Hazardous Waste Sites, Washington*, Silver Spring, MD: HMCRI, pp. 200–207.

von Lindern, I.H., and von Braun, M.C. (1988) Reconstructive analysis of lead exposures in a smelter community using geographic information system techniques. *Proceedings: Society for Occupational Environmental Health 1988 Annual Conference on Toxic Wastes and Public Health: The Impact of Superfund*, Washington, DC.

Wegner, G. (1976) *Shoshone Lead Health Project Summary Report*, Boise, ID: Idaho Department of Health and Welfare.

Woodward Clyde Consultants and TerraGraphics (1986) *Interim Site Characterization Report for the Bunker Hill Site*, Walnut Creek, CA: EPA Contract No. 68-01-6939.

Yankel, A.J., von Lindern, I.H., and Walter, S.D. (1977) The Silver Valley lead study: the relationship between childhood blood lead levels and environmental exposure. *Journal of the Air Pollution Control Association* 27: 763–767.

33

U.S. EPA Region 6 Comparative Risk Project: Evaluating Ecological Risk

DAVID A. PARRISH,
LAURA TOWNSEND,
JERRY SAUNDERS,
GERALD CARNEY, AND
CAROL LANGSTON

The U.S. Environmental Protection Agency (EPA) administers over twenty separate Federal laws dealing with environmental regulation and protection. The Region 6 office in Dallas conducted a project to compare the residual risk to human health, human welfare, and ecological systems associated with 22 environmental problem areas (Table 33-1). This chapter describes the evaluation of ecological risk resulting from certain environmental problems to ecoregions.

A work group of ecologists, GIS analysts, and other environmental professionals was formed to conduct the ecological evaluation. The work group reviewed ecologi-

Table 33-1: List of environmental problem areas for Region 6 Comparative Risk Project

Environmental problem area		Type of stressor
1.	Industrial wastewater discharges to oceans, lakes, and rivers	chemical
2.	Municipal wastewater discharges to oceans, lakes, and rivers	chemical
3.	Aggregated public and private drinking water supplies	chemical
4.	Nonpoint discharges to oceans, lakes, and rivers	chemical
5.	Physical degradation of water and wetland habitats	physical
6.	Aggregated groundwater contamination	chemical
7.	Storage tanks	chemical
8.	RCRA hazardous waste sites	chemical
9.	Hazardous waste sites—abandoned/Superfund sites	chemical
10.	Municipal solid waste sites	chemical
11.	Industrial solid waste sites	chemical
12.	Accidental chemical releases to the environment	chemical
13.	Application of pesticides	chemical
14.	Sulfur oxides and nitrogen oxides (incl. acid deposition)	chemical
15.	Ozone and carbon monoxide	chemical
16.	Airborne lead	chemical
17.	Particulate matter	chemical
18.	Hazardous/toxic air pollutants	chemical
19.	Indoor air pollutants other than radon	chemical
20.	Indoor radon	chemical
21.	Radiation other than radon	chemical
22.	Degradation of terrestrial ecosystems and habitats	physical

Table 33-2: Major ecoregions in U.S. EPA Region 6 (Omernik and Gallant, 1987)

Ecoregion name (number)	AR	LA	NM	OK	TX
Southern Rockies (21)			X		
Arizona/New Mexico Plateau (22)			X		
Arizona/New Mexico Mountains (23)			X		
Southern Deserts (24)			X		X
Western High Plains (25)			X		X
Southwestern Tablelands (26)				X	X
Central Great Plains (27)				X	X
Flint Hills (28)				X	
Central Oklahoma/Texas Plains (29)				X	X
Central Texas Plateau (30)					X
Southern Texas Plains (31)					X
Texas Blackland Prairies (32)					X
East Central Texas Plains (33)					X
Western Gulf Coastal Plains (34)		X			X
South Central Plains (35)	X	X			X
Ouachita Mountains (36)	X			X	
Arkansas Valley (37)	X			X	
Boston Mountains (38)	X			X	
Ozark Highlands (39)	X			X	
Central Irregular Plains (40)				X	
Southeastern Plains (65)		X			
Mississippi Alluvial Plains (73)	X	X			
Mississippi Valley Loess Plains (74)		X			
Southern Coastal Plains (75)		X			

cal assessments conducted previously for specific problem areas. Many problem areas did not have ecological assessments. Where ecological assessments had been conducted, different methodologies had been used. The work group developed an Ecological Risk Index (ERI) model using ecological concepts and a GIS framework to evaluate relative risks from diverse environmental problems.

GEOGRAPHIC FRAMEWORK

The 22 environmental problems impact ecological systems in a variety of ways (see Table 33-1, Type of Stressor). Some, such as discharges to streams, the application of pesticides, or emissions to the air, may inhibit or simulate biological responses in the environment. Others, like filling of wetlands or urbanization, alter habitat. The work group decided that the common denominator for comparing risks was area of impact (AI).

The work group selected ecoregions as the geographic framework for this project. Omernik and Gallant (1987) have divided EPA Region 6 into 24 distinct ecoregions (Table 33-2). Some of the reasons for using ecoregions were that they: provided scientifically described geographical and ecological units; served as a template for program data collection; allowed for application of GIS; and provided information about the general health of large ecological units recognizable in the landscape (Office of Planning and Analysis, 1990). A GIS data layer of the ecoregions was provided by the EPA Office of Research and Development Laboratory in Las Vegas, Nevada.

ECOLOGICAL FUNCTIONS

The work group made the assumption that ecological risk exists when environmental threat impairs the ability of an ecoregion to perform basic ecological functions. Seven basic ecological functions were identified from the literature (Rodale, 1972; Southwick, 1976) (Table 33-3). For five of the functions, spatial data sets or maps were selected as indicators of ecosystem vulnerability. GIS was used to calculate densities by ecoregion for three of the indicators.

ECOLOGICAL RISK INDEX

The work group developed an Ecological Risk Index (see the following). To account for differences in how each problem area affected ecological functions, a degree of impact (DI) value was assigned. The value ranged from 1 to 5. A high DI value (i.e., 5) indicated an impact that was judged by the work group to take 50 years or more before the ecosystem might recover. Examples are urbanization and Superfund sites. Lower DI values indicated lesser impacts. Work group members developed a rationale for the DI and the area of impact (AI) for each environmental problem area (Office of Planning and Analysis, 1990). Where possible these rationales were based on risk assessment procedures used by EPA program offices. Each program office compiled the area of impact for each environmental problem area by ecoregion.

The work group developed a vulnerability index (DV) for six of the ecological functions in Table 33-3. The DV also ranged from 1 to 5, with high values indicating a high degree of vulnerability to environmental threats. Arid areas of New Mexico with many endangered species and low stream densities were the most vulnerable for most of the ecological functions. The Mississippi delta area was most vulnerable for soil production and maintenance.

The Ecological Risk Index was then calculated as:

$$ERI = AI \times DI \times DV / AE \quad (33\text{-}1)$$

where:

ERI = Ecological Risk Index
DI = degree of impact
DV = degree of vulnerability
AI = area of impact
AE = area of ecoregion

Table 33-3: List of ecological function and vulnerability data used for Region 6 Comparative Risk Project

1. Distribution of water, minerals and nutrients via the hydrologic cycle.
   Vulnerability index: stream density.*
2. Oxygen production and carbon dioxide consumption.
   Vulnerability index: primary productivity.
3. Filtering and detoxifying of pollutants.
   Vulnerability index: none, assumed equal for all ecoregions.
4. Soil production and maintenance.
   Vulnerability index: erosion potential.
5. Production of aquatic organisms.
   Vulnerability index: density of aquatic endangered species.*
6. Production of terrestrial organisms.
   Vulnerability index: density of terrestrial endangered species.*
7. Conversion of energy (sunlight) into organic matter.
   Vulnerability index: function not used in final analysis.

(* evaluated with GIS)

The index values and the areas were loaded into a relational database. The ERI was then evaluated by environmental problem area and by ecoregion. If ecoregion-level data were not available for a problem area, the overall ERI was estimated using regionwide data. The problem areas were grouped into four categories based on ERI values (Table 33-4). The GIS allowed the work group

Table 33-4: Ecological risk rankings of environmental problem areas for Region 6 Comparative Risk Project

Category 1: (highest risk)

Degradation of terrestrial ecosystems and habitats
Application of pesticides
Physical degradation of water and wetland habitats
*Global warming
*Stratospheric ozone depletion

Category 2:

Nonpoint discharges to oceans, lakes, and rivers
Hazardous/toxic air pollutants

Category 3:

Ozone and carbon monoxide
Municipal wastewater discharges to oceans, lakes, and rivers
RCRA hazardous waste sites
Industrial wastewater discharges to oceans, lakes, and rivers
*Municipal solid waste sites
*Industrial solid waste sites
*Aggregated groundwater contamination
Accidental chemical releases to the environment

Category 4: (lowest risk)

Hazardous waste sites—abandoned/superfund sites
Particulate matter
Airborne lead
*Storage tanks

Problem areas for which no evidence was provided to indicate ecological harm:

Sulfur oxides and nitrogen oxides (including acid deposition)
Radiation other than radon

Problem areas for which the work group concluded there was negligible or no ecological risk (primarily human health concerns):

Aggregated public and private drinking water supplies
Indoor air pollutants other than radon
Indoor radon

(* means estimated risk)

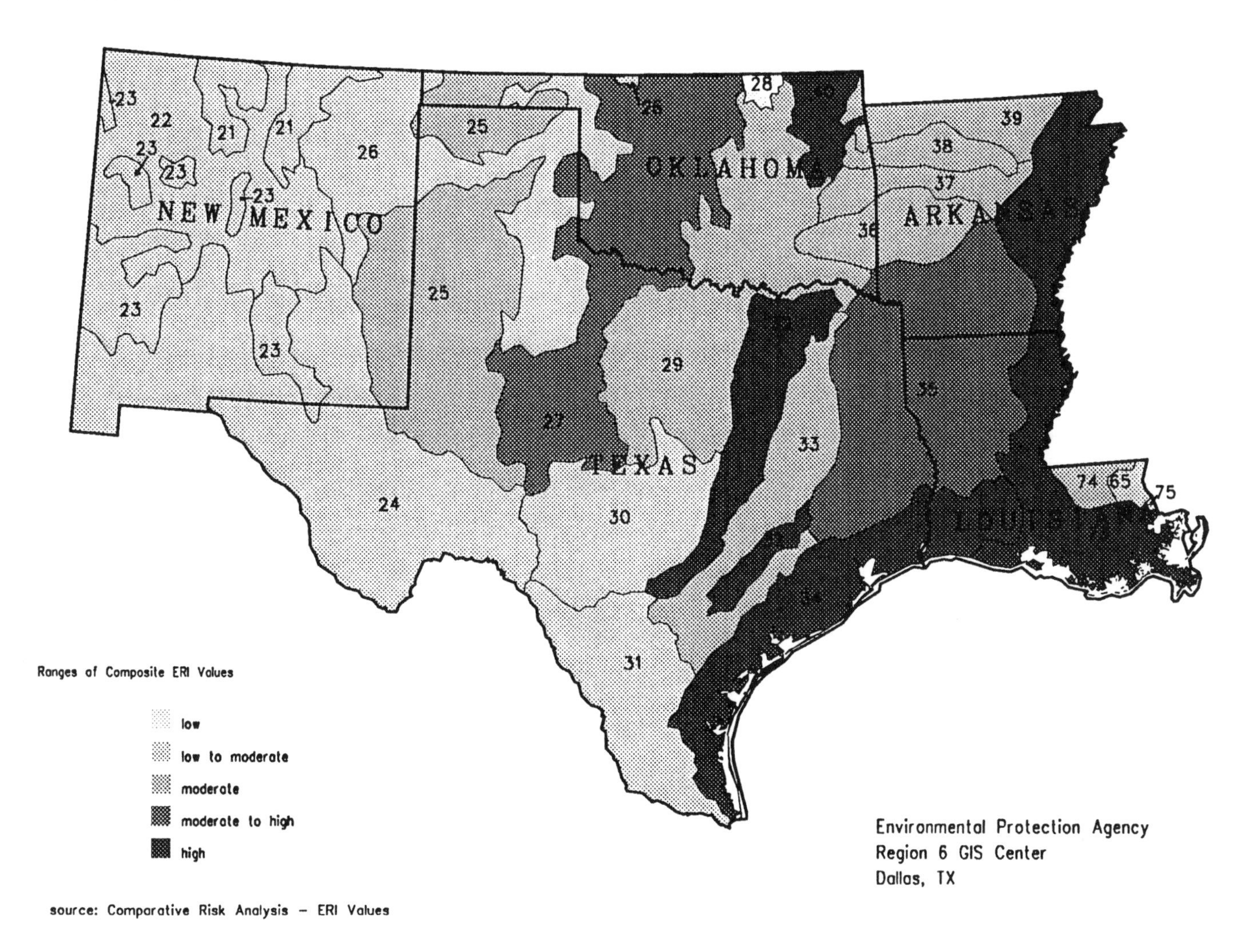

Figure 33-1: Ecoregional ERI rankings: distribution of ecoregion's ecological risk.

to display the results spatially and to identify ecoregions at highest risk (Figure 33-1). Although not evaluated with the ERI, two global-scale concerns, global warming and stratospheric ozone depletion, were included to highlight their potential risk to ecological systems in Region 6.

The highest-risk environmental problems identified by the ERI model were found throughout most ecoregions or impacted most ecoregions. These areas and problems became the focus of the Region's strategic planning process. For this project ecoregions were used to organize for data collection and evaluation. However, the ERI model was designed with the flexiblity to handle other geographic frameworks (e.g., watersheds or counties) and more objective measures of impact and ecosystem function. The Region has promoted the use of the ERI with GIS layers at larger scales by Texas for a statewide risk study.

GIS ISSUES AND RECOMMENDATIONS

Overall, this project was limited by the availability and the quality of existing databases. The application of GIS was limited by ready access to a GIS during the early phases of the study. Vulnerability indices were based on data sets or maps that had to be accumulated from the literature. EPA data had to be compiled manually due to the missing or unreliability of geographic coordinates in EPA data sets. The project treated ecoregions as homogeneous areas, since consistent data on sensitive habitats were not located. Scale was not an issue in this study. However, as the Region uses this methodology to focus on particular geographic areas, the availability of high-quality, large-scale data sets will become more of an issue. EPA has established a "Locational Data Policy" to compel programs to collect and store geographic coordinates of known quality in all information systems. The policy requires information on the accuracy and the collection methods of the coordinates. An accuracy goal of 25 m has been set for all data points.

Additionally, at the Federal level through the Federal Geographic Data Committee and at the Regional level through state program grants, EPA is promoting the development of medium- to large-scale data sets for key environmental information, such as endangered species habitat, soils, land use/land cover, and hydrography. The availability of these data should enhance future GIS-based ecological risk projects.

ACKNOWLEDGMENTS

The authors recognize the dedicated efforts of the Region 6 Comparative Risk Project Ecological Assessment Workgroup, who believed in the potential of GIS. Also, we acknowledge the computer support of Kelvin Moseman and Celia Morrison of Computer Sciences Corporation.

REFERENCES

Office of Planning and Analysis (1990) *Region 6 Comparative Risk Project, Appendix A: Ecological Report*, Dallas, TX: U.S. Environmental Protection Agency, Region 6.

Omernik, J.M., and Gallant, A.L. (1987) *Ecoregions of the South Central States*, Washington, DC: U.S. Environmental Protection Agency, USPO 795–479.

Rodale, R. (1972) *Ecology and Luxury Living May Not Mix*, Emmans, PA: Rodale Press.

Southwick, C. (1976) *Ecology and the Quality of Our Environment*. 2nd Edition, New York: Van Nostrand Company.

34

Models, Facts, and the Policy Process: The Political Ecology of Estimated Truth

JOHN LESLIE KING AND
KENNETH L. KRAEMER

"Before I draw nearer
to that stone to which you point,"
said Scrooge, "answer me one question.
Are these the shadows of things that Will be,
or are they the shadows of things that May be, only?"
—C. Dickens, *A Christmas Carol*

It is a truism that better information leads to better decisions. And in the ongoing effort to improve decision-making in public and private organizations, the effort to improve information available for decisions is paramount. We live in an era obsessed by the notion that knowledge is power. And technologies for managing information bring knowledge to the fore like never before possible. But do these technologies allow us to say, via transitivity, that information technology is power? The answer depends on how the technologies are put to use.

Recent advances in computer-assisted information management, and particularly those in the modeling of reality, have heightened our awareness of the need to refine not only our technologies but our techniques for using them. It is not enough to know how to build a powerful database, a sophisticated model, a glitzy display, unless the sole objective is to win the local science fair. If one wants to use these technologies to make changes in the world, it is necessary to know how they fit into the world. And more important, it is necessary to know that the world is a lot bigger than these technologies. How does the feather move the boulder?

This chapter provides a particular view on the role of modeling, facts, and knowledge generally, in the making of policy. The focus here is mainly on public policy, since any clear understanding of public policy formation will contain as a subset most of what is needed to understand policy-making in the private sector. This chapter is motivated by a tradition of research on model use generally (Gass, 1983), coupled with our research over 18 years into the uses of computers and telecommunications technologies in complex organizations. In particular, it draws on extensive studies of the role of computer-based models in local and national policy-making [1]. The models investigated are not, strictly speaking, "GIS based." Rather, they are mainly economic and program policy models. Nevertheless, these models offer a useful perspective on the role of modeling in policy-making that should be of interest to those attempting to apply GIS-based models to actual policy problems. In particular, this chapter explores some of the common myths of the role of knowledge in policy-making, and focuses specifically on the political character of information technologies, their use, and their outcomes. Effective use of technology in policy-making depends on making the technology purely political. The chapter concludes with some suggestions for making this happen.

THE "DATAWARS" PERSPECTIVE ON MODELING IN THE POLICY MAKING PROCESS

In 1977 we were asked by a group of researchers at a (then) West German national laboratory why models for policy-making support were routinely used in U.S. Federal government agencies. They asked the question because they had been developing such models for use in the German federal ministries, and had met with dismal failure. Even their most technically successful model, BAFPLAN, a microanalytic simulation model for predicting participation rates in the national college student assistance program, had met with only lukewarm acceptance. Gazing across the Atlantic, they noticed that models of this kind were in routine use in the United States and, in fact, appeared to be having an effect in the policy-making process. Why, they wondered, were the Americans more successful at such modeling?

Our German colleagues reasoned that the models used by the Americans were not likely to be better than those they had developed. Most of the models were similar in concept and construction, and the Germans knew their technology for modeling was as good as anyone's. The likely answer, it seemed, was in the *implementation* of the models. The Americans had apparently figured out a much more successful procedure for implementing models in the policy process. And indeed, Greenberger and colleagues (1976) had recently reported the results from a study that suggested that model implementation was often the most troubling phase of the policy modeling effort. Preliminary results of Brewer and Shubik's (1979) study of simulation in military planning suggested the same thing. And the management science literature generally was awash with discussion of the problems of model implementation (Schultz and Sleven, 1975; Fromm et al., 1975; GAO, 1976; House and McLeod, 1977; Radnor et al., 1970). If implementation was the issue, and the Americans had improved implementation procedures, the sensible thing for the Germans was to ask the Americans how their model implementation procedures worked and why they worked so well.

We commenced a study called IMPMOD, or Implementation of Models, which lasted from 1978 to 1981. We studied two broad classes of models: macroeconometric models used most often in economic policy-making by the Treasury Department, the Congressional Budget Office, the Office of Management and Budget, and the Federal Reserve; and microanalytic simulation models used mainly in assessment of complex policies for taxation and social welfare programs [2]. We also looked in some detail at the BAFPLAN experience, though this was not a formal part of the study. The study consisted of a series of detailed case studies of model development, implementation, and use, plus reanalysis of the Fromm, Hamilton, and Hamilton (1975) survey of Federal model use. The results of the study came mainly from the case studies.

We discovered, much to our surprise, that American success in model use came about largely *in spite* of the characteristics of the models or the plans for their implementation. In fact, by studying only the models and their implementation, we could discover very little of interest from the standpoint of model use. Rather, it was in studying the way the models were actually used in the policy process that we discovered why they were used so widely. Put simply, the models were used because they were effective weapons in ideological, partisan, and bureaucratic warfare over fundamental issues of public policy. Those models that were most successful, as measured by the extent of their use, were those that had proved most effective in the political battles over what kinds of economic and domestic policy should be followed, whether Democrats or Republicans should get the credit, and which bureaucratic agencies would receive the power and funds to implement the policies. Successful combatants in the policy debates had to have strong models of their own, and moreover, they usually needed copies of the opposition's models as well, in order to mount their offensive and defensive campaigns. Models in Federal policy-making were successful as a result of what we called "datawars"—the explicit use of model-based information in policy warfare.

These discoveries produced a result our German colleagues had not anticipated. The German experience with modeling reflected not so much the character of their models, or the nature of their implementation efforts, but the basic structure of their mechanisms of technocratic support for policy-making. While the German parliament, Chancellorship, and executive agencies were democratic and pluralistic, the "technical" side of public policy-making in Germany was highly restricted to long-established and highly independent entities such as the Bundesbank, the national banking authority, and the Bundesstatistischampt, or federal statistics office. The former was powerful and independent because it was established so by law (similar to the U.S. Federal Reserve). The latter was powerful and independent because of a long tradition of assuming that technical expertise is "nonpolitical," and therefore not to be questioned except on technical correctness. Both agencies saw it in their interest to maintain a tight hold over all technical supports for policy-making, as part of their "skill bureaucracy"—their bureaucratic hold over power by virtue of technical skill. However, neither saw any incentive to become embroiled in genuinely political fights involving their models and data, simply because they could not guarantee the outcomes of such fights. They could, they feared, end up losing not only the support of powerful interests in the government by taking the "wrong" side, but even worse, failure to prevail could undermine confidence in the technical "correctness" of the models and technologies themselves. As a result, the Germans had great difficulty getting models into sufficiently serious policy discussion to even warrant their use, much less a central role in policy debates. In the rough-and-tumble world of true democracy, "pure" models were as useless as "pure" science and "pure" technology.

THE INHERENT POLITICS OF MODELING

Since the early 1980s we have seen the power of models as genuine political instruments in the policy process. In fact, during our field work for the IMPMOD study, we discovered that even potent changes in leadership ideology at the top did not permanently disturb the pluralistic infrastructure of modeling down below. In fact, it made it more robust. When the Reagan Administration swept into office, carrying not only a conservative President but a conservative Senate with it, the word quickly went out that the "old" ways of thinking and doing business were doomed. Many predicted that the modeling efforts on the economic side, which were based mainly on post-Keynesian theory, and thus in conflict with supply-side "Reaganomics," would be eliminated from the Executive branch. In fact, even the staunchest Reagan aides soon discovered that they could not control the discourse of models because formal economic modeling was too well established in the fabric of policy debates.

A major thorn in the Administration's side was the Congressional Budget Office (CBO), which had been established in 1974 to provide an analytical counterweight to the President's hegemony in budget analysis capability. The CBO had grown greatly in stature and expertise in only 6 years, and had acquired all the top macroeconometric models. It depended critically on them for forecasting the balance between revenues and expenditures necessary for the House and Senate committees charged with ways and means and budgets to do their constitutionally mandated jobs. CBO also depended on the inputs from the microanalytic simulation models for taxation and welfare policy to calibrate their estimates of the actual revenues that would be generated under different taxation schemes, and the demand for welfare services under different scenarios. Even the conservative members of the Senate fell silent when they confronted the prospect of making tax and budget judgments with nothing more than ideology and guesswork to back them up. The models, as problematic as they might be, offered something. And what they offered, in particular, were numbers derived from established, well-understood processes of analysis. As Jodie Allen, Deputy Assistant Secretary of Labor, had explained, in the heat of policy battles, "some numbers beat no numbers every time."

The new conservatives had no numbers, and although they gained a few quick victories through the power of their stunning electoral mandate, they soon found themselves frustrated by the modeling results the opposition was consistently placing on the table to "prove" the disastrous consequences of conservative policies. The conservatives discovered that long-run success depended on having versions of the opposition's models to determine what kinds of offensives they would face, and if possible, a solid arsenal of offensive models of their own. The major models used during previous administrations remained in place in all the Executive agencies as well as the Fed and CBO because they were essential to political strategy. Also, they were the only game in town when it came to providing systematic estimates of likely consequences from different policy actions. The conservatives responded with some modest ammunition of their own, too. In relatively short time, several "supply-side" economic models were produced, including the controversial "Claremont Model," which was soon discredited by economists of both sides as incomplete and naive at best. More substantial modeling efforts incorporating conservative viewpoints were undertaken by the model-providing companies on the macroeconometric side, such as DRI, Chase Econometrics, and Wharton Economic Forecasting Associates, while both Mathematica and The Urban Institute found demand for their MATH and TRIM models had not slackened. Models had become a permanent part of the discourse of national policy-making.

Modeling is here to stay in national policy-making. The experience with economic and social welfare models has been repeated with modeling in other arenas such as agriculture, energy, education, transportation, and of course the military. The reasons for this success of modeling are less "rational" and "scientific" than many scientists and policy analysts care to think. Rather than providing a base of information and analysis that produces the "correct" answer, thereby causing consensus under the "truth wins" decision rule, models often serve to channel discussion in ways that provide offensive or defensive advantage to particular parties. To some, this is an unfortunate outcome, "politicizing" the use of models. To others, however, it is a desirable outcome because it shows that models can be incorporated and used in the inherently political process of policy-making in a democracy. In fact, it can be argued that efforts to keep modeling out of politics, or even the less intentional development of modeling infrastructure that biases model use away from intensely political application, is the surest route to marginalization of models as tools for policy analysis. If they are not useful as weapons in political debate, they will have little or no role in such debate.

A CONSTRUCTIVE ROLE FOR MODELING IN POLICY-MAKING

The question arises, at this point, whether models have any role in policy debate other than a purely political role as weapons. Or put differently, does the use of models contribute anything besides grist for the mill of political haggling? This is an important question, because if the role of model use is limited to political warfare, models will at best be the pawns in a seemingly endless modeling "arms race." There will surely be a role for purveyors of models in such a scenario, but one wonders whether there is anything particularly noble or valuable in modeling in such a situation. In this regard, we are optimistic. Our studies of modeling and computer-based information systems for decision-making in the IMPMOD and other projects convinces us that models do play an important, substantive, and constructive role in policy formation. This role is not in the corny form of the "answer machine" that provides policy-makers with the "truth" about a given situation. Rather, the models tend to have three singular and powerful influences on the policy process as a political form of game playing.

The first role of modeling is its role as a clarifier of issues in debate. Modeling is a systematic and formal process of analysis that requires specification and documentation of assumptions. Most complex policy problems require models with large substantial investment in infrastructure—technology, skilled people, and data resources. Every modeler must admit to constraints on each of these components of infrastructure, and every sensible modeler makes compromises along specific lines of reasoning to accommodate the constraints. These compromises make obvious the fact that models are incomplete efforts to describe real phenomena. The essential question for every critic of models is, what gets in and what is left out? The answer reveals the biases of the modelers: their assumptions, ideologies, world view, and so on. And these biases are the bases of serious political discourse. In a model, the modeler's biases are written in hard code and documentation, which then invites serious scrutiny. The critics can then question why certain variables are included versus excluded, or why this variable is treated exogenously versus endogenously, or why variables are weighted as they are. The model provides a systematic argument for and against various biases, by its very nature. The model becomes the Rosetta Stone by which policy analysts with different biases can speak a common language to debate critical assumptions. In the words of Dutton and Kraemer (1976), the model becomes a key focus of negotiation about what makes sense and what does not in the essential phase of setting ground rules for the debate. In the words of King and Star (1990), the models become "boundary objects" that bring together people from different social perspectives. Models are a way of defining common ground.

The second role of modeling is its role in enforcing a discipline of analysis and discourse. The singular power of modeling is consistency: A good model will return results that differ only in appropriate amounts given changes in particular inputs (e.g., changing the coefficient of an exogenous variable). A model is a kind of mechanical thing; willy-nilly tinkering with it in an effort to produce temporary political advantage will usually produce results that are wildly out of bounds. This constrains the ability of the modelers on all sides of a debate to engage in bluffing or deception, particularly when all sides have access to each others' models. Instead, each side must adhere faithfully to the technical realities of modeling, even if they disagree wildly over the biases inherent in different models. Again, this enforces discipline and attention to the underlying issues as well as to the essential but troublesome questions of how particular results might be obtained. Everyone knows that the Spirit of Modeling produces the shadows of what Might be, only. No one knows what Will be. But there can be better or poorer shadows, depending on the quality of the model. And quality of analysis and information therefore becomes a justifiably important issue in debate.

Finally, models do provide an interesting and powerful form of "advice." It is not remotely close to the "correct answer" envisioned by Simon (1960) and others, wherein the technological wonders of management science would give decision-makers answers to the questions of what they should do. Rather, it is advice on what not to do. The best models point out just how bad the results of a truly problematic policy might be, and this proves to be exceedingly valuable in the policy-making process. An example of such powerful use of modeling was uncovered in our IMPMOD study. The Treasury Department uses a microanalytic simulation model called the Personal Income Tax Model (PITM). This model uses a snapshot of 50,000 taxpayer returns gathered at a specific point in time, and "ages" that population of returns (e.g., raising/lowering incomes, altering exemption and deduction profiles, etc.) in a manner consistent with demographic projections under different scenarios. Proposed changes in tax policy are then run against the aged population to provide a picture of probable results (e.g., changes in revenues, tax incidence, etc.). The PITM might well be the most heavily used model for policy analysis in the world. In 1980 there were nearly 1,500 runs of this model—an average of more than six every working day. The majority of runs were requested by members of the administration or Congress interested in changing some aspect of tax legislation. They had discovered that the model was good at showing which proposals were likely to produce politically unpalatable results, such as proposals for new taxes that would cost more to administer than they would raise in revenue. Politicians of every political stripe routinely ran their proposed tax bills through the PITM, simply to avoid making fools of themselves.

A qualification is in order on the matter of modeling's contribution to the quality of decision outcomes. Major tax legislation was indeed passed during the Reagan Administration, and the PITM correctly forecasted decreases in revenue that would result, with serious consequences for the Federal budget deficit. But at that time the groundswell of ideological support for tax law changes was so strong that the PITM's forecasts were eclipsed by the larger political reality. Leaders from both parties were trying to make tax reform "their" issue, so the question was not whether there would be radical reforms, but what the reforms would look like. The expected magnitude of reform was so far from the baseline on which the PITM was built that it was arguable that the PITM's forecasts would not be very good anyway. The results of radical changes are unlikely to be predicted accurately by models based on the performance of variables under the status quo. But radical reforms are uncommon, so this pitfall of modeling is seldom encountered.

We concluded from our work that models do not prevent (or cause) any particular policy changes. They simply provide policy-makers with an idea of whether the proposed policies are likely to produce results in an "acceptable range." In this way, modeling helps diminish the number of policy proposals that look good superficially, but that can be shown by systematic analysis to have very serious downside consequences. This contribution, coupled with the "boundary object" role of models, and the enforcing of discipline that model use produces, has salutary effects on policy-making. One way of thinking about these effects is to assume that, where the facts really matter in a policy debate, modeling can provide a useful adjunct. The contribution is not from giving the right answer, but in helping sort out the nature of the debate, focusing disciplined attention on the issues, and setting boundaries likely to contain "sensible" outcomes. Models are not much use in times of ideological upheaval, simply because the decisions are based on beliefs rather than facts. Ideological policy-makers appeal to their own versions of facts, and dismiss the facts of others as falsehoods. In this way, the fundamental assumptions of policy modeling are upended. But no radical reforms can persist for long. Eventually, a new status quo emerges, and as it does so, sufficient policy stability to support modeling reemerges. Again, the utility of modeling as an adjunct for incremental policy making comes into the foreground.

A PERSPECTIVE ON "SCIENTIFIC MODELING"

At this point, the experienced reader might ask whether the kinds of models discussed above are not "social science" models, and thus quite different from the "scientific" models of physical processes of the environment toward which GIS-based information might be directed. The question requires a qualified response.

To start, we must ask if the two kinds of models are truly different, and if so, what consequences might such differences have in policy terms. It can be argued that the modeling of certain physical phenomena can be done with greater efficacy and precision than can modeling of some social phenomena. Finite element analytical simulations, for example, predict almost exactly how a structure will behave under particular stresses. In contrast, macroeconometric simulations of the national economy seldom predict anything very exactly. Thus, we could conclude that "scientific" models represent the "truth" more vividly, as demonstrated by their predictive power, and that they will be readily accepted into the policy process.

Curiously, however, the policy relevance of scientific models that are "always right" is already moot. To the extent that anything can be predicted with precision, as can the structural behavior of a particular bridge design, the effect of such precision is to move the models out of policy altogether and into the realm of engineering expertise. Few legislative bodies will tangle with the risks associated with bridge design, because failures are so noticeable and disastrous. Engineers stand ready to build bridges for almost any occasion, and they use models to do so. But their models lose their utility when one steps back into the question of whether a bridge should be built, or where it should be built, or how large it should be, or how it should be paid for. These questions force the modeler back into the "social" realm where the word "science" is something of a euphemism.

Engineering models have narrowly circumscribed utility, but policy models do not. In fact, policy models suffer from the peculiar dilemma that they are most useful when the truth is not altogether clear. Policy models are useful only as projections into the future, and the future is up for grabs. There is no policy utility whatsoever for a model that correctly predicts retrospectively how things actually came out. All politicians can do that. And there is little comfort to a politician in knowing that a given model has retrospectively predicted what is already known unless there can be a *guarantee* that predictions of the future will be accurate. Since no honest scientist is willing to make such a prediction, the "distance" between social science models and scientific models becomes one of degree and not of kind.

We believe the experiences with social science models are excellent bellwethers of modeling in the policy process, regardless of how "scientific" the models might be. The reason for this is simply the nature of the policy process. Models are useful when the fundamental questions of what might happen are unclear. Once everyone agrees on the facts, there is no further need for models to clarify the facts. Consider the ongoing efforts to model the economy. Everyone argues about what will happen, but no one argues for long with what is happening. The 1992 Presidential campaign offered a wonderful example, in which the incumbent president, George Bush, was forced out of a position of denial about the recession and into an active campaign to convince voters that his economic programs would do the most to overcome the recession. The models were turned from trying to show that the country was or was not in a recession, and toward the question of what policies would best get the country out of the recession.

This could be cited as an example of the fuzziness of social science models, but exactly the same thing can be said about "scientific" modeling of the environment. A good case in point is provided by the current rush of concern about depletion of the ozone layer. Models predicting serious depletion of the ozone layer due to CFCs and other chemicals were laughed at by a substantial majority of the atmospheric science community as well as policy-makers for many years. No amount of modeling changed the prevailing view. The response was always the same: More data are required before policy conclusions can be drawn. But resistance to the ozone depletion argument began to weaken once the ozone "hole" was found over Antarctica in the mid-1980s. And in 1992, when a significant "hole" was found over North America, politicians were suddenly of one voice to curtail the production and use of ozone-depleting chemicals. The issue of "whether" appeared to be solved, while the question of what to do about it moved to center stage, along with numerous, conflicting model-based predictions of what should be done.

Prediction is not proof. Even the seemingly airtight "laws" of thermodynamics are of little weight when the political bias is against them, a poignant lesson learned not long ago by the people of the United States, the administrators of NASA, and the friends and families of the seven astronauts aboard the *Challenger*'s final flight.

MODELING IN THE AGE OF GEOGRAPHIC INFORMATION SYSTEMS

We believe the experiences we witnessed in the use of macroeconometric and microanalytic simulation models in national policy-making will be recapitulated in the application of GIS to the policy process. This prediction is based on the abiding character of the policy process, which will not be changed by any kind of technological breakthrough. What differences, if any, might we see from GIS application that we did not see in our earlier studies? Two come to mind.

First, the peculiar character of geographic databases is an important element in the GIS's role in policy-making. As proponents of geographic database construction have maintained for many years, everything in government (and public policy) is one way or another tied to the Earth. And despite the dynamic character of the Earth, such as changing shorelines or riverbanks, and differences of opinion on what names to apply to particular areas, such as "wetlands," it is possible to create a fairly reliable record of land characteristics that everyone in a policy debate will agree on. This forms a powerful boundary object; one much more powerful than those created by the macroeconometric or microanalytic simulation models. For one thing, the basic elements of a geographic database are tangible and observable, while notions like the "money supply" or "propensity to participate in program" are not. Moreover, geographic facts tend to remain facts over time, which makes them by themselves boundary objects. Geographic databases will naturally draw together policy-makers of different perspectives simply because they agree on these basic facts, but they disagree on what to do with what is represented by those facts. This is a blessing and a curse for GIS use in policy making, at once drawing policy-makers to GIS as a support for their work, but on the other hand, making the GIS the battlefield of intense debate. Rising to the occasion offered by that battlefield role is a major challenge of GIS professionals.

Second, the very breadth of GIS application to policy problems makes GIS likely to be pulled into many different kinds of policy debates. Unlike the worlds of modeling for economic planning or welfare policy analysis, where idiosyncratic models served each policy need, GIS are likely to support a plethora of models for many policy needs. This too will be a challenge for GIS professionals, because different policy debates take on widely different characteristics. Debates about land use or environmental pollution abatement typically pit economic interests of industrialists and land developers against environmental groups interested in maintenance or restoration of areas and in limiting growth. Debates about policies to fight crime are of a different character altogether, focused on determining appropriate levels of law enforcement, deployment of law enforcement resources, and locations of jails and other correctional facilities. Debates about traffic and transportation management often pit advocates for various transportation "solutions" (slow growth, roads, fixed-rail rapid transit) against one another, and increasingly, they pit governments short on resources for essential transportation infrastructure against developers who want to develop land but not pay for the full costs of infrastructure required to support such development. GIS-based models can and will be used in each of these kinds of debates. Whether they are used constructively in the policy debate depends on the facility with which GIS professionals can adapt the systems to serve the arguments of different parties in the debates. If the models serve only particular interests, to the exclusion of others, the models will be challenged with the intent of destroying credibility in particular models or in modeling generally.

These challenges will not be met easily. They will definitely not be met through application of the naive notion that the "truth" will win in policy debates, and the corollary belief that any particular models contain the "truth." This notion is naive, but, more dangerous, it is insulting to policy-makers, including those who win as a result of using particular models in their arguments. Policy-makers do not like to think of themselves as dependent on their staffs, and they particularly dislike being dependent on techniques they cannot really control. A model that gives "true" answers might give the right true answers in one case, and the wrong true answers in another. And in any case, if everyone does agree on the truth of a matter, it ceases to be a policy issue and is remanded to administration for disposal.

GIS modelers, and modelers in general, have several tasks before them in the quest to achieve widespread and constructive use of models in the policy process. This brief list illustrates the most important:

1. Modelers should avoid believing or giving the impression that their models hold the "answers" for policy-makers. They hold, instead, the refined results of particular points of view. The difference is critical, and modelers must be sure policy-makers understand the difference.
2. Modelers should recognize the biases inherent in their own models, and that such biases are inevitable in any

serious model. Moreover, these biases are the bases of essential public policy debate, and model use can focus required attention on these biases while not depriving modeling per se of any of its power. The constructive consequences of such recognition are to encourage development of multiple models, each incorporating different perspectives, and thereby allowing for comparison of results from the different models/perspectives as part of debate.

3. Modelers should encourage those they work for to support broadening access to modeling by all parties in the debate. Efforts to keep different parties out only damage the credibility of modeling itself, suggesting that the models will not stand up to scrutiny by opponents. In such debates, models soon become little more than "fire once and forget" tools, good for one or two wins and soon abandoned.
4. Modelers should work to build the common infrastructure of modeling across all parties in debates, and in particular, the common data sets required for understanding the issues and the modeling techniques that improve simulation and forecasting accuracy.
5. Modelers should join those public policy debates that align with their own political beliefs whenever possible, bringing the skills and tools of modeling to their side of the debates.
6. Modelers should ensure that legitimate policy positions in debates that do not have modeling support obtain necessary support, either as an entitlement of participation in the policy process, or via third-party arrangements involving philanthropic or other organizations.

CONCLUSION

Success of GIS-based models in policy-making will depend on the skill and facility with which GIS professionals develop and adapt their models to serve the analytical needs of the largest number of combatants in policy warfare. At one level, this sounds a bit disingenuous—like the arms merchant willing to sell to both sides in a conflict. But two factors make it more noble than that.

One is the fact that these models are not destructive; at worst, they are ineffectual, and at best, they help to strengthen the quality of the policy making process. To the extent that the GIS professional is "arming" the combatants, the arms are used in the fight against ignorance, confusion, and obfuscation. And the arms of modeling are successful only when all the combatants have them.

The other ray of hope is the fact that modelers are, themselves, fully capable as individuals of putting their talents to work for causes in which they believe. Thus, modelers who favor policy A can work for the proponents of policy A, and likewise for policy B, C, D, and so on.

Indeed, it makes little sense to assume that there can be, or even should be, one modeling group supporting all sides in a conflict. This is not realistic politically, and in any case, a modeling entity thus constituted will forever be on shaky ground. The ability of modelers to criticize and compromise with respect to each others' work in policy debates depends on camaraderie in the *profession* of modeling. It does not assume common ground on the issues themselves. This is a great strength of the "datawars" view of modeling in the policy process.

NOTES

[1]This work is the legacy of the research conducted at the Public Policy Research Organization of the University of California, Irvine, since the early 1970s. For readily accessible accounts of this research, see: Kraemer and King (1976); Danziger, Dutton, Kling, and Kraemer (1982); Dutton and Kraemer (1986); King and Kraemer (1985); Kraemer, Dickhoven, Tierney, and King (1987); and Kraemer, King, Dunkle, and Lane (1989). This research was supported by the National Science Foundation and by the Gesellschaft für Mathematik und Datenverarbeitung (GMD) of the Federal Republic of Germany. Comments and queries may be directed to king@ics.uci.edu.

[2]For the macroeconometric model study we concentrated on the Data Resources Incorporated national model and its associated time-series databases. This model had been adopted for use by every Federal government entity engaged in economic analysis. For the microanalytic simulation study we concentrated on two versions of the Real Income Maintenance model, which was first built as part of the 1972 President's Commission on Income Maintenance, and was subsequently elaborated as the Transfer Income Model (TRIM) and the Micro Analysis of Transfer to Households (MATH) models supported by The Urban Institute and Mathematica Policy Research, respectively. Details can be found in King (1984a,b), Kraemer and King (1986), and Kraemer, Dickhoven, Tierney, and King (1987).

REFERENCES

Brewer, G.D., and Shubik, M. (1979) *The War Game*, Cambridge, MA: Harvard University Press.

Danziger, J.N., Dutton, W.H., Kling, R., and Kraemer, K.L. (1982) *Computers and Politics*, New York: Columbia University Press.

Dutton, W.H., and Kraemer, K.L. (1986) *Modeling as Negotiating*, Norwood, NJ: Ablex.

Fromm, G., Hamilton, W.L., and Hamilton, D.E. (1975)

Federally Supported Mathematical Models: Survey and Analysis, Washington, DC: U.S. Government Printing Office.

GAO (1976) *Improvement Needed in Managing Automated Decision Making by Computers Throughout the Federal Government*, Washington, DC: General Accounting Office.

Gass, S.I. (1983) Decision-aiding models: validation, assessment and related issues for policy analysis. *Operations Research* 31: 603–631.

Greenberger, M., Crenson, M.A., and Crissey, B.L. (1976) *Models in the Policy Process: Public Decision Making in the Computer Era*, New York: Russell Sage Foundation.

House, P.W., and McLeod, J. (1977) *Large-Scale Models for Policy Evaluation*, New York: John Wiley and Sons.

King, J.L. (1984a) Ideology and use of large-scale decision support systems in national economic policymaking. *Systems, Objectives, Solutions*, December.

King, J.L. (1984b) Successful implementation of large-scale decision support systems: computerized models in U.S. economic policy making. *Systems, Objectives, Solutions*, May.

King, J.L., and Kraemer, K.L. (1985) *The Dynamics of Computing*, New York: Columbia University Press.

King, J.L., and Star, S.L. (1990) Conceptual foundations for the development of organizational decision support systems. *Proceedings of the Twenty Third Hawaii International Conference on Systems Science*, Los Alamitos, CA: IEEE Society Press, pp. 143–151.

Kraemer, K.L., and King, J.L. (1976) *Computers and Local Government: A Review of Research*, New York: Praeger.

Kraemer, K.L., and King, J.L. (1986) Computerized models in national policymaking. *Operations Research* 34(4): 501–512.

Kraemer, K.L., Dickhoven, S., Tierney, S. Fallows, and King, J.L. (1987) *Datawars: The Politics of Modeling in Federal Policymaking*, New York: Columbia University Press.

Kraemer, K.L., King, J.L., Dunkle, D., and Lane, J.P. (1989) *Managing Information Systems: Change and Control in Organizational Computing*, New York: Columbia University Press.

Radnor, M., Rubenstein, A.H., and Tansik, D. (1970) Implementation in operative research and R&D in government and business organizations. *Operations Research* 18: 967–981.

Schultz, R.L., and Sleven, D.P. (1975) *Implementing Operations Research/Management Science*, New York: American Elsevier.

Simon, H.A. (1960) *The New Science of Management Decision*, New York: Harper.

V

Spatial Data

LOUIS T. STEYAERT

Many diverse types of spatial data are needed to monitor and understand dynamic terrestrial processes, as well as to develop the environmental simulation models that are needed for scientific assessment of environmental problems, including the effects of human interactions. Multidisciplinary data sets of land surface/subsurface characteristics are essential inputs to such models. The models require data on the multitemporal behavior of land surface properties, as well as the parameterization of spatially heterogeneous and complex landscape characteristics. Such spatial data sets are needed to support environmental simulation models that are cross-disciplinary and increasingly use an integrated/coupled systems approach to model spatially distributed processes across multiple time and space scales.

This rich set of spatial data requirements forges fundamental links between GIS, remote sensing, and environmental simulation models. Remote sensing technology is essential to the development of many spatial data sets for the study of environmental processes. GIS complements remote sensing by providing the framework for integrated spatial analysis of diverse data structures in order to help understand and parameterize land surface processes. GIS also has a role in developing and tailoring integrated spatial data sets, including remote-sensing-derived thematic layers, for input to models.

This section focuses on spatial data issues and resources, as aptly introduced in the section overview (Kemp, Chapter 35). The next four chapters provide examples of key spatial data sets available for environmental simulation modeling. Kineman (Chapter 36) uses an example of global ecosystems database development to help illustrate several important concepts for scientific database development and application. Loveland and Ohlen (Chapter 37) illustrate the role of remote sensing technology and describe experimental land data sets available for environmental monitoring and modeling. The status of digital soils databases for the United States is described by Lytle in Chapter 38. Finally, the importance of terrain in landscape processes and the topic of digital elevation models (DEM) are discussed by Hutchinson in Chapter 39.

The overview chapter by Kemp introduces spatial data issues involving data collection strategies, the use of cartographic and remote sensing products as sources of digital data, digital characteristics of spatial databases, and archival sources. Data collection issues include the choice of spatial model for discretizing geographic properties, for example, discretizing the real world by objects with clearly defined boundaries, discrete sampling of continuous surface properties with point data, or defining regions such as polygons with assumed internal homogeneity. These spatial models are transformed into various data structures such as the raster or vector formats. Paper maps and remotely sensed data are major sources for collecting digital data. However, the use of maps requires caution to consider standard cartographic practices properly, the effects of map scaling, coordinate systems, projections, and data capture methods to avoid potentially serious errors. Some of the remote sensing sources include satellite data and relatively high-resolution digital orthophotographs. Specialized image processing is usually needed to apply geometric corrections, radiometric calibrations, enhancements, and other corrections. Additional spatial database issues include file formats, tiling and edge-matching, and topological considerations. Several sources of spatial data are provided.

Chapter 36 (Kineman) describes the conceptual basis for the Global Ecosystems Database Project (GEDP) developed by the National Oceanic and Atmospheric Administration and the U.S. Environmental Protection Agency. Issues of the design of scientific databases and the importance of the GEDP in combination with GIS for characterizing and describing ecosystem processes are also discussed. The general data themes of the GEDP on CD-ROM include vegetation cover and ocean color from satellite sources, surface climatology from surface and satellite observations, various vegetation and ecological classifications from multiple sources, soil classifications from surveys and maps, and various boundary data. Design considerations included the challenge of providing users with sufficient information on the data sets (that is, the history, experimental design, accuracy, etc.) and a set of tools that permit the user to appropriately apply the dataset to new problems. The chapter makes an excellent point that the integrated analysis of such databases is essential for descriptive studies to characterize ecosystems and understand environmental processes. This illustrates the potential role of spatial data and GIS to complement environmental simulation modeling. The chapter concludes with discussion of the need for a GIS "tool box" approach to support the database, the role of the GEDP as an intermediate-scale dataset for global studies, the importance of compatible nested gridding of the data, and the essential requirement for peer review of database products.

In Chapter 37, Loveland and Ohlen describe the use of satellite remote sensing technology by the EROS Data Center (EDC) of the U.S. Geological Survey (USGS) to develop experimental land data sets. The focus is on the processing of daily 1 km Advanced Very High Resolution Radiometer (AVHRR) data available from the NOAA polar orbiting satellites. EDC produces four levels of prototype AVHRR data products that are designed for both biophysical and land cover data requirements of global change researchers. These data products include processed daily AVHRR scenes, composite images, time series sets, and land cover characteristics. Preprocessing options for daily AVHRR scenes over North America and other areas of the Earth include calibrations, corrections for illumination and atmospheric effects, and registration. Regional and continental data sets are composited to overcome data contamination problems due to clouds, haze, dust, or other attenuating factors. Time series image data sets are formed by geographic areas: conterminous United States, Alaska, Mexico, and Eurasia. To illustrate, the 14-day image composites for the conterminous United States are available on CD-ROM (1990–1991) and include ten channels of information: channels 1–5 (calibrated), the normalized difference vegetation index (NDVI), satellite zenith, solar zenith, relative solar/satellite zenith, and date of pixel observation. This set of time series images provides information on the dynamics of land surface processes. Loveland and Ohlen describe the use of these data to develop a land cover characteristics database for the conterminous United States. They discuss the land cover classification strategy to obtain seasonally distinct, homogeneous land cover regions. The land characteristics database concept is presented including database contents: land surface regions, land cover/vegetation descriptions, seasonal growing statistics, terrain summary statistics, summary climate parameters, and ecological components. The purpose of these various AVHRR products is to support diverse land data requirements for land and water management, climate modeling, and environmental monitoring.

The sources and status of digital soils databases available from the U.S. Department of Agriculture Soil Conservation Service (USDA/SCS) are described in Chapter 38 (Lytle). Three soil geographic databases which represent soil data at different spatial scales have been devised. These include the Soil Survey Geographic Data Base (SSURGO), the State Soil Geographic Data Base (STATSGO), and the National Soil Geographic Data Base (NATSGO). Each database includes digitized soil map unit delineations linked with attribute data for each map unit, providing information on the extent and properties of each soil. The use of GIS to manage and apply these databases by users is a major component of the SCS program. The SSURGO data are designed for relatively large-scale natural resource planning and management issues; scales for base maps range from 1:12,000 to 1:31,680. Although soil surveys in the U.S. are 90 percent completed, approximately 12 percent of the analog data are now in digital format. The STATSGO data are intended for multicounty, state, and regional applications. These data are based on generalizations of the more detailed SSURGO data, and generally use USGS 1:250,000 scale topographic maps as the base map. The STATSGO data will be available for the entire U.S., except Alaska, by late 1992. NATSGO is available for the conterminous U.S. and is intended for regional and national applications. The collective use of Major Land Resource Areas (MLRAs), state soils maps plus other information (topography, climate, water, and potential natural vegetation), and sampling survey results from the National Resource Inventory in the development of NATSGO are described. The interpretation of these databases and future activities are also discussed including the status of small-scale global digital soils data.

The development of a continent-wide digital elevation model (DEM) for Australia is described in Chapter 39 (Hutchinson). Such digital information on terrain is essential in environmental modeling because topography has a large role in determining hydrologic, climatic, and other characteristics of the landscape. The general procedure for calculating a regular grid DEM is based on analysis of surface specific point elevation data, contour line data, and streamline data. The basic algorithm involving an iterative finite difference interpolation approach is described. The method features computational efficiency, allowing it to be applied to large data sets, and it incorporates a drainage basin enforcement condition that removes spurious sinks in the DEM. An alternative technique for ensuring drainage enforcement and the removal of sinks based on the use of streamline data is discussed. This DEM interpolation procedure overcomes limitations in the automatic calculation of streams and ridges solely from contour data. The algorithm and data sources are used to develop a continent-wide DEM for for Australia with a resolution of 1/40th of a degree (approximately 2.5 km).

35

Spatial Databases: Sources and Issues

KAREN K. KEMP

Digital databases of scientific data have been widely available for over three decades. Established database management techniques, developed originally for business and military applications on mainframe computers, have provided many useful tools for the design and use of large numeric databases. Techniques for accessing and manipulating traditional tabular format data are widely taught in various university engineering, business, and science departments. However, relatively recently, databases have begun to proliferate in which the spatial characteristics of the stored entities are at least as important as their nonspatial characteristics. The genesis and increasing numbers of these databases have been spurred on by both the recent development of a wide range of commercially successful GIS and the trend towards research on global issues. The improved availability of digital spatial data will prove to be a bonanza for environmental modelers.

Spatial data perform many different functions in environmental models. Relative locations of entities are used directly as indicators of distance or separation. Location combined with single attributes such as elevation are used to devise basic parameters related to energy and flow. Interpretation of model results can often be aided by displaying the output on a perspective view of the land surface. Such visualization requires that each value in the model output is referenced to a location and its elevation. Perhaps the most common role of spatial data is as input to the process of land characterization, in which the spatial heterogeneity of a unit of the landscape is compressed into a single value. Such values may then be input to nonspatial, often probabilistic, mathematical models. Inadvertent false assumptions about the quality of spatial data can significantly impact the reliability of model output. An understanding of the special characteristics of spatial data and the problems related to the use of spatial databases is critical.

Geographic attributes are very different from other types of measurements. They are at least two dimensional, and sometimes three (e.g., geologic data) or four (time) dimensional. Also important are the relationships between objects in space. Which objects are closest together; which route would I follow to get from here to there; what objects can be found within 1/2 km of this place? Thus, while spatial databases have much in common with traditional tabular databases, they have additional traits that require that spatial data be stored and manipulated in unique, new ways. This chapter articulates some of the important issues related to the use of spatial data and spatial databases for environmental modeling and provides an introduction to topics examined in more detail in the following chapters. The first section of the chapter picks up and develops themes introduced earlier by Goodchild (Chapter 2), Nyerges (Chapter 8), and Goodchild (Chapter 9).

CHARACTERISTICS OF SPATIAL DATA COLLECTION

Discretizing space: spatial models

Environmental modelers seek to develop mathematical simulations of some portion of the natural environment. However, as Harvey points out, "In reality any system is infinitely complex and we can only analyze some system after we have abstracted from the real system" (Harvey, 1969, p. 448). Abstraction requires a logical and reproducible technique for selection from and simplification of the natural system. As well, the storage and manipulation of data in traditional computer implementations of mathematical models requires discrete numerical entities. Thus it is necessary to find some way of reducing the infinite complexity and continuous variation of nature to discrete, observable, and measurable units.

In the geographic context, this discretization of space is easiest if we are dealing with distinct objects, such as individual animals, stream channels, or buildings. These objects have clearly defined boundaries and can be uniquely identified and located as discrete points (e.g., animals), lines (e.g., streams) or areas (e.g., buildings). However, the situation is considerably more difficult when the objects we wish to observe vary continuously, such as elevation, soil temperature, or vegetation density. In this case, it is necessary to devise a discrete model of the spatially distributed phenomena that can be used as a proxy for complex reality (Chapter 2). One such model also uses points. Here we sample a continuously varying field at representative locations. This is the approach often used for measuring elevation or meteorological phenomena. This model requires an assumption that values between points can be determined from some smooth function of the known values. Note the fundamental conceptual difference between points as discrete objects and points as individual measurements of continuous phenomena.

An alternative model for measuring continuous phenomena divides space into a set of contiguous regions. The characteristics we are seeking to sample are measured, and representative values are assigned to each region. Here the regions may be defined before measurement (as in demographic census) or after measurement (as in soil and vegetation mapping). Unlike the point model described, it is assumed in this model that values do not vary within regions but that they change abruptly at region boundaries. Problems related to this spatial model include the transformation of natural heterogeneity to homogeneous regions, the abrupt change in value at region boundaries, and the effect of a priori definition of regions on the measured values (Openshaw, 1983; Haining, 1990).

It is clear that neither of these models, nor any of the others currently in use, can depict reality exactly. Nevertheless, such spatial models are a necessary first step towards developing mathematical models of spatially distributed processes. It is necessary to understand, however, that numerical manipulation of spatial data may seriously degenerate the relationship between spatial model and reality. Awareness of the fundamental assumptions made in the process of measuring and recording spatial phenomena is essential. For further discussion of the conceptual and statistical aspects of this issue, the reader is referred to the earlier chapters by Goodchild (2 and 9) and some of the extensive geographic literature available (for example, Peuquet, 1988; Gatrell, 1983; Haining, 1990).

Recording spatial phenomena: data structures

Once a scheme for sampling reality has been devised, the next step requires transformation of the conceptual model of space into a form that can be represented on the computer, the data structure. In current GIS, there are two common types of data structure, raster and vector. Raster data structures use a complete rectangular tesselation of space. Since every location in a given study area will be represented by a value in the raster, this data structure is frequently used for the representation of continuous phenomena like soils, vegetation, or electromagnetic radiation. Vector data structures use the points and lines of traditional cartography to represent spatial entities. Lines may be used to represent linear features like rivers and roads or to enclose areas such as vegetation or political zones. In this structure it is not necessary for spatial entities to cover the study area completely; some places may contain no entities of interest.

Any conceptual model of space can be transformed into any variation of these data structures, though certainly some are better matches. For example, a set of regions representing soil types can be represented in a vector data model by storing the boundaries of each soil polygon, identifying each polygon with a unique ID, and, in an attribute table, associating the values of various soil characteristics to each polygon ID. Alternatively, a grid can be laid over the soils map, and the predominant soil type falling in each grid cell can be recorded as cell values in a layer of a raster database. Assumptions about the relationship between the stored data and reality must be amended as a result of this transformation. Values of continuous phenomena stored as a grid of points (vector structure) imply a different sampling method than values stored as a raster. In the first case, it may be assumed that the data represent the exact values at the point locations, while in the second case, the data are representative of all the locations within the grid cell (i.e., average, maximum, mode).

Therefore, the most important characteristics of spatial data and the databases formed from it are the spatial model that has been used to discretize space and the related data structure. These will affect directly how the data are collected and how they can be used. Related to these issues is the problem of spatial data accuracy, a theme that underlies much of the current research in GIS. Goodchild introduced many important aspects of spatial data accuracy in Chapter 9, and Goodchild and Gopal (1989) provide an excellent review of recent research directions. The following sections emphasize some of these relevant issues.

MAPS AS SPATIAL DATA SOURCES

Digital spatial data obtained from map sources carry with them characteristics related to the scale and projection of the source map, and subsequent analysis is hindered by the results of traditional cartographic techniques (Fisher, 1991). Maps as analog spatial databases are shaped by spatial models and data structures. A spatial model would have influenced the collection of the original data, while the form of the graphic representation on the map represents the data structure.

Cartographic license

As a source of precise digital data, a map is often a relatively poor choice. Traditionally, in order to compile a map, the cartographer determined which information was relevant for inclusion and then prepared a map designed for visual interpretation. There are several common "cartographic license" techniques used to help readers decipher the map graphics. Two of these with particular impact on the use of maps as digital data sources are (1) the practice of slightly displacing overlapping objects, such as parallel roads and rivers, and (2) generalization, a technique used to reduce the amount of information that is displayed (Muller, 1991). While these are important and useful techniques for paper map interpretation, they cause indiscriminate locational errors in digital spatial databases derived from maps. The significance of these errors, however, will depend upon the applications in which the data are used.

The effect of map scale

The scale of the source map affects the amount of detail that can be captured in the digital database. Consider using a map drawn at a scale of 1:100,000 to extract the location of a river. By joining with straight lines a set of carefully chosen points located along the blue line of the river, we can produce a line that looks almost exactly like the one on the map. However, if we take this representation of the river and enlarge it to a scale of 1:25,000, it will look very angular and not at all like the same river we would find on a map of the area prepared at this larger scale. Now consider overlaying this 1:100,000 scale representation of the river with a line representing the boundary of the state, which by legal definition runs along the centerline of this river, digitized and reduced from a different source map at 1:24,000 scale. Major discrepancies between the location of the river and the location of the boundary will be evident. Which of these is the truth?

Map data often give a false impression of precision. Lines on maps are usually drawn at a minimum width of 0.5 mm or less. On a map having a scale of 1:25,000 such a line actually represents a band 12.5 m wide on the ground. This imprecision is considered entirely appropriate for indistinct boundaries such as soil groups or vegetation types where the change between classes is typically gradual. However, when that line is digitized, its location is stored to a multidecimal precision. This line will instantaneously gain a false precision in the order of cm on the ground. The inherent fuzziness of the mapped line is lost. Now consider the same 0.5 mm line on a map at 1:1,000,000 scale. Here it represents a band of width 500 m on the ground. Problems arising from integrating digital data obtained from large- and small-scale maps should be apparent.

Coordinate systems and map projections

There are many different ways to identify location. The most common of these are latitude and longitude, Universal Transverse Mercator (UTM) coordinates, and local cadastral systems. However, many other systems are in current use, and the one chosen for a specific spatial database is a critical and necessary attribute. Not infrequently, a researcher will receive a set of spatial data referenced to a coordinate system in which the location 0,0 is somewhere near a corner of the study area. When it becomes necessary to integrate this data set with another set obtained from a different source, the location of this local coordinate system with relation to the other must be determined. This generally requires mathematical transformation of the data along with some guesswork based on landmarks and other local knowledge.

For any but the largest-scale site maps, the question of map projection is critical. It is a basic principle of map making that when projected onto a flat map, objects on the Earth's surface are distorted in some way, either in size, shape, or relative location (Maling, 1980). When information is digitized from a map, the recorded locations will often be based on a rectangular coordinate system determined by the position of the map on the digitizing table (Star and Estes, 1990). In order to determine the true (Earth) locations of these digitized entities, it is necessary to devise the mathematical transformation required to convert these rectangular coordinates into the positions on the curved surface of the Earth that are represented on the map. Hundreds of different projections exist, though luckily only a few dozen are generally used. Mathematical formulae to convert map units into latitude and longitude are readily available for most common projections (Snyder, 1987). Such transformation functions are normally built in to commercial GIS, though the accuracy of the transformations are rarely disclosed.

Capturing mapped data

In order to enter mapped data into a digital database, it is necessary to convert the graphic points, lines, and areas into a representation based on a specific digital data structure. Usually, the lines and points representing features on maps are digitized directly, producing a vector model of the map. Here, an operator moves a hand-held cursor over the map, clicking a button whenever the cross-hairs on the cursor are centered over locations for which coordinates must be stored. Manual digitizing is often considered to be a tedious and error-prone activity, but with practice, maps can be digitized efficiently.

It is also possible to scan a map optically and produce a colored or grey-tone pixel-based representation similar to a satellite image. The colors or shades in the image must be related to the map legend, and manual interpretation of linear features like roads is necessary. Depending upon the quality of the original maps, the scanner, and the automatic editing capabilities of the software, considerable editing may be required to make meaningful databases out of scanned maps (see Jackson and Woodsford, 1991; Star and Estes, 1990, for details about digitizing and scanning). For large projects with sizeable budgets for data input, combination scan digitizing systems now exist in which digitized lines are extracted automatically from scanned images.

Choosing the appropriate method of capturing mapped data depends on the final data structure required and the size of the project. If a project requires the input of large quantities of mapped data, it may be preferable to contract the services of one of the many GIS service companies specializing in data conversion.

SPATIAL DATA FROM REMOTE SENSORS

Sensors on satellites and other high-altitude platforms are an increasingly important source of spatial data, particularly for projects on regional and global scales. In contrast to maps as data sources, these sensors record spectral characteristics of the Earth's surface without interpretative bias. Data are both collected and distributed in the raster data structure. Satellite sensor sampling strategies are neutral and exhaustive. As well, it is often possible to get data for a specific region on a number of different dates and, for currently active sensors, to get up-to-date data about the Earth's surface. Numerous different satellite sensors provide data collected from a range of spectral bands (wavelengths) and spatial resolutions (width of the rectangular portion of the Earth covered by one pixel). Data from the Landsat TM (Thematic Mapper—30 and 120 m resolution), Landsat MSS (Multi-Spectral Scanner—80 m resolution), NOAA AVHRR (Advanced Very High Resolution Radiometer—1.1 km resolution) and SPOT HRV (High Resolution Visible Range Instruments—10 and 20 m resolution) are particularly popular for environmental applications.

While now comparatively old, *The Manual of Remote Sensing* (Colwell, 1983) provides an excellent background on remote sensing theory. More recent textbooks include those by Lillesand and Kiefer (1987) and Sabins (1987). These and other similar sources provide detailed information on the different sensors, their spectral ranges, operational dates, spatial resolutions, and application fields.

Image processing

Before they can provide meaningful measurements to users, the raw, unbiased reflectance values received by the satellite sensors require considerable mathematical processing. Manipulations are required both to register the grid of pixels to specific locations on the Earth's surface and to transform the data into useful information. Understanding the various algorithms operating on the raw data is usually beyond the ability of nonspecialists, and the resulting spatial data must be accepted on faith. Frequently the result of this manipulation is classified data in which pixel values indicate classes of an attribute (e.g., nominal data such as vegetation type or land use classes) rather than interval or ratio data. Classification algorithms and techniques are among the most highly disputed in the discipline. Problems with classification of remotely sensed data are described in this publication by Loveland (Chapter 37) as well as by Star and Estes (1990). Methods for statistically estimating the error resulting from the classification of satellite images have been developed and do provide some quality assurance (Davis and Simonett, 1991).

The concurrent use of data from different sensors or even different flights of the same sensor requires registration of the images to ensure that the rectangular pieces of the ground surface represented by the pixels cover precisely the same ground locations in all images. It may be necessary to estimate new images from the original ones through a process of resampling in which a new grid having a different size cell or oriented at a different angle is used to estimate new pixel values from the original image. Just as with maps, it is important to consider the relationship between spatial models, data structures, and reality before such data are used. A good review of these issues related to the use of remotely sensed data in natural resource applications of GIS is provided by Trotter (1991).

Digital orthophotos

Digital orthophotos provide another source of digital data. These products are scanned airphotos that have been rectified to eliminate displacement caused by variable elevation of the ground surface and the tilt of the camera. Properly registered with other digital data sets, these images can be used directly as backdrops for vector data or to provide a basemap for onscreen digitizing.

CHARACTERISTICS OF SPATIAL DATABASES

As spatial data are collected digitally, they are usually entered into a database. Having highlighted a few of the unique characteristics of spatial data, we now broaden our attention to a consideration of some important issues related to spatial databases.

File format

Just as space can be discretized in a number of different ways, there are many different physical file formats used. The manner in which spatial data are related to nonspatial data and whether these are stored separately or together are distinguishing features. Different file structures permit varying strengths and efficiencies. The power and competitiveness of commercial GIS are determined by their proprietary file structures and associated algorithms.

When a database is constructed for a special purpose, it is virtually always designed to match the data and file structures specified by the commercial software chosen for or, more often, assigned to the project. Thus, many special purpose databases have very specific formats. If a researcher on another project using a different GIS elects to use this previously constructed database, it will be necessary to undertake what is often a time-consuming import/export process (Guptill, 1991). Fortunately, some help for this problem is beginning to appear in the form of standard file formats. Proprietary data structures and file formats can often be transformed into an intermediate common structure that can, in turn, be imported by other programs. Many of these intermediate transfer formats are commonly available simply as a result of the widespread use of particular software packages (e.g., AutoCAD's DXF file structure), though recently the development of universal standards has begun.

In the U.S., several agencies of the Federal government under the direction of the Federal Interagency Coordinating Committee on Digital Cartography (FICCDC), now the Federal Geographic Data Committee (FGDC), have devised an extensive and widely supported Spatial Data Transfer Standard (SDTS). This standard required a multiyear development process and has now been adopted by the National Institute of Standards and Technology as a Federal Information Processing Standard (FIPS 173). In future it will be the only format used for exchange of spatial data between Federal agencies. The great advantage of this Federally sanctioned standard is that all GIS vendors will be required to write functions for the import from, and possibly export to, the SDTS if they or their customers wish to do business with the Federal government. Excellent descriptions of this data standard have appeared recently in the literature [see the original proposal (DCDSTF, 1988) and a short, more recent summary by Tosta (1991)].

In Europe, several national and international proposals are being discussed for the exchange of spatial data. These include the National Transfer Format (NTF) of the United Kingdom (Guptill, 1991), the Authoritative Topographic Cartographic Information System (ATKIS) of Germany, and the European Transfer Format (EFT) of the Comité Européen des Résponsables de la Cartographie Officielle (Strand, 1991). Many other countries, associations of countries, and international data collection agencies are exploring similar methods for standardization of spatial data file formats.

Tiling and edge-matching

If the database coverage is regional or larger, it is normally necessary to divide the covered area into pieces, often called tiles. Tiles are comparable to map sheets and indeed are often defined by the same boundaries. Tiling may introduce some problems when a researcher needs to use data from two or more adjacent tiles. Edge-matched tiles ensure that objects that extend across adjacent tile boundaries are properly related. Without edge-matching, lines such as roads or property boundaries may not meet across the tile edges, gaps or overlap may occur due to inaccuracies in the projection transformations along the tile margins, and objects that extend across tile boundaries may appear to be two unrelated objects. In edge-matched tiled databases, no gaps or overlaps occur, and objects are given ID numbers that are unique across the entire database so that an object that extends across the boundary of adjacent tiles can be related. Using cleanly edge-matched tiles, some GIS can provide seamless graphic output and data query facilities that make the existence of the tile boundaries transparent to the user.

Topology

In the context of vector-structured spatial databases, topology is the explicit description, within the physical data structure, of the spatial relationships between stored entities. These relationships may include such things as the hydrologic sequence of the individual links of a river, or the identity of the neighbors of each region. By including topology, analyses involving spatial connectedness such as hydrologic modeling or impact assessment of a pollution plume can be directly computed. This is in direct contrast to digital design files or those in CAD databases, in which stored entities have no relationships. For example, in a cartographic database, the set of lines that together comprise the boundary of a region may not be explicitly related to the definition of that region. In order to know the spatial extent of a specific region, it is necessary to either plot the map or execute complex algorithms. Such nontopological databases are generally thought to be of limited use in sophisticated GIS applications although some GIS are capable of deriving the topological structure of such data.

AVAILABLE SPATIAL DATABASES

Many governments have created agencies specifically entrusted with responsibilities for collecting and maintaining spatial databases. Maintenance of archives of data collected from satellite sensors has always been a primary objective of the space agencies. In the U.S., the Federal EROS (Earth Resources Observation Systems) Data Center in Sioux Falls, SD, in cooperation with EOSAT (Earth Observation Satellite Company), the commercial operator of the Landsat satellites, maintains and distributes images from the archive of Landsat data. The EOS (Earth Observing System) Program, which will gather data from several new sensors on U.S., European and Japanese satellites over a 15 year period beginning in the mid 1990s, includes a major emphasis on developing systems for facilitating access to the data by the research community.

In contrast to the single-source databases of the space agencies, multithematic archives of spatial data require concerted cooperative efforts between several different government agencies. Examples of such databases exist at many different levels of government organization. At the state level, several examples of integrated natural resource databases exist [see, for example, North Carolina's (Siderelis, 1991) and Minnesota's (Robinette, 1991)]. The Australian Resources Information Systems (ARIS) is an example of a database depository for data at the national scale (Cocks et al., 1988). Internationally, the World Data Centers have existed since 1957 as centers for the exchange of scientific data (Hastings et al., 1991; Townshend, 1991). CORINE is the European Community's continent scale spatial database (Wiggins et al., 1987; Mounsey, 1991). UNEP has recently sponsored the development of the Global Resource Information Database (GRID) network, which seeks to provide environmental data on a global scale (UNEP, 1990).

An excellent review of the development of several different global databases of spatial data for scientific applications is provided by Mounsey and Tomlinson (1988). Tens of other excellent examples at each of these scales exist.

A 1990 survey by *GIS World* identified over 400 miscellaneous spatial databases available in the U.S. from government agencies and private companies (GIS World, 1990). Certainly this number has grown considerably since the survey was conducted. These range from county to global coverage and include themes as diverse as water well locations, administrative boundaries, and geology. Every one of these is unique in some way and would provide suitable input for many different environmental applications. The following chapters by Hutchinson, Lytle, Loveland and Ohlen, and Kineman (36 through 39) describe in detail four very different database projects. Lytle (Chapter 38) describes the development of digital soils databases from existing maps and tabular data. Hutchinson (Chapter 39) explains the procedures required to create reliable, continental-scale elevation data sets from dense point elevation data. Results of analyses based on this data set illustrate some of the critical issues related to the use of such databases. Loveland and Ohlen's description of the development of a land characterization database from remotely sensed and ancillary data (Chapter 37) critically examines the processes of classification and interpretation. Kineman's Chapter 36 discusses the development of the Global Ecosystems Database Project and argues that careful design should produce a product specifically suited to scientific research on global issues. These chapters are useful illustrations of the many issues that must be clearly recognized by environmental modelers as they integrate data from such sources into their mathematical models.

In addition to the environmental databases described in these chapters, it is useful also to mention briefly some that are becoming increasingly important as the spatial frameworks for many nonspatial databases.

National mapping agency databases—USGS products

As the agency responsible for the production of the traditional national topographic quadrangle maps, the U.S. Geological Survey has recognized its dawning role as the central organization responsible for the maintenance and distribution of current digital map data for the nation (Starr and Anderson, 1991). Products distributed by the Earth Science Information Centers (ESIC) of the USGS from the National Digital Cartographic Data Base (NDCDB) include topographic map line data (Digital Line Graphs, DLGs) and elevation data (Digital Elevation Models, DEMs).

DLGs contain points, lines, and polygon outlines digitized from the original separates used to print topographic maps at scales from 1:24,000 to 1:2,000,000. The data are topologically structured with adjacency and connectivity explicitly encoded. Coded attributes identify features according to the detailed SDTS standard feature types. Data categories include contours, boundaries, hydrography, transportation, and the U.S. Public Land Survey System.

DEMs are produced by digitizing contours on existing maps or by scanning stereomodels of high-altitude photographs. Data are supplied as a grid of point elevations. For the USGS standard 7.5 minute topographic quadrangle areas, elevations are provided in a grid with 30 meter ground spacing. Lower-resolution coverage is available in 3-arcsecond grids developed from the Defence Mapping Agency's 1:250,000 scale maps (USGS, undated).

Census geography – TIGER

Most developed countries have recently created, or are in the process of creating, detailed databases of the nation to be used for the storage, display, and analysis of census data (Rhind, 1991). The U.S. product, the TIGER/Line files, is a good example of these census geography databases. For a complete description of the genesis and future of TIGER, see Marx (1990).

TIGER stands for Topologically Integrated Geographic Encoding and Referencing System and is the U.S. Census Bureau's digital basemap database. It was designed solely for use by the Census Bureau, but, as a result of farsighted planning by its designers, it has become one of the most important and widely available digital spatial databases in the U.S. Based on the USGS 1:100,000 DLG files, this topologically structured database contains digital geographic information on roads, railroads, rivers; census collection geography such as boundaries of census tracts, blocks, political areas; feature names and classification codes; and within metropolitan areas it contains address ranges and ZIP codes for streets (U.S. Bureau of the Census, 1990). Coverage is available for the entire U.S. Coordinates are not precise, and considerable variation between TIGER geography and other digital street network products is observed when databases are integrated. While the Census Bureau does not have plans to keep this database updated between decennial censuses, many third-party vendors are developing related products with improved locational accuracy and updated detail (GIS World, 1991).

Digital Chart of the World

The U.S. Defense Mapping Agency (DMA), along with the military mapping agencies of Canada, the U.K., and Australia, have embarked on an ambitious project to produce a digital basemap of the entire world from the 270 1:1,000,000 scale Operational Navigation Charts (ONCs). The final product will be approximately 4 CD-ROMs containing 2 gigabytes of vector data on hypsography (contour lines), hydrography, land cover, landmarks, ocean features, aeronautical details, physiography, political boundaries, railroads, roads, utilities, and vegetation. Based on a 0.5 mm linewidth common on published maps, these data can be considered to have a minimum "resolution" of 500 m. DCW will be distributed with software to access, query, and display the data. The product is being developed under contract by Environmental Systems Research Institute (ESRI, the private company that developed and markets ARC/INFO) and when completed in 1992 will be available in the U.S. from the DMA.

FINDING THE SPATIAL DATA YOU NEED

Locating the data to meet one's needs is still a hit-or-miss task. In the U.S., a good source of information about earth science databases is an Earth Science Information Center (ESIC). There are more than 75 of these across the country, some operated by the USGS, others by other Federal or state agencies. In several cases, these ESICs can be found at University Map Libraries. These centers have many USGS products available for review and/or purchase as well as offering access to computer databases on aerial photography, satellite imagery, and assorted digital products listed in the USGS's Cartographic Catalog.

At the moment, even with ESIC personnel to assist one, the most practical way to find what one needs is by undertaking a time-consuming search of numerous digital and printed database indexes and pursuing the data through a network of knowledgeable individuals. The indexes should provide metadata about the databases, ideally providing bibliographic-style details such as date of collection, agency involved, location, file format, accuracy, etc. Unfortunately, no bibliographic standards exist for these indexes, some are collected for only a few years and quickly become out of date, and many important databases will not be listed anywhere. The assistance of an experienced spatial information specialist, such as a map librarian at a major research institution, can help direct one to the less accessible sources of information. Fortunately, the problem of locating spatial data has been recognized by national and international government agencies. Many have started taking on responsibilities for developing and maintaining standardized bibliographic metadatabases on spatial and other scientific data. In this direction, the USGS is developing the Global Land Information System (GLIS), which is to contain descriptive information about a wide range of data sets pertaining to the Earth's land surface at regional, continental, and global scales. Users will be able to query by theme or location, evaluate data sets, and place orders on-line. The first phase of operation of this system is planned for late 1992.

CONCLUSION

The future for the discriminating use of and easy access to the overwhelming amount of spatial data being compiled today is bright. The upsurge in research into GIS has brought to attention many of the critical issues about spatial data that have been the quiet research domain of geographers and others for decades. Federal and international agencies are accepting responsibilities for the collection, management, and dissemination of these data. Major efforts at devising appropriate systems for accessing these databases are being funded. The opportunity to access spatial databases on the entire range of topics relevant to environmental modelers is expanding exponentially.

REFERENCES

Broome, F.R., and Meixler, D.B. (1990) The TIGER data base structure. *Cartography and Geographic Information Systems* 17(1): 39–47.

Burrough, P.A. (1986) *Principles of Geographical Information Systems for Land Resources Assessment*, Oxford University Press.

Cocks, K.D., Walker, P.A., and Parvey, C.A. (1988) Evolution of a continental-scale geographical information system. *International Journal of Geographical Information Systems* 2(3): 263–80.

Colwell, R.H. (ed.) (1983) *The Manual of Remote Sensing*, Falls Church, VA: American Society of Photogrammetry.

Cooke, D. (1990) An overview of TIGER. In *The 1990 GIS Sourcebook*, Fort Collins, CO: GIS World Inc., pp. 321–324.

Davis, F.W., and Simonett, D.S. (1991) GIS and remote sensing. In Maguire, D.J., Goodchild, M.F., and Rhind, D.W. (eds.) *Geographical Information Systems: Principles and Applications*, London: Longman, Vol. 1, pp. 191–213.

DCDSTF (Digital Cartographic Data Standards Task Force) (1988) The proposed standard for digital cartographic data. *The American Cartographer* 15(1): 9–142, plus microfiche.

Fisher, P.F. (1991) Spatial data sources and data problems. In Maguire, D.J., Goodchild, M.F., and Rhind, D.W. (eds.) *Geographical Information Systems: Principles and Applications*, London: Longman, Vol. 1, pp. 175–189.

Gatrell, A.C. (1983) *Distance and Space: A Geographical Perspective*, Oxford: Oxford University Press.

GIS World (1990) *The 1990 GIS Sourcebook*, Fort Collins, CO: GIS World Inc.

GIS World (1991) *The 1991–92 International GIS Sourcebook*, Fort Collins, CO: GIS World Inc.

Goodchild, M.F., and Gopal, S. (1989) *Accuracy of Spatial Databases*, London: Taylor and Francis.

Guptill, S.C. (1991) Spatial data exchange and standardization. In Maguire, D.J., Goodchild, M.F., and Rhind, D.W. (eds.) *Geographical Information Systems: Principles and Applications*, London: Longman, Vol. 1, pp. 515–530.

Haining, R. (1990) *Spatial Data Analysis in the Social and Environmental Sciences*, Cambridge: Cambridge University Press.

Harvey, D. (1969) *Explanation in Geography*, London: Edward Arnold.

Hastings, D.A., Kineman, J.J., and Clark, D.M. (1991) Development and applications of global databases: considerable progress, but more collaboration needed. *International Journal of Geographical Information Systems* 5(1): 137–146.

Jackson, M.J., and Woodsford, P.A. (1991) GIS data capture hardware and software. In Maguire, D.J., Goodchild, M.F., and Rhind, D.W. (eds.) *Geographical Information Systems: Principles and Applications*, London: Longman, Vol. 1, pp. 239–49.

Lillesand, T.M., and Kiefer, R.W. (1987) *Remote Sensing and Image Interpretation*, John Wiley and Sons.

Maguire, D.J., Goodchild, M.F., and Rhind, D.W. (1991) *Geographical Information Systems: Principles and Applications*, London: Longman.

Maling, D.H. (1980) *Coordinate Systems and Map Projections*, London: George Philip and Son.

Marx, R.W. (1990) The TIGER system: yesterday, today and tomorrow. *Cartography and Geographic Information Systems* 17: 89–97.

Mounsey, H.M. (1991) Multisource, multinational environmental GIS: lessons learnt from CORINE. In Maguire, D.J., Goodchild, M.F., and Rhind, D.W. (eds.) *Geographical Information Systems: Principles and Applications*, London: Longman, Vol. 2, pp. 185–200.

Mounsey, H.M., and Tomlinson, R. (1988) *Building Databases for Global Science*, London: Taylor and Francis.

Muller, J-C. (1991) Generalization of spatial databases. In Maguire, D.J., Goodchild, M.F., and Rhind, D.W. (eds.) *Geographical Information Systems: Principles and Applications*, London: Longman, Vol. 1, pp. 457–75.

Openshaw, S. (1983) The modifiable areal unit problem. *Concepts and Techniques in Modern Geography, No. 38*, Norwich: GeoAbstracts.

Peuquet, D.J. (1988) Representations of geographic space: toward a conceptual synthesis. *Annals of the Association of American Geographers* 78: 375–94.

Rhind, D.W. (1991) Counting the people: the role of GIS. In Maguire, D.J., Goodchild, M.F., and Rhind, D.W. (eds.) *Geographical Information Systems: Principles and Applications*, London: Longman, Vol. 2, pp. 127–137.

Robinette, A. (1991) Land management applications of GIS in the State of Minnesota. In Maguire, D.J., Goodchild, M.F., and Rhind, D.W. (eds.) *Geographical Information Systems: Principles and Applications*, London: Longman, Vol. 2, pp. 275–283.

Sabins, F.F., Jr. (1987) *Remote Sensing: Principles and Interpretation, 2nd ed.*, New York: Freeman.

Siderelis, K.C. (1991)Land resource information systems. In Maguire, D.J., Goodchild, M.F., and Rhind, D.W. (eds.) *Geographical Information Systems: Principles and Applications*, London: Longman, Vol. 2, pp. 261–273.

Snyder, J.P. (1987) Map projections—a working manual.

U.S. Geological Survey Professional Paper 1395, Washington, DC: U.S. Government Printing Office.

Star, J., and Estes, J.E. (1990) *Geographic Information Systems: An Introduction*, Englewood Cliffs, NJ: Prentice-Hall.

Starr, L.E., and Anderson, K.E. (1991) A USGS perspective on GIS. In Maguire, D.J., Goodchild, M.F., and Rhind, D.W. (eds.) *Geographical Information Systems: Principles and Applications*, London: Longman, Vol. 2, pp. 11–22.

Strand, E.J. (1991) A profile of GIS standards. *The 1991–02 International GIS Sourcebook*, Fort Collins, CO: GIS World, Inc., pp. 417–421.

Tosta, Nancy (1991) SDTS: setting the standard. *GeoInfo Systems*, July/August, pp. 57–59.

Townshend, J.R.G. (1991) Environmental databases and GIS. In Maguire, D.J., Goodchild, M.F., and Rhind, D.W. (eds.) *Geographical Information Systems: Principles and Applications*, London: Longman, Vol. 2, pp. 201–16.

Trotter, C.M. (1991) Remotely-sensed data as an information source for geographical information systems in natural resource management: a review. *International Journal of Geographical Information Systems* 5(2): 225–239.

U.S. Bureau of the Census (1990) *TIGER Questions and Answers*, Washington, DC: U.S. Department of Commerce.

USGS (undated) *Digital Cartographic and Geographic Data*, Reston, VA: Department of the Interior, U.S. Geological Survey, National Cartographic Information Center.

Wiggins, J.C., Hartley, R.P., Higgins, M.J., and Whittaker, R.J. (1987) Computing aspects of a large geographic information system for the European Community. *International Journal of Geographical Information Systems* 1(1): 77–87.

36

What is a Scientific Database? Design Considerations for Global Characterization in the NOAA–EPA Global Ecosystems Database Project

JOHN J. KINEMAN

Although data for global change research are becoming increasingly available to researchers on a disciplinary and institutional basis, carefully managed data integration efforts are required to support specific project and programmatic missions within NOAA, EPA, and other agencies. A cooperative agreement between the NOAA National Geophysical Data Center (NGDC) and the EPA Environmental Research Laboratory—Corvallis (ERL-C) established the Global Ecosystems Database Project (GEDP) in 1990 to link database integration efforts at NGDC with ERL-C's research within the EPA Global Climate Research Program (Campbell et al., 1992). The project is a part of NGDC's Global Change Database Program. Its goal is to provide modern, global-, and continental-scale data (for the entire Earth's surface) needed by the global change research community. The primary focus of the GEDP is to develop an integrated, quality-controlled, global database (including time sequences and model outputs) for spatially distributed characterization (and modeling support) related to global change.

The GEDP is making advances in database structure and function, and helping to define the role of GIS in characterization and modeling. Results of this work include the distribution of CD-ROMs containing successive improvements of the database, critical reviews, and supporting materials. In 1991 a prototype database (NGDC, 1991) was assembled and distributed for review on CD-ROM (Prototype 1: March 1, 1991). The first public release (Release 1: April 1, 1992) incorporated additional data and revisions based on the first year review, and contained an integrated set of global 2–60 minute geographic (latitude/longitude) raster (grid) data and vector boundaries, including multiple digital data sets within six general themes:

1. Vegetation cover and ocean color from satellite;
2. Surface climatology from station, ship, and satellite;
3. Vegetation and ecological classifications from surveys, maps, and satellite;
4. Soil classifications from surveys and maps;
5. Elevation and terrain data from maps; and
6. Boundary (vector) data from maps.

Considerable expansion of the database is planned in the remaining 4 years of the project, including additional variables, outputs and models from the EPA research program, continental databases, and finer-resolution data as available. Data compiled by the project in 1991 were also contributed to the ISY Global Change Encyclopedia to be released in 1992 by the Canada Center for Remote Sensing (Cihlar et al., 1991). Further details about the project, database, and CD-ROM products can be obtained from NGDC. The history of these efforts is reported elsewhere (Ruttenberg et al., 1990; Kineman and Ruttenberg, 1991; Kineman and Clark, 1987; Clark and Kineman, 1988; Kineman et al., 1990; Hastings et al., 1991).

THE DESIGN OF DATA

The greatest problem in the distribution of environmental and ecological data sets is their usefulness outside the institutions or programs that created them, and within a rapidly expanding global change community. While review and publication standards provide effective quality assurance for research in general (including the production of data sets), they apply less directly to the distribution of data sets for subsequent uses, which is often a second or lower priority in research programs and funding. The result is that when data are removed from their original research context, we have inadequate mechanisms for evaluating their design or verifying their accuracy for given purposes. This issue of verifiability becomes critical in the context of large multidisciplinary system studies such as global change, which must rely on a common data pool.

To both the statistician and scientist all data are inseparable from experimental designs, that is, the particular design that originally determined things like sensing method, sampling scheme, precision, acceptable error, temporal compositing and time steps, correction algorithms, and so forth (Kineman, 1989). Such design criteria embody numerous decisions, including the particular phenomena that the data are expected to reveal. There can be great difficulty, therefore, in using precollected and predesigned data to test new ideas and/or search for new phenomena, unless design considerations can be accounted for in new applications.

For these reasons, increased distribution of data cannot meet the needs of the global change community without some common methods for redesign and verification. Not only is complete design information required for each data set, but improved analytical tools are needed for comparing data sets and combining or converting them into variables that are appropriate for each new study, carrying with them (ideally) statistical confidence levels (error estimates) based on the combined effect of their design parameters. This process requires sophisticated analytical capabilities operating on a uniformly structured and suitably designed database—all of which we are far from having on a community-wide basis.

Let us take the greenness index produced by NOAA as an example. Even if all physical parameters influencing the data, such as sensor calibrations, sun-angle effects, transmissivity, etc., could be corrected for, we still do not know what "greenness" means in ecological terms, or how to relate this measurement to other measures of biomass, growth, health, or even land-cover classification. These are research questions that require either intensive (and largely impractical) ground truth on a global scale, or analytical combinations with other data sets, each of which has similar problems. Add to this the problem of matching the scale of data with the scale of phenomena, and the scale of other data; and then consider that almost no global data set currently available comes with quantitative estimates of accuracy, even for the original parameter measured; and that current processing systems (GIS) have very poor methods for assessing and tracking accumulated error in the analytical process.

This is what I mean by "removed from the experimental context"—the suitability and accuracy of design criteria for such representation is not generally known or testable. This defeats the applied scientist who wishes to be purely descriptive in quantitative statistical terms, as well as modelers who wish to test theory—the former because accuracy has not been quantified, the latter because sampling designs may be inappropriate for the critical test.

These issues have traditionally been left to the individual researchers to sort out, often duplicating efforts considerably. But the multidisciplinary, community-wide goals of the global change program will require cooperation, which cannot be achieved without common ground. Our awareness of the issues can be used to define what a general database should be able to do, and how it can fit into the scientific process. That definition will then determine the important characteristics of the database and its operational methodology (that is, its form and function, borrowing terms from ecology).

CHARACTERIZATION AND THE SCIENTIFIC METHOD

Recognizing the need for a common methodological basis for database design, we may ask if there is a scientific approach that can encompass this exercise. "Ecological characterization" was defined in the early 1970s within the Coastal Ecosystems Project of the U.S. Fish and Wildlife Service as:

> a study to obtain and synthesize available environmental data and to provide an analysis of the functional relationships between the different components of an ecosystem and the dynamics of that system. It is simply a structured approach to combining information from physical, chemical, biological, and socioeconomic sciences into an understandable description of an ecosystem.—Watson, 1978; and subsequent publications of the Coastal Ecosystems Project, Office of Biological Services, Fish and Wildlife Service, U.S. Department of the Interior, Washington, DC 20240

Characterization, as thus conceived, is descriptive, goal-directed, and organized by conceptual models of the ecosystem, which provide the scientific context, experimental designs, and information priorities needed to focus the effort. To apply this concept to the global change scientific program, which is a goal-directed, multidisciplinary ecosystem study, characterization efforts must be developed as a complementary link to process-oriented research and modeling efforts (i.e., our present understanding of the Earth system). In the context of modern databases and GIS technology, we may think of a computer database designed for adaptive characterization, defined as an analytical process of integrating and redesigning existing multidisciplinary data for the purpose of representing important spatial patterns and temporal trends, along with capabilities for empirical testing, synthesis, and quantitative assessments of confidence based on experimental design information (i.e., metadata).

Environmental (and ecological) characterization thus defined differs from dynamic modeling in important ways. It is focused entirely on synthesis of existing data and process information. As such, its goal is to apply accepted integration and synthesis methods to data sets to provide empirically valid representations of the natural system, given a variety of applications. Though guided by modeling efforts, it must remain entirely complementary to them (i.e., as independent as possible to avoid circular testing). As an effort in its own right, such characterization is closely akin to comparative ecosystems analysis, sharing many of the difficulties and opportunities of that field (Cole et al., 1991). As a goal-directed activity, this approach can ideally support monitoring, assessment, prediction, model inputs and tests, exploration, education, resource studies, environmental management and planning, and other information applications. Figure 36-1 shows science divided into the complementary aspects of theory development and descriptive analysis, wherein characterization lies. These are extremes in the way we might choose to look at nature that are usually combined in any given study, but they correspond well with methodological differences between simulation modeling and characterization.

The diagram not only shows these as two necessary and complementary activities in science; it also presents them as equally important. Furthermore, they follow parallel methods in regard to the traditions of pure research (i.e., empirical science), and are thus equally "scientific." The most common routines of science can be traced in Figure 36-1, along with the critical interactions facilitated by the GIS/modeling linkage (center box in the diagram).

Both observation and analysis (left side of Figure 36-1) are theory laden according to most philosophers. But this influence may range from specific requirements of hypothesis testing to more general concepts of spatial representation. To the extent that the characterization effort can be flexible yet scientifically rigorous, the outcome may serve a variety of goals. For example, a purely observational database may support exploration and visualization, specific sampling designs for model tests, and data synthesis efforts to generate needed model inputs. It may alternatively incorporate the model outputs to characterize predicted states. It is clear that both characterization and modeling are required for prediction.

As implied in the diagram, GIS is the technology that is potentially most suited (given appropriate developments) to support the requirements of an operational scientific database. On the other hand, other tools, such as simulation techniques, are optimized for executing complex time-dependent mathematical or stochastic process equations often used in theory development and modeling. This distinction, though somewhat rough, is evident in different computer requirements (Constanza and Maxwell, 1991). GIS can operate on micro- to workstation-level computers, performing the necessary stepwise comparisons and combinations of data that are mostly limited by access times for large data volume rather than by computational demands. Simulation, on the other hand, tends to be limited by complex and often iterative computational demands that may require supercomputers. Global simulation is thus restricted to smaller databases on very coarse grids of 2.5 to 5 degree cells.

What makes each side of this comparison "scientific"

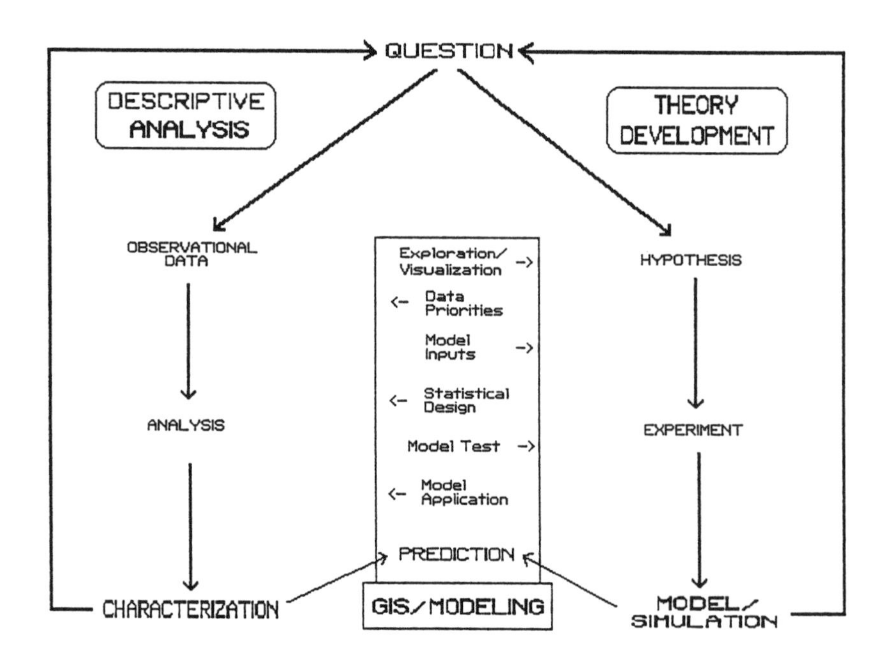

Figure 36-1: Science divided into its complementary aspects of theory development and descriptive analysis.

is, first of all, rigorous application of the scientific method of quantified empirical testing or confirmation. On the theory development side, this means reference to real world observations designed specifically for the needed test, or, in this case, reference to a pre-established observational database that can be easily adapted. On the descriptive analysis side, the empirical method is represented in analysis with respect to statistical design, where confidence limits are established from methodological considerations, statistical analysis, and intercomparison with other data (including ground truth where possible). Second, neither side can be acceptable scientifically, if it does not incorporate the best results of the other: Theory is lame without reference to accurate descriptions, and descriptions are meaningless without suitable theory for interpretation.

A GLOBAL CHARACTERIZATION DATABASE

Global change programs currently emphasize prediction as the primary scientific goal. Yet we actually understand very little about whole system behavior, certainly not enough to make operational predictions, or to assume that current models represent everything that is important. As one would expect, our awareness and concern about global issues is advancing at a much greater rate than our theoretical understanding. Whether or not we develop a predictive capability in the near future, the pressure for descriptive studies will continue to grow. It would be foolish indeed to delay in accurately assessing and monitoring the state of global systems, or accurately describing change. A study of global change cannot afford to overlook phenomena that are not yet a part of our models.

It seems reasonable to develop as fundamental a database as possible, given current theoretical knowledge about what it should represent. This database should be designed to characterize patterns and trends and to represent basic processes that, when suitably combined, may be converted into a variety of useful environmental indices and/or predictors. This, in turn, can serve as an empirical basis for testing theory (and modeling), a surrogate reality that can greatly reduce the need to collect specific data for each study. There is a compromise in this, to be sure, in that experimental designs will unavoidably be forced into pre-existing molds. This makes statistical considerations and flexibility (i.e., "redesign" issues) paramount in structuring a characterization database. Most data sets, particularly from remote sensing, require comparative information for interpretation. Developing that information will require an extensive global cooperative effort, new concepts in database design, and improved analytical tools.

As with all characterization, the GEDP is a focused effort: it is limited by project-specific needs, but it also hopes to have a more general impact by making advances in structure and methodology. Various constraints help control the scope of the GEDP and ensure its usefulness for characterization. One of these is certainly the research priorities of those groups supporting its development (i.e., the link to modeling, as has been discussed). The next three most defining aspects have been GIS structure, scale, and geographic sampling (i.e., projection and grid structures).

GIS structure and function

Characterization requires both a carefully structured database and appropriate analytical software. The scientific requirements may dictate the contents of the database; however, it is the operational needs of global change science that dictate structure and function. Specifically, multidisciplinary integration requires that there exist conventional ways of structuring and working with an integrated database that are acceptable across disciplines. The need to be able to "redesign" information, as has been discussed, requires that these conventions support rigorous statistical approaches to combining and comparing data sets, as well as for data quality and error assessment. While far from ideal, current GIS technology offers a start in this direction by suggesting conventions in spatial data structure as well as a "toolbox" of conventional processing functions that can be combined for a given analysis.

We can list the major elements of descriptive analysis by looking closely at the interactions between theory and description shown in Figure 36-1. This is done in Figure 36-2. These functions correlate well with GIS concepts, which imply a number of conventions that are already well established, such as geographical (vector and raster) object definitions, common geographical referencing, labeling and legend conventions, and others. Future research and development in GIS, if properly informed by the global change community, can provide systems that are increasingly well suited for environmental characterization. At the same time, links with different software tools that are equally well suited for theory development and dynamic modeling may be formed. The fundamentally different structure of these two activities, their different technical requirements, and the scientific need for a complementary relationship implies that they must develop as independent (though linked) activities.

Natural scale ranges

Aside from its general focus on the global environment, the scope of the GED effort is defined more than anything by scale. It is known that many environmental and ecological phenomena are scale dependent (Rosswall et al.,

DESCRIPTIVE ANALYSIS	
FUNCTION	SCIENTIFIC PURPOSE
Data Integration	Multi-thematic representation of variables, with a common analytical structure, with meta-data for verification and error analysis.
Exploratory Data Analysis	Visualization; exploration; Hypothesis formation
Inter-comparison and statistical analysis	Quality and error assessment; empirical testing (hypotheses, models, and characterizations); statistical modelling
Statistical design and error analysis	Experimental Design / re-design for various applications
Environmental Characterization & Data	Multi-disciplinary representation of phenomena; monitoring and assessment; description of patterns and trends; model inputs and tests

Figure 36-2: The major elements of descriptive analysis.

1988). We must assume (for practical reasons) that within any given study, certain "natural" groupings can be defined wherein data scales can be combined and between which phenomena can be interrelated. Figure 36-3 shows one such grouping for three very broad categories of research.

Two of these represent active areas of current research: site studies and global simulation modeling. However, between these two activities there is a gap in scale where relatively less work has been done. The Global Ecosystems Database attempts to fill this gap, which initially covered roughly one order of magnitude in scale, from

RESOURCE MANAGEMENT & SITE STUDIES		GLOBAL CHARACTERIZATION		GLOBAL SIMULATION MODELING	
10-100m	1-4km	4-16km	50-100km	100km	500km
Spot, Landsat	AVHRR LAC/GAC	AVHRR/ GVI	Climate data, classifications, etc.	Climate data, Classifications, etc.	GCM runs

Figure 36-3: Groupings of scales for environmental and ecological phenomena.

5–10 minutes (10 km) to 1 degree (100 km). The effort may eventually be expandable to include 1 km data for site and regional data.

Integration of data within this intermediate range of scales may provide links between global circulation models (GCMs) and more detailed resource management studies. As such linkages are formed, they will serve as a means for communicating information between studies at variously defined scale groupings. Because such scale divisions are not entirely arbitrary, but also relate to observable phenomena, it is likely that different phenomena of interest exist in all three scale ranges shown in Figure 36-3. The challenge to the database system is to develop techniques for relating scales within defined ranges at the level of data, and to develop methods for interrelating information between these natural scale groupings at the level of prediction and analysis. In this way, the problem of scale is reduced by defining the approximate boundaries of such scale groups, in terms of natural phenomena and the practicalities of observing.

Compatible sampling grids

Combining raster (gridded) data from dissimilar grids raises important scientific and statistical questions in regard to the original design of the database. For example, classifications based on various spatial statistics (e.g., spatial dominance) at one scale may not be directly convertible to the same statistic at another, since wholly different classes may be implied. Similar problems exist at the same scale between differently oriented grids (e.g., edge versus center cell registration). There are many interpolation and sampling methods for continuous and classed data, each appropriate for different kinds of data, depending on the underlying nature of spatial dependence that is assumed.

The critical issue is that certain kinds of grid changes cannot legitimately be accomplished after the database is produced, without additional research. Either registration or scale differences can mean that special interpolation or sampling may be required if two or more dissimilar data sets are to be used in combination, as is the intent of an integrated database. Ideally, software systems (i.e., GIS) should assist the user by providing estimates of scale combination errors, or at least warnings when dissimilar grids are compared (features not currently in existing GIS, even those that perform scale changes transparently). While this problem cannot be eliminated entirely without scale-independent source data (a virtual impossibility), it can be reduced by establishing certain conventions in the database structure.

There are several ways to establish geographic reference with gridded data. An independent geographic reference for each data cell (as with vector data) is one approach, but this virtually ensures later interpolation or sampling problems for unmatched grids. A common origin and grid registration eliminates the problem for data at the same scale. By further establishing a "nested grid" convention, whereby commonly registered grids are even multiples and divisions of each other, the problem is reduced further since cell edges are aligned when one scale overlays another.

Such a convention of "nested grids" is used in the GEDP for geographic (latitude/longitude) raster data as a means of reducing interpolation requirements in the final database. By accomplishing the more complicated resampling or interpolation problems during data integration, they can be fully documented and reviewed as part of the database. If this idea catches on, the original investigators may be convinced to provide data in the conventional grid structure, thus eliminating the need for secondary changes during integration. In this proposed convention, there are nested grid scales of 5 degrees, 1 degree, 30 minutes, 10 minutes, 5 minutes, 1 minute, and 30 seconds; all of which are integer multiples or divisions of each other.

With a commonly registered database structured in this manner, comparison is possible with relatively simple expansion or contraction procedures that preserve the original data values and are computationally simple. This does not reinterpret the data, but does allow direct overlay comparisons, given that one keeps track of the original scales and their appropriate meaning (which is rightfully part of the definition of the original variable itself). Scale integration beyond this point becomes an analysis problem that must be assessed by the user for each given application.

PEER REVIEW AND DISTRIBUTION

The purpose of establishing conventions in data structure and characterization methodology is to facilitate community-wide verifiability. In general, global ecosystems data are poorly developed and generally difficult to obtain and use in digital form. Furthermore, ecosystems data tend to be more subjective than many environmental data sets, involving varying classification methods (e.g., for vegetation, soils, land cover, etc.) and highly interpreted information. Ecological models are rather limited and diverse, and there is generally poor consensus within the "global change" community on data requirements because it is so multidisciplinary. This situation implies that considerable interaction with the scientific community is needed to obtain a consensus on data requirements. Also, experimentation is needed on methods for analyzing and exchanging data and their derivatives. And finally, a widely based peer review is needed to evaluate the usefulness and scientific value of global data.

At least four levels of external review are needed. These are to assess (1) quality, (2) structure and functionality,

(3) scientific applications, and (4) education, outreach, and policy applications. The GED effort addresses the first and second of these through the peer review process, and the third in relation to the research needs of ERL-C (which is partially funding the effort). The fourth level of review is being investigated through various projects at NGDC and the associated World Data Center-A.

The review process is intended to encourage data exchange, comparison of data, data quality assessment, data "publication" standards, GIS development, and the development of common analytical methods. NGDC plans annual public releases of the database and its updates, following each peer review period.

ACKNOWLEDGMENT

The Global Ecosystems Database Project is supported by EPA Contract No. DW13934786-01-0, "Co-developing data, tools, and methods for characterization and analysis of environmental system patterns to support EPA Global Climate Change Research and Modeling."

REFERENCES

Campbell, W.G., Kineman, J.J., Lozar, R.T., and Marks, D. (1992) A geographic database for modeling the role of the biosphere in climate change. Submitted to *Climate Change.*

Cihlar, J., Simard, R., Manore, M., et al. (1991) Global change encyclopedia: a project for the International Space Year. *Advances in Space Research* 11/3: (3)249–(3)253.

Clark, D.M., and Kineman, J.J. (1988) Global databases: a NOAA experience. In Mounsey, H., and Tomlinson, R. (eds.) *Building Databases for Global Science*, London: Taylor and Francis, pp. 216–232.

Cole, J., Lovett, G., and Findlay, S. (eds.) (1991) *Comparative Analyses of Ecosystems: Patterns, Mechanisms, and Theories*, New York: Springer-Verlag.

Costanza, R., and Maxwell, T. (1991) Spatial ecosystem modeling using parallel processors. *Ecological Modeling* 58: 159–183.

Hastings, D.A., Kineman, J.J., and Clark, D.M. (1991) Development and application of global databases: considerable progress, but more collaboration needed. *International Journal of Geographical Information Systems* 5/1: 137–146.

Kineman, J.J. (1989) Observable 'data'. *EOS* 70/17: 547.

Kineman, J.J., and Clark, D.M. (1987) Connecting global science through spatial data and information technology. *International Geographic Information Systems (IGIS) Symposium: The Research Agenda. Proceedings—Volume I*, Washington, DC: NASA, pp. 209–228.

Kineman, J.J., Clark, D.M., and Croze H. (1990) Data integration and modeling for global change: an international experiment. *Proceedings of International Conference and Workshop on Global Natural Resource Monitoring and Assessments: Preparing for the 21st Century, September, 1989, Venice, Italy*, Rockville, MD: American Society of Photogrammetry and Remote Sensing, pp. 660–669.

Kineman, J.J., and Ruttenberg, S. (1991) The IGBP multithematic database pilot project: status and plans. *Proceedings of the Eleventh Annual ESRI User Conference—Volume 1*, Redlands, CA: Environmental Systems Research Institute, Inc. pp. 103–109.

NGDC (1991) *Global Ecosystems Database, Version 0.1 (Beta-test): Database Documentation*, NGDC key to Geophysical Records Documentation No. 25, Boulder, CO: National Geophysical Data Center.

Rosswall, T., Woodmansee, R.G., and Risser, P.G. (eds.) (1988) *Scales and Global Change: Spatial and Temporal Variability in Biospheric and Geospheric Processes*, New York: J. Wiley, 355 pp.

Ruttenberg, S., Kineman, J.J., and Eastman, J.R. (1990) The Global Change Database Project (GCDP): a data project for the IGBP. *International Science: Special Edition for Global Change*, Paris: International Council of Scientific Unions.

Watson, J.F. (1978) Ecological characterization of the coastal ecosystems of the United States and its territories. *Proceedings: Energy/Environment '78*, Los Angeles: Society of Petroleum Industry Biologists, pp. 47–53.

37

Experimental AVHRR Land Data Sets for Environmental Monitoring and Modeling

THOMAS R. LOVELAND AND
DONALD O. OHLEN

Data needed for global environmental monitoring and modeling studies are commonly deficient in their information content, scale, consistency, or not available. This is in spite of the fact that near-global coverage of imagery from Earth-orbiting satellites such as Landsat have been available for nearly 20 years, and data from the National Oceanic and Atmospheric Administration (NOAA) Advanced Very High Resolution Radiometer (AVHRR) have been available for over 10 years.

It is well accepted that remote sensing methods are essential to address the data needs for global studies of land surface processes (Pinker, 1990). Remote sensing can potentially provide acceptable information regarding the spatial distribution, and with some sensors, even the multitemporal variation of a wide variety of land cover characteristics at the Earth's surface. There are several strategies for using remote sensing to estimate biospheric processes for large-area environmental studies.

One traditional approach involves the extrapolation of data from site-specific field investigations into a spatial context. Land cover maps are used to guide the spatial distribution of field data associated with specific land cover types into forms needed for models developed at these sites. Remote sensing is typically used to map land cover.

Another, more sophisticated approach, involves the development of process-level models at individual sites and using these to parameterize remote-sensing-driven algorithms for large-area analyses. The use of leaf area index as a parameter to estimate evapotranspiration or net primary production is an example of this widely used approach.

Finally, research on the development of biophysical remote sensing methods is another approach for collecting parameters needed for process models. Biophysical properties are estimated from canopy characteristics that are determined by inversion of radiative transfer equations.

The International Geosphere–Biosphere Program (IGBP) is promoting a concept for a global land cover database that contains several remotely sensed data elements that collectively cover these three remote-sensing–model-parameterization strategies, including (IGBP, 1990):

- Basic calibrated remote sensing data;
- Preprocessed data planes;
- Attribute data sets on baseline and change; and
- Parametric data sets on baseline and change.

The IGBP identified NOAA AVHRR as the prime source of data for this land data set.

The U.S. Geological Survey's (USGS) EROS Data Center (EDC) has begun to develop 1 km resolution AVHRR time series data sets (Loveland et al., 1991) that parallel the conceptual definition of IGBP's global land cover database (Sadowski, 1990). EDC produces four levels of prototype AVHRR data products that are designed for both the biophysical and land cover data requirements of global change researchers (Figure 37-1):

- Processed daily AVHRR scenes over most of North America, and other areas of the Earth;
- Composite images for specified time intervals for the conterminous United States, Alaska, Mexico, and Eurasia;
- Time series sets for the same geographic areas; and
- Land cover characteristics prototype database for the conterminous United States.

DAILY PROCESSED AVHRR SCENE PRODUCTS

AVHRR data are available as part of NOAA's operational meteorological satellite program. While designed largely for meteorological applications, substantial progress has

AVHRR Land Data Products Developed at the EROS Data Center:

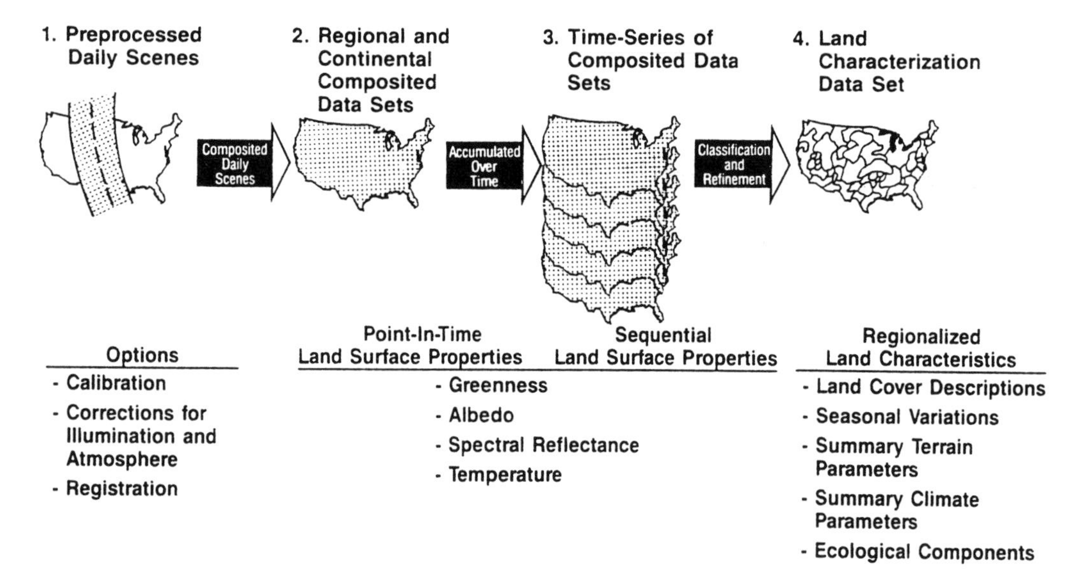

Figure 37-1: Four-stage sequence of U.S. Geological Survey EROS Data Center's prototype Advanced Very High Resolution Radiometer land data sets.

been made during the past decade in the application of the data for land characterization studies.

The AVHRR sensor orbits on a Television Infrared Observation Satellite platform. The current program configuration calls for two functional satellites in orbit at all times. Each satellite images the Earth twice per day, with one satellite acquiring early morning and early evening coverage, with the other collecting afternoon and late-night imagery. Key AVHRR imagery specifications include (Kidwell, 1991):

Image geometry	
Orbital width	2400 km
Resolution (nadir)	1.1 km square
Viewing look angle	56 degrees from nadir
Coverage interval	12 hours
Spectral channels (micrometers)	
Visible	0.58–0.68
Near infrared	0.72–1.1
Middle infrared	3.55–3.93
Thermal infrared	10.3–11.3
Thermal infrared	11.5–12.5

AVHRR offers a number of advantages for global change and environmental monitoring studies including (Harris, 1985):

- Appropriate resolution in relation to global coverage requirements;
- Data required to generate the types of land cover classes needed for global studies;
- Past and future continuity of data;
- Daily global coverage; and
- Reliable data for the analysis of land cover and vegetation dynamics on a continental scale.

The daily coverage, in particular, enhances the likelihood that cloud-free information can be obtained for specific periods, and makes it possible to monitor changes in surface processes over short periods, such as a growing season (Miller, Howard, and Moore, 1988; Tappan et al., 1989; Justice et al., 1985).

EDC has the capability to receive 1 km AVHRR data directly covering much of North America through its High Resolution Picture Transmission facility (Figure 37-2). EDC also accesses the tape recorded Local Area Coverage data stream as it is relayed to selected ground stations. This allows access to data over any part of the globe in which NOAA has scheduled tape recorded data acquisition. EDC provides processed AVHRR products for environmental monitoring and global change studies. Selected daily data are processed into single-date data sets that are:

- Calibrated using prelaunch calibration, or other community standard calibration coefficients for the visible and near-infrared (NIR) channels, and with sensor blackbody-based coefficients for the thermal channels;
- Geographically referenced to user-selected map projections, coordinate systems, and resolutions (Kelly and Hood, 1991);

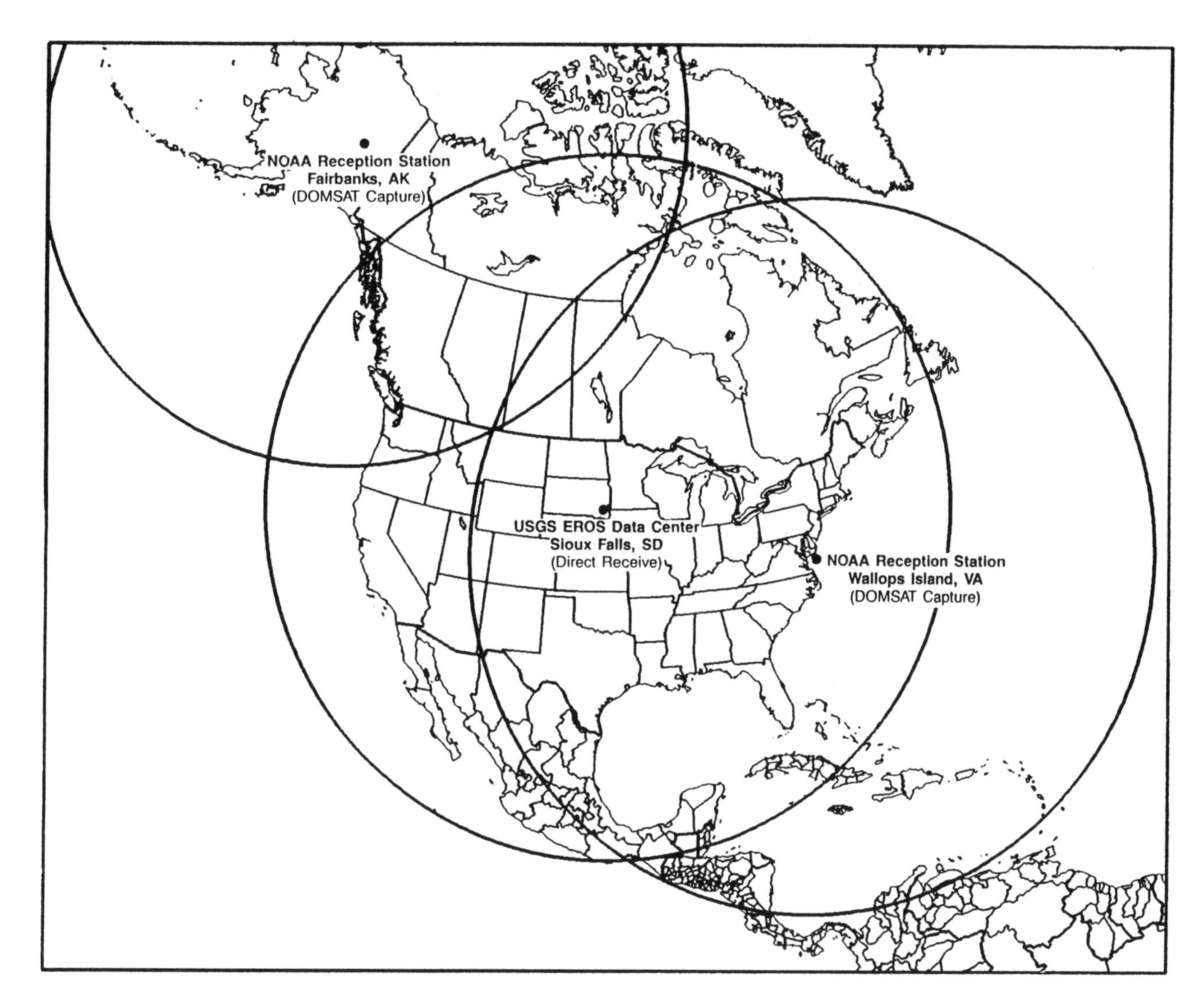

Figure 37-2: Direct reception coverage area for U.S. Geological Survey EROS Data Center's daily Advanced Very High Resolution Radiometer scene acquisitions.

- Enhanced with the generation of ancillary channels including normalized difference vegetation index (NDVI), and viewing geometry (solar angle, solar azimuth, and satellite zenith).

The NDVI transformation is calculated using the visible and NIR AVHRR channels (1 and 2, respectively) with the following formula:

$$\text{NDVI} = (\text{NIR-visible})/(\text{NIR+visible})$$

NDVI relates to standing green biomass and is an indicator of relative amounts of photosynthetic activity. For instance, studies have concluded that NDVI is correlated to several physical properties including leaf area index and primary production, and indirectly related to evapotranspiration, surface roughness, and stomatal resistance (Pinker, 1990; Spanner et al., 1990; Dorman and Sellers, 1989).

COMPOSITE AND TIME SERIES COMPOSITE DATA SETS

The daily processed AVHRR scenes have a fundamental limitation due to degradation of the satellite signal caused by clouds, haze, dust, or other attenuating factors. To overcome daily data contamination problems, composite data sets derived from multiple daily images are needed. The time period for compositing must still permit the monitoring of seasonal phenomena. For this reason, EDC produces 14-day AVHRR composite data sets from the 1 km AVHRR data for the conterminous United States (Plate 37-1).

The composite data sets are prepared to minimize the contaminated data that were collected over the composite period. Using a maximum NDVI compositing technique, the data corresponding to the highest NDVI value (or "greenest") per pixel are selected for the periodic composite data set. The maximum NDVI compositing procedure requires that each daily observation be georeferenced and that the NDVI be derived. On a per-pixel basis, each NDVI value for all daily observations in the composite period is examined and only the data corresponding to the date with the highest value are retained for each pixel. The maximum NDVI technique assumes that any contamination of the spectral signals will reduce the NDVI value from that of a "clean" signal. Therefore, the maximum NDVI represents the best possible pixel value for a particular location within each composite period (Holben, 1986).

The composite data sets assembled at EDC include ten data channels for each period (Eidenshink, 1992):

- Channels 1–5 (calibrated);
- NDVI;
- Satellite zenith;
- Solar zenith;
- Relative solar/satellite azimuth;
- Date of pixel observation.

The set of composite images, over time, form a time series for studying the dynamics of land surface processes. The spectral channel data and NDVI, in particular, may serve as indicators of landscape characteristics.

Attempts have also been made to use the visible and NIR channels as surrogates for solar albedo (Henderson-Sellers, 1980). However, standard methodologies are not available to convert satellite-sensed radiances routinely into information on surface albedo. The difficulties are related to the fact that satellites measure only the Earth atmosphere reflectance in narrow spectral intervals at fixed local time, and are further limited by the lack of internal calibration within the AVHRR sensor that monitors changes in the radiometric sensitivity of the reflective spectral channels over time (Townshend et al., 1991). This latter factor also limits the utility of NDVI as a quantitative measure of other surface parameters.

While there are limitations to AVHRR time series composite data sets, they still provide a valuable, unique means for monitoring dynamic events. The true value of AVHRR data arises primarily from their multitemporal use, since this allows land surfaces to be characterized using their seasonal or phenological variations.

LAND COVER CHARACTERISTICS DATABASE

As much as there is a need for biophysical land data, there is also an immediate requirement for reliable information on land cover at global scales. Currently, the time-varying physiological parameters used in land process models are more frequently taken from reports of field measurements in the literature and extrapolated to general biomes than measured directly through biophysical remote sensing methods (Dorman and Sellers, 1989). The opportunity exists to develop a global land cover database from the historical archive of AVHRR data. NDVI, particularly if viewed over time, clearly illustrates patterns of land cover, but the level of effort to produce reliable land cover data exceeds the capabilities of most researchers (Henderson-Sellers et al., 1986).

The fourth level of products in EDC's AVHRR land data series involves the development of prototype land cover characteristics databases that can be applied to environmental monitoring and modeling applications. An effort is currently underway to test a conceptual framework for land characteristics databases using the 1990 conterminous United States AVHRR time series data set (Figure 37-3).

Classification methods

The classification methods used in the prototype were based on several design considerations. The methods used had to be (Loveland, 1991):

- Applicable and repeatable over large (that is, continental or global) areas;
- Capable of measuring significant seasonal, ecological, and cultural variations in land cover;
- Applicable to very large data sets;
- Able to deal with data varying in quality; and
- Capable of producing flexible results that are not application specific.

A multitemporal, multisource classification approach using 1 km AVHRR/NDVI data from March–November 1990, augmented with terrain, ecoregions, and climate data was selected for use in the conterminous United States test (Loveland, 1991). This strategy relied heavily upon overlaying, exploring, and interrelating the disparate multisource data. The classification steps involved:

- Stratification of vegetated and nonvegetated land in order to concentrate land cover region definition on vegetated landscapes;
- Unsupervised classification methods to define 70 spectral–temporal ("seasonally distinct") classes (Plate 37-2);
- Evaluating, labeling, and characterizing the 70

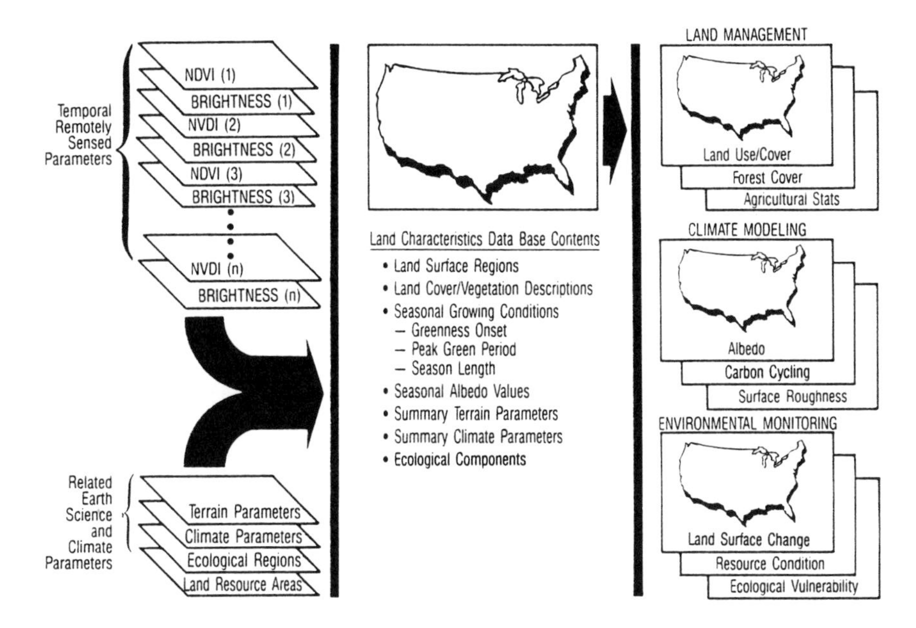

Figure 37-3: Conceptual strategy for large-area land characterization includes use of remote sensing and multisource data to create a spatial database that includes seasonally distinct land cover regions and associated attributes that can be tailored to a number of disparate applications.

classes using a combination of graphical, statistical, and visual analysis of the classified data; and

- Postclassification refinement in instances where the 70 classes were not uniquely associated with a single cover type.

This four-step process resulted in a final classification with 157 seasonally distinct land cover regions.

The conterminous United States land cover characteristics database comprises three components: (1) 157 seasonally distinct land cover regions, (2) descriptive and quantitative attributes for each region, and (3) the spatial databases used to produce the first two components.

Database components

Land cover regions. The seasonally distinct land cover regions are composed of relatively homogeneous land cover associations (for example, similar flora and physiognomic characteristics), which have unique phenology, including onset, peak, and seasonal duration of greenness (Figure 37-4 illustrates the seasonal elements of selected regions). These land cover regions were developed to have ecological meaning so the associated attributes can serve as indicators of surface processes.

Attributes. Detailed quantitative and descriptive attributes characterize each region. The attributes for each region can be likened to a spreadsheet and permit updating, calculating, or transforming the entries into new parameters or classes. This provides the flexibility to use the land characteristics database in many models without extensive modification of model inputs. Attributes included or planned are:

- Descriptions of vegetation composition and physiognomy;
- Quantitative seasonal characteristics including monthly NDVI and seasonal parameters (onset, peak, and duration of greenness);
- Site characteristics including topography, soils, climate, and ecosystem membership; and
- Translation tables linking the regions to common land cover classification schemes (such as UNESCO, USGS Anderson System, and the vegetation types used in the Simple Biosphere Model and the Biosphere Atmospheric Transfer Scheme).

As research evolves, other attributes will be added to the land characteristics database. For example, surface albedo, primary production, evapotranspiration, and leaf area index attributes could be added to the regions when consensus methods are reached for their calculation.

Spatial databases. These spatial data sets, all with 1 km resolution, were used either to develop or characterize the seasonally distinct land cover regions. Included in these data are:

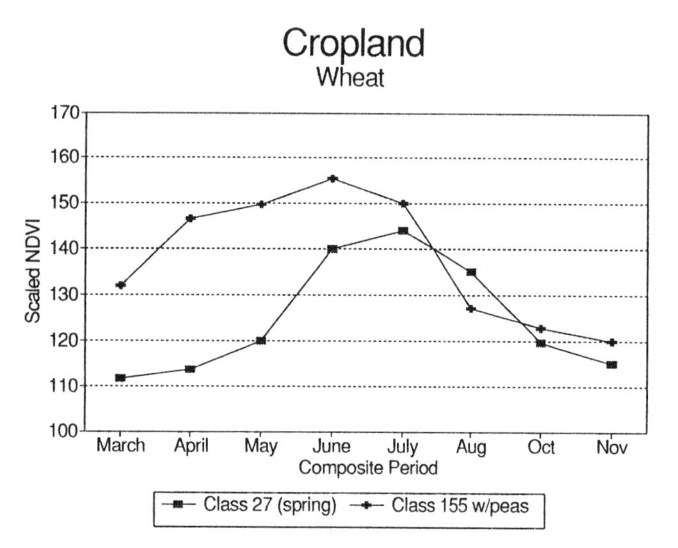

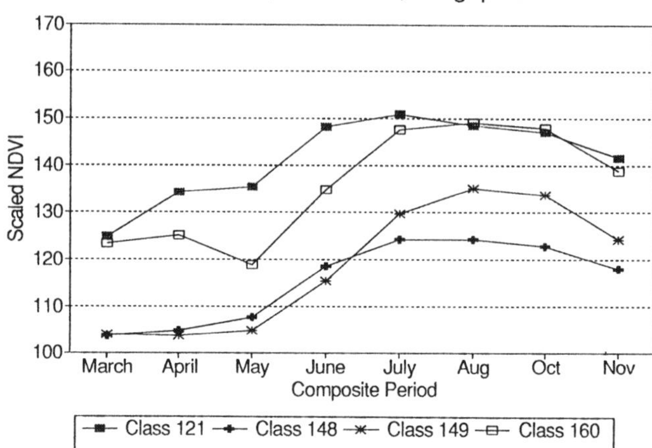

Figure 37-4: Normalized difference vegetation index seasonal profiles for selected land cover regions. The graphs illustrate the phenology and relative productivity of selected agricultural and forest types.

- 1990 time series AVHRR composite data;
- Digital elevation data;
- Climate station data including length of frost free period, average annual precipitation, average monthly precipitation, and monthly mean temperature;
- Ecoregions from the U.S. Environmental Protection Agency (Omernik, 1987);
- Major land resource areas developed by the U.S. Department of Agriculture/Soil Conservation Service (USDA/SCS, 1981); and
- Land use/land cover data from the USGS (Fegeas et al., 1983).

Comments

In general, homogeneous land cover regions are well defined if they are composed of relatively large, regular landscape patches. In areas such as the eastern United States, it was difficult to identify pure seasonally distinct land cover regions. In these cases, regions must be described according to the complexities of land cover that comprise the mosaic. Some cover types, such as wetlands and urban areas, cannot be reliably mapped using these procedures. Either their small size (particularly wetlands) or their complex cover conditions (urban lands) make classification very difficult.

PLANS

The four levels of AVHRR land data products developed at EDC are currently limited to selected areas of the globe. A global effort is desirable but not feasible until a systematic effort to collect the daily 1 km AVHRR coverage is undertaken. Fortunately, plans endorsed by IGBP and the National Aeronautics and Space Administration (NASA) call for the initiation of at least a 1-year effort to assemble such a data set (IGBP, 1991). NASA and EDC, through the Earth Observing System program, will initiate this effort in early 1992. While specific product development strategies are still undefined, it will be possible to develop a consistent four-stage land data products program for all of the Earth's land masses.

REFERENCES

Dorman, J.L., and Sellers, P.J. (1989) A global climatology of albedo, roughness length and stomatal resistance for atmospheric general circulation models as represented by the simple biosphere model (SiB). *Journal of Applied Meteorology* 28: 833–855.

Eidenshink, J.C. (1992) The 1992 conterminous U.S. AVHRR data set. *Photogrammetric Engineering and Remote Sensing* 58(6) (in press).

Fegeas, R.G., et al. (1983) USGS digital cartographic data standards: land use and land cover digital data. *U.S. Geological Survey Circular 895-E*, Washington, DC: U.S. Geological Survey.

Goward, S.N., Tucker, C.J., and Dye, D.G. (1985) North American vegetation patterns observed with the NOAA-7 Advanced Very High Resolution Radiometer. *Vegetatio* 64: 3–14.

Harris, R. (1985) Satellite remote sensing: low spatial resolution. *Progress in Physical Geography* 9(4): 600–606.

Henderson-Sellers, A. (1980) Albedo changes—surface

surveillance from satellites. *Climate Change* 2(3): 275–281.

Henderson-Sellers, A., Wilson, M.F., Thomas, G., and Dickinson, R. E. (1986) Current global land-surface datasets for use in climate-related studies. *NCAR Technical Note 272+STR*, Boulder, CO: National Center for Atmospheric Research, 110 pp.

Holben, B. (1986) Characteristics of maximum value composite images from temporal AVHRR data. *Monitoring the Grasslands of Semi-Arid Africa Using NOAA-AVHRR Data*, pp. 1417–1434.

IGBP (1990) Global change. *Report No. 12*, Stockholm, Sweden: IGBP Secretariat.

IGBP (1991) *Improved Global Data for Land Applications*, Paris: Universitś de Paris, 60 pp.

Justice, C.O., Townshend, J.R.G., Holben, B.N., and Tucker, C.J. (1985) Analysis of the phenology of global vegetation using meteorological satellite data. *International Journal of Remote Sensing* 6: 1271–1318.

Kelly, G., and Hood, J. (1991) AVHRR conterminous United States reference dataset. *Technical Papers, 56th ACSM-ASPRS Annual Convention, Baltimore, Maryland, March 1991*, Bethesda, MD: ASPRS/ACSM, Vol. 3, pp. 232–239.

Kidwell, K.B. (1991) *NOAA Polar Orbital Data Users Guide*, Washington, DC: NOAA National Climate Center.

Loveland, T.R. (1991) A strategy for large-area land characterization—the conterminous U.S. example. *Proceedings, U.S. Geological Survey Global Change Research Forum*, Reston, VA: U.S. Geological Survey, Circular 1086 (in press).

Loveland, T.R., Merchant, J.M., Ohlen, D.O., and Brown, J.F. (1991) Development of a land-cover characteristics database for the conterminous U.S. *Photogrammetric Engineering and Remote Sensing* 57(11): 1453–1463.

Miller, W.A., Howard, S.M., and Moore, D.G. (1988) Use of AVHRR data in an information system for fire management in the western United States. *Proceedings, 20th International Symposium on Remote Sensing of Environment, Nairobi, Kenya, December 1986*, Ann Arbor, MI: ERIM, Vol. 1, pp. 67–79.

Omernik, J.M. (1987) Ecoregions of the conterminous United States. *Annals of the Association of American Geographers* 77(1): 118–125.

Pinker, R.T. (1990) Satellites and our understanding of the surface energy balance. *Paleogeography, Paleoclimatology, Paleoecology* (Global and Planetary Change Section) 82: 321–342.

Sadowski, F.G. (1990) Prototype land data sets for studies of global change. *Proceedings, 10th International Geoscience and Remote Sensing Symposium, Washington, DC, May 1990*, New York, NY: IEEE, Vol. 2, p. 1235.

Spanner, M.A., Pierce, L.L., Running, S.W., and Peterson, D.L. (1990) The seasonality of AVHRR data of temperate coniferous forests: relationship with leaf area index. *Remote Sensing of Environment* 33: 97–112.

Tappan, G.G., Howard, S.M., Loveland, T.R., Tyler, D.J., and Moore, D.G. (1989) Seasonal vegetation monitoring with AVHRR data for grasshopper and locust control in West Africa. *Proceedings, 22nd International Symposium on Remote Sensing of Environment, Abidjan, Cote D'Ivoire, October 1988*, Ann Arbor, Michigan: ERIM, Vol. 1, pp. 221–234.

Townshend, J.R.G., and Justice, C.O. (1988) Selecting the spatial resolution of satellite sensors required for global monitoring of land transformations. *International Journal of Remote Sensing* 9(2): 187–236.

Townshend, J.R.G., Justice, C.O., Li, W., Gurney, C., and McManus, J. (1991) Global land cover classification by remote sensing: present capabilities and future possibilities. *Remote Sensing of Environment* 35: 243–255.

Townshend, J.R.G., Justice, C.O., and Kalb, V. (1987) Characterization and classification of South American land cover types. *International Journal of Remote Sensing* 8(8): 1189–1207.

USDA/SCS (1981) Land resource regions and major land resource areas of the United States. *Agriculture Handbook 296*, Washington, DC: U.S. Department of Agriculture/Soil Conservation Service.

38

Digital Soils Databases for the United States

DENNIS J. LYTLE

A soil survey is a field investigation supported by laboratory data resulting in a soil map displaying the geographic distribution of different kinds of soils. It lists the soil properties and defines, classifies, and interprets the soils for various uses. The soil survey data are available in digital form (spatial and attribute database) for limited areas and in the traditional published text format for most of the country. The SCS is in the process of converting many of these analog products into digital spatial databases.

People involved in environmental and earth sciences, natural resource management, and land use planning recognize soil surveys as an important source of detailed information about the landscape. However, they often have difficulty using soil data as presented in standard U.S. Department of Agriculture (USDA) Soil Conservation Service (SCS) soil survey publications. Often the maps are not in a digital form that permits easy integration with other resource information in a GIS for the purpose of analysis. As a result, the SCS, as part of its effort in the National Cooperative Soil Survey (NCSS), has established three digital soil geographic databases to improve the storage, manipulation and retrieval of soil map information. The three databases are:

1. Soil Survey Geographic Database (SSURGO)
2. State Soil Geographic Database (STATSGO)
3. National Soil Geographic Database (NATSGO)

These three soil geographic databases have been devised to represent soil data at different scales. Each database includes digitized soil map unit delineations linked with attribute data for each map unit, giving the extent and the properties of each of these soils. With these computerized databases, users will be able to retrieve, analyze, display, and store soils data more efficiently. Also, by using GIS they will be able to integrate soil data with other spatially referenced resource and demographic data.

DEVELOPMENT OF SOIL SURVEYS AND DESCRIPTION OF THE DATABASES

Soil data are somewhat unique in that spatial distribution and variability on the landscape generally occur in a predictable pattern that is related to landforms. During field studies a soil scientist visually identifies landscapes and their associated landforms. By relating the kinds of soils to specific landforms and investigating respective soil properties, soil map unit concepts are formed. With these concepts soil map units are delineated, and the properties of the soil components are described (USDA, 1951). Components of map units of each geographic database are generally phases of soil series. Miami silt loam with 0 to 2 percent slopes is an example of a soil series phase, where Miami is the soil series name and silt loam, 0 to 2 percent slopes are the phase portion of the larger range of properties for Miami Series. The full taxonomic classification of Miami Series is "fine-loamy, mixed, mesic Typic Hapludalfs" (USDA, 1990). Series are the lowest level in the U.S. system of soil classification (USDA, 1975). Phases of series permit the most precise interpretation by the data user. Interpretations are developed and displayed differently for each geographic database to be consistent with the level of detail expressed.

Included with each soil geographic database is an attribute data file for each map unit component. The Soil Interpretations Record (SIR) database (SCS, 1983) provide these attribute data. The SIR database contains data for more than 25 soil properties for the approximately 18,000 soil series recognized in the U.S. Information is given for each major layer (layer data) of the soil and for the soil as a whole (component data). This database also contains interpretations (interpretation data) for numerous uses. Table 38-1 provides a listing of these data.

SSURGO has been designed to be used primarily for large-scale natural resource planning and management by landowners and local government agencies. Soil maps in the SSURGO databases are prepared using field methods that require direct observation along traverses and determination of map unit composition by field transects. They

Table 38-1: Soil database data element

Component data	
Component name	Capability subclass: irrigated and nonirrigated
Kind of component	Prime farmland codes
Taxonomic classification	Crop yields: irrigated and nonirrigated:
Class determining phase	common crops
Soil interpretation record: number	Woodland limitations and ratings for:
Percent of map unit	erosion hazard
Slope: low or high	equipment limitations
Flooding: frequency, duration, months of the year	seeding mortality
Drainage	windthrow hazard
Water table: depth, kind, months of year	plant competition.
Ponding: depth, kind, months of year	Tree species: existing and potential
Hydric: yes or no	Ordination symbol
Bedrock: depth to, and hardness	Site index
Cemented pan: depth to, and hardness	Windbreaks: recommended trees to plant
Subsidence depth: total and initial	Wildlife habitat suitability
Hydrologic group	Range and woodland understory vegetation:
Potential frost action	plant species
Corrosion: steel and concrete	percentage composition
Capability class: irrigated and nonirrigated	potential production
Layer data	
USDA texture	Available water capacity
Layer depths	Bulk density
T & K erosion factors	Organic matter percent pH
Wind erodibility group	Cation exchange capacity
Rock fragment 3–10 in. and > 10 in.	Salinity
Percent of soil passing #4, 10, 200 sieves	Sodium absorption ratio
Percent clay	Gypsum
Liquid limit and plasticity index	Permeability
UNIFIED and AASHTO class	Shrinkswell
Interpretive data (Restrictive features and ratings for):	
Sanitary facilities:	sand
septic tank absorption	gravel
fields	topsoil
sewage lagoon areas	Water management:
sanitary landfills	pond and reservoir area
daily cover for landfills	embankments, dikes, and levees
Building site development:	excavated ponds, aquifers
shallow excavations	drainage
dwellings without basements	irrigation
dwellings with basements	terraces and diversions
small commercial buildings	grassed waterways
local roads and streets	Recreation development:
lawns	camp areas
landscaping	picnic areas
golf fairways	playgrounds
Construction material:	paths and trails
roadfill	

Table 38-2: Status of SSURGO analog and digital soil surveys

	Analog surveys	Digital conversion
Acres completed	1.5 billion	200 million
Total U.S. (excl. Alaska)	1.7 billion	1.7 billion
Percent of U.S. complete	90	12
Cost per acre	$2.00	$0.075
Total investment to date	$3.0 billion	$15 million
Remaining investment	$400 million	$112.5 million

are made using National Cooperative Soil Survey (NCSS) standards (SCS, 1983). The maps comprising the basis for this database are published at scales usually ranging from 1:12,000 to 1:31,680. The digital files are composed of line segments that are generated from maps that are either scanned or manually digitized. Map bases are U.S. Geological Survey (USGS) 7.5 minute quadrangles or orthophotographs. SSURGO is being developed so that most new surveys started will be in digital form when complete. A very large effort will be needed to convert the existing soil survey data to a digital form. Table 38-2 lists the status of soil surveys in the U.S. and the status of the conversion of these surveys to digital form. It also lists the estimated cost of converting these data. Most of the 200 million acres converted to a digital format has been completed under cooperative cost share agreements with state and local agencies. There is an extremely high demand to convert the remaining soil survey to digital form, especially in urbanizing areas and in areas where soil surveys are used as a basis for land taxation.

STATSGO is intended to be used primarily for multicounty, state, and regional natural resource planning, management, and monitoring. Soil maps for STATSGO are made by generalizing from more detailed SSURGO soil survey maps. Map unit composition for STATSGO was determined by transecting or sampling areas on the more detailed maps, usually on a paper copy of the maps, and expanding data to characterize the whole map unit. A few states that had complete or nearly complete detailed soil surveys used acreage data from the detailed soil surveys to determine composition. Where SSURGO maps were not available, other soil data and data about the geology, topography, vegetation, and climate were assembled and evaluated; soils of similar areas were studied, and a judgment of the probable classification and extent of the soils was made. The STATSGO soils maps comply with national standards (SCS, 1984). They were prepared using the USGS 1:250,000 scale topographic quadrangles as base maps. Line segment maps were both scanned and manually digitized (USDA, 1991). STATSGO is currently available for a number of states, most of which are in the northeastern U.S. It will be available for the entire U.S., excluding Alaska, by October 1992.

NATSGO is used for national and regional resource appraisal, planning, and monitoring. The boundaries of the Major Land Resource Areas (MLRAs) on the national MLRA map, vintage 1978, were used for the NATSGO spatial layer (USDA, 1981). These boundaries were modified and digitized in 1983. The MLRA boundaries were developed primarily from state general soil maps. Information on land use, elevation and topography, climate, water, and potential natural vegetation were also used to locate boundaries. Map unit composition for NATSGO was determined by sampling done as part of the SCS 1982 National Resource Inventory (SCS, 1979). Sample data were expanded for MLRAs. Sample design was statistically significant to state portions of MLRAs.

The NATSGO map was compiled and manually digitized at a scale of 1:2,000,000. The source map was published in Agriculture Handbook 296 at a scale of 1:7,500,000 (USDA, 1981). Digital versions in SCS are usually displayed at 1:5,000,000. NATSGO is available for the conterminous U.S.

The SSURGO, STATSGO, and NATSGO data can be ordered from the U.S. Department of Agriculture, Soil Conservation Service, National Cartographic Center, 501 Felix St., P.O. Box 6567, Fort Worth, Texas 76115.

INTERPRETATION OF THE DATABASES

As mentioned earlier, interpretations are made and displayed differently for each soil geographic database to be consistent with the level of detail expressed in the maps. SSURGO map units contain one to three components (Figure 38-1). In general, the more detailed the map and

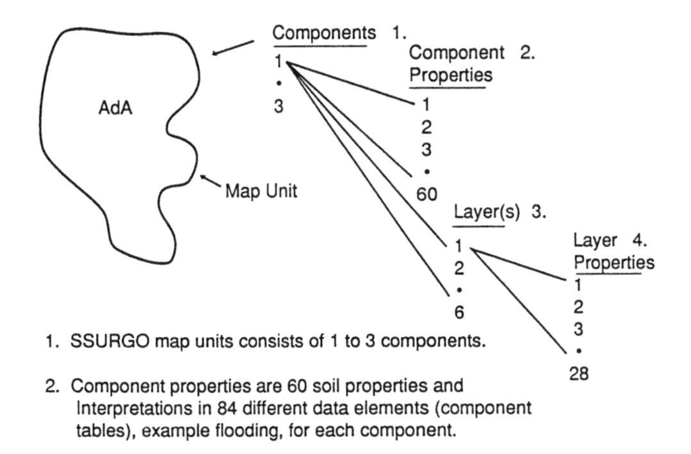

1. SSURGO map units consists of 1 to 3 components.
2. Component properties are 60 soil properties and Interpretations in 84 different data elements (component tables), example flooding, for each component.
3. From1-6 soil layers for each component.
4. Layer consists of 28 soil properties for each layer and 53 different data elements (layer table). For example, clay percent

Figure 38-1: SSURGO map unit.

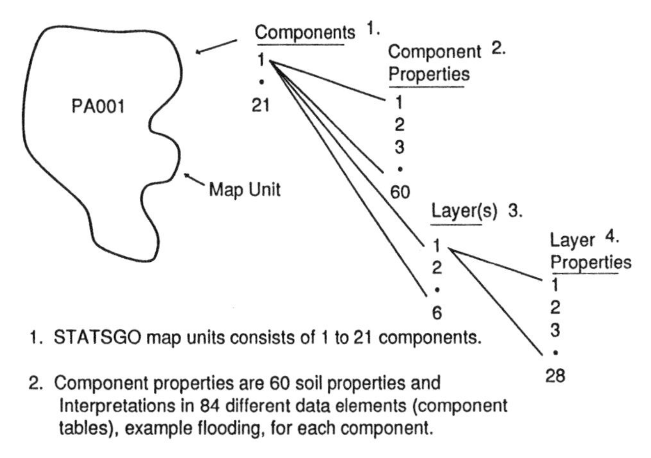

1. STATSGO map units consists of 1 to 21 components.
2. Component properties are 60 soil properties and Interpretations in 84 different data elements (component tables), example flooding, for each component.
3. From 1-6 soil layers for each component.
4. Layer is comprised of 28 soil properties for each layer and 53 different data elements (layer table). For example, clay percent.

Figure 38-2: STATSGO map unit.

the larger the map scale, the fewer the number of components per map unit. Data are given in the database for the percent of the area that each component represents. For instance, in a Lyman–Tunbridge complex, the Lyman component may make up 45 percent and the Tunbridge component may make up 35 percent, with the remaining 20 percent being inclusions of named and unnamed soils that have similar or dissimilar properties. The percentages of the individual included soils are often unknown and thus are not represented in the database. The percentages given in the database do not necessarily represent the composition for each map unit delineated, but represent the average composition of the map unit as it occurs throughout the survey area, usually a county. Interpretative maps of these data are usually made by presenting either the most limiting soil component rating or the dominant (largest percent) soil component. Thus, a good or fair or poor, or slight or moderate or severe designation is possible for a map unit.

GIS technology makes it possible to query the database to see if a particular component meets a criterion or has a good or fair or poor rating. The result is then accumulated into a category file. The second component is tested, and the result is accumulated. When all components in an area have been tested, the percentages of components are aggregated for each category by map unit ID (MUID). A map can then be generated for each category (good, fair, or poor) with a legend showing the percentage of map units (not delineations) that meet the criteria. Interpretations made and displayed in this way allow the user to make a more informed decision based on the percentage meeting the criteria. For example, instead of the entire map unit being rated poor because the most limiting component rated poor, as in the case of Lyman–Tunbridge, perhaps only 45 percent rates poor, 35 percent rates fair, and the rest has an unknown rating.

The last approach is the one taken in interpretation of STATSGO (Figure 38-2) map units. STATSGO map units may contain up to 21 components, making it impossible to decide on a dominant or most limiting component. For additional discussion of analysis of STATSGO data see Bliss and Reybold (1989).

Interpretations of NATSGO map units (Figures 38-3 and 38-4) are made and presented in a way very similar to STATSGO. There are differences though, because of the way the data were collected and stored. The spatial data for NATSGO, as stated earlier, is the MLRA map. The attribute data come from the 1982 National Resources Inventory (NRI). The NRI is conducted every 5 years by the SCS. A statistical sampling procedure is used to conduct the NRI. The Primary Sampling Units (PSUs) are selected according to a stratified random sampling procedure. The PSUs are typically 160 acres, although in some parts of the country other unit sizes are used. Some information is collected on the basis of the entire PSU; additional data are collected at three (or fewer) points within the PSU. The 1982 NRI sampled 352,786 PSUs and over one million points. The statistical design called for at least a 2 percent sample of non-Federal lands in the U.S., excluding Alaska. At each PSU sample point the soil component is linked to the MLRA.

An analysis of the soils data in NRI involves a query of each PSU point to see if it meets a preselected criterion. If it does, the area that the point represents (the expansion factor) is accumulated into a variable. If the point does not meet the criterion, then the area is accumulated into

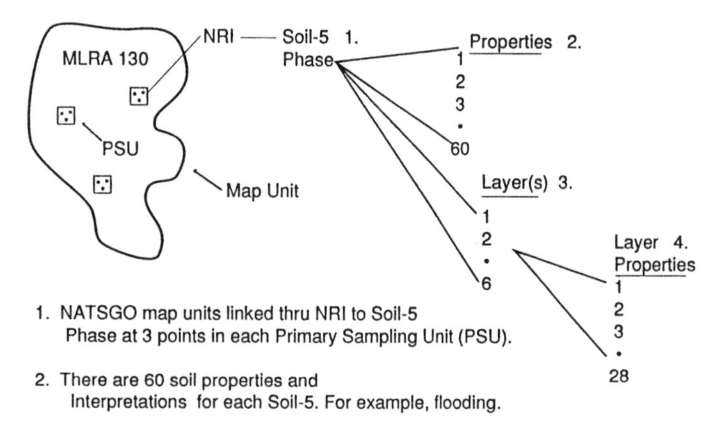

Figure 38-3: NATSGO map unit.

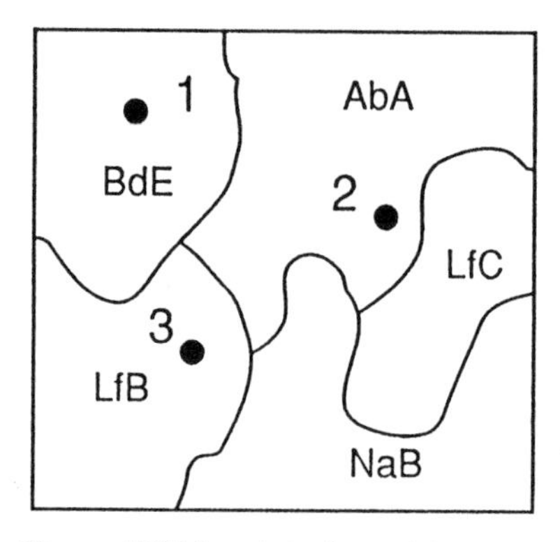

Figure 38-4: 1982 NRI primary sampling unit.

a second variable. When all the points have been processed, the sum of the two areas should always equal the total area of non-Federal land and water in the MLRA. The percentage of the MLRA that meets the criteria can then be calculated and displayed in map form. Because of the lack of information for Federal lands, analysis should be performed in a GIS with a mask eliminating federal lands. Any maps produced should exclude federal lands in order to avoid giving false impressions (Bliss, 1989).

To assure consistent, accurate interpretations, and to increase the use of soil geographic data, SCS is developing interfaces between the spatial and attribute data that incorporate the methods previously discussed. These interfaces have been developed for the SCS AT&T UNIX system 5, INFORMIX and GRASS software, and AT&T 6386 hardware. They are being developed by commercial vendors.

The methods previously discussed dealt with interpreting attributes to components of map units. GIS technology also offers methods to display the accuracy of the map line and methods to capture soil data in a spatial form that was previously not recorded because of the size of the area. GIS technology will revolutionize not only the way data are analyzed and displayed, but it will also affect the way data are gathered. Further GIS technology is driving the conversion of analog soils data to digital formats.

FUTURE ACTIVITIES

Soils digital spatial data are one of 10 layers of information that will initially become part of a National Geo-Data System (NGDS). In concept, NGDS will be a system of Federal digital databases independently owned and held, but linked to a central source for access by users. Responsibility for developing a spatial data transfer standard (SDTS) for digital soils data that will allow their inclusion in the NGDS lies with the Soils Subcommittee of the Federal Geographic Data Committee (FGDC). The SCS chairs the subcommittee, which is made up of Federal agencies that "collect or finance the collection of soils digital spatial data as part of their mission, or have direct application of these data through their legislated mandate" (FGDC, 1991).

A generic SDTS (USGS, 1990) provides a general framework for digital spatial data transfer. A Federal profile will be developed for data types such as polygonal, line, and point data. In addition to establishing a structure and format for the spatial data, SDTS requires a data dictionary and a statement of data quality for both the spatial and attribute data. The soil subcommittee will be involved in the development, review, and approval of SDTS for soils data and its submission as a Federal Information Processing Standard (FIPS).

Another activity that is underway is the joint development of a 1:30,000,000 soil map of the world involving the SCS, U.S. Forest Service, and the U.S. Geological Survey. This data set is built on the United Nations Food and Agriculture Organization soil map published in 1990, which has been converted to the U.S. Soil Taxonomy polygon by polygon. Interpretations will be possible from the taxonomic classification as well as from an attribute data set that will be developed from world soil pedon data. A similar effort is planned at the 1:5,000,000 scale, and possibly at the 1:1,000,000 scale for the world using the U.S. Soil Taxonomy.

Difficulties exist in developing attribute databases for these small-scale maps. Most of the data that are available are pedon laboratory data for sample points. These data provide estimates of water holding capacity, organic carbon, fertility, etc. The challenge is to find a meaningful way to expand these data to a larger geographic area. The most promising means is by classifying each pedon into the U.S. Soil Taxonomy and relating the pedon to a geographic area with the same classification; thus, the 1:5,000,000 and 1:1,000,000 maps mentioned earlier. Other means of extrapolating the pedon data must also be explored. The Soil and Terrain digital database (SOTER) is part of the United Nations Environment Program Global Assessment of Soil Degradation project to develop a global soil data set; however, efforts have not advanced because of funding limitations and because of problems with too much detail in the attribute information. The project is being coordinated by the International Soil Reference and Information Center. It has a nominal scale of 1:1,000,000 and a 10 to 20 year time frame for completion.

RECOMMENDATIONS

Standards for the conversion of analog data to a digital form must be made available to all those who are doing conversion. Cooperative efforts must be developed to demonstrate the proper use of soils data with other "layers" of natural resource data in GIS. A major focus of these cooperative efforts should be to develop methods to communicate the accuracy of maps that result from the combination of natural resource data of varying scales and accuracies. Also, very few resources are currently being devoted to development of global soil data sets. The technology and expertise exists in numerous U.S. institutions and organizations to develop these data sets, and several joint efforts have been proposed. These data will be critical for modeling activities associated with the Global Change Program. Development must start now if data are to be available for the day when models and systems can handle these larger-scale data sets. Indeed that day must be here now.

REFERENCES

Bliss, N.B. (1989) A natural resource database: techniques for linking the Major Land Resource Area Map, the 1982 National Resources Inventory and the Soil Interpretation Record Databases in a geographic information system. *Final report of work performed by the EROS Data Center, Sioux Falls, SD for the Soil Conservation Service, Washington, DC.*

Bliss, N.B., and Reybold, W.U. (1989) Small-scale digital soil maps for interpreting natural resources. *Journal of Soil and Water Conservation* (Jan-Feb): 30–34.

FGDC, Soils Subcommittee (1991) *Soils Digital Spatial Data (Exhibit A draft)*, Washington, DC: Federal Geographic Data Committee.

SCS (1979) *National Resources Inventory Instructions 1981–82*, Washington, DC: U.S. Department of Agriculture, Soil Conservation Service.

SCS (1983) *National Soils Handbook*, Washington, DC: U.S. Department of Agriculture, Soil Conservation Service.

SCS (1984) *National Instruction No. 430-302*, Washington, DC: U.S. Department of Agriculture, Soil Conservation Service.

USGS (1990) *Spatial Data Transfer Standard (draft)*, Washington, DC: U.S. Geological Survey.

USDA (1951) Soil survey manual. *Handbook 18*, Washington, DC: U.S. Department of Agriculture, 503 pp.

USDA (1975) Soil taxonomy: a basic system of soil classification for making and interpreting soil surveys. *Handbook 436*, Washington, DC: U.S. Department of Agriculture, Soil Conservation Service, 754 pp.

USDA (1981) Land resource regions and major land resource areas of the United States. *Handbook 296*, Washington, DC: U.S. Department of Agriculture, Soil Conservation Service, 156 pp.

USDA (1990) Soil series of the United States, including Puerto Rico and the U.S. Virgin Islands. Their taxonomic classification. *Miscellaneous Publication 1483*, Washington, DC: U.S. Department of Agriculture, Soil Conservation Service, 459 pp.

USDA (1991) State Soil Geographic Data Base (STATSGO) data users guide. *Miscellaneous Publication 1492*, Washington, DC: U.S. Department of Agriculture, Soil Conservation Service, 88 pp.

39

Development of a Continent-Wide DEM with Applications to Terrain and Climate Analysis

MICHAEL HUTCHINSON

Digital elevation models (DEMs) are commonly perceived as providing a suitable means for encoding the spatial distribution of topography and various dependent environmental quantities across the Earth's surface. In hydrological applications, topography can be used to quantify surface drainage and infiltration processes (Beven and Kirkby, 1979; Moore et al., 1988). In climate analysis, topography plays a critical role in methods for generating accurate continent-wide interpolations of various climate variables (Hutchinson and Bischof, 1983; Weiringa, 1986; Hutchinson, 1991). It must therefore be an important factor in the translation of broad-scale climate scenarios generated by general circulation models down to the fine resolutions required for environmental planning and management.

The proper use of DEMs poses a number of problems, not least the provision of both the elevation data and accompanying descriptions of relevant surface soil and vegetation properties, especially at the broad scales required for continental and global modeling. A procedure for calculating regular grid DEMs from surface specific point elevation data, contour line data, and stream line data has been developed by Hutchinson (1988,1989). Two important features of the method are its computational efficiency, so that it can be applied to the very large data sets required for continent-wide DEMs, and a drainage enforcement algorithm, which automatically removes spurious pits. This significantly improves the accuracy of the interpolated DEM and assists in the detection of data errors. It also simplifies subsequent hydrological analyses. The method has been used to generate a 1/40th degree (about 2.5 km) DEM for Australia from continent-wide coverages of point elevation and stream line data (Hutchinson and Dowling, 1991). This DEM improves on an existing 1/10th degree DEM (Moore and Simpson, 1982) in terms of spatial resolution, attention to drainage properties, and the incorporation of stream line data and additional point data. The primary point elevation data have also undergone extensive checking and revision.

The essential features of the DEM interpolation procedure and its application to the production of a DEM for Australia, a continent with extensive areas of low relief, are described. The results of applying the drainage analysis software of Jenson and Domingue (1988) are also briefly described, and an initial assessment is made of the utility of the DEM for terrain and climate analysis. This includes an assessment of the adequacy of the spatial resolution of the DEM. The DEM is found to respect known drainage divisions, making it a suitable candidate for supporting continent-wide hydrological analyses. The extension of the interpolation technique to determine stream lines and ridge lines from contour data automatically, typically at finer, regional scales, is also described.

THE DEM INTERPOLATION PROCEDURE

The DEM interpolation procedure incorporates several features designed to match the special features of elevation data and the intended hydrological and other modeling applications of the interpolated DEM. These elevation-specific features make most general-purpose interpolation algorithms inappropriate. They include:

1. Elevation data sets are very large. Continent-wide applications may require over a million data points. The usual forms of high-quality "global" interpolation methods, such as kriging and thin plate splines, in which every interpolated point depends explicitly on every data point, are computationally impracticable. On the other hand, local interpolation methods, which are usually based on partitioning space into small elements and fitting simple functions on each element, achieve computational efficiency at the expense of somewhat arbitrary restrictions on the form of the fitted function. Such techniques are sensitive to the positions of the data points, and spurious edge effects can be generated.

2. Point elevation data are normally measured with relatively small error, and the actual elevation surface is rarely very smooth. Both of these considerations make smooth trend surface interpolation inappropriate.
3. Most landscapes have many hill tops (local maxima) but relatively few sink points (local minima) so that most landscapes possess a connected drainage pattern. General-purpose interpolation techniques do not recognize this bias.
4. Elevation data are normally sampled nonrandomly in one of two ways—as surface-specific points (i.e., points on hill tops, ridges, and stream lines) or as contour lines. An elevation specific interpolator should take advantage of these sampling strategies. Thus, if all significant hill tops have been sampled, then the interpolated DEM should not permit local maxima away from data points. On the other hand, corners in elevation contours indicate the positions of stream lines and ridges.
5. Stream line data may also be available. These should be used to impose morphological constraints on the DEM so that elevations decrease monotonically down each stream line and there is a break in slope across each stream line.
6. In areas with low relief there may be few surface-specific point data and only widely spaced contour data lines. The interpolated DEM should be stable in such data-sparse areas and not generate spurious features unsupported by data. In particular, the drainage integrity of the DEM should be maintained in such areas.

The following subsections describe how the DEM interpolation procedure addresses each of these points.

The basic interpolation algorithm

An iterative finite difference interpolation technique, as described by Hutchinson (1989), has been adopted. The method has been designed to have the computational efficiency of a local method without sacrificing the continuity, rotation invariance, and general freedom from spurious features enjoyed by high-quality global interpolation methods. Importantly, because the fitted grid values are available at every stage during the iteration, it is relatively straightforward to monitor the drainage properties of the DEM and to impose appropriate morphological constraints via simple ordered chain conditions on the grid points. The interpolation problem is solved by minimizing a discretized rotation-invariant roughness penalty subject to the DEM having a specified small mean-square residual from the elevation data points. The roughness penalty is defined in terms of first- and second-order derivatives of the fitted grid (Hutchinson, 1989). The basic finite difference approach was initiated by Briggs (1974).

The method is essentially a discretized version of the thin plate spline technique (Wahba, 1990), for which the roughness penalty is usually linearized curvature of the fitted surface. In this case, however, the roughness penalty is modified, in a rotation-invariant manner, to allow the fitted DEM to follow the sharp changes in terrain associated with streams and ridges. Unmodified minimum curvature interpolation, though a good general purpose strategy, tends to maintain local trends too strongly and so generate spurious areas of overshoot and undershoot in regions with high relief (Hutchinson, 1989). The appropriate roughness penalty for elevation interpolation can be related to the statistical nature of the actual terrain (Frederiksen et al., 1985; Goodchild and Mark, 1987). A similar approach has been investigated by Mitas and Mitasova (1988).

The iteration technique employs a simple multigrid strategy that calculates grids at successively finer resolutions, starting with a coarse initial grid and successively halving the grid spacing until the final user-specified grid resolution is obtained. For each resolution, grid points are calculated by Gauss–Seidel iteration with overrelaxation (Golub and van Loan, 1983), subject to the specified roughness penalty, data residual, and ordered chain constraints. Iteration for each grid resolution terminates when the specified maximum number of iterations (normally 30) is reached. The computational cost of the procedure is therefore optimal in the sense that it is essentially proportional to the number of interpolated grid points. The computational efficiency is such that DEMs with over a million points may be easily interpolated on a modest UNIX workstation.

Drainage enforcement algorithm

The drainage enforcement algorithm has been described by Hutchinson (1989). It attempts to remove all sink points in the fitted DEM that have not been identified as such in input sink data, in recognition of the fact that sinks are usually quite rare in nature (Band, 1986; Goodchild and Mark, 1987). The action of the drainage enforcement algorithm is similar to the action taken by basin delineation programs (Jenson and Domingue, 1988) to overcome the problem of spurious sinks and is also related to the method of cartographic generalization suggested by Warntz (1975). The approach described here, which attempts to maintain a connected drainage pattern, provides a more secure physical basis for cartographic generalization than earlier partially lexicographic approaches. The drainage enforcement algorithm in fact provides the "judicious filtering" called for by Band (1986).

The essence of the algorithm is to recognize that each spurious sink is surrounded by a drainage divide containing at least one saddle point. If the sink is associated with

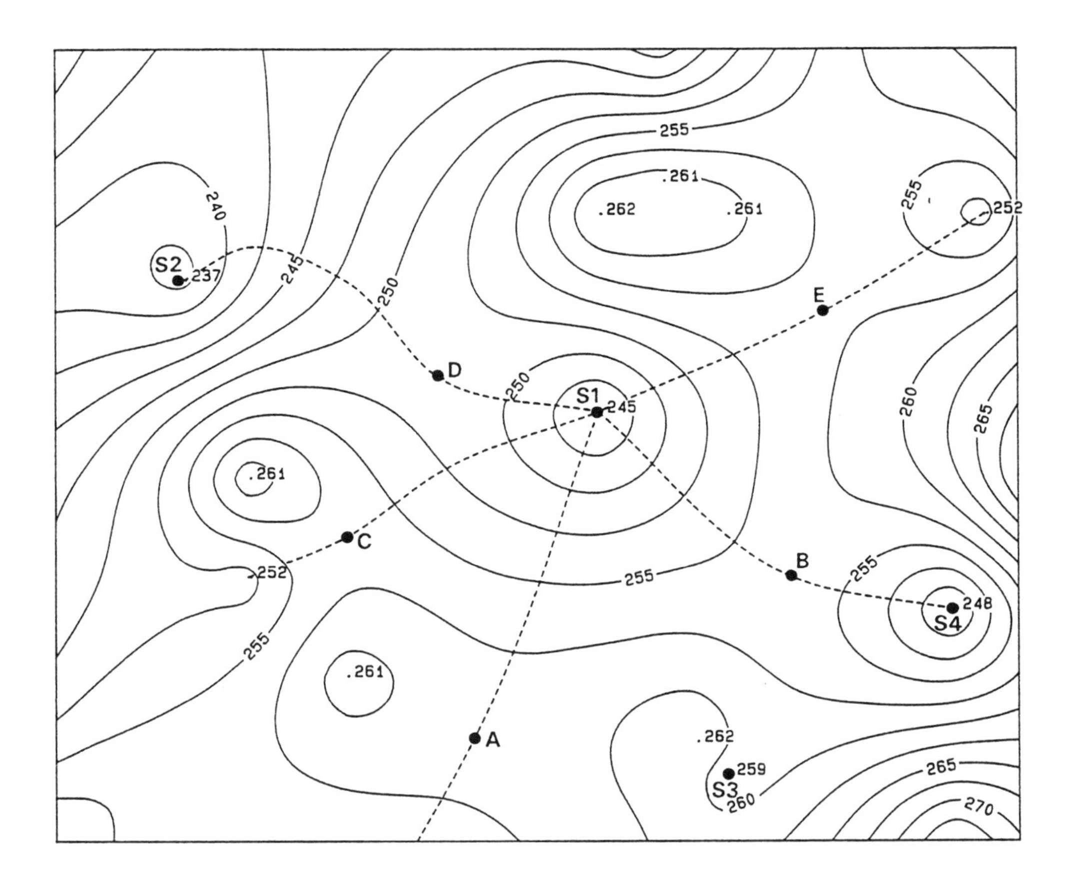

Figure 39-1: Saddle points A, B, C, D, E associated with the sink S1 via flow lines which are indicated by dashed lines. Additional sinks are denoted by S2, S3, S4. Data points are indicated by their heights in meters.

an elevation data point, then the lowest such saddle, provided it is not also associated with an elevation data point, identifies the region of the DEM that is modified to remove the sink. If, on the other hand, the lowest saddle point is associated with an elevation data point but the sink is not, then the sink and its immediate neighbors are raised above the height of the data point saddle. If neither the sink nor the lowest neighboring saddle is associated with elevation data points, then DEM points in the neighborhood of both the sink and the saddle are modified to ensure drainage. Finally, if both sink point and saddle point are associated with elevation data points, then a choice is made, depending on a user-supplied tolerance, between enforcing drainage and maintaining fidelity to the elevation data. This last situation arises in particular when generating coarse-resolution DEMs at continental scale.

The drainage enforcement algorithm proceeds concurrently with the iterative interpolation procedure described. For each grid resolution the DEM is periodically inspected for sinks and their accompanying lowest saddle points. This is illustrated in Figure 39-1, where the saddle points associated with the sink S1 are the points A, B, C, D, and E. Constraints effecting drainage clearance are then applied to the DEM by inserting ordered chain conditions that lead from each spurious sink, via the lowest associated saddle point, to a data point or existing ordered chain on the other side of the saddle. In the example of Figure 39-1, the lowest saddle associated with the sink S1 is the point D. Since this saddle point is not associated with a data point, and it leads to a sink S2 that is strictly lower than S1, an ordered chain is inserted from S1 to S2 via D, as shown in Figure 39-2. Each ordered chain in Figure 39-2 is made up of two flow lines leading from the lowest saddle associated with each sink in Figure 39-1.

The integration of the drainage enforcement algorithm into the interpolation process has made a significant difference to large parts of Figure 39-1 in a way that cannot be achieved by postprocessing methods that simply raise sinks to form a locally flat surface. In particular, derived parameters such as slope and aspect are much more realistic in the valleys associated with the inferred drainage lines than what would be calculated from Figure 39-1 either before or after simple postprocessing to remove the

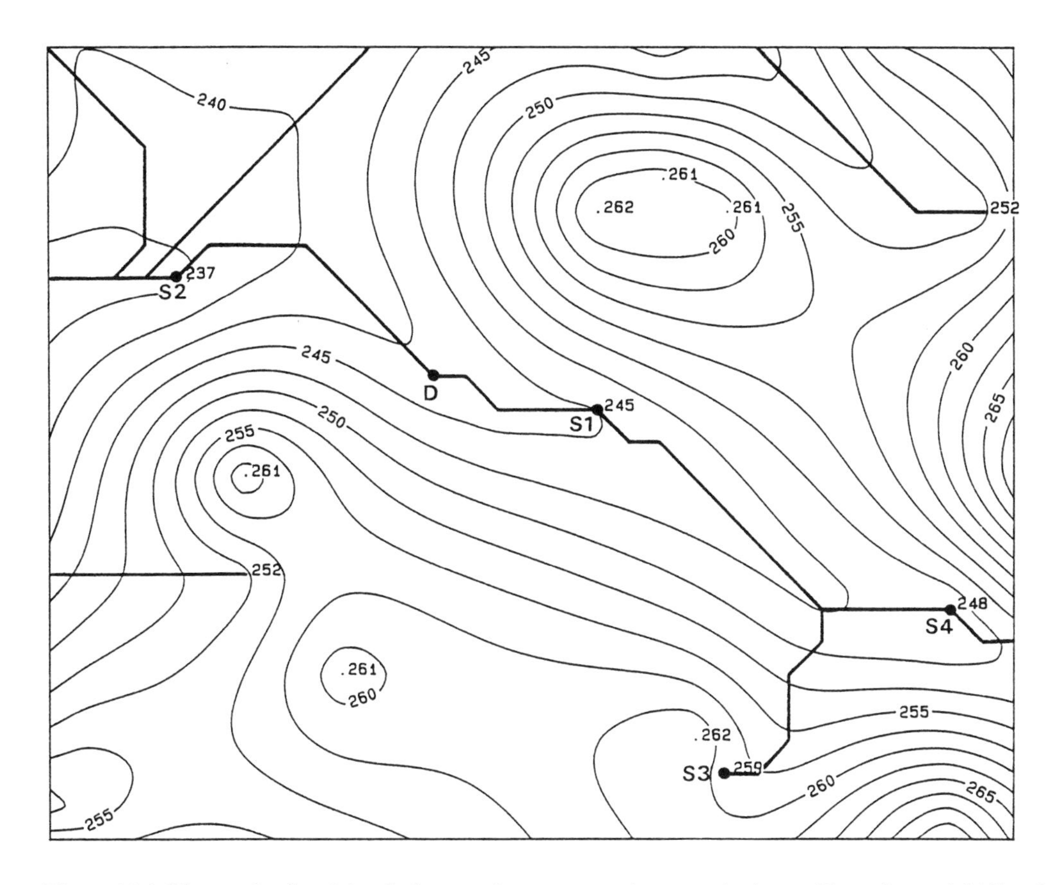

Figure 39-2: The result of applying drainage enforcement to the example shown. Piecewise straight lines indicate inferred drainage lines.

sinks. The action of the drainage enforcement algorithm is modified in practice by the systematic application of three user-supplied elevation tolerances. These allow the amount of drainage enforcement to be adjusted in relation to the accuracy and density of input elevation data, as well as the level of terrain generalization required. When the tolerances have been set appropriately, the remaining sink points are normally those associated with genuine sinks, with significant errors in elevation data or with areas where the input data are not of sufficient density to identify the drainage pattern of the fitted DEM reliably (Hutchinson, 1989).

Incorporation of stream line data

Drainage enforcement can also be obtained by incorporating stream line data as described by Hutchinson (1989). When such data are available, this can lead to more accurate placement of stream lines than can be achieved automatically by the procedure. Data stream lines can also be used to remove sinks that would not otherwise be removed. This is again of particular relevance when calculating generalized, coarse-resolution DEMs. Ordered chain conditions are associated with the stream lines, which ensure approximately linear descent between successive elevation data points down each stream line. They also ensure that each stream line acts as breakline for the interpolation conditions so that each stream line lies at the bottom of its accompanying valley (Hutchinson, 1989).

Automatic calculation of streams and ridges from contour data

Contours are a popular source of elevation data that implicitly encode a number of terrain features, including points on stream lines and ridges. Their main disadvantage is that they can significantly undersample the areas between contour lines, especially in areas of low relief. This has led most investigators to prefer contour-specific algorithms over general-purpose algorithms when interpolating contour data. Such algorithms usually use linear

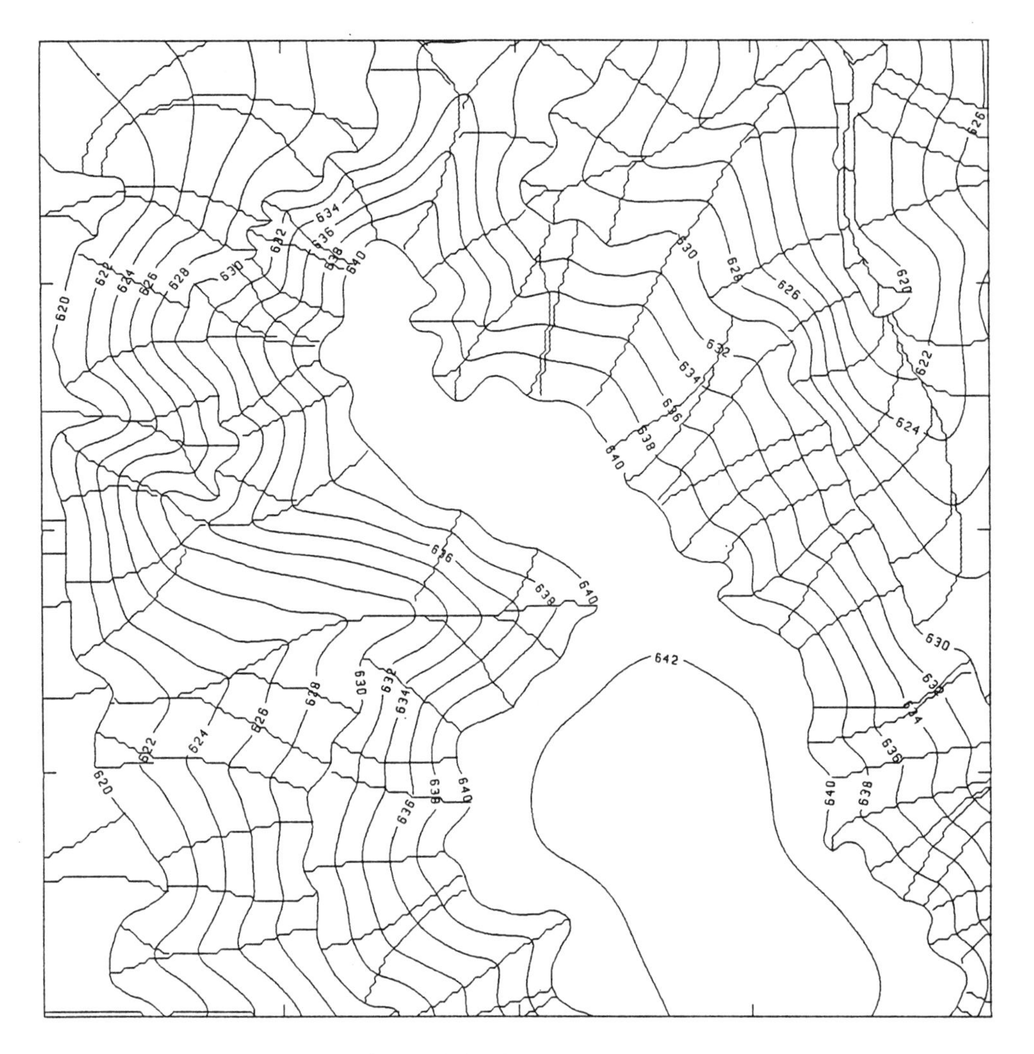

Figure 39-3: Minimum curvature interpolation from 10 m contours with stream lines and ridge lines calculated automatically.

interpolation along straight lines of steepest slope in an attempt to extend across the areas between the contour lines at least some of the fine structure contained in the data contours.

Three main weaknesses can be identified in such methods. These are mainly due to deficiencies in the techniques devised to interpolate linearly along lines of steepest slope, rather than any inherent defect in the concept itself. The first deficiency is that the orientation of each line of steepest slope is limited on computational grounds to one of a small number of possible orientations. This can lead to significant misplacement of search directions. The second deficiency is that search directions are usually straight, even though streams and ridges often exhibit significant curvature between data contour lines. The third and most fundamental deficiency is that the search directions are defined in relation to the DEM points to be interpolated rather than being attached directly to critical points on the data contours.

The procedure overcomes these deficiencies by first identifying the corner points, or points of locally maximum curvature, on all data contours. A preliminary minimum curvature DEM is then interpolated from the contour data. The surface defined by this DEM is systematically traversed in the direction of steepest ascent or descent on the concave side of the contour at each corner point. This means that the grid is traversed uphill along ridges and downhill along streams until the next data contour is encountered. This yields stable search directions. As shown in Figure 39-3, the resulting calculated streams and ridges are then curved paths in general that follow the natural shape of the landscape. Ordered chain conditions are associated with these paths, which maintain approximately linear interpolation between successive data points along each path (Hutchinson, 1988). The example in Figure 39-3, which shows contours at 2 m intervals, has been interpolated from data contours at 10 m intervals. An attractive feature of this process is that it yields, in conjunction with the contour lines, a systematic classification of the landscape into simple, connected, approximately planar, terrain elements, which are bounded by contour segments and flow line segments. These are similar to the elements calculated by Moore et al. (1988), but simpler because the flow lines consist only of inferred stream lines and ridge lines. They are also determined in a more stable manner, which incorporates both uphill and downhill searches, depending on the shape of the terrain. These elements can be used to simplify hydrological analyses. The example of Figure 39-3 illustrates this classification.

It should be noted, however, that DEM interpolation from contour data is generally better suited to finer scales where the contour data are less generalized and for which the corners are more reliable indicators of stream lines and ridges. At coarser scales it is often just as effective, and much less demanding on data storage, to digitize the main corner points of contour lines and use these as point elevation data.

A DIGITAL ELEVATION MODEL OF AUSTRALIA

An existing DEM of Australia, at a resolution of 1/10th degree (approximately 10 km), has been calculated by Moore and Simpson (1982) using the minimum curvature procedure of Briggs (1974) applied to a point elevation data set containing about 320,000 points. These data were measured during a continent-wide gravity survey conducted by the Bureau of Mineral Resources (Aniloff et al., 1976). Though detailed enough to detect significant geological structures, this DEM suffers from a number of acknowledged limitations from the point of view of hydrological analysis. Extreme heights are not well represented, and no stream line data were used in the generation of the DEM. In fact, the relatively coarse resolution of the DEM meant that it could only support quite generalized drainage structures. The point data were also known to contain a number of errors.

Hutchinson and Dowling (1991) have described a new DEM for Australia, at a resolution of 1/40th degree (approximately 2.5 km). A shaded relief view of a 1/20th degree version of this DEM is shown in Plate 39-1. It has been calculated using the elevation specific gridding technique described here. The data consisted of about 580,000 spot heights, the stream line network digitized from 1:2,500,000 scale maps and about 400 known sink points. The spot height data included 400,000 points obtained by Bureau of Mineral Resources ground survey, 19,000 trigonometric points and 83,000 bench marks, obtained from the Division of National Mapping, 10,000 points from the digitized coastline at 1:250,000 scale, and an additional 65,000 points digitized from 1:250,000 scale maps in areas where the continent-wide data sets were lacking in sufficient detail. Additional stream lines were digitized from the same maps to remove remaining drainage anomalies. The 400 sink data points were digitized to prevent drainage clearance of genuine depressions, although the drainage enforcement algorithm has generally been found to be sufficiently sensitive to ensure that genuine depressions are not cleared, whether or not they have been identified as such in the data (Hutchinson, 1989).

Numerous errors in all data sets were detected and corrected by examining remaining sinks in the interpolated DEM. The sensitivity of this morphological approach to error detection is indicated by noting that errors as small as 20 m were easily detected. Such errors would remain undetected as outliers by more conventional statistical methods. Positional errors in the stream line data derived from 1:2,500,000 scale maps were reduced to about 0.1 km by "rubber sheeting" the stream lines using a set of control points comprising one point for each 1/2 degree cell across Australia. The control points were digitized from 1:250,000 scale maps.

It is difficult to quote a single estimate for the overall elevation accuracy of the DEM since this is dependent on local relief and the resolution of the DEM as well as the accuracy of the input elevation and stream line data. In areas of low relief the elevation errors in the DEM approach the errors in the data of about 10 m. In areas of complex terrain the elevation errors may exceed 200 m. These can occur when the interpolation procedure has to resolve conflicts between maintaining fidelity to the data and maintaining descent down stream lines at the chosen resolution. All such conflicts were resolved in favor of the stream line data. Derived terrain parameters such as slope must be regarded as being highly generalized in such areas.

Plate 39-1 clearly shows the low elevation and relief of the Australian continent. Approximately 50% of the land surface is less than 250 m above sea level and 75% is less than 400 m above sea level. Over 90% is less than 600 m above sea level. Much of the western plateau region has been tectonically stable since the early Tertiary Period. The improvement in resolution of the 1/40th degree DEM is illustrated in Plate 39-2, which shows the island of Tasmania at 1/10th degree and 1/40th degree resolution, respectively. Tasmania represents approximately 1% of the land area of the Australian continent. The 1/10th degree DEM has difficulty in respecting even the coarsest drainage patterns.

Drainage basin analysis

A drainage basin analysis was performed on the DEM using the software of Jenson and Domingue (1988). The results of this analysis are discussed in detail by Hutchinson and Dowling (1991). Processing time for this analysis amounted to just a few hours on a VAX 8700 computer. Since large parts of central Australia do not drain to the coast, inland and coastal drainage basins were treated separately. The coastal drainage analysis was relatively straightforward, yielding broad drainage divisions that were in very close agreement with those mapped by the Australian Water Resources Council (1976).

The inland drainage analysis is new. This region comprises over 30% of the land surface. Its low relief and arid climate have meant that drainage is not well coordinated and drainage divisions have been poorly defined. The inland drainage analysis gave rise to 225 distinct closed basins. Each basin could be identified with a known lake, ephemeral lake, or salt lake, as depicted on 1:1,000,000 scale topographic maps. A method was developed to combine these basins in order to obtain the major interior drainage divisions. Each basin was combined with the adjacent basin associated with its lowest pour point, provided that the adjacent basin was lower in elevation and the height of the pour point above the bottom of the higher basin did not exceed a threshold of 30 m. The end result of this process was that each inland basin was associated, via a succession of descending basins, with a portion of the coastal drainage zone. This provided strong support for the generally accepted notion that the inland basins of Australia are the remnants of an external palaeodrainage system. The inland basins were grouped according to the portion of the coast with which they were associated, giving rise to seven major drainage divisions with inland drainage, as depicted in Figure 1 of Hutchinson and Dowling (1991). Departures of the new inland basin analysis from that of the Australian Water Resources Council (1976) were supported in all cases by detailed examination of available topographic maps, as well as by an analysis of the palaeodrainage systems by van de Graff et al. (1977). It can be reasonably inferred that the inland drainage basin analysis afforded by the 1/40th degree DEM is superior to the analysis accepted by the Australian Water Resources Council.

An assessment of scale

Hutchinson and Dowling (1991) examined the cumulative distributions of a number of hydrologically relevant terrain parameters derived from the 1/40th degree DEM for Murray–Darling basin and the Derwent River basin. The same parameters were also derived for the same basins from a 1/20th degree DEM to get an indication of the potential effects of DEM resolution or scale on subsequent hydrological analyses. The results are as follows:

1. Boundaries of the catchments, as calculated from the 1/40th and 1/20th degree DEMs, are in very close agreement. The calculated areas for the Murray–Darling agree to within 0.1%. The calculated areas for the much smaller Derwent basin agree to within 1%. Catchments would thus appear to be well defined by both the 1/40th and the 1/20th degree DEMs. However, differences between the two resolutions in the definition of some smaller catchments were detected.
2. Elevation is not always recognized as a hydrologically relevant terrain parameter. However, it is well known to play a dominant role in moderating climate (Hutchinson and Bischof, 1983; Hutchinson, 1991). Cumulative distributions of elevation for each catchment were also in close agreement for the two resolutions. It is thus reasonable to expect that differences between the two DEM resolutions would have little consequence for continent-wide climate assessments.
3. Differences between the two resolutions for cumulative values of slope were barely significant for the Murray–Darling basin, which is dominated by large areas of low relief, but plainly significant for the more rugged Derwent basin. Interestingly, the differences only become apparent for slopes exceeding 2%, the upper limit of the USDA soil class I.
4. Cumulative distributions of specific catchment area and wetness index show the greatest dependence on DEM resolution. As explained by Hutchinson and Dowling (1991), the differences in the cumulative curves indicate that most of the additional structure introduced by the 1/40th degree DEM occurs in the upper reaches of the catchments. For both catchments, over half of the 1/40th degree DEM points have at most one upstream neighbor.

CONCLUSION

The feasibility of calculating and working with a continent-wide DEM of sufficient spatial detail to support a variety of hydrological and other environmental analyses has been clearly demonstrated. The computational efficiencies of both the DEM interpolation procedure and the drainage analysis software mean that they are potentially applicable to any continent, provided only that sufficient data are available.

The DEM interpolation procedure, with its accompanying drainage enforcement algorithm, provides a sound basis for terrain generalization and displays significant advantages over conventional statistical techniques for extracting information from and detecting errors in spatially distributed topographic data. It is feasible that the drainage enforcement algorithm could also be used to good effect in filtering much finer-resolution elevation data, as afforded by high-resolution satellite imagery, in a way that maintains and probably enhances its drainage integrity.

It would appear that both the 1/40th degree DEM and

the coarser-resolution 1/20th degree DEM can adequately define much of the hydrologically relevant structure of Australia for continent-wide analyses. This includes catchment definition, the spatial distribution of climate, and the distribution of the lowest slope classes. The significance of differences in specific catchment area and wetness index have yet to be determined, since these parameters are inherently scale dependent. Ideally, methods for determining catchment response using DEMs should be able to take proper account of the spatial scale.

ACKNOWLEDGMENT

The assistance of Kate Ord and June McMahon in constructing Plates 39-1 and 39-2 is gratefully acknowledged.

REFERENCES

Aniloff, P., Barlow, B.C., Murray, A.S., Denham, D., and Sandford, R. (1976) Compilation and production of the 1976 Gravity Map of Australia. *BMR Journal of Australian Geology and Geophysics* 1: 273–276.

Australian Water Resources Council (1976) *Review of Australia's Water Resources 1975*, Canberra: Department of National Resources, Australian Government, 170 pp.

Band, L.E. (1986) Topographic partition of watersheds with digital elevation models. *Water Resources Research* 22(1): 15–24.

Beven, K.J., and Kirkby, M.J. (1979) A physically-based variable contributing area model of basin hydrology. *Hydrological Sciences Bulletin* 24: 43–69.

Briggs, I.C. (1974) Machine contouring using minimum curvature. *Geophysics* 39: 39–48.

Frederiksen, P., Jacobi, O., and Kubik, K. (1985) A review of current trends in terrain modelling. *ITC Journal* (1985) 101–106.

Golub, G.H., and van Loan, C.F. (1983) *Matrix Computations*, Baltimore: John Hopkins University Press.

Goodchild, M.F., and Mark, D.M. (1987) The fractal nature of geographic phenomena. *Annals, Association of American Geographers* 77(2): 265–278.

Hutchinson, M.F. (1988) Calculation of hydrologically sound digital elevation models. *Proceedings, Third International Symposium on Spatial Data Handling, Sydney*, Columbus: International Geographical Union, pp. 117–133.

Hutchinson, M.F. (1989) A new procedure for gridding elevation and stream line data with automatic removal of spurious pits. *Journal of Hydrology* 106: 211–232.

Hutchinson, M.F. (1991) The application of thin plate smoothing splines to continent-wide data assimilation. In Jasper, J.D. (ed) *Data Assimilation Systems*, BMRC Research Report No. 27, Melbourne: Bureau of Meteorology, pp. 104–113.

Hutchinson, M.F., and Bischof, R.J. (1983) A new method for estimating the spatial distribution of mean seasonal and annual rainfall applied to the Hunter Valley, New South Wales. *Australian Meteorological Magazine* 31: 179–184.

Hutchinson, M.F., and Dowling, T.I. (1991) A continental hydrological assessment of a new grid-based digital elevation model of Australia. *Hydrological Processes* 5: 45–58.

Jenson, S.K., and Domingue, J.O. (1988) Extracting topographic structure from digital elevation data for geographic information system analysis. *Photogrammetric Engineering and Remote Sensing* 54(11): 1593–1600.

Mitas, L., and Mitasova, H. (1988) General variational approach to the interpolation problem. *Computers and Mathematics with Applications* 16(12): 983–992.

Moore, I.D., O'Loughlin, E.M., and Burch, G.J. (1988) A contour-based topographic model for hydrological and ecological applications. *Earth Surface Processes and Landforms* 13: 305–320.

Moore, R.F., and Simpson, C.J. (1982) Computer manipulation of a digital terrain model (DTM) of Australia. *BMR Journal of Australian Geology and Geophysics* 7: 63–67.

van de Graff, W.J.E., Crowe, R.W.A., Bunting, J.A., and Jackson, M.J. (1977) Relict early Cainozoic drainages in arid Western Australia. *Zeitschrift fur Geomorphologie N.F.* 21: 379–400.

Wahba, G. (1990) Spline models for observational data. *CBMS-NSF Regional Conference Series in Applied Mathematics*, Philadelphia: Society for Industrial and Applied Mathematics, 169 pp.

Warntz, W. (1975) Stream ordering and contour mapping. *Journal of Hydrology* 25: 209–227.

Weiringa, J. (1986) Roughness-dependent geographical interpolation of surface wind speed averages. *Quarterly Journal of the Royal Meteorological Society* 112: 867–889.

VI

Spatial Statistics

DONALD E. MYERS

Spatial statistics and geographic information systems are natural partners, especially for the analysis and modeling of environmental data. The variety of problems that are addressed by spatial statistics and ways in which geographic information systems can aid in the manipulation of environmental data are amply illustrated in the chapters presented in this section. Environmental data sets have two characteristics that set them apart from many other kinds. First they are nearly always multivariate; that is, there is more than one variate or analyte of interest, and these are correlated in some sense. Second, each data value is associated with a location either by specific coordinates or by association with some area or volume. This positional association is also normally manifested in another way, namely through some form of spatial correlation. At least three perspectives on the way in which these two characteristics are utilized are presented in this collection. Subsequent sections of this introduction will identify where these occur in the various chapters and provide an overview.

SPATIAL STATISTICAL MODELS

The objective in constructing an environmental model may vary considerably, but a model is most often used to explain or predict. Cressie and Ver Hoef (Chapter 40) point out that environmental models not only need to be spatially based but should incorporate time and often need to incorporate the relationship between an organism and its environment. Spatial statistics seems to be a new field, but in fact it has its roots in classical statistics and in particular in the work of Fisher, Yates, and Whittle. Several examples of models are given, especially in the context of sampling design. In addition to an overall discussion of the relevance of spatial statistics to environmental modeling, several examples are given to illustrate this relevance. The first example uses a random function model and pertains to acid rain, that is, wet deposition of hydrogen ions in the Eastern U.S. resulting from the dispersal into the atmosphere of various sulfurous and nitrous pollutants produced by industry and transportation. This is linked to the identification and prediction of ecological effects. The methods used in this example are a part of what is known as geostatistics. A more complete overview and description of geostatistics is provided in Cressie's Chapter 41. An example using data for the percent vertical cover in a dolomite glade illustrates the use of a spatial lattice model. Spatial point pattern methods are illustrated using data from a census of longleaf pines in southern Georgia. In this model, time is incorporated into the birth, death, and growth process.

Geostatistics is based on the use of a random function model wherein the data are viewed as a nonrandom sample from one realization of the random function. It had its origins in applications to mining, hydrology, meteorology, and forestry, although the developments in the latter two areas proceeded in slightly different ways. Cressie describes the importance of the spatial correlation function, which is most often given in the form of a variogram. Difficulties pertaining to and methods for estimating the variogram are described. The random function model leads easily to a regression form for the estimator, known as kriging, for spatial prediction or estimation. An extension useful in the presence of a nonstationarity, known as universal kriging, includes the thin plate spline as a special case. It is shown that nonpoint data are easily accommodated by the model and the model is easily adaptable to simulation. The final section is devoted to multivariate geostatistics wherein both spatial and intervariable correlation are incorporated in the model. Matheron and coworkers derived a linear predictor for one variate utilizing the spatial correlation for that variate and the intercorrelation with other variates. The work of Myers, which provides the general setting, shows that the single-variate case is subsumed in the general case and provides the natural extension of the single-variate case.

Geostatistical methods explicitly incorporate the position coordinates into the random function model. Anselin (Chapter 46) models spatial correlation in a manner that does not explicitly utilize coordinates. A heuristic description might be that spatial correlation means that values at locations close together are more correlated than values for locations far apart. In geostatistics this concept is used to derive an estimator for the values at unsampled locations. Anselin uses the same concept to characterize clusters or patterns. In the geostatistical model, spatial correlation is quantified by the spatial correlation function, that is, the variogram or covariance. In the discrete space autoregressive model, spatial correlation is quantified by contiguity or weight matrices. For the latter, the data locations are considered to be a subset of a regular or irregular grid or lattice. An example is given using data from the Global Change Database (Chapter 36) from an area around the border between the Central African Republic, Sudan, and Zaire. Four variables, GREEN (greenness vegetation index), TEMP (temperature), ELEV (modal elevation), and PREC (precipitation), are used. One part of the analysis is concerned with examination of the clustering of GREEN versus the clustering of the other variables, and the second part of the analysis considers GREEN as the dependent variable in a spatial regression using the remaining as explanatory variables. Several statistics useful for testing for spatial correlation are described, and there is a comparison between the use of least squares and maximum likelihood for determining the coefficients in the spatial regression equation. Anselin has incorporated these methods into a software package called SPACESTAT.

Conventional inference techniques are based on an assumption that the sample is selected from a hypothetical population and the inferences pertain to the parameters or characteristics of this population. More realistically the sample is commonly selected from a finite universe. Overton (Chapter 47) describes the use of probability sampling, which was incorporated into the National Surface Water Survey (NSWS) and is an integral part of the current EPA initiative, EMAP. As an example consider the universe of lakes in the NSWS. A probability sample is a subset of this universe selected in such a manner that for each element of the sample the probability of its having been selected is known and this probability is positive. A representation of the universe, called a frame, is used to select the sample. For example, a frame might simply be a list of the elements of the universe or in a spatial context it could be a map. The Horwitz–Thompson theorem provides assurance that certain population parameter estimators based on probability samples are unbiased and determines the variances of these estimators. It is shown that in the case of model-based inferences the usual estimators must be replaced by weighted estimators in order to maintain consistency. In the design of EMAP, the use of probability samples is extended to spatial problems by first overlaying a triangular point grid on the map and then perturbing the grid in a random manner. This provides a sampling grid with the same configuration but randomly positioned. Any small region of fixed area is equally likely to contain a sample grid point.

APPLICATIONS

While most GIS include some routines or programs for the analysis of data, their principal thrust as yet is in the graphical manipulation and presentation of the data. In particular, GIS do not as yet really incorporate any spatial statistical components. Conversely the standard statistical packages include neither spatial statistical routines nor the capabilities of a GIS. The remaining chapters in this section on spatial statistics illustrate the use of a GIS as an aid to spatial statistical analyses or the use of spatial statistics to complement the use of a GIS. These are important because of the lack of an adequate interchange in the literature pertaining to these two fields. Only one of the papers explicitly uses a GIS in the analysis. In general all the papers begin with a geostatistical perspective, and that significantly affects the way in which the connection with a GIS is presented. Englund (Chapter 43) is concerned with the use of simulation, whereas the other chapters pertain to estimation and modeling.

The random function model in geostatistics can be thought of as a collection of realizations together with a probability assignment, although the data are a sample from only one realization. The kriging estimator is a smoother, and the variability exhibited by the data together with the estimated values is less than that of the data. Simulation of additional realizations is a method for reproducing the variability and hence is useful for designing sampling plans and for planning in general. The simulations can be conditioned to the data by kriging. There are several different algorithms commonly used for simulation; Englund has used the sequential Gaussian method. Reproducing or characterizing this variability is useful in determining the reliability of maps produced by kriging. Englund illustrates this by considering two "layers" in a GIS that have been produced by sampling different variates, followed by kriging. In this example neither variate is of interest alone, but certain characteristics of the intersection of the layers are of interest (for example, areas where both variates have high values). If these layers are produced only from the data and interpolated by kriging or some other method, there is still the question of the reliability of the resulting intersection map. Englund shows how to use simulation to quantify and characterize the reliability.

Ver Hoef (Chapter 45) considers three different methods to predict spatial-cover abundance for a glade in the Missouri Ozarks. The first approach is a classical regression with cover abundance as the response variable and shade as the explanatory variable. In the second method, abundance at one location is predicted using only the values for abundance at nearby locations and incorporating the spatial correlation (kriging). The third method is a combination of the other two. Residuals from the regression are used in kriging; that is, the method attempts to separate the dependence on the explanatory variable and the spatial dependence of abundance on itself. The three methods are compared by using cross-validation. Sequentially, each data value for abundance is deleted and estimated using only the remaining data. Then the mean-square estimation error is computed, and this statistic is used to discriminate between the methods. Note that the use of the term universal kriging, in this chapter, is not quite the same as the usage in the geostatistical literature. What is called universal kriging in this chapter has been called kriging with external drift in the geostatistical literature; it is related to but not the same as co-kriging. The form of universal kriging used herein is found to be superior to either regression or kriging alone when predicting cover abundance.

A similar approach is used by Jager and Overton (Chapter 42) in their study of spatial patterns for acid neutralizing capacity (ANC) for lakes in the Adirondacks of New York. The data are taken from the National Lake Survey and constitute a probability sample. The lakes are stratified according to alkalinity levels (low, medium, and high). Using elevation and pH as explanatory variables, two regression equations are obtained for LANC, where LANC is the base 10 logarithm of (ANC + 150); one regression is used for the low and medium alkalinity levels and one for the high. In each case LANC is regressed on precipitation pH and elevation. Using the regression residuals, sample variograms for LANC were computed and modeled. The authors then infer spatial patterns for LANC using spatial patterns of the explanatory variables and the regression equations.

Rhodes and Myers (Chapter 44) consider a different application of geostatistics to lake survey data. The Eastern Lake Survey–Phase I provided data on a number of variates thought to be relevant to predicting the acidification of lakes as related to effects on marine life in the lakes. Sampled lakes had been selected as a probability sample with one water sample (6.2 liters) taken from each lake irrespective of surface area (lakes smaller than 5 hectares were excluded from the sample). In a geostatistical analysis each data value is either associated with a point or is the spatial average value over an area or volume. Intralake variability is important both in the estimation and modeling of the variogram as well as in the subsequent kriging step. The limitation of one sample location per lake prevents direct estimation of this intralake variability and has a significant impact on the set of sample locations in a kriging neighborhood. Short-range variability will contribute to a nugget effect in the sample variogram when there is a lack of sample pairs for short distances. The magnitude of the nugget can at least be used as a proxy for estimating the intralake variability. The GIS package GRASS 3.1 was used to select pseudo sample locations in some of the larger lakes. This was done by using the GIS to overlay a grid on a map of the lakes and thus to identify grid points within a given lake. To simulate variability within the lakes, pseudo sample locations within a given lake were assigned a value obtained from the lake sample value by adding a random multiple of the square root of the nugget of the variogram. The effect of incorporating these additional sample locations, and thus simulating intralake variability, is evaluated in several ways.

THE FUTURE

While there is clearly interest in merging spatial statistics into GIS, there is little consensus on what techniques or routines should be included or how they would be used. Some interfaces have already been built, including one between the statistical package S+ and the GIS GRASS. Although S+ is not always viewed as a spatial statistics package per se, it is possible to incorporate spatial statistical functions, as well as to take advantage of other graphical features. It is to be hoped that such connections will encourage statistics departments to implement and teach GIS software. Papers on spatial statistics are appearing more often in the statistical literature and at statistics meetings.

One of the reasons for the delay in merging these tools is that GIS is as yet relatively unknown in the statistical community at large and in much of the spatial statistics community in particular. This is in part attributable to the development of GIS in the context of geography and particularly an emphasis on vector-based GIS and hence on vector data. Many spatial statistical methods would more naturally relate to raster data. One aspect of the way statistics functions in an academic setting is pertinent to these problems. Younger statistics faculty need to establish themselves in order to obtain promotion and tenure. This frequently inveighs against significant collaboration in other disciplines such as working in the interface between spatial statistics and GIS. Similarly, an established statistician whose research has been limited to mathematical statistics does not have much incentive to become involved in interdisciplinary work. These trends deserve some attention in the statistical community.

A second possible reason is that spatial statistics, particularly geostatistics and probability sampling, are relatively unknown in the GIS community. This is in part attributable to the origins of GIS, which were principally in geography and closely related fields rather than those giving rise to geostatistics. This is reflected in the initial emphasis on vector-based GIS. As the technology and the software begin to blur the distinction between vector- and raster-based systems, the base for applications will increase. Environmental modeling needs will accelerate this trend.

40

Spatial Statistical Analysis of Environmental and Ecological Data

NOEL CRESSIE AND
JAY M. VER HOEF

The *environment* is the surroundings of an organism. *Ecology* is the study of an organism's relationship to its environment. In order to keep research goals well focused, care should be taken to maintain the distinction between the meaning of the two words. Data coming from both the organism and its environment are often spatial in nature. A GIS is well suited for storage, analysis, and display of such data, but little software development has taken place to allow standard spatial statistical analyses to be carried out.

In a complex world, a model is a powerful tool for distilling the phenomenon under study down to its essential features. Those features depend closely on the scientific or engineering questions being addressed. A well-fit model has considerable explanatory and predictive capabilities, but, by its very nature, there will be residual uncertainty made up of everything that the model cannot explain or predict. Statistics, the science of uncertainty, allows the modeler to take the physical, chemical, or biological features of the phenomenon into account, but within a framework that recognizes the limitations of the model. The result is a statistical model.

The statistical model itself is usually based on simplifying assumptions that may or may not be appropriate. Here, sensitive lack-of-fit diagnostics play an important role in identifying inappropriate assumptions. It will be argued subsequently that *classical* statistical assumptions should often be discarded in the environmental context, giving rise to more realistic models and new statistical (including diagnostic) methods. In the future, spatial statistics and time series are areas with which every serious environmental and ecological modeler will need some familiarity.

It is our belief that only careful statistical (space–time) modeling can give the clearest answers to important environmental issues that threaten to do irreparable damage to our planet. Such models incorporate uncertainty due to inter alia incomplete sampling, the measurement process, and imprecise spatial and temporal coordinates of observations. Without a statistical approach, it is often impossible to know whether a degradation or improvement is real or whether it is due to chance fluctuations.

This chapter will present an overview of spatial statistics and its relationship to environmental and ecological modeling, with particular consideration given to the role of the GIS. Imagine three layers: The bottom layer is made up of organisms and their environment, linked by ecology. Data from the bottom layer are directed into the middle layer, a statistical-analysis "engine" that includes a GIS component. The top layer is made up of statistical models. It is through the GIS, in particular, that *spatial* statistical models can interact with organisms and their environment. Perhaps the strongest theme of our chapter is that the potential for the GIS to be an effective conduit (in both directions) between the top and bottom layers has yet to be realized.

The next section gives a historical review of statistics for spatial data and contrasts it with the more classical approach that models variability with independent and identically distributed (IID) errors. Through three very different examples, the following section demonstrates the challenge that faces GIS software to facilitate spatial statistical modeling and inference. The final section presents two contrasting approaches to spatial statistical inference, along with future opportunities and conclusions.

HISTORICAL REVIEW OF SPATIAL STATISTICS

Much of the statistics taught to scientists and engineers has its roots in the methods advocated by Fisher (1925), which were developed for agricultural and genetics experiments. Statistics and its applications have grown enormously in recent years, although at times it has been difficult for the new developments (in, e.g. ecology and environmental engineering) to fit into the old framework of blocking, randomization, and replication within a well-designed experiment. In the most extreme case, there is just one experimental unit, whence one uses the term *observational study* to describe the investigation. (On occasions, it may be possible to obtain replication when the observational study is one of several like studies, but those occasions are rare.)

It may seem then that statistical methods would have little to say about an observational study. However, statistics is a vital, evolving discipline that is highly adaptive to new scientific and engineering problems. At the very least, statistical design and analysis forces one to think about important factors, control data, sources of variability, and statistical models; it reveals new structure and so prompts conjectures about that structure; and it is a medium for productive multidisciplinary communication. It may even be possible to substitute observational studies in the field with well-designed experiments in the laboratory. Indeed, in the *computer* laboratory, large simulation experiments have been a very effective tool.

Spatial statistics is a relatively new development within statistics; ironically, Fisher was well aware of potential complications from the spatial component in agricultural experiments. In Fisher (1935), he wrote: "After choosing the area we usually have no guidance beyond the widely verifiable fact that patches in close proximity are commonly more alike, as judged by the yield of crops, than those which are far apart." Fisher went to great lengths to remove spatial dependence by randomizing treatment assignments. Thus, as well as controlling for bias, randomization also neutralizes the effect of spatial correlation (Yates, 1938), although it does not neutralize the effect of spatial correlation at spatial scales larger or smaller than the plot dimensions of the experiment. Fairfield Smith (1938) was concerned with choosing plot dimensions so that any increase in plot size would yield little decrease in error variance. Although his analysis was empirical, the very formulation of the problem recognizes the presence of spatial correlation. Statistical models for such phenomena did not begin to appear until much later (Whittle, 1954).

Spatial statistics is a tool that is immediately applicable to environmental problems. In an observational study, each problem has its own source of randomness that depends on the current state of knowledge (or lack of it) for the phenomenon under study. This approach will be contrasted with an approach based on probability sampling whereby external randomness is imposed on the problem. There, we conclude that certain types of probability sampling can be useful adjuncts to spatial statistics.

Bayesian statistics (e.g., DeGroot, 1970) provides a technology that is well suited to situations where there are few or no replications, such as in environmental problems. When the ratio (number of data)/(number of parameters) is small, say less than ten, it is hard to see how one can do without the Bayesian technology. Later in this chapter, Bayesian statistics is used to predict 300 parameters from 300 data. The Bayesian approach allows the prior distribution of parameters to be updated, via Bayes's formula, yielding a posterior distribution. Then, any inference is based on this posterior distribution.

Another illustration of the use of Bayesian statistics is in choosing between competing theories. For the purpose of illustration, suppose there are just two such theories, H_0 and H_1. In Bayesian statistics, prior opinion yields the odds ratio, $\Pr(H_0)/\Pr(H_1)$. In the light of data $\mathbf{Z}$, this ratio can be updated:

$$\{\Pr(H_0 \mid \mathbf{Z})/\Pr(H_1 \mid \mathbf{Z})\} = \{f_0(\mathbf{Z})/f_1(\mathbf{Z})\}\{\Pr(H_0)/\Pr(H_1)\},$$

where f_0 and f_1 are the joint densities of the data $\mathbf{Z}$ under H_0 and H_1, respectively. It is this probability that is usually of more scientific interest than the P value, which is the probability of a sample "more extreme than" $\mathbf{Z}$ given H_0.

In conclusion, we believe that there is much promise in using environmental and ecological models that incorporate *both* spatial and Bayesian statistics. These models are well suited to solving current challenges, such as:

1. Distinguishing between perturbations resulting from natural phenomena and those caused by human practices. Greenhouse warming and the depletion of protective ozone are two phenomena of considerable concern to this planet.
2. Choosing indicators (or response variables). They need to be sensitive to results, spatially and temporally stable, of aesthetic or economic relevance, and easy to measure.
3. Dynamic modeling in space. Causative components can be incorporated into the statistical model, allowing the results of various natural scenarios and policy decisions to be simulated. Moreover, the need for complicated spatial structures can often be obviated by including the time dimension.
4. Developing remedial policies that reverse the effects of past practices. Continuous monitoring, assessment, and modeling give feedback on which policies are effective. (Models can help the process but, in the absence of politicians with vision and leadership to champion the policies, our work means very little.)

SPATIAL STATISTICAL MODELING

This section will start with some general comments about spatial statistical modeling and how it currently relates to GIS. It will be seen that even summary statistics from a GIS are rather crude; to illustrate what could be done, three very different spatial statistical examples are presented.

A GIS is a computer hardware and software system designed to *collect* (input), *manage* (storage, retrieval), *analyze* (aggregate, estimate, optimize, simulate), and *display* (tables, maps, dynamic graphics) spatially referenced data. The statistician is interested in all four aspects, although this chapter is mostly concerned with analysis.

As a discipline within statistics matures, it goes through the three stages of description, indication, and inference. The last two are model based; the difference between them is that indication has no measure of precision associated with an estimator or predictor. Interestingly, the evolution of the three stages is not one way. Vague statistical models are at the back of all effective descriptive statistics. It is worth emphasizing then that even at the description stage, a GIS should produce statistics that cater to the spatial models that may be fit. Currently, this is *not* happening, and attempts to link GIS with standard statistical packages could spell disaster. Since most statistical packages have *no* spatial modeling capabilities (geostatistical packages are an exception to this), attempts at inference are in danger of being just plain wrong. Proper use of spatial statistical methods can solve the pseudoreplication problem (Hurlbert, 1984) with which ecologists have been struggling for some time, but those methods are not currently available in a GIS. Outside the GIS environment, the state of spatial statistical modeling is very healthy and experiencing considerable growth. Its successful coupling with GIS must truly be a team effort.

Basic summary statistics such as histograms, scatter plots, means, standard deviations, etc. are available from most GIS, and basic image analysis operations such as smoothing, edge detection, area and perimeter statistics, etc. are available from several GIS. One of the key features of a GIS is the ability to overlay different databases that are each derived from the same geographic region. It is surprising then that multivariate statistical methods are not featured in the GIS. A start has been made by Haslett et al. (1991) to adapt dynamic graphical multivariate methods to a spatial setting. This technology is not hard to use and would fit very nicely into a GIS. Other methods of data analysis for various types of spatial data are given by Cressie (1991, Sections 2.2, 6.2, and 8.2).

The most general spatio-temporal model considered here is:

$$\{\mathbf{Z}(\mathbf{s}; t) : \mathbf{s} \in D(t), t \in T\}; \tag{40-1}$$

where s and t index the location and time of the random observation $\mathbf{Z}$ (which could be real-valued, a random vector, or even a random set). The model (40-1) achieves a great deal of generality by allowing index sets $\{D(t) : t \in T\}$ and T to be possibly random sets themselves; see the following. A particular case of (40-1) is the purely spatial model

$$\{\mathbf{Z}(\mathbf{s}) : \mathbf{s} \in D\}; \tag{40-2}$$

where D is a (possibly random) index set in the Euclidean space $\mathbf{R}^2$ or $\mathbf{R}^3$.

In the three subsections that follow, D will take different forms. The first section deals with geostatistical data where D is a continuous region, the second deals with lattice data where D is a finite or countable set of sites, and the third deals with point patterns where D is a random set, specifically, a spatial point process. These three cases cover the majority of spatial statistical models.

Geostatistical data

In (40-2), geostatistical data occur when D is a fixed region that contains a rectangle of nonzero volume. Thus, there are an uncountable number of possible observations from $\{Z(\mathbf{s}) : \mathbf{s} \in D\}$, the underlying spatial process (or random field). Suppose

$$Z(\mathbf{s}) = \mu(\mathbf{s}) + \delta(\mathbf{s})\,; \mathbf{s} \in D, \tag{40-3}$$

where $\mu(\cdot)$ is the deterministic mean structure that captures the large-scale variation including possible physical, chemical, or biological components. The stochastic term $\delta(\cdot)$ has zero mean, by definition, and contains the small-scale spatial dependence; it may also have components that relate to the substantive problem. Ideally, the spatial-process model helps us gain an understanding of the phenomenon under study, model parameters are interpretable in terms of that phenomenon, and we are able to design optimal sampling plans that tell us the "best" places to take observations.

In this subsection, we shall concentrate on the problem of sampling design, since it arises so often in environmental problems. Some obvious initial questions to ask are:

1. What is to be measured, how should one measure it, and how much will it cost to measure it?
2. Are there baseline data, in the form of local topographical and meteorological data, historical data, and "found" data from previous studies?
3. In the absence of reliable baseline data, will a pilot study be necessary to assess the underlying spatial and temporal variability?
4. Are replicates needed?
5. What is the spatial support (or level of aggregation) of the measurements?

Optimal spatial (statistical) design finds sites $\{\mathbf{s}_1, \mathbf{s}_2, \ldots, \mathbf{s}_n\}$ and the number of sites n that optimize a prespecified (statistical) criterion. Sometimes monitoring sites are already in place, and one wishes to know how to add (or delete) sites optimally.

Consider the optimal design of an acid-deposition network. In the United States, acid deposition results mainly from the atmospheric alteration of sulfur and nitrogen air pollutants produced by industrial processes, combustion, and transportation sources. Total acid deposition includes acid compounds in both wet and dry form. Dry deposition is the removal of gaseous pollutants, aerosols, and large particles from the air by direct contact with the Earth (NAPAP, 1988). Since dry deposition is difficult to monitor, and attempts at any such monitoring are relatively new, we focus on wet deposition here.

It is generally accepted that an important factor in the relatively recent increase of acid deposition is the emission of industrial byproducts into the atmosphere; the consequences for aquatic and terrestrial ecosystems are potentially disastrous. Most fish populations in freshwater lakes are very sensitive to changes in pH (EIFAC, 1969). More fundamentally, such changes could also adversely affect most other aquatic organisms and plants, resulting in a disruption of the food chain. Acid deposition has also been closely connected with forest decline (Pitelka and Raynal, 1989) in both Europe and the United States.

Wet deposition, or acid precipitation as it is commonly called, is defined as the hydrogen ion concentration in all forms of water that condenses from the atmosphere and falls to the ground. Measurement of the total annual amount of hydrogen ion is the end result of a very complicated process beginning with the release of pollutants into the atmosphere. They might remain there for up to several days and, depending on a variety of meteorological conditions (e.g., cold fronts or wind currents), they may be transported large distances. While in the atmosphere, the pollutants are chemically altered, then redeposited on the ground via rain, snow, or fog.

A model for the spatial distribution of total yearly hydrogen ion (H^+), measured on the Utility Acid Precipitation Study Program (UAPSP) network in 1982 and 1983, was developed by Cressie et al. (1990) and Cressie (1991, Section 4.6.1). We present their results for the 1982 data, including implications of the fitted model for network design. Nineteen monitoring sites yielded data in the form of latitudes, longitudes, and annual acid deposition (in μmole H^+ / cm^2). Through various exploratory methods of data analysis, the following random field model was fit to the yearly acid deposition over the eastern half of the U.S.:

$$Z(\mathbf{s}) = \beta_0 + \beta_1 x + \beta_2 y + \beta_3 x^2 + \beta_4 xy + \beta_5 y^2 + \delta(\mathbf{s}), \qquad (40\text{-}4)$$

where the coordinates of $\mathbf{s} = (x, y)'$ are expressed in radians and $\delta(\cdot)$ is a zero mean, second-order stationary random process (whose variogram is given by Cressie, 1991, p. 261).

Suppose one wishes to add a site at $\mathbf{s}_0$. Let $\lambda_1, \ldots, \lambda_n$ minimize the mean-squared prediction error,

$$E(Z(\mathbf{s}_0) - l_1 Z(\mathbf{s}_1) - \ldots - l_n Z(\mathbf{s}_n))^2$$

with respect to $l_1, \ldots, l_n$, subject to the unbiasedness condition,

$$E(Z(\mathbf{s}_0)) = E(l_1 Z(\mathbf{s}_1) + \ldots + l_n Z(\mathbf{s}_n)).$$

Define the so-called kriging variance,

$$\sigma_k^2(\mathbf{s}_0) = E\left(Z(\mathbf{s}_0) - \lambda_1 Z(\mathbf{s}_1) \ldots - \lambda_n Z(\mathbf{s}_n)\right)^2, \qquad (40\text{-}5)$$

which is the minimized mean-squared prediction error.

Let $S \equiv \{\mathbf{s}_1, \ldots, \mathbf{s}_n\}$ denote the existing network and let $S_P \equiv \{\mathbf{s}_{n+1}, \ldots, \mathbf{s}_{n+m}\}$ denote $m \geq 2$ potential new sites from which one will be chosen. Define $S_{+i} \equiv S \cup \{\mathbf{s}_i\}$, $i = n+1, \ldots, n+m$, to be augmented networks. Then S_{+j} is preferred if it predicts best (on the average) the remaining $m-1$ sites in S_P.

Let $\sigma_k^2(\mathbf{s}_0; S_{+i})$ denote the kriging variance for predicting the acid-deposition level at $\mathbf{s}_0$ using the augmented network S_{+i}, where $i = n+1, \ldots, n+m$. For illustration, define the objective function

$$V(\mathbf{s}_j) \equiv \sum_{\substack{i=n+1 \\ i \neq j}}^{n+m} \sigma_k^2(\mathbf{s}_i; S_{+j})/(m-1); \quad j = n+1, \ldots, n+m. \qquad (40\text{-}6)$$

Then the site in S_P that achieves min $\{V(\mathbf{s}_j) : j = n+1, \ldots, n+m\}$ will be declared the optimal site to add. (Other criteria are considered in Cressie et al., 1990, that put more emphasis on "hot spots" and regions of large kriging variance.)

Eleven potential sites (Minneapolis, Minnesota; Des Moines, Iowa; Jefferson City, Missouri; Madison, Wisconsin; Springfield, Illinois; Altoona, Pennsylvania; Charlottesville, Virginia; Charleston, West Virginia; Baltimore, Maryland; Trenton, New Jersey; and Knoxville, Tennessee) were chosen to improve geographic coverage of the existing network (of 19 sites). From among these eleven sites, Baltimore was chosen as the optimal site to add. Its associated average kriging variance, given by (40-6), was 2.56 (μmoles $H^+/cm^2)^2$, compared to Minneapolis's 2.59 (the second smallest value); Charlottesville had the largest value of 2.77.

Several comments are worth making. First, in principle, it is easy to add or delete multiple sites in the same way as given. However, in practice, the computational burden grows exponentially with the multiplicity. Sacks and Schiller (1988) suggest the annealing algorithm as one possible way to deal with this problem. Second, it is obvious that if the criterion (40.6) is changed, the optimal site(s) may change. For example, Zimmerman (1991) applied a variogram-estimation criterion to the same set of potential sites given. He found the optimal additional site to be Charleston, West Virginia. (This choice yields important information on spatial dependence at small distances.)

Finally, with little prior knowledge of the process, or faced with conflicting criteria, a *regular* network of moni-

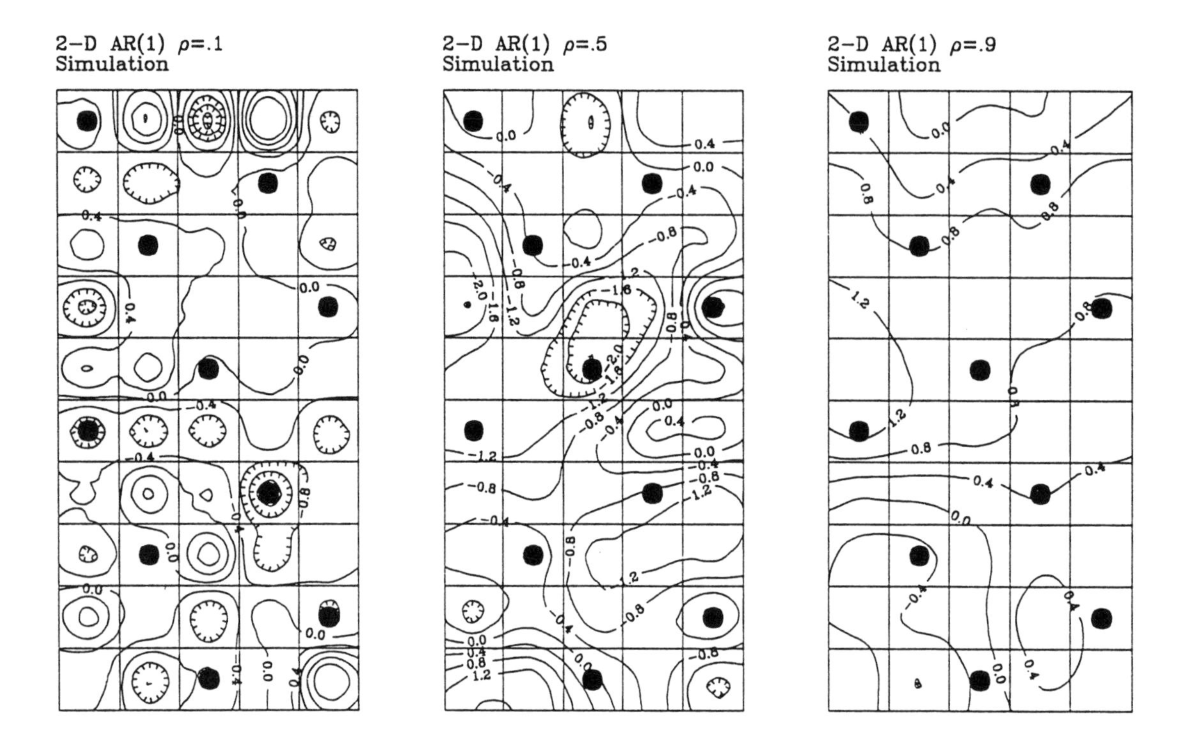

Figure 40-1: Simulations of standard Gaussian (normal) data from an AR(1) model with different levels of spatial dependence.

toring sites can be better than a completely random or stratified random network. For example, suppose that the objective is to predict the average of a two-dimensional (2D) process from spatial data obtained through simple random sampling (SRS), stratified random sampling (STS), and a fixed, regular network design (FXD). For illustration purposes, suppose that the process is a realization of a 2D AR(1) model on a grid of 10 × 5 contiguous plots. That is, $\{Z(i, j) : i = 1, \ldots, 10; j = 1, \ldots, 5\}$ has the following properties:

$$E(Z(i, j)) = \mu,$$

$$\mathrm{var}(Z(i, j)) = \sigma^2,$$

$$\mathrm{cov}(Z(i, j), Z(k, m)) = \sigma^2 \rho^h;\ 0 \le \rho \le 1;$$

where $h \equiv \{(i-k)^2 + (j-m)^2\}^{1/2}$.

Several simulations of standard Gaussian (normal) data from a 2D AR(1) model with different values for ρ are given in Figure 40-1. (The isopleth lines were interpolated between the 10 × 5 data generated and are given as a qualitative summary of the effect of increasing ρ.) The $\{Z(i, j)\}$ can be viewed as a discretization of a geostatistical process. The important feature of this example that makes it geostatistical is that only 10 of the 50 plots are sampled, from which the 50-plot average is to be predicted.

In order to compare SRS, STS, and FXD, a comparison criterion is needed. For any realization of the 2D AR(1) model, there will be an average $\overline{Z}$ (possibly different from μ) over all 50 plots. We wish to predict $\overline{Z}$ with a sample mean, denoted as $\hat{\overline{Z}}$, of 10 plots. The squared-error loss is $(\overline{Z} - \hat{\overline{Z}})^2$, and a measure of risk can be obtained from simulation as follows. Data were generated 1,000 times from a standard Gaussian 2D AR(1) model for various values of ρ, as in Figure 40-1. For each simulation, the value $\overline{Z}$ was calculated, and an estimate $\hat{\overline{Z}}$ was obtained using SRS, STS, and FXD. For SRS, 10 of 50 plots were chosen randomly. For STS, one value was chosen randomly from each row. The FXD design was fixed for each simulation and is shown as the solid circles in Figure 40-1. For a fixed value of ρ and for each sampling method, $(\overline{Z} - \hat{\overline{Z}})^2$ was calculated and then averaged over the 1,000 simulations. The results are given in Figure 40-2, which shows that the fixed, regular network (FXD) has the smallest risk for all values of ρ. In a later section, it is further recommended that the fixed, regular network be given a random starting point.

An important example of a regular network of environmental monitoring sites is Tier 1 of the Environmental Monitoring and Assessment Program (EMAP) initiated by the United States Environmental Protection Agency in 1988 (Messer et al., 1991). Its basic goal is to monitor the vital signs (status, extent, changes, and trends) of U.S.

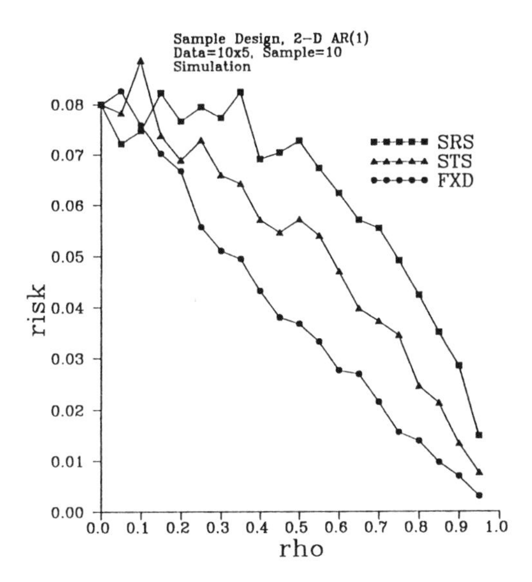

Figure 40-2: Averages over 1,000 simulations for calculated risk as a function of spatial dependence, for each sampling scheme.

natural resources on a regional and national basis with known confidence. From it, one can gauge more clearly the effect of an environmental insult or the efficacy of new Earth-friendly programs. Some external randomness is imposed on the starting point of the regular network, resulting in a probability sample. This provides a validity and robustness to the estimation of distribution functions, means, and regression relationships. However, to carry out more complex analyses, such as those found in spatial statistics, a modeling approach is needed. For further discussion, see the following.

The simulations given previously indicate that, when possible, a good network design will stratify regionally so that, within any region, spatial stationarity is achieved. Then the best design within a region is a regular one. For the Tier 1 EMAP design, stratification is too difficult to implement due to conflicting design criteria from different ecological resources. Thus, a regular sampling design at the national level is specifi

Lattice data

In (40-2), lattice data occur when D is a finite (or countable) set of sites. Of all the possible spatial structures that (40-2) can generate, a data set whose spatial locations are a regular lattice is the closest analogue to a time series observed at equally spaced time points. Lattice data can often be formed by aggregation over a region, as will be the case for the example to be given. The random-field model (40-3) is also applicable to these types of problems, although the models for spatial dependence are often different from those chosen for analyzing geostatistical data. The model featured in this subsection will be a Markov random field (e.g., Besag, 1974). Since the design question does not arise in this context, we shall concentrate on fitting models and estimating (or predicting) parameters.

We examine a set of data consisting of percent vertical cover in a dolomite glade in the Ozarks of southeast Missouri. Along a transect 30 meters in length, photographs were taken through the vegetation against a white background, and image analysis was used on the photographs to calculate percent vertical cover $Z(i)$; $i = 1, \ldots, 300$, for each 10 cm segment along the line. (For a full description of the method of data collection, see Ver Hoef et al., 1989.) The data were transformed to $Y(i) \equiv \log[Z(i) + 1]$, due to skewness of the $Z(i)$ toward large values. The transformed data are shown in Figure 40-3a. The bimodal nature of the histogram of $\{Y(i) : i = 1, \ldots, 300\}$ shown in Figure 40-4 indicates that the values are a mixture of two populations, one with a mean of about 1.5 (in a patch), and the other with a mean of about 0.6 (out of a patch). The objective is to estimate the true parameter value, $\mu = 0.6$ or 1.5, for *each* of the 300 segments and then to

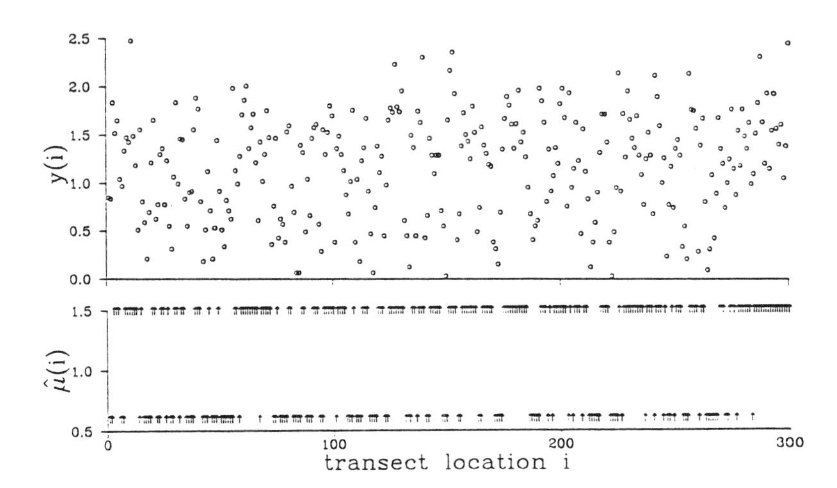

Figure 40-3: Transformed percent vertical cover along a 30 m transect across a dolomite glade in the Ozarks of southeast Missouri.

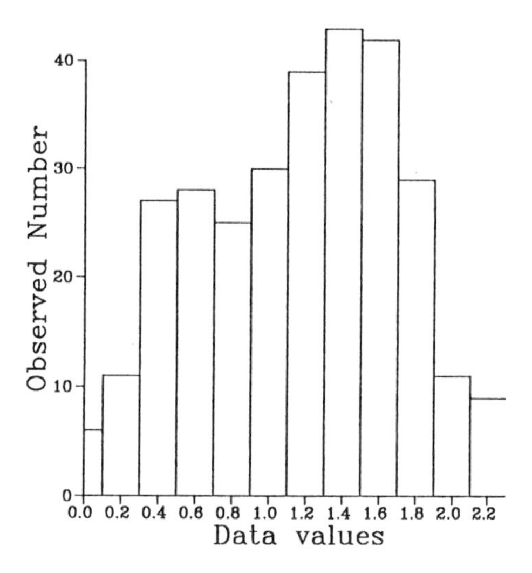

Figure 40-4: A histogram of the values shown in Figure 40-3.

determine the spatial pattern of patches. Since the number of parameters equals the number of data points, a Bayesian approach is taken.

Assume the following model. Conditional on the means, the random variables $\{Y(i) : i = 1, \ldots, 300\}$ are independently distributed as Gaussian (normal) random variables:

$$g(Y(i) \mid \mu(i)) \sim \text{Gau}\,(\mu(i), \sigma^2); \quad i = 1, \ldots, 300:$$

Then the joint conditional density is

$$f(\mathbf{Y} \mid \mu) = \prod_{i=1}^{300} g(Y(i) \mid \mu(i)),$$

where $\mathbf{Y} \equiv (Y(1), \ldots, Y(300))'$ and $\mu \equiv (\mu(1), \ldots, \mu(300))'$. Now let μ have a prior distribution

$$\pi(\mu; a, b, \beta) \propto$$

$$\exp\left\{\sum_{i=1}^{n}\sum_{j=1}^{n} \beta_{ij}\left[\frac{[a-\mu(i)][a-\mu(j)]}{(a-b)^2} + \frac{[b-\mu(i)][b-\mu(j)]}{(a-b)^2}\right]\right\}$$

where

$$\beta_{ij} = \beta_{ji} \equiv \begin{cases} \beta & \text{if } |i-j| = 1 \\ 0 & \text{otherwise} \end{cases},$$

$a = 0.6$, $b = 1.5$, and $n = 300$ in this example. This prior is a generalization of the one given by Greig et al. (1989) for 0-1 values of $\mu(\cdot)$. The parameter β controls the amount of "smoothing" and, based on spatial exploratory data analysis, was chosen to be 0.1 here. The prior distribution $\pi(\cdot)$ is a Markov random field; that is, conditional probabilities depend only on local neighborhoods.

With the model $f(\cdot)$ and the prior $\pi(\cdot)$, the posterior distribution is

$$p(\mu \mid \mathbf{Y}) \propto f(\mathbf{Y} \mid \mu)\pi(\mu).$$

Now, given $\mathbf{Y}$, we want to find the μ that achieves the maximum probability from the posterior distribution $p(\cdot)$; this is called the maximum a posteriori (MAP) estimate of μ (e.g., Geman and Geman, 1984). A direct procedure would be to maximize over all $2^{300} > 10^{90}$ possibilities for μ, but this is computationally prohibitive. To perform the maximization, we used simulated annealing instead (Geman and Geman, 1984). The predicted values $\{\hat{\mu}(i) : i = 1, \ldots, 300\}$ are given in Figure 40-3b. What we have done here is essentially to extract the "signal" μ from the "signal + noise" $\mathbf{Y}$. The vector μ is now useful for ecological interpretation (e.g., for determining the average patch size), which was not possible from the raw data. A full description of this method is given in Ver Hoef (1991).

The example given is one-dimensional, but the same principles apply in higher dimensions (simulated annealing is "borrowed" from statistical image analysis in two dimensions). Consider an irregular lattice of small areas, such as one might find in a cancer atlas. Assuming squared-error loss, Cressie (1992) uses Bayesian spatial methods to smooth observed incidence rates of sudden infant death syndrome.

Spatial point patterns

In (40-2), spatial point patterns occur when D is a random set, specifically a spatial point process. Thus, the particular realization is a spatial point pattern consisting of the location of events (in some sample window A). A spatial point process is any random mechanism for locating events $s_1, s_2, \ldots$, $\mathbf{R}^2$ or $\mathbf{R}^3$. By fitting models to patterns, we hope to gain some understanding of the generating mechanism and ascribe scientific interpretations to model parameters. Diggle (1983) is a readable reference with biological and ecological applications.

Rathbun and Cressie (1990) consider a spatial point pattern of trees evolving over time. They model the pattern according to a space–time survival point process, where events are born at some random location and time and then live and grow for a random length of time. By taking a reductionist approach, they consider three components: the birth process, the growth process, and the death process.

A census of longleaf pines in a portion of the Wade Tract (southern Georgia, USA) was taken in 1979. The study area was followed over time, allowing the following definitions to be made (Platt et al., 1988):

Adults: trees at least 30 cm diameter at breast height (dbh) in 1979.

Subadults: trees between 10 and 30 cm dbh in 1979.

Juveniles: trees between 2 and 10 cm dbh in 1979.
Recruits: trees at least 2 cm dbh in 1987, but less than 2 cm in 1979.

A thorough statistical analysis recognizes a *multivariate* point pattern and looks for dependencies between various tree classes. Data taken at later times allow growth modeling of the survivors in a 9-year period, as well as a modeling of the death process of nonsurvivors.

From their models, Rathbun and Cressie (1990) were able to ascertain the following relations:

Birth process: Births tend to occur close to disturbance paths. Births tend to occur in forest openings, isolated from subadult and adult trees. Thus, trees will tend to spend their lifetimes in single-aged cohorts.
Growth process: Interactions within the same or larger size classes are of the competitive kind, particularly among juveniles and subadults.
Death process: Interactions of juveniles with subadults reduce survivorship of juveniles, but there is little effect of competitive interactions of subadults with adults. Finally, adult mortality appears to be clustered.

DISCUSSION

The use of statistical models for solving environmental and ecological problems does not (or at least should not) come from any compulsion to use mathematics for its own sake. The complexity of the physical, chemical, and biological processes as well as almost unpredictable modifications of ecosystems by humans, almost demands a statistical approach. The paradigms of Fisher (1925) are helpful but not always adequate for the statistical analysis of spatio-temporal data. However, a statistical model for a single, deterministic (but unobserved) surface may not seem adequate either. In the next section, we compare two ways to define a probability structure on the surface that will allow statistical inference to proceed.

Spatial probability sampling

Suppose that measurements, both actual and potential, are denoted

$$\{z(\mathbf{s}) : \mathbf{s} \in D\}, \qquad (40\text{-}7)$$

where $\mathbf{s}$ is a spatial location vector on a transect, in the plane, on the surface of a sphere, or in three-dimensional space. The index set D gives the extent of the region of interest and $\mathbf{s}$ varies continuously over it. Suppose that data $\mathbf{z} \equiv (z(\mathbf{s}_1), \dots, z(\mathbf{s}_n))'$ are observed at known sites $\{\mathbf{s}_1, \dots, \mathbf{s}_n\}$. The model-based approach to inference (prediction of unobserved values or estimation of population parameters) assumes that (40-7) is a single realization from a random process $\{Z(\mathbf{s}) : \mathbf{s} \in D\}$. What is the source of probability in this model for (40-7)?

When there are infinitely many possible response surfaces from which to choose, a standard tactic in science is to deal with these statistically. The applied scientist or engineer does not come to the problem context free. The accumulated knowledge of his or her training, revelations in the latest journal articles, and experience with data of a similar type all lead to a probability distribution for (40-7). One might expect there to be general consensus from within the profession on this "prior" distribution; nevertheless, it represents a personal quantification of lack of knowledge about (40-7). Objectivity is an important part of the scientific process, but it is not the whole story. Good scientists build good prior distributions, in the sense that they assimilate well past and present knowledge into accurate predictions about future observations.

The Bayesian approach allows new information (data) to be incorporated into the prior distribution via Bayes's formula, yielding a posterior distribution. In the context of random spatial processes, suppose the new information is $\mathbf{Z} \equiv (Z(\mathbf{s}_1), \dots, Z(\mathbf{s}_n))'$. Then the posterior distribution is the conditional distribution of the process $Z(\cdot)$ *given* $\mathbf{Z} = \mathbf{z}$. Now, the posterior could be called a *new prior,* and any future information would update it in the same way to yield a new posterior, and so forth. Any inference should be based on the *latest* posterior distribution.

Thus, the source of probability is lack of knowledge, and, because some people have more knowledge about certain things than others, probabilities of particular events depend very much on the individual who assigns them. For example, suppose you are playing poker. Your (rational) decision regarding how many and what cards to discard depends on the probabilities of certain cards coming from a well-shuffled deck of cards. Knowing that the other players have cards but not knowing what they are does not influence those probabilities. However, among poker players, it is well known that extra knowledge of any kind (from the discarded cards, the other hands, or the remainder of the deck) will change the probabilities of obtaining certain hands. The same is true for spatial processes, where the extra knowledge about the process comes from spatial continuity (nearby observations tend to be more alike than those far apart).

The model-based approach described is not universally accepted in spatial statistics. When studying one particular oilfield or one particular aquifer, it has been argued (e.g., Matheron, 1965; Journel, 1985) that there is no randomness in (40-7) (apart from measurement error). Then the claim is that inference should be carried out assuming that $\mathbf{z}$ is a (probability) sample from a fixed but unknown $z(\cdot)$. Not surprisingly, a similar debate has occurred in classical sampling theory between those advocating design-based methods and those advocating model-based methods (see, e.g., the article by Hansen et al., 1983, and the discussion subsequent to it).

We have demonstrated that it is more efficient to sample stationary processes regularly than completely randomly and that, at least initially, as the degree of spatial dependence (i.e., spatial continuity) increases, so too does the advantage of regular sampling. Intuitively, for an isotropic process, the optimal grid will be equilateral triangular, and the regions of closest proximity to grid points will be hexagons. For a square grid, points in the corners of a square are farther from the center point than any other points on the edge and so harder to predict. This intuition is confirmed by simulation studies (e.g., Olea, 1984; Yfantis et al., 1987). Note, however, that there are occasions when certain locations will *have* to be sampled, such as centers of large population, and conversely when certain locations should not (or cannot) be sampled.

Can a probability sample still be taken if the grid is regular? From a model-based point of view, stationary processes are stationary no matter where one starts looking at them. Thus, a random *starting point* of the grid should make no difference to the eventual estimation of model parameters. Design-based inference depending on this externally created probability structure has some advantages: The procedure prevents biases in the site-selection process. For some purposes, the assumption of stationarity can be dispensed with. For simple quantities like distribution functions, their moments, and regression relations, inference can be free of model assumptions. However, that strength can also be a weakness if one stops with design-based inference, since, without a model, the range of questions that can be asked and answered are severely limited. Previous sections illustrated the enormous scope of model-based inference.

Armed with good model diagnostics and flexible spatial models, the statistician can make substantial contributions to environmental science and ecology. There is also room for prudent probability sampling, for example, stratification to obtain regions within which the process is (approximately) stationary followed by regular (probability) sampling within each region.

Future opportunities

Scientific problems regarding the environment are hard, partly due to patchiness of data quality and quantity and partly due to the inherent variability of regional and global processes. Modern (and future) statistical methods, brought to bear with the force of new technologies such as GIS, can make the most efficient use of the environmental dollar. Areas in which we expect to see considerable future statistical activity include:

- Spatio-temporal models of correlation that include causative components.
- Probability sampling designs that recognize and exploit spatial continuity.
- Resampling (bootstrapping) methods that yield valid inferences in the presence of spatial dependence.
- Bayesian statistics, where scientists and engineers are able to express their prior information when constructing a statistical model.

Not only are there opportunities for exciting research programs, there is also the chance for statisticians to become involved in the (re)education process at all levels, from beginning service courses in statistics through advanced Ph.D.-level statistics courses, and in short courses at both statistical and environmental gatherings.

CONCLUSIONS

This planet is feeling the effects of its burgeoning population. The population, and hence the need for resources, does not show any tendency to decrease or even level off. We are witnessing deforestation, reduced biodiversity, stratospheric ozone depletion, and acid deposition. We may be entering an age of global warming, desertification, and sea-level rise.

Scientists and engineers are doing their best to understand, quantify, and predict the effects of global change. The time scale of global change is beyond that of a presidency, and its spatial scale is beyond that of any one nation. Thus, long-term monitoring networks are essential, as are remote sensing programs and international environmental agreements.

Without baseline data to serve as the "control," and without sound ecological models, one can say almost nothing about the environmental impact of anthropogenic or natural phenomena.

Since the scales are often global, it is important to move towards standardization of sampling design, field and laboratory methods of data collection, and statistical analyses of data. Those analyses should not neutralize or ignore the spatio-temporal continuity in environmental phenomena. Inappropriate randomization can lead to a substantial loss of statistical efficiency, and, worse yet, statistical methods based on classical (IID) assumptions may be invalid. Through proper spatial statistical modeling, both validity and efficiency can be attained. The GIS has the potential to provide a powerful interface between environmental/ecological processes and spatio-temporal (statistical) models.

ACKNOWLEDGMENTS

This research was funded in part by NATO, the National Park Service, and the National Science Foundation under grant number DMS-9001862, shared by Statistics and Probability and Geography and Regional Science. The paper was written while the first author was visiting the Department of Statistics, Stanford University, during the fall of 1991. The comments of Scott Overton on an earlier draft are much appreciated. Responsibility for any errors of commission or omission remain with the authors alone.

REFERENCES

Besag, J. E. (1974) Spatial interaction and the statistical analysis of lattice systems. *Journal of the Royal Statistical Society B* 36: 192–225.

Cressie, N. (1991) *Statistics for Spatial Data*, New York: Wiley.

Cressie, N. (1992) Smoothing regional maps using empirical Bayes predictors. *Geographical Analysis* 24: 75–95.

Cressie, N., Gotway, C. A., and Grondona, M. O. (1990) Spatial prediction from networks. *Chemometrics and Intelligent Laboratory Systems* 7: 251–271.

DeGroot, M. H. (1970) *Optimal Statistical Decisions*, New York: McGraw-Hill.

Diggle, P. J. (1983) *Statistical Analysis of Spatial Point Patterns*, New York: Academic Press.

European Inland Fisheries Advisory Commission (EIFAC) (1969) Water quality criteria for European freshwater fish—extreme pH values and inland fisheries. *Water Research* 3: 593–611.

Fairfield Smith, H. (1938) An empirical law describing heterogeneity in the yields of agricultural crops. *Journal of Agricultural Science* (Cambridge) 28: 1–23.

Fisher, R. A. (1925) *Statistical Methods for Research Workers*, Edinburgh: Oliver and Boyd.

Fisher, R. A. (1935) *The Design of Experiments*, Edinburgh: Oliver and Boyd.

Geman, S., and Geman, D. (1984) Stochastic relaxation, Gibbs distributions and the Bayesian restoration of images. *IEEE Transactions on Pattern Analysis and Machine Intelligence* PAMI-6: 721–741.

Greig, D. M., Porteous, B. T., and Seheult, A. H. (1989) Exact maximum a posteriori estimation for binary images. *Journal of the Royal Statistical Society B* 51: 271–279.

Hansen, M. H., Madow, W. G., and Tepping, B. J. (1983) An evaluation of model-dependent and probability-sampling inferences in sample surveys. *Journal of the American Statistical Association* 78: 776–793.

Haslett, J., Bradley, R., Craig, P., Unwin, A., and Wills, G. (1991) Dynamic graphics for exploring spatial data with application to locating global and local anomalies. *American Statistician* 45: 234–242.

Hurlbert, S. H. (1984) Pseudoreplication and the design of ecological field experiments. *Ecological Monographs* 54: 187–211.

Journel, A. G. (1985) The deterministic side of geostatistics. *Journal of the International Association for Mathematical Geology* 17: 1–14.

Matheron, G. (1965) *La Theorie des Variables Regionalisees et ses Applications*, Paris: Masson.

Messer, J. J., Linthurst, R. A. and Overton, W. S. (1991) An EPA program for monitoring ecological status and trends. *Environmental Monitoring and Assessment* 17: 67–78.

National Acid Precipitation Assessment Program (NAPAP) (1988) *Interim Assessment. The Causes and Effects of Acidic Deposition, Volumes I–IV*, Washington, DC: U.S. Government Printing Office.

Olea, R. A. (1984) Sampling design optimization for spatial functions. *Journal of the International Association for Mathematical Geology* 16: 369–392.

Pitelka, L. F., and Raynal, D. J. (1989) Forest decline and acidic deposition. *Ecology* 70: 2–10.

Platt, W. J., Evans, G. W., and Rathbun, S. L. (1988) The population dynamics of a long-lived conifer (*Pinus Palustris*). *American Naturalist* 131: 491–525.

Rathbun, S. L., and Cressie, N. (1990) A space–time survival point process for a longleaf pine forest in southern Georgia. *Statistical Laboratory Preprint No. 90-13*, Ames, IA: Iowa State University.

Sacks, J., and Schiller, S. (1988) Spatial designs. In Gupta, S. S., and Berger, J. O. (eds.), *Statistical Decision Theory and Related Topics IV*, New York: Springer, Vol 2., pp. 385–399.

Ver Hoef, J. M. (1991) Statistical Analysis of Spatial Pattern in Ecological Data. *Unpublished Ph.D. Dissertation*, Ames, IA: Iowa State University.

Ver Hoef, J. M., Glenn-Lewin, D. C. and Werger, M. J. A. (1989) Relationship between horizontal pattern and vertical structure in a chalk grassland. *Vegetatio* 83: 147–155.

Whittle, P. (1954) On stationary processes in the plane. *Biometrika* 41: 434–449.

Yates, F. (1938) The comparative advantages of systematic and randomized arrangements in the design of agricultural and biological experiments. *Biometrika* 30: 444–466.

Yfantis, E. A., Flatman, G. T., and Behar, J. V. (1987) Efficiency of kriging estimation for square, triangular, and hexagonal grids. *Mathematical Geology* 19: 183–205.

Zimmerman, D. L. (1991) Design criteria for semivariogram model discrimination and parameter estimation. *Technical Report No. 191*, Iowa City, IA: Department of Statistics, University of Iowa.

41

Geostatistics: A Tool for Environmental Modelers

NOEL CRESSIE

The prefix "geo" in geostatistics originally implied statistics pertaining to the Earth (Matheron, 1963; see also Hart, 1954, who used the term differently from Matheron, in a geographical context). However, more recently, geostatistics has been used in a variety of disciplines ranging from agriculture to zoology. (Within meteorology, a virtually identical theory called objective analysis was developed by Gandin, 1963.) Its flexibility in being able to incorporate the known action of physical, chemical, and biological processes along with uncertainty represented by spatial heterogeneity, makes it an attractive tool for environmental modelers.

In the sections that follow, the basic ideas behind a geostatistical analysis (including the variogram, kriging, splines, conditional simulation, and change of support) are presented. A separate section considers spatial prediction of multivariate data. A sample of environmental applications is given in the last section.

THEORY AND METHODS OF GEOSTATISTICS

Geostatistics is mostly concerned with spatial prediction, but there are other important areas, such as model selection, effect of aggregation, and spatial sampling design, that offer fruitful open problems. The emphasis in this section will be on a spatial-prediction method known as *kriging*. Matheron (1963) coined the term in honor of D. G. Krige, a South African mining engineer (see Cressie, 1990, for an account of the origins of kriging).

The variogram

First, a measure of the (second-order) spatial dependence exhibited by the spatial data is needed. A model-based parameter (which is a function) known as the variogram is defined here; its estimate provides such a measure. Statisticians are used to dealing with the autocovariance function. It is demonstrated here that the class of processes with a variogram contains the class of processes with an autocovariance function, and that kriging can be carried out on a wider class of processes than the one traditionally used in statistics.

Let $\{Z(\mathbf{s}) : \mathbf{s} \in D \subset \mathbf{R}^d\}$ be a real-valued stochastic process defined on a domain D of the d-dimensional space $\mathbf{R}^d$, and suppose that differences of variables lagged $\mathbf{h}$ apart vary in a way that depends only on $\mathbf{h}$. Specifically, suppose

$$\text{var}[Z(\mathbf{s}+\mathbf{h}) - Z(\mathbf{s})] = 2\gamma(\mathbf{h}) \quad \text{for all } \mathbf{s}, \mathbf{s}+\mathbf{h} \in D; \tag{41-1}$$

typically the spatial index $\mathbf{s}$ is two or three dimensional (i.e., $d = 2$ or 3). The quantity $2\gamma(\cdot)$, which is a function only of the *difference* between the spatial locations $\mathbf{s}$ and $\mathbf{s} + \mathbf{h}$, has been called the *variogram* by Matheron (1963), although earlier appearances in the scientific literature can be found. It has been called a *structure function* by Yaglom (1957) in probability and by Gandin (1963) in meteorology, and a *mean-squared difference* by Jowett (1952) in time series. Kolmogorov (1941) introduced it in physics to study the local structure of turbulence in a fluid. Nevertheless, it has been Matheron's mining terminology that has persisted. The variogram must satisfy the conditional negative semidefiniteness condition,

$$\sum_{i=1}^{k}\sum_{j=1}^{k} a_i a_j 2\gamma(\mathbf{s}_i - \mathbf{s}_j) \leq 0$$

for any finite number of spatial locations $\{\mathbf{s}_i : i = 1, \dots, k\}$, and real numbers $\{a_i : i = 1, \dots, k\}$ satisfying

$$\sum_{i=1}^{k} a_i = 0.$$

When $2\gamma(\mathbf{h})$ can be written as $2\gamma^0(\|\mathbf{h}\|)$, for $\mathbf{h} \in \mathbf{R}^d$, the variogram is said to be *isotropic*; otherwise it is said to be anisotropic, in which case the process Z is also referred to as anisotropic.

Variogram models that depend on only a few parameters θ can be used as summaries of the spatial dependence and as an important component of optimal linear prediction (kriging). Three basic isotropic models, given here in terms of the semivariogram (half the variogram), are:

Linear model (valid in $\mathbf{R}^d, d \geq 1$)

$$\gamma(\mathbf{h}; \theta) = \begin{cases} 0 & \mathbf{h} = \mathbf{0} \\ c_0 + b_l \|\mathbf{h}\| & \mathbf{h} \neq \mathbf{0}. \end{cases}$$

where $\theta = (c_0, b_l)', c_0 \geq 0, b_l \geq 0$;

Spherical model (valid in $\mathbf{R}^1$, $\mathbf{R}^2$, and $\mathbf{R}^3$)

$$\gamma(\mathbf{h}; \theta) = \begin{cases} 0 & \mathbf{h} = \mathbf{0} \\ c_0 + c_s \left[\frac{3}{2}(\|\mathbf{h}\|/a_s) - \frac{1}{2}(\|\mathbf{h}\|/a_s)^3 \right] & 0 < \|\mathbf{h}\| \leq a_s \\ c_0 + c_s & \|\mathbf{h}\| \geq a_s, \end{cases}$$

where $\theta = (c_0, c_s, a_s)', c_0 \geq 0, c_s \geq 0, a_s \geq 0$;

Exponential model (valid in $\mathbf{R}^d, d \geq 1$)

$$\gamma(\mathbf{h}; \theta) = \begin{cases} 0 & \mathbf{h} = \mathbf{0} \\ c_0 + c_e[1 - \exp(-\|\mathbf{h}\|/a_e)] & \mathbf{h} \neq \mathbf{0}, \end{cases}$$

where $\theta = (c_0, c_e, a_e)', c_0 \geq 0, c_e \geq 0, a_e \geq 0$.

Another semivariogram model is the *rational quadratic model* (valid in $\mathbf{R}^d, d \geq 1$):

$$\gamma(\mathbf{h}; \theta) = \begin{cases} 0 & \mathbf{h} = \mathbf{0} \\ c_0 + \dfrac{c_r \|\mathbf{h}\|^2}{1 + \|\mathbf{h}\|^2 / a_r} & \mathbf{h} \neq \mathbf{0} \end{cases}$$

where $\theta = (c_0, c_w, a_r)', c_0 \geq 0, c_r \geq 0, a_r \geq 0$.

A semivariogram model that exhibits negative correlations caused by periodicity of the process is the *wave* (or *hole-effect*) model (valid in $\mathbf{R}^1$, $\mathbf{R}^2$, and $\mathbf{R}^3$):

$$\gamma(\mathbf{h}; \theta) = \begin{cases} 0 & \mathbf{h} = \mathbf{0} \\ c_0 + c_w \dfrac{1 - a_w \sin(\|\mathbf{h}\|/a_w)}{\|\mathbf{h}\|} & \mathbf{h} \neq \mathbf{0}, \end{cases}$$

where $\theta = (c_0, c_w, a_w)'\ c_0 \geq 0, c_w \geq 0, a_w \geq 0$.

A further condition that a variogram model must satisfy is (Matheron, 1971)

$$2\gamma(\mathbf{h}) / \|\mathbf{h}\|^2 \to 0, \quad \text{as} \quad \|\mathbf{h}\| \to \infty.$$

In fact, the *power* semivariogram model,

$$\gamma(\mathbf{h}; \theta) = \begin{cases} 0 & \mathbf{h} = \mathbf{0} \\ c_0 + b_p \|\mathbf{h}\|^\lambda & \mathbf{h} \neq \mathbf{0}, \end{cases}$$

where $\theta = (c_0, b_p, \lambda)', c_0 \geq 0, b_p \geq 0, 0 \leq \lambda < 2$,

is a valid semivariogram model in $\mathbf{R}^d, d \geq 1$.

When the process Z is anisotropic, the variogram is no longer purely a function of distance between two spatial locations. Anisotropies are caused by the underlying physical process evolving differentially in space. Sometimes the anisotropy can be corrected by an invertible linear transformation of the lag vector $\mathbf{h}$. That is,

$$2\gamma(\mathbf{h}) = 2\gamma^0(\|A\mathbf{h}\|), \qquad \mathbf{h} \in \mathbf{R}^d,$$

where A is a $d \times d$ matrix and $2\gamma^0$ is a function of only one variable.

Replacing Eq. (41-1) with the stronger assumption

$$\text{cov}(Z(\mathbf{s}+\mathbf{h}), Z(\mathbf{s})) = C(\mathbf{h}), \text{ for all } \mathbf{s}, \mathbf{s}+\mathbf{h} \in D \quad (41\text{-}2)$$

and specifying the mean function to be constant, that is,

$$E(Z(\mathbf{s})) = \mu, \text{ for all } \mathbf{s} \in D \qquad (41\text{-}3)$$

defines the class of *second-order* (or wide-sense) *stationary* processes in D, with (auto) covariance function $C(\cdot)$. Time-series analysts often assume Eq. (41-2) and work with the quantity $\rho(\cdot) \equiv C(\cdot)/C(\mathbf{0})$. Conditions (41-1) and (41-3) define the class of *intrinsically stationary* processes, which is now shown to contain the class of second-order stationary processes.

Assuming only Eq. (41-2),

$$\gamma(\mathbf{h}) = C(\mathbf{0}) - C(\mathbf{h}); \qquad (41\text{-}4)$$

that is, the semivariogram is related very simply to the covariance function. An example of a process for which $2\gamma(\cdot)$ exists but $C(\cdot)$ does not is a one-dimensional standard Wiener process $\{W(t) : t \geq 0\}$. Here, $2\gamma(h) = |h| (-\infty < h < \infty)$, but $\text{cov}(W(t), W(u)) = \min(t, u)$, which is not a function of $|t - u|$. An analogous result is true in $\mathbf{R}^d$ (Cressie, 1991, p. 302). Thus, the class of intrinsically stationary processes *strictly* contains the class of second-order stationary processes.

Now consider estimation of the variogram from data $\{Z(\mathbf{s}_i) : i = 1, \ldots, n\}$. Suppose these are observations on an intrinsically stationary process [i.e., a process that satisfies Eqs. (41-1) and (41-3)], taken at the n spatial locations $\{\mathbf{s}_i : i = 1, \ldots, n\}$. Because of assumption (41-3), $\text{var}[Z(\mathbf{s}+\mathbf{h}) - Z(\mathbf{s})] = E(Z(\mathbf{s}+\mathbf{h}) - Z(\mathbf{s}))^2$. Hence, a natural method-of-moments estimator of the variogram $2\gamma(\mathbf{h})$ is

$$2\hat{\gamma}(\mathbf{h}) \equiv \sum_{N(\mathbf{h})} [Z(\mathbf{s}_i) - Z(\mathbf{s}_j)]^2 / |N(\mathbf{h})|, \quad \mathbf{h} \in \mathbf{R}^d, \quad (41\text{-}5)$$

where the average (41-5) is taken over $N(\mathbf{h}) = \{(\mathbf{s}_i, \mathbf{s}_j) : \mathbf{s}_i - \mathbf{s}_j = \mathbf{h}\}$, and $|N(\mathbf{h})|$ is the number of distinct elements in $N(\mathbf{h})$. For irregularly spaced data, $N(\mathbf{h})$ is usually modified to $\{(\mathbf{s}_i, \mathbf{s}_j) : \mathbf{s}_i - \mathbf{s}_j \in T(\mathbf{h})\}$, where $T(\mathbf{h})$ is a tolerance region of $\mathbf{R}^d$ surrounding $\mathbf{h}$. Other estimators, more robust than Eq. (41-5), are given in Cressie and Hawkins (1980) and Cressie (1991, Sec. 2.4). Parametric models, $2\gamma(\cdot;\theta)$, can be fit to the estimator (41-5) by various means; as a compromise between efficiency and simplicity, Cressie (1985) advocates minimizing a weighted sum of squares

$$\sum_{k=1}^{K} \left\{\frac{2\hat{\gamma}(\mathbf{h}(k))}{2\gamma(\mathbf{h}(k);\theta)} - 1\right\}^2 |N(\mathbf{h}(k))|$$

with respect to variogram model parameters θ. The sequence $\mathbf{h}(1), \ldots, \mathbf{h}(K)$ denotes the "lags" at which an estimator (41-5) was obtained, and that satisfy range and replication conditions such as those given by Journel and Huijbregts (1978, p. 194, Eq. III.42). Zimmerman and Zimmerman (1991) summarize and compare several methods of variogram-parameter estimation based on simulated Gaussian data. They find that minimizing the weighted sum of squares usually performs well, and never does poorly, against such competitors as maximum likelihood estimation (both ordinary and restricted) and minimum norm quadratic unbiased estimation.

Kriging

For the purposes of this section, assume that the variogram is known; in practice, variogram parameters are estimated from the spatial data. Suppose it is desired to predict $Z(\mathbf{s}_0)$ at some unsampled spatial location $\mathbf{s}_0$ using a linear function of the data $\mathbf{Z} \equiv (Z(\mathbf{s}_1), \ldots, Z(\mathbf{s}_n))'$:

$$\hat{Z}(\mathbf{s}_0) = \sum_{i=1}^{n} \lambda_i Z(\mathbf{s}_i). \quad (41\text{-}6)$$

It is sensible to look for coefficients $\{\lambda_i : i = 1, \ldots, n\}$ for which Eq. (41-6) is uniformly unbiased and that minimize the mean-squared prediction error $E(Z(\mathbf{s}_0) - \hat{Z}(\mathbf{s}_0))^2$. More generally, one could try to minimize $E(L[Z(\mathbf{s}_0), p(\mathbf{Z})])$ with respect to predictor $p(\mathbf{Z})$, where L is a given loss function. For example, the loss function proposed by Zellner (1986),

$$L[Z(\mathbf{s}_0), p(\mathbf{Z})] = b(\exp\{a[Z(\mathbf{s}_0) - p(\mathbf{Z})]\} - a[Z(\mathbf{s}_0) - p(\mathbf{Z})] - 1), \quad a \in \mathbf{R}, b > 0.$$

allows overprediction to incur a different loss than underprediction. Minimizing mean-squared prediction error results from using

$$L[Z(\mathbf{s}_0), p(\mathbf{Z})] = b[Z(\mathbf{s}_0) - p(\mathbf{Z})]^2, \quad b > 0;$$

which is the squared-error loss function. In all that is to follow, squared-error loss is used.

The uniform unbiasedness condition imposed on Eq. (41-6) is $E(\hat{Z}(\mathbf{s}_0)) = \mu = E(Z(\mathbf{s}_0))$, for all $\mu \in \mathbf{R}$, which is equivalent to

$$\sum_{i=1}^{n} \lambda_i = 1. \quad (41\text{-}7)$$

If the process is second-order stationary and Eq. (41-7) is assumed,

$$E\left(Z(\mathbf{s}_0) - \sum_{i=1}^{n} \lambda_i Z(\mathbf{s}_i)\right)^2 = \quad (41\text{-}8)$$
$$C(\mathbf{0}) - 2\sum_{i=1}^{n} \lambda_i C(\mathbf{s}_i - \mathbf{s}_0) + \sum_{i=1}^{n}\sum_{j=1}^{n} \lambda_i\lambda_j C(\mathbf{s}_i - \mathbf{s}_j).$$

If the process is intrinsically stationary (a weaker assumption) and (41-7) is assumed,

$$E\left(Z(\mathbf{s}_0) - \sum_{i=1}^{n} \lambda_i Z(\mathbf{s}_i)\right)^2 = \quad (41\text{-}9)$$
$$2\sum_{i=1}^{n} \lambda_i \gamma(\mathbf{s}_i - \mathbf{s}_0) - \sum_{i=1}^{n}\sum_{j=1}^{n} \lambda_i\lambda_j \gamma(\mathbf{s}_i - \mathbf{s}_j).$$

Using differential calculus and the method of Lagrange multipliers, optimal coefficients $\lambda = (\lambda_1, \ldots, \lambda_n)'$ can be found that minimize Eq. (41-9) subject to Eq. (41-7); they are

$$\lambda = \Gamma^{-1}\left[\gamma + \frac{(1 - \mathbf{1}'\Gamma^{-1}\gamma)\mathbf{1}}{\mathbf{1}'\Gamma^{-1}\mathbf{1}}\right], \quad (41\text{-}10)$$

and the minimized value of Eq. (41-9) (kriging variance) is

$$\sigma_k^2(\mathbf{s}_0) = \gamma'\Gamma^{-1}\gamma - \frac{(1 - \mathbf{1}'\Gamma^{-1}\gamma)^2}{\mathbf{1}'\Gamma^{-1}\mathbf{1}}. \quad (41\text{-}11)$$

In Eqs. (41-10) and (41-11), $\gamma = [\gamma(\mathbf{s}_1 - \mathbf{s}_0), \ldots, \gamma(\mathbf{s}_n - \mathbf{s}_0)]'$, $\mathbf{1} = (1, \ldots, 1)'$, and Γ is the $n \times n$ symmetric matrix with $(i, j)^{\text{th}}$ element $\gamma(\mathbf{s}_i - \mathbf{s}_j)$, which is assumed to be invertible.

The kriging predictor given by Eqs. (41-6) and (41-10) is appropriate if the process Z contains no measurement error. If measurement error is present, then a "noiseless version" of Z should be predicted; Cressie (1988) has details on when and how this should be implemented.

Thus far, kriging has been derived under the assumption of a constant mean. More realistically, assume

$$Z(\mathbf{s}) = \mu(\mathbf{s}) + \delta(\mathbf{s}), \quad \mathbf{s} \in D, \tag{41-12}$$

where $E(Z(\mathbf{s})) = \mu(\mathbf{s})$, for $\mathbf{s} \in D$, and $\delta(\cdot)$ is a zero-mean, intrinsically stationary stochastic process with $\text{var}[\delta(\mathbf{s} + \mathbf{h}) - \delta(\mathbf{s})] = \text{var}[Z(\mathbf{s} + \mathbf{h}) - Z(\mathbf{s})] = 2\gamma(\mathbf{h}), \mathbf{h} \in \mathbf{R}^d$. In Eq. (41-12), the "large-scale variation" $\mu(\cdot)$ and the "small-scale variation" $\delta(\cdot)$ are modeled as deterministic and stochastic processes, respectively, but with no unique way of identifying either of them. What is one person's mean structure could be another person's correlation structure. Often this problem is resolved in a substantive application by relying on scientific or habitual reasons for determining the mean structure.

Suppose $\mu(\mathbf{s}) = \mathbf{x}(\mathbf{s})'\beta$, a linear combination of variables that could include trend-surface terms or other explanatory variables thought to influence the behavior of the large-scale variation. Thus,

$$Z(\mathbf{s}) = \sum_{j=0}^{p} x_j(\mathbf{s})\beta_j + \delta(\mathbf{s}), \quad \mathbf{s} \in D, \tag{41-13}$$

where $\beta \equiv (\beta_0, \ldots, \beta_p)'$ are unknown parameters and $\delta(\cdot)$ is intrinsically stationary [i.e., satisfies Eqs.(41-1) and (41-3)] with zero mean. Although the model has changed, the problem of predicting $Z(\mathbf{s}_0)$ using an unbiased linear predictor (41-6) remains. The uniform unbiasedness condition is now equivalent to the condition

$$\lambda' X = \mathbf{x}_0', \tag{41-14}$$

where $\mathbf{x}_0 \equiv (x_0(\mathbf{s}_0), \ldots, x_p(\mathbf{s}_0))'$ and X is an $n \times (p+1)$ matrix whose $(i, j)^{\text{th}}$ element is $x_{j-1}(\mathbf{s}_i)$. Then, provided Eq. (41-7) is implied by Eq. (41-14), minimizing the mean-squared prediction error subject to Eq. (41-14) yields the *universal kriging predictor*

$$\hat{Z}_U(\mathbf{s}_0) = \lambda_U' \mathbf{Z}, \tag{41-15}$$

where

$$\lambda_U = \Gamma^{-1}\left[\gamma + X\left(X'\Gamma^{-1}X\right)^{-1}\left(\mathbf{x}_0 - X'\Gamma^{-1}\gamma\right)\right]; \tag{41-16}$$

the (universal) kriging variance is

$$\sigma_k^2(\mathbf{s}_0) = \gamma'\Gamma^{-1}\gamma - \left(X'\Gamma^{-1}\gamma - \mathbf{x}_0\right)'\left(X'\Gamma^{-1}X\right)^{-1}\left(X'\Gamma^{-1}\gamma - \mathbf{x}_0\right). \tag{41-17}$$

Another way to write the equations (41-14) and (41-15) is

$$\hat{Z}_U(\mathbf{s}_0) = \mathbf{v}_1'\gamma + \mathbf{v}_2'\mathbf{x}_0, \tag{41-18}$$

where $\mathbf{v}_1$ (an $n \times 1$ vector) and $\mathbf{v}_2$ [a $(p + 1) \times 1$ vector] solve

$$\begin{aligned} \Gamma\mathbf{v}_1 + X\mathbf{v}_2 &= \mathbf{Z} \\ X'\mathbf{v}_1 &= \mathbf{0}. \end{aligned} \tag{41-19}$$

Equations (41-18) and (41-19) are known as the dual-kriging equations, since the predictor is now expressed as a linear combination of the elements of $(\gamma', \mathbf{x}'_0)$. From Eq. (41-19), it is clear that thin-plate spline smoothing is equivalent in form to universal kriging (see Watson, 1984, where the relationship between the two prediction techniques is reviewed). Kriging has the advantage that in practice the data are first used to estimate the variogram, so adapting to the quality and quantity of spatial dependence in the data. Furthermore, kriging produces a mean-squared prediction error, given by Eq. (41-17), that quantifies the degree of uncertainty in the predictor. Cressie (1991, Sec. 5.9) presents these two faces of spatial prediction along with 12 others, including disjunctive kriging and inverse-distance-squared weighting.

Conditional simulation of spatial data

Simulation of spatial data $\{Z(\mathbf{s}_i) : i = 1, \ldots, N\}$ with given means $\{\mu(\mathbf{s}_i) : i = 1, \ldots, N\}$ and covariates $\{C(\mathbf{s}_i, \mathbf{s}_j) : 1 \leq i \leq j \leq N\}$ can be carried out in a number of ways, depending on the size of N and the sparseness of Σ_N, the $N \times N$ symmetric matrix whose $(i, j)^{\text{th}}$ element is $C(\mathbf{s}_i, \mathbf{s}_j)$. One way is to use the Cholesky decomposition $\Sigma_N = L_N L'_N$, where L_N is a lower-triangular $N \times N$ matrix (e.g., Golub and Van Loan, 1983, pp. 86–90). Then $\mathbf{Z}_N \equiv (Z(\mathbf{s}_1), \ldots, Z(\mathbf{s}_N))'$ can be simulated by

$$\mathbf{Z}_N = \mu_N + L_N\varepsilon_N \tag{41-20}$$

where $\mu_N \equiv (\mu(\mathbf{s}_1), \ldots, \mu(\mathbf{s}_N))'$, and ε_N is an $N \times 1$ vector of simulated independent and identically distributed random variables, each with zero mean and unit variance. Other methods, including polynomial approximations, Fourier transforms, and turning bands, are presented and compared in Cressie (1991, Sec. 3.6).

Now consider the simulation of values of $\{Z(\mathbf{s}) : \mathbf{s} \in D\}$ *conditional on* observed values $\mathbf{Z}_n$. Call this conditionally simulated process $\{W(\mathbf{s}) : \mathbf{s} \in D\}$, and suppose $\{V(\mathbf{s}) : \mathbf{s} \in D\}$ is an unconditionally simulated process with the same first and second moments as $\{Z(\mathbf{s}) : \mathbf{s} \in D\}$. For example, Eq. (41-20) might be used to simulate $\mathbf{V}_N \equiv (V(\mathbf{s}_1), \ldots, V(\mathbf{s}_N))'$, where $N \geq n$.

Consider conditional simulation at an arbitrary location $\mathbf{s}_{n+1}$ in D (Journel, 1974). Now write

$$\Sigma_{n+1} = \begin{bmatrix} \Sigma_n & \mathbf{c}_n \\ \mathbf{c}_n' & C\left(\mathbf{s}_{n+1}, \mathbf{s}_{n+1}\right) \end{bmatrix}$$

and notice that the two terms of the decomposition

$$Z\left(\mathbf{s}_{n+1}\right) = \mathbf{c}_n'\Sigma_n^{-1}\mathbf{Z}_n + \left[Z\left(\mathbf{s}_{n+1}\right) - \mathbf{c}_n'\Sigma_n^{-1}\mathbf{Z}_n\right] \tag{41-21}$$

are uncorrelated. Hence, the conditional simulation

$$W\left(\mathbf{s}_{n+1}\right) = \mathbf{c}_n'\Sigma_n^{-1}\mathbf{Z}_n + \left[V\left(\mathbf{s}_{n+1}\right) - \mathbf{c}_n'\Sigma_n^{-1}\mathbf{V}_n\right], \quad \mathbf{s}_{n+1} \in D, \tag{41-22}$$

has the same first two moments, unconditionally, as the process $\{Z(\mathbf{s}) : \mathbf{s} \in D\}$ and $W(\mathbf{s}_i) = Z(\mathbf{s}_i), i = 1, \ldots, n$. That is, unconditional simulation of sample paths of V yields, through Eq. (41-22), conditionally simulated sample paths of W.

It is apparent from Eqs. (41-20) and (41-21) that when the ε_is are Gaussian, so too is the process $\{W(\mathbf{s}) : \mathbf{s} \in D\}$. However, this may not reflect the reality of the conditional process when the original process $\{Z(\mathbf{s}) : \mathbf{s} \in D\}$ is "far from" Gaussian, even though the first two moments match and the two processes agree at the data locations. There is clearly a danger in using conditional simulation uncritically.

Change of support

The change-of-support problem remains a major challenge to geostatisticians. Although data typically come as $\mathbf{Z} = (Z(\mathbf{s}_1), \ldots, Z(\mathbf{s}_n))'$, suppose that prediction is required for $Z(B) \equiv (1/|B|)\int_B Z(\mathbf{u})\, d\mathbf{u}$. Kriging adapts very easily to accommodate the change from point support $\mathbf{s}_0$ to block support B. For example, in Eqs. (41-10) and (41-11), γ is modified to $\gamma(B) \equiv [(1/|B|)\int_B \gamma(\mathbf{s}_1 - \mathbf{u})\, d\mathbf{u}, \ldots, (1/|B|)\int_B \gamma(\mathbf{s}_n - \mathbf{u})\, d\mathbf{u}]'$, and in Eq. (41-11) $\sigma_k^2(B)$ has the extra term $(-1/|B|^2)\int_B\int_B \gamma(\mathbf{u} - \mathbf{v})d\mathbf{u}\, d\mathbf{v}$. But in mining applications and emission compliance, for example, the quantity of greatest interest is the conditional distribution $\mathrm{pr}(Z(B) > z|\mathbf{Z})$. Both disjunctive kriging (Matheron, 1976) and indicator kriging (Journel, 1983) attempt to estimate this quantity based on bivariate distributional properties of the (possibly transformed) process. The problem is important enough to pursue beyond these initial approaches.

MULTIVARIATE SPATIAL PREDICTION

The definition of the variogram for a univariate process was given in a previous section. Now let $\{\mathbf{Z}(\mathbf{s}) : \mathbf{s} \in D\}$, where $\mathbf{Z}(\mathbf{s}) \equiv (Z_1(\mathbf{s}), \ldots, Z_k(\mathbf{s}))'$, be a multivariate spatial process. Suppose that each component process possesses a variogram:

$$2\gamma_{jj}(\mathbf{h}) \equiv \mathrm{var}[Z_j(\mathbf{s} + \mathbf{h}) - Z_j(\mathbf{s})], \quad \mathbf{h} \in \mathbf{R}^d, \; j = 1, \ldots, k.$$

There are two ways to generalize this notion to account for crossdependence between $Z_j(\cdot)$ and $Z_{j'}(\cdot)$. The most natural one for multivariate spatial prediction (cokriging) is seen below to be

$$2\gamma_{jj'}(\mathbf{h}) \equiv \mathrm{var}[Z_j(\mathbf{s} + \mathbf{h}) - Z_{j'}(\mathbf{s})], \quad \mathbf{h} \in \mathbf{R}^d. \tag{41-23}$$

which, apart from a mean correction, is the quantity proposed by Clark, Basinger, and Harper (1989).

The other generalization,

$$2\nu_{jj'}(\mathbf{h}) \equiv \mathrm{cov}(Z_j(\mathbf{s}+\mathbf{h}) - Z_j(\mathbf{s}), Z_{j'}(\mathbf{s}+\mathbf{h}) - Z_{j'}(\mathbf{s})), \quad \mathbf{h} \in \mathbf{R}^d, \tag{41-24}$$

can only be used for cokriging under special conditions on the matrix covariance function (41-26) (Journel and Huijbregts, 1978, p. 326; Myers, 1982). It has, however, been the generalization traditionally recommended (e.g., Myers, 1982, 1984, 1988, 1991; Wackernagel, 1988). Relationships between the $\{\gamma_{jj'}(\cdot)\}$, $\{\nu_{jj'}(\cdot)\}$, and cross-covariances are given by Myers (1991) and Ver Hoef and Cressie (1991).

Let

$$E(\mathbf{Z}(\mathbf{s})) = \mu; \quad \mathbf{s} \in D \tag{41-25}$$

$$\mathrm{cov}(\mathbf{Z}(\mathbf{s}), \mathbf{Z}(\mathbf{u})) = C(\mathbf{s}, \mathbf{u}), \quad \mathbf{s}, \mathbf{u} \in D, \tag{41-26}$$

where $\mu \equiv (\mu_1, \ldots, \mu_k)'$ and $C(\mathbf{s}, \mathbf{u})$ is a $k \times k$ matrix (not necessarily symmetric). The cokriging predictor of $Z_1(\mathbf{s}_0)$ is a linear combination of all the available data values of all the k variables

$$\hat{Z}_1(\mathbf{s}_0) = \sum_{i=1}^{n}\sum_{j=1}^{k} \lambda_{ji} Z_j(\mathbf{s}_i). \tag{41-27}$$

Notice that Eq. (41-27) assumes that all components of $\mathbf{Z}(\mathbf{s}_i)$ are available for each i. Should this not be the case, a straightforward modification to (41-27) through (41-31) is possible (e.g., Journel and Huijbregts, 1978, p. 325; Myers, 1984).

Asking for a predictor that is uniformly unbiased, that is, $E(\hat{Z}_1(\mathbf{s}_0)) = \mu_1$ for all μ, yields the necessary and sufficient condition

$$\sum_{i=1}^{n} \lambda_{1i} = 1, \; \sum_{i=1}^{n} \lambda_{ji} = 0; \quad j = 2, \ldots, k. \tag{41-28}$$

Therefore, the best linear unbiased predictor (or cokriging predictor) is obtained by minimizing

$$E\left(Z_1(\mathbf{s}_0) - \sum_{i=1}^{n}\sum_{j=1}^{k} \lambda_{ji} Z_j(\mathbf{s}_i)\right)^2, \tag{41-29}$$

subject to the constraints (41-28). In principle, this problem is no more difficult than ordinary kriging, except there are a greater number of Lagrange multipliers $m_1, m_2, \ldots, m_k$ needed for the constraints in Eq. (41-28). In terms of covariances and cross-covariances, the cokriging equations are easily seen to be

$$\sum_{i=1}^{n}\sum_{j=1}^{k} \lambda_{ji} C_{jj'}(\mathbf{s}_i, \mathbf{s}_{i'}) - m_{j'} = C_{1j'}(\mathbf{s}_0, \mathbf{s}_{i'}); \tag{41-30}$$

$$i' = 1, \dots, n, \; j' = 1, \dots, k,$$

combined with the linear equations (41-28). The analogous equations for predicting *all* elements of $\mathbf{Z}(\mathbf{s}_0)$ can be found in Myers (1982). The minimum mean-squared prediction error, or (co)kriging variance is

$$\sigma_k^2(\mathbf{s}_0) = C_{11}(\mathbf{s}_0, \mathbf{s}_0) - \sum_{i=1}^{n} \sum_{j=1}^{k} \lambda_{ji} C_{1j}(\mathbf{s}_0, \mathbf{s}_i) + m_1. \quad (41\text{-}31)$$

When predicting all elements of $\mathbf{Z}(\mathbf{s}_0)$, the appropriate analogue of Eq. (41-31) is a mean-squared prediction error *matrix*, which is given in Ver Hoef and Cressie (1991).

The covariance-matrix function $C(\mathbf{s}, \mathbf{u})$ usually has to be estimated from the data. The assumption $C(\mathbf{s}, \mathbf{u}) = C^*(\mathbf{s} - \mathbf{u})$ allows estimation of $C^*(\cdot)$ from $\mathbf{Z}(\mathbf{s}_1), \dots, \mathbf{Z}(\mathbf{s}_n)$.

It is also possible to formulate cokriging in terms of variograms and cross-variograms. Substituting $\{-\gamma_{jj'}(\cdot)\}$ for $\{C^*_{jj'}(\cdot)\}$ in Eqs. (41-30) and (41-31) yields the appropriate cokriging equations. To see why, consider the simple case of $k = 2$ and notice the following algebraic identity:

$$\begin{aligned}\left[Z_1(\mathbf{s}_0) - \hat{Z}_1(\mathbf{s}_0)\right]^2 = &-\sum_{i=1}^{n} \sum_{j=1}^{n} \lambda_{1i} \lambda_{1j} \left[Z_1(\mathbf{s}_i) - Z_1(\mathbf{s}_j)\right]^2/2 \\ &+2\sum_{i=1}^{n} \lambda_{1i} \left[Z_1(\mathbf{s}_0) - Z_1(\mathbf{s}_i)\right]^2/2 \\ &+2\sum_{k=1}^{n} \lambda_{2k} \left[Z_1(\mathbf{s}_0) - Z_2(\mathbf{s}_k)\right]^2/2 \\ &-\sum_{i=1}^{n} \sum_{k=1}^{n} \lambda_{1i} \lambda_{2k} \left[Z_1(\mathbf{s}_i) - Z_2(\mathbf{s}_k)\right]^2/2 \\ &-\sum_{k=1}^{n} \sum_{l=1}^{n} \lambda_{2k} \lambda_{2l} \left[Z_2(\mathbf{s}_k) - Z_2(\mathbf{s}_l)\right]^2/2\end{aligned}$$

After taking expectations, one obtains the desired result.

However, although substitution of $\{-\upsilon_{jj'}(\cdot)\}$ for $\{C^*_{jj'}(\cdot)\}$ in Eqs. (41-30) and (41-31) is recommended in much of the geostatistics literature, it is only appropriate should $C^*(\cdot)$ be a symmetric matrix. To provide models for $\{\upsilon_{jj'}(\cdot)\}$, but to have a restrictive condition on $\{C^*_{jj'}(\cdot)\}$ that usually cannot be checked, is self-defeating; the rationale behind working with cross-variograms is to finesse the need for cross-covariances. Further, if the condition can be checked and does not hold, the cokriging equations in terms of $\{\upsilon_{jj'}(\cdot)\}$ are wrong; an example can be found in Ver Hoef and Cressie (1993).

Building valid, flexible models for $\{C^*_{jj'}(\cdot)\}$ or $\{\gamma_{jj'}(\cdot)\}$ and fitting them to available data requires further research. An open problem is to find necessary and sufficient conditions to ensure valid models for $\{\gamma_{jj'}(\cdot)\}$. In environmental studies, an important special case is where $Z_j(\mathbf{s}_i)$ is actually $Z(\mathbf{s}_i; t_j)$, an observation on the space–time process $Z(\cdot\,;\cdot)$. For a matrix formulation of cokriging, the reader is referred to Myers (1982, 1984, 1988, 1991) and Ver Hoef and Cressie (1993). Myers (1982, 1984, 1988) does *not* use the cross-variograms $2\gamma_{jj'}$; upon replacement of $-\gamma_{jj'}(\cdot)$ for $C^*_{jj'}(\cdot)$ in the cokriging equations based on $\{C^*_{jj'}(\cdot)\}$, Cressie (1991, Sec. 3.2.3) and Myers (1991) show that an equivalent set of cokriging equations results. Equations given in terms of $\{\upsilon_{jj'}(\cdot)\}$ should be avoided, in general, for reasons given earlier.

Finally, the problem of simultaneously predicting $\mathbf{Z}(\mathbf{s}_0)$ (or, for that matter, some subset of variables at possibly different locations) begs the question of which *multivariate* criterion should be minimized. Ver Hoef and Cressie (1991) show that the mean-squared prediction error matrix $E\{(\mathbf{Z}(\mathbf{s}_0) - \mathbf{p}(\mathbf{Z}; \mathbf{s}_0))\,(\mathbf{Z}(\mathbf{s}_0) - \mathbf{p}(\mathbf{Z}; \mathbf{s}_0))'\}$ is minimized (in a special sense) by the cokriging predictors $\hat{Z}_1(\mathbf{s}_0), \dots, \hat{Z}_k(\mathbf{s}_0)$, subject to $E(\mathbf{Z}(\mathbf{s}_0)) = E(\hat{\mathbf{Z}}(\mathbf{s}_0))$. This allows accurate joint prediction regions to be constructed for all the variables or any subset of them.

APPLICATIONS AND CONCLUDING REMARKS

Environmental data are inherently spatial, in a domain where the spatial index is frequently continuous. Not surprisingly, then, geostatistics is having an increasingly important role to play in environmental modeling efforts. For the spatial statistician, this leads to a whole new set of exciting problems, and standard geostatistics has to be augmented with new methods. In the references that follow, I have tried to give a certain amount of coverage to such methods. They represent a sampling, and no attempt has been made at completeness.

Journel (1984, 1988) has written informative expositions of the use of geostatistics in environmental problems. Istok and Cooper (1988) demonstrated how to predict groundwater contaminant concentrations using geostatistics, and Myers (1989) implemented it to assess the movement of a multipollutant plume. Geostatistical studies of acid deposition have been carried out inter alia by Eynon and Switzer (1983), Bilonick (1985, 1988), Le and Petkau (1988), and Cressie, Gotway, and Grondona (1990).

Little mention has been made in this chapter of space–time geostatistics. In obvious notation, the generic problem is to predict $\{Z(\mathbf{s}; t_0) : \mathbf{s} \in D(t_0)\}$ from data

$$\{(Z(\mathbf{s}_{1,i}; t_i), \dots, Z(\mathbf{s}_{n_i,i}; t_i)) : i = 1, \dots, m\}; \qquad (41\text{-}32)$$

where $t_1 < t_2 < \dots < t_m \le t_0$. The data are assumed to be an incomplete sampling of the space–time random process

$$\{Z(\mathbf{s}; t) : \mathbf{s} \in D(t);\, t \in T\}, \qquad (41\text{-}33)$$

where the domain $D(t)$ may vary with time. Most commonly, $D(t) \equiv D$ and usually $T = \{1, 2, \dots\}$, which enables Eq. (41-33) to be viewed as a time series of spatial processes, each process occurring at equally spaced time points. In order to estimate model parameters with acceptable precision, there has to be repeatability across space or time. For example, Switzer (1989) averages over time to obtain estimates of *nonstationary* (in space) covariances. These are then used to predict $Z(s_0; t_i)$ from contemporaneous data.

This presentation of geostatistics has followed a statistical modeling approach; much more can be found in Part I of Cressie (1991). There is another approach, relying on probabilities obtained from the sampling scheme, that Journel (1985) and Isaaks and Srivastava (1989) sometimes prefer. However, its limitations are sorely felt when one wishes to go beyond marginal and bivariate distribution (or moment) estimation, to kriging. Provided good diagnostics are used to verify the goodness of fit of a spatial statistical model, the modeling approach advocated in this article is extremely powerful.

ACKNOWLEDGMENTS

This research was partially supported by the National Science Foundation under grant DMS 9001862. Part is taken almost verbatim from a section of the article "Geostatistical Analysis of Spatial Data" by N. Cressie, which appeared in the National Research Council volume *Spatial Statistics and Digital Image Analysis* (1991, National Academy Press, Washington, DC). This article was written while the author was visiting the Department of Statistics at Stanford University during the fall of 1991. The comments of Don Myers on an earlier draft are much appreciated. Responsibility for any errors of commission or omission remain with the author alone.

REFERENCES

Bilonick, R. A. (1985) The space–time distribution of sulfate deposition in the northeastern United States. *Atmospheric Environment* 19: 1829–1845.

Bilonick, R. A. (1988) Monthly hydrogen ion deposition maps for the Northeastern U.S. from July 1982 to September 1984. *Atmospheric Environment* 22: 1909–1924.

Clark, I., Basinger, K. L., and Harper, W. V. (1989) MUCK: A novel approach to co-kriging. *Proceedings of the Conference on Geostatistical, Sensitivity, and Uncertainty Methods for Groundwater Flow and Radionuclide Transport Modeling,* Columbus, Ohio: Battelle Press, pp. 473–493.

Cressie, N. (1985) Fitting variogram models by weighted least squares. *Journal of the International Association for Mathematical Geology* 17: 563–586.

Cressie, N. (1988) Spatial prediction and ordinary kriging. *Mathematical Geology* 20: 405–421.

Cressie, N. (1990) The origins of kriging. *Mathematical Geology* 22: 239–252.

Cressie, N. (1991) *Statistics for Spatial Data*, New York: Wiley.

Cressie, N., Gotway, C. A. and Grondona, M. O. (1990) Spatial prediction from networks. *Chemometrics and Intelligent Laboratory Systems* 7: 251–271.

Cressie, N., and Hawkins, D. M. (1980) Robust estimation of the variogram, I. *Journal of the International Association for Mathematical Geology* 12: 115–125.

Eynon, B. P., and Switzer, P. (1983) The variability of rainfall acidity. *Canadian Journal of Statistics* 11: 11–24.

Gandin, L. S. (1963) *Objective Analysis of Meteorological Fields*, Leningrad: GIMIZ.

Golub, G. H., and Van Loan, C. F. (1983) *Matrix Computations,* Baltimore, MD: The Johns Hopkins University Press.

Hart, J. F. (1954) Central tendency in areal distributions. *Economic Geography* 30: 48–59.

Isaaks, E. H., and Srivastava, R. M. (1989) *An Introduction to Applied Geostatistics*, Oxford: Oxford University Press.

Istok, J. D., and Cooper, R. M. (1988) Geostatistics applied to groundwater pollution. III: Global estimates. *Journal of Environmental Engineering* 114: 915–928.

Journel, A. G. (1974) Geostatistics for conditional simulation of ore bodies. *Economic Geology* 69: 673–687.

Journel, A. G. (1983) Non-parametric estimation of spatial distributions. *Journal of the International Association for Mathematical Geology* 15: 445–468.

Journel, A. G. (1984) New ways of assessing spatial distribution of pollutants. In G. Schweitzer (ed.) *Environmental Sampling for Hazardous Wastes,* Washington, DC: American Chemical Society, pp. 109–118.

Journel, A. G. (1985) The deterministic side of geostatistics. *Journal of the International Association for Mathematical Geology* 17: 1–14.

Journel, A. G. (1988) Nonparametric geostatistics for risk and additional sampling assessment. In L. H. Kieth (ed.) *Principles of Environmental Sampling*, Washington, DC: American Chemical Society, pp.45–72.

Journel, A. G., and Huijbregts, C. (1978) *Mining Geostatistics,* London: Academic Press.

Jowett, G. H. (1952) The accuracy of systematic sampling from conveyor belts. *Applied Statistics* 1: 50–59.

Kolmogorov, A. N. (1941) The local structure of turbulence in an incompressible fluid at very large Reynolds numbers. *Doklady Akademii Nauk SSR* 30: 301–305.

Le, D. N., and Petkau, A. J. (1988) The variability of rainfall acidity revisited. *Canadian Journal of Statistics* 16: 15–38.

Matheron, G. (1963) Principles of geostatistics. *Economic Geology* 58: 1246–1266.

Matheron, G. (1971) *The Theory of Regionalized Variables and its Applications*, Fontainebleau: Cahiers du Centre de Morphologie Mathematique.

Matheron, G. (1976) A simple substitute for conditional expectation: The disjunctive kriging. In Guarascio, M., David, M. and Huijbregts, C. (eds.) *Advanced Geostatistics in the Mining Industry*, Dordrecht: Reidel, pp. 221–236.

Myers, D. E. (1982) Matrix formulation of co-kriging. *Journal of the International Association for Mathematical Geology* 14: 249–257.

Myers, D. E. (1984) Cokriging—new developments. In Verly, G., David, M., Journel, A., and Marechal, A. (eds.) *Geostatistics for Natural Resources Characterization*, Part 1, Dordrecht: Reidel, pp. 295–305.

Myers, D. E. (1988) Multivariate geostatistics for environmental monitoring. *Sciences de la Terre, Serie Informatique Geologique, Nancy* 27: 411–427.

Myers, D. E. (1989) Borden field data and multivariate geostatistics. In Ports, M. A. (ed.) Hydraulic Engineering, *American Society of Civil Engineering*, pp. 795–800.

Myers, D. E. (1991) Pseudo-cross variograms, positive-definiteness, and cokriging. *Mathematical Geology* 23: 805–816.

Switzer, P. (1989) Non-stationary spatial covariances estimated from monitoring data. In Armstrong, M. (ed.) *Geostatistics*, Vol. 1, Dordrecht: Kluwer, pp. 127–138.

Ver Hoef, J. M. and Cressie, N. (1991) Multivariable spatial prediction. *Mathematical Geology* (in press).

Wackernagel, H. (1988) Geostatistical techniques for interpreting multivariate spatial information. In Chung, C. F., Fabbri, A. G., and Sinding-Larsen, R. (eds.) *Quantitative Analysis of Mineral and Energy Resources*, Dordrecht: Reidel, pp. 393–409.

Watson, G. S. (1984) Smoothing and interpolation by kriging with splines. *Journal of the International Association for Mathematical Geology* 16: 601–615.

Yaglom, A. M. (1957) Some classes of random fields in n-dimensional space, related to stationary random processes. *Theory of Probability and its Applications* 2: 273–320.

Zellner, A. (1986) Bayesian estimation and prediction using asymmetric loss functions. *Journal of the American Statistical Association* 81: 446–451.

Zimmerman, D. L., and Zimmerman, M. B. (1991) A Monte Carlo comparison of semivariogram estimators and corresponding ordinary kriging predictors. *Technometrics* 33: 77–91.

42

Explanatory Models for Ecological Response Surfaces

HENRIETTE I. JAGER AND
W. SCOTT OVERTON

It is often the spatial patterns in environmental and ecological variables that arouse interest and demand explanation. For environmental response variables, the causal influences of interacting environmental factors produce the patterns of interest. Ecological response variables by definition involve living organisms and are at least one step removed from spatial patterns in the physical environment. The spatial organization of ecological variables, such as species abundances, is often viewed as a collection of individual species responses to variation in the physical environment (Gleason, 1926), although competition and other ecosystem interactions may also influence spatial arrangement.

The response of ecological variables to spatial environmental gradients can be direct, or it can be biologically integrated through interactive responses of the ecosystem. Physical factors such as soil and geologic and physiographic structure provide the base physical environment and substrate. Climatic factors such as precipitation, temperature, constituents of atmospheric deposition, and solar radiation represent important exogenous driving variables. Complex ecosystems provide yet another dimension of the environment of particular ecological processes, and also interact with the physical environment, modifying it and ameliorating its effects. The influence of these factors is seldom simple, and predictive models must be tailored to the ecological problem at hand. Quite often, simple explanatory variables suffice for complex relations. For example, the environmental cue that notifies temperate populations of spring's arrival is widely modeled as the "degree-day," a cumulative measure of temperature that ignores temperatures below a given threshold. This simple measure mimics the way in which many organisms physiologically integrate ambient temperatures, as evidenced by the success of the model.

Other variables operate as complex environmental gradients (Whitaker, 1970) that reflect a suite of primitive gradients that covary in a predictable way in nature. Populations apparently respond to these complex gradients, which act as surrogates for their causal constituents. For example, elevation itself does not cause changes in species composition, but temperature, soils, and rainfall are covarying environmental factors that influence species composition along an elevational gradient. Atmospheric pressure and density so closely follow elevation that their effects can almost be considered to be caused by elevation. The issue of whether variables in a model are causal is complicated by the number of links (indirect effects) and the scale of interest. To illustrate, a mobile population is likely to cue on any convenient signal that a potential habitat falls into its elevation range (e.g., the presence of familiar prey species or nest building materials). Such a proximate cue is likely to be an integrated index of elevation rather than a direct measurement of environmental variables. Similarly, explanatory models may use variables that integrate complex patterns of causal factors, rather than the base causal factors themselves. Elevation is useful because we are familiar with the environmental factors associated with elevation, and because elevation is an integrative factor that carries certain causal complexes with a high degree of reliability.

Understanding the spatial organization of ecological systems is a fundamental part of ecosystem study. While discovering the causal relationships of this organization is an important goal, our purpose of spatial description on a regional scale is best met by use of explanatory variables that are somewhat removed from the mechanistic causal level. Regional-level understanding is best obtained from explanatory variables that reflect spatial gradients at the regional scale and from categorical variables that describe the discrete constituents of (statistical) populations, such as lakes. The scale on which we are concerned with spatial pattern is quite different from the scale of study of ecosystems; our scale is more the scale that Whitaker (1970) addressed in his treatment of gradients than in his treatment of ecological processes.

In this chapter, we use a regression model to predict lake acid neutralizing capacity (ANC) based on environmental predictor variables over a large region. These predictions are used to produce model-based population estimates. Two key features of our modeling approach are that it honors the spatial context and the design of the sample data. The spatial context of the data is brought into the analysis of model residuals through the interpretation of residual maps and semivariograms. The sampling design is taken into account by including stratification variables from the design in the model. This ensures that the model applies to a real population of lakes (the target population), rather than of whatever hypothetical population the sample is a random sample.

OVERVIEW

Environmental predictors of spatial pattern

Faced with a spatially distributed environmental response variable, our goal is to construct a response surface for spatial pattern that consists of a regression model involving appropriate explanatory variables. There are several qualities of interest to us in identifying a model. First, we seek robust model relations that describe the regional-scale manifestations of environmental processes. This implies that we do not consider spatial processes to contribute to the causal determination of pattern in the response variables at the scale being analyzed here. Second, the explanatory variables of interest are spatially extensive (known at each location of interest). When used in the manner proposed, these extensive data can provide enhanced resolution of spatial patterns, as well as enhanced population inferences.

In addition, we account for the relevant sampling design features of the data on which the analyses are based, preserving the informative ties with the well-defined populations being sampled. The case study presented here involves a survey of lakes in which the sample was drawn from a list (sampling frame) of lakes in upstate New York that were represented on 1:250,000 scale maps. The Eastern Lake Survey (ELS) was a synoptic survey with a single index sample taken during the period of minimum within-lake variability (fall) from each lake. The target population of lakes was defined by eliminating frame lakes that met one or more of the following criteria: (1) the point identified as a lake on the 1:250,000-scale map appeared not to be a lake on more detailed maps; (2) proximity of the lake to intense urban, industrial, or agricultural land use; or (3) lake surface area smaller than 4 hectares. In the context of the sampling design, we demonstrate the utility of frame data (data available for all lakes in the list frame) in generating improved population estimates, as well as in generating better description of spatial pattern. Our approach is consistent with the model-based approach to population description common to finite sampling, as reported by Royall and Cumberland (1981), for example. The model that we describe here exploits the more extensive information in the explanatory variables to improve inferences relative to the response variable.

The approach described here can be used at many scales in environmental science, but we will concentrate on regional-level studies. This implies that we are interested in multivariate relationships that explain spatial patterns over a relatively large region. In two nonstationary kriging techniques, universal kriging and kriging with intrinsic random functions, the pattern model is usually local and parameter estimates for the drift are generally not of interest. In contrast, residual kriging proceeds from a global model of predicted drift. While kriged residuals were not actually added to our regression estimates, the iterative residual kriging framework was useful to us in our spatial analysis of residuals.

This case study involves a response variable, acid neutralizing capacity (ANC), for the finite population of lakes. ANC is measured by titration and is used here in μeq/L. This variable measures the ability of the lake to buffer acidic inputs. Based on sample data from the ELS (Linthurst et al., 1986), and frame data on the entire population, we predicted lake ANC for population lakes that were not included in the sample. The predictions combine the large-scale spatial patterns inherent in the suite of explanatory variables to provide a multivariate pattern model for ANC, expressed as a spatial response surface. This pattern can be supplemented by local deviations that are obtained by kriging interpolation. In this paper, we restrict our attention to predictions of spatial pattern rather than on the kriging portion.

The proposed methods have important implications for investigators conducting regional surveys of environmental or ecological resources. There are emerging analytic methods that will utilize both the predicted mean surface and the spatial model of the residuals. Facsimile populations constructed to have exactly the location of real population units and patterns of mean and variability closely similar to those of the real population are proving very useful in assessment of sampling designs. This comparison of sampling designs can be conducted using simulated (facsimile) populations both before and after conduct of the sample. Those aspects of the present effort directed towards improving the model for residuals will be useful in constructing realistic facsimile populations.

In the spatial analysis of residuals described here, our regression or response surface model can be viewed, from the perspective of geostatistics, as a model of the deterministic portion of a stochastic process. This case study is different from usual geostatistical analysis in several respects. The sample is a probability sample of an explicit stratified population of lakes, with the sampling intensity varying among strata. This fact causes us to approach several parts of the analysis differently than if the sample were unstratified. Additionally, predictions are made for the specific locations at which population units occur, rather than for all points in the domain. The spatial distribution of the frame lakes (the full set of "lakes" represented on the maps used in sample selection) is therefore automatically considered, with no assumption that lake density is spatially uniform. Finally, the use of environmental predictor variables to model spatial pattern departs from the common use of geostatistical models to represent the drift in that it will not be expressed solely as a function of location. Our use of the term "pattern" corresponds to the term "external drift" used by Ahmed and DeMarsily (1987). This approach to modeling pattern is preferred because ecologists, unlike geologists, generally have access to spatial data that can help to explain large-scale patterns of interest.

The spatial analysis of residuals used here evolved from the iterative residual analysis procedure outlined by Neuman and Jacobsen (1984). Details of the methods used can be found in Jager et al. (in preparation). Beginning with a regression analysis that assumes uncorrelated and homogeneous errors, the program alternately estimates semivariogram model parameters from the resulting residuals, constructs the variance–covariance (VC) matrix of the errors, and reestimates the regression coefficients by weighted least squares (as defined by Graybill, 1983, p. 177), until the estimates stabilize. We have modified the earlier analyses by accounting for heterogeneity of variance of the residuals among strata prior to spatial analysis of residuals.

Maps were produced using SURFER (Golden Software) and point semivariograms were used in the spatial analysis of residuals. Our interpretation of these data is that the volumetric support of the ANC measurement is 6.2 L based on the size of the sample bottle used. We did not consider the measurement as a lake average because "each lake is represented by an index chemistry, rather than, for example, mean chemistry or some other integration over time and space" (Linthurst et al., 1986, p. 3).

Residual analysis–general considerations

Regression analysis and geostatistics share the same basic model,

$$\mathbf{Z} = \mathbf{Xb} + \mathbf{R} \tag{42-1}$$

but while regression focuses on the model and parameter estimates **b**, geostatistics focuses on the correlation structure of the residuals, **R**. These are opposite sides of the same coin. In regression, unknown properties of the residuals are a hindrance to analysis. In geostatistics, the presence of a deterministic spatial pattern is a hindrance to obtaining a valid semivariogram model. The usual ordinary least-squares regression procedure (OLS) assumes that the (residual) errors are homogeneous and uncorrelated. Both heterogeneous and correlated errors can be taken into account by using generalized least squares (GLS), a procedure that explicitly accounts for the variance–covariance structure of the residuals. In the presence of heterogeneous and correlated errors, estimates by OLS will be less efficient than estimates by GLS, but still unbiased. Loss of efficiency is modest unless correlation is high, although the efficiency lost due to heterogeneity of errors may be more serious. Thus analysis by OLS can be used to generate the residuals that provide the initial basis for residual analysis. Iteration and reanalysis by GLS is not expected to generate a greatly different set of residuals, and the theoretical gain in efficiency from GLS will not be fully achieved because the variance–covariance matrix must be estimated from the observed residuals.

Analysis by either OLS or GLS can eliminate spatial pattern and better provide for effective kriging. If the residuals satisfy weak stationarity assumptions, then the empirical semivariogram will describe the VC structure of the sampled lakes. The residuals will not fully satisfy the kriging assumptions if a deterministic trend or spatial pattern remains. If the error structure is confounded with unexplained pattern, then an experimental semivariogram constructed from the residuals (the residual semivariogram) will exhibit several characteristic features (Starks and Fang, 1982; Neuman and Jacobsen, 1984). These include anisotropy (the correlation depends on direction, as well as distance) and a shape that increases parabolically or worse. While the appearance of these features does not prove that patterns are present, the presence of these features can be used as a diagnostic tool to guide modeling decisions (e.g., the direction of anisotropy may suggest an explanatory factor). Some guidance for determining the range of reasonable semivariogram parameters can also be obtained from independent estimates of small-scale variability (the nugget) and maximum or large-scale variability (the sill). Alternating least-squares and spatial analysis of residuals thus seems a natural approach to investigating spatial pattern.

ADIRONDACK CASE STUDY

The U.S. Environmental Protection Agency conducted a synoptic survey of lakes in the eastern U.S. during the fall of 1984 (Linthurst et al., 1986; Blick et al., 1987). A probability sample of lakes was drawn from the 1:250,000 scale map population. The sample was stratified by previously drawn contours that identified regions in which surface waters would be expected to have low, medium, or high alkalinity on the basis of geology, soils, and other information. Figure 42-1 shows the region of upstate NY with which we are concerned, the sample lakes, and the boundaries of three strata. Many chemical and some physical measurements were made on each of the lakes in-

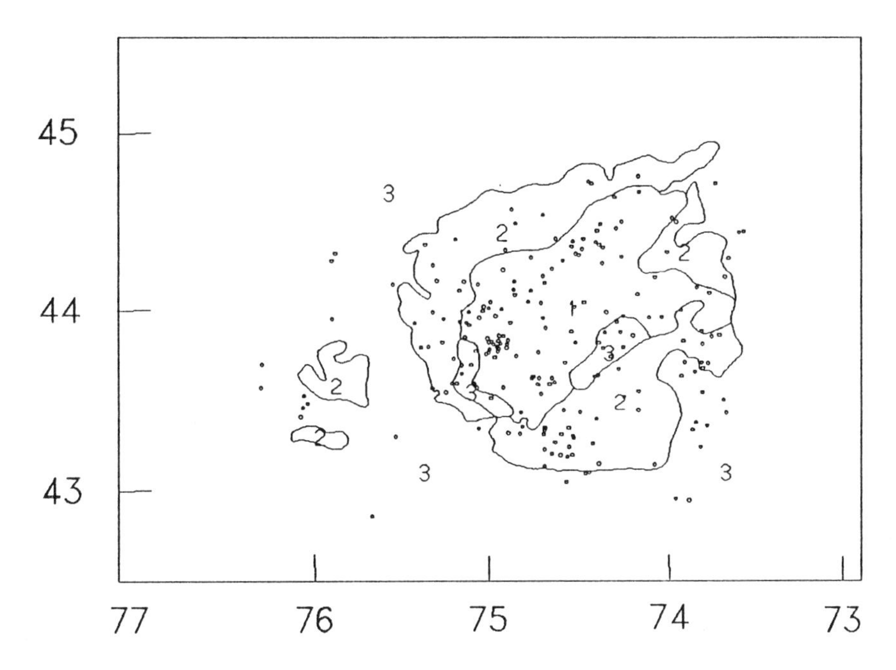

Figure 42-1: Map of sample lakes and stratum boundaries.

cluded in the sample, including ANC. The base properties of the population/sample in Subregion I-A (containing the Adirondacks) that is addressed by our analysis are provided in Table 42-1. Strata are alkalinity classes. $\hat{N}$ is the estimated size of the target population, as specified by the survey, and n is the number of target lakes in the sample.

The sample weights vary among the strata; weights can be thought of as the number of population lakes represented by each sample lake. Although the differences in weights in Table 42-1 are not large, the strata should still be analyzed separately. The differences between the sample size of 208 and 155, and between the population size of 1684 and the estimated 1290, are due to nontarget lakes contained in the frame population. $\hat{N}$ is the estimated size of the target population, and all design-based estimates apply to this population. In our exercise of predicting ANC for the nonsample lakes, we are unable to identify the lakes in the frame population that are nontarget; thus our projections are for the full frame population of 1684 lakes. This is a fundamental limitation to this model-based method; any subpopulation must be identifiable on the frame in order to make inferences in this manner about that subpopulation.

Data from probability samples sometimes pose analytic problems by virtue of design features that invalidate conventional statistical analyses. In this example, there is only one general restriction and that is associated with the strata, and with the differential sampling rate among the strata. It is necessary to conduct the regression analyses by stratum, and to maintain the stratum distinctions if the fitted models are different. If the regression relations are equal for different strata, then those strata can be pooled for analysis and prediction. In certain cases, regressions weighted by the sampling weights are appropriate. Likewise, the sample weights should be taken into account when kriging and an extension of the method proposed

Table 42-1: Characteristics of the sampling design for subregion I-A of the Eastern Lake Survey (Linthurst et al., 1986, p. 36)

Stratum	Frame size	Sample size	Weight	n	$\hat{N}$	SE($\hat{N}$)
1	711	75	9.633	57	549.08	33.08
2	542	65	8.338	51	425.54	26.13
3	431	68	6.719	47	315.79	22.14
Total	1684	208		155	1290.11	

here can be used for kriging from a variable-probability sample.

Regression analysis

Several environmental and topographic variables were considered as possible predictors of ANC. These were elevation (m), pH of precipitation, precipitation (cm per year), and watershed slope (%). Elevation and slope were obtained for the lake locations (both those sampled and predicted) from the TOPOCOM digitized elevation maps from 1:250,000 scale maps. The pH of precipitation and precipitation amount were obtained from relatively large-scale maps (Olsen and Slavich, 1986) converted to GIS coverage. In addition, the stratum to which each lake belongs was included as an indicator variable. All interactions between these variables were evaluated for inclusion in the model. Other potential explanatory variables, such as lake type, lake depth, and lake size, were not available simply because they were measured only on the sample lakes. However, inclusion of these variables in the regression analysis provides evidence of their potential value, in case the frame data were to be made available.

Logarithmic transformation of ANC was made to make the distribution more symmetric. The new variable was defined as LANC = $\log_{10}(\text{ANC} + 150)$, to account also for possible negative values of ANC. LANC was regressed on the suite of explanatory variables using generalized least squares (GLS). The parameters obtained by GLS are not significantly different, in this case, from those obtained by ordinary least squares (OLS). The regression equation for ANC for lakes belonging to the two lower ANC strata (strata nos. 1 and 2 in Table 42-1) is shown in Equation (42-2). Equation (42-3) gives the equation for lakes in the high ANC stratum no. 3. The regression parameters were estimated simultaneously by using dummy variables to indicate stratum membership for all stratum-specific parameters and the final model explained 67% of the variation in LANC.

$$\text{LANC} = -9.08 - 0.0012\ \text{elevation} + 2.78\ \text{pH} \quad (42\text{-}2)$$

$$\text{LANC} = 6.77 - 0.0012\ \text{elevation} - 0.85\ \text{pH} \quad (42\text{-}3)$$

These prediction equations produce a spatial pattern for LANC by virtue of the spatial pattern of the explanatory variables, elevation, and pH. Part of the pattern in ANC has been "explained" by the association with these two spatially patterned explanatory variables. Figure 42-2 illustrates the spatial pattern apparent in the observed data, and Figure 42-3 the smoother patterns in the fitted regression. Figure 42-4 presents the spatial patterns observed in the predicted values generated on the frame population; visual enhancement derives from the great amount of information present in the explanatory variables known on the frame. The known spatial patterns in these explanatory variables are interpreted via the regression equations to generate spatial patterns in LANC (the response surface). The surfaces in these figures were generated by SURFER.

Residual analysis – case study

Given that we have generated a response surface reflecting the pattern predicted by the explanatory variables, we now wish to explore spatial patterns in the residuals, both in terms of means of residuals, and in terms of their variance/covariance structure. The spatial properties of residuals are used here to identify a regression model that leaves no evidence of unexplained pattern and to improve the precision of the regression analysis. In some cases, this structure can also be used to improve the response surface predictions by adding kriged residuals. In this analysis, a semivariogram model for residuals was used to estimate the covariance structure of the errors and to increase the precision of the GLS regression parameter estimates, but no kriged residuals were added to the predictions because the results of kriging cross-validation were poor.

Residual analysis begins with the residuals produced by the GLS regression models, Eqs. (42-2) and (42-3). Directional semivariograms for both LANC and the residuals were examined for anisotropy and other features that might result from unexplained spatial pattern (Figure 42-5a,b). The residual semivariograms do not differ markedly and are therefore more isotropic. We interpret this as verification that the regression model has removed much of the gradient-like pattern in LANC. Note also that the relative nugget is larger in the residual semivariograms.

Table 42-2: Summary of LANC residual error for each stratum

Stratum	*n*	Residual error mean square	Root-residual error mean square
1	57	0.0115	0.1043
2	51	0.0205	0.1431
3	47	0.0607	0.2464
Pooled	155	0.02938	0.1714

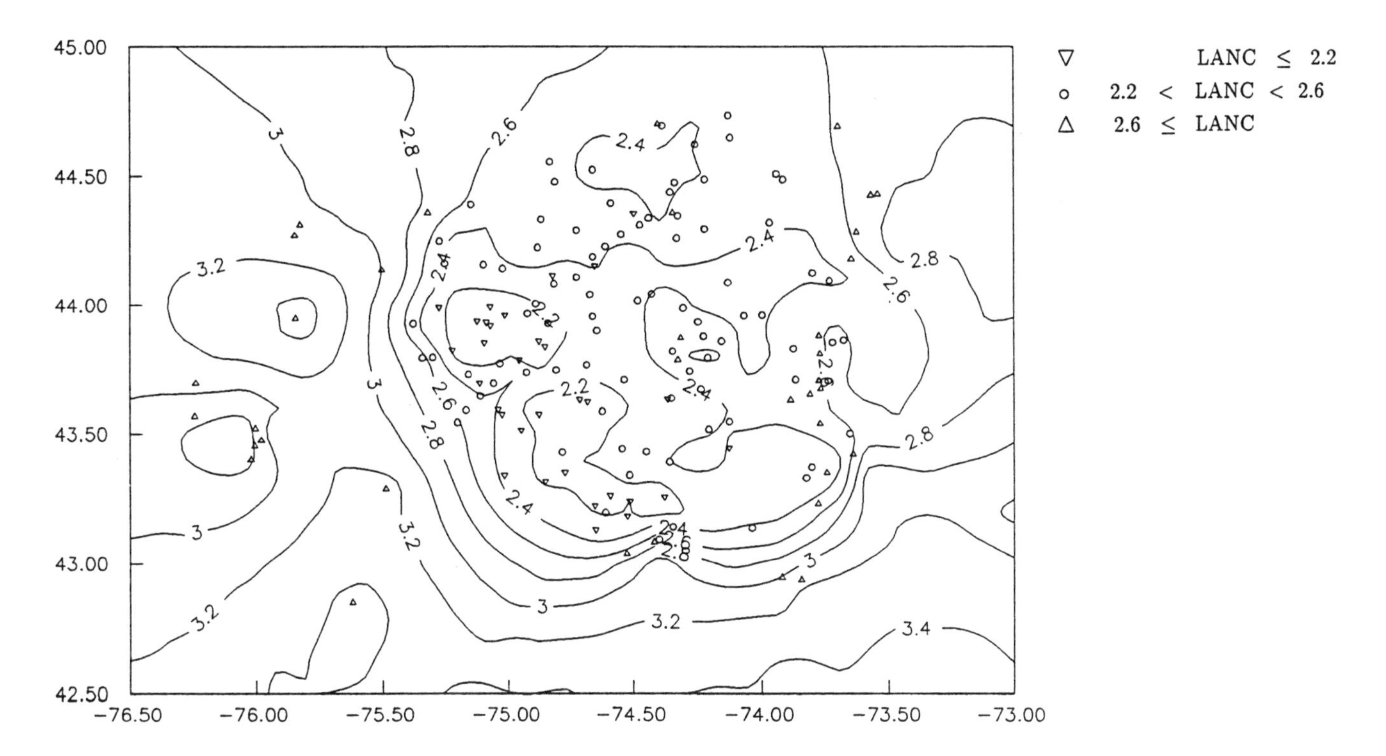

Figure 42-2: Map of observed LANC values on the sample.

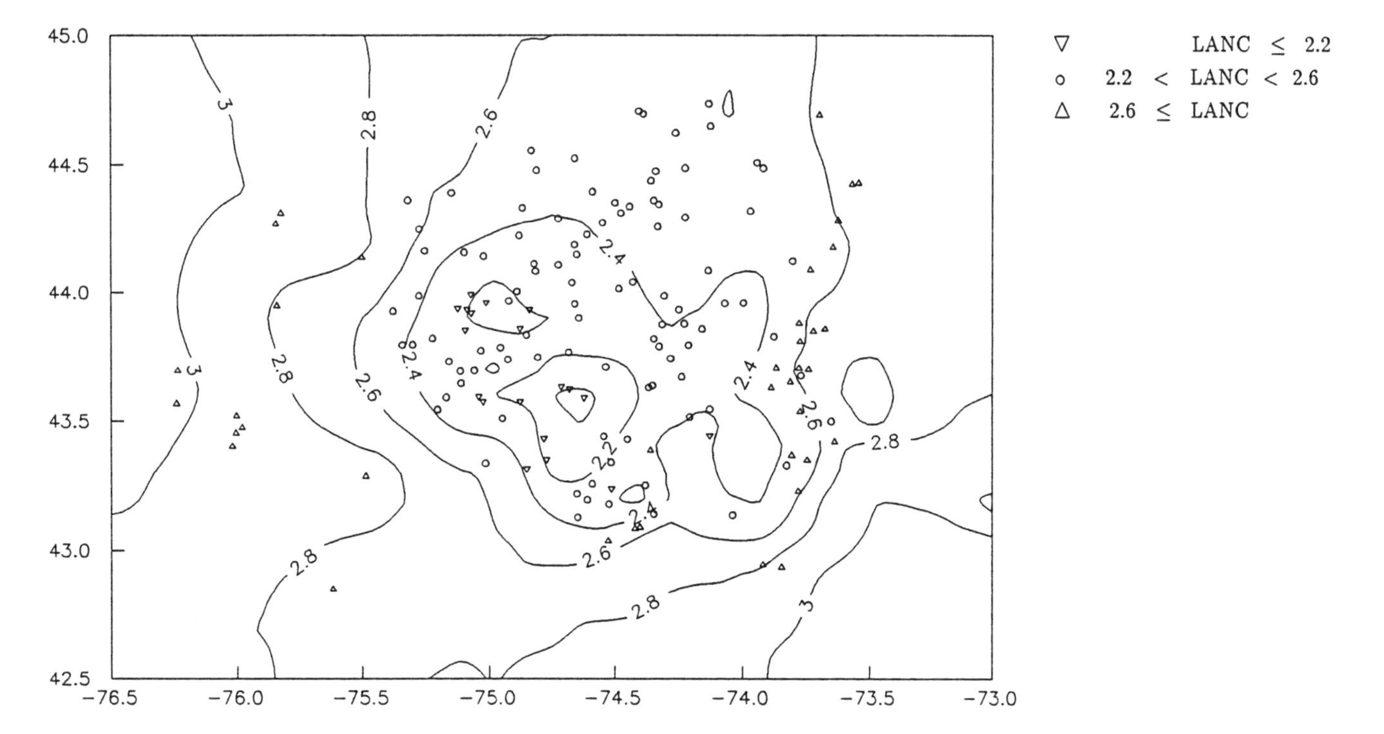

Figure 42-3: Map of predicted LANC values on the sample.

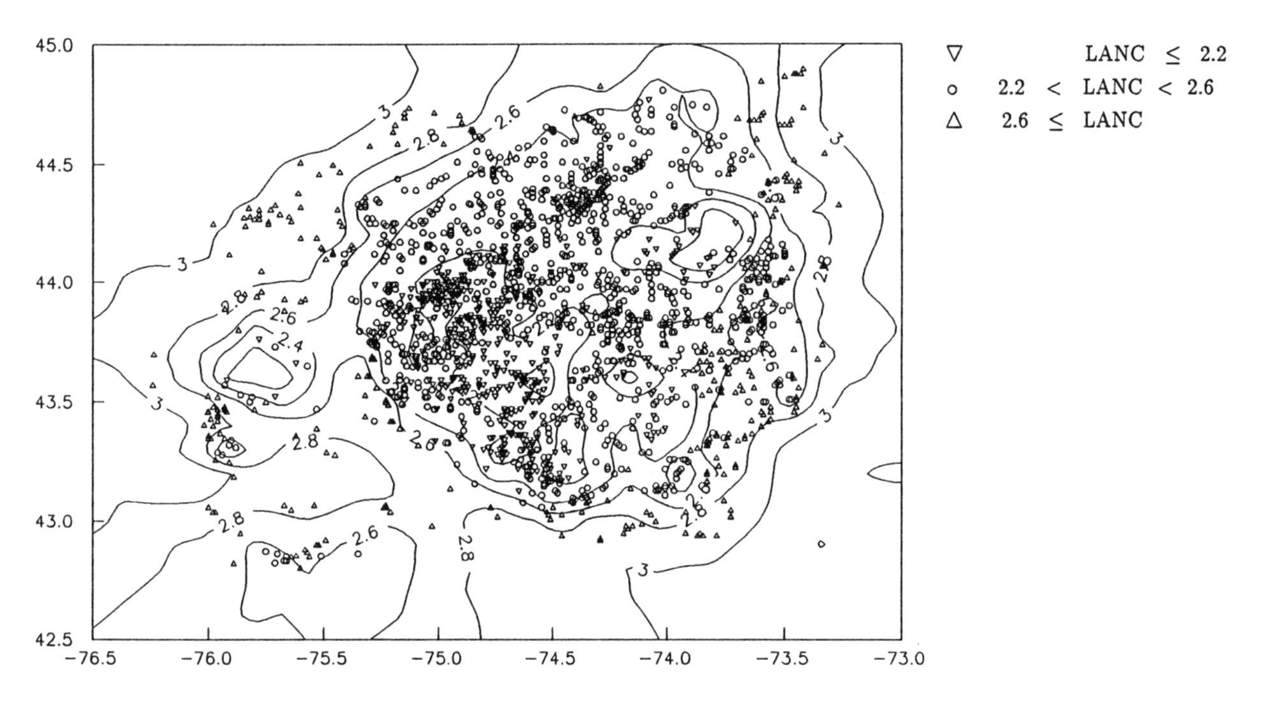

Figure 42-4: Map of predicted LANC values on the frame population.

We interpret this to mean that the model has partially explained the apparent spatial autocorrelation in LANC.

As the three strata are spatially defined, it is natural to examine the magnitudes of residual variance for evidence of homogeneity, by stratum, as in Table 42-2. Given evidence of heterogeneity of variance among strata, it is easy to eliminate these differences simply by dividing the residuals from the various strata by their respective standard deviations. Use of standardized residuals is preferred so that all strata can be analyzed together, spatially and otherwise, for further pattern. Poststratification based on the observed pattern is possible, with the limitation that scaling requires these classes (clusters) to be identified on the frame (classes must be identified in terms of frame variables).

Further increase in the nugget after scaling leaves very little residual spatial autocorrelation (Figure 42-6). Attempts to fit a parametric model for residual semivariance by maximum likelihood estimation revealed no significant difference between the nugget and the sill parameter. In addition, the predictions made using the fitted semivariogram in cross-validation showed virtually no correlation with the measured values. It is surprising that there is not a remnant of small-scale autocorrelation reflecting high-resolution spatial processes below the resolution of the regionally defined explanatory variables. In our example, such processes could affect two lakes very close together. The issue is clouded by the simple fact that correlation among residuals, due to the regression estimation process, will generate spatial autocorrelation when the explanatory variables are spatially patterned. Thus, we would not expect total elimination of autocorrelation in the residual surface, even if the model were perfect, having uncorrelated errors. In this case study, we conclude that there is a virtual absence of autocorrelation following the pattern fit, which suggests that there is no evidence of remaining spatial pattern in these data.

Patterns in the residuals exposed by a spatial analysis of residuals contribute to the general understanding of the population. The VC matrix can be thought of as a partitioned matrix with blocks of locations (sample units) that share the same variance on the diagonal. Recall that differences in variance among strata have been eliminated by scaling, so that all remaining variation will be on subpopulations that are not identified by the strata. The off-diagonal blocks define the covariances between sample units from different subpopulations (in this case, strata).

IMPLICATIONS FOR NATIONAL MONITORING PROGRAMS

Model-based estimation – case study

Population statistics for the population of Adirondack lakes obtained using our pattern model are given in Table

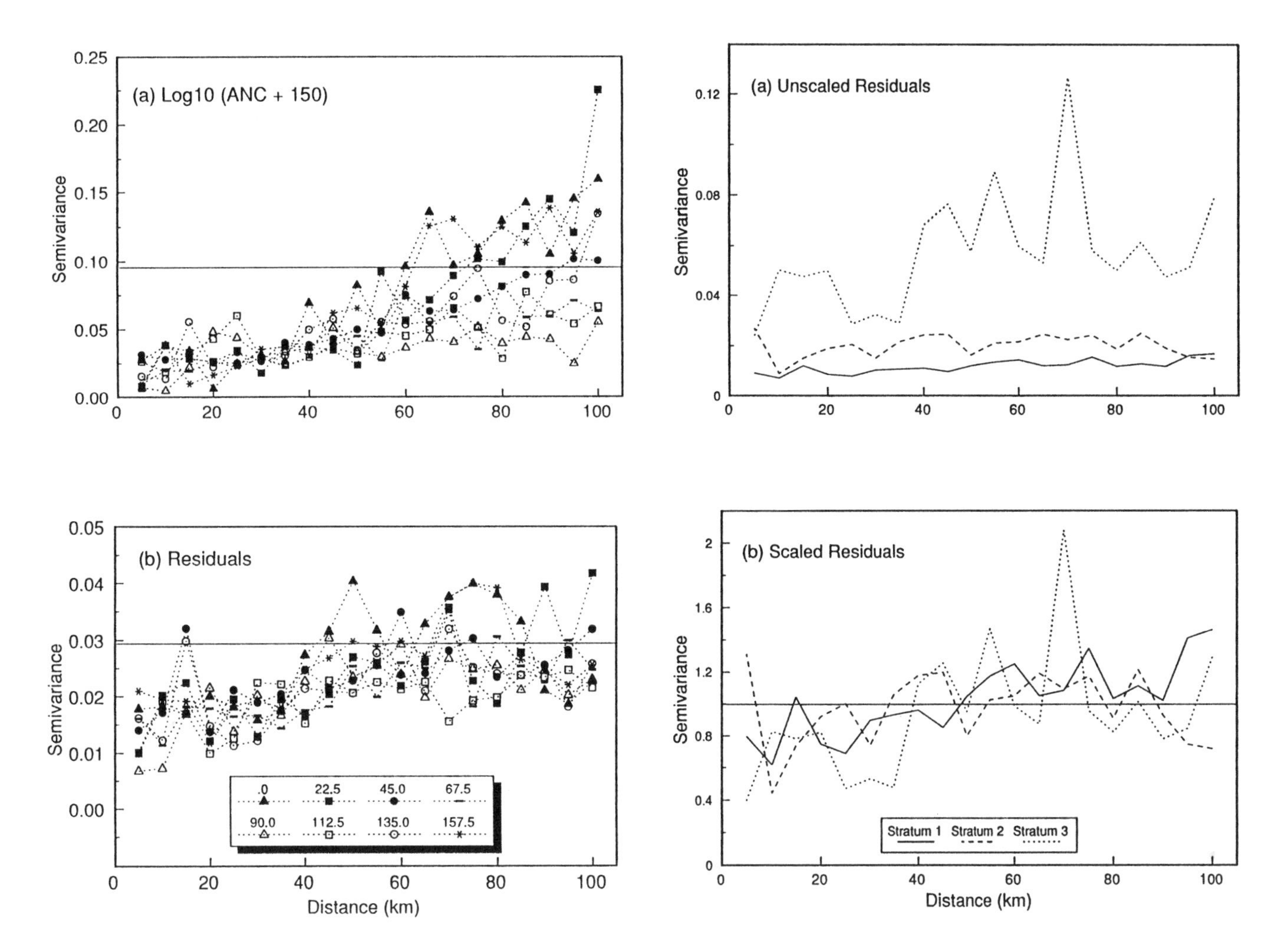

Figure 42-5: Directional semivariograms for (a) LANC and (b) scaled residuals.

Figure 42-6: Semivariograms of (a) unscaled and (b) scaled residuals.

42-3, in contrast to statistics obtained from the design-based methods employed in the Eastern Lake Survey. Although there is a potential inherent increase in precision from use of the model-based method, the inability to identify nontarget lakes in the frame population prevents us from taking advantage of this potential in generating population statistics. The model-based results of this table should be considered only as indicating the potential of the method.

The methods of projection to obtain Table 42-3 are straightforward. For model-based analyses, a predicted value is made for each lake in the frame but not in the sample; these are combined with the sample values, and the full set of 1649 lake predictions is simply analyzed for the population characteristics. The backtransformed model-based estimates for ANC in Table 42-3 are generated by backtransforming the predicted and observed values of LANC: ANC = $10^{\text{LANC}} - 150$. The two alternative model-based estimates will be discussed here. Design-based estimates are produced by expansion of the sample values to population estimates. This is performed via standard probability sample estimation formulae, such as provided by Overton (Chapter 47). In this framework, LANC and ANC are simply treated as two variables on the sample. In Table 42-3, $\hat{N}$ is the estimated population size, $\hat{T}_y$ estimates the total ANC over all lakes projected to be in the population, and $\hat{N}_{50\ \mu eq/L}$ is the estimated number of lakes with ANC $\leq 50\,\mu$eq/L.

Several features of the estimates are notable. First, $\hat{N}$ is different for the two estimation methods: While the

Table 42-3: Comparison of model- and design-based population estimates for Adirondack lakes (Region I-A)

	LANC		ANC (μeq/L)			
				Model-based		
Statistic	Design-based	Model-based	Design-based	Back-transformed	Bias-corrected	Non-transformed
$\hat{T}_y$	3,146	3,882	278,325	199,716	227,864	319,698
Mean	2.439	2.354	215.74	121.11	138.18	193.87
Std. devn.	0.286	0.246	413.34	194.29	206.98	315.55
Median	2.399	2.315	100.54	56.44	69.18	75.99
$\hat{N}$	1290.1	1649	1290.1	1649	1649	649
$\hat{N}_{50\,\mu eq/L}$	487.6	774	487.6	774	691	664

sample provides an estimate of the size of the target population for the design-based method, the target lakes are not identified on the frame used in producing model-based estimates. Comparing the estimates $\hat{N}$ of and $\hat{N}_{50}$ suggests that a large proportion of the nontarget lakes have low predicted ANC, but we have no way to confirm this from the data. Second, the discrepancy in population sizes (1649 vs. 1684) is accounted for by 35 lakes from the high-ANC stratum in the far west portion of the state that were included in the original Lake Survey frame but that are not included in our analysis because data on the predictor variables are lacking. Excluding sample lakes west of 76 degrees from the design-based estimates has little effect on the gap between the design-based and model-based estimates. Third, for LANC, there is reasonably good agreement between the estimates that are less sensitive to N, the population size (the mean, standard deviation, and median), and poor agreement for those estimates influenced by population size ($\hat{N}_{50}$ and $\hat{T}_y$). Agreement of $\hat{T}_y$ would be greatly improved if it were possible to remove the nontarget lakes from the model-based estimates. Finally, agreement is poor for all estimates of the parameters for ANC. The model-based estimates (generated from the backtransformed predictions of LANC) appear to be badly biased. This bias can be reduced by adding a bias-correction term involving the estimated prediction variances, s^2, from the regression: $E[\text{ANC}] = \exp\{k\text{LANC} + \frac{1}{2}k^2s^2\} - 150$, where $k = \ln(10)$. This correction does narrow the gap between the model-based and design-based estimates, but not by much. Alternatively, the analysis can be conducted without the log transformation. Preliminary results of a reanalysis without transforming ANC showed better agreement with the design-based estimates, especially for mean ANC. Resolution of the remaining differences hinges on identification of the nontarget lakes.

This discussion of the results of Table 42-3 suggests the need for a well-defined strategy for the use of models in population estimation. If model-based estimation is based on transformation, followed by back-transformation of predictions, then the resultant substantial biases must be dealt with in some manner. Bias correction may not be satisfactory; it generally seems preferable not to transform and to accept the loss of efficiency in return for consistency and near-unbiasedness. Loss of efficiency can be reduced by weighted regression, without incurring bias in predictions, but simple unweighted regressions of the natural variable may still be preferable. However, if the transformed variable represents an alternate measure of the population that is in some sense more satisfying from a scientific standpoint, then perhaps the criteria should be consistency and near-unbiasedness for the parameters of this alternate population, rather than for the original one. In Table 42-3, one should ask if LANC or ANC is more relevant.

Model-based estimation

Model-based methods for predicting survey variables on the unsampled population units can be valuable for regional surveys of environmental resources. A multivariate regression model such as the regression model described here can be used to predict unknown population values from relationships developed from the survey sample. The uncertainty associated with estimates of population parameters can be reduced considerably through the use of information contained in explanatory variables and imparted through a model. Additional information about the location of each resource unit has general potential of further reducing the uncertainty through use of interpolation models. The pattern in the fitted surface is thus generated by the known patterns in the explanatory variables, so that these known patterns have been used to strengthen spatial inferences from the sample. The situation found in this case study would seem to be highly desirable, from the point of view of ecological response surfaces, with virtually all apparent spatial patterns accounted for by the explanatory variables.

The spatial analysis of residuals used in our model-building procedure (evaluation of residual maps and spatial autocorrelation) was useful to us as a model diagnostic tool. This case study demonstrates several roles played by geostatistics in model building via the exploration of spatial pattern in the residuals. These roles include (1) the selection of predictor variables, (2) identification of regions with homogeneous residuals, and (3) the capability to increase model precision through GLS pattern estimation.

Application of this model approach to developing regional analysis of spatially distributed environmental data depends on the general availability of extensive data for explanatory variables. In effect, these extensive data must be available for the entire frame of population units. Digitized GIS coverages of spatially extensive environmental data are very valuable for a regression-based approach, but must be in a specific form suitable to the needs of any particular study. Identification of needs and possibilities, and provision of specific explanatory variables as part of the frame materials, would seem to be an integral part of any comprehensive monitoring program.

The validity of model-based estimates depends on the appropriateness of the assumptions made. In the Adirondack case study we have already raised one issue, whether data require transformation to meet model assumptions, and this is probably just the tip of the iceberg. Basing the estimates on a probability sample ensures that an assumption-free (design-based) methodology is always available. Then model-based methods can be used to enhance inferences without the general validity of those inferences depending on the model assumption (see Overton, Chapter 47).

This case study illustrated several important points for future environmental surveys. First, we demonstrated the potential of model-based estimation for the description of spatial pattern in environmental variables. Second, we learned the importance of adequate investment in frame materials as the inadequacies in ELS frame data prevented us from producing population estimates for the target population of lakes. Finally, the value of analyzing residuals in a spatial context was underscored by the insights gained during our model-building procedure.

ACKNOWLEDGMENTS

We are especially indebted to George Weaver from Oregon State University for his help in conducting analyses and addressing some fruitful questions about the use of data transformations. This research was supported by Cooperative Agreement CR816721 between the U.S. Environmental Protection Agency and Oregon State University and by Interagency Agreement DW89934074-2 with the U.S. Department of Energy under contract DE-AC05-84OR21400 with Martin Marietta Energy Systems, Inc. This is Environmental Sciences Division publication no. 3839.

REFERENCES

Ahmed, S., and DeMarsily, G. (1987) Comparison of geostatistical methods for estimating transmissivity using data on transmissivity and specific capacity. *Water Resources Research* 23(9): 1717–1737.

Blick, D.J., Messer, J.J., Landers, D.H., and Overton, W.S. (1987) Statistical basis for the design and interpretation of the National Stream Survey, Phase I: lakes and streams. *Lake and Reservoir Management* 3: 470–475.

Gleason, H.A. (1926) The individualistic concept of the plant association. *Bulletin of the Tory Botanical Club* 53: 7–26.

Graybill, F.A. (1983) *Matrices with Applications in Statistics*, Belmont, CA: The Wadsworth Statistics/Probability Series, Wadsworth Publishing Co., Inc.

Jager, H.I., Sale, M.J., and Schmoyer, R.L. (1990) Cokriging to assess regional stream quality in the Southern Blue Ridge province. *Water Resources Research* 26(7): 1401–1412.

Linthurst, R.A., Landers, D.H., Eilers, J.M., Brakken, D.F., Overton, W.S., Meier, E.P., and Crowe, R.E. (1986) Characteristics of lakes in the Eastern United States. Vol. I, Population description and physico-chemical relationships. *EPA/600/4-86/007a*.

Neuman, S.P., and Jacobson, E.A. (1984) Analysis of nonintrinsic spatial variability by residual kriging with application to regional groundwater levels. *Mathematical Geology* 16(5): 499–521.

Olsen, A.R., and Slavich, A.L. (1986) Acid precipitation in North America: 1984 annual data summary from Acid Deposition Data Base. *EPA 600/4-86-033*, Research Triangle Park, NC: U.S. Environmental Protection Agency.

Overton, W.S. (1989) Calibration methodology for the double sample of the National Lake Survey Phase II sample. *Technical Report 130*, Corvallis, OR: Department of Statistics, Oregon State University.

Royall, R.M., and Cumberland, W.G. (1981) An empirical study of the ratio estimator and estimators of its variance. *Journal, American Statisical Association* 76: 66–77.

Starks, T.H., and Fang, J.H. (1982) The effect of drift on the experimental semivariogram. *Mathematical Geology* 14(4): 309–319.

Whitaker, R.H. (1970) *Communities and Ecosystems*, New York: The Macmillan Co., 158 pp.

43

Spatial Simulation: Environmental Applications

EVAN J. ENGLUND

Spatial simulation is a technique that has great potential as a tool for dealing with the various problems associated with spatial uncertainty. Although there were antecedents in time series, spatial simulations for earth science applications date from work in geostatistics in the early 1970s (Journel, 1974). Compared to geostatistical interpolation (kriging), there have been relatively few published applications of spatial simulation. These have occurred primarily in the mining and petroleum industries, and in hydrology. It is noteworthy that the recent geostatistics text by Isaaks and Srivastava (1989) does not mention the technique. This lack of attention may have been due to the amount of computer time required—typically one or two orders of magnitude greater than that for kriging and contouring the same area. Fortunately, improved algorithms and fast desktop computers have combined to make the method feasible (each simulation shown below would require about 2 minutes on a 486-25 PC).

In this chapter, two spatial data layers are created by contouring irregularly spaced sample data sets. The layers are then combined in a typical GIS logical operation, and spatial simulation is used to evaluate the resulting combined uncertainty. The two sample sets are drawn from much larger "exhaustive" sets that are examined to assess the validity of the simulation results.

KRIGING, SPATIAL SIMULATION, AND CONDITIONAL SIMULATION

Kriging is a linear unbiased spatial interpolation method that provides minimum mean-squared error estimates at unsampled locations. While kriged surfaces, like other regressions, may be good estimators, they are also unrealistically smooth and continuous.

Spatial simulations, conversely, fill in realistic-looking detail, but are poor estimators. Both kriging and simulation are controlled by variogram models, which quantify the spatial variability of data. Spatial simulation refers to the generation of spatial data through the use of a random number generator, constrained to honor a specified variogram model. Conditional simulation forces the simulated data also to honor a pre-existing set of sample data.

A series of conditional simulations generated with different random seeds from the same initial data and variogram model can be thought of as equally likely possibilities that might explain the available observed data. A kriged map, providing the "best" estimates at each location, is essentially equivalent to the average of a very large number of simulations. Conditional simulations can be used in a number of ways, from simple displays of the uncertainty of data, to much more complex sensitivity analyses, where simulations are used to provide variable input to deterministic models.

AN EXAMPLE

The intersection of two kriged maps

Consider two variables, V_1 and V_2, which have been measured in separate sampling campaigns over the same area. Each has an irregular network of sample locations, as shown in Figure 43-1. From each data set, variograms were computed and modeled. As Figure 43-2 suggests, fitting variogram models is not always simple. Kriged estimates of each variable were then made on a regular grid of nearly 20,000 points (Figure 43-3). The kriging process has taken the original sparse measured data and generated complete areal coverage of estimated values. Now imagine a hypothetical decision-maker who has determined that V_1 and V_2 have a negative synergism; though neither variable is of interest by itself, when both have high values there is cause for concern. Given the two kriged maps as data layers in a GIS environment, it would be a simple matter to combine them and generate a new classification map

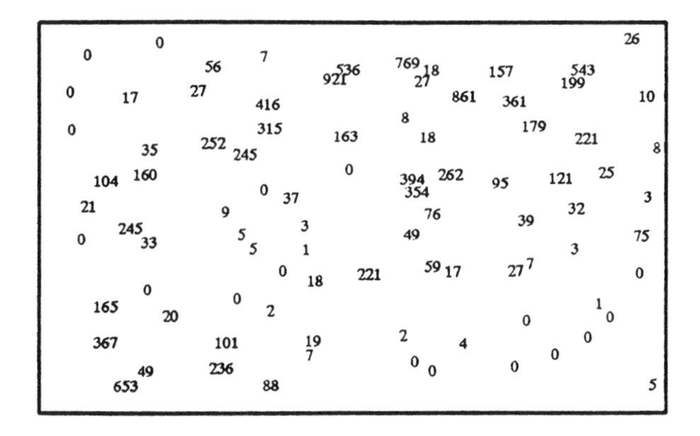

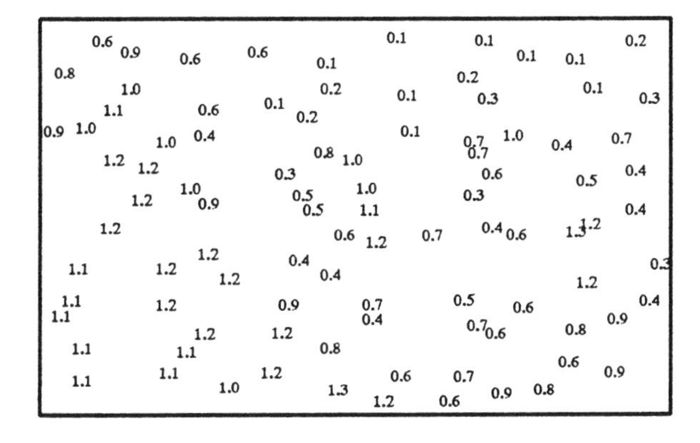

Figure 43-1: Sample locations and measured values for variables V_1 (top) and V_2 (bottom).

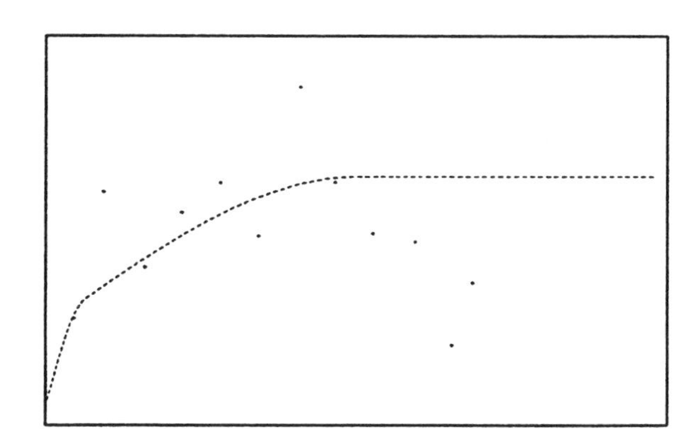

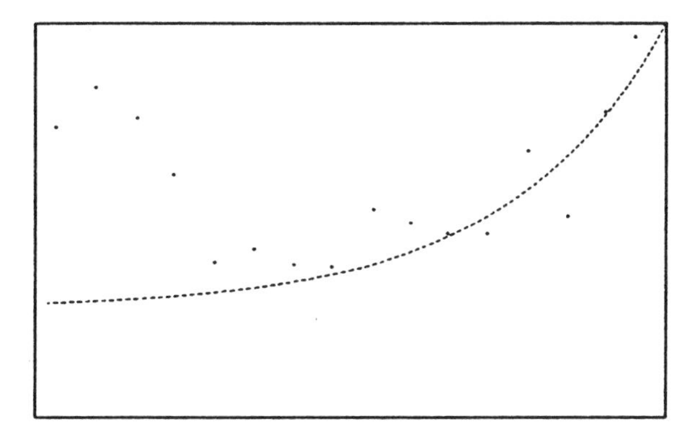

Figure 43-2: Example directional variograms for variables V_1 (top) and V_2 (bottom). Observed values are shown as points; fitted models as lines.

showing where both V_1 and V_2 exceed specified limits (Figure 43-4).

How good is this classification map? Is it likely to be nearly perfect, mediocre, or totally wrong? How can the uncertainty be conveyed to the decision-maker in an understandable and useful form?

One way to look at spatial uncertainty is with kriging errors. Although Isaaks and Srivastava (1989) rightly caution against the use of the kriging standard deviation for estimating confidence intervals, it is nevertheless often the only measure of spatial error available. A simple example of an error zone for false negatives based on kriging standard deviations is shown in Figure 43-5. Expected error rates within this zone range from 50% near the inner boundary to 16% or less at the outer boundary. A similar zone could be created for false positives.

This type of error display does not provide the decision-maker with a realistic picture. It reinforces the oversmoothing effect of kriging, and also fails to account for spatial autocorrelation of errors. If, say, 20% of the points in the error zone were actually misclassified, there is no way to tell whether they are scattered randomly through the zone, or whether they are strongly clustered to form occasional bulges in the boundary.

Intersections of simulated maps

Figure 43-6 shows three conditional simulations of V_1 that honor both the original data in Figure 43-1 and the variogram model in Figure 43-2. The only different parameter was the seed for the random number generator. Figure 43-7 shows three comparable simulations of V_2.

While the kriged maps of the two variables appear to be generally similar in terms of the smoothness and complexity of the estimation surfaces, the simulations present a markedly different picture. The V_2 simulations retain much of the appearance of the kriged map. The simulated surfaces are rougher, but still fairly continuous, and the shaded bands of the kriged map are generally present in

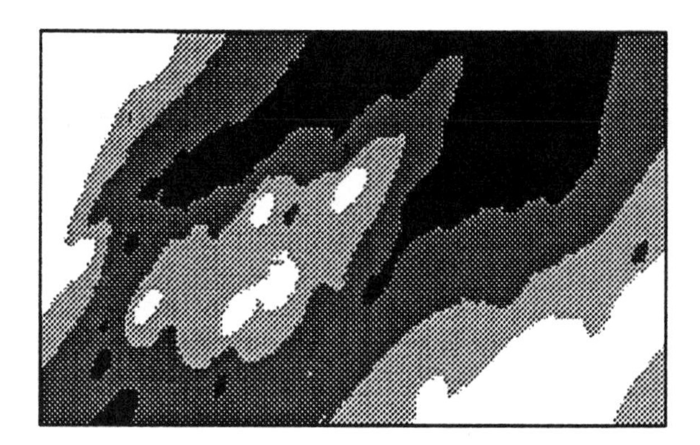

Figure 43-3: Maps of kriged estimates of variables V_1 (top) and V_2 (bottom). Darker shades indicate higher estimated values.

Figure 43-4: Classification based on two kriged maps. Dark area indicates where both V_1 and V_2 are estimated to be high.

Figure 43-5: Zone with highest probability of false negatives is shown in lighter shade.

the same places. By contrast, the V_1 simulations bear little resemblance to the kriged map. The simulations are very discontinuous, and many areas that are high on one simulation are low on another (e.g., compare the lower right-hand corners of the maps in Figure 43-6).

These maps give us an intuitive grasp of the uncertainty of the estimates. If reality were to turn out to be like any of these simulations, it is obvious that the V_2 kriged map would be a much better estimator than the V_1 kriged map.

Note that the three V_1 simulations, while they differ considerably in detail, nevertheless share a common distinctive appearance we could call "texture." The same is true for the V_2 simulations. This is basically a graphical illustration of the information contained in the variogram models, which describes the variability at all spatial scales, and controls the simulation process.

Interestingly, geostatistical simulation is a method that contains fractal simulation as a subset, but is more flexible. According to Burrough (1983), the slope of a variogram model plotted on log-log axes can be translated directly into a fractal (Hausdorff-Besicovitch) dimension. A linear variogram model on log-log axes would therefore indicate constant fractal dimension, and lead to simulations with similar textures at all scales. Neither variogram model used in this example is linear on the log-log plot, resulting in different textures at different scales.

The simulations by themselves give us an idea of the spatial uncertainty of each variable. By combining pairs of simulations in the same way we combined the kriged maps, we can look at the uncertainty in our original classification. Figure 43-8 shows the resulting three simulated classification maps. The decision-maker looking at a series of such maps and realizing that reality might be like any one of them will have a much better idea whether or not the current information is adequate for the decision.

Figure 43-9 shows the true distributions of V_1 and V_2,

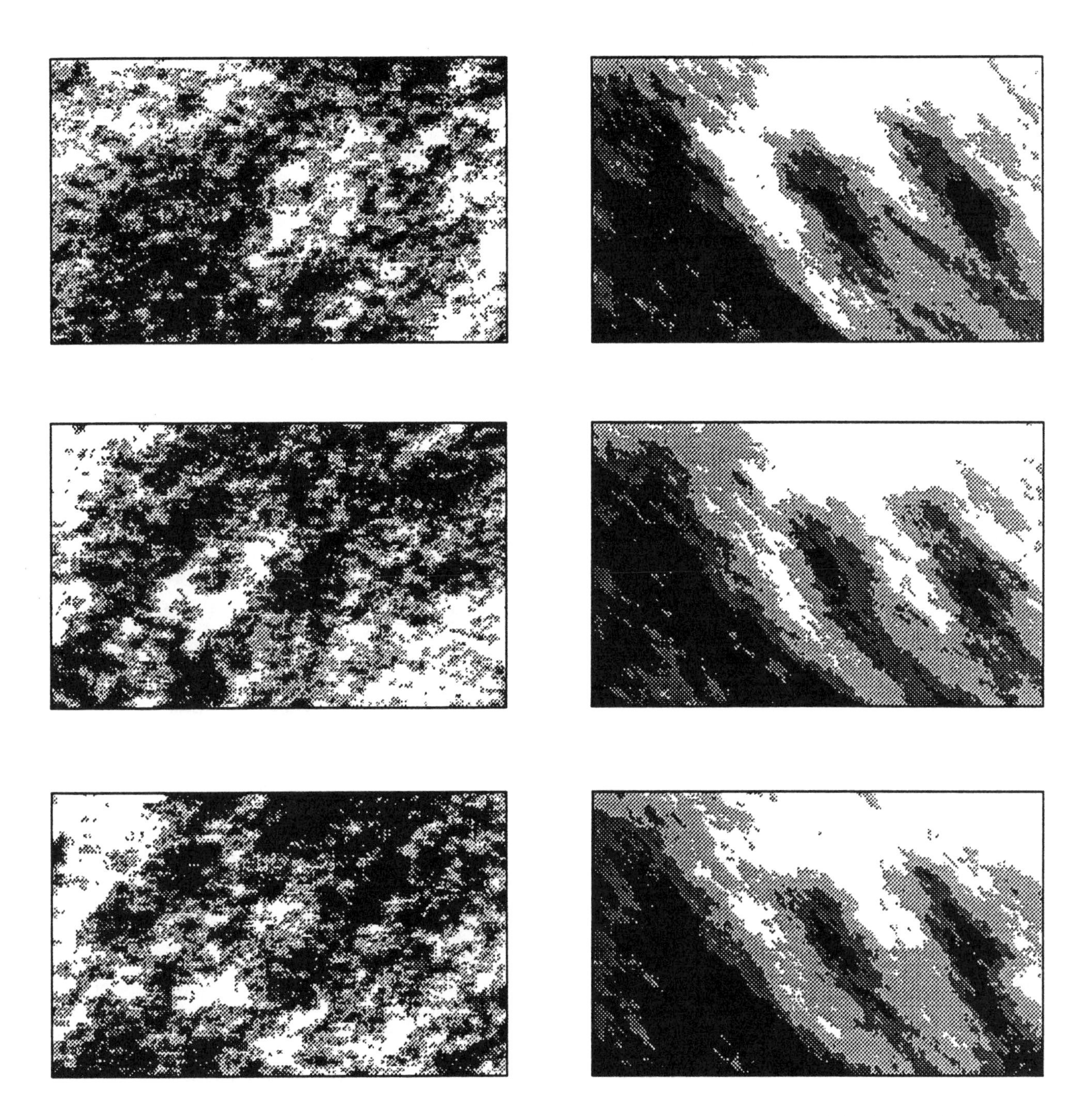

Figure 43-6: Three conditional simulations of V_1.

Figure 43-7: Three conditional simulations of V_2.

and the true classification map based on their intersection. Note that the simulated maps do not perfectly represent all of the features observed in the actual maps; for example, the simulations of V_2 appear to be more variable than the real thing. In Figure 43-2, the type of model that was fitted to the experimental V_2 variogram is one that implies that the data represent a continuous but relatively rough surface. The experimental variogram, however, is so noisy that many alternate models could be considered equally valid. An investigator with prior knowledge that the measured phenomenon has a continuous but relatively smooth surface could have used a more appropriate model and produced more realistic simulations.

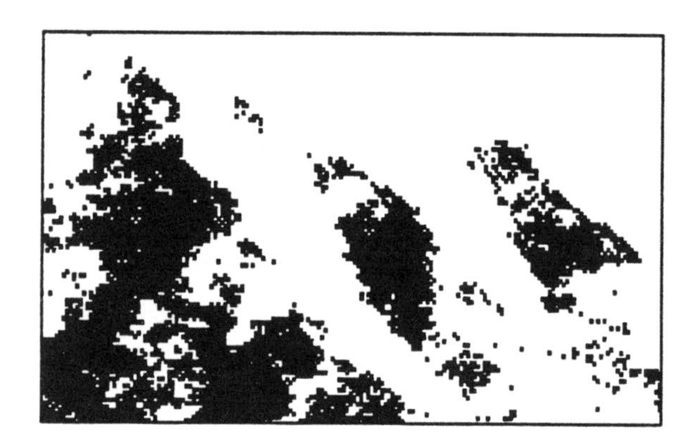

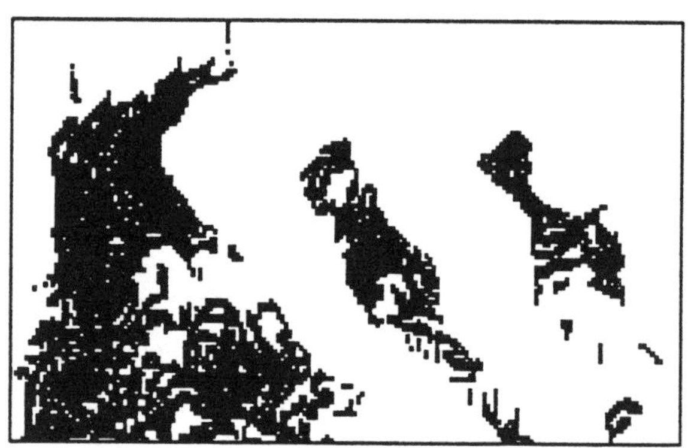

Figure 43-8: Three simulated classification maps. Each is based on a pair of conditional simulations. Dark areas indicate where both V_1 and V_2 were simulated to be high.

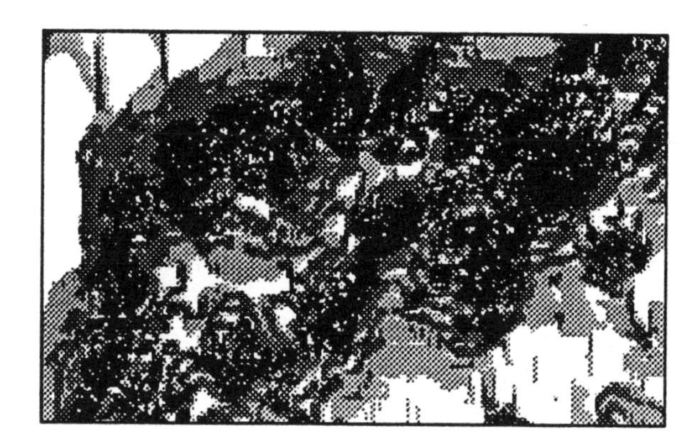

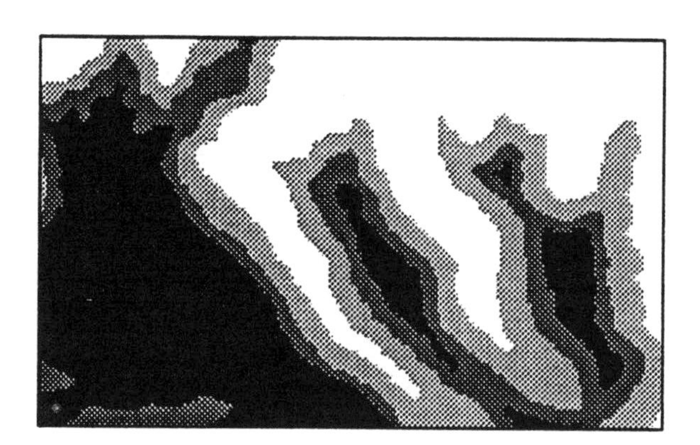

Figure 43-9: Actual maps of V_1 (top), V_2 (center), and classification (bottom).

QUANTIFYING THE CONSEQUENCES OF UNCERTAINTY

When visualization of uncertainty alone is not sufficient for the decision-maker, the simulations may be used to make quantitative estimates of errors and their consequences. In the example given, a sensitivity analysis could be conducted by assuming that decisions were to be made from the existing kriged classification map. Then, for any one of the simulations, the number of false positives and false negatives that would occur if that simulation were reality can be computed. Repeating this process on, say, 100 simulations will give both the expected numbers of errors, as well as best case and worst case numbers. Neptune et al. (1990) describe the development of a table of maximum allowable error rates for remediation decisions for different levels of contaminant concentration at a Superfund site. These maximum rates could easily be compared to the expected rates from the simulations.

If the costs associated with incorrect decisions can be estimated, an objective function can be used to determine the optimum (minimum total cost) sampling design. Although it is difficult to estimate such costs in environmental situations such as site remediation, simple "loss-function" models (Englund, 1990) may provide a reasonable approach. A site model is first generated by conditional simulation and used as the basis for a Monte Carlo resampling scheme. A number of additional samples are drawn from the model; new kriged estimates are made; and the benefits due to better decisions are compared to the costs of obtaining the additional data. The process is repeated with different sampling schemes until the most cost-effective one is found. While such an approach is computer intensive, it is possible on any desktop computer system capable of serious GIS applications.

DISCUSSION

Both kriging and conditional simulation require the same inputs—sample measurements with spatial coordinates and a variogram model. Thus, there is no inherent reason why conditional simulation should not be used as routinely for uncertainty analysis as kriging is used for interpolation. Kriging is becoming more frequently used in GIS applications, and at least one standard GIS package (ARC-INFO) is adding a kriging option. It is unlikely, however, that conditional simulation will become available in the GIS environment until a substantial demand has been established. As happened in the case of kriging, this is likely to require the gradual accumulation of case studies in the literature, as well as wider availability of textbooks, training, and user-friendly software.

SPATIAL SIMULATION SOFTWARE

A reader interested in experimenting with spatial simulation has few options apart from expensive commercial geostatistics packages. The following are three sources of public domain simulation software:

TUBA, a nonconditional simulation program, is distributed by SEASOFT, P.O. Box 53124, Albuquerque, NM 87153.
COSIM, a conditional (by cokriging) simulation program, is available from a bulletin board service operated by the Computer Oriented Geological Society (COGS), P.O. Box 1317, Denver, CO 80201. The BBS number is 303 740 9493.
GSLIB, Geostatistical Software Library and User's Guide, Oxford University Press, 1992, includes several different conditioneal simulation programs.

ACKNOWLEDGMENT

The information in this document has been funded wholly by the US Environmental Protection Agency. It has been subjected to Agency review and approved for publication. Mention of trade names does not constitute endorsement or recommendation for use.

REFERENCES

Burrough, P.A. (1983) Multiscale sources of spatial variation in soil. I. Application of fractal concepts to nested levels of soil variation. *Journal of Soil Science* 34: 577–597.
Englund, E.J. (1990) A variance of geostatisticians. *Mathematical Geology* 22(4): 417–455.
Isaaks, E.H., and Srivastava, R.M. (1989) *Applied Geostatistics*, Oxford University Press.
Journel, A.G. (1974) Geostatistics for conditional simulation of ore bodies. *Economic Geology* 69(5): 673–687.
Neptune, D. (1990) Quantitative decision making in Superfund: a data quality objectives case study. *Hazardous Materials Control* May–June: 19–27.

ADDITIONAL READING

Armstrong, M. (ed.) (1989) *Geostatistics: Proceedings of the Third International Geostatistics Congress*, Volumes 1–3 (contains several papers on conditional simulation).
Dimitrakopoulos, R. (1990) Conditional simulation of intrinsic random functions of order *k*. *Mathematical Geology* 22(3): 361–380.
Easley, D.H., Borgman, L.E., and Weber, D. (1991) Monitoring well placement using conditional simulation of hydraulic head. *Mathematical Geology* 23(8): 1059–1080.
Gomez-Hernandez, J.J., and Srivastava, R.M. (1990) ISIM3D: an ANSI-C three-dimensional multiple indicator simulation program. *Computers & Geosciences* 16(4): 395–439.
Palmer, M.W. (1988) Fractal geometry: a tool for describing spatial patterns of plant communities. *Vegetatio* 75: 91–102.

44

GRASS Used in the Geostatistical Analysis of Lakewater Data from the Eastern Lake Survey, Phase I

HANNAH RASMUSSEN RHODES AND
DONALD E. MYERS

This case study describes how GRASS was used in the geostatistical analysis of the ELS-I data. The objective was to obtain estimates of the spatial distributions of ion concentrations for chemical analytes most relevant to lake acidification processes. In applying geostatistical techniques the support of the data (e.g., the area associated with a particular lake sample) has a crucial effect. The original objective of the ELS-I sampling did not require incorporating the size of the lakes nor require more than one sample per lake. In order to examine the influence of lake size on the kriging analysis, pseudo sample positions within the large lakes (surface area greater than 500 ha) sampled during the ELS-I were compiled using GRASS. These pseudo sample positions allow for incorporating the differing sample supports into the analysis.

ELS-I DATA

The data analyzed were obtained from the ELS-I initiated in 1983 by the U.S. Environmental Protection Agency (EPA). The ELS-I was designed to describe the chemical status of lakes in areas of the eastern U.S. containing the majority of low-alkalinity lakes. These were defined as having acid neutralizing capacity (ANC) $\leq$ 400 μeq l^{-1}. Complete details on statistical design, lake selection, analytical methodologies, field methods, and quality-assurance protocols are given elsewhere (Best et al., 1986; Linthurst et al., 1986; Overton et al., 1986; Overton, 1986; Kanciruk et al., 1986); so only the key design features are presented here.

A stratified design was used wherein sample lakes were selected statistically from each stratum. The first level of stratification included three regions of the eastern U.S. (Northeast, Upper Midwest, Southeast) expected to be the most susceptible to change as a result of acidic deposition. The second stratification level was obtained by dividing each region into subregions exhibiting geographic homogeneity with respect to water quality, physiography, vegetation, climate, and soils. Finally, each subregion was further subdivided into three alkalinity map classes (ANC < 100, 100 – 200, and > 200 μeq l^{-1}). The population of lakes was screened to exclude lakes surrounded by or adjacent to anthropogenic activities such as intense urban, industrial, or agricultural land use. The survey was conducted in the Fall of 1984 with one water sample per lake collected from a depth of 1.5 m at the deepest part of the lake. 26 chemical analytes related to surface water acidification and several physical attributes were measured for each lake.

The geostatistical analysis was restricted to the Northeast, identified as region 1, which is divided into five subregions: Adirondacks (1A), Poconos/Catskills (1B), Central New England (1C), Southern New England (1D), and Maine (1E). Several chemical analytes from the PC version of the ELS-I data were examined, but due to space limitations only the results pertaining to ANC in subregion 1A are presented in this case study. Figure 44-1 shows the locations of the lakes sampled in subregion 1A.

COMPILATION OF PSEUDO SAMPLE POSITIONS USING GRASS

Consider the layout in Figure 44-2, where a lake of interest is located near one large lake and four smaller lakes. Let l_e denote the lake to be estimated, l_i sampled lake i, x the original sample location, and d_i the distance between l_e and the sample location for l_i. When the kriging estimate and variance are calculated, the number of samples in the search neighborhood must be specified. Suppose a search neighborhood encompassing four samples is chosen. Since $d_1 > d_2 > d_3 > d_4 > d_5$ the value corresponding to the large lake, l_1, will not be included in the kriging

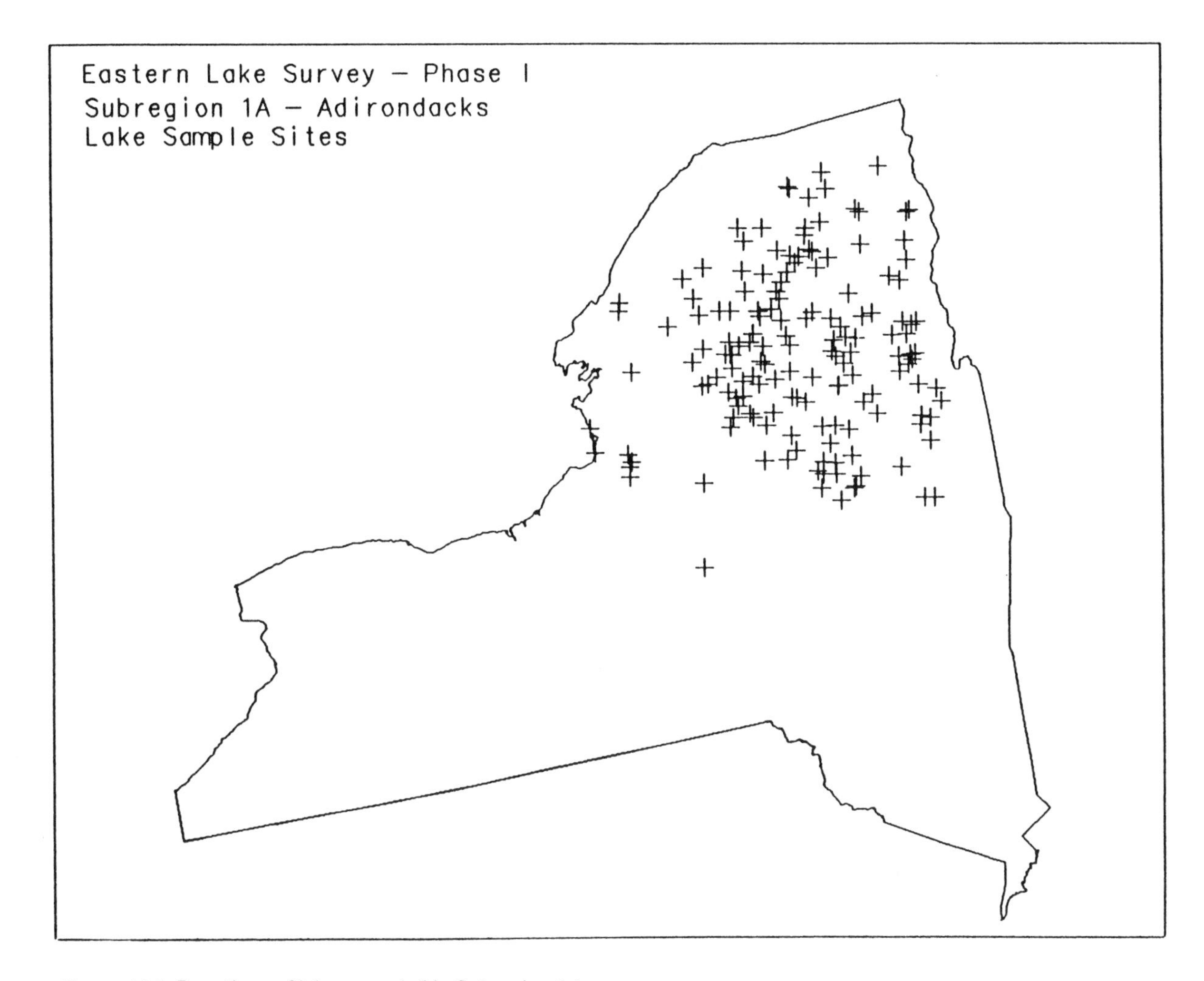

Figure 44-1: Locations of lakes sampled in Subregion 1A.

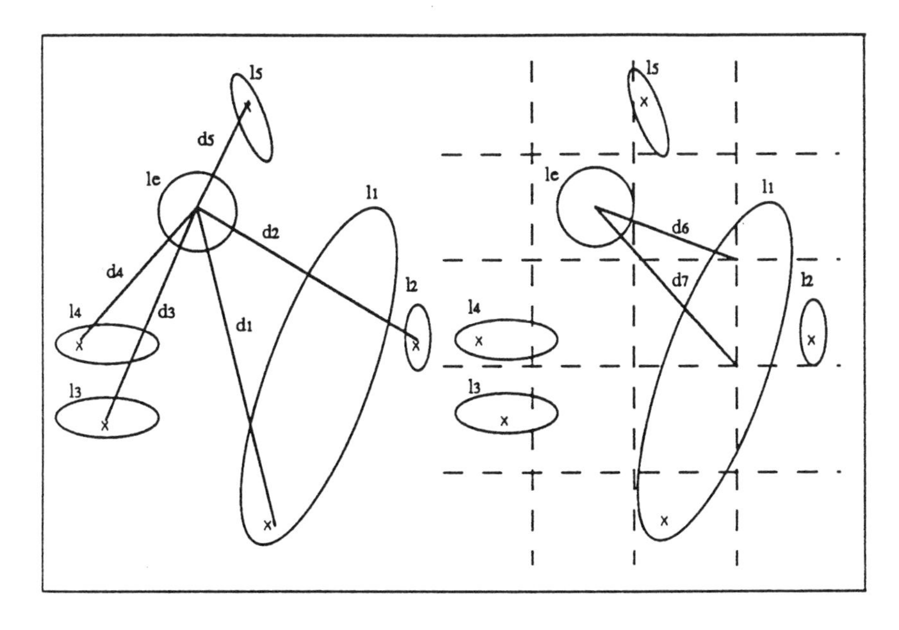

Figure 44-2: Theoretical example.

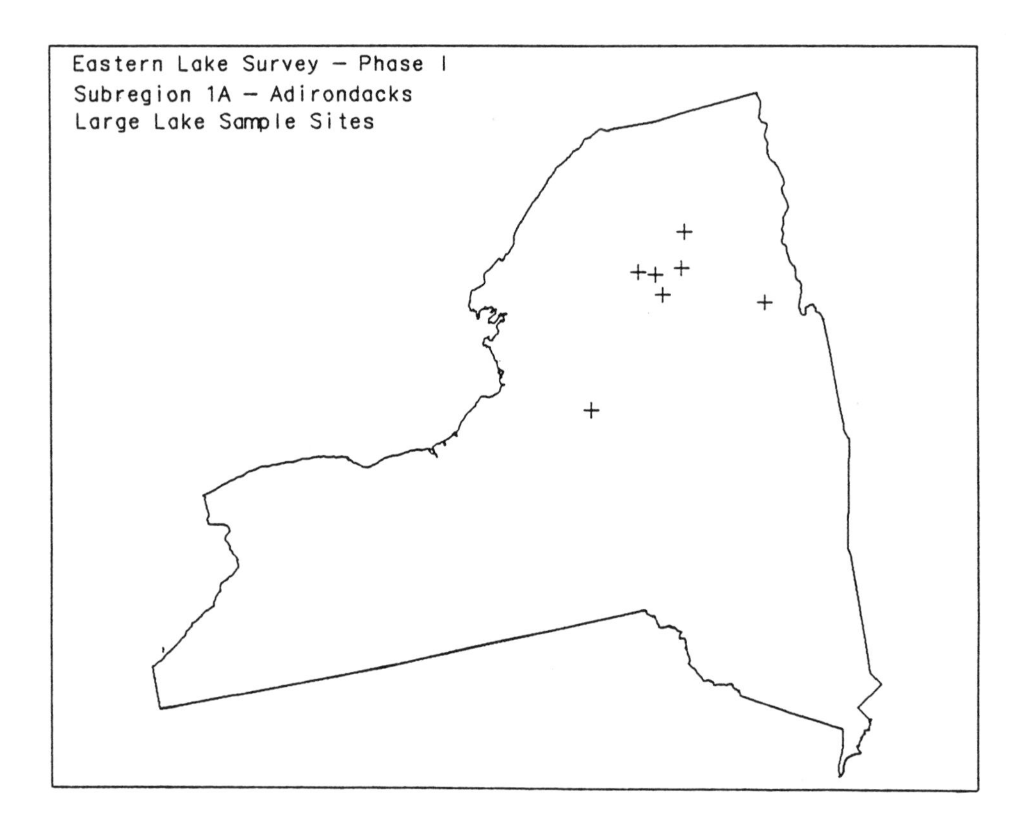

Figure 44-3: Location of large lakes sampled in Subregion 1A.

estimation of the value for l_e. Consider overlaying a grid and adding pseudo sample positions at the gridpoints lying within the boundary of l_1. Now $d_1 > d_2 > d_3 > d_7 > d_4 > d_6 > d_5$; so the values corresponding to lakes l_2 and l_3 will no longer influence the kriging estimate for l_e, but the value corresponding to l_1 will be included.

To obtain outlines of the large lakes sampled during the ELS-I, hydrological digital line graph (DLG) data displaying small-scale (1:2,000,000) streams and water bodies in the northeastern U.S. were purchased from the U.S. Geological Survey. It should be pointed out that the ground coordinate system for the DLG data is the Albers Equal-Area Conic projection, whereas the ELS-I sample locations have coordinates given in latitude and longitude.

The GRASS 3.1 programs used in the compilation process will be outlined here. First, a default window specifying the geographic coverage of the DLG data was created, and the *import.to.vect* program was used to convert the DLG data into the GRASS vector format. Windows for the five subregions of region 1 were defined with the *window* command. The locations for the lakes sampled in region 1 were converted from latitude and longitude to Albers coordinates in order to have a common coordinate reference. Then site files of the Albers coordinates for the large lakes sampled were created. *Dvect* was used to display the outlines of lakes in the northeastern U.S., and the large lakes sampled were marked by overlaying the site files using *d.sites*. Figure 44-3 shows the location of the large lakes sampled in subregion 1A. The *window* command was used to zoom in on the large lakes sampled, and *Dgrid* was used to overlay a 2000 m sized grid. A 1000 m sized grid was also overlaid in order to examine the effect of grid size. Pseudo sample positions were then obtained at each of the grid points falling within the boundary of the large lake with *Dwhere*. Figure 44-4 depicts one of the large lakes sampled in subregion 1A with the original sample location marked and a 2000 m sized grid overlaid. Table 44-1 lists the number of pseudo sample positions added to the large lakes sampled in subregion 1A with a 2000 m and a 1000 m sized grid overlaid. The lake name, surface area, and ANC value are also included.

The last step in the compilation process consisted of converting the Albers coordinates for the pseudo sample positions to latitude and longitude, adding them to the ELS-I data files, and assigning values to the pseudo sample positions. Three approaches for attributing values were considered: (1) original analyte value, (2) original analyte value + rand$*\sqrt{\text{nugget}}$, and (3) original analyte

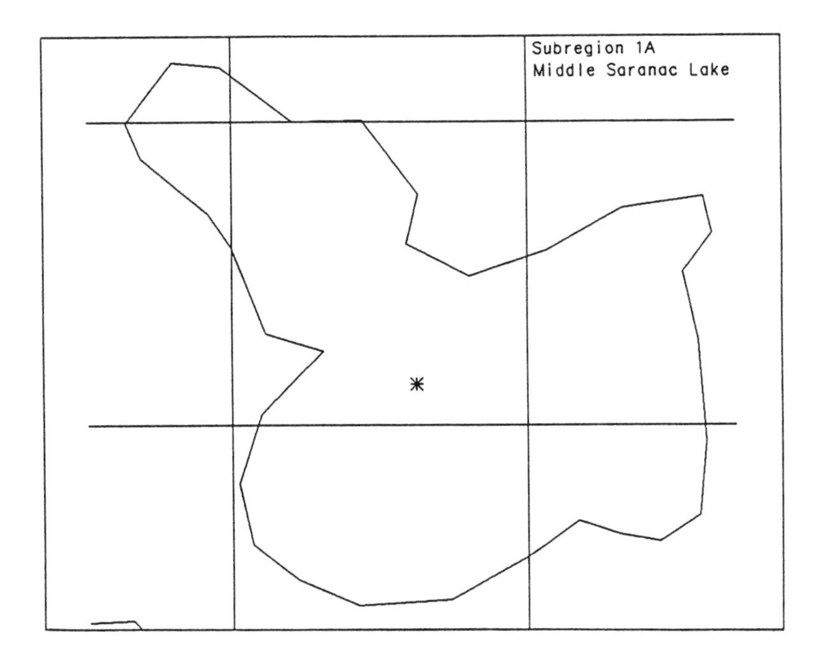

Figure 44-4: Middle Saranac Lake.

value + rand*std.dev., where rand is a random number between –1.0 and 1.0. The first approach treats the sampled lakes as perfectly mixed, whereas the latter two incorporate spatial variability within the sampled lakes. Three augmented data sets were thus created with varying values assigned to the pseudo sample positions obtained with the 2000 m sized grid. The first approach was used for the pseudo sample positions from the 1000 m sized grid. Table 44-2 lists various statistics for ANC computed from the original and the four augmented data sets.

GEOSTATISTICAL ANALYSIS

Kriging is a geostatistical estimation (interpolation) technique. The early theoretical work was done by Matheron, and the theory has been developed extensively (Journel and Huijbregts, 1978; Myers et al., 1982; Myers, 1991). A discussion of the kriging theory is beyond the scope of this case study. Examples of the development of the kriging theory and applications of its use to acid precipitation data can be found in Eynon and Switzer (1983), Finkelstein (1984), Bilonick (1985), Bilonick (1988), Venkatram (1988), Guertin et al., (1988), Marcotte (1989), or Seo et al. (1990 a, b).

Before the kriging analysis was performed, the degree of drift in the chemical analyte values was evaluated with the BLUEPACK-3D RECCO program. BLUEPACK-3D is a geostatistics software package developed jointly by the Centre de Geostatistique et de Morphologie Mathematique in Fontainebleau, France, and the BRGM (French Geological Survey). The actual kriging analysis

Table 44-1: Large lakes information

Lake ID	Lake name	Surface area	2000 m grid	1000 m grid	ANC μeq l^{-1}
1A1-044	Long Lake	1619	8	29	84.2
1A1-042	Lows Lake	1019	3	14	54.6
1A3-030	Delta Reservoir	958	3	10	1515.8
1A1-043	Little Tupper Lake	923	3	9	64.7
1A2-035	Brant Lake	595	1	6	381.4
1A1-074	Middle Saranac Lake	567	2	8	178.0
1A1-056	Forked Lake	514	3	12	56.7

Table 44-2: Statistics for ANC

Statistic	Original	2000 m grid	2000 m (nugget)	2000 m (std. dev.)	1000 m grid
Mean	251.910	255.846	252.261	259.005	255.702
Std. dev.	448.297	453.018	452.717	461.541	451.423
Minimum	−34.200	−34.200	−97.404	−281.370	−34.200
25th %tile	27.650	38.150	25.219	21.950	54.600
Median	115.700	107.400	111.850	117.750	84.200
75th %tile	234.725	231.200	232.850	273.350	212.750
Maximum	3226.800	3226.800	3226.800	3226.800	3226.800

was performed using Geo-EAS (Geostatistical Environmental Assessment Software), which is a collection of interactive software tools for performing geostatistical analyses. Geo-EAS was produced under the sponsorship of the EPA Environmental Monitoring Systems Laboratory at Las Vegas, Nevada.

First PREVAR was run to generate the pair comparison file containing the distances and relative directions between pairs of sample points. Variogram values were calculated with VARIO utilizing the pair comparison file. Pair distance intervals (lags) were specified with increments equal to the maximum interpair distance divided by 48. Omnidirectional, as well as four directional variograms, were generated in order to determine possible anisotropies. Variogram models were fitted visually to the sample variograms and the cross-validation program XVALID was run to test the models. In cross-validation each data location is suppressed one at a time, and an estimate is obtained using the other data locations, thus producing an estimation error for each sample location. In this manner XVALID was used to determine the number of sample points in the kriging search neighborhood yielding the smallest average estimation error. The actual kriged estimates were produced by KRIGE, and CONREC was executed to generate contour maps from the gridded data file produced by KRIGE.

It should be pointed out that when the variograms were computed at the alkalinity map class stratification level, the number of data pairs per lag was very small even for as few as five lags, making the variograms unsuitable for modeling. Therefore the subsequent kriging analysis was performed at the subregion stratification level. First the kriging analysis was performed on the original data set. No drift was evident in the ANC values, and the directional variograms revealed no significant anisotropies. A Gaussian model with a nugget of 25,000, a sill of 3,000,000, and a range of 6.9 was fitted to the omnidirectional variogram. Figure 44-5 shows the plot of the variogram values versus distance with the model overlaid. The kriging estimates were produced with the region divided up into 100 cells, and a kriging search neighborhood encompassing 14 points was used. 16 contouring levels ranging from −50 to

Table 44-3: Kriging estimates for ANC

Statistic	Original	2000 m grid	2000 m (nugget)	2000 m (std. dev.)	1000 m grid
Mean	251.366	255.458	251.728	257.695	246.495
Std. dev.	310.064	344.329	342.941	343.712	353.122
Minimum	4.790	4.790	2.408	4.790	4.790
25th %tile	74.259	70.247	59.996	74.761	58.315
Median	154.980	141.985	140.045	139.550	111.775
75th %tile	267.158	262.825	265.930	257.595	250.540
Maximum	1416.100	2079.700	2016.500	2105.900	1581.100

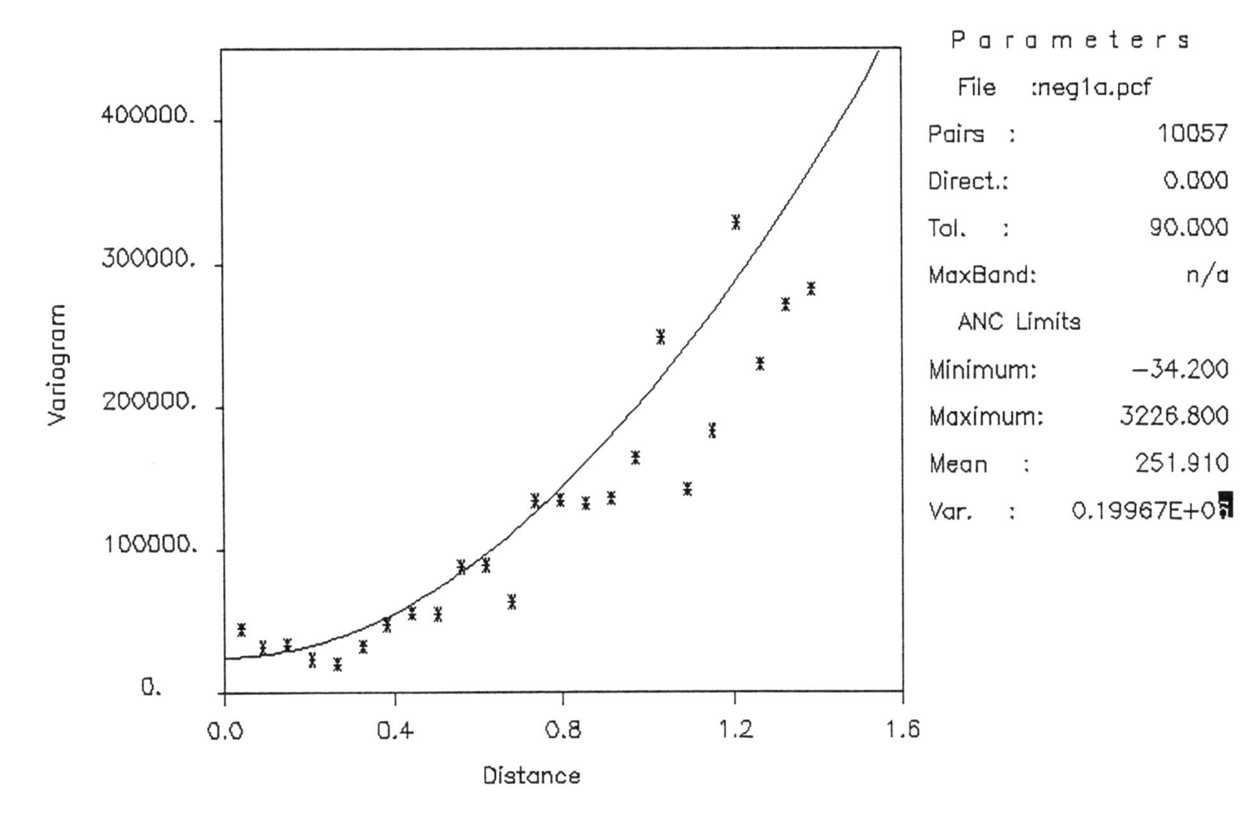

Figure 44-5: Variogram for ANC.

$2200 \mu eq\, l^{-1}$ were specified. The kriged estimates for ANC are shown in Figure 44-6 and contours for the kriging standard deviations are depicted in Figure 44-7.

In order to examine the effects of lake size on the kriging estimates, kriging analyses were performed again on the four augmented data sets following the steps outlined and using the model fitted to the variogram from the original data set. Cross-validation was then performed on the four augmented data sets in order to compare the various estimates. Table 44-3 lists statistics for the kriging estimates for ANC and Table 44-4 lists statistics for the difference between the kriging estimates for ANC and the actual ANC values. Figures 44-8 through 44-13 show the contour maps for the kriged ANC values and the kriging

Table 44-4: Difference between kriging estimates for ANC and actual ANC

Statistic	Original	2000 m grid	2000 m (nugget)	2000 m (std. dev.)	1000 m grid
Mean	–0.544	–0.388	–0.532	–1.310	–9.207
Std. dev.	335.024	295.203	297.010	308.807	259.017
Minimum	–2501.800	–2501.800	–2501.800	–2501.800	–2501.800
25th %tile	–66.355	–63.937	–66.807	–83.542	–20.340
Median	24.213	14.795	18.228	22.286	0.000
75th %tile	114.470	81.794	103.311	113.495	45.737
Maximum	837.880	860.480	852.330	861.640	744.220

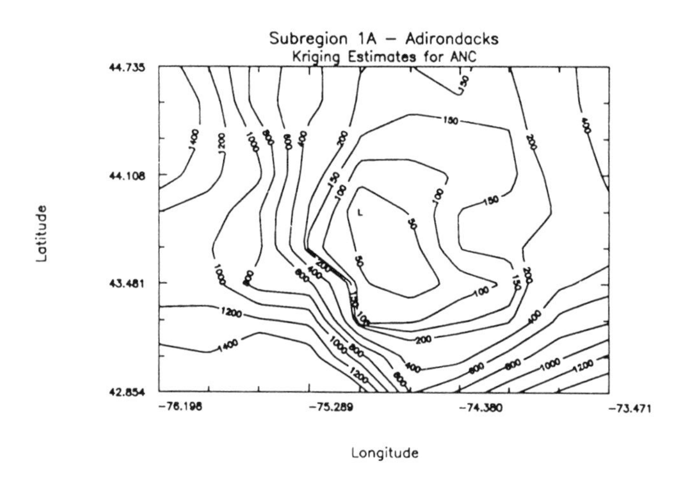

Figure 44-6: Kriging estimates for ANC.

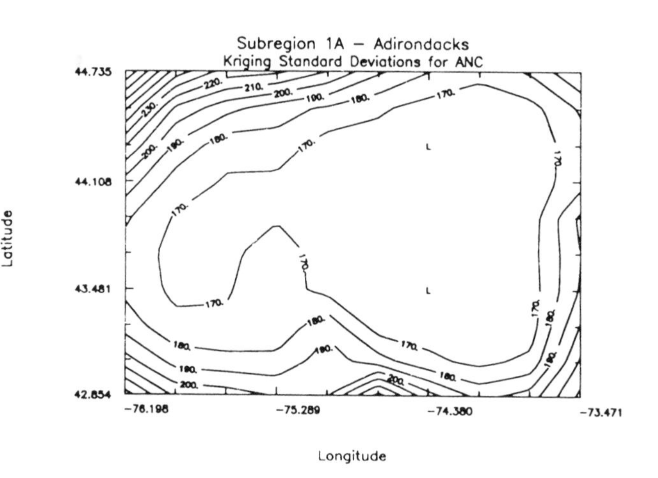

Figure 44-7: Kriging standard deviations for ANC.

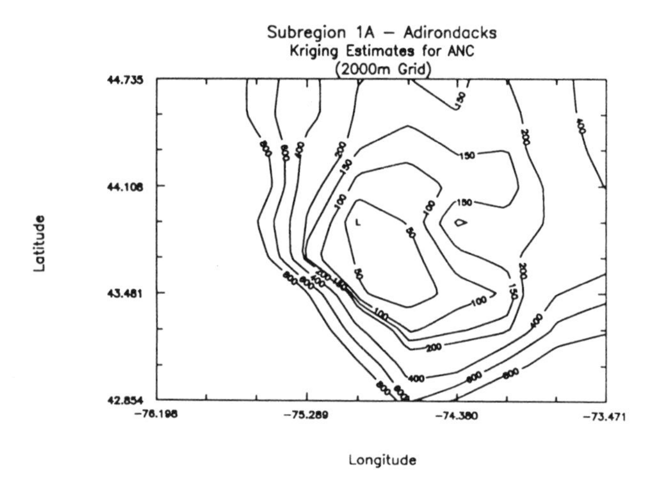

Figure 44-8: Kriging estimates for ANC, 2000 m grid.

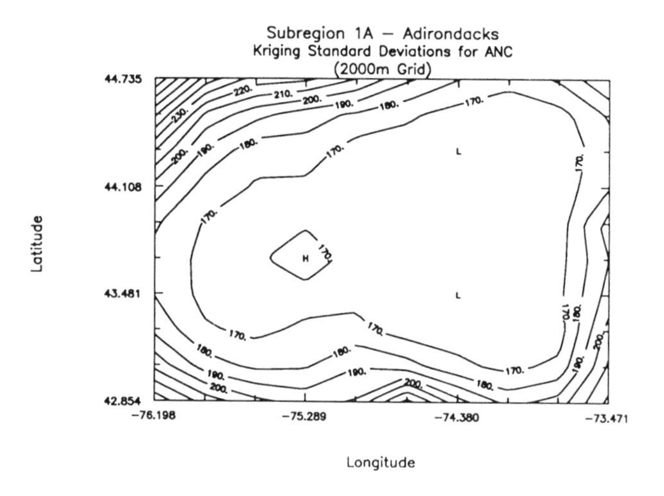

Figure 44-9: Kriging standard deviations for ANC, 2000 m grid.

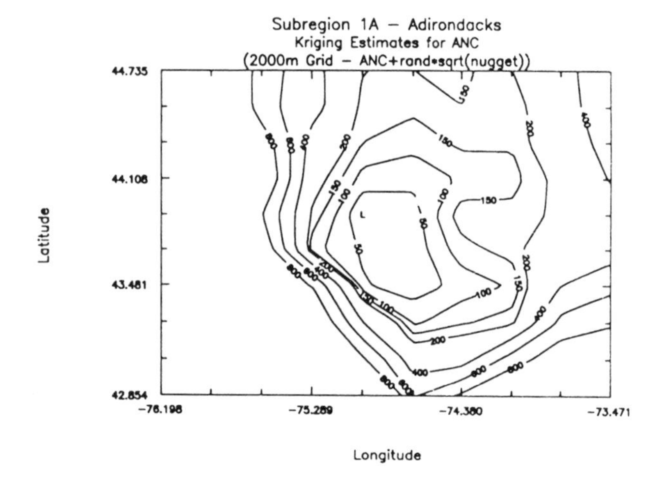

Figure 44-10: Kriging estimates for ANC, 2000 m grid (Nugget).

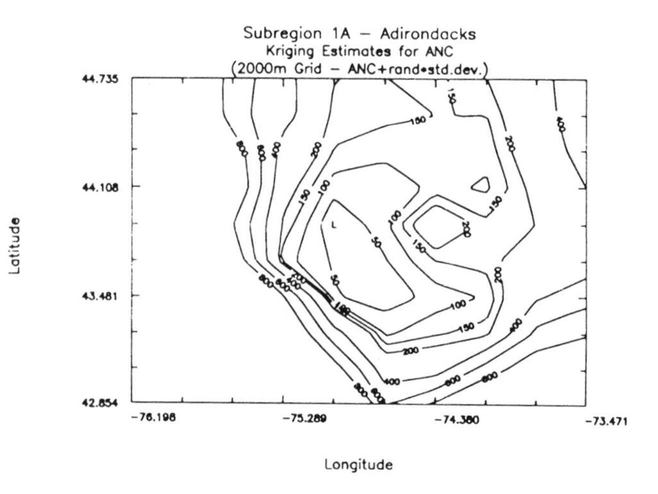

Figure 44-11: Kriging estimates for ANC, 2000 m grid (std. dev.).

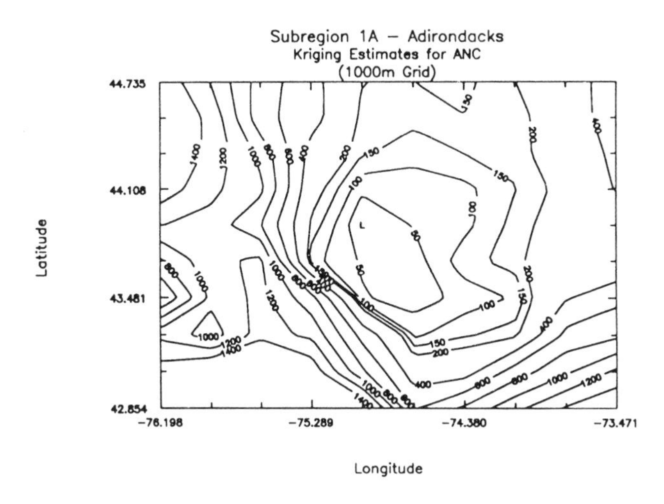

Figure 44-12: Kriging estimates for ANC, 1000 m grid.

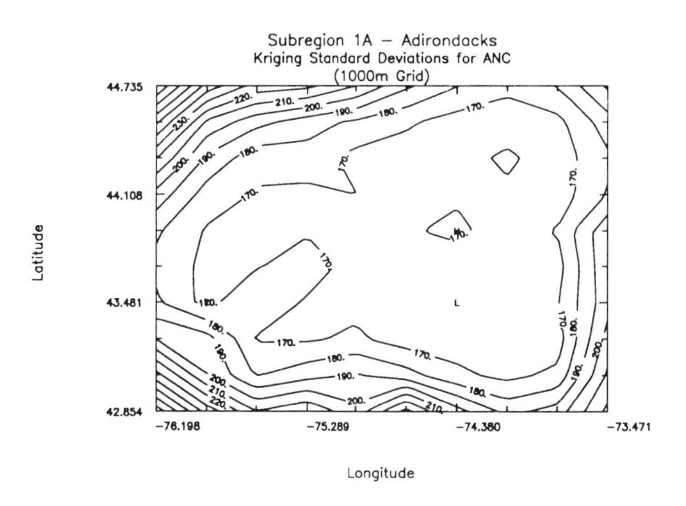

Figure 44-13: Kriging standard deviations for ANC, 1000 m grid.

standard deviations. The plots of the kriging standard deviations for the three augmented data sets obtained from the 2000 m sized grid were virtually identical and were therefore not included.

SUMMARY AND CONCLUSIONS

To examine the influence of lake size on the kriging analysis, GRASS was used to identify and assign pseudo sample positions within the large lakes sampled during the ELS-I. The inclusion of the pseudo sample positions had significant effects on the kriging analysis, as evidenced by the changes in the contour maps for the kriged ANC values. The results are sensitive to the size of the grid used to identify the pseudo sample positions and the values assigned to the pseudo sample positions.

NOTICE

Although the research described in this chapter has been funded wholly or in part by the EPA, it has not been subjected to Agency review and therefore does not reflect the views of the Agency, and no official endorsement should be inferred.

REFERENCES

Best, M.D., Creelman, L.W., Drouse, S.K., and Chaloud, D.J. (1986) National Surface Water Survey, Eastern Lake Survey - Phase I, quality assurance report. *Technical Report EPA/600/4-86/011*, Las Vegas, NV: U.S. Environmental Protection Agency.

Bilonick, R.A. (1985) The space–time distribution of sulfate deposition in the Northeastern United States. *Atmospheric Environment* 19(11): 1829–1845.

Bilonick, R.A. (1988) Monthly hydrogen ion deposition maps for the Northeastern U.S. from July 1982 to September 1984. *Atmospheric Environment* 22(9): 1909–1924.

Eynon, B.P., and Switzer, P. (1983) The variability of rainfall acidity. *The Canadian Journal of Statistics* 11(1): 11–24.

Finkelstein, P.L. (1984) The spatial analysis of acid precipitation data. *Journal of Climate and Applied Meteorology* 23: 52–62.

Guertin, K., Villeneuve, J., Deschenes, S., and Jacques. G. (1988) The choice of working variables in the geostatistical estimation of the spatial distribution of ion concentration from acid precipitation. *Atmospheric Environment* 22(12): 2787–2801.

Journel, A.G., and Huijbregts, Ch. J. (1978) *Mining Geostatistics*, Orlando, FL: Academic Press Inc.

Kanciruk, P., Gentry, M.J., McCord, R.A., Hook, L.A., Eilers, J.M., and Best, M.D. (1986) National Surface Water Survey, Eastern Lake Survey – Phase I, data base dictionary. *Technical Report ORNL/TM-10153*, Oak Ridge, TN: Environmental Science Division, Oak Ridge National Laboratory.

Linthurst, R.A., Landers, D.H., Eilers, J.M., Brakke, D.F., Overton, W.S., Meier, E.P., and Crowe, R.E. (1986) Eastern Lake Survey – Phase I, characteristics of lakes in the Eastern United States. Volume I: population descriptions and physico-chemical relationships. *Technical Report EPA/600/4-86/007a*, Las Vegas, NV: U.S. Environmental Protection Agency.

Marcotte, D. (1989) Spatial estimation of frequency distribution of acid rain data using Bigaussian kriging. *Statistical Applications in the Earth Sciences*, pp. 287–296.

Myers, D.E. (1991) Interpolation and estimation with spatially located data. *Chemometrics and Intelligent Laboratory Systems* 11: 209–228.

Myers, D.E., Begovich, C.L., Butz, T.R., and Kane, V.E. (1982) Variogram models for regional groundwater geochemical data. *Mathematical Geology* 14(6): 629–644.

Overton, W.S. (1986) Working draft, analysis plan for the EPA Eastern Lake Survey, March 18, 1985. *Technical Report 113*, Corvallis, OR: Dept. of Statistics, Oregon State University.

Overton, W.S, Kanciruk, P., Hook, L.A., Eilers, J.M., Landers, D.H., Blick, D.J., Jr., Brakke, D.F., Linthurst, R.A., and DeHaan, M.D. (1986) Eastern Lake Survey—Phase I. characteristics of lakes in the Eastern United States. Volume II: lakes sampled and descriptive statistics for physical and chemical variables. *Technical Report EPA/600/4-86/007b*, Washington, DC: U.S. Environmental Protection Agency.

Seo, D., Krajewski, W.F., and Bowles, D.S. (1990a) Stochastic interpolation of rainfall data from rain gages and radar using cokriging 1. Design of experiments. *Water Resources Research* 26(3): 469–477.

Seo, D., Krajewski, W.F., and Bowles, D.S. (1990b) Stochastic interpolation of rainfall data from rain gages and radar using cokriging 2. Results. *Water Resources Research* 26(5): 915–924.

Venkatram, Akula (1988) On the use of kriging in the spatial analysis of acid precipitation data. *Atmospheric Environment* 22(9): 1963–1975.

45

Universal Kriging for Ecological Data

JAY M. VER HOEF

The general problem considered here is spatial prediction for a raster data set, where we have one or more complete layers of explanatory variables, and another sparsely filled layer for a response variable (Figure 45-1). The goal is to predict the response variable for the remaining cells in the grid. Three approaches will be compared: linear regression, kriging, and universal kriging.

The type of data depicted in Figure 45-1 occurs frequently in ecology. Often, explanatory variables such as slope, aspect, soil type, geological bedrock type, etc. are easy to obtain and may be in electronic form currently. However, data collection of the response variable for each cell may be too expensive or destructive. For example, to collect data on the biomass of a plant species is time consuming and would be totally destructive if collected on every cell in a grid. Thus, it would be very useful to predict biomass spatially for all of the cells in the raster grid, based on a smaller subset or using explanatory variables.

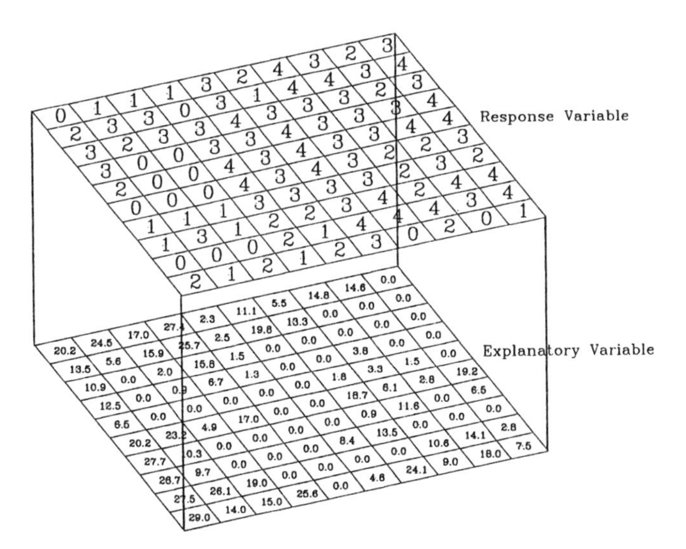

Figure 45-1: Example of raster data where the explanatory variable fills the grid but the response variable only sparsely fills the grid.

Since ecology is the relationship of organisms to their environment (McIntosh, 1985), historically ecologists have been concerned with the response of organisms to environmental (explanatory) factors (Whittaker, 1975, p. 111), which is the niche concept (Pianka, 1978, p. 237). Linear and nonlinear models of organismal response to explanatory, environmental factors can be built, and in general this approach will be called regression. Regression is appealing for spatial prediction because it uses information on the relationship of organisms to their environment, that is, the explanatory variables.

Optimal linear spatial prediction, or kriging, is another approach for predicting the response variable at unsampled locations. Kriging has been described in many places (for example, Matheron, 1963; Journel and Huijbregts, 1978; Cressie, 1991). For spatial prediction, kriging can be thought of simply as an optimally weighted average of the surrounding data for that same response variable. Kriging is less satisfactory to ecologists since it does not use relationships of the organism to its environment; rather, it only uses the idea that locations close together seem to be more similar.

The approach I will feature is universal kriging (Matheron, 1969; Huijbregts and Matheron, 1971; Cressie 1991). Universal kriging combines aspects of regression and kriging; so it retains the usefulness of relating an organism to its environment, together with the idea that locations in close proximity tend to be more similar. Some of the models given here are also described by Cressie in this volume, but there are notational differences so that comparisons can be made between regression, kriging, and universal kriging.

PREDICTION MODELS

Denote the response variable as a vector. For example, in Figure 45-1,

$$\mathbf{z} = \begin{bmatrix} z(\mathbf{s}_1') \\ z(\mathbf{s}_2') \\ \cdot \\ \cdot \\ \cdot \\ z(\mathbf{s}_n') \end{bmatrix} = \begin{bmatrix} z(1,3) \\ z(1,9) \\ \cdot \\ \cdot \\ \cdot \\ z(10,10) \end{bmatrix} = \begin{bmatrix} 1 \\ 2 \\ \cdot \\ \cdot \\ \cdot \\ 1 \end{bmatrix}$$

where $\mathbf{s}_t=(i,j)'$ is the spatial location; the ith row and jth column in the raster grid, indexed by t; $t=1,...,n$. The columns of **X** contain explanatory variables. For example, in Figure 45-1,

$$\mathbf{X} = \begin{bmatrix} [\mathbf{x}(\mathbf{s}_1')]' \\ [\mathbf{x}(\mathbf{s}_2')]' \\ \cdot \\ \cdot \\ \cdot \\ [\mathbf{x}(\mathbf{s}_n')]' \end{bmatrix} = \begin{bmatrix} 1 & x(1,3) \\ 1 & x(1,9) \\ \cdot & \cdot \\ \cdot & \cdot \\ \cdot & \cdot \\ 1 & x(10,10) \end{bmatrix} = \begin{bmatrix} 1 & 17.0 \\ 1 & 14.6 \\ \cdot & \cdot \\ \cdot & \cdot \\ \cdot & \cdot \\ 1 & 7.5 \end{bmatrix}.$$

The vector β contains parameters. For the example here, $\beta=[\beta_0,\beta_1]'$, where β_0 is a parameter for an overall mean and β_1 is a coefficient for the explanatory variable. Then, a linear regression model can be written as,

$$\mathbf{z} = \mathbf{X}\beta + \delta, \quad (45\text{-}1)$$

where δ is assumed to be a zero mean vector of independent random errors (var$[\delta]=\sigma^2\mathbf{I}$; **I** is the identity matrix). After selecting a regression model, with the appropriate explanatory variables included in **X**, the parameters β can be estimated with ordinary least squares, $\hat{\beta} = (\mathbf{X}'\mathbf{X})^{-1}\mathbf{X}'\mathbf{z}$. Let $\mathbf{x}(\mathbf{s}_0')$ be a vector containing the values of the explanatory variables at some location to be predicted, $\mathbf{s}_0$. For example, from Figure 45-1, $\mathbf{x}(\mathbf{s}_0')$ might be $\mathbf{x}(1,1) = [1,20.2]'$. Then the regression predictor $\hat{p}_r(\mathbf{s}_0')$ for the response variable z(1,1) at $\mathbf{s}_0$ is simply

$$\hat{p}_r(\mathbf{s}_0') = [\,\mathbf{x}(\mathbf{s}_0')]\,'\hat{\beta} = [\,\mathbf{x}(\mathbf{s}_0')]\,'(\mathbf{X}'\mathbf{X})^{-1}\mathbf{X}'\mathbf{z}. \quad (45\text{-}2)$$

The advantage of statistical models is that prediction variances can also be obtained:

$$\sigma^2_{\hat{p}_r(\mathbf{s}_0')} = \sigma^2 + \sigma^2[\,\mathbf{x}(\mathbf{s}_0')]\,'(\mathbf{X}'\mathbf{X})^{-1}\mathbf{x}(\mathbf{s}_0'), \quad (45\text{-}3)$$

where σ^2 is estimated by,

$$\hat{\sigma}^2 = \frac{(\mathbf{z}-\mathbf{X}\hat{\beta})\,'(\mathbf{z}-\mathbf{X}\hat{\beta})}{n-\text{rank}(\mathbf{X})}.$$

Notice that in regression, prediction depends primarily on the explanatory variables at $\mathbf{s}_0$.

Kriging, on the other hand, uses only the surrounding data for the response variable; denote the kriging predictor $\hat{p}_k(\mathbf{s}_0') = [\,\mathbf{a}(\mathbf{s}_0')]\,'\mathbf{z}$. In terms of the linear model (45-2), kriging usually assumes a constant mean,

$$\mathbf{z} = \mathbf{1}\mu + \delta, \quad (45\text{-}4)$$

where δ is a vector whose statistical dependence is described by a variogram model. That is, Var$[\delta(i,j)-\delta(u,v)]=2\gamma(h)$, where $h=[(i-u)^2+(j-v)^2]^{0.5}$; in words, the variance of the difference between $\delta(i,j)$ and $\delta(u,v)$ depends only on their distance apart. The function $2\gamma(h)$ is the variogram, and $\gamma(h)$ is called the semivariogram. The vector $[\mathbf{a}(\mathbf{s}_0')]$ contains the optimal weights for **z**, where optimal is defined by minimizing the mean-squared prediction error,

$$E[\hat{p}(\mathbf{s}_0') - z(\mathbf{s}_0')]^2;$$

where kriging uses a linear combination of the data; that is, minimize

$$E\{[\,\mathbf{a}(\mathbf{s}_0')]\,'\mathbf{z} - z(\mathbf{s}_0')\}^2.$$

A sufficient condition for the predictor to be uniformly unbiased over all β is $\mathbf{1}'[\mathbf{a}(\mathbf{s}_0')]=1$. For more complete treatments of kriging, see, for example, Journel and Huijbregts (1978) or Cressie (1991). The kriging equations may be written,

$$\begin{bmatrix} \Gamma & \mathbf{1} \\ \mathbf{1}' & 0 \end{bmatrix} \begin{bmatrix} [\mathbf{a}(\mathbf{s}_0')] \\ \lambda \end{bmatrix} = \begin{bmatrix} \gamma(\mathbf{s}_0') \\ 1 \end{bmatrix}, \quad (45\text{-}5)$$

where Γ is a matrix of all pairwise semivariogram values for the data **z**, and $\gamma(\mathbf{s}_0')$ is a vector of semivariogram values between the data **z** and the location to be predicted, $z(\mathbf{s}_0')$, and **1** is a vector of ones. A Lagrange multiplier, λ, occurs due to the unbiasedness conditions. Variograms indicate the spatial dependency in the data and must be estimated from the data. See Journel and Huijbregts (1978) or Cressie (1991) for various variogram models and Cressie (1985, 1991) for methods to estimate the variogram. The optimal weights are obtained by solving the kriging equations (45-5) for $[\mathbf{a}(\mathbf{s}_0')]$,

$$\hat{p}_k(\mathbf{s}_0') = [\mathbf{a}(\mathbf{s}_0')]'\mathbf{z} = \left([\gamma(\mathbf{s}_0')]' + \frac{1-[\gamma(\mathbf{s}_0')]'\Gamma^{-1}\mathbf{1}}{\mathbf{1}'\Gamma^{-1}\mathbf{1}}\mathbf{1}'\right)\Gamma^{-1}\mathbf{z}. \quad (45\text{-}6)$$

The prediction variance is given by,

$$\sigma^2_{\hat{p}_k}(\mathbf{s}_0') = [\gamma(\mathbf{s}_0')]'\Gamma^{-1}\gamma(\mathbf{s}_0') - \frac{[\mathbf{1}'\Gamma^{-1}\gamma(\mathbf{s}_0')-1]^2}{\mathbf{1}'\Gamma^{-1}\mathbf{1}} \quad (45\text{-}7)$$

Notice that kriging does not use information on explanatory variables.

Universal kriging is a generalization of kriging and regression; a good description is given by Cressie (1991).

For universal kriging, we assume the linear model (45-1), but in contrast to regression we also assume that δ has spatial dependence that is described by a variogram model, rather than independence among all $\delta(\cdot)$. This is related to the idea of kriging the residuals from regression. For example, let

$$\hat{\delta} = (\mathbf{z} - \mathbf{X}\hat{\beta})$$

be the residuals from regression. Then a predictor might be

$$[\mathbf{x}(s_0')]'\hat{\beta} + \hat{\delta}'[\mathbf{a}(s_0')],$$

where $\mathbf{a}(s_0')$ comes from the kriging equations using as the "data." This could be thought of as a kriging-adjusted regression predictor. However, it turns out that the optimal solution is obtained by minimizing the mean-squared prediction error and is given by the universal kriging equations. Let

$$\hat{p}_U(s_0) = [\mathbf{b}(s_0')]'\mathbf{z}$$

be the universal kriging predictor. The vector of optimal weights can be obtained by solving the universal kriging equations for $\mathbf{b}(s_0')$,

$$\begin{bmatrix} \mathbf{T} & \mathbf{X} \\ \mathbf{X}' & \mathbf{0} \end{bmatrix} \begin{bmatrix} [\mathbf{b}(s_0')] \\ \lambda \end{bmatrix} = \begin{bmatrix} \tau(s_0') \\ \mathbf{x}(s_0') \end{bmatrix}, \qquad (45\text{-}8)$$

where $\mathbf{X}$ and $\mathbf{x}(s_0')$ are as defined for the linear model (45-1), $\mathbf{T}$ is a matrix of all pairwise semivariogram values for the residuals, δ, in the linear model (45-1), and $\tau(s_0')$ is a vector of the semivariogram values between the residuals of the data and the residual at the location to be predicted. As a practical matter, the variogram is estimated from the residuals

$$\hat{\delta} = \varepsilon = (\mathbf{z} - \mathbf{X}\hat{\beta})$$

and the semivariogram is used in the universal kriging equations. If $\mathbf{X}=\mathbf{1}$ and $\mathbf{x}(s_0')=1$ (no explanatory variables), then the universal kriging equations are identical to the kriging equations; if all $\delta(\cdot)$ are independent ($\mathrm{Var}[\delta]=\sigma^2\mathbf{I}$), universal kriging is identical to regression prediction. The universal kriging predictor is

$$\hat{p}_U(s_0) = [\mathbf{b}(s_0')]'\mathbf{z} = [\tau(s_0')]'\mathbf{T}^{-1}\mathbf{z} + \left\{[\mathbf{x}(s_0')]' - [\tau(s_0')]'\mathbf{T}^{-1}\mathbf{X}\right\}(\mathbf{X}'\mathbf{T}^{-1}\mathbf{X})^{-1}\mathbf{X}'\mathbf{T}^{-1}\mathbf{z}. \qquad (45\text{-}9)$$

The prediction variance is given by

$$\sigma^2_{\hat{p}U}(s_0') = [\tau(s_0')]'\mathbf{T}^{-1}\tau(s_0') - [\mathbf{X}'\mathbf{T}^{-1}\tau(s_0') - \mathbf{x}(s_0')]'(\mathbf{X}'\mathbf{T}^{-1}\mathbf{X})^{-1}[\mathbf{X}'\mathbf{T}^{-1}\tau(s_0') - \mathbf{x}(s_0')]. \qquad (45\text{-}10)$$

CASE STUDY

To compare the prediction methods, regression, kriging, and universal kriging, a data set was used which contained all response variables (Figure 45-2). These data come from a glade in the Ozark area of southeastern Missouri. The response variable is a log-transformed cover value of an herbaceous plant species, *Coreopsis lanceolata*, and the explanatory variable is the square root of the total basal area of all woody plant species. Ecologically, the explanatory variable indicates a shading factor. Each cell in the grid represents a real square 7×7 m plot.

In this case, we know all of the values of the response and explanatory variables. To compare methods, the response value at each location was removed, one at a time, and predicted back with each method. Next, the difference between the true value and predicted value was squared, and then averaged over all 100 sites. This is called cross-validation for the mean-square prediction error (MSPE):

$$\mathrm{MSPE}(m) = \frac{1}{100}\sum_{t=1}^{100}\left[\hat{p}_m(s_t') - z(s_t')\right]^2, \qquad (45\text{-}11)$$

where m is the prediction method r, k, or u for regression, kriging, or universal kriging, respectively; and $\hat{p}_m(s_t)$ is the predictor for location s_t, where $z(s_t)$ was removed during cross-validation.

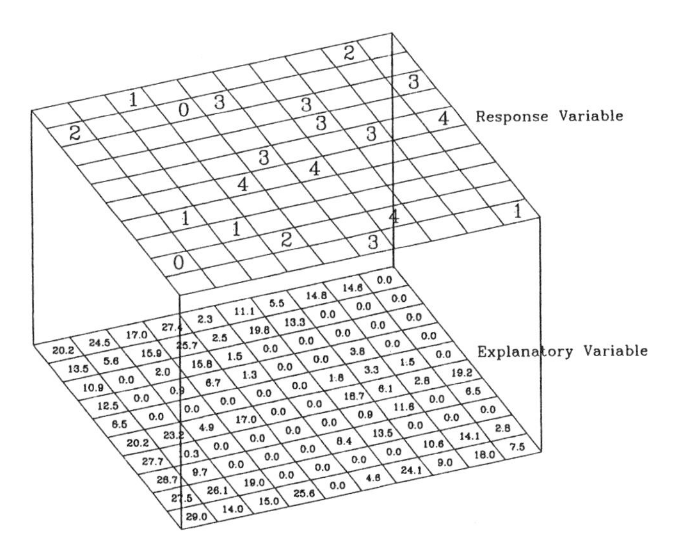

Figure 45-2: Data set used for case study. The response variable is log-transformed cover values for Coreopsis lanceolata in 7×7 m contiguous plots, and the explanatory variable is the square root of basal area for all woody species in the plot.

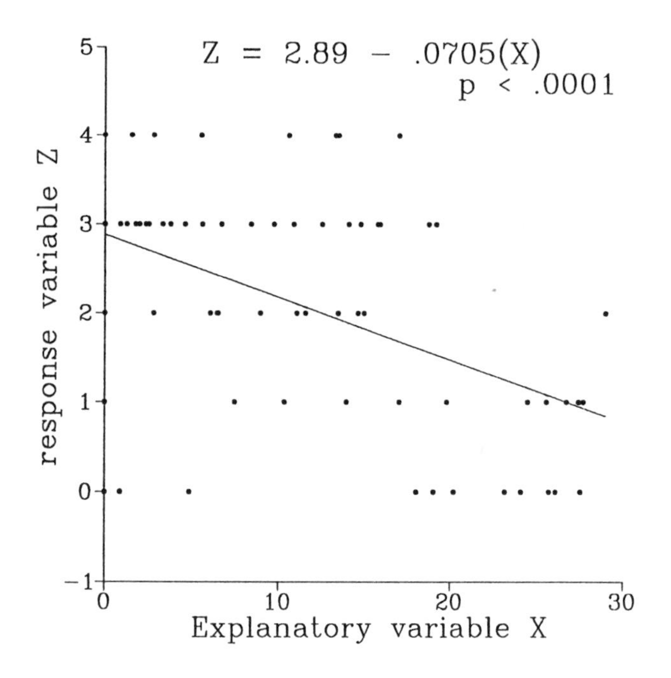

Figure 45-3: Linear regression model for all data in case study.

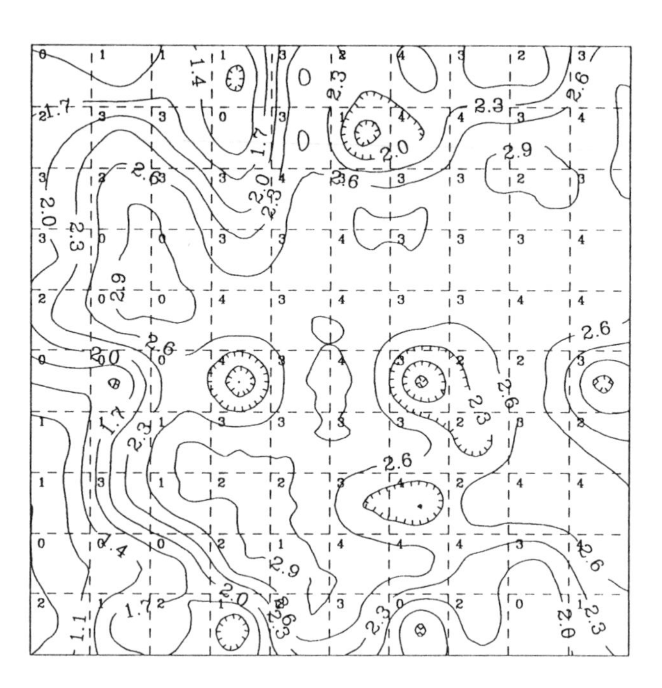

Figure 45-4: Regression predictions during cross-validation, given as isopleth lines. The plots are outlined by the dashed line, with the true value in the upper left-hand corner.

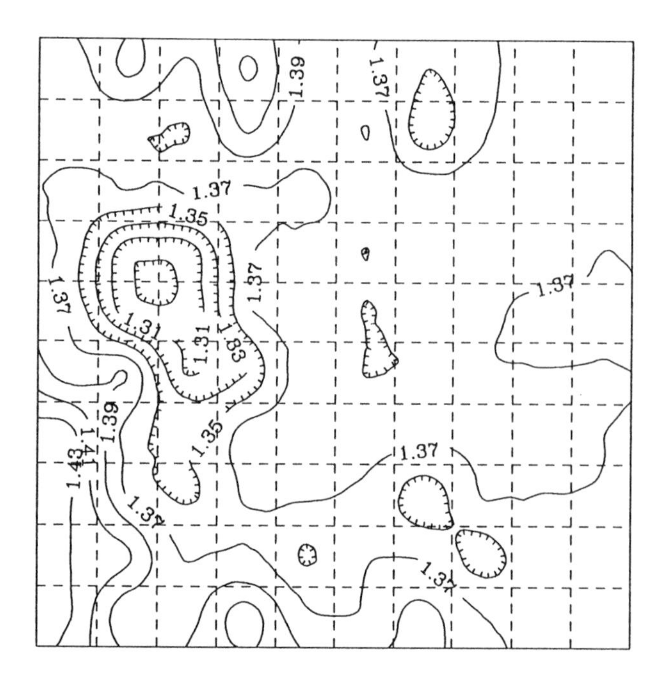

Figure 45-5: Regression prediction variances during cross-validation, given as isopleth lines. The plots are outlined by dashed lines.

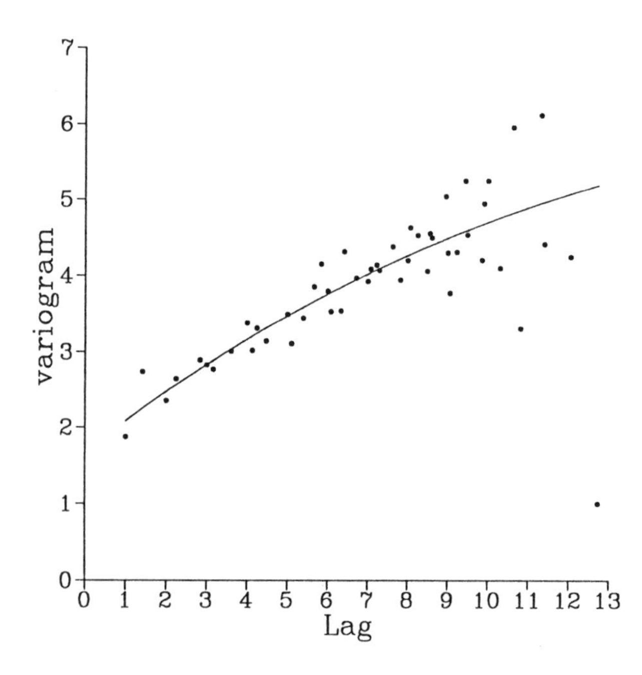

Figure 45-6: Empirical variogram (solid circles) and fit variogram (curve) for case study data.

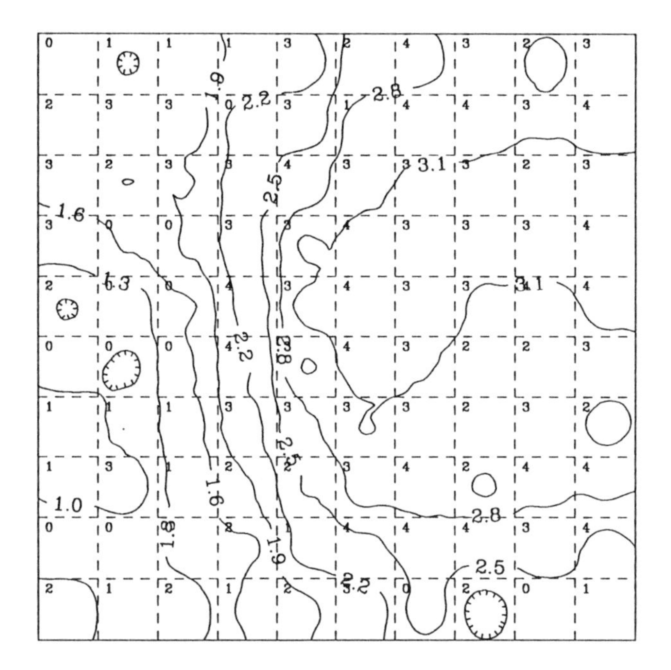

Figure 45-7: Kriging predictions during cross-validation, given as isopleth lines. The plots are outlined by the dashed line, with the true value in the upper left-hand corner.

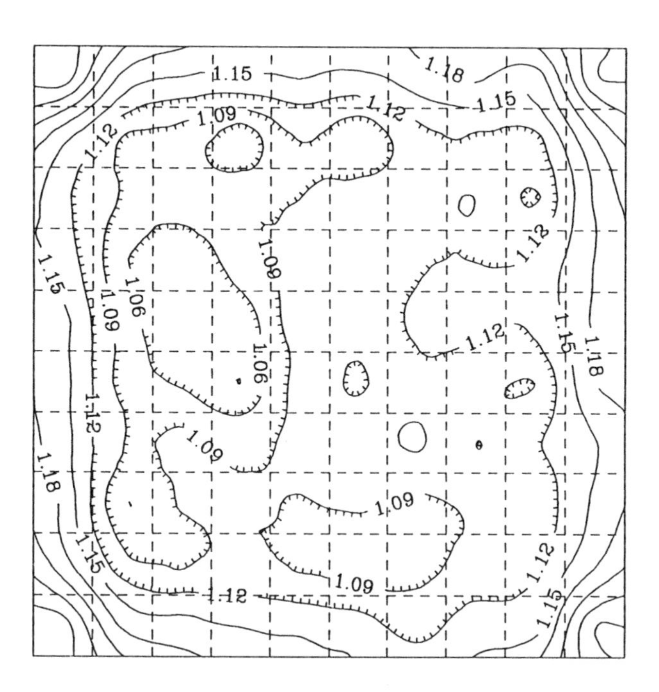

Figure 45-8: Kriging prediction variances during cross-validation, given as isopleth lines. The plots are outlined by the dashed line.

The regression model for all one hundred locations is given in Figure 45-3, and it shows that the response variable has a significant linear relationship to the explanatory variable. Each time a datum was removed during cross-validation, a new linear regression equation was fit and used in (45-2). Figure 45-4 shows the predicted values, using (45-2), as isopleth lines for all 100 locations, along with the true value in the upper left-hand corner of each cell. Figure 45-5 shows the prediction variances (45-3) calculated at each location during cross-validation. From (45-11) for regression, MSPE(r)=1.37.

Prediction using kriging depends on the variogram. The empirical variogram was calculated using the classical formula,

$$2\hat{\gamma}(h) = \frac{1}{n(H)} \sum_{H} [z(i,j) - z(u,v)]^2; \qquad (45\text{-}12)$$

where H is the set of pairs $\{[z(i,j), z(u,v)]; [(i-j)^2+(u-v)^2]^{0.5}=h\}$ and $n(H)$ is the number of distinct pairs in H. These values (45-12) are given as solid circles in Figure 45-6 for all of the data. An exponential variogram model was fit to the empirical values (45-12) using the SAS® nonlinear weighted least-squares procedure. The exponential model is

$$2\gamma(h) = c_0 + c_1[1 - \exp(-c_2 h)] \qquad (45\text{-}13)$$

and the estimated parameters were c_0=1.61, c_1=5.09, and c_2=0.092. The fitted model is given as a curve in Figure 45-6. Each time a datum was removed during cross-validation, a new variogram was calculated and used in the kriging predictor (45-6). Figure 45-7 shows the predicted values, using (45-6), as isopleth lines for all 100 locations, along with the true values in the upper left-hand corner of each cell. Figure 45-8 shows the prediction variances (45-7) calculated at each location during cross-validation. Notice that prediction variances for kriging are smaller toward the center and larger toward the edges. This result is well-known to kriging practitioners. From Eq. (45-11) for kriging, MSPE(k)=1.15, which is smaller than MSPE(r), making kriging a better predictor than regression for these data.

To use universal kriging, the empirical variogram (45-12) was calculated for the residuals from regression. These residuals are given as isopleth lines in Figure 45-9. The empirical variogram (solid circles) and the fitted exponential variogram (curve) for the residuals are given in Figure 45-10. For the residuals, the fitted parameters (45-13) were c_0=0.14, c_1=2.60, c_2=1.15. Each time a datum was removed during cross-validation, a new regression equation was estimated and a new variogram was calculated on the residuals and used in the universal kriging predictor (45-9). Figure 45-11 shows the predicted values, using Eq.

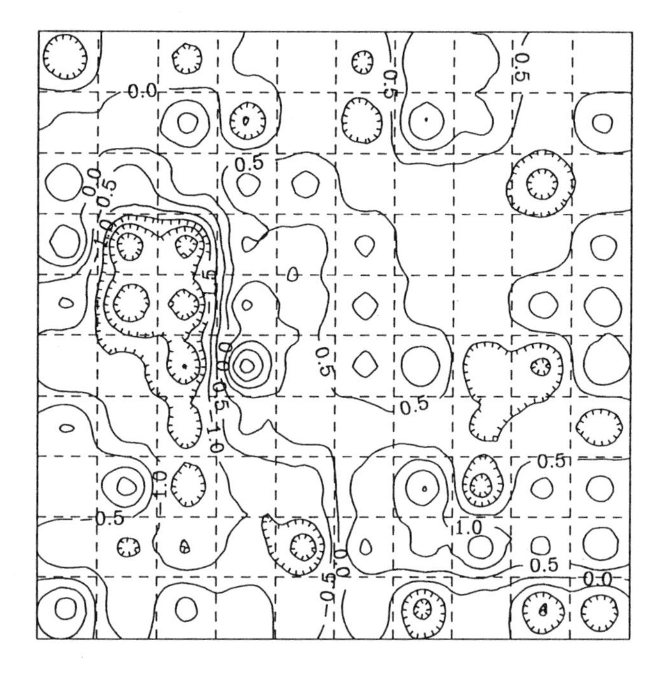

Figure 45-9: Regression residuals, given as isopleth lines. The plots are outlined by the dashed line.

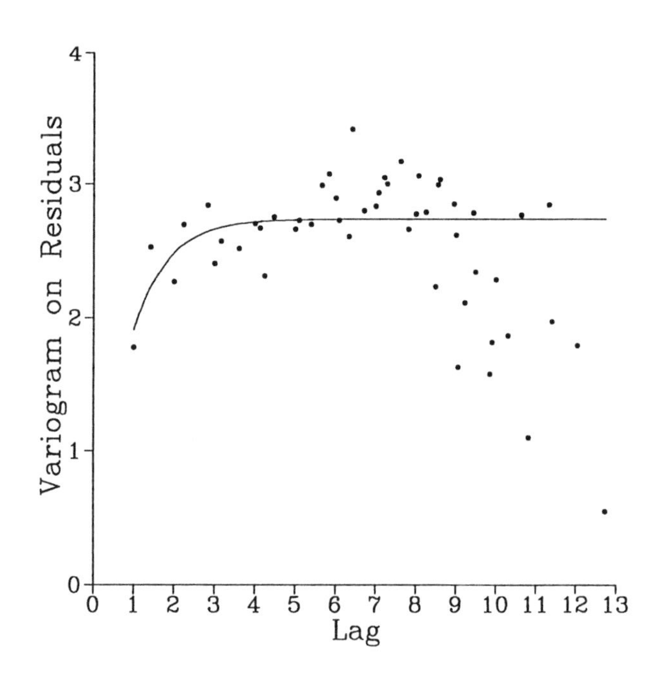

Figure 45-10: Empirical varigram (solid circles) and fit variogram (curve) for residuals in Figure 45-9.

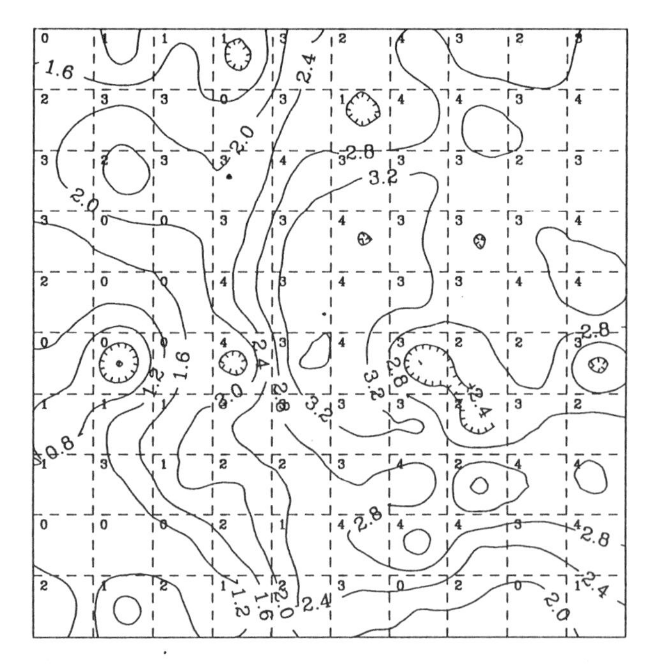

Figure 45-11: Universal kriging predictions during cross-validation, given as isopleth lines. The plots are outlined by the dashed line, with the true value in the upper left-hand corner.

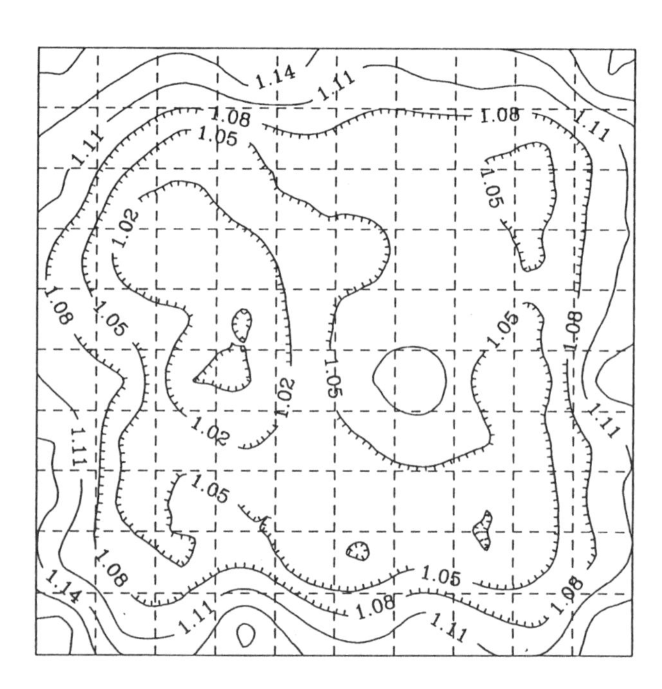

Figure 45-12: Universal kriging prediction variances during cross-validation, given as isopleth lines. The plots are outlined by the dashed line.

(45-9), as isopleth lines for all 100 locations, along with the true value in the upper left-hand corner of each cell. Figure 45-12 shows the prediction variances (45-10) calculated at each location during cross-validation. From Eq. (45-11) for universal kriging, MSPE(u)=0.98, making universal kriging a better predictor than both regression and kriging for these data.

By comparing Figures 45-4 and 45-7, it is clear that regression prediction differs from kriging. And although here kriging is a better predictor than regression, there appears to be an interesting ecological relationship between the shading factor and the spatial abundance of *Coreopsis lanceolata*. Universal kriging makes use of that relationship in providing the best predictive model.

CONCLUSIONS

Universal kriging is ideally suited for predicting missing values of a response variable in a GIS when there are explanatory variables. It has the appealing property of using two sources of information. First, as in regression, it uses the relationship between the response variable and explanatory variable, which is appealing to an ecologist. Second, it uses the fact that the residuals may have spatial information left in them, such as residuals that are closer together may tend to be more similar. Hence, universal kriging provides the best predictive model of these three methods, and it also incorporates some understanding of ecological relationships through regression.

Other data sets have been analyzed, ranging from the case where no explanatory variables affected the response variable, to the case where there was no apparent spatial dependence in the residuals after the effect of the explanatory variables. In all cases, universal kriging did as well as or better than regression or kriging.

Computer packages that perform universal kriging are scarce, to be found only in some specialized geostatistical software. It is hoped that in the future the joining of environmental modeling and the GIS framework will contain universal kriging.

ACKNOWLEDGMENTS

This research was funded in part by the following: Federal Aid in Wildlife Restoration and the Alaska Department of Fish and Game, the Whitehall Foundation, the North Atlantic Treaty Organization, the National Park Service, and the National Science Foundation, grant number DMS-9001862.

REFERENCES

Cressie, N. (1985) Fitting variogram models by weighted least squares. *Journal of the International Association for Mathematical Geology* 17: 563–586.

Cressie, N. (1991) *Statistics for Spatial Data*, New York: John Wiley and Sons.

Huijbregts, C.J., and Matheron, G. (1971) Universal kriging (an optimal method for estimating and contouring in trend surface analysis). In McGerrigle, J.I. (ed.) *Proceedings of Ninth International Symposium on Techniques for Decision-Making in the Mineral Industry. The Canadian Institute of Mining and Metallurgy, Special Volume* 12: 159–169.

Journel, A.G., and Huijbregts, C.J. (1978) *Mining Geostatistics*, London: Academic Press.

Matheron, G. (1963) Principles of geostatistics. *Economic Geology* 58: 1246–1266.

Matheron, G. (1969) *Le Krigeage Universal*, Fontainebleau: Cahiers du Centre de Morphologie Mathematique, No. 1.

McIntosh, R.P. (1985) *The Background of Ecology, Concept and Theory*, Cambridge: Cambridge University Press.

Pianka, E.R. (1978) *Evolutionary Ecology*, New York: Harper and Row.

Whittaker, R.H. (1975) *Communities and Ecosystems*, New York: MacMillan Publishing Company.

46

Discrete Space Autoregressive Models

LUC ANSELIN

What is special about spatial data?

Empirical work in environmental modeling is often based on data for which the location of the observations is an important attribute. This type of data can be referred to as *spatial data*. Clearly, it is ideally suited for incorporation into a geographic information system (GIS). The observations consist of values for a cross-section of spatial units, such as points, regular grids, or irregular polygons, either for one time period (a so-called pure cross-section) or for several time periods (a time series of cross-sections, or panel data). Following Cressie (1991), the statistical analysis of spatial data can be approached from three related but distinct perspectives, respectively, focusing on *geostatistical data*, *point patterns*, and *lattice data*. In this chapter, I will focus on the latter, for which the units of observation form a set of discrete locations, either associated with a regular lattice or grid structure (as in a raster-based GIS), or with an irregular lattice or polygon structure (as in a vector-based GIS). The distinctive characteristic of the statistical analysis of such spatial data is that the spatial association between values observed at different locations (spatial dependence) and the systematic variation of phenomena by location (spatial heterogeneity) become the major focus of inquiry. This viewpoint is taken in a large body of the spatial statistics and spatial econometrics literatures, as reviewed in, for example, Cliff and Ord (1981), Upton and Fingleton (1985), Anselin (1988a), Griffith (1988a), Ripley (1988), and Haining (1990). Due to constraints on the length of this chapter, I will limit the discussion to the treatment of *spatial dependence*, or *spatial autocorrelation*. Spatial heterogeneity (or nonstationarity) will not be covered (for reviews or relevant issues, see Anselin, 1988a, 1990a).

The remainder of this chapter consists of three sections. First, I introduce the notion of spatial autocorrelation more formally and outline a number of tests for its presence in a data set or GIS data layer. In the second section, I review some methodological issues that arise when regression analysis is carried out for cross-sectional data, in which spatial dependence is present. I refer to this as *spatial regression analysis*. This is the problem encountered when a formal association between different data layers in a GIS is tested in the form of a regression model, where one dependent variable (data layer) is related to a number of explanatory variables (data layers). The presence of spatial dependence in the data may in many instances invalidate the properties of standard regression analysis. Consequently, specialized diagnostic tests and estimation methods will need to be implemented. A number of these tests as well as estimation techniques that are based on the maximum likelihood (ML) principle are reviewed. The description in these first two sections is based in part on materials in Anselin (1991) and O'Loughlin and Anselin (1992). The treatment of spatial regression is closed with a brief discussion of an alternative approach to this problem, based on techniques that are formally similar to those developed in time series analysis. In the concluding section, I outline some ideas on the integration between a GIS and spatial regression analysis.

Throughout, I will illustrate the various methods with a simple empirical example that uses a small subset of the Global Change Database (NOAA, 1990). The sample dataset consists of 7×7 square raster grids with 10 arcminute spacing, for a total of 49 observations, roughly situated around the border between the Central African Republic, Sudan, and Zaire. Four variables are used: GREEN, TEMP, ELEV, and PREC. Their definitions are given in Table 46-1, and the actual data (as well as the results of some transformations, to be discussed below) are listed in Table 46-2. Their spatial layout is graphically illustrated in Figure 46-1. It should be noted that the

Table 46-1: List of variables

GREEN	greenness vegetation index, based on AVHRR
TEMP	temperature in 1/10 degrees celsius
ELEV	modal elevation in meters
PREC	precipitation (mm per year)

Table 46-2: Data for empirical illustrations

N	GREEN	W_GREEN	TEMP	ELEV	PREC	RESID
1	126	131.00	277	730	107	–4.27
2	130	128.33	274	670	106	2.64
3	129	127.33	271	610	104	4.07
4	126	126.33	271	610	104	1.07
5	125	125.00	276	610	89	–2.46
6	125	126.33	281	610	74	–4.99
7	122	128.50	281	670	74	–14.18
8	132	129.33	271	730	114	–0.51
9	130	130.75	268	670	115	1.36
10	126	130.75	265	610	116	1.23
11	125	129.00	265	580	116	3.33
12	124	128.00	266	610	109	–3.20
13	132	126.00	268	640	102	0.22
14	132	128.33	268	670	102	–2.88
15	132	135.67	264	790	121	–9.88
16	135	135.25	261	700	124	1.05
17	139	136.50	258	670	127	6.78
18	140	136.50	258	640	127	10.88
19	130	129.50	257	610	129	4.00
20	123	126.25	255	670	130	–10.58
21	131	121.67	255	670	130	–2.58
22	140	140.67	264	730	121	4.31
23	140	141.00	261	700	124	6.05
24	145	143.25	258	700	127	9.69
25	152	140.25	258	700	127	16.69
26	131	132.00	257	670	129	–1.19
27	112	118.25	255	730	130	–27.77
28	110	116.67	255	730	130	–29.77
29	150	144.33	263	760	123	11.24
30	144	143.00	261	760	125	4.33
31	142	145.75	258	730	128	4.07
32	145	142.25	258	730	128	7.07
33	134	129.50	257	730	136	–1.03
34	109	121.25	256	730	144	–23.12
35	107	112.00	256	760	144	–28.22
36	149	145.67	262	760	124	9.79
37	140	149.25	260	760	127	0.36
38	149	142.50	258	760	129	8.45
39	141	144.75	258	760	129	0.45
40	133	138.75	257	760	143	–1.76
41	132	124.50	256	760	157	3.02
42	117	121.67	256	760	157	–11.98
43	147	152.00	262	760	124	7.79
44	155	144.67	260	760	127	15.36
45	147	152.00	258	760	129	6.45
46	152	145.33	258	760	129	11.45
47	148	141.33	257	760	143	13.24
48	139	135.33	256	790	157	6.92
49	126	128.00	256	760	157	–2.98

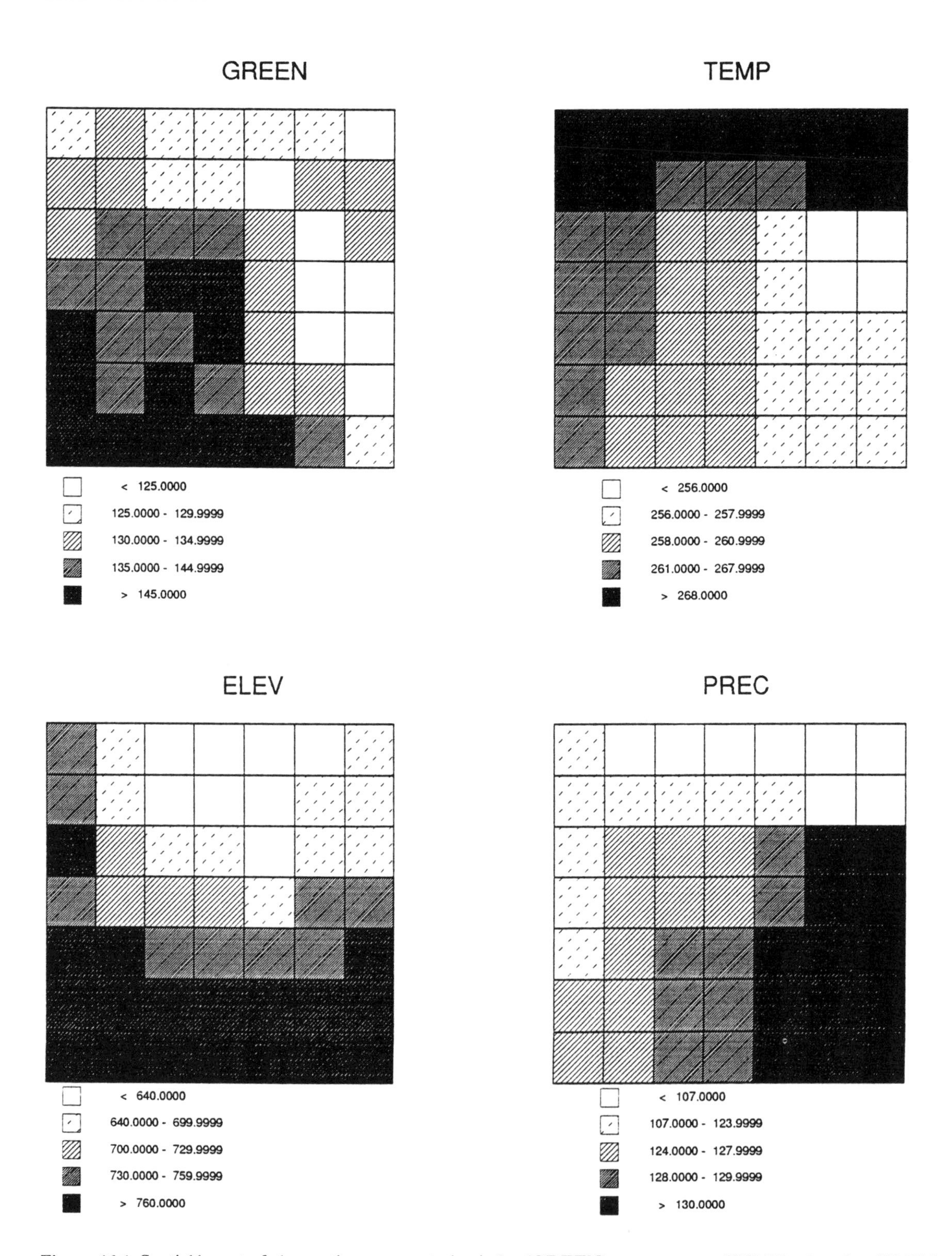

Figure 46-1: Spatial layout of observations on vegetation index (GREEN), temperature (TEMP), elevation (ELEV), and precipitation (PREC).

example is purely illustrative and is not intended to make a contribution to the substantive understanding of the interaction between the included variables. All computations are carried out by means of the *SpaceStat* software package for the analysis of spatial data (Anselin, 1992).

SPATIAL AUTOCORRELATION

Concepts

A basic tenet underlying the methodology of discrete space autoregressive modeling (but also much of spatial statistics and geostatistics) is the proposition that values at close-by locations are more correlated than values at locations that are far apart. This inverse relation between value association and distance has also been called the *first law of geography* (Tobler, 1979). The resulting clustering in space of similar magnitudes implies that values at one location are in part determined by the values at neighboring locations. In other words, the spatial clustering or positive spatial autocorrelation results in a loss of information relative to observations that would be independent. Much of the methodology of spatial statistics deals with detecting significant patterns of spatial clustering and with correcting for the resulting loss of information.

A crucial aspect of defining spatial autocorrelation is the determination of the relevant "neighborhood," that is, those locations surrounding a given data point that would be considered to "influence" the observation at that data point. Unfortunately, the determination of the set of "neighbors" is not without a certain degree of arbitrariness. Formally, the membership of observations in the neighborhood set for each location is expressed by means of a square contiguity or spatial weights matrix (W), of dimension equal to the number of observations (N), in which each row and matching column correspond to an observation pair. The elements w_{ij} of this matrix take on a nonzero value (typically, 1, for a binary matrix) when locations (observations) i and j are considered to be neighbors, and a zero value otherwise. By convention, the diagonal elements of the weights matrix, w_{ii}, are set to zero. Commonly, the weights matrix reflects simple contiguity, that is, the fact that two areal units (grid cells, polygons) share a common border. Contiguity is also often defined as a function of the distance between spatial units of observation, or their centroids (or other "meaningful" points that are not centroids, but are uniquely associated with each polygon). In this sense, two units are considered to be contiguous if the distance between them falls within a chosen range. This is similar to the approach taken to compute a semivariogram in geostatistics, where the distance class between observations is used to estimate the relation between value association (covariance, correlation) and distance. In essence, the spatial weights matrix summarizes the topology of the data set in graph-theoretic terms (nodes and links). Higher-order contiguity is defined in a recursive manner, in the sense that a unit is considered to be contiguous of a higher order to a given location if it is first-order contiguous to a unit that is contiguous of the next lower order (and not already contiguous of a lower order, to avoid circularity). For example, units that are considered to be second-order contiguous to a location are first-order contiguous to the first-order contiguous ones. Higher-order contiguity thus results in bands of observations around a given location being included in the neighborhood set, at increasing distances.

Clearly, a large number of weights matrices can be derived for the same spatial layout or map. The fact that the choice of the weights matrix is an important determinant of the results of any spatial statistical analysis should always be kept in mind. As a practical matter, it is often useful to experiment with a few different matrices, unless there is a compelling reason on theoretical grounds to consider only a single one. For example, for the grid cells of the data set used in the empirical examples (shown in Figure 46-1) three different types of first-order contiguity can be considered, depending on whether the commonality of edges (rook criterion), of vertices (bishop criterion), or of both is taken into account (queen criterion). In Figure 46-2, the binary weights matrix is given for the rook criterion applied to the 49 observations in the 7×7 square grid. Note the characteristic sparseness of the matrix (many more zero elements than nonzero elements) and the band pattern exhibited by the ones.

In general, the spatial weights matrix does not have to be binary, but can take on any value that reflects the "potential" interaction between spatial units i and j, for example, based on inverse distance or inverse distance squared, as in a gravity model of spatial interaction, or computed from the relative length of the shared border (for an extensive discussion, see Cliff and Ord, 1981). For ease of interpretation, the weights matrix is often standardized such that the elements of a row sum to one. In other words, each element w_{ij} of W is divided by its row sum, $\Sigma_j w_{ij}$ (with the summation over all columns j). Since the row sums in row i (for w_{ij}) and row j (for w_{ji}) are not necessarily the same, the resulting row-standardized matrix is no longer symmetric, even though the original binary contiguity matrix was. For such a square $N \times N$ row-standardized weights matrix, W, the product with an $N \times 1$ vector of observations has an intuitive interpretation. If the observations are represented as a vector x, the resulting product would be Wx. Since each element of the product Wx equals $\Sigma_j w_{ij} x_j$, Wx is in fact a vector of weighted averages of neighboring values. This operation and the associated variable are often referred to as a *spatial lag*, similar to the terminology used in time series analysis. However, strictly speaking, the analogy with a lag in time series analysis is not valid, since the latter is an actual shift in time, whereas in a spatial lag there is no shift, only an averaging of neighboring values.

To illustrate this concept, consider the first row of the weights matrix in Figure 46-2. Following the rook criterion, the nonzero elements in the second and eighth column of the first row indicate that observations 2 and 8 are first-order contiguous to the first spatial unit (the cells to the right and below the upper left corner cell in the maps of Figure 46-1). For example, in order to obtain a spatial lag for the first observation on the variable GREEN (the greenness vegetation index), the observations associated with its neighbors (respectively, 130 and 132) are averaged (since two contiguities results in each receiving the weight 0.50), yielding a value of 131 (shown as the first observation on the variable W_GREEN in Table 46-2).

```
0100000100000000000000000000000000000000000000000
1010000010000000000000000000000000000000000000000
0101000001000000000000000000000000000000000000000
0010100000100000000000000000000000000000000000000
0001010000010000000000000000000000000000000000000
0000101000001000000000000000000000000000000000000
0000010000000100000000000000000000000000000000000
1000000010000010000000000000000000000000000000000
0100000101000001000000000000000000000000000000000
0010000010100000100000000000000000000000000000000
0001000001010000010000000000000000000000000000000
0000100000101000001000000000000000000000000000000
0000010000010100000100000000000000000000000000000
0000001000001000000010000000000000000000000000000
0000000100000001000001000000000000000000000000000
0000000010000010100000100000000000000000000000000
0000000001000001010000010000000000000000000000000
0000000000100000101000001000000000000000000000000
0000000000010000010100000100000000000000000000000
0000000000001000001010000010000000000000000000000
0000000000000100000100000001000000000000000000000
0000000000000010000000100000100000000000000000000
0000000000000001000001010000010000000000000000000
0000000000000000100000101000001000000000000000000
0000000000000000010000010100000100000000000000000
0000000000000000001000001010000010000000000000000
0000000000000000000100000101000001000000000000000
0000000000000000000010000010000000100000000000000
0000000000000000000001000000010000010000000000000
0000000000000000000000100000101000001000000000000
0000000000000000000000010000010100000100000000000
0000000000000000000000001000001010000010000000000
0000000000000000000000000100000101000001000000000
0000000000000000000000000010000010100000100000000
0000000000000000000000000001000001000000010000000
0000000000000000000000000000100000001000001000000
0000000000000000000000000000010000010100000100000
0000000000000000000000000000001000001010000010000
0000000000000000000000000000000100000101000001000
0000000000000000000000000000000010000010100000100
0000000000000000000000000000000001000001010000010
0000000000000000000000000000000000100000100000001
0000000000000000000000000000000000010000000100000
0000000000000000000000000000000000001000001010000
0000000000000000000000000000000000000100000101000
0000000000000000000000000000000000000010000010100
0000000000000000000000000000000000000001000001010
0000000000000000000000000000000000000000100000101
0000000000000000000000000000000000000000010000010
```

Figure 46-2: Spatial weights matrix based on rook criterion (row and column sequence corresponds to data points).

Measures of spatial autocorrelation

When a spatial layout of values is observed as in Figure 46-1, the visual interpretation of pattern or spatial clustering only provides a qualitative and intuitive insight. In order to quantify this information, a number of measures of spatial autocorrelation have been suggested. The common approach in these measures is to assess the extent to which the observed spatial arrangement of values is indeed special, or instead could be the result of a random assignment. Measures of spatial autocorrelation are derived from the null hypothesis that space does not matter, or that the assignment of values to particular locations is irrelevant. Under this null hypothesis, it is the values only that provide insight, and all allocations of these values to locations are equally likely. In contrast, under the alternative hypothesis of *spatial autocorrelation* (spatial dependence, spatial association), the interest focuses on instances where large values are surrounded by other large values in neighboring locations, or small values are surrounded by small values, or large values are surrounded by small values (and vice versa). The former two are referred to as positive spatial autocorrelation, the latter as negative spatial autocorrelation. Whereas positive spatial autocorrelation implies a spatial clustering of similar values, negative spatial autocorrelation is a qualitatively different concept, in that it implies a checkerboard pattern of values.

Tests for spatial autocorrelation in cross-sectional observations on a single variable (a single data layer) are based on the magnitude of an indicator that combines the value observed at each location with the values at neighboring locations. In essence, the tests are measures of the similarity between association in value (covariance or correlation) and association in space (contiguity). Spatial autocorrelation is taken to be present when a so-called spatial autocorrelation statistic computed for a particular map pattern takes on an extreme value, compared to what would be expected under the null hypothesis of no spatial association. Clearly, exactly what is considered to be extreme depends on the distribution of the test statistic under the null hypothesis, and on the chosen level of the Type I error (or, the critical value for a given significance level). Analytically, the distribution under the null hypothesis can be based on an underlying uncorrelated normal distribution or on a so-called randomization assumption. For the latter, normality is not needed, but each observation is assumed to occur equally likely at all locations (or, space does not matter).

Probably the two most commonly used measures for spatial autocorrelation are Moran's I statistic and Geary's c statistic (see Cliff and Ord, 1973, for an extensive discussion). Both indicate the degree of spatial association as summarized for the whole data set. While Moran's I is based on cross products to measure value association, Geary's c uses squared differences. Formally, Moran's I for N observations on a variable x (with observations x_i at location i) is expressed as:

$$I = (N/S_0)\Sigma_i\Sigma_j w_{ij}(x_i - \mu)(x_j - \mu)/\Sigma_i(x_i - \mu)^2$$

where μ is the mean of the x variable, w_{ij} are the elements of the spatial weights matrix, and S_0 is the sum of the elements of the weights matrix: $S_0 = \Sigma_{ij}w_{ij}$. Geary's c statistic is expressed as:

$$c = (N-1)/2S_0 \{\Sigma_i\Sigma_j w_{ij}(x_i - x_j)^2 / \Sigma_i(x_i - \mu)^2\}$$

in the same notation as previously.

Inference for these statistics is carried out by computing a so-called z value or standard deviate, which is obtained by subtracting the expected value and dividing by the standard deviation of the statistic under the null hypothesis, either normality or randomization (the expression for the moments of the statistic are given in Cliff and Ord, 1973, 1981). The resulting z values can then be compared to a table of standard normal variates to assess significance. A value of Moran's I that is larger than its theoretical mean of $-1/(N-1)$, or, equivalently, a positive associated z value, points to positive spatial autocorrelation. In contrast, for Geary's c, positive spatial autocorrelation is indicated by a value smaller than its mean of 1, or by a negative z value. Both Moran's I and Geary's c have been shown to be special cases of a general cross-product statistic that indicates the association between two matrices of similarity for a set of objects. This general framework and the associated gamma statistic have found widespread adoption to measure spatial autocorrelation in contexts where the standard assumptions do not hold (for technical details and applications, see Hubert et al., 1981, 1985; Hubert and Golledge, 1982; Hubert, 1985, 1987).

A slightly different approach can be based on the so-called distance statistics recently suggested by Getis and Ord (1992). These statistics are computed by defining a set of neighbors for each location as those observations that fall within a critical distance (d) from the location. Consequently, for different distance measures, a different set of neighbors will be found. This can be formally expressed in a set of symmetric binary weights matrices, $W(d)$, with elements $w_{ij}(d)$ indexed by distance d. For each distance d, the elements $w_{ij}(d)$ of the corresponding weights matrix $W(d)$ equal one if the distance $d_{ij} < d$. A summary measure of spatial association for a given distance d is the general statistic, $G(d)$, defined as:

$$G(d) = \Sigma_i\Sigma_j w_{ij}(d)x_i x_j / \Sigma_i\Sigma_j x_i x_j$$

in the same notation as used previously. As shown in Getis and Ord (1992), this measure is only applicable for positive variables. Similar to the approach taken for Moran's I and Geary's c, a standard z value can be computed from the $G(d)$, its mean and standard deviation, and significance can be assessed from a standard normal table (see Getis and Ord, 1992, for technical details). As argued by Getis and Ord (1992), the interpretation of spatial association given by the $G(d)$ measure is different from Moran's I or Geary's c. A positive and significant value indicates spatial clustering of high values, while a negative and significant value indicates spatial clustering of small values. This is in contrast to the "traditional" interpretation of spatial association, where both instances would be considered to be positive spatial autocorrelation.

An additional feature of the G statistics is the ability to compute a measure of spatial association for each individual spatial unit. For each observation, i, the G_i and G_i^* statistics indicate the extent to which that location is surrounded by high values or low values for the variable under consideration. Formally, the G_i measure for observation i can be expressed as:

$$G_i(d) = \Sigma_j w_{ij}(d)x_j / \Sigma_j x_j$$

with the summation in j exclusive of i, and the other notation as given. The G_i^* measure is formally identical:

$$G_i^*(d) = \Sigma_j w_{ij}(d)x_j / \Sigma_j x_j$$

except that the summation over j is now inclusive of i (i.e., includes the value of x for the observation for which the statistic is computed). In other words, the G_i statistic should be interpreted as a measure of clustering of like values around a particular observation, irrespective of the value at that location. In contrast, the G_i^* statistic includes the value at the location within the measure of clustering.

To illustrate these concepts, Moran's I, Geary's c, and the G statistic are listed in Table 46-3, for the four variables in the data set, and using first order contiguity according to the rook criterion. Note that the weights matrix for these statistics is row-standardized for Moran's I and Geary's c, but unstandardized (symmetric, as in Figure 46-2) for G. Also, for I and c, the computations are in deviations from the mean, while for G, the raw values are used. Both Moran's I and Geary's c provide a strong indication of significant positive spatial autocorrelation in all four variables. Using the G statistic, the evidence is much weaker, and only for PREC is strong positive asso-

Table 46-3: Measures of spatial association (z values in parentheses, based on randomization assumption)

Variable	I	c	G
GREEN	0.731	0.257	0.0720
	(7.01)	(−7.03)	(1.06)
TEMP	0.791	0.173	0.0712
	(7.65)	(−7.88)	(−0.64)
ELEV	0.842	0.159	0.0714
	(8.00)	(−7.92)	(−0.06)
PREC	0.839	0.136	0.0733
	(8.16)	(−8.27)	(2.70)

ciation indicated, that is, a spatial clustering of large values. Even though it may seem contradictory, the lack of significance of the G statistic for the other variables can be consistent with strong spatial autocorrelation indicated by the I and c statistics. This is because the G statistic does and the I and c do not distinguish between clusters of high values and low values: Both are considered to be evidence of spatial clustering for the latter, but only one of them is for the former. Also, the G statistic may well have poor power to detect spatial clustering of small values (Cressie, personal communication). A closer look at which locations show the strongest degree of spatial clustering around them is given in Tables 46-4 and 46-5, and graphically illustrated in Figure 46-3 (for G_i^* only). For both positive and negative values, the five locations with the most extreme G_i and G_i^* statistics are listed in the tables. For the positive extremes, there is very little spatial overlap between the variable GREEN and the other variables. In contrast, for the negative extremes, four of the five locations listed for GREEN also appear for TEMP, and three of them appear in the positive extreme list for PREC. In the next section, the variable GREEN will be considered as the dependent variable in a regression model. A close scrutiny of the results for the G_i and G_i^* statistics indicates the extent to which the spatial pattern of clustering for the dependent variable is the same as for the explanatory variables. If it is, then the inclusion of the explanatory variables in the model is likely to eliminate the indication of spatial autocorrelation. If, on the other hand, the spatial patterns of dependent and explanatory variables do not coincide, as is the case in this example, then there is no reason to expect that the inclusion of explanatory variables would eliminate the spatial autocorrelation. I turn to this issue next.

SPATIAL REGRESSION ANALYSIS

Regression and spatial autoregression

In linear regression analysis, the objective is to explain the variance of a variable, y (the dependent variable), by a linear combination of observations on explanatory variables, X. In matrix notation, the standard regression model can be expressed as:

$$y = X\beta + \varepsilon$$

where y is an $N \times 1$ column vector of observations on the

Table 46-4: Observations with extreme G_i statistics (z values in parentheses)

	GREEN	TEMP	ELEV	PREC
Positive values				
	45 (2.84)	6 (3.64)	47 (1.98)	41 (3.23)
	37 (2.77)	5 (2.86)	37 (1.95)	42 (3.01)
	43 (2.29)	7 (2.83)	49 (1.72)	48 (2.98)
	31 (2.16)	2 (2.64)	48 (1.71)	49 (2.79)
	39 (1.98)	1 (2.29)	38 (1.69)	34 (2.13)
Negative values				
	35 (−3.54)	27 (−1.91)	12 (−3.56)	6 (−3.93)
	27 (−2.89)	34 (−1.83)	11 (−3.38)	7 (−3.20)
	28 (−2.75)	41 (−1.75)	4 (−3.36)	5 (−2.96)
	34 (−2.31)	28 (−1.74)	5 (−3.05)	13 (−2.40)
	42 (−1.90)	35 (−1.65)	10 (−2.74)	14 (−2.23)

Table 46-5: Observations with extreme G_i* statistics (z values in parentheses)

	GREEN	TEMP	ELEV	PREC
Positive values				
	45 (3.03)	6 (4.27)	47 (2.18)	41 (3.64)
	37 (2.75)	7 (3.70)	37 (2.16)	42 (3.47)
	43 (2.53)	5 (3.38)	41 (2.16)	48 (3.44)
	44 (2.41)	2 (3.09)	49 (1.94)	49 (3.29)
	46 (2.36)	1 (3.03)	38 (1.92)	34 (2.41)
Negative values				
	35 (–4.06)	27 (–2.16)	12 (–3.86)	6 (–4.53)
	27 (–3.34)	34 (–2.03)	11 (–3.86)	7 (–4.02)
	28 (–3.31)	28 (–2.02)	4 (–3.68)	5 (–3.46)
	34 (–2.98)	41 (–1.96)	5 (–3.42)	13 (–2.68)
	42 (–2.34)	35 (–1.87)	10 (–3.14)	14 (–2.53)

dependent variable, X is an $N \times K$ matrix of observations on the explanatory variables (typically including a vector of ones as the first column, to incorporate a constant term, or intercept), with an associated $K \times 1$ vector of regression coefficients, and ε is an $N \times 1$ vector of random error terms. The latter are unobserved, but are crucial to understand the probabilistic model of regression analysis. Similarly, the population regression coefficients are unobserved, and need to be estimated from the observations on y and X. By far the most common method by which this estimation is carried out is ordinary least squares (OLS). In this approach, the sum of squared residuals is minimized. When a number of assumptions are satisfied, such as uncorrelated error terms with fixed variance, the OLS estimates are optimal, in the sense that they are unbiased and achieve the smallest variance (most efficient) of all linear estimates (this is referred to as BLUE, for best linear unbiased estimates). Other requirements for OLS to be BLUE are that the relationship between y and X is linear, that all relevant variables are included among the X (no misspecification), and that the X are uncorrelated with the error terms (for a more technical discussion of the properties of OLS, see, e.g., Spanos, 1986; Greene, 1990; and, in a spatial context, Anselin, 1988a). If any of the various assumptions are not satisfied, some of the optimality properties of OLS may no longer hold. The extent to which the assumptions are valid for a particular sample can be checked by so-called diagnostic tests, which I will discuss in more detail.

A special case of a regression model is where the explanatory variable consists of a spatial lag. In other words, the dependent variable is regressed on a spatial lag of itself, hence the term spatial *auto*regression (note the formal analogy to the terminology used in time series analysis). As pointed out, the definition of a spatial lag is not unique and depends on the choice of a spatial weights matrix. Formally, a first-order spatial autoregression can be expressed as:

$$y = \alpha + \rho Wy + \varepsilon$$

where the constant term α is included for convenience (otherwise, the y would have to be expressed in deviations from its mean to ensure that the mean of the error term ε is zero), the Wy is the spatial lag for y, and ρ is referred to as the spatial autoregressive coefficient. As before, the ε is an $N \times 1$ vector of random errors. However, in contrast to what holds for the classical regression model or for an autoregressive model in time series, these error terms are no longer uncorrelated with the explanatory variable (the spatial lag, Wy). As a result, OLS is no longer a proper estimation method, and it will yield biased and inefficient estimates (for technical details, see Anselin, 1988a). The alternative is to use a maximum likelihood (ML) estimator. This will yield efficient estimates that are optimal in an asymptotic sense (i.e., for infinitely large samples). ML estimation for a spatial autoregressive model is based on the assumption of normality, which is often not satisfied in practice. Also, its asymptotic properties do not necessarily yield very good results for small samples.

The estimate for ρ can clearly be considered as an indication of spatial autocorrelation, for example, as an alternative to the use of the I, c, or G statistics as mentioned. Since a spatial autoregressive process uses only information on the spatial dependence in the data, it is in many ways similar to the kriging method of geostatistics. Similar to kriging, a spatial autoregressive process can be used for spatial interpolation, when the elements in the spatial weights matrix are expressed in function of distance.

A generalization of the spatial autoregressive model is a so-called mixed regressive–spatial-autoregressive specification. In addition to the spatial lag (the autoregressive part), the set of explanatory variables now also include

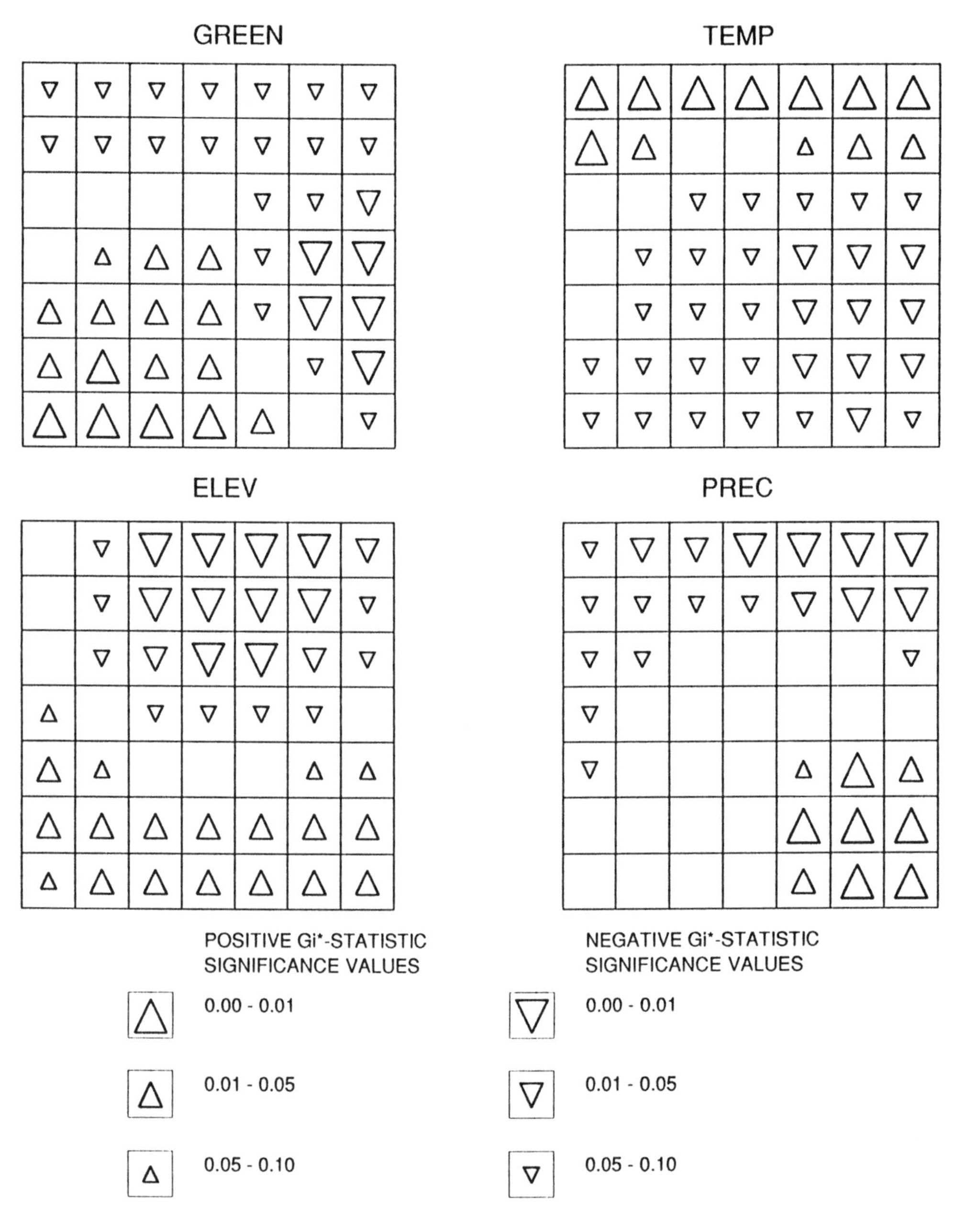

Figure 46-3: Most extreme G_i* statistics for vegetation index (GREEN), temperature (TEMP), elevation (ELEV), and precipitation (PREC), using first order contiguity.

other variables (the regressive part). Formally, a mixed regressive–spatial-autoregressive model can be expressed as:

$$y = \rho Wy + X\beta + \varepsilon$$

in the same notation as previously. As for the spatial autoregressive model, the correlation of the spatial lag Wy (as one of the explanatory variables) with the error term invalidates the optimality of OLS as an estimator for this model, and the ML approach needs to be used instead. Note that this is in contrast to what holds in time series analysis (for technical details, see Ord, 1975; Anselin, 1988a; Haining, 1990).

The estimates of the coefficients in a mixed regressive–spatial-autoregressive model can be interpreted in several ways. The inclusion of Wy in addition to other explanatory variables allows one to assess the degree of spatial dependence in the model, while controlling for the effect of other explanatory variables. Hence, the main interest is in the spatial effect. Alternatively, the inclusion of Wy allows one to assess the significance of the other (nonspatial) explanatory variables, after the spatial dependence is controlled for. Formally, this can be represented as:

$$y - \rho Wy = X\beta + \varepsilon$$

or, as a regression of the spatially filtered dependent variable on the remaining explanatory variables (see Getis, 1990, for a more extensive discussion of spatial filtering in regression analysis). Also, the model can be viewed in its nonlinear form, when the main interest is in understanding the mean of the process:

$$y = (I - \rho W)^{-1}X\beta + (I - \rho W)^{-1}\varepsilon$$

and

$$E[y] = (I - \rho W)^{-1}X\beta$$

Substantive versus error dependence

The inclusion of the spatial lag term Wy in the mixed regressive–spatial-autoregressive model is a formal expression for the spatial dependence in the data. I refer to this form of dependence as *substantive spatial dependence*, since it pertains to the variable of interest (y) and is an expression of a spatial process, such as a spatial diffusion process. Another form of spatial dependence occurs when the error term follows a spatial autoregressive process. Clearly, these error terms are no longer uncorrelated, and as a result OLS estimates will not be efficient. Formally, a model with *spatial error dependence* can be expressed as:

$$y = X\beta + \varepsilon$$

$$\varepsilon = \lambda W\varepsilon + \zeta$$

that is, as the combination of a standard regression model and a spatial autoregressive model in the error term ε. The error term ζ is assumed to be well behaved (uncorrelated, with fixed variance). In order to stress the difference with the substantive spatial dependence, the autoregressive parameter in the error dependence models is expressed by the symbol λ (rather than ρ). In contrast to its interpretation in the substantive model, the coefficient λ is considered to be a *nuisance* parameter, usually of little interest in and of itself, but necessary to correct for or filter out the dependence. Note that since the mean of the error term ε is zero, irrespective of the value of λ, the mean of y is not affected by the spatial error dependence.

For both substantive and error spatial dependence some of the assumptions of the standard regression model are violated when this dependence is ignored. When a substantive spatial effect is present, but the spatial lag term Wy is omitted from the model, the estimates for the regression coefficients β will be biased. This is but a special case of the standard problem associated with omitted variables. When spatial error dependence is present, but ignored, the consequences are not quite as grave. While the OLS estimates are no longer efficient, they remain unbiased. In other words, if the main focus is on prediction or interpolation, OLS is fine, since it yields unbiased estimates. However, if the interest lies in statistical inference, that is, in generalization based on the "significance" of the estimates (t tests and F tests) and on a measure of fit (R^2), OLS will be unreliable. Again, this is but a special case of the problem associated with nonspherical (correlated) errors in a regression model.

The standard approach towards detecting the presence of spatial dependence in a regression model is to apply diagnostic tests. This is complicated by the great similarity between substantive dependence and error dependence. In formal terms, the spatial error model is equivalent to a mixed regressive–spatial-autoregressive model of a special kind, referred to in the literature as the spatial Durbin or common factor specification (for technical details, see Burridge, 1981; Bivand, 1984; Anselin, 1988a). Indeed, the spatial autoregressive error can also be expressed as:

$$\varepsilon = (I - \lambda W)^{-1}\zeta$$

that is, the product of the inverse of an $N \times N$ matrix and a white noise term. This expression can be substituted into the standard regression specification, such that:

$$y = X\beta + (I - \lambda W)^{-1}\zeta$$

After some simple matrix manipulatons, this yields:

$$y = \lambda Wy + X\beta - \lambda(WX\beta) + \zeta$$

or, a mixed regressive–spatial-autoregressive model with spatial lags for the dependent as well as for the explanatory variables, but in the parameter λ to indicate its relation to the error dependence model. The formal equivalence between this model and the spatial error model is only satisfied if a set of nonlinear constraints on the coefficients is satisfied. Specifically, the negative of the product of λ (the coefficient of Wy) with each β (coefficients of X) should equal $-\lambda\beta$ (the coefficients of WX). In the spatial econometric literature, this is termed the common factor hypothesis, in analogy with a similar approach in time series analysis (Hendry and Mizon, 1978).

The implications of the common factor model are twofold. On the one hand, it will be very difficult to distinguish substantive spatial dependence from error spatial dependence in a diagnostic test, since the latter implies a special form of the former. Also, once a spatial error specification has been chosen for a regression model, the common factor constraints need to be satisfied, or else this specification cannot be valid.

Tests for spatial dependence in a regression model

In light of the previous discussion, it should be clear that there is a need for two types of diagnostic tests for spatial dependence in a regression model: those geared at detecting substantive dependence and those developed to detect error dependence. Both types should be based on a model estimated without spatial dependence. Although it is possible to carry out a significance test on the spatial autoregressive coefficient in a model that includes substantive or error spatial dependence, for example, by means of an asymptotic t test or likelihood ratio test, it is much simpler to take a model without spatial effects as the point of departure. Indeed, OLS can be applied to the latter, while for the spatial models a nonlinear ML procedure would need to be followed.

Of the two types of diagnostics, tests for spatial error dependence have received most attention in the literature. The best known approach is an application of Moran's I to the residuals of an OLS regression. This test was first suggested by Cliff and Ord (1972). In matrix notation, the statistic takes on the form:

$$I = e'We \; / \; e'e$$

where e is an $N \times 1$ vector of regression residuals, W is a spatial weights matrix, and $e'e$ is the sum of squared residuals. The usual correcting factor of N/S_0 (as in Moran's I test for spatial autocorrelation listed previously) is not needed here since the weights matrix should be row standardized in order to satisfy stationarity requirements (i.e., to keep the implied spatial process from becoming explosive). This test is formally equivalent to the familiar Durbin–Watson test for serial correlation in the time domain (Durbin and Watson, 1950). Similarly, its exact distribution depends on the particular explanatory variables in the model. In addition, the elements of the spatial weights matrix are important as well. In practice, the statistic is converted to a z value that is then compared to a standard normal distribution (see Cliff and Ord, 1972, 1981, for technical details). It should be kept in mind that this is an asymptotic approximation, which may not be well satisfied in small samples. Moreover, as Anselin and Rey (1991a,b) have shown, Moran's I for regression residuals is very sensitive to the presence of other forms of specification error, such as non-normality and heteroskedasticity. Also, the test is not able to discriminate properly between spatial error dependence and substantive spatial dependence.

An alternative test is based on the Lagrange multiplier principle, and was first suggested by Burridge (1980). It is very similar in expression to Moran's I and is also computed from the OLS regression residuals. However, some additional expressions in matrix traces are needed to achieve an asymptotic χ^2 distribution (with one degree of freedom) under the null hypothesis of no spatial dependence (H_0: $\lambda = 0$). The test is:

$$\text{LM(err)} = \{e'We/\sigma^2\}^2/\text{tr}[W'W + W^2]$$

where tr stands for the matrix trace operator (i.e., the sum of the diagonal elements), and σ^2 is a maximum likelihood estimate for the error variance, $\sigma^2 = e'e/N$. Note that this is not the standard unbiased result provided by most regression packages, for which the sum of square residuals, $e'e$ would be divided by the degrees of freedom (N–K).

A test for substantive spatial dependence (i.e., an omitted spatial lag) can also be based on the Lagrange multiplier principle and was suggested in Anselin (1988b). Its form is slightly more complex, but again only requires the results of an OLS regression. The test is:

$$\text{LM(lag)} = \{e'Wy/\sigma^2\}^2 \;/\; \{(WXb)'MWXb/\sigma^2 + \text{tr}[W'W + W^2]\}$$

where Wy is the spatial lag, WXb is a spatial lag for the predicted values (Xb), and M is a familiar projection matrix, $M = I - X(X'X)^{-1}X'$. The other notation is as previously used. The LM(lag) test is also distributed as χ^2 with one degree of freedom under the null hypothesis of no spatial dependence (H_0: $\rho = 0$).

From a large number of Monte Carlo simulation experiments in which the performance of Moran's I was compared to the two Lagrange multiplier tests, Anselin and Rey (1991a,b) concluded that Moran's I does not provide a sensitive indication of the form of spatial dependence (the alternative hypothesis). Even though this test is by far the most familiar of the three, it should actually be used with great caution. In contrast, the two LM tests provide a good sense of which alternative is the most appropriate one when one of the two is significant, and the other is not. Moreover, when both are significant (not an uncommon occurrence), the one with the largest value (the "most" significant) tends to point to the correct alternative.

As an illustration of the three approaches, the results of an OLS regression of the variable GREEN on TEMP, ELEV, and PREC are given in the first column of Table 46-6 (referred to as model 1 in the table). The complete list of residuals is also presented in Table 46-2, as the variable RESID. A graphical representation in Figure 46-4 gives a clear intuitive sense of spatial clustering. More formally, as shown at the bottom of Table 46-6, all three tests for spatial dependence are highly significant: Moran's I yields a z value of 6.64, LM(err) is 30.41 and LM(lag) is 35.23. Since the latter two are distributed as χ^2 with one degree of freedom, the critical level for $\alpha = 0.01$ is 6.63. The fact that LM(lag) is more extreme than LM(err) would indicate that the correct source of spatial dependence is an omitted spatial lag variable.

Table 46-6: Regression and spatial autoregression

VARIABLE	Model 1[a]	Model 2[b]	Model 3[c]	Model 4[d]
W_GREEN		0.887	0.866	
		(16.01)	(13.89)	
CONSTANT	364.82	15.01	84.32	209.70
	(2.43)	(7.43)	(1.08)	(1.03)
TEMP	–0.933		–0.245	–0.298
	(–1.87)		(–0.97)	(–0.45)
ELEV	0.103		0.013	–0.005
	(2.95)		(0.73)	(-0.12)
PREC	–0.480		–0.091	0.043
	(–2.20)		(–0.82)	(0.15)
λ				0.892
				(16.59)
Measures of fit				
R^2:	0.19			
Lik:	–184.56	–161.45	–160.78	–160.80
AIC:	377.11	326.91	331.56	331.60
Diagnostics for spatial effects				
$Z(I)$	6.64			
LM(ERR)	30.41			
LM(LAG)	35.23			

[a]Model 1: OLS estimation of regressive model.
[b]Model 2: ML estimation of pure spatial autoregressive model.
[c]Model 3: ML estimation of mixed regressive–spatial-autoregressive model.
[d]Model 4: ML estimation of regressive model with spatial error dependence.

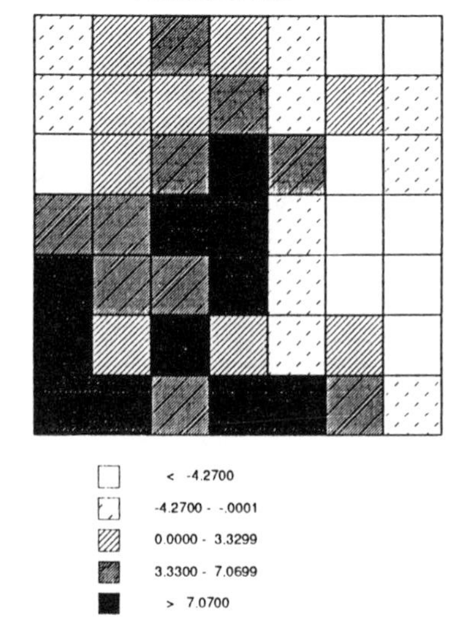

Figure 46-4: Spatial layout of regression residuals.

Estimation of spatial regression models

Due to the presence of spatial dependence, the usual framework for estimation that is based on a random sample of independent observations cannot be used for spatial regression models. Instead, an approach is taken in which the dependence is formally incorporated in the joint probability density of the observations. These dependent random variates are then transformed into uncorrelated ones and a new joint distribution is derived by means of the so-called Jacobian (for technical details, see Ord, 1975; Anselin, 1988a; Griffith, 1988a). For spatial regression models, this Jacobian turns out to take the form:

$$J = \det(I - \rho W)$$

where det stands for determinant of a matrix, I is an $N \times N$ identity matrix, W is a spatial weights matrix, and ρ is a spatial autoregressive coefficient.

Estimation of spatial regression models is typically carried out by means of a maximum likelihood approach, in which the probability of the joint distribution (likelihood) of all observations is maximized with respect to a number of relevant parameters. For the mixed regressive–spatial-autoregressive model,

$$y = \rho Wy + X\beta + \varepsilon$$

the likelihood is obtained by assuming a normal distribution for the error term , with mean zero and variance σ^2. The log (logarithm) of the likelihood for this model is:

$$L = -(N/2)\ln(2\pi) - (N/2)\ln\sigma^2 + \ln[\det(A)] - (1/2\sigma^2)(Ay-X\beta)'(Ay-X\beta)$$

where N is the number of observations, and for notational simplicity, the expresson $I-\rho W$ is replaced by A. The parameters with respect to which this likelihood needs to be maximized are ρ, β, and σ^2. It turns out that the estimates for the regressive coefficients β, conditional upon the value for ρ, can be found as:

$$b = b_0 - \rho b_L$$

where b_0 and b_L are OLS regression coefficients in a regression of X on y and Wy respectively. Similarly, the error variance σ^2 can be estimated as:

$$\sigma^2 = (1/N)(e_0 - \rho e_L)'(e_0 - \rho e_L)$$

where e_0 and e_L are the residual vectors in the regressions for b_0 and b_L. Substitution of these results into the log-likelihood function yields a so-called concentrated likelihood, which only needs to be maximized with respect to the parameter ρ:

$$LC = -(N/2)\ln[(e_0 - \rho e_L)'(e_0 - \rho e_L)/N] + \ln[\det(A)]$$

Inference in this model is based on an asymptotic variance matrix, obtained from the second derivatives of the likelihood function (for technical details, see Ord, 1975; Anselin, 1988a). Maximum likelihood is not the only approach possible to estimate a mixed regressive–spatial-autoregressive model. Alternatives are the use of instrumental variables and an application of the bootstrap (for details, see Anselin, 1990b).

The results of an ML estimation of a pure spatial autoregressive and a mixed regressive–spatial-autoregressive model are listed in the second and third columns of Table 46-6. In both instances there is very strong evidence for positive spatial clustering. In the mixed model, the introduction of the spatial lag also alters the qualitative interpretation of the original regression model: None of the original regressors is any longer significant. This is an example of a situation where after controlling for spatial dependence, only white noise is left. The original impression of an association between explanatory and dependent variables (as suggested in the first column of Table 46-6) disappears when spatial dependence is introduced. Another indication of the importance of spatial dependence in this example is given by the considerable improvement in fit. This is shown by the values for the log-likelihood and AIC, listed at the bottom of each column. The AIC (Akaike information criterion) is a measure of fit that includes a tradeoff between fit and parsimony, similar to the adjusted R^2 used in OLS. The log-likelihood (as a measure of fit) is corrected for the number of explanatory variables that are included. The lower the AIC, the "better" the model (for details, see Anselin, 1988a). In our example, both spatial models have a considerable higher log-likelihood than the OLS model: –161.45 and –160.78 versus –184.56. However, even though the mixed autoregression has a higher likelihood than the pure form, the latter may be considered to be a better model since it achieves a lower AIC: 326.91 versus 331.56. This was to be expected, since none of the regressive coefficients turned out to be significant (as also would be indicated by a simple likelihood ratio test on their joint significance).

A regression with spatial error autocorrelation is a special case of a model with a nonspherical error variance. If the autoregressive coefficient λ were known, then the model could be estimated by means of generalized least squares (GLS). It is easy to see from the spatial Durbin or common factor specification given that this model can also be expressed as:

$$y - \lambda Wy = (X - \lambda WX)\beta + \zeta$$

If we set $y^* = y - \lambda Wy$ and $X^* = X - \lambda WX$, and since ζ is a well-behaved error term in the usual sense, this is a standard regression in the spatially transformed variables y^* and X^*. The problem is that λ is typically not known, and needs to be estimated together with the other parameters (the β and σ^2). Hence, GLS is really not practical, and instead a so-called estimated or feasible generalized least squares approach is taken, conditional upon a consistent estimate for λ. In contrast to the situation for timewise serial error correlation, this is not straightforward. In fact, it turns out that the only practical strategy is a maximum likelihood approach. Assuming normality for the error term and using the concept of a Jacobian for this model as well, the log-likelihood can be found to be:

$$L = -(N/2)\ln(2\pi) - (1/2)\ln\sigma^2 + \ln[\det(B)] - (1/2^2)(y-X\beta)'B'B(y-X\beta)$$

with $B = I - \lambda W$, for notational simplicity (the other notation is as previously used). In contrast to the case for the mixed regressive–spatial autoregressive model, there is less computational advantage in using a concentrated likelihood that is a simple function of the coefficient λ. In practice, rather than maximize the likelihood jointly for all parameters, an iterative procedure can be followed, where the ß coefficients are estimated conditional upon the λ, and the λ are estimated conditional upon the ß, and the procedure switches back and forth until convergence is achieved (this is sometimes referred to as EGLS or FGLS). As for the mixed regressive–spatial-autoregressive specification, inference is based on an estimated asymptotic variance matrix (for technical details, see Anselin, 1988a).

The results for a ML estimation of the regression model with spatial autoregressive errors are given in the last column of Table 46-6. Again, the spatial autoregressive coefficient is highly significant and positive. Also, the qualitative interpretation of the other regression coefficients changes after the spatial effect is incorporated (they are no longer significant). Compared to the substantive model, the error model has a slightly lower fit, based both on likelihood (–160.80 versus –160.78) as well as on AIC (331.60 versus 331.56).

Spatial time series

An alternative approach towards the specification and estimation of regression models with dependence in space (and time) can be based on principles developed in the analysis of time series. Specifically, the perspective pioneered by Box and Jenkins (1976) has been extended to so-called SARMA and STARMA (or even STARIMA) models, where the acronym SARMA stands for space autoregressive moving average, in analogy with the ARIMA terminology for time series. STARMA (for space time ARMA) models take on the general form:

$$y_{i,t} = \varepsilon_{i,t} + \Sigma_s \Sigma_k \alpha_{s,k} W^{(s)} y_{i,t-k} + \Sigma_s \Sigma_k ß_{s,k} W^{(s)} \varepsilon_{i,t-k}$$

In this expression, i stands for the locational coordinate (sometimes made explicit in terms of longitude and latitude) and t for the time period. The $W^{(s)}$ matrix is a spatial lag operator for order of contiguity s, similar to the time lag operation (or backshift operator) $t-k$. The first sum of lagged terms, in the dependent variable y, makes up the autoregressive part (AR), while the second sum of lagged terms, in the error term, makes up the moving average part (MA). For each of these, the sum is over the order of the spatial dependence or contiguity, s, and the order of the time lag, k.

The specification of the lag orders for a STARMA model from observed data is no trivial matter. Similar to the approach taken for ARIMA models in time series analysis, the autocorrelation and partial autocorrelation coefficients are computed and investigated for distinctive patterns in function of the lag length. A number of factors complicate this approach for spatial data. Foremost among these is the difficulty to obtain stationarity in space, let alone in space–time. Also, as Hooper and Hewings (1981) have shown, the types of space–time processes for which the autocorrelation and partial autocorrelation functions are well defined and useful in the identification process is very limited. Nevertheless, there is a considerable body of literature on STARIMA modeling, with applications in areas such as epidemiology, marine fisheries, water resources, and regional forecasting. A detailed treatment of this approach is beyond the scope of the current chapter. For overviews of applications and a discussion of technical issues pertaining to the STARIMA model and its extensions (such as STARMAX models, which also include other explanatory variables, the X), the interested reader is referred to Bennett (1979), Pfeifer and Deutsch (1980a,b,c), and Stoffer (1986).

CONCLUSION

Integration of spatial regression modeling and GIS

As is well known, most commercial GIS implementations are rather limited in what they offer in terms of statistical tools for the analysis of spatial data. It should therefore come as no surprise that spatial regression modeling is mostly absent in the practice of GIS analysis. The integration of spatial data analysis within a GIS is discussed at greater length in Anselin and Getis (1992) and Goodchild et al. (1992). In essence there are three ways in which this can be implemented:

- A complete integration of all spatial analysis functions within the GIS software;
- An efficient link between separately developed modules for spatial data analysis and the database in a GIS, including an effective exploitation of the "spatial" information (i.e., the topology of the data);
- A completely separate function for spatial data analysis and GIS as a database, with standard import and export functions to move the data between the two.

So far, most applications fall within the last category, due to the difficulties of accessing proprietary data formats in commercial GIS and the limited facilities of current macro languages. Several examples of the joint use of GIS and statistical packages can be found in the volume edited by Allen, Green, and Zubrow (1990). Some examples of a closer integration between GIS and statistical software are discussed in Goodchild et al. (1992).

By and large, however, the regression techniques that are implemented are nonspatial. As pointed out in Anselin and Griffith (1988) and Haining (1990), one of the problems is the lack of software to carry out the complex and nonlinear estimation and inference for spatial process models. A number of recent advances have been made in software development for spatial regression analysis in the form of libraries of macro routines for commercial statistical packages (e.g., Griffith, 1988b; Griffith et al., 1990; Bivand, 1990, 1991). A self-contained spatial data analysis software package, *SpaceStat*, is introduced in Anselin (1992). So far, however, the linkage between this software and a GIS is very limited, but progress is being made (e.g., the work of Bivand, 1990, in which data from ARC/INFO is imported into a SYSTAT module for spatial regression analysis). The investigation of ways to integrate spatial data analysis and GIS effectively is one of the main research questions on the agenda of the NCGIA Research Initiative on Spatial Analysis and GIS (NCGIA Initiative 14).

ACKNOWLEDGMENTS

The research on which this chapter is based was supported in part by Grant SES-8721875 from the National Science Foundation and by the National Center for Geographic Information and Analysis (NSF Grant SES-8810917). Research assistance with the empirical illustrations and graphics from Rusty Dodson and Sheri Hudak is greatly appreciated. Useful comments on an earlier draft were made by Noel Cressie and Michael Goodchild.

REFERENCES

Allen, K., Green, S. and Zubrow, E. (1990) *Interpreting Space: GIS and Archaeology*, London: Taylor and Francis.

Anselin, L. (1988a) *Spatial Econometrics: Methods and Models*, Dordrecht: Kluwer Academic.

Anselin, L. (1988b) Lagrange Multiplier test diagnostics for spatial dependence and spatial heterogeneity. *Geographical Analysis* 20: 1–17.

Anselin, L. (1990a) Spatial dependence and spatial structural instability in applied regression analysis. *Journal of Regional Science* 30: 185–207.

Anselin, L. (1990b) Some robust approaches to testing and estimation in spatial econometrics. *Regional Science and Urban Economics* 20: 141–163.

Anselin, L. (1992) *SpaceStat, A Program for the Statistical Analysis of Spatial Data*, Santa Barbara, CA: National Center for Geographic Information and Analysis S92-1.

Anselin, L., and Getis, A. (1992) Spatial statistical analysis and geographic information systems. *The Annals of Regional Science* 26(1): 19–33.

Anselin, L., and Griffith, D. (1988) Do spatial effects really matter in regression analysis? *Papers, Regional Science Association* 65: 11–34.

Anselin, L., and Rey, S. (1991a) Properties of tests for spatial dependence in linear regression models. *Geographical Analysis* 23: 112–131.

Anselin, L., and Rey, S. (1991b) *The Performance of Tests for Spatial Dependence in a Linear Regression*, Santa Barbara, CA: National Center for Geographic Information and Analysis, Technical Paper 91-13.

Bennett, R. (1979) *Spatial Time Series*, London: Pion.

Bivand, R. (1984) Regression modelling with spatial dependence: an application of some class selection and estimation methods. *Geographical Analysis* 16: 25–37.

Bivand, R. (1990) Spatial statistics: front-end inference support for GIS. *Proceedings, Third Scandinavian Research Conference on Geographical Information Systems, Helsingor, Denmark*, pp. 244–254.

Bivand, R. (1991) SYSTAT-compatible software for modelling spatial dependence among observations. *Paper Presented at the 7th European Colloquium on Theoretical and Quantitative Geography, Stockholm, Sweden*.

Box, G., and Jenkins, G. (1976) *Time Series Analysis, Forecasting and Control*, San Francisco: Holden Day.

Burridge, P. (1980) On the Cliff-Ord test for spatial correlation. *Journal of the Royal Statistical Society B* 42: 107–108.

Burridge, P. (1981) Testing for a common factor in a spatial autoregressive model. *Environment and Planning A* 13: 795–800.

Cliff, A., and Ord, J.K. (1972) Testing for spatial autocorrelation among regression residuals. *Geographical Analysis* 4: 267–284.

Cliff, A., and Ord, J.K. (1973) *Spatial Autocorrelation*, London: Pion.

Cliff, A., and Ord, J.K. (1981) *Spatial Processes, Models and Applications*, London: Pion.

Cressie, N. (1991) *Statistics for Spatial Data*, New York: Wiley.

Durbin, J., and Watson, G.S. (1950) Testing for serial correlation in least squares regression I. *Biometrika* 37: 409–428.

Getis, A. (1990) Screening for spatial dependence in regression analysis. *Papers, Regional Science Association* 69: 69–81.

Getis, A., and Ord, J.K. (1992) The analysis of spatial association by use of distance statistics. *Geographical Analysis* 24(3): 189–206.

Goodchild, M.F., Haining, R.P., and Wise, S. (1992) Integrating GIS and spatial data analysis: problems and possibilities. *International Journal of Geographical Information Systems* 6(5): 407–423.

Greene, W. (1990) *Econometric Analysis*, New York: Macmillan.

Griffith, D. (1988a) *Advanced Spatial Statistics*, Dordrecht: Kluwer Academic.

Griffith, D. (1988b) Estimating spatial autoregressive model parameters with commercial statistical packages. *Geographical Analysis* 20: 176–186.

Griffith, D., Lewis, R., Li, B., Vasiliev, I., Knight, S., and

Yang, X. (1990) Developing Minitab software for spatial statistical analysis: a tool for education and research. *The Operational Geographer* 8: 28–33.

Haining, R. (1990) *Spatial Data Analysis in the Social and Environmental Sciences*, Cambridge: Cambridge University Press.

Hendry, D., and Mizon, G. (1978) Serial correlation as a convenient simplification, not a nuisance: a comment on a study of the demand for money by the Bank of England. *Economic Journal* 88: 549–563.

Hooper, P., and Hewings, G. (1981) Some properties of space–time processes. *Geographical Analysis* 13: 203–223.

Hubert, L. (1985) Combinatorial data analysis: association and partial association. *Psychometrica* 50: 449–467.

Hubert, L. (1987) *Assignment Methods in Combinatorial Data Analysis*, New York: Marcel Dekker.

Hubert, L., and Golledge, R.G. (1982) Measuring association between spatially defined variables: Tjostheim's index and some extensions. *Geographical Analysis* 13: 273–278.

Hubert, L., Golledge, R.G., and Costanzo, C.M. (1981) Generalized procedures for evaluating spatial autocorrelation. *Geographical Analysis* 13: 224–233.

Hubert, L., Golledge, R.G., Costanzo, C.M., and Gale, N. (1985) Measuring association between spatially defined variables: an alternative procedure. *Geographical Analysis* 17: 36–46.

National Oceanic and Athmospheric Administration (1990) *Global Change Database Project, Pilot Project for Africa*.

O'Loughlin, J., and Anselin, L. (1992) Geography of international conflict and cooperation: theory and methods. In M.D. Ward (ed.), *The New Geopolitics*, London: Gordon and Breach, pp. 11–38.

Ord, J.K. (1975) Estimation methods for models of spatial interaction. *Journal of the American Statistical Association* 70: 120–126.

Pfeifer, P., and Deutsch, S. (1980a) A STARIMA model-building procedure with applications to description and regional forecasting. *Transactions of the Institute of British Geographers* 5: 330–349.

Pfeifer, P., and Deutsch, S. (1980b) A three-stage iterative procedure for space–time modeling. *Technometrics* 22: 35–47.

Pfeifer, P., and Deutsch, S. (1980c) Identification and interpretation of first order space–time ARIMA models. *Technometrics* 22: 397–408.

Ripley, B. (1988) *Statistical Inference for Spatial Processes*, Cambridge: Cambridge University Press.

Spanos, A. (1986) *Statistical Foundations of Econometric Modelling*, Cambridge: Cambridge University Press.

Stoffer, D. (1986) Estimation and identification of space-time ARMAX models in the presence of missing data. *Journal of the American Statistical Association* 81: 762–772.

Tobler, W. (1979) Cellular geography. In S. Gale and G. Olsson (eds.), *Philosophy in Geography*, Dordrecht: Reidel, pp. 379–386.

Upton, G., and Fingleton, B. (1985) *Spatial Data Analysis by Example*, New York: Wiley.

47

Probability Sampling and Population Inference in Monitoring Programs

W. SCOTT OVERTON

A fundamental difference between probability sampling and conventional statistics is that "sampling" deals with real, tangible populations, whereas "conventional statistics" usually deals with hypothetical populations that have no real-world realization. The focus here is on real finite populations as well as on populations representing real-world continuous responses in two-dimensional space.

A general perception of the issue of sampling from a real collection of objects is given by Stuart (1962), and the urn sampling problem is a special case of the general simple random sample (SRS). However, this perception is not easily obtained from conventional statistics where the inference problem is usually communicated as one of hypothesizing a distribution (infinite population) from which the data are assumed to be an independent random sample. Then the theory of independent and identically distributed random variables provides the basis of conventional statistical inference.

This theory may be inappropriate for a problem involving a real (meaning real-world) population. If the objectives of investigation identify a real population as the target of interest, it may not be acceptable to substitute a hypothetical population for the real one, unless there is a modeling process by which a specific hypothetical population can be identified. Statistical habits and practices that are normal and acceptable in conventional statistics can lead to unacceptable inferences in this context.

Too often the objective is lost, and inferences are made about a hypothetical population rather than the real one. Most important, the necessary protocols for inferences about real populations may be neglected, because they are not identified in conventional statistics. The recent paper by de Gruijter and ter Braak (1990) addresses this general issue. However, they present a somewhat different orientation than is given in this chapter. Here the emphasis is on the distinction between real populations and hypothetical ones, and on the effect of that distinction in a sampling context.

Superpopulations are hypothetical populations that are especially focused on real-world populations (Cassel et al., 1977). It is often hypothesized that the (usually finite) real-world population was derived by some process from the superpopulation; this provides an example of a model process by which the real population is a realization of the superpopulation. Alternatively, a superpopulation may be "generated" from the real population by the sampling design. Identification of the process may or may not be the basis for model-based inference; it may or may not involve the transfer of objective parameters from the real population to the superpopulation. These considerations are at the root of the issue of whether superpopulation inference is appropriate for finite population objectives.

Superpopulation inference is indistinguishable from conventional inference, and this may lull one into believing that any sample will lead to rigorous inference, just by invoking a superpopulation. But the superpopulation to which the inferences actually apply can be drastically changed by the sampling protocol, resulting in rigorous model-based inference about the wrong population parameter (see the section on model-based inference to follow). In reality, an appropriate superpopulation reflecting properties of the real population that is the subject of survey objectives must be carefully preserved by the sampling and analytic protocol. There is no guarantee that the parameter is equivalent to the objective parameter, just because a superpopulation (model) approach is taken.

The appropriate statistical criterion for providing that inferences are being made about the intended population and parameter is consistency (consistency is here used in the sense of Fisher consistency, rather than in the more common usage as convergence in probability; Fisher, 1956, p. 143). We wish to employ a sampling design that guarantees the availability of estimating methods that will generate consistent characterization of the intended property of the intended real population, not of some substitute, hypothetical or not. The design protocol that most straightforwardly and unambiguously does this is probability selection, and a sample that is derived from such a protocol is called a probability sample. Probability sampling has a well-defined theory of inference, providing "design-based" properties— properties that derive from the design protocol, not from assumptions or models.

After consistency, efficiency is the foremost criterion of inference, referring to precision, relative to some reference precision. Use of model-based methods is predicated on enhanced precision, hence greater efficiency. Choice among probability designs will also be made on this criterion, but we find that in multiple-objective surveys, a design that increases precision for one objective will often decrease precision for another. Model-based inference, being objective-specific, does not present this dilemma. But greater efficiency without consistency is simply unacceptable, and consistency is the prime criterion.

Understanding when to use methods like probability sampling is not complete without understanding when not to use them. There is a substantial literature on the issue of achieving "representativeness" in the context of model-based inference. This position is briefly identified. The special circumstances for which an "index" sample is appropriate are also identified, with indices briefly discussed as providing an alternate concept of representativeness.

The context of this chapter is provided by EPA's current initiative EMAP (Messer et al., 1991) and the recently conducted National Surface Water Surveys (NSWS, Linthurst et al., 1986; Kaufmann et al., 1988). EMAP is devoted to monitoring a variety of ecological resources throughout the United States with respect to general concerns of well-being and ecosystem health. The NSWS focused on evidence of acidic deposition in areas of the United States that were deemed potentially susceptible to acidification. A probability sample was used for selecting lakes from the universe of lakes and stream reaches from the universe of stream reaches. In both, the universes (populations) were carefully and explicitly defined by the frame materials. Estimators were consistent for population attributes; Horvitz–Thompson estimation (Horvitz and Thompson, 1952) was used throughout. The proposed plan for EMAP is similar in concept, but more complex.

PROBABILITY SAMPLES

The common model for probability sampling begins with a universe, **U**. For finite sampling the universe is a finite point set; u indicates an element of the set. The strong definition of a probability sample (p sample) leading to the classical sampling model is: strong definition: a probability sample is a subset of a universe selected in such a manner that:

- The probability p_s of obtaining any subset of that universe is known, and
- The inclusion probability π_u of any element u of the universe is positive.

The inclusion probability of any element is the probability with which the element is included in the sample. To say that this probability is positive is to say that the element can be included in the sample. This definition of a p sample establishes the finite sample space and a field of probability, allowing complete probabilistic analysis. This space is the foundation of finite sampling theory. The probability field for a particular design completely determines the properties of that design.

A useful alternate model for a p sample is the representation of a design in terms of the inclusion probabilities, defined on the universe. If all of the inclusion probabilities are known, then this model is equivalent to that defined previously, in terms of probabilities of the samples. Inclusion probabilities are defined by operations on the sample space. For example, the first-order inclusion probability of element u is given by:

$$\pi_u = \sum_{\{S:\, u \in S\}} p_s$$

with the second-order inclusion probability defined by:

$$\pi_{uv} = \sum_{\{S:\, u, v \in S\}} p_s$$

and the higher-order generalization simply specifies the set for summation.

This strong definition of a p sample is unnecessarily restrictive. Rigorous inferences and assessments can be made with far less information about the design, as shown by examination of the Horvitz–Thompson theorem. Define a population, **Y**, as a function on the universe, $\mathbf{Y}: \mathbf{U} \rightarrow Y$, where Y is the range of the function (population), and where the value assigned to element u is y_u.

Horvitz–Thompson theorem:
If $S \subset \mathbf{U}$ is selected such that $\pi_u > 0$, for all $u \in \mathbf{U}$, then $\hat{T}_y = \sum_S y_u/\pi_u$ is unbiased for $T_y = \sum_{\mathbf{U}} y_u$, and

$$V(\hat{T}_y) = \sum_{\mathbf{U}} y_u^2 \left(\frac{1-\pi_u}{\pi_u}\right) + \sum_{\mathbf{U}} \sum_{\mathbf{U}-u} \frac{y_u}{\pi_u} \frac{y_v}{\pi_v} \left(\pi_{uv} - \pi_u \pi_v\right)$$

(summations are specified in terms of the set of population units to be summed). If n is fixed, the variance can be expressed alternately,

$$V(\hat{T}_y) = \frac{1}{2} \sum_{\mathbf{U}} \sum_{\mathbf{U}-u} \left(\pi_u \pi_v - \pi_{uv}\right) \left(\frac{y_u}{\pi_u} - \frac{y_v}{\pi_v}\right)^2 .$$

Horvitz–Thompson estimation (HTE) is simply application of the theorem. Horvitz–Thompson inference (HTI) combines estimation and variance estimation based on the theorem, providing generalization of design-based inference and a strong intuitive base for the practice of sampling. HTI requires knowledge of only the first- and second-order inclusion probabilities, and often these need to be known only on the sampled units. Further, there are good sampling designs for which reliable design-based variance estimators do not exist, and it is desirable that these designs be classed as probability designs. For example, the only use of the second-order inclusion probabilities is variance estimation. If variance cannot be estimated even when the π_{uv}s are known (as in systematic sampling), but the design is otherwise satisfactory, then the criterion of rigor should not exclude that design.

Really required for rigorous inference are:

- The capacity for Horvitz–Thompson estimation, or its equivalent;
- Precise and unambiguous identification of the target population; and
- A provably adequate protocol/algorithm for assessment of precision.

The capacity for HTE is provided by any sample that meets the postulates of the theorem. Equivalent capacity to that provided by the theorem is needed for the point and areal samples used for continuous universes. For these cases, there is a direct extension to probability estimation, as we will identify later.

Needed, then, is a weak definition of a probability sample that will provide these properties for designs that fit the definition. Further, for a p sample to retain its identity as a p sample in future application or review, essential information must be archived. Given these criteria, the following weak definition of a p sample will suffice for a rigorous monitoring program: weak definition: a probability sample is a subset of the universe selected by an explicit protocol so that π_u is:

- Known for each element of the realized sample, and
- Positive for each element of the universe.

Our prescription for rigorous monitoring inference is then for a probability sample taken according to this weak definition, with the specific selection protocol and details of selection archived for future reference and use. Archiving the essential documentation is necessary so that the sample can be used as a probability sample in later years in trend assessment, or in generating inferences for newly identified subpopulations of concern.

Careful documentation of the exact protocol and details of selection is necessary to allow analysis or simulation of the monitoring study as part of either theoretical or empirical verification. In effect, this information will allow an investigator to complete the specification of the design according to the strong definition of a probability sample if he so wishes. Archiving this documentation will also provide for application of new analytic protocols as they are developed in the future, and will allow students and other users to review the basis of the data set.

POPULATIONS

The universe/population pair often is referred to just as the "population," it being understood that a population must be defined on an underlying universe. Several kinds of populations are identified. Finite populations are defined on a finite point set; these are "populations" of objects, like lakes. Extensive and continuous populations are defined on a (continuous) spatial region. The differences in these kinds of universe/population pairs usually will require somewhat different sampling designs and somewhat different representation. The parameter of interest for a population defined on a discrete universe often is the population total,

$$T_y = \sum_{\mathrm{U}} y_u;$$

the analogous parameter on a continuous universe is the integral over the universe,

$$T_y = \int_{\mathrm{U}} y(u)\, du.$$

The term "population" is also often used in the biological sense, as for animal populations, or populations of flora. This usage should easily be distinguished by context from the statistical usage defined, but the term "resource" will be used here for ecological "populations" to minimize the potential ambiguity.

Examples of populations of resources

Lakes. The universe is a finite point set, each point representing a lake. Populations defined on the universe represent the attributes of interest.

Individual waterbody. The universe is the areal extent of the waterbody. Populations may be continuous, as for concentrations of nutrients, or extensive, as for attributes of flora or benthic fauna.

Wetlands. The universe is a spatial region occupied by wetlands. Populations are generally extensive and highly structured, like attributes of plant communities.

FRAMES

A frame is a representation of the universe/population that is used in selecting a sample. Though a great variety of frames exists, list frames and map frames capture the general concept. List frames are simply lists of the units of the universe, constructed in such a manner that a real world unit is unambiguously identifiable from its list representation. A map frame is a map on which the positions of the objects of the universe are represented; the lakes on 1:250,000 USGS topographic maps provided the frame for the National Lake Survey (Linthurst et al., 1986). Descriptive information about the units of the frame is referred to as "frame data" or "frame materials." Frame data provide the basis for stratification, variable probability selection, and other specific details of the sample design. Frame data also provide the basis for model-based inference.

Sometimes the nature of a population is changed by the way the sampling frame is structured. For example, a common frame structure for a continuous spatial universe imposes a spatial partition of the universe into areal sampling units. This frame converts a continuous universe into a discrete one. It is important to convert the population by a compatible operation to preserve the parameter T_y.

POPULATION DESCRIPTION AND ANALYSES

The end result of a sampling strategy is a set of descriptive or analytic statistics that are prescribed by the objectives and the design. For most cases, distribution functions will serve as the generic parameter/descriptor, with other statistics being implied by the empirical distribution function. Population comparisons are also readily made in terms of distribution functions. These functions provide the strongest conceptual image of the concept of consistency as well; the real population has a fixed and knowable distribution function (Figure 47-1). The criterion of consistency is simply that the sampling effort produces an estimator of this function, not that of a subpopulation or of a superpopulation having a different function (Figure 47-2). Consistency for other parameters is similarly addressed.

Unfortunately, inference can be shifted away from the intended parameters either by an inappropriate sample or by choice of an inappropriate statistic. Biased sample selection (in the usage of Stuart, 1962) is due to (1) unknown preferential selection of units from some subpopulations at the expense of others, or (2) exclusion of certain subpopulations from selection. A p sample, by definition, suffers from neither of these faults, and HTE provides unbiased estimators of all linear parameters. Consistency is then extended by Generalized HTE to include nonlinear functions of linear parameters, as will be indicated here.

It is important to recognize that variable probability sampling does not constitute biased selection, if the inclusion probabilities are known. It is ignorance of the preferential selection that creates a problem, preventing application of HTI. Restriction of the sample to a subpopulation, however, does not prevent use of HTI for the subpopulation.

In concert with the estimated distribution function, several standard unbiased estimators of linear parameters are provided, all by the HT theorem:

$$\hat{T} = \sum_S y/\pi,$$

$$\hat{\bar{Y}} = \hat{T}_y/N,$$

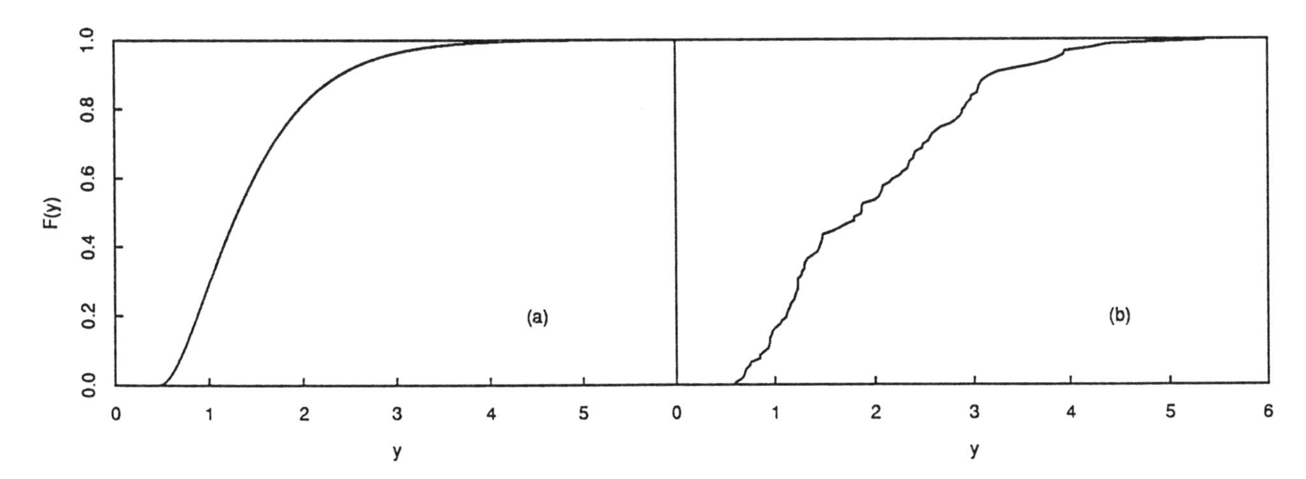

Figure 47-1: A real population (b) and a hypothetical population (a) each represented by its distribution function. When the real population is the objective of study, it may be unacceptable to substitute a hypothetical one.

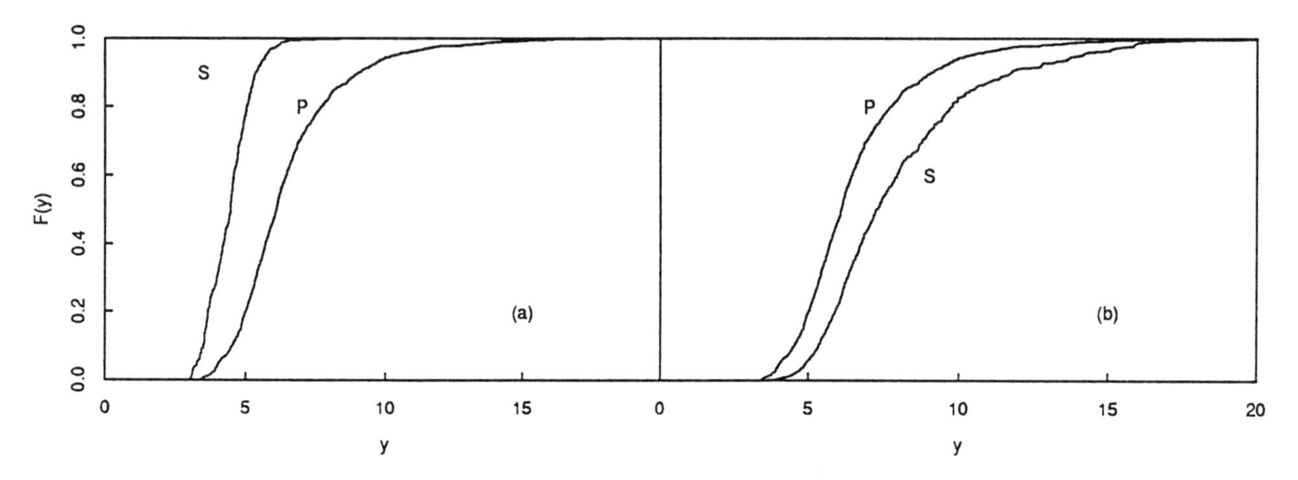

Figure 47-2: Distribution functions of: (a) a real population, P, and a real subpopulation, S, of that population; (b) a real population, P, and a superpopulation, S, which was generated by interaction of the sampling design and the real population.

$$\hat{N}(y') = \sum_{S_{y'}} 1/\pi,$$

$$\hat{F}(y') = \frac{1}{N} \sum_{S_{y'}} 1/\pi,$$

where $S_{y'}$ is the subset of S such that $y_u \leq y'$ (the formulas are simplified by suppressing the subscripts in y_u and π_u; all instances of y and π are to be read as having that subscript; summation is always over a designated set of units).

Subpopulation estimation is simply accomplished by identifying the subset of the sample that is in the subpopulation. Let S_a be that subset of the sample belonging to subpopulation A. Then estimators of subpopulation parameters are:

$$\hat{T}_{ya} = \sum_{S_a} y/\pi$$

and

$$\hat{N}_a = \sum_{S_a} 1/\pi.$$

Other estimators are consistent for nonlinear parameters, such as:

$$\hat{V}_y = \frac{1}{N-1} \left\{ \sum_{S} y^2/\pi - \hat{T}_y^2/\hat{N} \right\}, \text{ when N is known,}$$

$$\hat{V}_y = \frac{1}{\hat{N}-1} \left\{ \sum_{S} y^2/\pi - \hat{T}_y^2/\hat{N} \right\}, \text{ when N is unknown,}$$

where

$$\hat{N} = \sum_{S} 1/\pi$$

and

$$V_y = \frac{1}{N-1} \left\{ \sum_{U} y^2 - T_y^2/N \right\}.$$

The mean of subpopulation A also requires a nonlinear estimator,

$$\hat{\bar{Y}}_a = \hat{T}_{ya}/\hat{N}_a.$$

Examination of these estimators will reveal that many will be simpler and more familiar if all the πs are equal. The variable probability forms should be used in general, but in practice there are many reasons for which one should attempt to generate equal probability designs. One reason for this is the common need to generate inferences for subpopulations, involving nonlinear statistics. Further, estimators of nonlinear parameters are also nonlinear statistics.

The ease of generating descriptions of arbitrary subpopulations persists, even with variable probabilities, and the general flexibility provided by these methods makes HTI a very powerful tool for monitoring studies. The only requirement is that a p sample, by the weak definition, be provided by the design. Other design features influence the ease of maintaining flexibility in the objectives and population definitions, but these are not of immediate concern here.

VARIANCE ESTIMATION FOR DISCRETE POPULATIONS

Rigorous monitoring programs must have the capacity to assess precision and confidence. In the interest of space, this topic is not developed thoroughly here, but a few words will establish the status of the subject.

The HT theorem provides two general forms of variance for the HTE of linear parameters. Each of the two forms leads to a variance estimator, each requiring knowledge of the first- and second-order inclusion probabilities in the sample, and each unbiased for $V(\hat{T}_y)$ for any design for which all the second-order inclusion probabilities are positive. If some π_{uv} are zero, as in the case of a systematic sample, then there does not exist a general consistent variance estimator in the context of HTI. Systematic designs are of particular interest, especially in spatial sampling; it is therefore critical to any rigorous monitoring program to provide another way to assess precision.

Our specification of a rigorous program has provided the basis for such an assessment. Having prescribed a p sample and archival of the exact protocol and selection materials, it is feasible to conduct computer simulations of a design that will faithfully reflect the design and determine its properties as applied to a hypothetical population. In current practice, when efficient spatial designs are used, variance estimation by probability methods is often unsatisfactory. If a hypothetical population can be constructed that closely resembles the real population, then the standard error of parameter estimators for the target case can be determined empirically. The term "facsimile population" refers to such a hypothetical population, and the term "facsimile population bootstrap" refers to a simulation protocol based on this hypothetical population. This protocol shows great promise, and has the extra benefit that all of the difficult aspects are involved with generating greater understanding of the structure of the real population. This is a model-based method, focused on design-based inference.

The information required for empirical assessment is also required for theoretical probability assessment. If the theoretical basis for assessment is currently lacking, but developed in the future, then it can be applied retrospectively, so long as the identity of the sample as a probability sample has been archived. Meantime, given this information, empirical assessment can fill the gap, and additionally provide verification of current theory, as well as stimulation of development of new insights and understanding.

MODEL-BASED INFERENCE

The use of predictive models to enhance population inference is well established (Cochran, 1977; Royall and Cumberland, 1981). The basic idea is very powerful; the sample provides data from which to fit the predictive model, $\hat{y} = \hat{g}(\mathbf{x})$, where $\mathbf{x}$ is a vector of predictor variables (frame data) that are known on the universe. Then the values of $\hat{y}$ are generated on the universe, and a common model-based estimator of T_y is:

$$\tilde{T}_y = \sum_S y_u + \sum_{U-S} \hat{y}_u.$$

Then
$$V(\tilde{T}_y) = E_m\left[\sum_{U-S}[\hat{y}_u - g(\mathbf{x})]\right]^2,$$

where E_m indicates expectation with respect to the model. An essential distinction is made by noting that for linear models and Gauss–Markov estimation,

$$E_m(\hat{y}_u) + g(\mathbf{x}_u),$$

so that
$$E_m(\tilde{T}_y) = T_y - \sum_{U-S}[y_u - g(\mathbf{x}_u)].$$

Model unbiasedness does not imply unbiasedness for the finite population parameter, and this model-based estimator is biased. It is still often advantageous to use such an estimator, and appropriate to use MSE rather than variance in assessing precision. We note here that other estimators are also used for the same objective, particularly when variable probability sampling has been employed.

In many circumstances we are interested in the regression relations of the finite population and will use regression analyses in describing those relations, rather than in estimating population totals, as discussed above. Now the issues that arise are the appropriate estimators of the model parameters, and the constraints on the design that are required to ensure consistency of the regression estimators for the regression parameters. For linear models, the usual analysis is prescribed by the Gauss–Markov theorem, and this theorem says nothing, directly, about how the sample is selected. However, it is easy to demonstrate that a variable probability sample requires estimators to be weighted in the manner of Horvitz–Thompson estimators, in order to estimate the parameters of the real population. If the design is a probability sample with variable probability, letting → mean "is consistent for," several identities that follow Fisher consistency help to see the patterns. It is noted that the weighted statistics are HT estimators of the indicated parameters; being linear, they are also unbiased.

unweighted	weighted
$\sum_S y \rightarrow \sum_U \pi y$	$\sum_S u/\pi \rightarrow \sum_U y$
$\sum_S y^2 \rightarrow \sum_U \pi y^2$	$\sum_S y^2/\pi \rightarrow \sum_U y^2$
$\sum_S yx \rightarrow \sum_U \pi yx$	$\sum_S yx/\pi \rightarrow \sum_U yx$

Extending the Fisher consistency criterion to nonlinear parameters, we see that:

$$b_{\text{untwd}} = \frac{\sum_S xy - \left(\sum_S y\right)\left(\sum_S x\right)/n}{\sum_S x^2 - \left(\sum_S x\right)^2/n} \rightarrow \frac{\sum_U \pi xy - \left(\sum_U \pi y\right)\left(\sum_U \pi x\right)/n}{\sum_U \pi x^2 - \left(\sum_U \pi x\right)^2/n}.$$

That is, the unweighted sample regression coefficient is consistent for the regression coefficient of the superpopulation that was generated by this sampling design on this population, rather than for the parameter of the real population. Figure 47-3b illustrates a case in which the difference is noticeable, but not great; note that the population mean is also different. It is easy to extend the consistent estimator to nonlinear parameters by Fisher's method. Letting $w_u = 1/\pi_u$,

$$b_{\text{wtd}} = \frac{\sum_S wxy - \left(\sum_S wy\right)\left(\sum_S wx\right)/\hat{N}}{\sum_S wx^2 - \left(\sum_S wx\right)^2/\hat{N}} \rightarrow \frac{\sum_U xy - \left(\sum_U y\right)\left(\sum_U x\right)/N}{\sum_U x^2 - \left(\sum_U x\right)^2/N}.$$

In model-based inference, one might opt to use unweighted sample regressions, and hence estimates of the superpopulation parameters. However, it is important to understand the distinction and to be consistent (in the procedural sense) throughout the analysis. In other circumstances, it is clearly indicated that the regression parameter of the real population is required by the objectives, and the weighted estimator must be used. Terminology is here a bit of a problem; these weighted nonlinear estimators are not really Horvitz–Thompson estimators, but rather a generalization of HTE for nonlinear parameters. I will refer to such estimators as GHTE, for generalized HTE.

The need to weight conventional statistics by the probabilities is widely overlooked, but easily seen to be necessary in order to maintain consistency. This complication is eliminated by maintaining an equiprobable sample, and is just one of several reasons for doing so. A stratified design with other than proportional allocation will also present variable probabilities for any population spanning more than one stratum. This case arises frequently, and it is difficult to justify unweighted analysis. Even if an equiprobable sample is used, then the practitioner must still be alert for cases in which a specific subpopulation has been eliminated by virtue of inclusion probabilities identically zero. Careful specification of the target population, as set by the sampling protocol, is essential.

Royall and Cumberland (1981) have argued that other methods (other than p samples) for providing representativeness can provide improved model-based inference. They balance the sample over the recognized factors of influence that are provided by the frame materials, a process analogous to established practice in experimental design. Then model-based methods utilize the information in the frame data to generate inferences for the population. In fact, if the X population is known and can be used as a design variable, and if the relation between the response variable and the X variable is sufficiently close, then the Royall and Cumberland (RC) position is

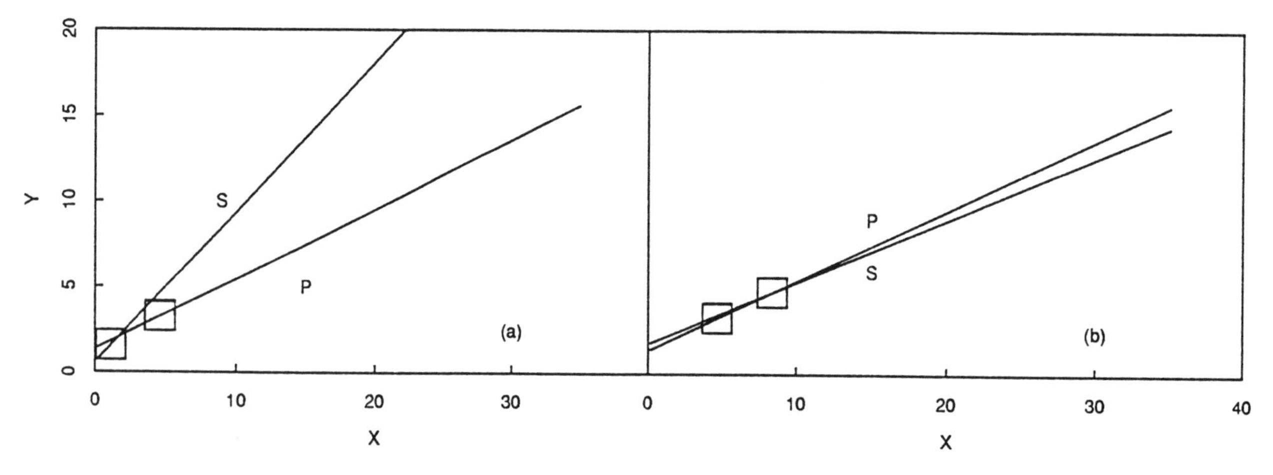

Figure 47-3: Regression analyses are affected by the sampling design. In (a), the subpopulation regression is seen as very different than that of the full population. In (b), the superpopulation regression is seen as somewhat different from that of the real population. Population means are indicated by the boxes.

not only tenable, but can lead to very precise characterization of the real population. Their perspective clearly does not demand a *p* sample.

But their perspective also does not suffer from application to a *p* sample. Think of this design problem as one of establishing a partition on the universe in such a manner that each cell of the partition is balanced with respect to the design variable, *X*. Then select one cell at random, as the sample, and use one of the model-based methods. Of course one might impose additional restrictions and rule out some cells as unacceptable samples, but if each element of the universe is contained in at least one cell, then this is still a *p* sample. In fact, a simple systematic sample, applied to a universe ordered on *X*, will provide a very similar balanced design. Similarly, a systematic sample in space will provide balance on space, as well as on any spatially patterned frame attribute; model-based methods are very useful in deriving inferences from such a sample.

The comments that accompany the cited RC paper will provide a good sample of opinions on the subject, but the RC position does not seem to be in opposition to the position taken here; certainly, there is no disagreement here with the enormous utility of the balanced design they advocate. Note, however, that a multiple-objective survey will be difficult to balance in a highly favorable manner on all objectives. More important, good balancing requires more prior knowledge of population structure than will often be available. In most monitoring cases, a systematic *p* sample will achieve nearly all that can be achieved by intricate balancing.

To summarize, if the probability sample is equiprobable, then the usual OLS estimators are consistent for real population parameters. If the sample is variable probability, then GHTE is required for consistency for the real population parameters; as the regression parameters are nonlinear, GHTE will be biased, but consistent. The need to use HTI in this case is often overlooked, and in fact, OLS regression estimators can be quite badly biased (and not consistent) when variable probability sampling designs are used. Generalized least squares (GLS) does not help in this case.

There is a literature on the issue of ignorability (c.f. Smith, 1976) on identification of design features that can be ignored in analysis. A more practical perspective is that of analytic surveys, and the modification of analysis because of a particular design feature (c.f. Kish, 1965). The foregoing treatment is an example of that perspective; analysis must take design features into account. An even more practical approach is to choose the design in recognition of the needed analyses. Monitoring programs may be directed primarily towards description, but inevitably there will arise the desire to analyze and compare. Designs that minimize difficulties of intended analysis should be used when feasible.

SPATIAL SAMPLING

The usual population characterization from a *p* sample treats the population as a collection of values, whether finite or infinite, on the unordered set of points of the universe. Representation in terms of distribution functions characterizes the population statistically, but does not reflect the structure of the universe nor the patterns of the population on that structure.

Continuous and extensive populations are defined on a spatially ordered universe, but finite populations also usually have a spatial component that may be of interest. For example, a population of lakes may be described in the population mode, or it can be described in terms of spatial pattern, in a manner to reflect the position of those lakes on a map. Currently, recognition of the importance of spatial pattern is so universal that the objectives of any survey are likely to include a spatial component.

Whatever the form of the frame, from list to map, if the sample is to reflect spatial pattern, then it should be selected in a manner to impose spatial restriction. Many writers, including Yates (1949), Matérn (1960), Olea (1984), Koop (1990), and Overton and Stehman (1990) have addressed various aspects of the spatial sampling issue. Many designs have been examined. The most precise are regular grids, of which the most precise are triangular grids. Some investigators advocate square grids, rather than triangular, because of computational advantages, and some advocate a design with some irregularity in order to estimate variance better.

Randomization, so as to provide a probability sample, is easily accomplished with any regular grid. The method used in EMAP (White et al., in press; Overton et al., in preparation), with the EMAP triangular grid, is as follows: (1) establish the base grid position, and the tesselation dual of that grid position, and (2) perturb the sample points from the centers to a common random position in the tesselation cells. This protocol provides a sampling grid in exactly the configuration of the base grid, but randomly positioned so that any small region of fixed area is equally likely to contain a sample grid point (Figure 47-4). This criterion defines an equiprobable point sample, from which several kinds of *p* samples can be generated, depending on the nature of the resource. Points sample areas with knowable probability.

For sample selection, the EMAP design employs an areal sample centered on each grid point, in addition to the grid points. This areal sample is nominally set to approximately 40 km^2, but can be whatever size is desired for a particular resource, and is nominally in the shape of a regular hexagon. A 40-hex samples resource "points" with inclusion probability 1/16, being the product of its area and the density of the grid. Any point in the United States has probability 1/16 of being contained in one of the 40-hex areal samples. Areas sample points with known probability.

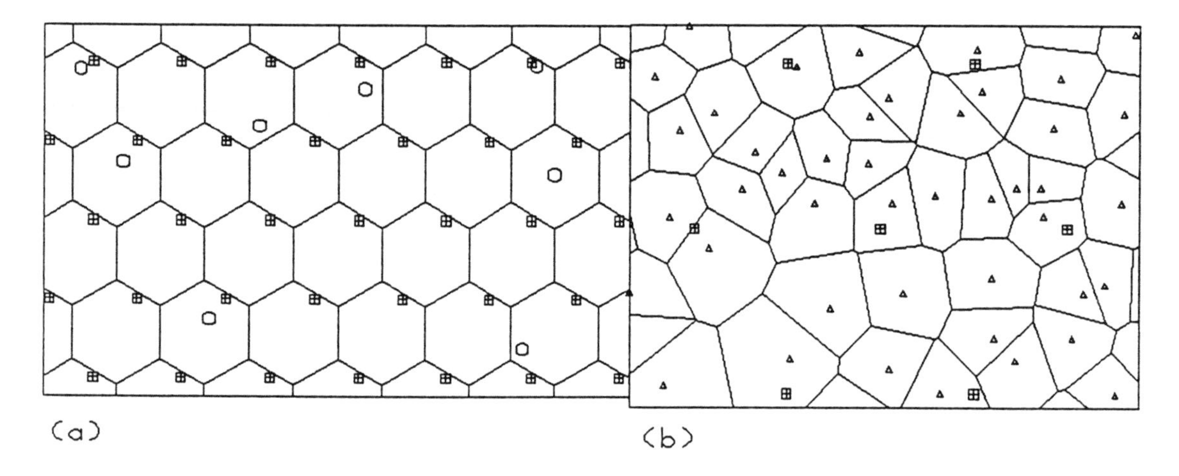

Figure 47-4: The triangular grid samples resources in several ways. In (a), the grid has been shifted to a random position in the base tesselation. Small areas of equal size are equally likely to contain a grid point after such a shift; this is the basis of an equiprobable sample. In (b), given a tesselation generated by the Thiessen polygons of a set of ecological objects, like lakes or trees, a cell of the tesselation will contain a point of the randomly located grid with probability proportional to the area of the cell.

Samples from populations of discrete units, like small lakes, may be sampled by identifying each unit with an explicit unambiguous position point, like the centroid of the lake. Then lakes can be sampled by a point-to-point rule of association, such as selecting the lake whose position point is closest to the grid point. Lakes can also be sampled by selecting all lakes whose position point falls into the areal sample or all lakes whose areal extent contains a grid point. The point-to-point rule involves identifying the area around each population unit that is closer to that unit than to any other—this is the Thiessen polygon of that unit. The probability of selection of any unit is then proportional to the area of its Thiessen polygon. It is appropriate to amend the rule to provide that no unit will be selected if it is greater than a particular distance from the grid point. The standard EMAP rule is that the unit must be the closest unit to the sample point that lies in the areal sample centered on the point; now the inclusion probability of the selected unit is proportional to the intersection of its Thiessen polygon and an identical areal sample centered on the unit position point.

Populations of discrete units may also be represented by spatial areas, rather than by points. This provides the option of selecting a unit if the grid point falls into the designated area of the unit. This method is appropriate for intermediate-sized objects (units). If the object area is large, relative to the sampling grid, or configured just so, then more than one grid point can simultaneously fall into the area; this is an inconvenience, but not a fatal problem. However, on the scale of EMAP, resource units that are large enough to contain more than one grid point are too large for representation as a member of a population of discrete units. Such units should be classed as extensive and characterized either as individuals or in the aggregate with others of the same class.

Continuous universes (with either extensive or continuous populations) are sampled either by (1) points (grid points or clusters of points centered on the grid points) with values of the population measured at the points, or (2) areal samples, being the portion of the areal sample (40-hex) centered on the grid point that is occupied by that resource. Such areal samples are composites of the points comprising them, and so have the same weights, 16, as would be used for any point. The appropriate values for areal samples are integrals of population values over the sample.

To summarize, areas are sampled by grid points falling into them, with inclusion probability proportional to the area, and points are sampled by being contained in the areal sample centered on the grid points, with inclusion probability (1/16) the product of the area of the areal sample and the density of the grid. Point-to-point rules of association always involve identification of an area within which the grid point will select an object associated with that area.

Inferences for discrete populations sampled in this manner are appropriately made by HTI, and generation of the usual population estimates is straightforward. Spatial analysis when the sample is variable probability poses some issues, and this topic is currently being investigated, as are other questions regarding variance estimation for samples taken systematically in space.

Inferences for extensive populations have no counterpart to the Horvitz–Thompson theorem. However, probability estimation follows basic principles, providing consistent estimators of the real population attributes. Investigation of the variance estimation problem for extensive resources is currently active, but empirical evidence indicates that the approximation formulae devised for variance of discrete resources under population randomization are satisfactory for many situations. For populations that are spatially highly autocorrelated at the scale of the sampling grid, the assumption of randomization will not be satisfactory, and it will be necessary to use another algorithm for variance estimation. The several approximations proposed to date are not generally satisfactory, and the "facsimile population bootstrap" is planned for this use.

INDEX SAMPLES

Many cases arise in sampling in which it can be recognized that the particular circumstances will be better served by a nonprobability sample than by one that qualifies as a *p* sample. A good example is provided by those circumstances suitable for an index sample. Consider a sample of lakes, and the general case that multiple observations on a lake will require reducing the number of lakes that can be sampled, so that observations are prescribed at one location and time on each sample lake. Hypothesize that the spatial and temporal patterns of the attribute of interest, on a single lake, are very pronounced, but that lakes are well organized, and that the patterns are well known. In such a case, an index sample, taken at a prescribed time and position of a lake, may be a much better representation of the nature of the lake than a random (probability) sample.

A familiar example in which an index sample is always used is determining body temperature of humans. Several index positions are used in different circumstances, but the "normal" value of each and interpretation of deviation from normal is well established. A *p* sample of size 1 would lead to a greatly inferior indicator of well being, even if one could be obtained without sacrificing the patient.

This issue is not restricted to cases involving a single observation; the same considerations apply when two or three observations are prescribed. Further, there are other cases in which a single observation will be made per sampling unit and a p sample is more appropriate than an index sample. It all revolves around the issue of how best to measure the state of the unit being observed under the circumstances. Then this state measurement becomes a variable on the sample of units (like lakes).

In reality, an index sample is a judgment sample for which the criteria of judgment have been formally identified as part of the observational protocol and are objectively applied in each case. Without an explicit protocol, a judgment sample has no identity; the objectivity that derives from the protocol imposes a level of rigor.

SUMMARY

Probability sampling with Horvitz–Thompson estimation is the basis of rigorous inference that derives its properties from the sampling protocol. The *p* sample provides explicit identity of the universe/population being sampled. Generalized Horvitz–Thompson estimation provides unbiased estimators of linear parameters and consistent estimators of all parameters, with consistency defined in the Fisher (1956) manner. A general body of design theory provides the capacity to tailor the design to specific circumstances.

Model-based inference can often improve precision of a *p* sample, sometimes greatly. When used without a *p* sample, the model must completely support the inferences. As model-based inferences are directed towards parameters of a superpopulation, care must be taken to ensure that the superpopulation parameters are identical to the real population parameters of concern. A *p* sample will provide the basis for establishing this identity of the superpopulation in terms of the real population. Applied to data taken by a *p* sample, model-based inference has the capability of consistent inference relative to the real population, but care must be taken to provide this property.

The usual model-based estimates are of the parameters of that superpopulation generated by the interaction of the design and the real population, as illustrated in Figure 47-3b. When the parameters of the real population are the objective, model analysis applied to a *p* sample must take into account those design features that influence design-based inference; otherwise, inference can be for parameters other than those intended. GHTE provides this capability for parameters of the finite population.

Monitoring programs will produce data that are used for a variety of analyses other than simple estimation of population parameters, becoming analytic surveys, in the language of sampling. Certain analyses may only be inconvenient when certain design features must be taken into account, but other analyses may be invalid. The several forms of variable probability sampling create quite difficult problems in analytic methodology like analysis of contingency tables and in graphical representation as by scatter plots. As a consequence, monitoring programs that require extensive and diverse analysis via a variety of models should minimize such design features. An equiprobable design will eliminate most of the essential difficulties of analysis, and will provide other advantages for a multipurpose survey. But the basic constraint that the design be a *p* sample should be relaxed only in very special circumstances. An equiprobable *p* sample will usually add little cost or inconvenience to a monitoring program, and will provide inexpensive insurance against poor inferences.

ACKNOWLEDGMENTS

I should like to thank Steve Stehman for critical review of this manuscript and for many in-depth discussions of its contents. Don Searles and the editors of this volume also provided helpful reviews. Thanks go to George Weaver for generating the figures and for empirically verifying several results. This work was supported by Cooperative Agreement CR816721 between the U.S. Environmental Protection Agency and Oregon State University.

REFERENCES

Cassel, C.M., Sarndal, C.E., and Wretman, J.H. (1977) *Foundations of Inference in Survey Sampling*, Wiley: New York.

Cochran, W.G. (1977) *Sampling Techniques* (3rd Edition), Wiley: New York.

de Gruijter, J.J., and ter Braak, C.F.J. (1990) Model-free estimation from spatial samples: a reappraisal of classical sampling theory. *Mathematical Geology* 22(4): 407–415.

Fisher, R.A. (1956) *Statistical Methods and Scientific Inference*, Oliver and Boyd.

Horvitz, D.G., and Thompson, D.J. (1952) A generalization of sampling without replacement from a finite universe. *Journal of the American Statistical Association* 47: 663–685.

Kaufmann, P.R., Herlihy, A.T., Elwood, J.W., Mitch, M.E., Overton, W.S., Sale, M.J., Messer, J.J., Cougan, K.A., Peck, D.V., Reckhow, K.H., Kinney, A.J., Christie, S.J., Brown, D.D., Hagley, C.A., and Jager, H.I. (1988) Chemical characteristics of streams in the Mid-Atlantic and Southeastern United States. Volume I: population descriptions and physico-chemical relationships. *EPA/600/3-88/021a*, Washington, DC: U.S. Environmental Protection Agency.

Kish, L. (1965) *Survey Sampling*, Wiley: New York.

Koop, J.C. (1990) Systematic sampling of two-dimensional surfaces and related problems. *Commun. Statist.-Theory Meth.* 19(5): 1701–1750.

Linthurst, R.A., Landers, D.H., Eilers, J.M., Brakke, D.F., Overton, W.S., Meier, E.P., and Crowe, R.C. (1986) Characteristics of lakes in the Eastern United States. Volume I: population descriptions and physico-chemical relationships. *EPA/600/4-86/007a*, Washington, DC: U.S. Environmental Protection Agency, 136 pp.

Matérn, B. (1960) *Spatial Variation*, New York: Springer-Verlag.

Messer, J.J, Linthurst, R.A., and Overton, W.S. (1991) An EPA program for monitoring ecological status and trends. *Environmental Monitoring and Assessment* 17: 67–78.

Olea, R.A. (1984) Systematic sampling of spatial functions. *Series on Spatial Functions, No. 7*, Lawrence, KS: Kansas Geological Survey, University of Kansas, 57 pp.

Overton, W.S., and Stehman, S.V. (1990) Statistical properties of designs for sampling continuous functions in two dimensions using a triangular grid. *Technical Report 143*, Corvallis, OR: Department of Statistics, Oregon State University, 36 pp.

Overton, W.S., White, D., and Kimerling, A.J. (in preparation) Statistical, geographic, and cartographic considerations of a sampling grid for environmental monitoring.

Royall, R.M., and Cumberland, W.G. (1981) An empirical study of the ratio estimator and estimators of its variance. *Journal of the American Statistical Association* 76: 66–77.

Smith, T.M.F. (1976) The foundations of survey sampling: a review. *Journal of the Royal Statistical Association* 139(2): 183–204.

Stuart, A. (1962) *Basic Ideas of Scientific Sampling*, Hafner: New York.

White, D., Kimerling, A.J., and Overton, W.S. (1992) Cartographic and geometric components of a global sampling design for environmental monitoring. *Cartography and Geographic Information Systems.* 19(1): 5–22.

Yates, F. (1949) *Sampling Methods for Censuses and Surveys*, Hafner: New York.

Epilog

MICHAEL P. CRANE AND
MICHAEL F. GOODCHILD

The preceding chapters are based upon papers selected from the invited presentations at the First International Conference/Workshop on Integrating GIS and Environmental Modeling held in Boulder, Colorado, during September, 1991. The conference was conceived about 2 years prior to the event by a handful of scientists convinced of the need for an explicit forum to review the role of scientific GIS in modeling the major components of the terrestrial environment. Because the meeting was the first to focus on this topic, the organizers had only their unsubstantiated convictions as a basis for estimating how much interest would be shown in a conference of this nature. The final registration count passed six hundred people, exceeding even the wildest expectations of the organizers.

Attendees came from 20 countries and brought to the conference an effective mixture of professional backgrounds drawn from the broad spectrum of environmental modeling and geoprocessing specialties. One of the primary objectives of the meeting was to involve this diverse mix of people in a discussion of points raised in the plenary sessions, by helping the organizers and speakers to identify and explore key issues in the adaptive use of GIS and allied technologies for modeling environmental problems, and by reflecting on prospects for the future. This was accomplished in a sequence of five working group sessions, each of which followed a major plenary session and thus formed an integral component of the conference agenda.

To promote participation and facilitate working group discussions, the large conference assembly was partitioned into ten breakout groups. Lack of prior knowledge about participants' backgrounds and specific interests necessitated their initial allocation to the breakout groups on an alphabetical basis. Participants were encouraged to move between groups in subsequent sessions as their interests and those of the groups developed. In some instances as many as fifty people comprised a particular breakout session. Although the sizes of the breakout sessions were typically more than the desired number for optimum group dynamics, each of the ten groups developed and pursued their individual programs with considerable enthusiasm. Moderators and rapporteurs were used in each workgroup to keep discussions moving and to ensure that everyone had an opportunity to speak, and to keep records of the discussions. Each working group was asked to develop a set of general conclusions and recommendations, and these are presented in the following. Discussions evolved in a process unique to each group, and this resulted in uneven treatment of the issues. Nevertheless, most of the working groups covered a broad range of topics during the course of their discussions, and this resulted in a fair amount of redundancy. For this reason the conclusions and recommendations are presented by topic rather than by working group.

RECOMMENDATIONS FROM THE GROUPS

Systems issues

Almost all of the working groups expressed a need for vendors to provide open architecture in order to facilitate systems integration. This applies to hardware components, operating systems, and applications software. To study the magnitude and complexity of environmental problems adequately, researchers need to use a variety of tools that interface smoothly with each other and enable the transfer of data between systems without having to go through torturous and time-consuming data format conversions. Many researchers have become very skilled at writing small FORTRAN or C programs to reformat data between packages, but this kind of activity ultimately reflects a lack of concern for coordination and integration on the part of system designers.

Another suggestion called for the establishment of an unbiased entity to perform systems benchmarking and evaluation on behalf of the scientific community. Systems would be evaluated in terms of functionality, ease of use, and ability to solve real-world problems. The evaluations would be conducted in an objective manner and result in a public and hopefully constructive document.

Data issues

Data issues generated the greatest response from the working groups. In particular, digital spatial data, its availability, format, standards, quality, and lineage were major concerns identified by all ten working groups. Attendees expressed frustration at the lack of adequate cataloging mechanisms to indicate what spatial data are actually available, where they are stored and in what format, and how they may be obtained. The employment of inconsistent standards in data collection was viewed as a significant complication in the ability of users to integrate data derived from diverse sources and perform meaningful analyses. A related problem is the paucity of information that accompanies data concerning its quality and specific processing history or lineage. Without this information users have no basis for performing error estimations or determining data reliability. While there are no quick or easy solutions to these problems, a number of useful suggestions were made by the working groups.

Several of the groups recommended the creation of a mechanism for the dissemination and exchange of information (metadata) about existing and proposed data sources, modeling algorithms, and public domain geoprocessing software. Strong support was voiced for the formation of an international electronic bulletin board as a means for both formal and informal communication about data. One group went so far as to name the system GRABIT, an acronym for Geospatial Resource Access Bulletin and Information Transfer, and several people from this group are exploring the feasibility of the proposal. Other groups proposed the establishment of a centralized storehouse/distributor for both data and software.

The question of data quality and data lineage will be addressed for agencies of the U.S. government with the eventual implementation of the Spatial Data Transfer Standard (SDTS). In addition to facilitating data sharing among government agencies, SDTS gives the user a mechanism for documenting data quality and lineage characteristics. A similar standard is being considered for adoption by the European Economic Community and in other countries. When these standards are in place, there will be increased pressure on private industry to comply with the standard and to provide software support. Better yet, working group participants would like to have an automated data quality/lineage tracking system as an integral component of their geoprocessing software, to operate in the background much like an audit trail.

Data costs were not identified as a prominent issue except by two groups, one of which proposed the concept of a public data library. This facility would verify and document data quality, catalog it, and distribute it free to users on the appropriate media by request. A slightly different idea was suggested by the second group—the library would contain a digital catalog and related metadata and provide access to a networked data system.

Another reason that data costs were not considered a bigger issue is because data sharing is becoming more widespread. Attendees felt that the sharing of data will be commonplace when data sources are better cataloged, and data quality and lineage are better documented.

GIS issues

The available software for doing spatial analysis generally falls short of what researchers need and, as a result, a lot of time is spent developing code and probably duplicating effort. The establishment of a repository for GIS functions and related software code was suggested as a possible solution, where a registry of functions would be maintained in the public domain to prevent redundant development activity. The intent would be for the repository to contain both public domain and proprietary vendor software.

A number of specific analytical/functional capabilities were identified as needed in GIS software, and these include enhanced interpolation routines, spatial statistics, time series integration, dynamic simulation and stochastic modeling, true three-dimensional modeling and mensuration, and linkages to CAD packages. In short, users want an integrated super GIS that provides at a minimum a basic set of tools for environmental modeling and error analysis. Some of these capabilities are in the process of being incorporated into public domain GIS software packages; a good example is the GRASS software developed by the U.S. Army Corps of Engineers, which is currently being interfaced to the S+ statistical package. Scientific users of GIS would also like access to source code, to be able to integrate new capabilities, and while they understand the basic dimensions of the software marketplace, are frequently frustrated by the relatively closed nature of much commercial software.

Several groups saw potential in object-oriented programming techniques in the development of the next generation of geoprocessing tools, and felt that such techniques could lead to a fertile rethinking and reorganization of many models.

Modeling and analysis issues

A strong recommendation was made for the creation of a taxonomic catalog of environmental models, to provide a better understanding of each model's value and its ability to interface with GIS. This would also help to improve communication between modelers and people working in the GIS realm. Somewhat related to this is the recurring idea of a clearing house that would catalog and evaluate environmental models, and distribute them on request. A standard spatial database subroutine package is needed, to enhance communication between the different components of the integrated GIS/modeling system.

PROSPECTS FOR THE FUTURE

The single most important accomplishment of the conference was the initiation of meaningful dialog between the geoprocessing and environmental modeling communities. One of the attendees compared the conference to a high school dance; at the beginning, the boys and girls stood on opposite sides of the dance floor, but by the end the ice had been well and truly broken. As conference attendees began to talk with one another, many realized they were discussing similar concepts but were using different sets of terms.

There was a strong recommendation from all ten working groups for a second conference on GIS and environmental modeling to address outstanding issues. Several groups felt that the next conference should be more specialized and focus on a few issues in greater depth. Topics proposed for consideration include lineage tracking requirements and implementation, error estimation, the problem of scale in models, types of models, integrating/interfacing programs, existing and desired standards and their implications, and new data structures. Whereas the first conference had been organized around the various areas of environmental modeling and had allowed generic issues to emerge in discussion, the second could reverse priorities by bring the issues to the fore, with discussion groups focusing on each area of modeling.

A number of the working groups felt that vendors should be more actively involved in the next conference—but not in a commercial way! Having vendors attend and actively participate in the scientific program would enable issues to be addressed more thoroughly, and would, it was hoped, provide a better understanding of where their development activities are headed and why.

CLOSING COMMENTS

As Jack Eddy argued so clearly in his introductory perspective (Chapter 1), it is time to think about putting the global environment into intensive care. In the 900 days that it took the World Commission on Environment and Development to produce the Bruntland Report, a drought in Africa killed up to 1 million people; the explosion at Chernobyl caused environmental damage throughout Europe; the leak at a chemical factory in Bhopal, India, killed 2,000 and injured 200,000 more; and a chemical fire in Switzerland caused toxic materials to be transported by the Rhine as far as the Netherlands.

By contrast, our scientific knowledge of the Earth system accumulates at an agonizingly slow pace. The computerized environmental models that capture that knowledge and deliver it to policy-makers in a predictive environment are still painfully primitive, and as this book has shown are still poorly integrated and lacking in many of the tools for visualization and presentation that would make them more influential in guiding policy and public opinion. Ten years from now we will look back on the GIS tools of the early 1990s as we now look back on those of the late 1970s—as quaint artifacts of an earlier age. Fifteen years ago virtually none of the GIS-based integration described in this book would have been possible, and software tools that we now think of as essential were not even conceived.

In a recent overview of the role of GIS in the environmental sciences, Townshend (1991) demonstrates "the power of GIS for improving the quality of environmental databases and expanding their applications." He goes on to say, however, that:

> the mere existence of these benefits does not in itself ensure the widespread adoption of GIS. Some large and valuable data sets and complex systems for their management already exist to support environmental science and its applications. Converting existing data sets to forms appropriate to GIS and achieving the transition from traditional to modern methods of spatial information handling will be a complex, expensive and time-consuming process: moreover, it will have to compete for resources with many other more glamorous activities. However, without the necessary investment in the infrastructure of information handling, our ability to make best use of our data in addressing fundamental environmental questions will be profoundly damaged.

We hope that this book has made a small contribution to the continuing dialog between environmental science and environmental policy, by exploring the usefulness of GIS as a tool to support environmental modeling and to analyze and present the results of modeling in useful ways, and by identifying some of the ways in which that usefulness can be improved in the future.

REFERENCES

Townshend, J.R.G. (1991) Environmental databases and GIS. In Maguire, D.J., Goodchild, M.F., and Rhind, D.W. (eds.) *Geographical Information Systems: Principles and Applications*, London: Longman, Vol. 2, pp. 201–215.

INDEX